ARCHITECTURE AND URBAN SPACE

PLEA'91 ARCHITECTURE AND URBAN SPACE

Sponsors

Comission of the European Communities
Junta de Andalucia
Sociedad Estatal EXPO'92
Universidad de Sevilla
and
Fundación Sevillana de Electricidad

ARCHITECTURE
AND
URBAN SPACE

Proceedings of the Ninth International PLEA Conference,
Seville, Spain, September 24–27, 1991

Edited by

SERVANDO ALVAREZ
Department of Energy and Fluid Mechanics Engineering,
University of Seville, Spain

JAIME LOPEZ DE ASIAIN
Seminar of Bioclimatic Architecture,
University of Seville, Spain

SIMOS YANNAS
Architectural Association Graduate School, London, U.K.

E. DE OLIVEIRA FERNANDES
University of Porto, Portugal

Published on behalf of
The International PLEA Organisation
by

KLUWER ACADEMIC PUBLISHERS
DORDRECHT / BOSTON / LONDON

ISBN 0-7923-1418-2

Published by Kluwer Academic Publishers,
P.O. Box 17, 3300 AA Dordrecht, The Netherlands.

Kluwer Academic Publishers incorporates
the publishing programmes of
D. Reidel, Martinus Nijhoff, Dr W. Junk and MTP Press.

Sold and distributed in the U.S.A. and Canada
by Kluwer Academic Publishers,
101 Philip Drive, Norwell, MA 02061, U.S.A.

In all other countries, sold and distributed
by Kluwer Academic Publishers Group,
P.O. Box 322, 3300 AH Dordrecht, The Netherlands.

Printed on acid-free paper

Printed in the Netherlands

PREFACE

These are the Proceedings of the PLEA'91 meeting, held in Seville under the joint responsability of the International Steering Committee of PLEA - Passive and Low Energy Architecture - and of a local organizing committee hosted by several research groups of the University of Seville.

This is the ninth annual PLEA Conference organized since 1982 in response to the permanent commitment of PLEA; to become close to the regional problems to contribute for the widening and the enhancement of the understanding of the problems associated with the built environment.

According to its general principles and objectives, PLEA aims to establish the highest standard of research but, also, to promote professionalism in building science and construction technology and in architecture and urbanism. While recognizing that the adoption of an "environment-conscious design" will help significantly towards solving some of our current environmental problems, PLEA advocates that the environment-conscious design is not just a means for avoiding ecological disaster, but a criterion of architectural quality and a creative meeting point between science, technology and architecture.

For that purpose, PLEA acts as an international interdisciplinary and wide open forum through many different activities such as conferences, workshops, architectural competions and exhibitions, scientific and technical publications and consultancy. In this year of 1991, besides this annual conference, PLEA International has also organized a PLEA Study Tour hosted by the authorities of the state of Arizona (USA). Through a very original programme, an obvious but somehow forgotten relationship was established between culture and the intelligence of the first human settlements and the technological requirements of the cities of the future.

PLEA 91 is organized in Seville at the eve of the World Fair, EXPO'92, that will take place at the capital of Andalucie from April through October 1992. That makes PLEA'91 full of added meaning and symbolism.

The World Fair 92 is devoted to highlighting the progress of the Mankind presenting a comprehensive "state of the art" of the technological progress and pointing out which are the major expectations and concerns for the future. At the same time it establishes a close link with other historical events that enlarged the frontiers of the World about 500 years ago. It also aims at suggesting a parallel between the development of new technologies and the benefits that resulted from knew knowledge in the 1500's and now.

The main objective of PLEA is the development of the principles of bioclimatic design through the integration of natural and innovative techniques. It therefore aims to respond to the fullfilment of some of the basic aspirations of Man, namely, in what regards a better built, healthier and more comfortable environment, within and around buildings.

The EXPO'92 is a major case study of the application of PLEA principles. The outdoor spaces have been studied and arranged in such a way that, through the use of low energy techniques the outdoor environment is made more acceptable for the visitors during the warm period the EXPO'92 is open. The extended use of passive and low energy cooling techniques makes the EXPO'92 site an outstanding demonstration of what can be done as an environment-conscious design.

Holding the sessions in the EXPO site and being able to contact with the different methods and technologies adopted in the urban space of the World Fair during the conference, PLEA'91 attempts to underline the relevance of those features.

Nowadays, energy in buildings is no longer a mere microscale issue. When about 40% of the energy used in the OECD countries is in buildings, mainly for comfort purposes and when the use of energy is responsible for some of the most crucial environmental problems that our World is faced with it is evident that every possible intervention in order to achieve a better use of energy in buildings converges with the most important objective of our civilization; to assure the survival of our way of living while providing the means to favour the access of a permanently growing share of the world population to the benefits of progress and development.

Architecture has not to be dominated by energy. But concepts of integration of a better climate information and of new advances in science and technology must be adopted in a wider way by the architectural profession. That is why PLEA'91 puts a focus on architecture and the urban space but does not leave apart the advances regarding design tools and passive technologies.

A great deal of effort has to be put on education, training and transfer of technology to the architectural profession.

However, there is still a large room for more research and for better tools to ease the learning process and application of those concepts. Topics as passive cooling, indoor air quality and comfort in environments with free-floating temperature are to receive much more effort. These will also be guidelines for future PLEA activities.

CONTENTS

Chapter 2: Projects

Part 4: DESIGN AIDS

Part 5: PASSIVE AND LOW ENERGY TECHNOLOGIES

Chapter 1: Progress in Model Development and Simulation

Chapter 2: Components Performance

Chapter 3: Cooling

Part 6: BUILDING ENERGY ANALYSIS

Chapter 1: Climatic Data

Part 7: EDUCATION AND TECHNOLOGY TRANSFER

Part 8: LATE PAPERS

Part 1

ARCHITECTURE AND URBAN SPACE

ARCHITECTURE AND URBAN SPACE

ANTONIO CRUZ/ANTONIO ORTIZ
Architects
C/. Santas Patronas, 36. 41001 Sevilla
SPAIN
Tph. (34.5) 4221674
Fax. (34.5) 4561689

ABSTRACT. Going through the analisys of several building recently projected and built, we present the relationship between architecture and urban space. This relationship must be considered as a consequence of the site as well as a new fact that structure the form of the site itself.

SEVILLE RAILWAY STATION.

In spite of its relative proximity to the center of the citiy, the areas selected for Seville's new central railway station had hardly been urbanized. In the area where the station has been consturcted, various factors, among them the existence of an old abandoned station, had allowed a huge pcket of land to remain virtually untouched.

The station will act as the origin for the reconversion of the entire, yet undeveloped, sector in an area with a high level of urban activity, This process will also be strengthened by the new streets being contructed on the railway's new layout, linking presently adjacent, but separated, parts of the city.

We, therefore thought it necessary beforehend to carry out the complete layout of the whole blovck where the construction of the station was foreseen. To this effect, three-storey buildings of a 25-meter average width were planned, occupying almost the entire perimeter of the block. In doing this, various goals were taken into account. First, to guarantee the station suitable and homogeneous surroundings in order to give the passengers a presumably much more ordered foreground than the yet unconstructed areas, the General Plan proposes around the block. On the other hand, far from isolating the station from the rest of the city -since the perimetral buildings will be constructed with the same materials as those used for the station it will relate to it and, in fact, extend it. The projects for some of these buildings have already been drawn up: the railroad company offices, the city bus stop canopy and a buildng for oficces and dwellings promoted by the Twon Hall. Thus, the greater part

S. Alvarez et al. (eds.), Architecture and Urban Space, 3–5.

of this large block's perimeter will share certain characteristics with the station, providing an adequate frame for it and constituting a large sector of the city, characterized by the presence of the station.

In the block's interior are two extensive, lateral areas, which will house both the parking lots and the traffic generated by the station and sorrounding buildings. Foreseen in front of the station are the main passenger aproaches, the taxi stop, and the bus stop canopy for both city and tour buses.

ADAPTATION OF THE CANDELARIA STATION TO THE MARITIME MUSEUM.

What it knwon in Cadiz as the "Baluarte de la Candelaria" is a group of buildings of military origin, constructed over different periods of time. Once modified, this buildings become the home of tne new Maritime Museum.

The buildings were in an abandoned state and of various types. Two of them separated the bastion from the rest of the city and were of similar character and construction. A third part, a roofed battlement, had been constucted during the last century on the limits of the city's sea defence walls to protect the artillery and gave on to a ramp leading to the roof. This whole complex outlined a courtyard contining a fourth -independent- building, the guards' quarters.

Merely restoring these different structures would not have been sufficient to satisfy the solicited requirements. As we were dealing with a midification of the builings'use as radical as this one, we realized thar connecting them by a sheltered walkway colud result in a sharp change in the courtyard's character, something desirable designed to link the vaious parts, resulting in a sheltered trail extending the full length of the Museum that never loses contact with the inner garden. These operations allow us to arrive at a new order that is superimposed on the apparently arbitrary layout of the buildings.

The character of these coverings has allowed them to be juxtaposed to the existing buildings, causing the vegetation and transparencies to become the themes of the new image the Museum should offer to the city.

SPORTS COMPLEX IN MADRID.

The Sports Complex of Madrid is located to the East of the city limits, bounded on the East by a strecth of the traffic ring M40 and on the West by the end of lo Arcentales Avenue and the newly formed, yet unnamed, streets.

Right from the initial porposals for the competition, we opted for a highly unitary solution, where the compound consituted a preliminary design that determined the shape of the various buildings to be subsequently erected.

A significant gesture of this desire is the creation of a continuous circular traffic ring that links up and structures the whole complex, providing acces to the vaious sports areas. Two parking zones are structured on this ring: one on the ring itself, and the other further West, for more frequent an intensive use.

Areas are marked out within the zone defined by the perimeter: to the North, with a terraced garden treatment, are the open air sports facilities for training: to the East, a green belt; and in a central location, a third an main one, a roughly square platform which holds the stadium and is connected to the differente sports pavilions. The general services for all the pavilions in the complex are housed under this platform in the area in front of the stadium.

The services are organized on two floors and are illuminated through courtyards. Dressing rooms, indoor practice areas, press, judges and VIP areas, etc. will facilitate the organization of large sporting events.

HUELVA BUS STATION.

The Huelva bus station has been planned on 33º-angle , wedge-shaped lot with a radius of 200 meters.

The attainment of a unified image has been the basic intntion in the developement of the whole complex. A large, flat, uninterrupted ceililng seven meters high shelters both the bus platforms and the inside hall area where passenger services and transport company offices area locted. On the lot's corners, outside this roof, are there small structuresof a lesses height: a gas station, additional docks and a commercial area. Their assimilation to the lot's perimetral wall makex their role as auxiliary elements evident.

The shape chosen for this large roof, which in the beginning could have almost been understood as a gesture inscribed in the lot's triangle, allowed us besides to resolve a complex and varied program such as this one.

All passenger services have been located on the ground floor around the hall. The floor upstairs will house transport company offices and personnel dessing rooms and lounges. Continuity is maintained along the passenger hall and platform area. Buses circulate around the platforms, avoiding interference with pdestrians. These platforms encircle an open garden, intending the nearness between vegetation and passenger to be one of the project's more attractive features.

SEARCHING FOR THE BIOCLIMATIC CITY

Jeffrey Cook, Regents Professor
School of Architecture
Arizona State University
Tempe, Arizona 85287-1605
USA

1. BIOCLIMATIC DESIGN

1.1 The Bioclimatic Dwelling

The Genesis of bioclimatic design is the dwelling. The passive single family house that extends its heating and cooling needs through natural means, and expands its symbolic relationships with the natural world through design is called **bioclimatic**. Perhaps an expression of the ideal goal is found in Palladio, (1508-1580) Preface to Book Four:

> "..., if we consider this beautiful machine of the world, with how many wonderful ornaments it has filled, and how the heavens, by their continual revolutions, change the seasons according as nature requires, and their motion preserves itself by the sweetest harmony of temperature; we cannot doubt, but that the little temples we make, ought to resemble this very great one, ..."

Thus human designs should attempt to imitate the dynamic natural balance of the global system.

In the minds and imaginations of many, the bioclimatic dwelling is an ideal that has a reality. Whether based on well tested vernacular or indigenous examples, or on up to date professionally designed versions using modern building materials and components, the idea is of a single building for human use, designed to suitably interact with its climate. Such traditional examples as the snow igloo of the Arctic can be examined in terms of their comfort and human accommodation. Simultaneously the morphology of their forms and processes can be used to read backwards: to project a description of their cold climate location that is a quite accurate reflection of real climatic conditions. Thus the design criteria of successful indigenous dwellings become a model for the creation of new building typologies using modern methods.

1.2 The Bioclimatic Settlement in Books

All bioclimatic researchers propose to extend their ideologies from the scale of the individual building to the larger scale of multiple buildings, and ultimately the city. But proportionally a very small part of their books is devoted to this much larger and more complex design problem of the city. Thus Konya (1980) in Design Primer for Hot Climates devotes only about four pages out of 132 to planning the multibuilding settlement,

S. Alvarez et al. (eds.), Architecture and Urban Space, 7–16.

Koenigsberger, Ingersoll, Mayhew, Szakolay in their 1973 Manual of Tropical Housing and Building also integrate the design issues of settlements within the respective climates. They devote some six pages out of 320 to settlement planning. Victor Olgyay (1963) in Design with Climate devotes a separate chapter (XII) to Examples in Four Regions. These 24 pages out of 190 appear to be among the least noticed of this generic work. All four Olgyay examples propose very low density solutions considering total land use even though neighborhood clusters for instance in the hot arid zone are compact with a high density. A rare exception is La Cittá del Sole (Solar City) 1985 edited by Sergio Los and Natasha Pulitzer that documents some 20 invited papers presented at a conference in Triste.

Generally authors who deal with temperate or cooler climates and thus with passive heating tend not to include any land planning or settlement design criteria, except as they affect sun angles and solar access. Edward Mazria (1979) in his 688 page "professional edition" of The Passive Solar Energy Book never gets beyond the single building. Inside Out by Brown, Reynolds, and Ubbelohde, a 1982 classroom text of "design proceedings for passive environmental technologies," is similar as are the several much more technical research and design manuals authored at least in part by Douglas Balcomb. Similarly, Donald Watson and Kenneth Labs in their 280 page Climatic Design, Energy Efficient Building Principles and Practices, 1983, never get beyond the envelope.

These authors and many others seem to follow the argument and the inverse of Alberti who in Book V Chapter XIV of his 1452 work of Ten Books on Architecture wrote:

> "...a House is a little City. We are therefore in the building of it, to have an Eye almost to everything that relates to the Building of a City; that it be healthy, furnished with all Manner of Necessaries, not deficient in any of the Conveniences that conduce to the Repose, Tranquility or Delicacy of life..."

This is Alberti's introduction to the discussion of the siting of a house in the country and the importance of site choice as it relates to climate. The implication in the field of design is that the **city is simply a large bioclimatic villa**, an extended passive house.

However, there are fundamental differences between a city and a house based on their intensions. A dwelling serves as shelter for the stabilty of the individual and his family unit. Thus its primary goals include comfort. The city is first of all a machine of socialization where ideas of comfort are overshadowed by security, commerce and interaction.

2. A METHOD OF ENQUIRY

Many bioclimatic researchers use a method of matching criteria with cases that could be open to challenge. Their process is one of establishing logical performance parameters and then searching through a field of candidates to find a well fitting match. Such was the method of Knowles in his 1978 book Energy and Form in looking at the passive solar aspects of a prehistoric cliff dwellers in the southwest of the USA. He only looked at ruins that face south, although Anasazi cliff dwellings have a variety of orientations. Obviously a more convincing method is to statistically analyze the whole universe of examples in a set, to determine whether criteria have a fit to more than selected models.

For this study, examples of city plans and city squares have been chosen primarily from a well documented book Beautiful Squares by Fritz Rothstein, published by Edition Leipzig in 1967 in an English edition. Since by definition this book intended to examine aesthetically pleasing cities, not bioclimatic designs, it represents a random unprejudiced set. But it might be inferred that cities whose design was so conscious as to develop their beauty, might also have been conscious in their response to climate.

In addition, samples were chosen only from the pre-industrial period. The argument is that industrial cities have their prime goals in the production, exchange and consumption of goods. Such goals typically are not climate sensitive and indeed may encourage defiance of climate if it can increase production, exchange or consumption. The city designed as an active and consumptive machine is not likely to respond as a passive organism, or even as life enhancing. Alternatively pre-industrial societies evolved over time to a slow maturity based on internal sustainability and thus became repository to local wisdom. Whether rural or urban, scarce resources encouraged thrift and conservation. Thus while maximizing personal comfort may not have been a top priority, the stabilization and enhancement of conditions for human life were certainly shared purposes. Perhaps these goals might also be found in cities that deliberately shaped themselves into recognizably beautiful places.

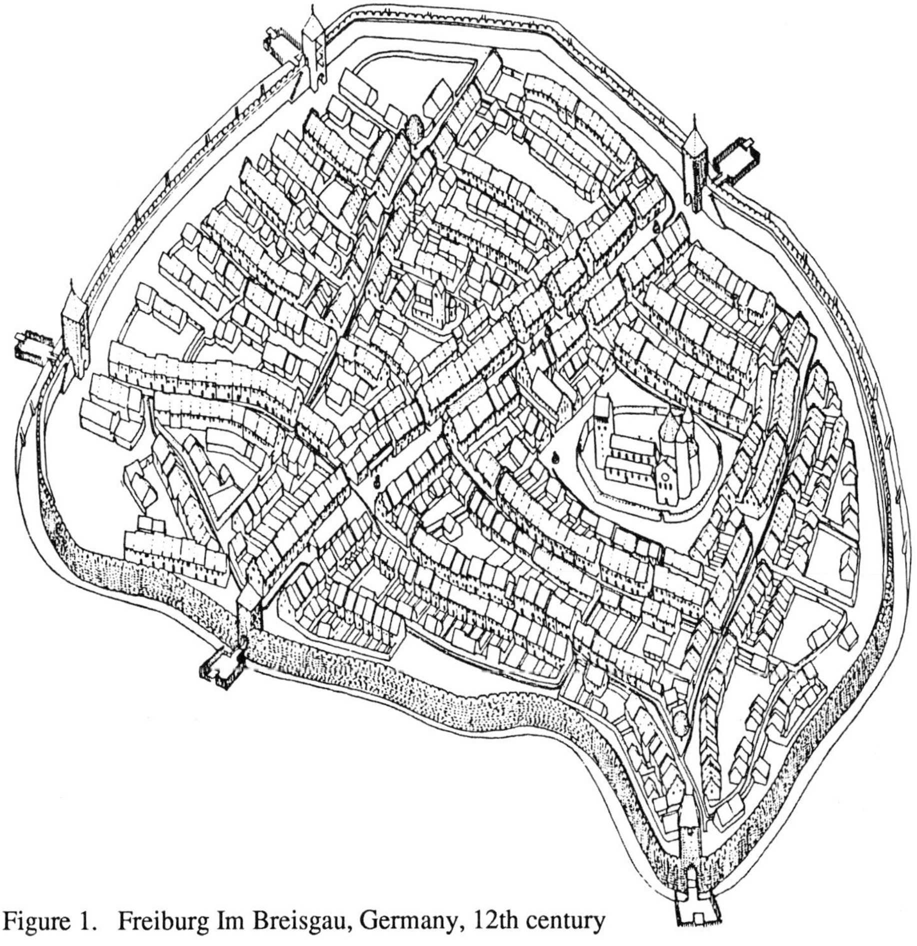

Figure 1. Freiburg Im Breisgau, Germany, 12th century

3. THE DESIGN OF CITIES

City designs usually are considered as having two origins. Either they are founded or 'new towns' with a preconceived and typically rationalized or regularized pattern, or they are organic, having evolved through the logic of their growth to express the ambitions of their citizens. Both types could be revealing of bioclimatic principles.

3.1 Freiburg Im Breisgau, Germany, 12th century 48° N lat, 8° E

In a dramatic view of Freiburg in Breisgau, Germany, (Figure 1) we can see the spatial organization of an organic medieval market town as it evolved in Germany from the twelfth century onward. It was founded in 1120 at the junction of trade routes between the North Sea and the Mediterranean, and the Rhine and the Danube. The surrounding fortified wall and gates of this new free trading city conform to local topographic conditions. The fortified church to one side represents the ultimate security whose defensive importance declined as the city walls became stronger. The market is linear in a broad street connecting the principal gates. Specializations in that market occur along its length.

Houses provide a continuous wall to the narrow streets that thus provide some climatic protection outdoors. Rear yards behind houses are generous to allow gardens and in some cases barns for animals. Orientation of buildings is always to the circulation pattern of the street, not in response to any climatic element. The street pattern is determined by trade routes. From a morphological point of view there is no evidence of bioclimatic criteria.

3.2 Lunenburg, Nova Scotia, Canada, 1753 45° N lat, 64° W

Lunenburg, Nova Scotia on the Atlantic coast of Eastern Canada (Figure 2) has a maritime climate. It is temperate, wet and changeable, with major storms coming off the ocean from the Southeast and winter storms from the Northeast.

The gridiron plan of Lunenburg is characteristic of preplanned new towns from ancient to modern times. In this case British Army Officers from nearby Halifax (1749) had surveyed the coast and chosen this site. The land is a neck between two well protected harbors. Originally the town had a wooden palisade wall as a protection to the West against Indian raids. By looking at the plan there is no evidence of bioclimatic sensitivity.

In fact, this is one of the best built examples of deliberate bioclimatic planning. The town street grid drapes over a steep hill that faces north/south. Only those parts of the town facing south and at the top of the hill were developed. The north facing side is so exposed that extensions to the Back Harbor have only been partially built in the late twentieth century.

The town faces South, addressing its Front Harbor and its prosperity. The South harbor which is also larger and more accessible became the port for the development of a major North Atlantic fishery. The central three blocks of the plan were reserved for civic buildings and an open green. In addition, the homes of leading citizens as represented by significant historic houses also are found along the middle of the slope facing south.

The bioclimatic importance of the plan is reinforced by the planning of the houses and their individual designs that step up the hill. House lots were small and the houses were built out to the street line. This provided wind protection along the street and a large enough

Figure 2. Lunenburg, Nova Scotia, Canada, 1753

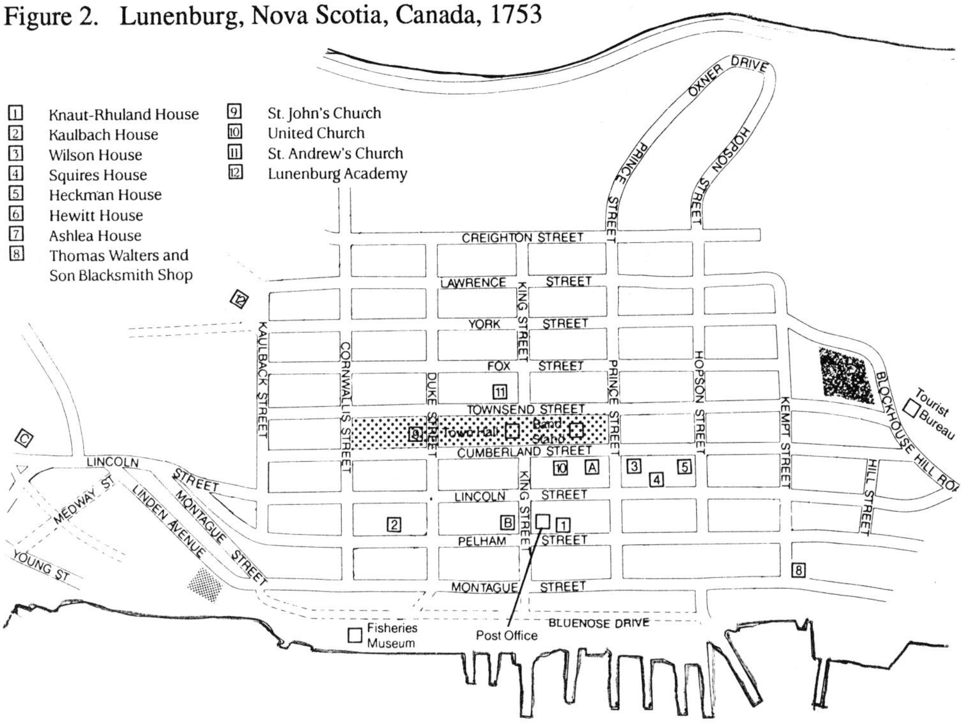

space in the rear yard for an extensive garden. Typically there were cellar kitchens entered at ground level on the lower side of the site, with storage behind. The main living or formal parlor was upstairs with direct access from natural grade on the upper site of the lot. The finest room was always on this main level and on the southwest corner, certainly the room with the finest exposure and the best view. (In this climate mornings may be foggy and afternoons clear.)

At Lunenburg preconceived knowledge about response to climate was reinforced by the three dimensional constructional development in a variety of house styles as the town evolved. Superficially the bioclimatic aspects of the town are hidden or unexplained and the town is known for its picturesque qualities and for its historic importance in the fishery. (Lunenburg was on the professional tour of PLEA 90 Halifax).

3.3 Rangoon, Burma, 1755 — 16° N. lat, 96° E

As part of the consolidation of Upper and Lower Burma with the help of the British, Rangoon (Figure 3) was founded as the new capital by King Alaungpaya on the site of the harbor town of Dagon. The strict gridiron plan of the 18th century along the North shore of an arm of the Irrawaddy delta is typically of preplanned cities. The broad streets and the small blocks may demonstrate a bioclimatic sensitivity to the humid warm tropics. Such a plan allows maximum exposure to breezes for ventilative cooling, and the plan which extends along the delta shore provides a maximum exposure. However, broad streets also can encourage public life and ceremonial activities: military parades or religious events.

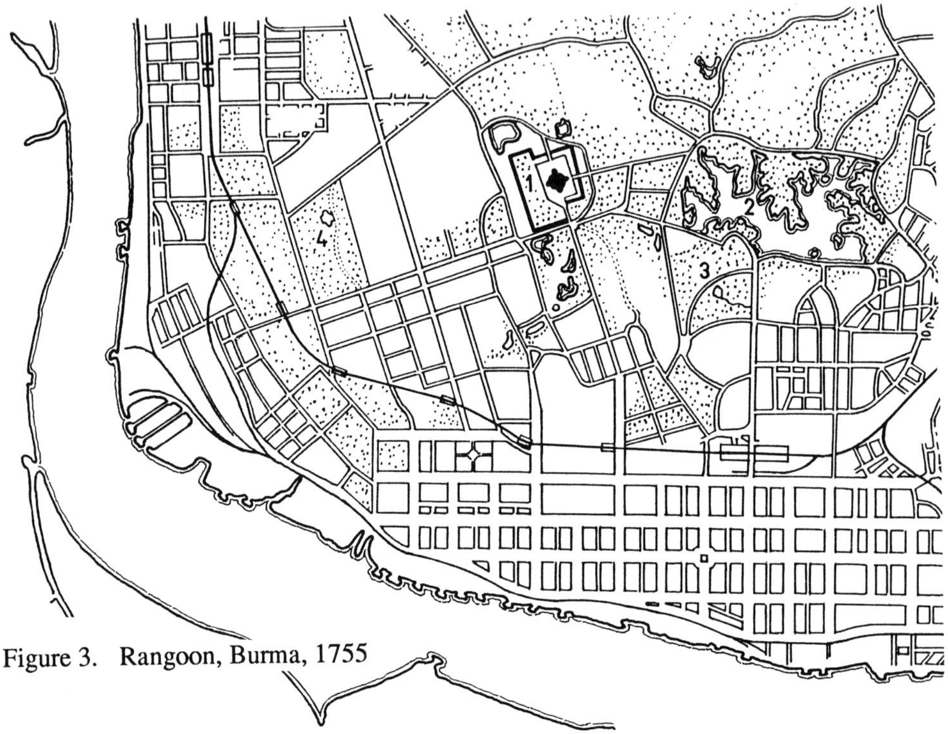

Figure 3. Rangoon, Burma, 1755

Behind the city rests its major monument. The great golden bell pagoda Shwe Dagon built on a natural hill some 170 feet high, received its present stupa form in 1768. It rises 350 feet above a stepped platform which is 680 x 900 feet and contains many smaller shrines and prayer halls. The general openness of the platform would allow the slightest breeze. But the absence of any shade, of any vegetation and the absence of fountains or ponds provides environmental conditions that exaggerate thermal discomfort.

The Royal Lake (3), the Victoria Memorial Park and Zoo (4), and the Government Building (5) are sited and planned in a picturesque way that give no indication of bioclimatic consideration. Thus while the main plan may be exemplary planning in tune with climate it is uncertain whether that was the prime determinate. Other elements of the capital display other criteria than bioclimatic.

3.4 Marakesh, Morocco — 30° N lat, 10° N

One of the most popular images of a bioclimatically successful city is Marakesh, Morocco. (Not mentioned is how *tabiya* or rammed red earth and stone walls restrict the height to two stories and provides much of the consistency of the city.) This often copied illustration at its original size (Figure 4) from page 204, Manual of Tropical Housing and Building by Koenigsberger, et al is accompanied by a technical discussion about the narrow self-shading streets and the cool air trapping of courtyards. Without challenging the correctness of this comfort analysis one must also acknowledge how thoroughly this consistent complex urban design form also satisfies social and religious criteria. It maximizes

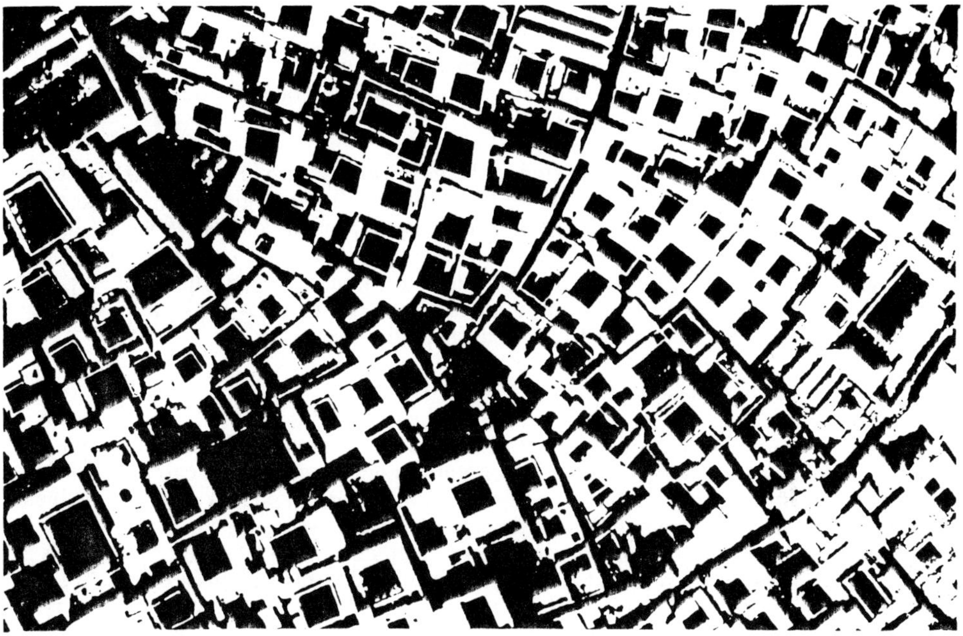

Figure 4. Marakesh, Morocco

domestic privacy and restricted public access in the neighborhoods. Its pedestrian dimensions also eliminate it from automobile intrustions.

Indeed the medieval and pedestrian qualities of Marakesh are remarkably similar to the same qualities in medieval Northern Europe. The principal differences are the form of the dwelling, and the circulating pattern of the streets. But the density and tightness with their pedestrian efficiency are surprisingly alike. Thus it may be argued that other criteria may have been the prime generators of the city form at Marakesh, and that its incidental bioclimatic suitability only secondarily reinforced or insured that its form has continued.

Bioclimatic researchers only find what they are looking for. Sometimes they find the finest examples, but perhaps solutions that were initiated by other goals.

4. THE DESIGN OF THE MICROCLIMATE OF CITIES

We are all aware of distinct changes in local climate over relatively short distances that occur in nature. Such changes in microclimate also happen in the built environment. Indeed the built environment can exaggerate natural microclimatic variations or can destroy the subtle changes that make each place in the natural world special and unique.

4.1 Natural Microclimatic Variation

All of us have had personal experiences that underline the importance of natural microclimatic variation. In my case the summer radio announcements of high temperatures at the Phoenix airport come with the satisfaction that at my house in the desert, 7 miles away, the temperatures in my car port are at least 10° F (6° C) lower. Yet on winter nights when there is a threat of a night freeze, my location is at least 5° F (3° C) warmer. Indeed weather patterns at all airports are exaggerated versions of local weather. The absence of trees and the paved runways are built interventions that demonstrate how climate can be adversely modified.

While working on a design team proposing the development of the southwest quadrant in the expansion of a British New Town, Basildon, a climatologist was included as a consultant. He proposed that housing built near the top of the meadows would have an average annual temperature 3-5° F (2-3° C) lower than housing built near the bottom. Essentially this natural variation could represent a substantial annual fuel expenditure for heating. Although only one criterium, this case suggests how early planning decisions, even before the building is conceived begin to shape the energy budget for its life.

4.2 Design Microclimatic Variation

The deliberate use of bodies of water, or of large areas of vegetation are among the most effective human interventions to provide climatic modification. Typically their use can reduce maximum conditions as well as raise minimum conditions. In contrast, the heavy use of black asphalt can exaggerate both the maximum and minimum conditions. Thus one of the most interesting bioclimatic components of a city can be the design of the streets and avenues. Will avenues of trees and attendant vegetation dominate the paved areas? Or will all the humanly hardened surfaces lack vegetation relief and suffer from heightened temperatures by day and have lower night time temperatures together with maximum exposure?

A particular case is Phoenix, Arizona where on a summer day it is easy to measure a 15° F (9° C) difference between a paved parking lot adjacent to an irrigated orange grove. Desert climates provide the maximum opportunity for bioclimatic design on an urban scale.

Thus designed microclimatic modification should be a deliberate goal of any city planner or urban designer. Perhaps the best current example of urban bioclimatic design is the Seville Expo where public spaces were conceived from the beginning in terms of minimizing summer thermal stress outdoors.

5. THE DESIGN OF THE COMPONENTS OF THE CITY

Although cities themselves are really more than huge buildings the same cannot be said for the components of a city. Thus streets and avenues, covered markets, covered streets, and squares and plazas while large in scale are potentially simple in both their diagrammatic configuration and in their vioclimatic performance.

5.1 Plaza Major, Madrid, Spain, 1617-1619, 1672, 1791-1853 — 40° N lat., 4° W

Among ceremonial urban squares and plazas of Europe, the Plaza Major (Figure 5) is among the most impressive for its clarity and directness. The plaza was cut from the surrounding tangle of streets and buildings with mathematical precision. Its bold simplicity is emphasized by the apparent casualness and intensity of the urban fabric surrounding it which seem the products of chance and not art. Plaza Major was intended to be the spacious public setting of an imperial court, and that is where major historic events as well as the pageantry and spectacles of the capital have unfolded to up to 50,000 people.

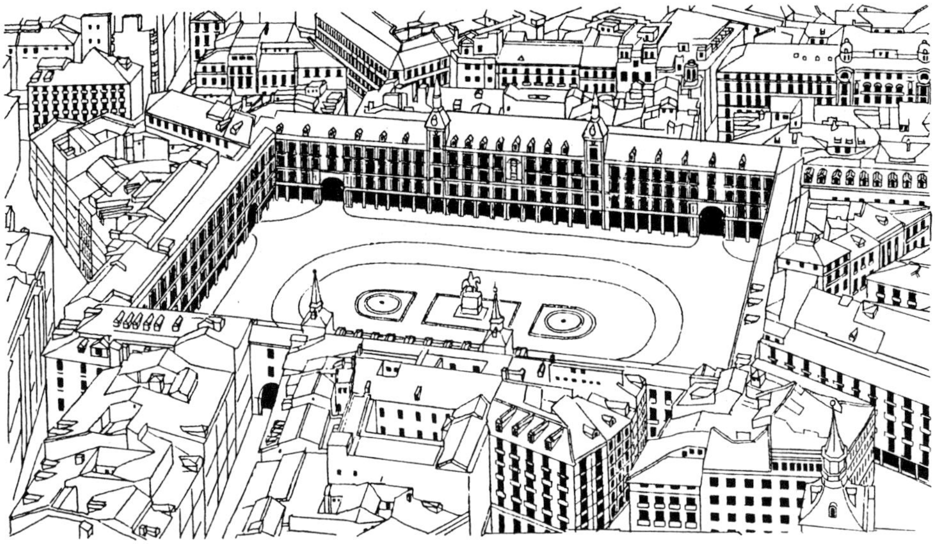

Figure 5. Plaza Major, Madrid, Spain,

Built as a rectangle approximately 300 x 600 feet (90 x 180 m) oriented to the cardinal points the Plaza Major was not intended as a bioclimatic design. But as with all constructions in nature it has bioclimatic ramifications. As the sun moves, as the wind blows, and the temperatures change the human environmental conditions are in constant flux, sometimes favoring one corner in the sun, sometimes another side in the shade, other times the openness of the center. But the opportunity for people to find their most optimum environment within the plaza is in conflict with the purpose and tradition of a space as a ceremonial outdoor room. However, much could be learned by monitoring natural conditions when people are free to find their own most comfortable location.

5.2 Piazza and Piazzetta di San Marco, Venice, Italy 45° N lat., 12° E

Books have been written on the virtues of St. Marks Square in Venice (Figure 6). Napoleon called it "the greatest Salon of Europe, the living room of the world. Most authors begin with the two distinguished architectural jewels that decorate the square: The exotic Basilica of St. Marks with its Byzantine splendor, and the freestanding Companile, the severe 325 foot (1000 m) high brick bell tower that serves as a spatial pivot between the Piazza and the Piazzetta. Many discuss the symbolic union of land and sea, of church and state, of merchant and noble, and the meeting of East and West. Others concentrate on the quality of the light and on the dynamics of the space, especially in the trapezoid Piazza that is 576 feet (175 m) long and has a width that varies from 180 feet to 270 feet (55 x 82 m).

In all its hypnotic effect on individuals and its magnetic social success, no one asks whether they are comfortable. Researchers have yet to examine St. Marks from a bioclimatic point of view. Here is one of the most admired urban spaces in the world in which one of its potential major qualities has not been explored. Yet if a simple rectangular plaza such as Madrid's Plaza Major can provide broad variations of human comfort, what is to be said about the environmental richness of varied experience in such a large and complex designed outdoor space?

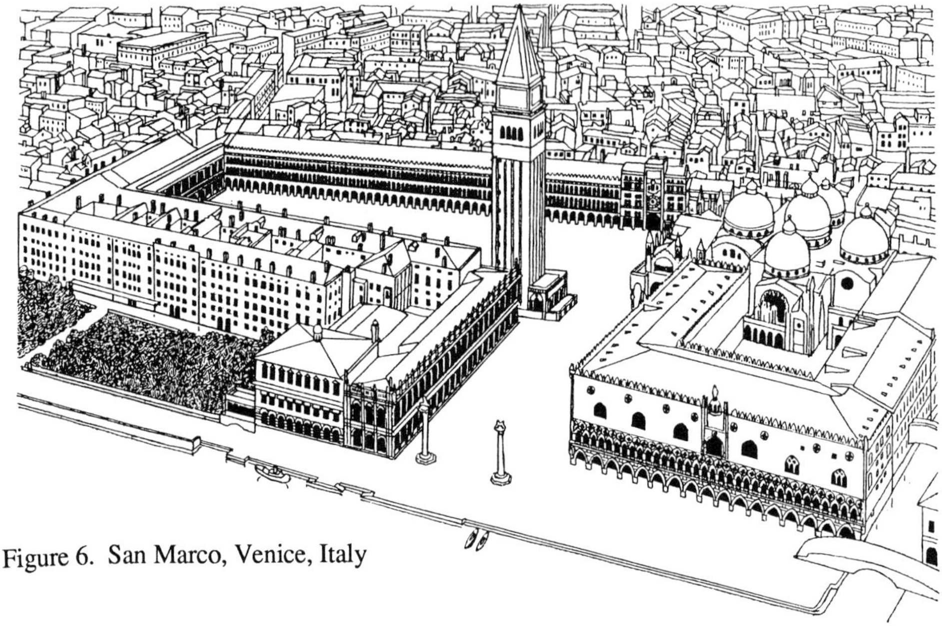

Figure 6. San Marco, Venice, Italy

6. SUMMARY

The purpose of this review has been both exploratory and provocative. The intension is to challenge a change, not just in the scale of thinking, but the scope of examination when looking at cities bioclimatically. Both need a larger dimension.

In general it can be concluded that the concept of the bioclimatic city is an underdeveloped research idea, that has had little attention. The analogy of expanding the principles of the house to the scale of the city is inadequate, not just for technical reasons but also because the criteria for cities are different than for houses. The primary goals for cities have always been social, cultural or materialistic, but not primarily ones of shelter or comfort which are the first needs in any dwelling. In addition, living with nature as a prime value system for urban conception is a prerequisite, as demonstrated in my paper Prehistoric Pueblos and Plazas of the Anasazi, also in these proceedings. But this standard of cosmic relationship has seldom characterized the urbanity of city dwellers.

Nevertheless, there are many public functions of the city that lend themselves to bioclimatic design. Historically, the public square, the open shopping street and more recently the covered street are critical components of urban life that have demonstrated beneficial and climate modification. Many of these urban components, although large in size, are not necessarily complex in the form of enclosure and thus may easily lend themselves to bioclimatic design. Such generalized recent prototypes as shopping malls and the landscaped pedestrian spaces of the Seville Expo demonstrate both the technique of bioclimatic modification, as well as the richness of aesthetic opportunities in such designs.

Perhaps the next generation of passive researchers will exercise as much thoroughness in examining urban prototypes and public outdoor spaces, as the past generation has devoted to the closed rooms of private one family houses. The world would be the richer.

Part 2

URBAN SPACE

URBAN DESIGN FOR HOT HUMID AND HOT DRY REGIONS

BARUCH GIVONI

Graduate School of Architecture and Urban Planning
University of California, Los Angeles, USA

ABSTRACT. The paper discusses the design objectives and the urban design elements which affect and can modify the urban micro-climate in hot humid and in hot dry regions. The design elements discussed are: Location of towns in a region; Density of the built-up area; Orientation and width of streets; Building design details affecting the comfort of people outdoors, and the Design details of public and private "green" areas. Urban design details are suggested for hot humid and for hot dry regions. The paper is based on a recent book of the author which was published by the World Meteorological Organization [1].

1. INTRODUCTION

Many features of the physical structure of the city can affect the urban climate. As the structure of a city can be controlled by urban planning and design it is possible to **modify** the urban climate through design. With such modifications it is possible to improve the comfort of the inhabitants outdoors and indoors, and to reduce the energy demand of the buildings for heating in winter and for cooling in summer.

The general effects of the following physical features of the urban area on its climate will be discussed in this paper:

- Location of a town within a region
- Density of the urban area
- Orientation and width of the streets
- Heights, and relative heights, of buildings
- Special design details of the buildings which affect the **outdoor** conditions.
- Urban Parks and private green areas

A special discussion is devoted to the climatic effect of internal patios, or courtyards, as this feature is common in many hot regions and, in the opinion of the author, its climatic impact is often misunderstood.

S. Alvarez et al. (eds.), Architecture and Urban Space, 19–31.

Climatic Impact of Courtyards and Attached Open Spaces.

Internal courtyards and attached open spaces, enclosed by walls, are very common in many hot regions, mainly in developing countries, and can be found both in single family houses and in townhouses. Houses with internal courtyards are considered by many persons as the most "appropriate" in hot regions, especially in deserts. This notion is based on the fact that such buildings are very common in such regions and were so since centuries, and millennia.

However, when analyzed from the thermal comfort and the energy consumption aspects it is found that their performance depends greatly on the detailed "treatment" of the courtyard. With some design details courtyards can provide pleasant outdoor environment and also improve the **indoor** comfort. With other details courtyards may elevate the indoor temperature and cause poor ventilation in the rooms located on the leeward side.

It should be realized that courtyard houses were developed over centuries in response to different needs and limitations, such as availability of certain materials and building technology, security, privacy, etc. Climate was only one of these factors and not necessarily the most important one. Present social needs often are different from those which existed when the various features of the buildings were evolved. Contemporary materials and technologies, as well as available scientific knowledge, enable today solutions not available before. The feature of the internal courtyard should therefore be analyzed in view of present needs and technology.

Many household activities can often take place outdoors in hot climate more comfortably than indoors. Such activities may include washing, cooking, playing and even sleeping. Therefore, such outdoor spaces could form an integral part of the house.

In hot dry regions care of open land is expensive because plants need water, which is scarce and expensive. Private care of open land is more likely in protected areas, demarcated by walls and fences. In an **enclosed** open space it is also possible to modify to some extent the climate outside the building, e.g. in a fenced yard or an internal courtyard patio.

Attached Open Spaces

Attached open spaces can be surrounded by high walls, completely isolated from the "public" outdoors. However, from the thermal aspects, they do not penetrate and indent the envelope of the building and thus do not increase the surface area of its envelope. When adequately treated, such attached open spaces (yards, balconies, etc.) can modify the climatic conditions of the ambient environment, next to the "skin" of the building. Such modifications may include:

- Shading the building walls adjacent to the open spaces, either by an overhand or by vegetation.
- Insulation the walls, e.g. by dense and high shrubs, creating semi-dead air space next to the walls.
- Increasing the humidity and lowering the temperature in the

semi-enclosed air space within the "walled" open space. Such modification of the climate can be achieved only if the open space is separated from the "general" environment by high walls and is shaded by a roof, or tree canopies, above it.

Internal courtyards

Internal courtyards **maximize** the thermal interaction between the building and the outdoor environment, introducing the outdoors into the heart of the building's core. It is commonly assumed that such internal patios help in maintaining cooled indoor temperatures. However, the actual climatic effects of an internal patio depends greatly on its design details and its "treatment".

An "untreated" internal patio with bare soil or with hard floor (concrete tiles, etc.) increase the rate of the indoor temperature rise during the daytime hours. This is due to the effective increase in the surface area of the building's envelope.

Also the ventilation of the rooms on the leeward side of the courtyard are poorer than in a building where inlet and outlet opening direct the air flow across the whole building. This is caused by the fact that a low pressure zone exist in the courtyard, as exist on the leeward to the building as a whole. The rooms behind the courtyard are thus surrounded by low pressure on all sides and therefore can not have effective cross ventilation.

Consequently an unshaded internal patio, without vegetation or other sources of shading and evaporative cooling is likely to cause higher indoor discomfort than in a building without a patio with the same floor area. The patio itself can provide, of course, an area with a more pleasant climate than that of the indoor's during the evening and night hours, in addition to the complete privacy in the open space.

However, with different treatments, a patio can have much more favorable impact on the indoor climate, as well as having more pleasant "outdoor" climate than the exposed environment in a hot-dry climate.

HOT-HUMID REGIONS

The major climatic characteristic of hot-humid regions, from the human comfort and urban and building design viewpoints, is the combination of high temperature and high humidity, cloudiness, and high rainfall. Most of the countries in the hot-humid areas are developing countries. The vast majority of people cannot afford air-conditioning. Therefore thermal stress (and its impact on health and productivity) should be minimized primarily by appropriate urban and building design details which do not involve high cost.

The main urban design objectives in hot-humid regions are:

- Minimizing thermal discomfort from the combination of high temperature and humidity.
- Minimizing the hazards of floods.

- Providing rain protection and shade for pedestrians.
- Enabling good natural ventilation of the urban space and for the buildings.
- Encouraging, by urban planning regulations, building types which maximize natural ventilation.

Locational Considerations in Hot-Humid Regions

A major factor determining the level of thermal stress in a hot-humid region is the street level air speed. Windward slopes are preferable to leeward slopes from the ventilation viewpoint. Valley slopes may experience downslope air currents during windless nights, thus providing more comfortable conditions as compared with the valley floor area (Lyons 1984, Givoni 1989). Seashore and large lakeshore areas may benefit from the daytime sea and nighttime land breezes.

Flat areas along rivers or near outlets of rivers may be subjected to severe floods when heavy rains, often in far away catchment areas, raise the water level above the "normal" flood level. Many towns in hot-humid coastal regions are in areas prone to hurricane (or cyclone as they are called in Southeast Asia) hazards. The storms are accompanied by very high rainfall. Often such areas are attractive for naturally irrigated agriculture because of the fertility of the land, or for providing opportunities for fishing, coastal trade, etc. This economic attractiveness should be weighted against the risks to life and property losses during cyclonic floods.

Minimizing Flood Hazards by Urban Design Features

During rains the city itself increases the excess of water run-off. The ability of the ground in an urban area to absorb water is reduced, as a result of the coverage of the land by buildings, roads, parking areas, etc. This factor increases the risks from floods in low lying flat urban areas. The risk of floods from excess rain water can be minimized by details of urban design. The following design details can be applied to achieve this goal:

- Increasing rain absorption in soil in the urban area, by the type of covering, thus reducing run-off.
- Preserving land features of natural drainage such as interconnected valley systems.

The network of valleys which naturally drain off the rain water from the town's area can serve as urban parks and other open spaces. With proper planting they can be re-used for such functions after the rain water subsides. Thus they can serve socio-functional needs of the people together with this climate related function.

Layout (Orientation) of the Street's Network

In hot-humid regions the main climatic objectives concerning the street layout would be to provide good ventilation conditions for pedestrians in the streets and good potential for ventilation of the buildings along the streets. The "ideal" design solutions for these two objectives are not the same and a compromise may be called for. Another objective is to provide shade over sidewalks in streets with large concentrations of pedestrians. Such shade can be provided by trees along sidewalks, as well as by special details of the buildings.

The best ventilation **within the streets** and the sidewalks is achieved when the streets are parallel to the direction of the prevailing winds during the afternoon hours (when the urban temperature reaches its maximum). However, when streets are parallel to the wind direction and the buildings along them are very close together and face the street, the **ventilation potential of the buildings** is compromised. The reason is that with this orientation, all the walls of the building are in "suction" zones. Effective indoor cross-ventilation can occur in a building only when at least one of its walls (and windows) are in a "pressure" zone.

Streets perpendicular to the wind direction, with closely spaced long buildings along them, as is common in many town in Developing Countries, may block the wind in the whole urban area and do not contribute at all to the urban ventilation. A good street layout from the urban ventilation aspect in a hot-humid region is when wide main avenues are oriented at an oblique angle to the prevailing winds (e.g. between 30 and 60 degrees). This orientation enables penetration of the wind into the heart of the town. The buildings along such avenues are exposed to different air pressures on their front and back facades. The upwind wall is at a pressure zone while the downwind wall is at a suction zone, thus providing the potential for natural ventilation.

Urban Density and Building Heights in Hot-Humid Regions

Urban density is among the major factors which determine the urban ventilation conditions, as well as the urban temperature. Generally speaking, the higher the density of the buildings in a given area the poorer will be its ventilation conditions. However, for a given density level there can be different urban ventilation conditions, depending on the particular configuration of buildings by which this density is obtained.

An urban configuration to be avoided as much as possible in hot humid regions is that of high long buildings, of about the same height, perpendicular to the prevailing wind direction. This configuration blocks the wind and creates poor ventilation conditions both in the streets and for the buildings. An urban profile of variable building height, where buildings of different heights are placed next to each other, and when the long facades of the buildings are oblique to the wind, enhances urban ventilation.

Characteristics of Different Building Types in Hot-Humid Regions

As most countries in the hot-humid areas are developing countries, the special social problems and needs of the population should be a critical factor in choosing a given building type. However, as this paper is limited to dealing with the climatic aspects, the social impacts of the different building types will not be discussed here, except for some comments when seemed appropriate.

Detached Single Family Houses

By definition, detached houses are exposed to the outdoor air on all sides. This feature provides good potential for natural ventilation. For a given thermal resistance of the envelope the expected indoor comfort during the daytime hours would not be worse than in more compact buildings. Furthermore, during the evening and night hours, when the outdoor wind usually weakens or subsides, a detached house will cool down faster than other types of buildings, providing better comfort. This building type is the least sensitive to the orientation of the building from the ventilation aspect.

In summary, a detached house is the most suitable in a hot-humid climate. This feature should be taken into account by urban planners, especially since this type is also more suitable socially and functionally to the needs of large low income families.

Townhouses (Row Buildings)

Townhouses are comprised of a set of several single family units, attached to each other on their sidewalls, thus forming a "row" of dwelling units. The units can range in height from a single story up to three stories. In hot-humid regions townhouses provide the next best building type for mass housing, from the climatic aspect, after detached houses. However, townhouses are much more sensitive than single family houses, from the ventilation aspect, to their orientation with respect to the wind direction.

In **multi-storied** townhouses, internal staircases can form natural shafts for vertical air flow. When these shafts are large enough, they transform the whole space of the house into an integrated space from the air flow aspect.

Multi-Story Apartment Buildings

Multi-storied apartment buildings may present difficulties in providing cross ventilation, unless they are planned with staircases (or elevators in buildings with too many stories for walkup) serving only two units per floor. Such buildings are also much more sensitive than single family houses, from the ventilation aspect, to their orientation with respect to the wind direction. The worst conditions would be in buildings with internal, double-loaded corridors. Therefore such apartment buildings should be avoided in hot-humid regions, in spite of being the least expensive, especially for buildings with elevators.

When the corridor is external and open (like a balcony), it is theoretically possible to have cross ventilation in the apartments along it. In order to have such ventilation, doors and windows in the walls along the corridor should be left open. This, of course, severely interferes with the visual and acoustical privacy of the occupants

When a staircase serves only two units per floor, each

apartment is exposed to at least two external walls. With adequate internal organization of the indoor space it is then possible to insure good ventilation conditions to all the dwelling units. To achieve this goal, the building should be so oriented that the wind will be either perpendicular or oblique with an angle of not more than 60 degrees from the perpendicular position.

Highrise "Tower" Buildings

This type of building, which requires sophisticated structure, elevators and other mechanical systems, is applicable only to relatively high income people and therefore can not solve the mass housing problems of the developing countries. However, it has some interesting features from the climatic aspect in hot-humid regions and therefore deserves attention.

When high narrow buildings are placed relatively far apart, they do not reduce the air speed near the ground level. In fact such buildings can increase appreciably the low level air speed, thus improving the ventilation potential for lower buildings between them and in the streets. The occupants of the high stories enjoy lower temperature, as well as lower humidity (as the vapor is generated at the ground level). This is in addition to the better ventilation potential and the view offered from the high stories.

Providing high standard, although expensive, housing in a city can also increase its attractiveness for professionals and high income population groups, an attraction which is an asset for towns of developing countries.

Design Details of Open "Green" Spaces in Hot-Humid Regions

The climatic objectives in planning public urban open "green" areas (parks, playing grounds, etc.) in hot-humid regions should include, inter alia:

- Providing shade for the users of these areas.
- Minimizing blockage of the wind within the open spaces and in the built-up areas around the open spaces.
- Minimizing damage of floods in the more sensitive urban areas.

Because of the high precipitation in hot-humid regions, local plants do not need irrigation most of the year. Therefore, it is possible to maintain "green" areas with less maintenance work and expenses than it would be in arid regions.

The impact of plants on human comfort in hot-humid areas can be a mixed bag. The shading provided by trees is always welcomed. However, the blockage of the wind and the contribution to the humidity level by evaporation from the leaves increases human discomfort.

Trees with a high trunk and wide canopy are the most effective plants in providing usable shade. If densely placed, they of course block the wind. Therefore, the best strategy with such trees is to have them only at the spots where their shade will be utilized, such as near benches, children play areas, along pathways, etc.

High shrubs block the wind and "contribute" to the humidity level without provision of shade. Therefore, their introduction should be minimal in parks in hot-humid regions, although they have aesthetic and functional uses as area dividers, etc. A combination of grasses, low flower beds and shade trees with high trunks is thus the most appropriate plants combination in parks in this climate.

Low lying areas prone to floods can be grassed and planted with trees which can withstand flooding. If the vegetation can withstand a given height of water for a few days, then such areas can be utilized as flood controls during and immediately after rain storms. During the periods between the storms, such areas can again be useful public green spaces for recreation, grazing, etc.

HOT DRY REGIONS

Hot dry regions are found in the sub-tropical latitudes, approximately between 15 and 35 degrees north and south of the Equator, in central and western Asia, the Middle East, Africa, North and South America, and in central and north Australia. These regions are characterized mainly by their aridity, high summer daytime temperatures, large diurnal temperature range and high solar radiation. In some arid regions typical daytime summer temperatures are in the 32-36°C range while in hotter regions they may experience temperatures well above 40 and up to 45° C and even higher.

Urban design should aim in hot-dry regions, from the climatic point of view, to achieve the following objectives:

- Choosing locations with the most favorable climate, mainly lower summer temperatures.
- Minimizing urban temperature elevation above the surrounding level (the "heat island").
- Enabling good ventilation potential for the buildings, especially in the evenings, and in the open spaces.
- Providing shade in summer in the streets, over sidewalks and in open spaces.
- Providing "solar access" in winter for the buildings and in open spaces, especially in regions with cold winters.
- Minimizing the dust level.

Town Location within a Region

In choosing a location for a town or a neighborhood in a hot dry region one should look for places with lower summer temperatures and good ventilation conditions, mainly in the evenings and at night. Areas with lower "potential" for dust are, of course, preferable to dusty areas.

A common opinion is that wind is not important, or even undesirable, in hot dry regions because the outdoor air temperature is above skin temperature and the wind increases the convective heat gain of the body. However, it is the personal experience of the author that, even with air temperature of about 40°C, a light wind actually reduces discomfort by reducing the wetness of the skin, especially when one stays outdoors and is exposed to the sun.

Urban Density Considerations in Hot-Dry Regions

The density of the built up area (land coverage by buildings) in hot-dry climate may have both positive and negative effects on human comfort outdoors and on the indoor climate of the buildings, depending on the details of the urban physical configuration. With proper design of the individual dwelling units a building can be cross-ventilated even when openings are provided only in the northern and southern walls. In this case, the distance between the buildings along the east/west axis can be can be minimized, thus minimizing also solar radiation in summer on the sensitive eastern and western walls and windows (when they are provided, or even eliminated altogether. The effect of distance between buildings along the north/south axis, however, may reduce the possibilities of utilizing solar energy for heating.

Untreated urban land in a hot-dry region is often a source of dust. The lack of rain in hot-dry regions and the high cost of "imported" water limit the capability of cities to landscape open land. In developing countries most of the population cannot afford the expenses of planting and maintaining private open areas between buildings. This consideration suggest higher urban densities than in other types of climate, e.g. a hot-humid ones.

Density and Urban Air Temperature in Hot-Dry Regions

The temperature modifications by a city are expressed mainly in the "heat island" phenomenon, especially during calm and clear nights, when the urban air temperature is usually higher than in the surrounding open country. However, it is possible to infer from theoretical considerations, as well as from actual measurements in and around buildings, that in hot-dry regions it might be possible to plan cities so that the ambient daytime air temperatures would be lower than in the surrounding country. High density of buildings and white roofs can ensure that a large fraction of the solar radiation will be reflected upwards. The longwave radiant loss from the large roof's area can then exceeds the absorbed solar radiation even on a clear day in mid-summer. Under these conditions, the average temperature of the roof surfaces will be lower than the average regional air temperature. If the city is large enough, and built densely enough, it can be assumed that it will be possible to achieve a daytime air temperature at the street level that is lower than in the surrounding areas.

Urban Density and Ventilation Potential in Hot Dry Regions.

Hot dry regions commonly have strong winds during daytime but the winds usually subside in the evenings. Furthermore, during the daytime hours building ventilation is not desirable, while in the evening it is essential for indoor comfort. Therefore the main concern in hot dry regions is how to provide the potential for evening and night ventilation for the buildings.

In high density neighborhoods, where streets are narrow, distances between buildings small and buildings are of about the same height, there is a sharp drop in the wind speed below the roofs' level. Below the roofs the wind is very light while above the roofs it is much stronger. Under these conditions it is difficult to provide good indoor ventilation, especially when the ambient wind speed is very low. This point is of particular importance for the rooms where the inhabitants sleep.

When the same family occupies the whole vertical section of the building, it is possible to make use of "wind catchers" to catch the wind blowing above the roofs level and to direct it downward. Such "devices" can be an integral part of the functional design of the building, such as stairwells leading to the (flat) roof, or to a second story extending only over part of the roof. The open part of the roof can be used as a private open space.

In neighborhoods with high rise apartment buildings, higher overall urban densities can be maintained with adequate potential for natural ventilation. Reasonable distances between the building blocks should be insured, and long buildings should not form "walls" perpendicular to the wind direction. The wind can then "negotiate" between the buildings so that the **potential** for natural ventilation exists. The extent to which an individual apartment will actually have adequate ventilation would depend on its design details.

Orientation and Width of Streets

In a hot dry climate the main objectives related to the street's layout are to provide maximum shade in summer for pedestrians and minimum solar exposure of the buildings along the streets. Narrow streets provide better shading by buildings for pedestrians on sidewalks than wide streets.

A north-south orientation of a street in high density city usually results in an east-west orientation of buildings along and parallel to the street, which will cause unfavorable solar exposure for these buildings. From the solar exposure viewpoint an east-west street orientation is preferable.

Protection against sun for pedestrians on the sidewalks can be provided by buildings with overhanging roofs, or colonnades in which the ground floor is set back from the edge of the road, with the upper stories jutting out, supported by pillars (or other means).

Reducing Dust by Town Planning

The main factors affecting the frequency, intensity and range of local dust storms are ground cover and the wind speed near ground level. Both these factors can be affected by neighborhood design features. As a planning policy, the built up area density should avoid un-built areas that the private citizen and the municipality cannot realistically keep up as landscaped areas. Public areas within the neighborhood borders should also be limited in size to areas which can be well kept. The land should be kept in its natural condition as far as possible, so that the natural ground cover of desert plants will limit the generation of dust.

"Green" Parks in Hot Dry Regions

Urban public and private green areas, to the extent that they can be kept properly (mainly irrigated), are a great asset in a hot dry region. Parks and playgrounds should provide ample shade and protection from dust in summer, as well as protection from cold winds in winter. Large lawns and flower beds without shade contribute little to the recreation possibilities of the inhabitants - ordinary citizens, elderly, and children alike - to rest, relax, or play on a hot sunny day.

It is not realistic to even try to modify the urban climate in hot-dry regions by a lot of plants. However, it is possible to modify the micro-climate next to the houses by small-scale planting around the buildings or within internal courtyards. This can provide outdoor spaces with more pleasant climate that the inhabitants can enjoy.

Building Type's Considerations in Hot-Dry Climate

Courtyards in Hot-Dry climate

Often houses with internal courtyard are considered the most "appropriate" for hot dry regions. This notion is based on the fact that such buildings are very common in desert regions and were so since many centuries. However, their performance depends greatly on the detailed "treatment" of the courtyard. With some details courtyards can provide pleasant environments while with other details they may elevate the indoor temperature and cause poor ventilation in the rooms on their leeward side.

Internal courtyards enlarge the surface area of the walls, a factor leading to a faster rate of daytime heating of the building's interior. To counteract this effect care should be taken to keep the daytime temperature of the courtyards itself as low as possible. This, of course would also enable use of the courtyard with more comfort.

To this end the courtyard should be shaded, either by architectural means or by vegetation. Early shading by vegetation can be provided by a pergola with climbing vines. Later, trees with wide canopy and high trunk, rising above the roof, can provide the shade. It is desirable that the vegetation shade will create a semi barrier between the courtyard air space and the ambient air above it. This will enable to cool the courtyard air by evaporation without too much mixing it with the warmer air above, mixing which usually is generated by the wind flowing over the roof.

Effective evaporative cooling of the courtyard air can be accomplished only when very large surface area of water has contact with flowing air. This can be provided by several means such as a fountain sprinkling very fine drops into a pond, water running down courtyard walls with large area, sprinkling the leaves canopy of trees in the courtyard, etc.

Single Family Detached Houses

Single family detached houses have the highest envelope surface area among the various building types. When they are built around an internal courtyard the envelope surface area is further enlarged. Consequently, in hot-dry regions, the rate of temperature rise during the daytime hours in single family houses is the fastest. From this viewpoint this building type may exhibit the highest indoor discomfort and cooling requirement in summer and the highest heating load in winter, as compared to building of other types, **unless properly oriented and ventilated.**

On the other hand, single family houses are less sensitive to orientation with respect to the sun and the wind direction. Therefore, when the topography or other factor do not enable optimal orientation of the buildings, both with respect to the sun and the wind direction, detached single family houses may be the most appropriate type in hot-dry regions in spite of their larger envelope surface area.

Townhouses (Row Buildings)

Each dwelling unit in a townhouse, except the end ones, has only two main external walls. Thus the exposed area of the building envelope is smaller in comparison with that of a single-family detached house. In a hot dry climate this factor reduces the rate of temperature rise during the daytime hours while, if the building is designed for effective cross ventilation, its cooling rate in the evenings may not be affected by the fact that the dwelling unit has only two external walls. Townhouses are more sensitive than detached houses to their design details and orientation, both with respect to the sun and the wind direction. In hot-dry climate a North-South orientation for the external walls will minimize the exposure of the townhouse to the sun in summer and maximize its potential for solar heating in the winter.

Multi-Story Apartment Buildings

Multi-story apartment buildings have the smallest surface area of the envelope of all building types. Therefore their rate of heat gain in summer and of heat loss in winter is also the lowest. On the other hand their ventilation and the inhabitants comfort depends on their design details and orientation.

Double-loaded corridor buildings have poor ventilation and poor solar exposure of at least one half of the apartments. Therefore they are not suitable at all in all types of hot climates.

Single-loaded corridor buildings may be appropriate in hot-dry climate if they are so oriented that the corridor is open in summer and is on the leeward side of the building. This orientation enables cross ventilation of the apartments when door and windows in the corridor wall are left open. However, the privacy of the occupants is ccmpromised when the dwelling is ventilated.

Direct access multi story apartment buildings provide direct access to several dwellings at each floor, usually from two to four units, from a staircase or an elevator. When only two apartments are accessed at each floor from the staircase all the dwelling unit can have, with appropriate orientation, cross ventilation in summer and potential for solar heating in winter. In hot-dry regions the appropriate orientation of the long facades would be to South and North, to minimize solar exposure in summer and to maximize it in winter. However, in many hot-dry regions the dominant wind direction in the evenings, when the buildings should be ventilated, is from the west. In such cases the optimal orientation would be NNW and SSE.

With three direct access dwelling units the ventilation and/or solar access conditions of at least one third of the apartments is compromised. With more than three dwelling units per staircase a significant number of apartments would have poor ventilation and solar exposure conditions. Therefore such buildings are not recommended for hot-dry regions.

References

1. Givoni, B. (1989) 'Urban Design In Different Climates'. World Meteorological Organization (WMO/TD, No. 346).

Chapter 1

Climatology and Urban Space

WIND IN URBAN SPACES
DEVELOPMENT AND APPLICATION OF AN EVALUATION METHOD

A. FERNANDEZ, S. de SCHILLER and J. M. EVANS
Research Centre Habitat and Energy,
Faculty of Architecture, Design and Urban Planning,
University of Buenos Aires.
Casilla de Correo 1765, Correo Central,
(1000) Capital Federal, Argentina.

ABSTRACT. This paper describes the application of a four stage method to evaluate wind problems arround proposed new buildings in an existing urban setting and establish specific design recommendations, based on a case study. Following a classification of the principal wind effects in urban spaces, the method used to evaluate wind problems includes the preparation and analysis of a meteorological data base; on-site measurements arround existing buildings and the study of wind effects using a wind tunnel. Finally conclusions and design recommendations are presented to improve the wind regime arround a new building planned for the Buenos Aires University Campus.

1. INTRODUCTION

Architects and urban planners cannot easily respond to problems caused by wind due to lack of understanding of the complex interaction between buildings and wind, a critical component of thermal comfort in urban spaces.

This paper describes the development and application of a method to evaluate wind problems arround buildings and establish specific design recommendations, based on a case study in Buenos Aires.

The Buenos Aires Metropolitan Region centred arround the Federal Capital, latitude 34'S, is located in the temperate humid zone on the River Plate where the presence of the 30 km wide estuary produces a characteristic coastal microclimate with mild winters and oppresive warm humid periods in summer. Thus winter wind proteccion and exposure to summer breezes are desirable bioclimatic requirements for outdoor spaces in this region.

This research and evaluation of wind problems includes a study and classification of the principal wind effects in urban spaces; the preparation and analysis of a meteorological data base; on-site measurements arround existing buildings in the Buenos Aires University Campus and the study of wind effects in a wind tunnel.

S. Alvarez et al. (eds.), Architecture and Urban Space, 35–40.

2. ANALYSIS OF WIND AND CLIMATE

Two meteorological stations are situated in the Federal Capital; one in the centre of the urban area (Central Observatory) and the other close to the River Plate (Aeroparque). The data published by the National Meteorological Service [1] confirms the presence of a special micro-climate near the river with lower average temperatures in summer, higher temperatures in winter and higher wind speeds. In general, the urban area has an even distribution of wind direction and annual average wind speeds between 9 and 13 km/hour.

A meteorological data base was prepared to analyse the wind regime in relation with other climatic variables. Using this base it was possible to relate wind direction with frequency, velocity distribution, average air temperature, etc. Graphs were prepared to summarise the following relationships for each month:

1). Variation of average wind speeds at different hours of the day.
2). Average distribution of wind frequencies for each hour.
3). Average velocities for each direction.
4). Frequency distribution of wind directions.
5). Average air temperature corresponding to each wind direction.

According to the Central Observatory data, the required design strategies for wind protection and use of cooling breezes in Buenos Aires are:

1). Wind protection: The highest wind frequencies correspond with the relatively warmer north winds and colder south winds. The latter have higher average velocities and correspond with temperatures 4'C lower than those measured during periods of north winds.

2). Catching summer breezes: The south-easterly winds are the most frequent, but are associated with occasional storms from the Atlantic. East winds are the most favourable for summer cooling as the average velocity of 11 km/hour is not excessive and the frequency is relatively high. North winds coincide with elevated average air temperatures that are 6'C higher than southerly winds.

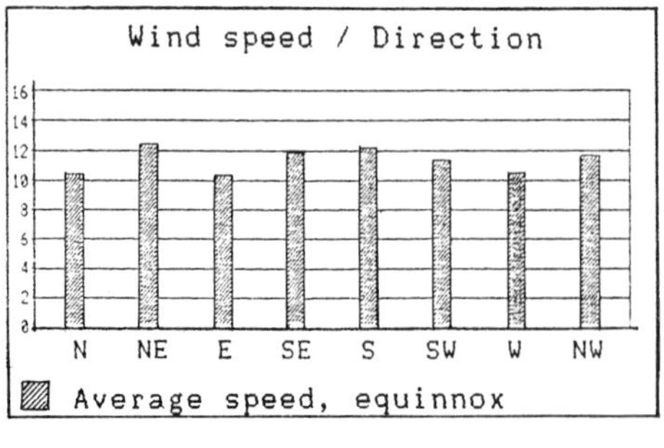

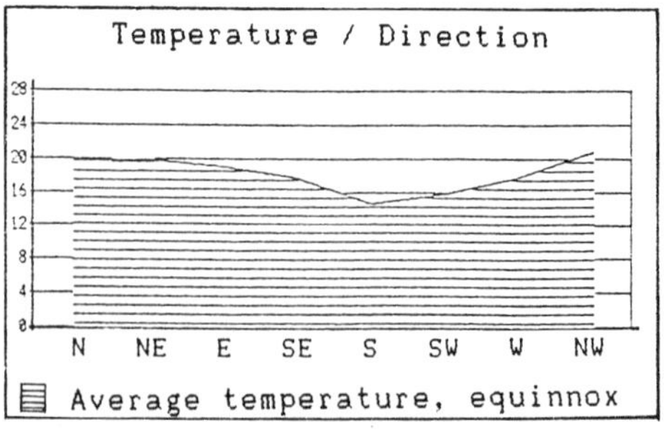

Figure 1. Relation between wind characteristics and other climatic variables, using data from the Central Observatory.

3. WIND IN URBAN SPACES

Wind patterns in urban areas depend on the morphological characteristics of the buildings and spaces between and arround them.

Buenos Aires has a generally irregular urban profile as a result of successive planning control measures, economic variations and individual development initiatives. The insertion of new buildings, which differ in form and height from the surrounding urban tissue, may transform existing comfortable outdoor spaces into unpleasant wind-swept areas.

A classification of urban spaces, using four basic elements (streets, squares, edges and intersections) was developed to analyse the interaction with wind phenomena. The variables considered include the shape and dimension of the space as well as the characteristics of the borders (form, continuity and proportions). Figure 2 summarises some of the most important wind phenomena in outdoor spaces.

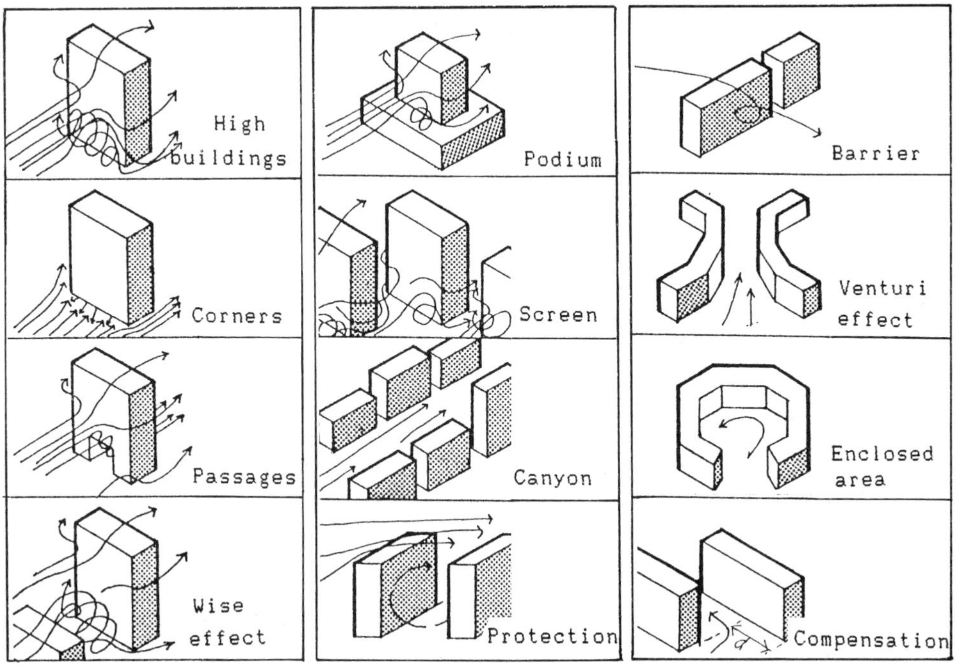

Figure 2. Wind effects in outdoor spaces.

The method used in the case study presented aims to develop guidelines and design recommendations to avoid undesirable wind effects and indicate possible architectural and urban design solutions. These are based on the need to control air movement and achieve desirable levels of direct sunlight to favour thermal comfort in urban spaces, over and above the minimum requirements of existing town planning regulations.

4. METHOD OF WIND EVALUATION

The method evaluates wind effects in outdoor urban spaces using the following four stage sequence:

1). Analysis of the general climatic and eolic characteristics of the locality, using graphs to summarise and visualise data base information: temperature, wind speed, frequency and direction. Sun path diagrammes are also used to compare orientations and determine desirable proportions of urban spaces.

2). Survey of wind conditions "in situ": a simple and economic manual instrument was developed to measure outdoor wind speeds. This has been used to study typical wind velocities and directions in a series of different urban spaces.

3). Comparative study and evaluation of wind effects and protected areas using the low speed wind tunnel, designed and constructed in the Laboratory of Bioclimatic Studies (LBS) [2]. The measured wind patterns can be compared with the simulations in the tunnel using scale models, and alternative new building forms and positions can be tested. The particular characteristics of the LBS open jet tunnel facilitate demonstrations and the filming of smoke tests [3].

4). Conclusions and design recommendations are developed to propose urban and architectural solutions that avoid undesirable wind effects, identify problem areas and indicate precautionary measures such as tree planting or protection of building entrances.

This evaluation proceedure can be applied in any locality with available meteorological data and provides a solid base for the development of design guidelines and recommendations.

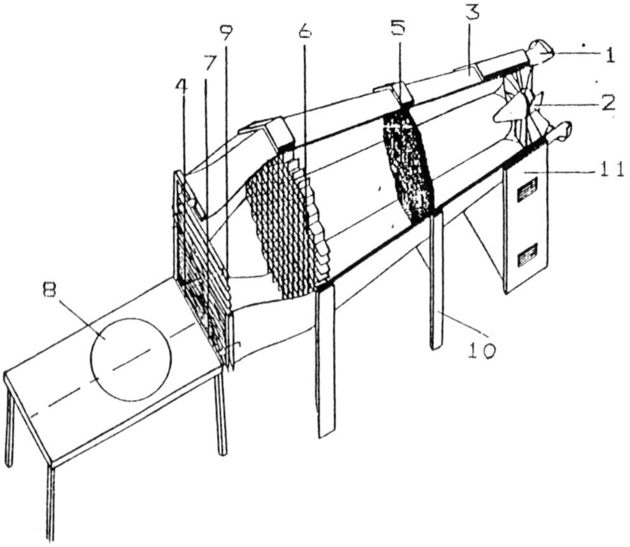

1. Fan housing 0.7 m.
2. Fan & screen.
3. Expansion cone.
4. Exit 0.8 x 0.8 m.
5. Screen.
6. Honeycomb.
7. Wind profiler.
8. Turntable
9. Working table
10. Supports.
11. 1 HP motor.

Figure 3. The open jet wind tunnel of the Laboratory of Bioclimatic Studies.

5. CASE STUDY

To test the evaluation method and verify the utility and accuracy of the new wind tunnel, the insertion of a new building block on a restricted site in the open and unprotected campus, adjacent to the existing building of the Faculty of Architecture, was studied to advise the university architects (Secretarla del Habitat) on possible wind problems and alternative protection measures. This study was undertaken at an early stage in the development of the project, allowing the findings to be incorporated in the design requirements of the future building.

The wide River Plate Estuary with low roughness creates a wind speed profile with high velocities at low level, as well as creating on-shore and off-shore breezes.

Models of the buildings, situated close to the River Plate Estuary, were tested in the wind tunnel for eight wind directions. In all of these tests, problems of wind acceleration were found in the space between the existing Faculty building and the proposed new building. These effects reduced the extent of existing protected areas in the wind shadow of the Faculty building.

The placing of a new lower building in front of the higher existing one generates higher wind speeds in a area that already suffers from wind exposure. This is particularly evident with wind from the south-west, called "Pampero" (wind from the Pampa), which has relatively low frequency but high average speeds. In this case, the slow circular wind turbulance behind the lower building combines with the circular flow windward of the higher building.

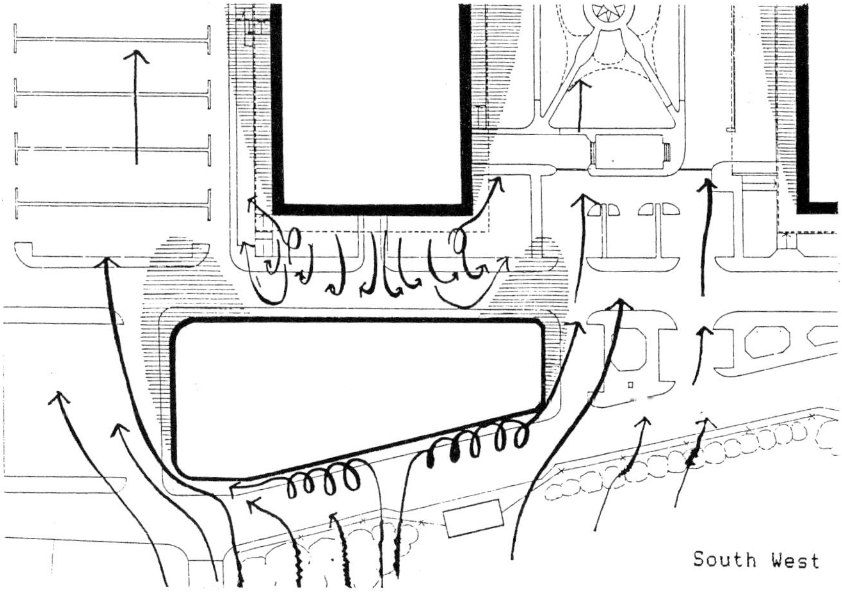

Figure 4. Wind effects in the University Campus with the new building.

As a result of the analysis of wind effects and shading of direct sunlight, design strategies were proposed to improve outdoor conditions arround the buildings studied. The main problem was to control or reduce wind speed in winter. In summer, light breezes can improve thermal comfort but in areas of pedestrian circulation additional movement is not essential; solar protection has a higher priority.

Wind-breaks formed by trees and other vegetation provide an ideal solution in this case, as the restricted site does not allow significant variations of building form or location. The systematic use of well placed barriers can reduce average wind speeds by 40%, with variations ranging between 20% and 80%. This wind protection also reduces heat losses from buildings in winter due to lower infiltration and reduced surface transmitance. Another proposal to reduce wind speeds at pedestrian level consisted of variations in roof slope to divert air movement, though tests showed that this was less effective.

Other recommendations include the protection of entrances using double doors to reduce cold draughts inside the building.

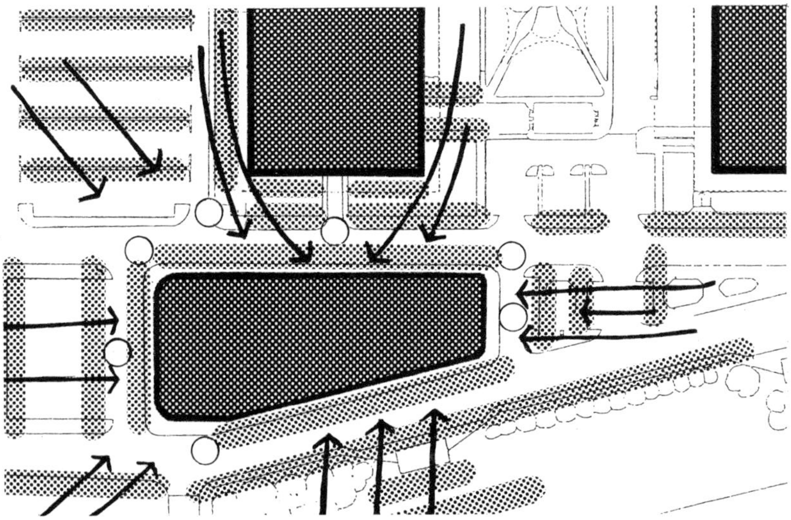

Figure 5. Proposed location of wind-breaks in the University Campus.

6. REFERENCES

[1] Servicio Meteorologico Nacional (1986) 'Estadisticas Meteorologicas 1971-1980' SMN, Buenos Aires.

[2] J. M. Evans, S. de Schiller et al. (1988) 'Equipamiento del Laboratorio de Estudios Bioambientales', Proceedings of the XII ASADES Workshop, Salta, Argentina.

[3] S. de Schiller, J. M. Evans and A. Fernandez (1990) 'Design for wind protection using a wind tunnel' in A. A. M. Sayigh (Ed) 'Energy and Environment: Into the 1990s' Proceedings of the 1st WorldRenwable Energy Congress, Pergamon Press, Oxford.

Reduction of cooling loads and CO_2 emissions through the use of vegetation in Italian urban areas

G. BARBERA (*), G. PECORELLA (**), G. SILVESTRINI (**)

(*) Istituto di Coltivazioni Arboree, Un. of Palermo, Viale delle Scienze, Palermo, Italy
(**) Cnr Ieren, Via C. Rampolla 8, 90142 Palermo, Italy

ABSTRACT: With the rapid diffusion of air conditioning in Italy, different options should be considered in order to reduce the impact of this application on the electric system. In this paper the effect on air temperature reduction due to an increase of green areas in the context of two italian towns has been examined. The savings on cooling load of a typical office module could decrease by 8-11% with an increase of the vegetated area of 10%. In terms of carbon dioxide recuctions this contribution is almost ten times higher than the amount stored through photosynthesis.

1. Introduction

In Italy the use of air conditioning has increased over the last years at the very rapid growth rate of 20%. By year 2000 the energy consumption for this application should increase to 13 TWh (one fourth concentrated in the residential sector) and by 2010 the demand should reach 27 TWh (50% concentrated in the residential sector), (Enel 1988). This trend will have also consequences on the rapid increase of the summer peak load of the electric system, that in this decade could in many areas exceed the winter peak load.

But the demand of power could increase even more in the first decades of next century if the global warming scenarios elaborated by climatologists will prove to be correct. It has been calculated that in the italian residential sector each 1 °C degree increase of the air temperature could imply a growth of the summer peak load of 3000 MW by 2020-2030 (Silvestrini, 1990).

At the city level also the heat island effect, an increase in air temperature due to the complex interactions between the urban strucure and the atmosphere, should be taken in account. In summer days the presence of green areas in some towns can also create an "oasis effect" with temperatures lower than the rural areas, but in many towns the daily average summer heat island is of 3-5 °C (Landsberg, 1978).

In order to reduce the electric demand, a significant attention should be therefore paid in building design, low energy cooling technologies and interventions in town planning. Different approaches have been suggested to reduce the urban temperature, like increasing the vegetative cover, choosing light colors for buildings and streets, controlling the urban thermal mass (Thurow, 1983).

S. Alvarez et al. (eds.), Architecture and Urban Space, 41–46.

2. Microclimatic effect of vegetation in urban area

We will analyze the possible contribution of an increase of the urban vegetated area in reducing the cooling loads of buildings, an option that has been extensively documented in literature (Alessandro, 1987).

The role of trees on buildings cooling can be useful in three ways, shading, reduction of wind speed, evapotranspiration. The first two aspects are important especially in presence of low level houses, while the reduction of air temperature through evapotranspiration can generate an "oasis effect" that can be percieved on a larger scale near green areas.

Experimental results have shown very strong cooling energy reductions for buildings protected by trees, that can be as large as 50% (Parker, 1981). Also the effect due to the presence of large urban parks or green belts has been measured, with 3-7 °C lower temperature compared to built areas (Alessandro, 1987). Oke (1972) has made an interesting study of evapotranspiration cooling rates in Montreal and the results are likely to be representative of a variety of mid latitude towns in summer. In fig. 1 the reduction af the air temperatures due to an increase of the vegetated area has been derived from a simulation model. It is important to underline the fact that the effect is not linear and that the greater reductions appear for the first 20-30 percent of the city area covered with trees.

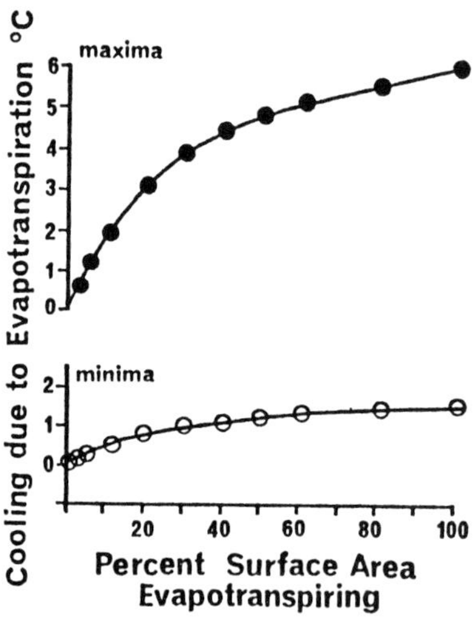

Fig. 1 Simulated effect on cooling of varying the area of a city covered by vegetated surfaces, (Oke, 1972)

In order to give quantitative data on the effect of an increase of green area in towns, in this paper a very simplified model has been used, following an approach proposed in Akbari (1987) that has permitted to evaluate the potential on cooling energy savings for the U.S. building stock. Considering the plantation of 100 million trees and the effect of an increase of the urban albedo, 44 TWh/year could be saved avoiding the emission of 66 million tons of CO_2 /year (Akbari, 1988).

3. Calculation of the air temperature reduction

The elements that need to be defined for the calculation of the air temperature reduction due to an increase of vegetated area are two : h, the mixing height that defines the air volume over the town of well mixed air (with the assumption that thermal turbulence is sufficient to cause an adiabatic lapse rate) and E_{tp}, the potential evapotranspiration of the vegetation. Defined the volume of mixed air over the town and the air change rate due to wind velocity it is then possible to calculate the temperature reduction due to the heat absorbed by the evaporative cooling.

The mixing height is generally low by night and increases after sunrise reaching 500-2000 meters in the afternoon. In presence of wind the value of h is greatly reduced.

The relation that is used has been proposed by Leahey and Friend (1971) and allows to estimate the contribution of artificial heat sources. The calculation of h derives from a simplified solution of a steady-state energy equation of a column of stable rural air moving to an urban area and can be expressed as follows:

$$h^2 = \frac{2}{c\,\rho\,\alpha} \int_{x1}^{x} [H + Q_{sr} - \sigma(T_0^4 - T_{0c}^4)]\, dx$$

where:

c = specific heat of air (kJ/kg °C)
ρ = air density (kg/mc)
u = wind speed (m/sec)
α = difference between urban and rural lapse rates (K/m)
H = artificial heat release rate (kW/m^2)
Q_{sr} = sensible heat flow in the rural region (kW/m^2)
σ = Stefan-Boltzmann constant ($5.7\ 10^{-11}\ kW/m^2\ K^4$)
T_o = ground surface urban temperature (K)
T_{oc} = ground surface temperature in the rural area (K)
x = distance from the transition point (m)
x1 = starting transition point of heat island (m)

This relation has been used to calculate an average mixing height over the urban area using simplified assumptions. The value of the artificial heat has been calculated in relation to an estimate of the energy consumption of the towns analyzed, giving an hourly profile with a peak in the afternoon, while the urban-rural temperature difference has been derived form the population value and corrected in relation to wind velocity. For the parameters not available, reference data reported by Akbari and Leahey have been used in order to obtain an estimate of the average h values.

To evaluate the amount of evapotranspiration a relation based on empirical results of water losses from different crops has been used (Jensen, 1963).

$$E_{tp} = (0.0252\,T + 0.078)\,R \qquad \text{(cm of water/day)}$$

where:

T = ambient air temperature (°C)
R = solar and sky radiation (cm of evaporated water/day)

In urban parks and city trees the E_{tp} value should be even higher than the one calculated with this relation (Clark, 1989). It is however important to note that this value represents the potential evapotranspiration, that is the theoretical value achieavable assuming an ample supply of water. Since we are evaluating the potential cooling effect of the vegetated area we assume that the trees are well watered. In many cases subsurface recharge through ground water flow and losses from water piper may occur, especially in italian towns where the losses from the water distribution system may be as great as 20- 40%.

4. Calculation of modified air temperature profiles for Milan and Rome

Assuming that the evaporative cooled air is uniformly mixed into a volume defined by the town area and the average mixing height, it is possible to calculate the reduction in air temperature at different hours. In reality the air will be cooler near the green areas, but in order to get a real sense of the temperature distribution, 2 or 3 dimensional atmospheric diffusion models should be used. The calculation of new temperature and relative humidity values, using the equations previously described, has been made for two italian towns, Milan and Rome. These locations have been selected from an analysis of the Test Reference Years of different italian towns in relation of their high percentage of summer hours with relative humidity values lower than 60% (Silvestrini, 1988).

In the italian context even a large effort to increase the actual vegetated percentage (usually between 5-15%) could hardly succeed in adding through new street trees and parks more than 10% of the built area. This value however is not impossible to achieve. The new Milan administration has for example decided to make a strong effort in increasing the vegetated area by adding 770 green hectars (partly using old abandoned industrial sites), a value the represents nearly 10% of the urban built area.

The figure 2 shows the average reduction in air temperature in July due to an increase of the green urban area by 5 and 10% of the town area. The calculated values should be considered just as first quantitative estimates of the reduction achieveable, considering the crude assumptions used.

5. Cooling savings in an office module

In order to verify the effect of an increase in vegetation on the reduction of cooling loads of a typical office module, a calculation has been made using the simulation model SMP based on the finite difference method (Butera, 1984).

An office of 120 m^3 in a multistorey building, with a floor area of 40 m^2, a thermal integrity that satisfies the italian building code has been considered. The glazed area is of 6 m^2 in the south exposure and 2 m^2 in the north facade, the only two external walls. Venetian blinds are used to protect from solar radiation. An air change rate of 1.5 ach during the working hours (8 a.m. - 18 p. m.) has been considered.

Calculations of the cooling loads have been made for the base case (using the hourly data of the Test Reference Year) and using the new temperature and humidity values altered by the presence of an increase of the vegetated area equal to 10% of the town area. The savings obtained for the specific enclosures analyzed are of 8% in Rome and of 11% in Milan.

In the calculations the cooling reduction due to the tree shading has conservatively not been considered due to the usual italian urban building typology. This contribution can however be relevant for low rise buildings.

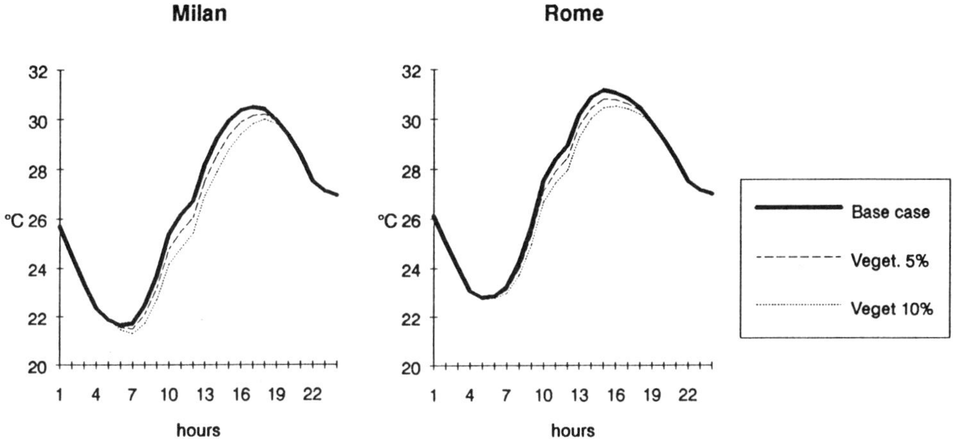

Fig. 2 Average hourly air temperatures for the month of July calculated for the climates of Rome and Milan supposing an increase of the urban tree canopy density of 5% and 10% of the built area

6. Conclusions

A rough evaluation of the effect of a 10% increase of the vegetated area can be made by an extrapolation of the savings calculated in the previous paragraph. Assuming from the national estimate of the cooling demand in 2000 that the consumption for this application should be about 600 and 800 GWh in Milan and in Rome, the savings in the two towns could respectively be of 66 GWhand 64 GWh. By 2010 the savings could double due to the increased diffusion of air conditioning.

In terms of CO_2 emissions, considering an average production of 0,69 kg CO_2 per kWh generated, by 2000 the increase in green area could avoid the emission of 40-50 thousand tons/year in each town. This amount should be added to the CO_2 absorbed by the trees during the growth, 10 tons/year per hectar as an average value in the italian context. This direct accumulation should lead, for example, to 8000 tons/year stored in Milan considering the 10% proposed increase in the vegetated area.

It is interesting to note that in the long term (2010-2020) the indirect effect in reducing the CO_2 emissions by saving the cooling energy demand will be nearly 10 times greater than the effect in accumulating carbon through photosynthesis.

7. Bibliography

Akbari H., Huang Y., Rosenfeld A. Taha H., (1987) "The potential of vegetation in reducing summer cooling loads in residential buildings", LBL-21291, Lawrence Berkeley Laboratory

Akbari H., Huang Y., Martien P., L. Rainer L., Rosenfeld, A. Taha H., (1988), "The impact of summer heat islands on cooling energy consumption and CO_2 emissions", LBL-25179, Lawrence Berkeley Laboratory, Berkeley, USA

Alessandro S., Barbera G., Silvestrini G., (1987), "Stato dell'arte delle ricerche concernenti l'interazione energetica tra vegetazione ed ambiente costruito", Quaderno CNR-IEREN n. 13, Palermo

Butera F., Farruggia S., Rizzo G., Silvestrini G. (1984), "Il codice di calcolo SMP per la simulazione del comportamento termico di moduli edilizi solari passivi e convenzionali: algoritmi usati e logica funzionale", Quaderno CNR-IEREN n. 1, Palermo

Clark J., Kjeigren, R., (1989) "Conceptual and management considerations for the development of urban tree plantings", Journal of Arboricolture, Vol. 15, n. 10

Enel (1988), "Programmi 1988-1992 e previsioni di copertura dei fabbisogni elettrici fino al 2000", Roma

Jensen M., Haise H., (1963), "Estimating evapotranspiration from solar radiation", Journal of the irrigation and drainage division, Pr. of the American Society of Civil Engineers, n. 89

Landsberg H. (1978), The urban climate, Academic Press

Leahey D., Friend J. (1971), "A model for predicting the depth of the mixing layer over an urban heat island with applications to New York city", J. of Applied Meteorology, n. 10, Dec.

Oke T., Yap D., Fuggle R. (1972), "Determination of urban sensible heat fluxes", International Geography, Ed. Adams and Helleiner, University of Toronto Press

Parker J. (1981), "Use of landscaping for energy conservation", STAR project 78 - 102, Florida State University System, Tallahassee, USA

Silvestrini G., Barbera G., Alessandro S, Orioli S., (1988), Ruolo del verde urbano nel migliorare il comfort termico e ridurre i consumi di climatizzazione estiva, Proceedings of 43th conference of ATI, Ancona

Silvestrini G.(1990) "Possibile impatto dell'incremento dell'effetto serra sulla penetrazione dei condizionatori d'aria in Italia", Energie Alternative - HTE, n. 66, Milano

Thurow C. (1983), "Improving street climate through urban design", American Planning Association, Chicago, USA

ANALYSIS OF THERMAL EFFECTS OF URBANIZATION

I. A. CAMILLONI and V.R. BARROS
Department of Atmospheric Sciences
University of Buenos Aires
University City
(1428) Buenos Aires
Argentina

ABSTRACT. In this paper we propose a model to estimate the mean temperature there would be in a city if the urban heat island were not present. We suppose that over homogeneous terrain and without anthropogenic influences the mean temperature over a certain period (5 years, in this case) depends only on the geographic coordinates: latitude (LAT), longitude (LON) and height (H) and relations between them. We select 29 non-urban meteorological stations within a limited plain and homogeneous area in Argentina.Using a multiple regression analysis method (Stepwise) we find that the most important parameters to determine the mean temperature in that region are latitude (LAT) and the product (LAT x H). We compare the mean temperature observed over a 5-year period with the one we obtain with the model for the selected stations. The model provides a reliable way to compute the magnitude of the urban heat island as the value of the error (0.3°C) is lower than the typical urban excess temperature estimated in 1.5 to 2°C.

1. Introduction

Urbanization processes produce many changes in the surface nature and in the properties of the local atmosphere. This kind of processes involves transformations in radiative, thermic, hydric and aerodynamic characteristics and produces changes in the turbulent fluxes of momentum, heat and water vapor.

Temperature is one of meteorological variables more often used in characterizing the climate of an urban area, because of its marked sensibility to urbanization and industrialization processes producing urban heat islands, etc.

The urban heat island is a physical phenomenon which refers to the differential heating registered in urban areas compared with its rural surrroundings.

Oke (1979) links the internal characteristics of the urban heat island to factors associated to land-use and nature and density of urban building. The magnitude of the heat island is quantified estimating the urban-rural differences of many parameters: instantaneous temperature, minimum temperature, maximum temperature, mean temperature over a certain period, etc (Oke (1973); Landsberg (1975); Oke (1979); Landsberg (1981); Oke (1982); WMO (1986)).

In this paper a method to determine the intensity of the urban heat island (ΔT_{u-r}) is presented as:

$$\Delta T_{u-r} = T(\text{city}) - T(\text{without city}) \tag{1}$$

where T(city) means the temperature observed within the city and T(without city) is the temperature

S. Alvarez et al. (eds.), Architecture and Urban Space, 47–52.

there would be in the area if the city were not present.

2. Data

The observations of 29 non-urban meteorological stations located in the surroundings of Rosario City and Paraná City are analyzed in the period 1959-1983 (see Fig. 1). Both cities are built up on the left and right side of Paraná River respectively. The region surface is almost plain with a mean height of 170 meters above sea level. The geographic coordinates of the stations selected are presented in Table 1.

Table 1. Gegraphical coordinates of the stations analyzed.

N°	latitude(S)	longitude(W)	height(m)	N°	latitude(S)	longitude(W)	height(m)
1	26°35'	62°51'	190	16	32°37'	60°11'	29
2	26°49'	60°27'	92	17	32°38'	62°41'	130
3	27°28'	58°49'	60	18	32°55'	60°47'	25
4	27°45'	57°38'	74	19	33°00'	58°37'	21
5	28°26'	58°55'	70	20	33°35'	59°24'	5
6	29°53'	61°57'	88	21	29°08'	59°16'	36
7	29°54'	63°41'	341	22	29°11'	59°42'	53
8	30°16'	57°39'	54	23	31°12'	57°55'	37
9	31°11'	61°33'	100	24	29°41'	57°09'	70
10	31°26'	60°56'	38	25	32°52'	58°31'	14
11	31°37'	60°41'	18	26	27°05'	61°07'	102
12	31°47'	60°29'	78	27	31°42'	60°49'	18
13	31°50'	60°31'	110	28	27°27'	58°46'	62
14	31°51'	59°05'	43	29	32°33'	60°51'	26
15	31°56'	60°25'	105				

The mean temperatures over each 5-year period (1959-1963, 1964-1968, 1969-1973, 1974-1978 and 1979-1983) are calculated using the three daily observations at each station at 08:00, 14:00 and 20:00 AST.

3. Model proposed

It is supposed that in absence of anthropogenic influences and over homogeneous terrain the mean temperature over a certain period (5 years, in this case) depends only on the geographic coordinates: latitude (lat), longitude (lon) and height (h):

$$T' = T'(\text{lat, lon, h}) + \text{error} \quad (2)$$

The model proposed is:

$$T' = T_0 + a_1\,LAT + a_2\,LON + a_3\,H + a_4\,(LAT)^2 + a_5\,(LON)^2 + a_6\,(H)^2 + a_7\,(LAT \times LON) + a_8\,(LAT \times H) + a_9\,(LON \times H) \quad (3)$$

where the variables LAT, LON and H are the normalized variables of lat, lon and h as:

$$f = (f_{max} - f) / (f_{max} - f_{min}) \tag{4}$$

The Stepwise regression procedure (Draper and Smith (1966)) analyzes the importance of each variable in the regression and provides the values of the coefficients a_1,............,a_9. The importance criterion is based on the proportion of the sume of squares (SSR) reduced.

The square regression coefficient r^2 and the mean residual value (MR) can be used to verify the adjustment of the data to the model. The mean residual value is defined as:

$$MR = \left(\sum_{i=1}^{N} (T_i - T'_i)^2 / N \right)^{1/2} \tag{5}$$

where T is the observed temperature, T ' is the model estimation and N is the number of cases.

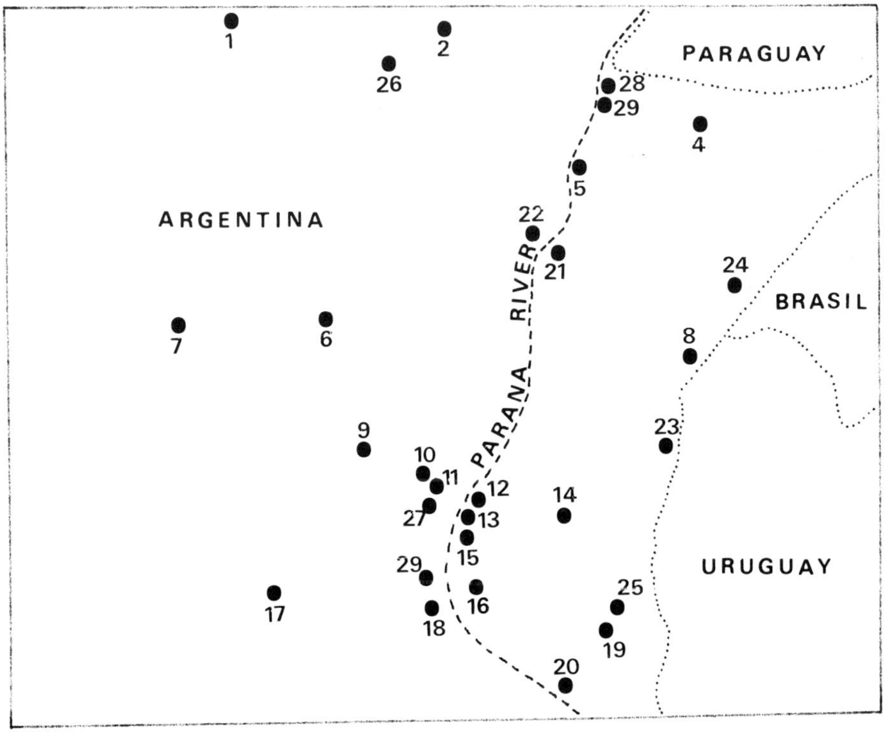

Figure 1. Region of validity of the model and stations used to propose it (Argentina).

4. Experiments

The objectives of the experiments are: 1) study the importance of each geographical variable for an eventual reformulation of the model (3), and 2) verify the model for different quantities of input information.

4.1. STUDY OF THE VARIABLE IMPORTANCE

The information available in the periods 1964-68 (19 stations: 1-4, 6-7, 9, 11, 13-14, 16, 18, 21-29) and 1969-73 (23 stations: 1-2, 4-7, 10-14, 16-23, 26-29) is considered.

The results of applying Stepwise procedure are presented in Table 2.

Table 2: Results of applying Stepwise procedure.(model (3)). Periods 1964-68 and 1969-73.

variable	regression coefficients (a_i) 1964-1968 T_0=22.06091 r^2 = 0.980	1969-1973 T_0=23.85116 r^2 = 0.990	proportion of SSR reduced 1964-1968	1969-1973
LAT	-2.76358	-7.41909	0.928	0.958
LON	3.38381	1.84416	0.008	0.000
H	4.52287	-2.12634	0.000	0.000
(LAT)2	-1.73946	1.33842	0.000	0.001
(LON)2	-1.95772	4.67815	0.000	0.001
(H)2	-2.49374	0.75731	0.001	0.000
LAT x LON	-2.49775	0.75731	0.001	0.000
LAT x H	-7.07883	0.96654	0.020	0.015
LON x H	-9.43555	5.43054	0.001	0.005

The most important variables are LAT and LAT x H. Then, model (3) is reformulated as:

$$T' = T_0 + b_1 \text{ LAT} + b_2 \text{ LAT x H} \qquad (6)$$

Parameters T_0, b_1 and b_2 can be calculated using Stepwise. The results are presentd in Table 3.

Table 3: Results of applying Stepwise procedure (model (6)). Periods 1964-68 and 1969-73.

variable	regression coefficients (b_i) 1964-1968 T_0=23.06842 r^2 = 0.973	1969-1973 T_0=23.24701 r^2 = 0.986	proportion of SSR reduced 1964-1968	1969-1973
LAT	-5.36500	-5.44585	0.928	0.958
LAT x H	-2.14600	-1.82166	0.200	0.015

4.2. MINIMUM INPUT DATA NECESSARY

The information available in the period 1969-73 (23 stations) is considered. Different experiments are done to analyze the minimum input data necessary to apply the model in the region considered.

Experiment No.1. The information of period 1969-73 is divided in two groups of 12 stations each: **A** (2,4,5-6, 10, 13-14, 16, 19, 22-23, 29) and **B** (1, 5, 7 11, 12, 17-18, 20-21, 26-28).

Data of group A are used to calculate the model parameters (T_o, b_1, b_2) and data of group B are used to verify the model.
Experiment No. 2. We change the groups of experiment No. 1 (A is used to verify the model and B is used to calculate the model parameters).
Experiment No.3. In this case the information of the period 1969-73 is separated in three groups of 8 stations each.: **A** (1, 11-12,17, 19, 22-23, 28), **B** (4, 6, 10, 13-14, 18, 20, 26) and **C** (2, 5, 7, 16, 21, 23, 27, 29). Model parameters are found with group A and group (B + C) is used to verify it.
Experiment No. 4. Group B of experiment No.3 is used to determine the model parameters and group (A+C) is used to verify it.
Experiment No. 5. Group C of experiment No.3 is used to find the model parameters and group (A+B) to verify it.
The results of experiments are presented in Table 4.

Table 4. Results of applying model (6) for different quantities of input data in period 1969-73

experiment No.	model T ' (°C)	r^2	MR (°C)
1	23.27546-5.50538LAT-1.42695LATxH	0.966	A:0.26 B:0.27
2	23.17719-5.39628LAT-1.80015LATxH	0.978	A:0.27 B:0.24
3	23.09884-5.36487LAT-0.68474LATxH	0.980	A:0.24 (B+C):0.30
4	23.25376-5.74322LAT-0.46505LATxH	0.986	B:0.18 (A+C):0.34
5	23.37280-5.36639LAT-2.66877LATxH	0.986	C:0.23 (A+B):0.31

5. Model application

The model is applied for the rest of the periods: 1959-63, 1974-78 and 1979-83. The results are presented in Table 5. Table 6 represents the mean temperature there would be in Rosario City (32° 55'S, 60°47'W) and Paraná City (31° 46'S, 60° 29'W) obtained with the model.

Table 5. Results of applying model (6) in the rest of the periods

period	model T '(°C)	r^2	MR(°C)	No. of stations
1959-63	23.17089-5.00998LAT-3.50789LATxH	0.929	0.29	2-4,6-8,10-12,18,21-22,24, 27
1974-78	23.28460-5.91435LAT-2.29667LATxH	0.982	0.23	2.4-7,13-14,18-20,22,24,27-29
1979-83	23.03724-4.54030LAT-5.60395LATxH	0.974	0.21	2,4,6,9,14,18-19,22,24,27-29

Table 6. Mean temperature (°C) in Rosario City and Paraná City

period	Rosario	Paraná
1959-63	18.6	19.0
1964-68	18.2	18.8
1969-73	18.4	18.8
1974-78	18.0	18.2
1979-83	18.7	18.5

6. Concluding remarks

- We find the model is reliable and useful to estimate the magnitude of T (Without city) in the region of analysis as the value of the error (MR) is of 0.3°C and the typical urban excess temperature is of 1.5-2°C. (Landsberg, 1981).
- When we applied the model with the information of 8 meteorological stations instead of 23, not significant differences appeared.
- The model is independent of the stations used to find its parameters (T_o, b_1 and b_2).
- The model can be applied to determine mean temperature series in any point of the region of analysis.

7. References

DRAPER, N. and SMITH, H. (1966). Applied Regression Analysis. John Wiley & Sons. 407 pp.
LANDSBERG, H. (1975). Atmospheric changes in a growing community. Inst. Fluid Dynamic Appl. Math. Tech. Note N° BN823. Univ. of Maryland, 54 pp.
LANDSBERG, H. (1981). The Urban Climate. Academic Press.
OKE, T. (1973). City size and the urban heat island. Atmos. Environ. 7. 769-779.
OKE, T. (1979). Review of Urban Climatology. WMO. Tech. Note N°169. 100 pp.
OKE, T. (1982). The energetic basis of the urban heat island. Quart. J. Roy. Meteor. Soc. 108, 1-24.
WMO (1986). Proc. of the Tech. Conf. on Urban Climatology and Its Applications with Special regard to Tropical Areas. Mexico, 1984. WMO N°652, 534 pp.

FACING ENVIRONMENTAL ISSUES IN OUR CITIES: THE CASE OF LIMA, PERU

LILIANA O. BELTRAN[1]
Facultad de Arquitectura
Universidad Nacional de Ingenieria
Av. José Gálvez 795
Lima 13 - PERU

GLADYS L. LOPEZ
Department of Architecture
370 Wurster Hall
University of California
Berkeley, CA 94720 - U.S.A.

ABSTRACT. This paper examines the transformation of Lima from an oasis to an urban heat island. The changes made in its urban evolution that intensify the urban heat island effect are analyzed: population, building density, percentage of green areas, and the building materials used. Analysis of the climate determines the overheated periods and the climatic elements that could be modified to improve pedestrian thermal comfort. Strategies to do this are discussed: those that affect climate directly such as increasing shading and wind exposure, and those that affect climate indirectly by increasing albedo. From the variety of shading types, those that fit Lima's existing urban residential patterns and its codes, trees and gazebos were selected. The efficiency of these shading types in reducing air temperature was tested using the simulation program URBMET. The study concludes by reinforcing the use of shading elements as the main strategy to reduce summer heat in Lima.

1. Introduction

Historically, cities have been one of humanity's great glories, but they have had a negative impact upon earth's ecological life support systems. Now the humaneness of cities is at risk ecologically, creating lethal places to aged and ill people. Fortunately, the concerns about man's influence on his environment are growing and we are facing a major adjustment in our understanding of the city's role in our ecosystem. The most obvious climatic difference between urban and rural areas is their temperature. In large urban areas, people notice the presence of pollutants whose permanence depends on the physical and geographic characteristics of the city in question.

2. From Oasis to Urban Heat Island

2.1 LOCATION

This study focuses on Lima, whose geographic coordinates are 12°02'36"S and 77°01'42"W and whose elevation is 154 meters above sea level. Lima is located on the west coast of Peru. Because of its closeness to the equator, the Peruvian coast should have a tropical climate and vegetation as do the coast of Ecuador and Colombia. But the cold winds that come from the Andes and from the Pacific Ocean are warmed over a hot soil and absorb humidity, forming diffused clouds and low fog that in winter can last several weeks. The dry coastal climate and the

[1] Presently working at the Windows and Daylighting Program, Lawrence Berkeley Laboratory, Mailstop 90-3111, Berkeley, CA 94720 - U.S.A.

S. Alvarez et al. (eds.), Architecture and Urban Space, 53–58.

irregularity of most of the rivers' flow have produced huge desert areas from a fertile soil. Thus, the coast is a long desert strip alternating with oases made by rivers that descend from the Andes. Lima, which lies on the banks of the Rimac River close to the foothills of the Andes, was considered an oasis along with other Peruvian coastal cities.

2.2 CLIMATE

Lima has two main seasons: a three-month summer (January-March) and a nine-month winter (April-December) that is mild, cloudy and misty. Maximum temperatures usually occur in February (26.1°C) or March (26.2°C) while the minimum occurs in September (14.7°C). The annual variation of the maximum daily temperature is 7.8°C while that of the minimum is 5.5°C (see Figure 1). During the day the temperature swing is relatively small, ranging from 3.5°C in August to 5.8°C in February. The small day to night temperature differences and the two marked seasons are the main characteristics of Lima. The annual maximum relative humidity remains almost constant year-round, ranging from 95% to 96.8%, while the minimum ranges from 61.2% in February to 74.4% in September. The number of maximum hours of sun occurs during January to April ranging from 55% to 65%, and the minimum occurs from July to September at 13%. Precipitation is very scarce year-round, the maximum average is 2.95 mm in August.

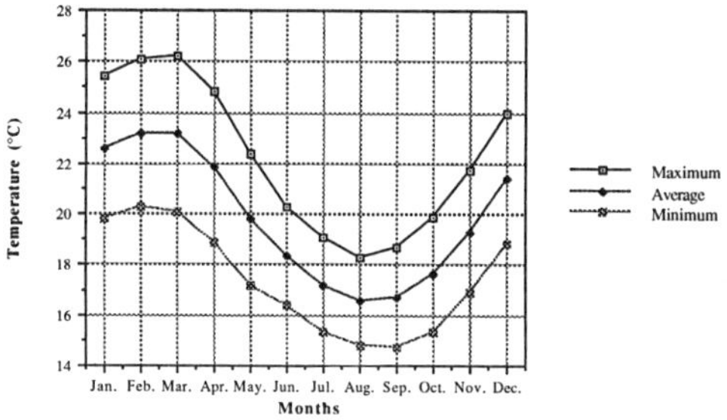

Figure 1. Lima's Average Air Temperature

The summer temperature-humidity lines of Lima's climate are plotted on an Olgyay's Bioclimatic Chart for Lima which indicates how far they are from the comfort zone (Figure 2). Therefore, Lima needs a cooling strategy to improve its thermal comfort during the summer season.

During daytime hours in summer there are overheated periods, while in winter most of the time the temperature remains in the comfort zone (Figure 2). During overheated periods, the high humidity increases the thermal discomfort that affects most pedestrians who are not protected from the sun and the heat of summer. Shading becomes necessary not only during overheated periods but also during comfort periods, since the comfort zone in Olgyay's chart is defined as the range of temperature where there is absence of discomfort in an average person in shade.

Lima, like many other large cities, has experienced through the years the increase of its temperature that meteorologists call the urban heat island effect, i.e., that urban areas are hotter than suburban and rural areas. Jauregui (1986) has also referred to Lima as an example of an urban heat island.

In most cities that use oil or coal as fuel in their power plants, the main reason to reduce the urban heat island effect is to reduce energy consumption and to avoid pollution. In Lima, where hydroelectric power plants are used, the goal is to achieve pedestrian thermal comfort and to ameliorate future overheating by global warming.

Lima's inhabitants are experiencing another problem of the big cities: pollution, which is affecting their health. The primary source of Lima's pollution is carbon monoxide produced by cars. Lima owes much of its air pollution problem to its transportation patterns, urban form and its location between ocean and mountains. In summer, the intense light promotes the formation of photochemical smog (ozone) and in winter the prolonged periods of inversion trap pollutants near the ground for days, deteriorating the air quality of Lima.

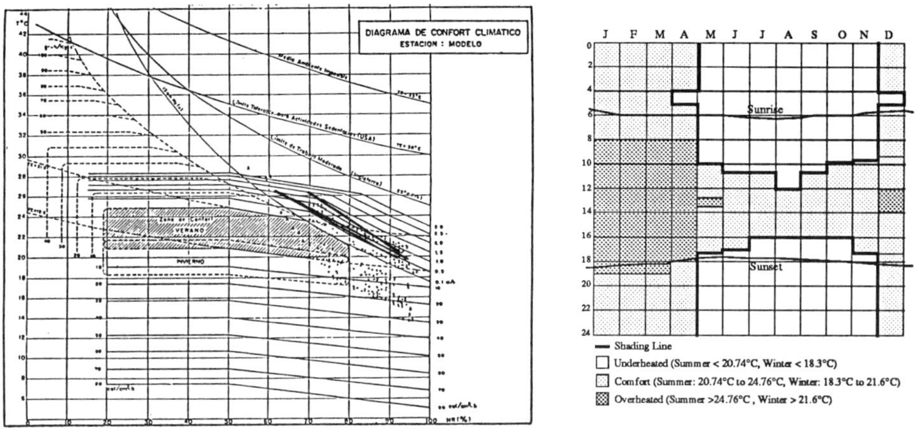

Figure 2. Lima's Climate on a Bioclimatic Chart and a Yearly Shading Chart (with the Use of Average Temperature Data Only).

2.3 PHYSICAL CONFIGURATION

The change from the oasis effect to the urban heat island effect in Lima was the result of the changes produced through the years in the city's size, physical configuration, building materials and life style.

In 1535, Lima had 70 inhabitants in an area of 215 Ha. The streets were straight and orthogonal. The blocks were square, oriented almost NE-SW. The lots allowed spacious gardens and patios. There was a uniform height of the buildings with the exception of the public buildings and churches. The building materials were adobe, stucco, and wood in balconies. People spent more time inside their homes than outside. By the beginning of the twentieth century, Lima had changed its image: there were 139,409 inhabitants in an area of 800 Ha. The lots were subdivided in many parts, reducing the sizes of patios and gardens. The shape and orientation of the blocks were quite variable. New buildings (industrial, military and public) and new technologies (electricity, telephone, cable car, automobile, concrete) were present. In spite of Lima's modernization, rapid expansion did not begin until 1920. The expansion of the city was produced mainly by the opening of development axes: avenues, boulevards and the creation of large urbanizations where a new concept was included in the street design: setback, usually with exterior gardens. The proliferation of these urbanizations, in these years, gave Lima the name of "the garden city".

In the 50's, industrial development started attracting people from all over the country and the immigration rate began to increase quickly. The demographic explosion produced the creation of new squatters. To avoid lot subdivisions the government established the minimum dimensions of a lot. At that time, Lima's image was of a metropolis.

In the 60's, 70's, and 80's, the explosive growth of Lima and its urban consolidation was accomplished mainly by the development of single family housing in previously agricultural areas. In 1989, the metropolis, with 7'068,000 inhabitants, occupied 296,441 Ha. and almost completely covered the valleys of the Rimac, Chillon and Lurin rivers with the exception of a few agricultural areas.

Lima has experienced a remarkable change in its morphology. One notices the presence of other urban centers, besides Lima's downtown, with high buildings and traffic concentration.

Peripheral suburbs, depending on their location, are occupied by squatter settlements with scant green areas or by large residences of the wealthy. The Spanish plazas are no longer used, although multifamily housing complexes still use plazas in their civic and recreational areas. The block shape is generally rectangular and its orientation is variable. The lots are rectangular and their width has decreased due to the increasing cost of living. The presence of trees in backyards is decreasing as a consequence of the lot's minimum area; the backyard is becoming too small to allow trees and visual extension at the same time. The percentage of green areas in the city has decreased because these areas do not have a landscape treatment or have been invaded. Brick and concrete are the main construction materials for safety reasons. Streets and avenues are designed to meet the automobile's needs which sometimes differ from the needs of pedestrians. The number of cars has increased the area of parking lots in commercial, institutional and multifamily housing buildings. Due to the economic crisis and small house dimensions at the present time, outdoor spaces are being more heavily used as business, social and recreational places.

3. Strategies Proposed to Improve the Pedestrian Comfort Level Outdoors

3.1. SHADING

During overheated periods, outdoor spaces in shade are desirable to reduce thermal stress. In cities at low latitudes, where horizontal surfaces receive the maximum annual insolation, shade over horizontal surfaces is necessary to reduce the amount of solar radiation reaching the ground and the amount of long wave radiation.

Figure 3 shows the shading effects in a typical residential street produced by surrounding two-story buildings. During summer around noon, the sun remains almost vertical, moving from a slightly South location (Dec. 22) to the North (Mar. 21). Shading is scarce in all directions during the time when it is needed. Sidewalks have no shadows cast over them and pedestrians are exposed to high solar radiation. Therefore, other street elements need to be proposed to reduce this problem.

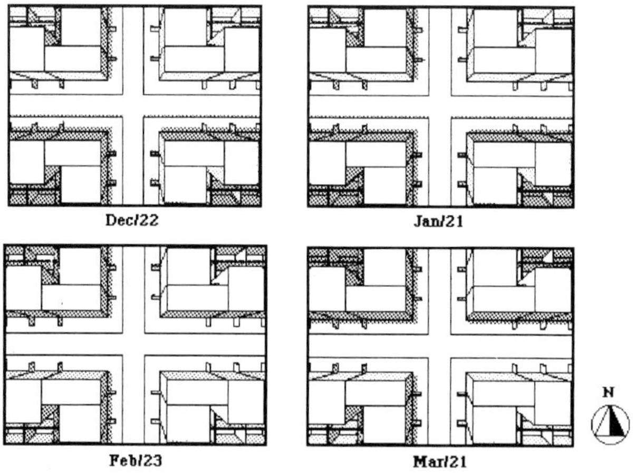

Figure 3. Summer Shading Studies at 1 pm. of a Typical Residential Street of Lima.

Trees affect urban climates and building energy use, by reducing the solar radiation reaching the ground and the buildings, and by lowering temperature and wind speed. The modification of each of these elements depends on foliage density and shape, as well as the tree's dimensions. But any tree, even one bereft of leaves, can have a noticeable impact on microclimate and on cooling loads. Urban trees also help to reduce pollution and mitigate the greenhouse effect. In Lima, tree planting will be easy and inexpensive to implement, as there already exists a 10 year-old program that organizes people by neighborhood to do landscaping in their parks. These groups whose objective is to make Lima beautiful and to improve its air quality, receive free trees once a year. Through appropriate landscaping in open spaces, they could achieve their goal and also reduce the city's temperature.

Gazebos and brightly colored canvas awnings were commonly used in the past, especially along commercial streets and sidewalk cafes. But this custom has been lost, and awnings are barely used nowadays. If gazebos are built of heavy materials such as concrete, they may not help to decrease temperature.

Arcades were mostly used around main plazas, but they can be used in commercial streets. Although they are a good way to protect against the summer sun, they can not be implemented in residential streets because of code restrictions.

The simulation program URBMET (Taha 1988) was used to calculate the air temperature under different types of shading on a typical February day. Shades with different transmissivity were modeled: 80%, 60%, and 40%, and that under trees. For the latter, the model canopy consisted of 684 trees uniformly distributed on a site of 307 by 150 meters. The foliage cover was 30% and the foliage transmissivity for shortwave radiation was 36%. The canopy height was 5 meters with a stem height of 2 meters. Figure 4 shows the temperature in February and the reduced temperature due to the different types of shade at 2 pm. The first case reduces the temperature by 2.7℃, the second one by 5.2℃, and the third by 6.8℃. The fourth one (trees) achieves a reduction 5℃. In all four cases the temperatures have been reduced sufficiently to lie within the comfort zone.

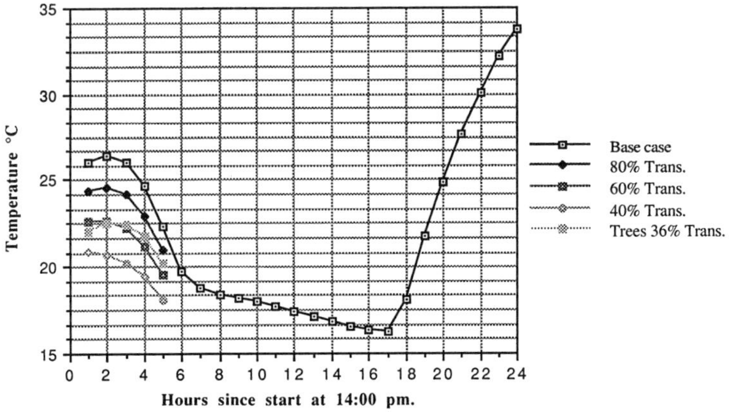

Figure 4. Simulation of Lima's Temperature Under Shading of Different Transmissivity using the computer model URBMET.

3.2 MATERIALS

Concrete is used as the main construction material: 80% of exterior walls of buildings are made of bricks and concrete blocks and 60% of the roofs are made of concrete. The large amount of concrete exposed to the sun has increased the heat storage capacity of the city.

To reduce this heat, it is necessary to use construction materials of low heat capacity, and increase the albedo of the city. Therefore, it is more effective to change the materials and color of horizontal, east and west surfaces. The color of the city's surfaces determine whether they absorb or reflect solar energy. The albedo of Lima has been defined mostly by the colors of the concrete and brick buildings which gives to the city a gray look. To change the albedo of roofs and paved areas could be expensive while building walls can be easily implemented by routine maintenance.

In 1985 a new code established the obligatory use of bright and traditional colors in public buildings. The main reason for this code was the psychological effect on the citizens, but it can also help to reduce heat if it is implemented in other buildings as well.

3.3 WIND

Urban ventilation offers a great potential to modify air temperature and pollution concentration in a hot-humid city,. It is also true that the speed and turbulence spectrum of a regional wind is modified by the urban structure. Lima's urban structure has reduced the wind speed in less than 2m./sec. in the downtown. The cooling capabilities of wind in the downtown is very small while in foothill and seashore areas is greater, due in the latter to the sea-breeze. Due to the huge extension of the metropolis and the fact that wind velocity at street level can be suppressed or increased by urban elements, we can state that the urban elements in downtown and seashore areas should not block SW and S winds.

A good street layout from the urban ventilation aspect, in a hot-humid region is when wide main avenues are oriented at an oblique angle to the prevailing winds. Givoni (1989). In the of case Lima, a good street layout that would provide wind ventilation for pedestrian and buildings has to be oriented along the SW and S or between the two directions.

4. Conclusions

This study presented solutions to reduce the urban heat island effect, improve air quality and ameliorate the quality of life utilizing design responses. These strategies can be incorporated in the urban and building codes of Lima. It was demonstrated that shading is an effective strategy that can reduce air temperature. Trees and gazebos can be easily implemented in Lima without breaking its social and cultural aspects in order to ameliorate its thermal pedestrian comfort in summer.

We expect that this paper calls the attention of researchers, planners and designers from around the world to find local solutions to improve the quality of life in urban areas. We --as planners, urban designers, and architects-- have the responsibility of the morphology as well as the quality of life in our cities. Our practice should be more responsive to these issues trying to maintain the harmony between man and nature.

5. References

Beltran, Liliana (1986) "Covered Courtyard for the Andean Cities", in the 11th National Passive Solar Conference Proceedings, American Solar Energy Society, Colorado, pp. 503-508.

Jauregui, Ernesto (1986) 'Tropical Urban Climates: Review and Assessment', in Urban Climatology and its Applications with Special Regard to Tropical Areas, World Meteorological Organization, Mexico, pp. 26-45.

Givoni, Baruch. (1989) Urban Design in Different Climates, World Meteorological Organization, Los Angeles.

Ortiz de Zevallos, Luis. (1978) Lima su Evolucion Creadora, Colegio de Arquitectos del Peru, Lima, Perú.

Quezada-Pacheco, Julio M. (1987) Estudio del Confort Climatico en Lima Metropolitana y Callao, Universidad Nacional Agraria La Molina, Lima, Peru.

Taha, Haider (1988) Site-Specific Heat Island Simulation: Model Development and Application to Micro-Climate Conditions, Master's Thesis, University of California Berkeley, California.

Chapter 2

Urban Design Concepts and Methods

SOLAR RIGHTS AND PLANNING CODES.

G. CASABIANCA, J. M. EVANS and S. de SCHILLER.
Research Centre Habitat and Energy,
Faculty of Architecture, Design and Urban Planning,
University of Buenos Aires.
Casilla de Correo 1765, Correo Central,
(1000) Capital Federal, Republica Argentina.

ABSTRACT. This paper examines the possibility to introduce solar rights in planning codes in latitudes between 22' and 55' S, the range found in continental Argentina. Six zones are identified according to the availability of direct sunlight, the angles of winter sun and the heating requirements which define the supply and demand for partial solar heating. The conclusions show that in latitudes up to 38'S, adequate levels of solar access can be achieved in typical suburban development without reducing densities or changing building forms. Between 38'S and 47'S, solar access is still possible though increasingly restrictive. Further south, solar access is difficult to achieve in the winter solstice in conventional development using existing planning legislation. Recommendations are given for achieving appropriate levels of solar access at different latitudes.

1. INTRODUCTION

Solar radiation provides a vital alternative energy source, especially apt for the thermal conditioning of buildings, reducing dependance on conventional energy and controlling negative environmental impacts.

To encourage the use of solar energy, it is essential to protect access to direct solar radiation, incorporating the "Right to Sunlight" in Urban Planning Codes and Building Regulations, the main legal instruments to limit individual rights in order to achieve benefits for the common good. However, they do not reflect the special requirements needed to achieve solar access. This study therefore examines the inplementation of solar access in planning and building codes.

To protect solar access, it is necessary to control building height so that reasonable levels of solar radiation can be received on neighbouring plots and excessive shadows cast by other obstacles are avoided.

Solar access, factor of vital importance in the development of planning and urban development policies, relates to densities, land subdivisions, building forms and urban tissue. Control of how and when adjacent buildings cast shadows on each other is required. Before

S. Alvarez et al. (eds.), Architecture and Urban Space, 61–66.

establishing these controls, standards should be fixed to obtain appropiate solar access levels for each case, based on sunlight requirements and site conditions that determine effective availability of solar radiation.

2. OBJECTIVE

This study aims to verify the feasibility of the "solar envelope" [1] as a control mechanism to protect solar access in urban and suburban areas in different regions of Argentina.

The results of this research demonstrate the possibility and alternatives to overcome problems of solar access rights in planning codes which control urban development, according to conditions and requirements in different regions of the country.

These alternatives apply conventional building controls such as maximum building heights, plot and building densities, building lines and set-backs, etc, to protect access to direct sunlight in new and existing urban and suburban areas. Building codes for different land use zones can thus provide a method to achieve and regulate solar access.

3. ESTABLISHING SOLAR ACCESS REGIONS

This study was developed in two stages: The first analyses different factors related to solar energy use in Argentina; the second examines the application of the solar envelope concept in different regions of the country.

The feasibility of solar energy use is directly related to climatic conditions and availability of solar radiation in each locality.

Argentina has a wide range of climatic zones determined by altitude and maritime influence as well as variation in latitude, from 22'S in the extreme north to 55'S in the south, excluding the Antarctic Sector.

A study of solar availability, heating requirements of buildings and the possibility of solar envelope application in 113 localities in different regions of the country, identified 6 zones shown in Figure 1 with the specific characteristics listed in Table 1.

Table 1. Potential Solar Use Zones and Characteristics.

Zone	Supply	Demand	Latitude	Winter Solar Altitude.
1	High	High	22'-30'	Very high
2	High	Low	22'-30'	Very high
3	High	High	30'-38'	High
4	Medium	Medium	30'-38'	High
5	Medium	High	38'-47'	Medium to low
6	Low	Very High	47'-55'	Low

Zones 1 and 2 do not present problems from the point of view of solar access and solar envelope application due to high angles of winter sun in these low latitudes which favour the reception of solar radiation in urban areas resulting from short shadows during most hours of the day. Zone 2, however, has very short winters and mild conditions which reduce the demand for solar heating.

In zones 3 and 4, conditions to apply the solar envelope are relatively easy to achieve. Angular elevation of winter sun is still high and it is possible to apply the solar envelope as a measure of building control without unduely restricting densities in urban areas. Zone 3, close to the Andes, has a higher heating demand due to colder winters at higher altitudes. Solar radiation intensity is also higher for the same reason.

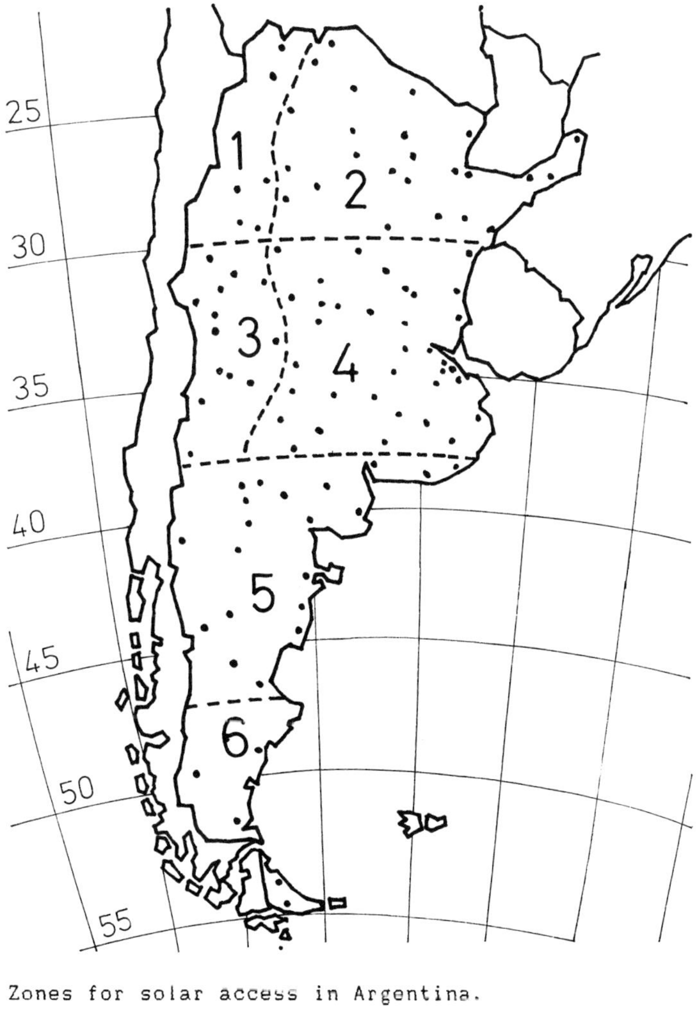

Figure 1. Zones for solar access in Argentina.

Zone 5, between 38'S and 47'S, has medium solar radiation intensity levels in winter combined with high winter heating demand. The lower solar altitude at these latitudes, limits urban densities to achieve solar access. For this reason, the application of the solar envelope in this zone is more demanding and requires careful study.

In Zone 6, south of latitude 47'S, solar access is desirable due to the high demand for heating, but mid-winter solar radiation is weak and cloud cover is high. The application of the envelope during the winter solstice (declination 23 45') also becomes very restrictive as low solar altitudes require wide spaces between buildings and reduced densities. The solar envelope has therefore been applied using 15th May -28 July (declination 18 45') as the critical dates for solar access.

4. VARIABLES FOR SOLAR ACCESS

The following factors are considered when evaluating solar envelope applications in each region identified (Figure 2):

- different situations of solar access according to plot size and proportions, building lines and set-backs from the plot boundaries.
- selection of alternative orientations.
- determination of the period required for solar access, according to the actual availability of solar radiation and applications of solar energy: conventional sunlight, passive solar systems, etc..
- selection of the date for verifying direct sunlight, considering the quantity and quality of incident solar radiation in different latitudes. For the initial study, the winter solstice was chosen to test solar access as this date combines the lowest solar altitudes with high heating requirements.
- definition of the limiting planes slope to determine levels of solar access (hours of potential direct sunlight) in different orientations and latitudes.

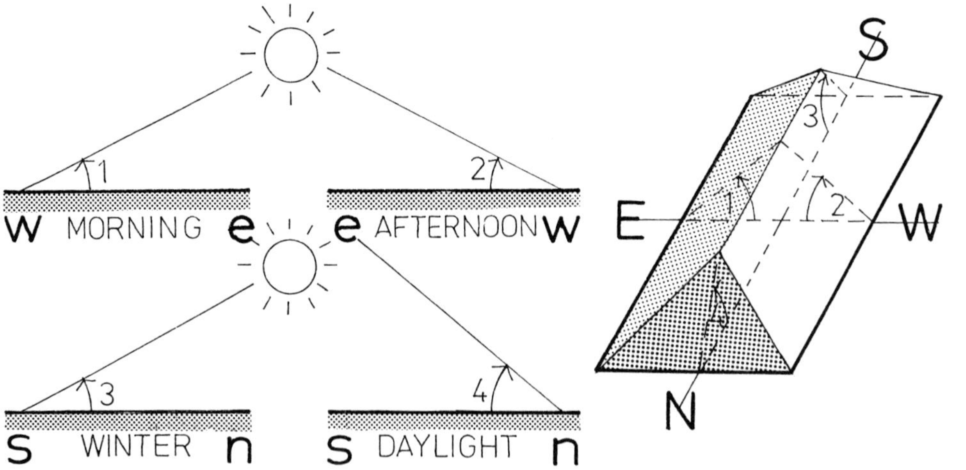

Figure 2. Application of the solar envelope

5. RESULTS

The solar envelope is defined as "a volume whose geometric form indicates the maximum permited building limits of a plot to ensure certain levels of direct sunlight and daylight in adjacent plots" [1]. The form of the solar envelope indicates the maximum dimensional limits of the building in each plot.

With a specially developed computer program, maximum plot coverage, plot ratios, building heights, residential densities, etc. can be determined according to selected levels of direct sunlight. The program allows additional inputs such as maximum building depth related to building type, fence height at site boundaries, building lines and set-backs, etc..

The results of this comparison can be directly applied in conventional building codes or adjusted to ensure campatibility between sunlight requirements and other criteria for urban development.

6. CONCLUSIONS

The general conclusions refer to the factors considered when applying the solar envelope; they include the following points:

1. The periods of solar access considered, ranging from 2 to 6 hours, are feasible in the majority of the zones studied (1 to 5) using existing urban land subdivisions and conventional building types.

2. Urban grid orientation has a direct influence on the volume enclosed by the envelope. Plots with frontages on streets running N-S have considerably lower densities than those on streets running E-W. Street grids with intermediate orientations (NE/SW - NW/SE +/- 15') produce favourable densities with similar volumes in similar sized plots allowing common values for plot ratios and urban densities in plots with different orientations (Figure 3).

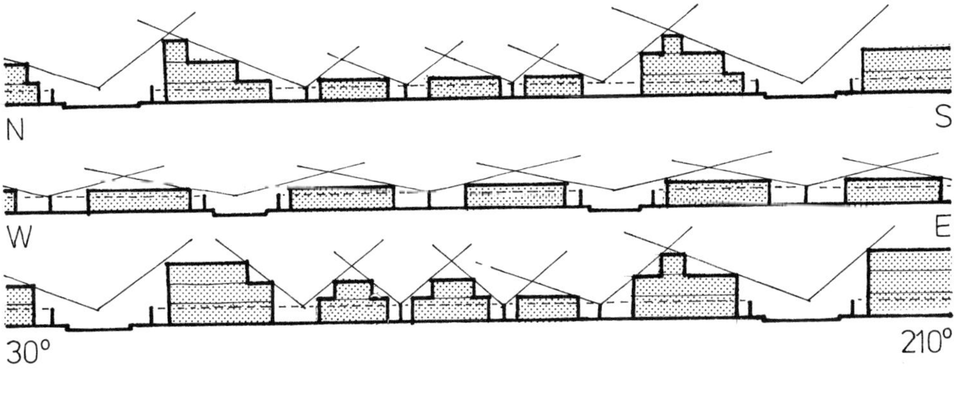

Figure 3. Variation of the volume of the solar envelope with orientation.

3. According to the plot ratios and densities obtained, buildings in rows or terraced housing between party walls give the highest densities. In the case of free standing buildings with facades requiring sun and daylight on all sides, large plots are needed with frontages of 16-20 m and buildings set back at least 3 m from the plot boundaries.

4. In general, the unification of building plots forming larger areas allows a considerable increase in the maximum permited floor area. However, larger plots may require additional controls to avoid excessively irregular street profiles.

5. The application of the solar envelope is especially apt for residential zones where solar gains can be used and the resulting medium and low densities are fully compatible with conventional building typologies. High density residential or mixed residential and comercial development require plots with larger dimensions (40m x 50m) together with favourable orientation (30' - 45' E or W of N); in this case the maximum solar access will be 4 hours.

6. In the case of mixed use zones, it is essential to adopt common criteria for the whole zone, avoiding variations for different uses, to achieve a uniform urban development which allows equal solar access opportunities for all plots.

7. In existing urban areas, it is essential to study the specific morphology of existing building development before establishing solar access criteria.

8. To promote and encourage the application of the solar envelope, it is important to adopt simple and uniform criteria for each zone according to the proposed and existing uses and the orientation of the urban grid. These criteria will also depend on the availability of solar radiation and latitude.

9. The variety of factors and criteria to be considered is very wide. During the development of this study, certain criteria were selected; the results obtained are indicative of application possibilities in Argentina and are related to the aims of this research. The solar envelope concept allows a wide range of criteria to be applied in relation to the specific planning objectives and the relative importance of solar access. The resulting urban tissue will correspond to the chosen criteria.

The results obtained show that the solar envelope is a feasible control mechanism to guide urban development if suitable criteria for solar access and implementation are chosen. The limits imposed by the envelope can be incorporated in planning codes using conventional indicators such as plot ratios, building heights, and building lines to regulate urban development without altering the characteristics of the urban tissue in the majority of the localities in Argentina.

7. REFERENCES

[1] Knowles, Ralph L. (1981) Sun Rhythm Form, The MIT Press, Cambridge, Massachusetts.

Decision project following informative system for environmental quality control

Prof.Arch.G CATERINA
Dipartimento di Configurazione e
Attuazione dell'Architettura
via Tarsia 31
80135 Napoli
ITALY

ABSTRACT . When urban context needs rehabilitation , the procedures for designing appropriate solutions could be various , according to different viewpoints selected in analysing the existing built . The methods of controlling environmental qualities of the ambient, could be identified employing a system of parameters based on the knowledge and classified according to their influence to the performance of the built at the variation of the external environmental factors.

The selected case-study , a small urban context in southern Italy , Santa Maria della Lobra - has been employed for verifying the proposed method and the parameters have been applied in order to control environmental qualities in urban design .

1. Introduction

Within the urban design the environment is one of the most provoking themes that the actual century has proposed.Broken the borders of the architecture at structural scale and/or at urban scale the cities,the villages,the industries,the countrysides, were conceived as a big circulation of air,water,nourishment and energetic flow."Let's plan the planet,the technique gives us the facilities",asserted Le Corbusier in the thirthies and in the sixthies it was added:

S. Alvarez et al. (eds.), Architecture and Urban Space, 67–73.

the human knowledge will furnish us the way to follow. But in the seventies the environment appeared in all his complexity: the needs are not so easy to be enumerated as Le Corbusier thought,the human environment is full of images,words,numbers,and objects in relation with them; in addition the systems of values that permitted collective plans,like the ones of Bernini in Rome and Haussmann in Paris,are not recognizable anymore.The abolition of unificatory principle,of hegemonic historical values has imposed to keep a constant relation with the concrete forms of our daily life in which are expressed the needs really perceived to recognize the common dominant forms and the diversities.The environment is not a collection of objects and people,it includes them,but it precedes them: it contains images,but itself is an inhabitated image.According to Prigogine,if we decide,we can isolate a crystal,but the city and the housing and work cells,separated from their environment,dye swiftly.Referred to the project of the environment,these assumptions hinder to act according to principles that we presume to be absolute and capable to influence and change , for their intimate nature,the envirommental quality.

The acting is in recognizing the complexity of the system and in definyng the relations of the project with the built interface.The decisional criterians are based on the knowledge and therefore the setting up of an informative system for the control of the environmental quality can indicate a different operative pathway for the decisional processes.

2.Experimental research

It was made an experimental research to define:

a) the lecture system of the urban environment in relation to the qualities of the environment itself;

b) the physic parameters that express the dominant shapes;

c) the characteristics of the different shapes;

d) the performances of the built at the variation of the external environmental factors.

For the experimental purposes it was chosen a small urban context in southern Italy:S.Maria della Lobra,that for its homogeneus disposition and for the presence of peculiar environmental values,allowed to study and verify a model of decisional approach to the theme of the environmental rehabilitation though:

1) the identification of the layout in the urban settlement in relation with the environmental characters of the area.
2) The main factors related to the modifications of the environmental quality.
3) The energetic variables in relation to the space variability.

References

-G.Ciribini (1990) La normativa dell'impatto ambientale,Alinea ed, FI-
G.Caterina (1989) Tecnologia del recupero edilizio,UTET ed.TO

We thank the architects G.Giordano and M.Contò for thein collaboration to the experimental project related to S.Maria della Lobra

INDICE

A – LETTURA DEI CARATTERI AMBIENTALI | A11 – INFRASTRUTTURA A RETE

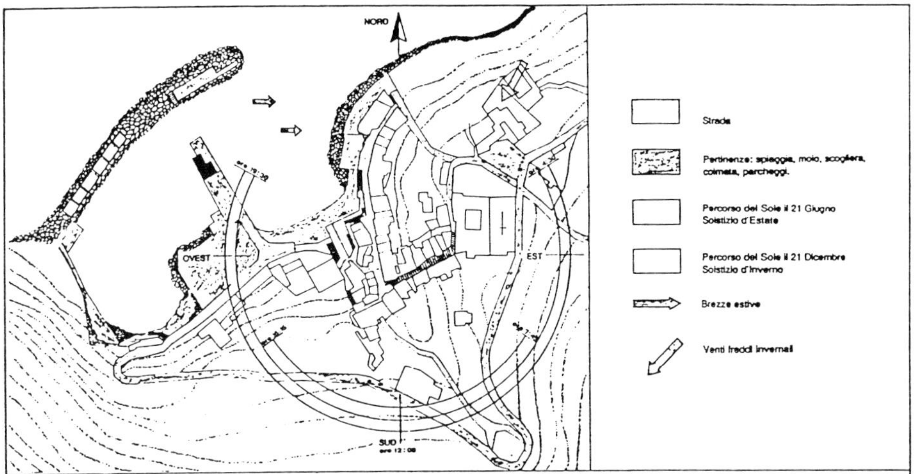

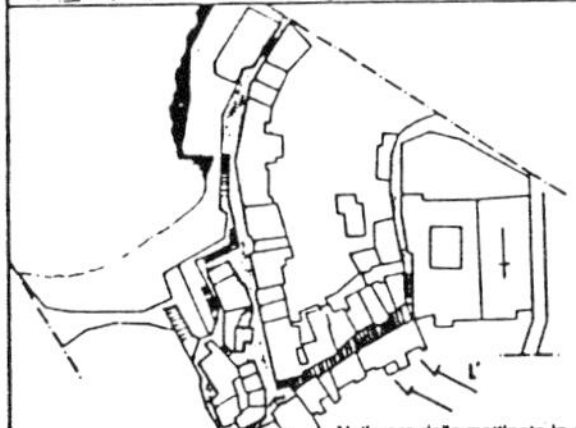

Nelle ore della mattinata la strada pedonale è ombreggiata dal costruito

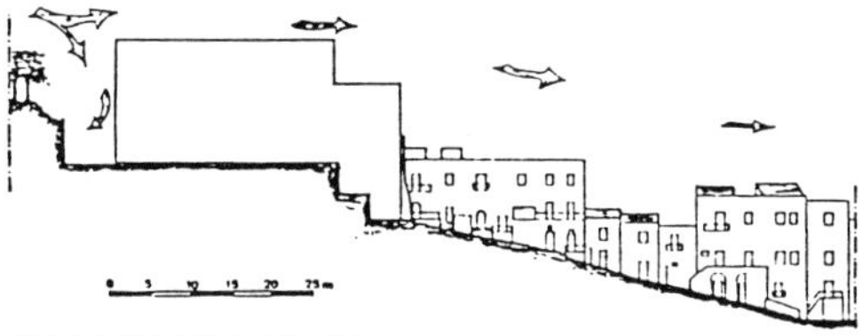

Sezione AA' - Tratto di strada che il Monastero ripara dai venti freddi provenienti da NordEst

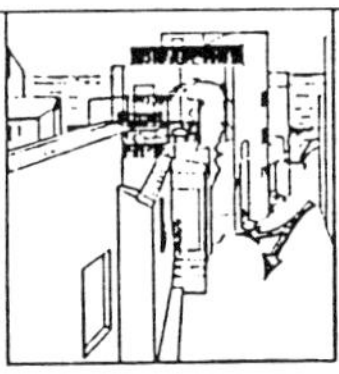

Veduta A - Strada che incanala le fresche brezze del mare

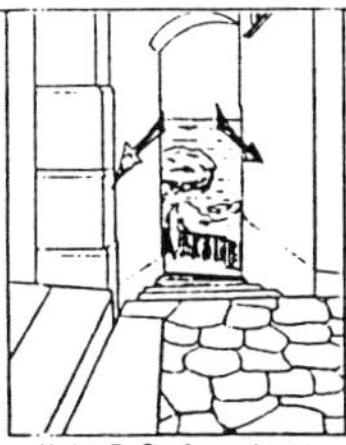

Veduta B - Stradina gradonata che porta al mare incanalando le fresche brezze estive

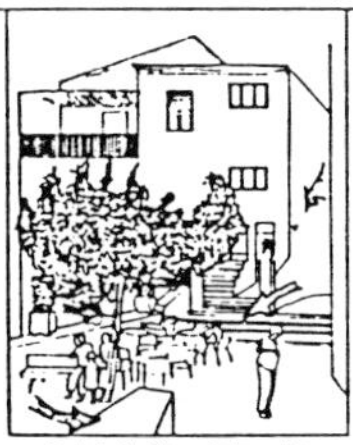

Veduta C - Piazzetta investita in Inverno dal freddo Maestro e rinfrescata in Estate dalle brezze di mare

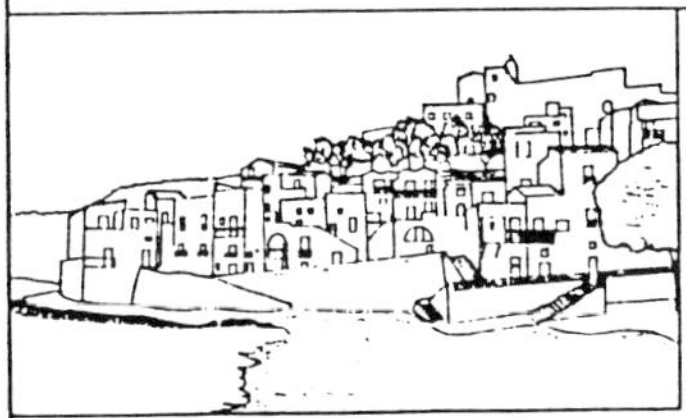

Piccole gradonate di raccordo del dislivello tra il costruito e la spiaggia

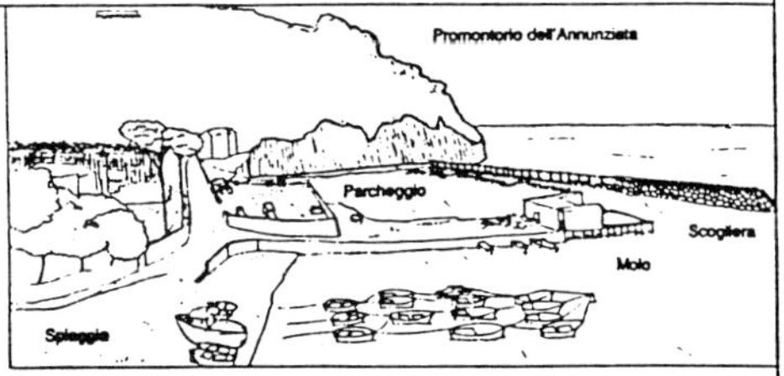

Veduta del piccolo porto - La spiaggia e il molo sono riparati dalla scogliera e dal promontorio dell'Annunziata. Si crea un piccolo porto tranquillo usato dai pescatori e dai turisti.

C = NORME PRESCRITTIVE	C3 = PRESTAZIONI: CONTROLLO DEL FLUSSO LUMINOSO

UNITA' AMBIENTALI	R.I.	Non vi è nessuna prescrizione.	
	D.M. '75	Art. 5. Tutti i locali degli alloggi, eccettuati quelli destinati a servizi igienici, disimpegni, corridoi, vani scala e ripostigli debbono fruire di illuminazione naturale diretta adeguata alla destinazione d'uso. Per ciascun locale di abitazione, l'ampiezza della finestra deve essere proporzionata in modo da assicurare un valore di fattore luce diurna medio non inferiore al 2 % e comunque la superficie finestrata apribile non dovrà essere inferiore ad 1/8 della superficie del pavimento.	D.M. '75 Vani abitabili dotati di sup. fin. > 1/8 sup. pavimento
	R.E.	Art. 34 Parte C. Riporta l'Art. 5 del D.M. '75, inoltre stabilisce che ... che la superficie finestrata apribile non deve essere mai inferiore a mq 2, nel caso che il locale sia provvisto di una sola apertura.	R.E. sup. finestre > 2 mq
ELEMENTI AMBIENTALI	D.M. '75	Art. 2 Comma 4. Le stanze da letto ... debbono essere provviste di finestre apribili.	D.M. '75 L = dotate di fin. apribile
	D.M. '75	Art. 2 Comma 4. Le stanze ... di soggiorno devono essere provviste di finestre apribili.	D.M. '75 S = dotate di fin. apribile
	R.I.	Art. 18. Ogni fabbricato, destinato ad abitazione deve avere un numero sufficiente di latrine costruite in modo da ricevere aria e luce dall'esterno e senza comunicazione con la cucina e le altre stanze di abitazione.	
	R.E.	Art. 34 b. ... I bagni possono ricevere aria e luce dall'esterno anche da spazi equiparabili a cavedi. Il rapporto tra la superficie delle finestre e quella dei pavimenti non deve essere inferiore ad 1/12 e in nessun caso la superficie della finestra può essere inferiore a mq 0,80.	R.E. sup. finestra / sup. pavim. > 1/12 sup. finestra > 0,80 mq

ORGANISMO EDILIZIO ABITATIVO	CLASSI AMBIENTALI	ELEMENTI AMBIEN	ELEMENTI AMBIENTALI
			GIUDIZIO DI VALORE
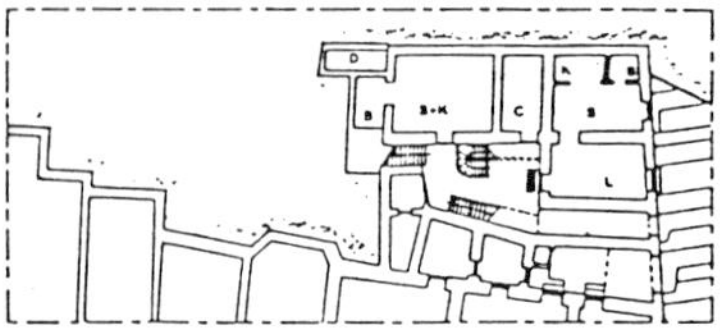 Pianta piano terra quota 2.00 m.	ALLOGGIO 1	SPAZI CHIUSI	S + K = D.M. sup finestra = 1,78 < 3,15 mq = sup pavimento / 8 — insufficiente illuminazione R.E. 1,78 < 2 mq — insufficiente illuminazione B = sup finestra = 0 — illuminazione nulla
	ALLOGGIO 2	SPAZI CHIUSI	S = D.M. sup finestra = 2,80 > 1,79 mq = sup pavimento / 8 — sufficiente illuminazione R.E. 2,80 > 2 mq — sufficiente illuminazione K = sup finestra = 0 — illuminazione nulla B = sup finestra = 0 — illuminazione nulla L = sup finestra = 3,30 > 1,54 mq = sup pavimento / 8 — illuminazione più che suffic.
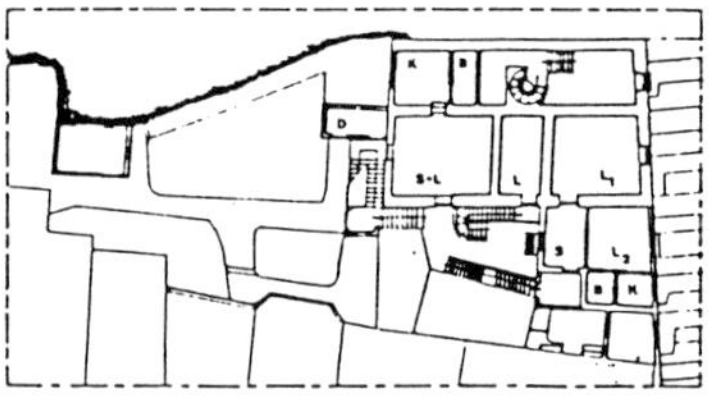 Pianta piano primo quota 6.60 m.	ALLOGGIO 3	SPAZI CHIUSI	S + L = sup finestra = 3 > 2,54 mq = sup pavimento / 8 — sufficiente illuminazione K = sup finestra = 1,02 < 1,19 mq = sup pavimento / 8 — insufficiente illuminazione sup finestra = 1,02 < 1,19 mq — insufficiente illuminazione B = sup finestra = 0 — illuminazione nulla
	ALLOGGIO 4	SPAZI CHIUSI	S = sup finestra = 2,4 > 1,54 mq = sup pavimento / 8 — sufficiente illuminazione sup finestra = 2,4 > 2 mq L2 = sup finestra = 1,4 > 1,3 mq = sup pavimento / 8 — sufficiente illuminazione sup finestra = 1,4 < 2 mq L1 = sup finestra = 1,4 < 2,12 mq = sup pavimento / 8 — insufficiente illuminazione sup finestra = 1,4 < 2 mq K = sup finestra = 10,5 mq (veranda) — ottima illuminazione B = piccole aperture — scarsa illuminazione
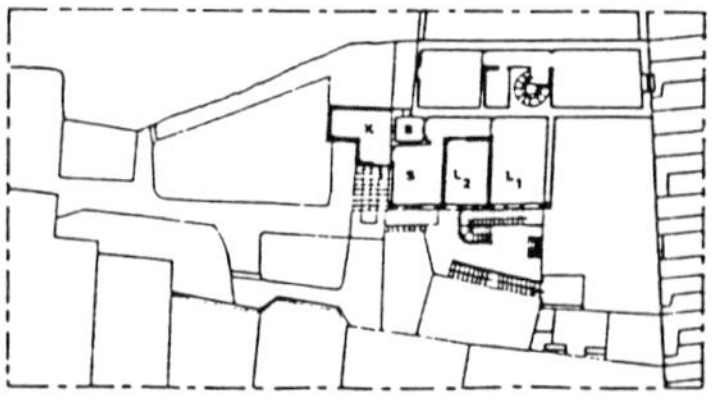 Pianta piano secondo quota 9.20 m.	ALLOGGIO 5	SPAZI CHIUSI	S = sup finestra = 0,7 < 0,85 mq = sup pavimento / 8 — insufficiente illuminazione sup finestra = 0,7 < 2 mq L1 = sup finestra = 1,88 < 2,92 mq = sup pavimento / 8 — insufficiente illuminazione sup finestra = 1,88 < 2 mq L2 = sup finestra = 1,88 > 1,5 mq = sup pavimento / 8 — sufficiente illuminazione sup finestra = 1,88 < 2 mq K = sup finestra = 0,8 > 0,44 = sup pavimento / 8 — sufficiente illuminazione

D - DISTRIBUZIONE DEGLI SPAZI INTERNI

D1 - SOLUZIONE A. TRASFORMAZIONE DISTRIBUZIONE INTERNA (PIANO TERRA)

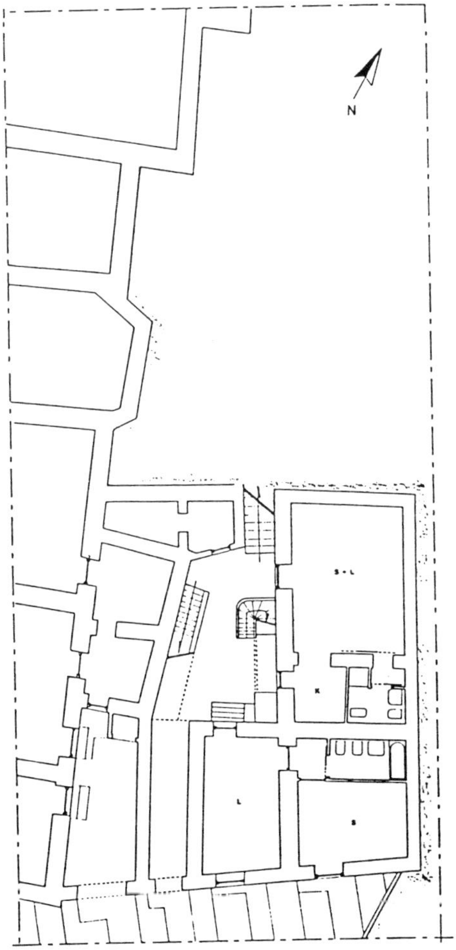

scala 1:100

piano terra - progetto

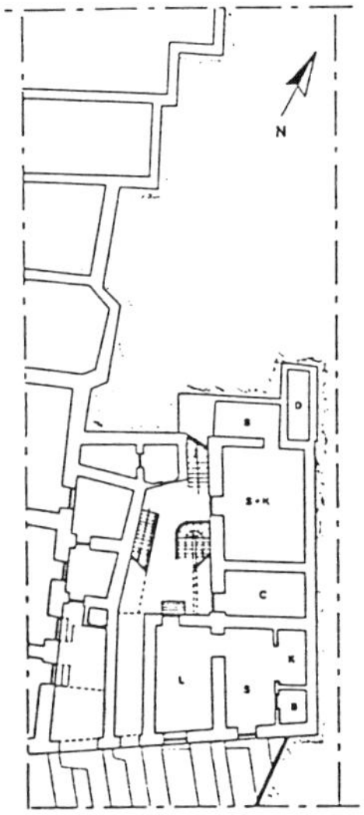

scala 1:200

piano terra - situazione attuale

ALLOGGIO n. 1

Situazione attuale

B: bagno con entrata dal soggiorno
Sup. tot. = 30,50 mq; insufficiente per due persone

Progetto

L'alloggio è stato trasformato in un monolocale di 41 mq per due persone.

ALLOGGIO n. 2

Situazione attuale

B: bagno con entrata diretta dal soggiorno
S: 12,28 mq; insufficiente per due persone.

Progetto

L'alloggio è stato trasformato in un appartamento per due persone.

CONTROL TECHNIQUES OF ENVIRONMENTAL QUALITY IN URBAN SPACE: RESEARCH AND CASE - STUDY.

GIAMPIERO CUPPINI, ANGELO MINGOZZI, GIOVANNI BACCI (*)
University of Bologna
Institute of Architecture and Urban Planning
viale Risorgimento, 2 (40136) Bologna
Italy

ABSTRACT. Urban quality control, considered on different dimensional scales, involves consideration of various factors some of these are closely related to the enviromental condition of the place: energy conservation and human comfort (psichological, igrothermic, acustic, visual).
The purpose of this study - partially financed by CNR - is to work out a methodology, parameters and techniques to optimize enviromental quality control in urban space referred to rehabilitation in particular.
A partial result of the research is described in this paper: a method to obtain static images by inserting the CAD drawing of the designed work into a picture of real enviroment. These images are useful instruments for checking and evaluating the visual impact of the designed work in terms of human visual and psychological comfort.
A case-study has been carried out to check the proposed method and the selected parameters and techniques.The selected case-study is about retrofit of a small "piazza" and annex new university building in the historical center of Bologna.

1. Objectives and results

The purpose of the study, financed by CNR (**), partially presented in this paper concerns methodologies, parameters and techniques to optimize enviromental quality control in urban space, referred to rehabilitation in particular.
Urban quality control, considered on different dimensional scales, involves consideration of various factors some of these are closely related to the environmental conditions of the place: human comfort (psichological, igrothermic, acustic, visual) and energy conservation.
This article briefly describes one result of the research relating to instruments for control the effects of interventions in urban areas in therms of visual and psychological comfort of the user.
In particular, we refer to the use of operative and methodological instruments for representing the visual impact of the designed work in the landscape "scenography".
This involves a study of urban landscape perception on the basis of visual and psychological categories.
Operators in the building process, concerned with the environmental compatibility of the designed work, can usefully refer to visual impact representations on the condition that any simulation is preminarily checked for dimensions, perspective and chromatic rendering.
Control methods are defined and tested in this study. The procedure involves the juxtaposition of a CAD "tridimensional" pattern of the designed work to pictures of real environment with the support of data processing technology.
The "visual fulcrum" of each image is chosen by mediating two different requirements: to get images with as many environment features as possible, and to give priority to the most usual views.

S. Alvarez et al. (eds.), Architecture and Urban Space, 75–80.

Any simulation is composed of two images: the present situation and the representation of the designed work. If the simulation is dimensionally, perspectively and chromatically correct, two comparable scenaries will be available for each view, thus enabling the operator to evaluate the visual-perspective impact of the designed work on the landscape.
The study and the analysis of methodology and operative instruments enabled us to pick out weak points, potentiality and precision now achievable with this representation technique.

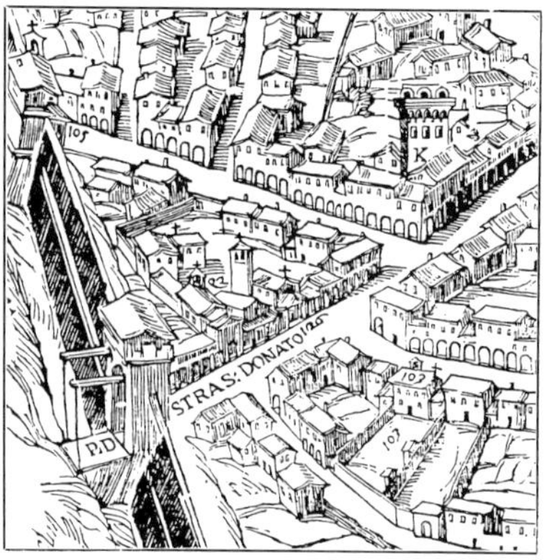

2. Methodology

The methodology developed in this research can be subdivided into five phases corresponding to eight partial checks and one final check (see block diagram of picture 1).

2.1. METHODOLOGY PHASES

2.1.1. *Design of simulation.* This phase involves choices - possibly modifiable in phases "2" and "3" - concerning the number of views and related "visual fulcra". The choice is based on the work designs and, if avaible, on the intervisibility chart, landscape units chart and existing visual and perceptive conditions chart.

2.1.2. *Execution of the CAD "tridimensional" pattern of the designed work.* This phase can involve a preliminary check of the choices made in the previous phase, by simulating the perspective views having the relative position of the change point related to the target point.

2.1.3. ***Survey of the existing state.*** This fase involves the use of the available technical instruments. It is divided into three subphases corresponding to three different checking phases:

- survey of flatness and height of the change point, target point and of at least one object included in the "visual fulcrum";
- pictures. It is essential to use "normal" objectives, i.e. having a perspective view similar to the one of human eyes; in some cases, (like in the case-study described at paragraph 3), however, the use of special objectives is advisable);
- computer processing of immages. This sub phase is the critical point of the process at the moment since image digitizing is mainly responsible for loss in simulation quality. The problem will be solved in the future by using still cameras. These instruments are already available but they produce images of poor quality at the present.

2.1.4. ***Rendering and pattern insertion.*** Pattern rendering, by using digitized images of real material too; insertion of the CAD "tridimentional" pattern into the chosen scenery with control of simulation perspective and dimension.

2.1.5. ***Output.*** Paper, slide, video and magnetic media are currently available as output supports.

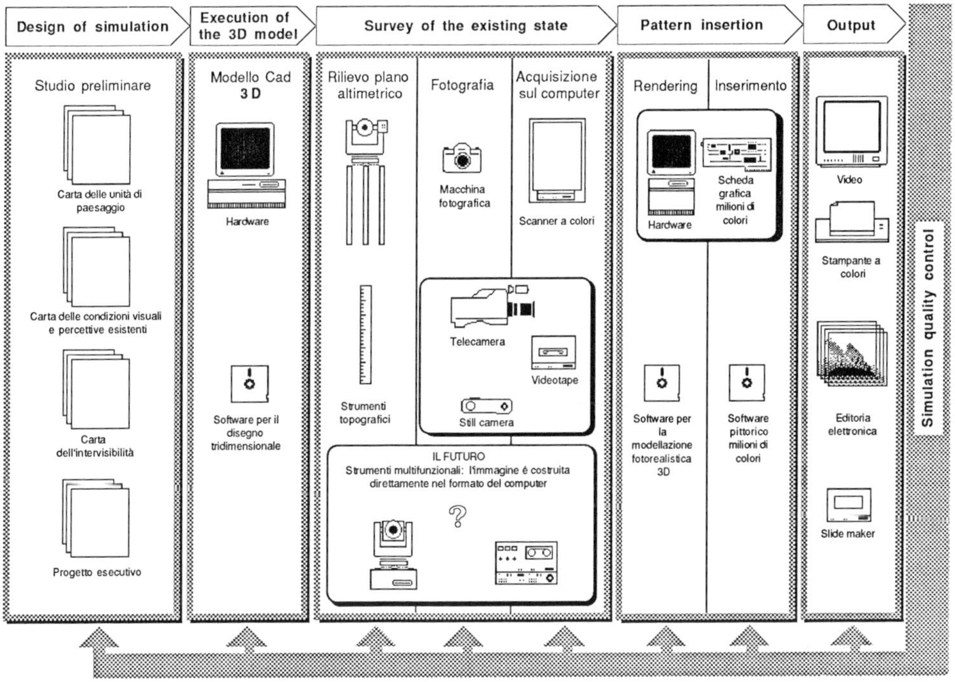

Figure 1. Methodology : block diagram

3. A case-study

A case-study has been carried out to check the proposed method and the selected parameters and techniques. The selected case-study is about retrofit of a small "piazza" and annex new university building in the historical center of Bologna.

The case study concerns the anlysis of the interactions between environment and designed work and vice versa, both relating to the igrothermic and to the visual well-being . In this paper we exclusively refer to aspects related to the visual impact of the designed building on urban enviroment.

The designed building includes three large halls with about threehunred places each. It is located in the cross-road area between Via Zamboni and Via Belle Arti to fill a gap of the urban structure caused by a demolition in the twenties.

Simulation is intended as an instrument for evaluations and for the achievement of the project objective, i.e. the mending of the urban fabric in the cross-road area.

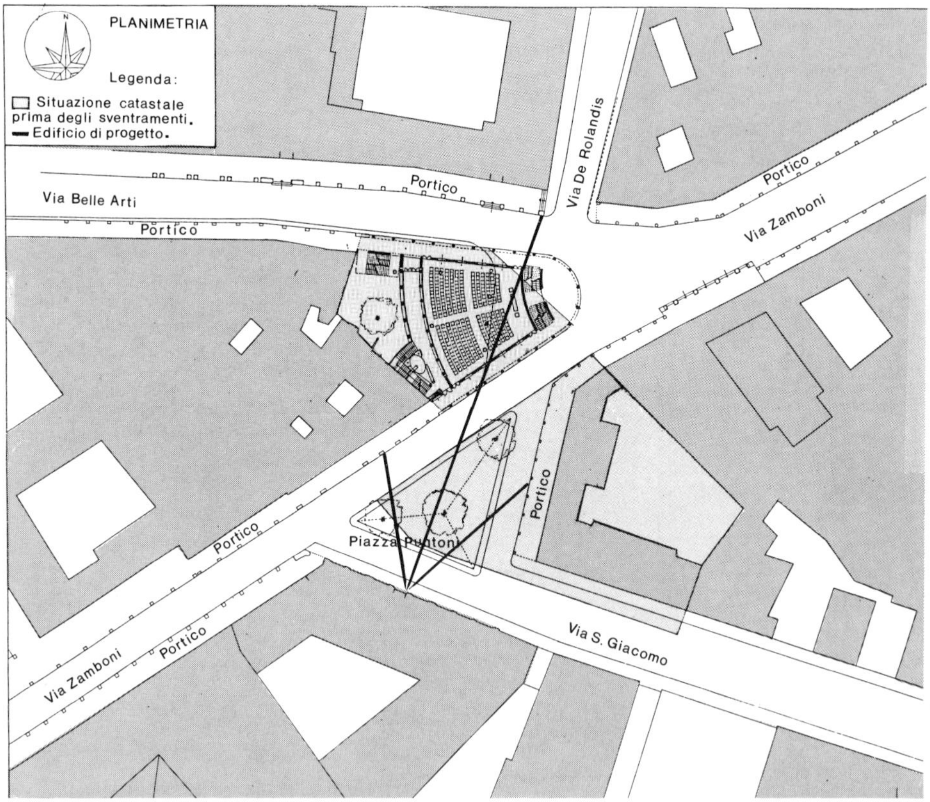

Figure 2. Case study: planimetry

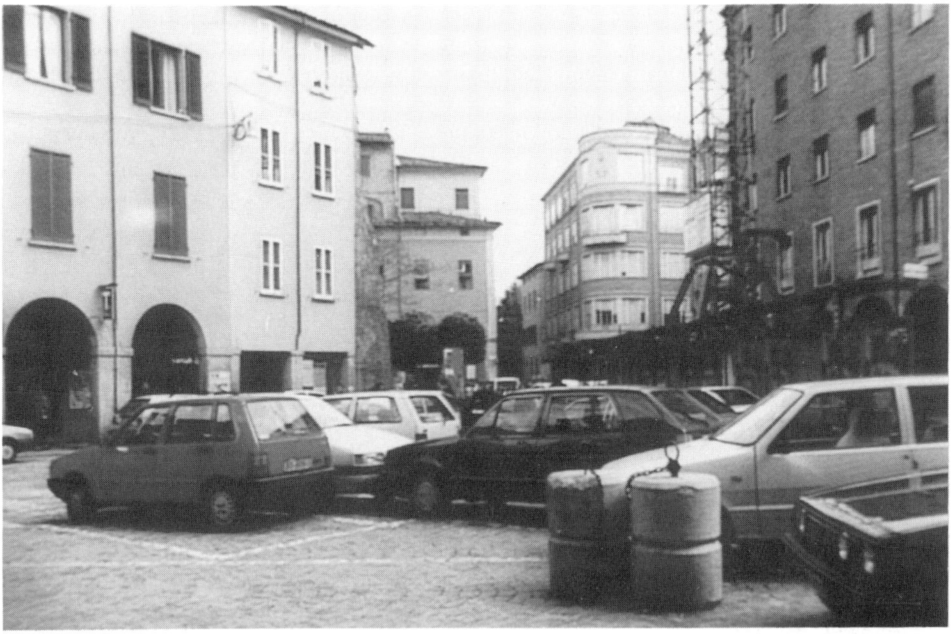

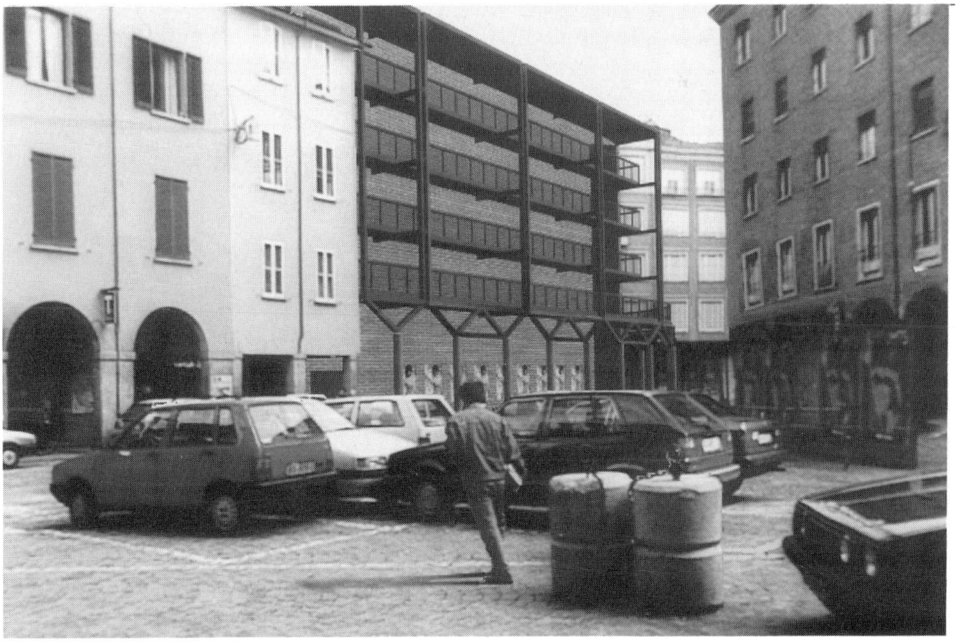

Figure 3. Case study: visual impact

Figure 2 shows the planimetry of the intervention area with indication of the "visual fulcrum", the change point and the target point of one simulation represented in figure 3.
It should be noted that the image contains disturbing elements such as the crane, the fance and some cars which were graphically eliminated, thus restoring the real situation and making the images more significant.
Figure 4 shows a visual impact simulation including a fance which marks the building overall dimensions on the ground.

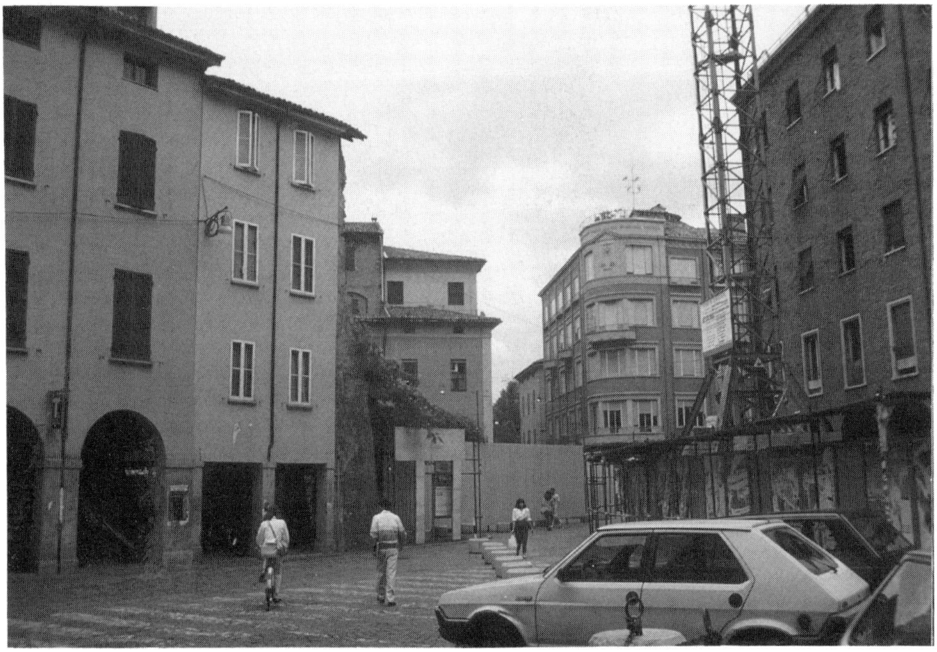

Figure 4. Case study: impact simulation with fence.

4. References

1. CUPPINI, G. "For a rehabilitation of the Bologna University district", Saecularia nona n° 3, Grafis edizioni, Bologna, (1990).
2. ONETO, G. Valutazione di impatto sul paesaggio, Pirola editore, Milano, (1988).
3. EBERHARD, D. "The importance of visual impact", Cadence vol. 5, n°5 (1990).

Note:

(*) A. Mingozzi ha curato i paragrafi 1 e 2; G. Cuppini ha curato il paragrafo 3; G. Bacci ha realizzato le immagini grafiche.

(**) I contenuti del presente articolo sono tratti da parte dei risultati della ricerca "Strumenti per il controllo di qualità dell' ambiente costruito" svolta, con la direzione scientifica di G. Cuppini, da A. Mingozzi nell' ambito della borsa di studio CNR bando n° 201.15.1 e dalle esperienze maturate da G. Bacci nell'ambito del Laboratorio di "Tecniche di Visualizzazione e Rappresentazione".

INTEGRATION OF URBAN EDGES AND ARCHITECTURE

G. INTINI
School of Architecture and Building Engineering
University of Liverpool
P.O. Box 147
Liverpool L69 3BX

Urban growth integration is either due to useful necessity, familiarity and aesthetics or due to the removal or adaptation of disruptive obstacles. Hence the study of disruptive or integrated forms of urban edges provides references for strategic intervention of urban therapy. However, the intention of every strategy should be intended as an attempt to expand a range of possibilities inherent in the nature of design rooted in a multiplicity of factors and systems determined by social, economic, technological and environmental parameters. Within this context art and science have a prominent and determining presence. A creative framework generates from the contrast between the stimulus of art abounding of emotions, desires, individuality, seeing, foreseeing and celebrating, and the impetus of science analysing the deficiency of knowledge with indifferent, impartial, rarely tolerant, observation of the idea behind the facts. Therefore the initial reactions to the projects have been intuitive followed by analysis to clarify the scopes and means of each project. Synthesis of analysed observations blend the various components offering a base for humanistic celebration to induce everyone's participation in order to accelerate integration. These processes allow patron and consultants active participation in a direct and personal dialogue. Parallel to this, a dialogue between the historical memory of the site takes place while the society observing and controlling bodies participate.

With this method, urban edges filtered through anthropological analysis highlight the dependency of historical urban growth on the notions of defensible space, need for shelter and its topologies, access to water and fertile land, transportation, climate and geomancy. Consequently permanent urban edges are identifiable in symbiotically integrated cases where the uniqueness of the edge celebrates urban presence.

Venice multi-functional canals typify edges instrumental to urban growth. Regretfully their use as a sewage route is ecologically unbalanced and thus requires intervention. However, it is the building design that integrated the canal edge allowing water penetration by raising living accommodation to a piano mobile which resolved the servicing and constructional problems together with the delicate relationship between public, semi-public and private spaces. Furthermore this solution has absorbed a variety of architectural styles and changes of use.

Architectural homogeneity as a direct result of the natural habitat is also described by Rodi Garganico (Italy) edged on a sea cliff initially for defence reasons. A different condition is found in

S. Alvarez et al. (eds.), Architecture and Urban Space, 81–85.

Locorotondo (Italy) also related to its valley initially for defence but with the advantage of access to fertile land and a suitable microclimate. Present conditions have an intensively cultivated valley and the town defence walls adapted into housing and artisan workshops surrounded by the original peripheral road relieving traffic from the town centre. Visually this edge is reinforced further by the man-made terracing of the valley. This manipulation was not left to chance but is a clear commitment for urban presence that can only be visible from the valley which automatically became a permanent edge. Permanency is also found in the rossian description of housing as a permanent element of the city and in the physical remains of the renaissance ideal town as identified in Palmanova and Ferrara. The latter adapted silted river beds as roads and the encircling defence ramparts as container for the first direct application of urban planning by an architect.

Another particular aspect of a permanent edge is the relationship of the urban setting and building design. Venice open spaces and buildings are designed to relate to a water edge without parapets. This solution adapted to the copeline is suitable for Liverpool Pier Head (England) dominated by remarkable buildings designed to front and be seen by the river Mersey. However, the exposure to harsh winds limits Pier Head suitability as a space for civic activities with neutral urban use. Where this condition is taken to interesting anthropological consequences is "il Campo" in Siena. Il Campo is not merely a civic space but a neutral area where the "contrade" rivalry is filtered via social interactions and events such as "il Palio". Especially during this event the piazza accommodates in public the release of the physical , social and psychological barriers between the contrade. Each "contrada" behaves as a secretive self-contained ministate, organised to worship their own horse and jockey hoping to win the coveted religious relic of the palio. Thus "il Campo" filters these passions but it is also a management device for the City to interact with visitors, and as an urban infrastructure to collect rainwater and avoid flooding the lower section of Siena. "Il Campo" geometry channels towards the lower section of the piazza openings of the drainage system which for further protection are lower than the raised portion of the piazza at this lower end. Both these social and technological considerations have informed the design for the piazzas of Mesquita, Pomarao and Minas de S. Domingos (Alentejo) in Portugal. These proposals are part of a regional strategy of renewal and are intended to bring civic cohesiveness to villages suffering abandonment due to the absence of infrastructures and stagnated economy although presenting unique architectural and urban forms set in natural and man-made environment of suggestive beauty. Essentially the strategy intends to integrate the existing socio-cultural activities with the proposed social and economical programmes together with the evaluation and preservation of the existing patrimony and cultural values. Hence the piazzas have the purpose to allow social interaction with visitors; to resolve the disjointed village centres by removing or integrating existing urban edges and collect rain water in tanks under the piazzas. Ad hoc solutions allow individual design variety. In Mesquita the central unmanaged area is adapted into a multi-functional market - entertainment - relax piazza, whereas Minas de S. Domingos has the two portions of the village cut by a road brought together by a centrally placed piazza incorporating the existing church and new commercial spaces to concentrate tourism away from housing. The solution for Pomarao allows the river to penetrate the village through a new marina, tourist and local centre and workshops surrounding the piazza while reinstating the original river edge. Thus with these intentions the edge is also reconsidered for new use and development potential, enhanced further by the application of solar passive energy and low technology application in order to recuperate the unique housing stock.

Conceptually the above strategy is also suitable for air space above urban railways and motorways but with the added requirement to reduce their impact as noise, environmental and visual polluters. To achieve this goal, the proposal to use air space above the railway penetrating Bari city centre in Italy suggests the construction of a superdeck linked to two peripheral wings. Positioned on the superdeck would be the new School of Architecture and an open house to allow a better integration of Bari with its interland. The superdeck would also act as a service spine and join with a landscaped route the two portions of the city separated by the railway. In addition traffic congestion would be reduced whereas movement would be possible within a walking radius and public transportation. Still in Bari a similar intervention over the motorway bordering the new stadium suggests the application of solar energy, supporting sport facilities and improved user friendly pedestrian links. Highlighting this proposal is the old dilemma of monumental architecture with limited use and the shortage of sport facilities available to citizens. To improve on this situation a sports for all strategy accounts for the new stadium as a civic flagship on the metropolitan level; the existing stadiums as intensive and back-up use at city level and intermediary facilities sited in neglected urban areas plus sports halls connected to educational institutions. Coordination and maintenance could be undertaken by area laboratories sited strategically and computer linked in order to engage in programmes of social integration of sport through information, motivation and assistance while attempting to contain the risks of vandalism and degradation. Hence a strategic location is the servicing edge positioned above the motorway closely connected with the new stadium. Housed on the deck would be an area laboratory and a research centre for sport injuries, performance and dieting designed using passive solar energy. Indeed such a design solution is also suitable for other sections of motorways and railways using floating energy sub-station to harness solar energy.

Historically the idea of bridging over urban edges has examples such as Pontevecchio in Florence and the Rialto bridge in Venice, and based on these references a bridge containing commercial kiosks and connected to urban sport facilities would link through a pedestrian route, the river front and shopping area of Liverpool. The bridged over Strand, a typical urban motorway, would be intensively landscaped and act as a route for cable technology to reach the office buildings. The improved pedestrian flow through an existing landscaped area suggests the interpretation of such areas as changeable, flexible spaces as applied for the competition concept to redesign Piazza d'Aosta facing Milan Central Station following the completion of the third underground line. Piazza d'Aosta is surrounded by architectural landmarks, intensive traffic inclusive of railway, underground, air terminal, buses, trains, taxis, cars, bicycles and pedestrians. The area contains shops, offices, hotels, restaurants, workshops, services and housing and is subjected to a varied daily influx of people throughout the year. Hence the concept proposes the piazza as a multi-functional pedestrian peninsula intensively landscaped and connected with the surrounding areas, edged out by traffic, via underground links which will house services and shops. Positioned on the piazza would be spaces for relaxation, sport, a multi-functional space surrounded by commercial kiosks, a new, prominent entrance to the underground station and a day nursery with children play areas to benefit the families living and working in the area. This arrangement would suit the flexibility required for urban areas subjected to varied use. In addition the intensive man made landscaped spine extending from the piazza towards the historical city centre would have a salubrious effect on the quality of life of Milan polluted city centre.

This salubrious conception of the city has interesting opportunities but especially when it is nature that intervenes as is the case of the disused Liverpool - Ainsdale railway line now a green corridor 20 miles long. Originally an aggressive and disruptive urban edge the loop is now an healthy space suggesting creative implications if linked with other green areas and established human links within the city. A web of these routes enclosing areas in need of renewal, would act as regenerating catalysts for urban archeology, research, relaxation and environmental friendly people route. A similar case is the area surrounding and linking the Portuguese villages described above, since it includes 20 km of abandoned mining area which nature is finding difficult to reappropriate due to the extensive damage done by mining. Despite this the area is ideal for tourism, film making, scientific study, industrial and mining archeology and a research programme to introduce planting. A museum of mining housed in a derelict warehouse adds permanency to the strategy whereas interpretation booths scattered around the mining provide referential material to the visitors. Such man made interventions are also necessary when confronted with the renewal of villages engulfed by the traffic reaching for the city centre from the suburban areas.

Wavertree, Liverpool 15, is in such a situation and the renewal of the High Street together with other aspects of the village is part of a research study to halt its decline. With the simple intervention of a relief road, located by the park and designed to absorb the High Street vehicular traffic without isolating the park from the village, the High Street becomes a renewal space capable of changes in land use and adapting new urban space with multi-functional activities free from vehicular presence around the existing landmarks and civic buildings.

The manipulation of traffic to reduce its noise impact and visual presence was determining in the design of the Bilal Cultural Centre in Liverpool and the Europan 2 Competition of Essling, Vienna in Austria.

The Bilal Cultural Centre implements 3 priority programmes, housing (175 units) a Mosque and the Islamic Centre. Urban analysis suggests that it is the location of the Mosque which determines the urban location of the 3 priorities. Phasing is also incorporated in the architecture which is self-contained and independent of the other interventions. Localised services and parking areas at the perimeter of the site allows the majority of the area to be pedestrian friendly and intensively landscaped.

Essling, a village included in Vienna's metropolitan area, is a major development area for the city's housing shortage. 550 housing units of 90 sqm. each and the integration of the village centre with the listed parkland and new development of commercial and leisure on 130,000 sqm. of land. The design includes three short roads to reach the housing units allowing two central landscaped pedestrian routes directly linking to the park, the village while connecting each dwelling. Additionally a relief road would liberate the village centre from intensive traffic thus affording the opportunity of creating two piazza's of a different nature. One piazza will be a direct link to the parkland and the new housing. The nature of its architecture is symbiotic with its natural surrounding while presenting a fabric scaled to enter the village without disruption. The architecture is of its time, fit for its purpose hence not based on style but an ad hoc response to the programme.

Traffic is also removed from an area of metropolitan scale such as the site of the Tokyo International Forum Competition. One of the concepts proposed is the use of the building to buffer the inner area of the site in order to eliminate any vehicular presence and create a peaceful and user friendly area. The relationship of architecture and urban scale is on two levels, one responsive to the existing urban scale and the other symbolical in order to respond to the programme in a progressive and individual architectural presence.

The above account describes urban edges to be relevant to urban conglomerates only when they have specific useful functions. Their application should be absorbed within a design approach open to suggestions, flexible and informed by the whole city. Consequent benefits should stabilise and use urban resources to offer urban dwellers better quality of life. Urban resources are identified in the architecture, the ecosystem and the people. Their relationship offer unquantifiable opportunity in the design of architecture capable to harness the environment without limiting its artistic celebrative presence. Hopefully this would, supported by continuous political commitment, eradicate urban isolation and alienation, while allowing new urban programmes informed by humanistic considerations, both collective and individual, and rejecting shortsighted commercialism, political opportunism and tacky uninspired architectural and planning styling.

Note. All the projects described are the work of Studio Intini based in Noci (Italy) and Liverpool (England). G. Intini is also a lecturer and design tutor at Liverpool University School of Architecture and Building Engineering.

G. Intini, School of Architecture and Building Engineering, University of Liverpool, P.O. Box 147, Liverpool. L69 3BX (051 794 2591)
Studio Intini, Via Pentimi 24, 70015 Noci Bari. (080 8978710)
Studio Intini, 16 Salisbury Terrace, Liverpool. L15 9HD (051 733 3343)

AN "ENVIRONMENT - SITE - URBAN SPACE - ARCHITECTURE" CARD INDEX.

Jean BOUILLOT
Architect
3, rue de l'Agent Bailly
75009 PARIS
France.

The aim of that card index is to collect for analysis and comparisons a sufficient quantity of case studies, selected among modern or (and) traditional built situations, so as to give on the one hand a good basic information and efficient design tools to the different space practisers, on the other hand a good general idea of the spatial conditions, through their different scales, in a geographical entity; for that last reason, that card index can be done at different and successive scales: planetary, national, regional, micro-regional.

All the cards are set up on the same basis and each one gathers both climatological and space informations.
The climatological informations are gathered on the top of the card and give datas about latitude, altitude, prevailing winds (direction), temperatures (average daily maxi & mini), bright sunshine (average monthly), precipitations (average), relative humidity (average maxi & mini).
The space informations are given following four scales (environment, site urban space, architecture), and each one of these four scales is introduced by a color slide showing the best standard in the place, followed by a bioclimatic analysis (bioclimatic chains for the two first ones, bioclimatic charts for the two others), and by the bioclimatic principles which proceed of the analysis. These space informations can be read vertically following the four scales (all the slides, all the bioclimatic analysis, all the bioclimatic principles), or horizontally at each scale (slide, bioclimatic analysis and principle).

The filing of the card index complies four geographical scales according to the climatic zone (cold, temperate, hot arid, hot humid), the continentality (insular, littoral, continental, mountaneer), the topography (plain, slope, valley, table-land) and the land-use mode (natural, rural, urban, industrial).

That card index can introduce different manipulations and analysis: proceeding vertically, the four scales for one case study, or horizontally, the same scale for several case studies, or by theme (urban-mountaneer sites, etc,), all the datas on each case being input from the beginning.

S. Alvarez et al. (eds.), Architecture and Urban Space, 87–92.

I COLD RUSSIA USSR	3 2 CONTINENTAL LITTORAL CARELIA	A PLAIN ONEGA LAKE	b RURAL KIJI	I 3-2 A b

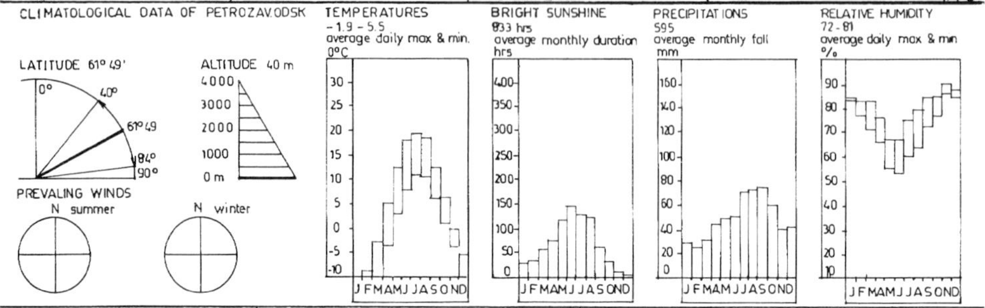

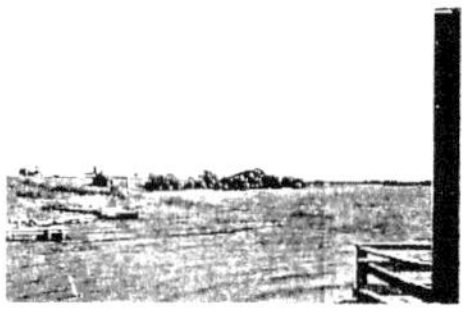

ENVIRONMENT BIOGRAMME

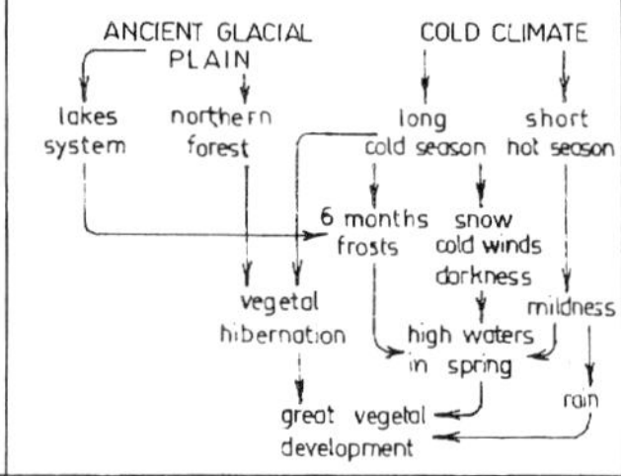

ENVIRONMENT BIOCLIMATIC PRINCIPLE

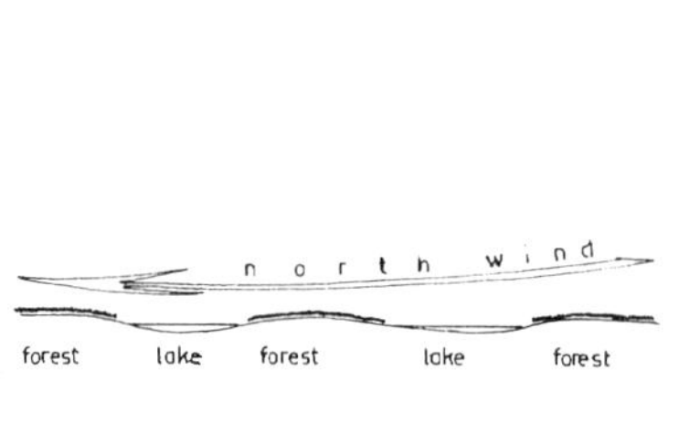

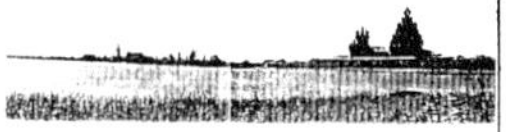

SITE BIOGRAMME

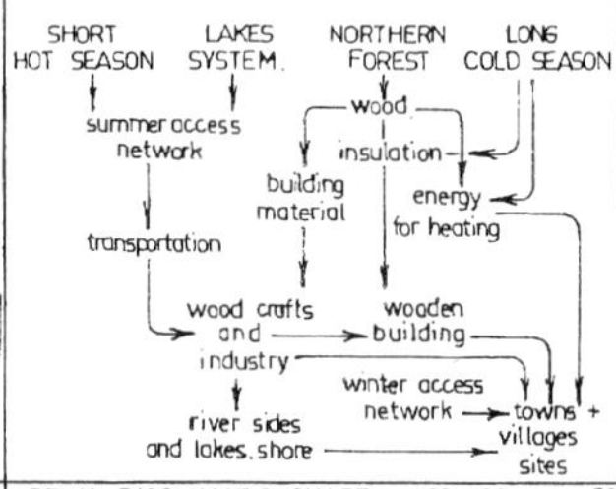

SITE BIOCLIMATIC PRINCIPLE:

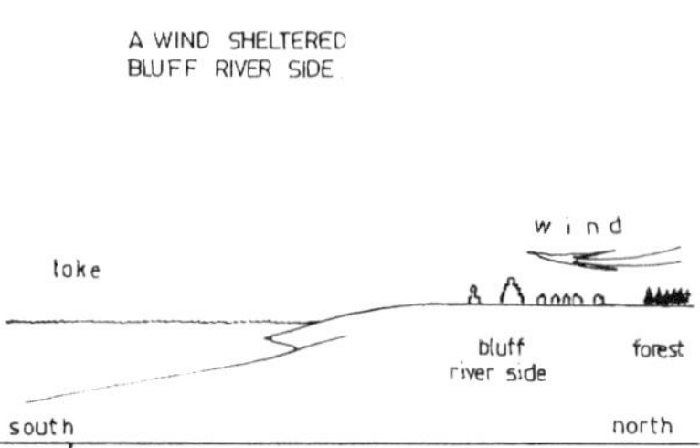

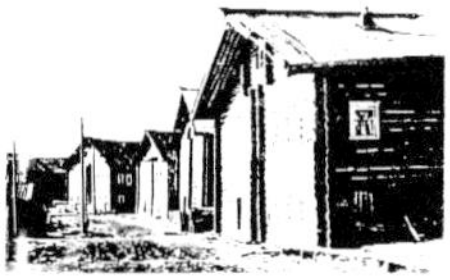

URBAN BIOCLIMATIC CHART

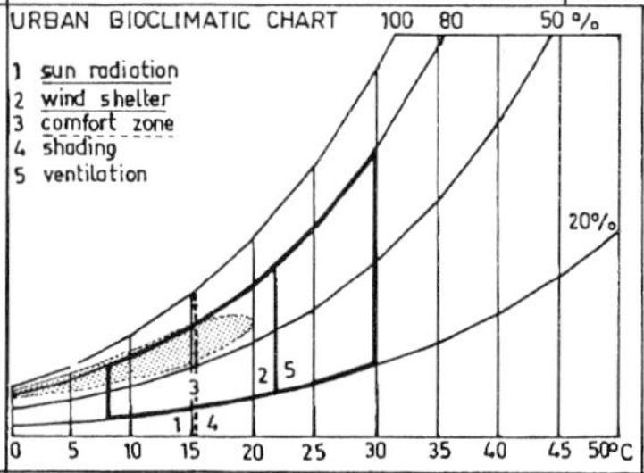

URBAN SPACE BIOCLIMATIC PRINCIPLE

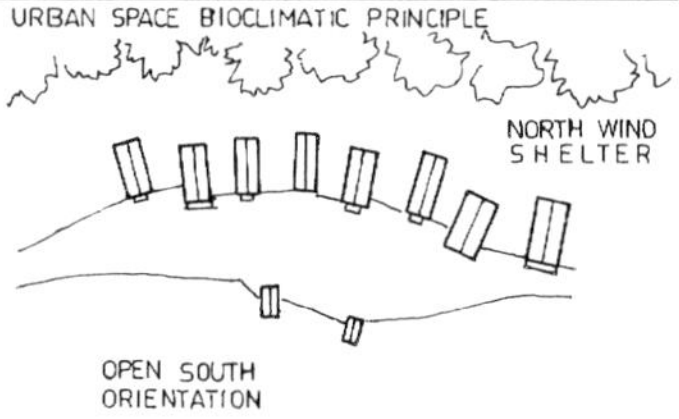

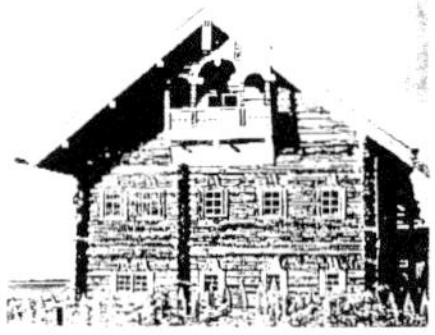

BUILDING BIOCLIMATIC CHART

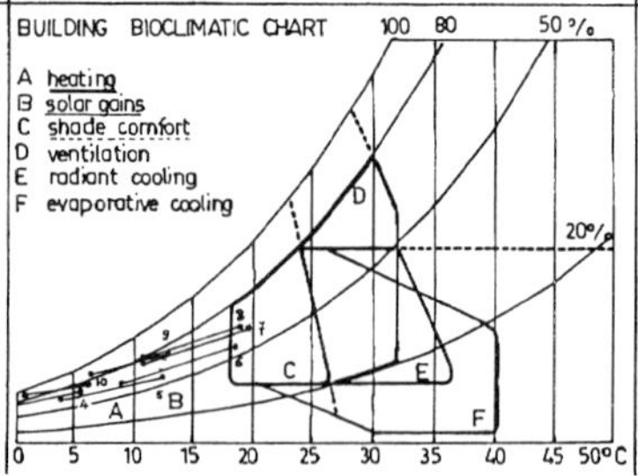

BUILDING BIOCLIMATIC PRINCIPLE

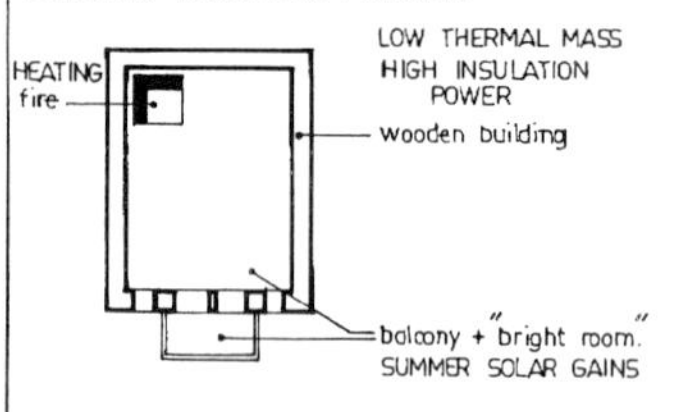

II TEMPERE Europe France	3 CONTINENTAL Bourgogne Cote d'or	BC VERSANT - VALLEE Montagne	b RURAL Corpoyer la Chapelle	II 3 BC b

DONNEES CLIMATOLOGIQUES DE BAIGNEUX LES JUIFS (61-80)

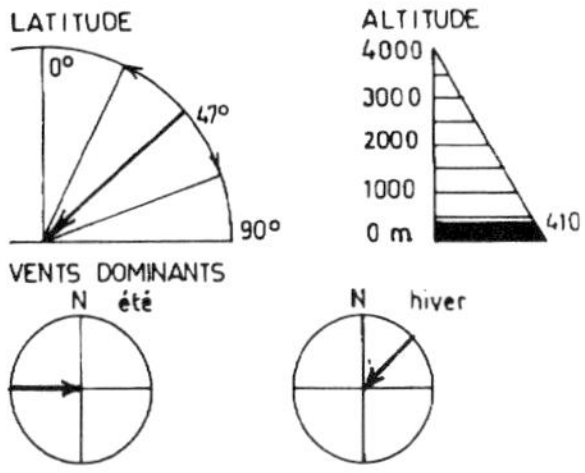

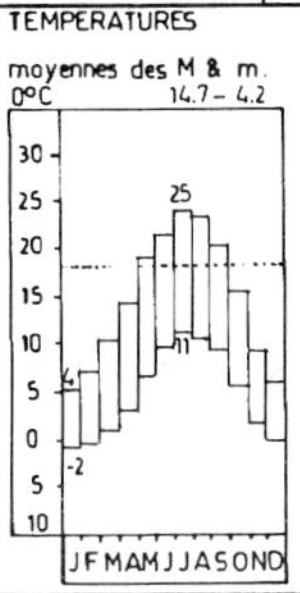

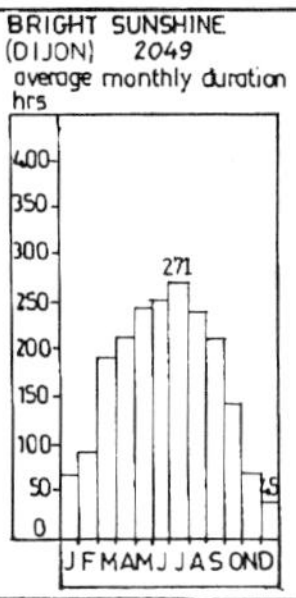

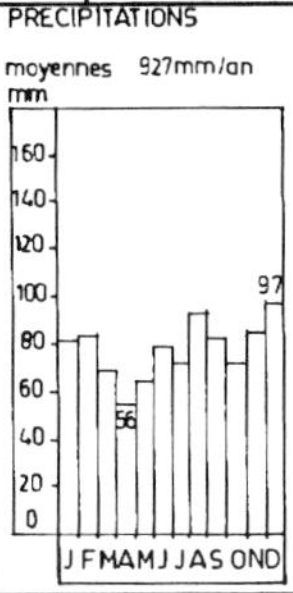

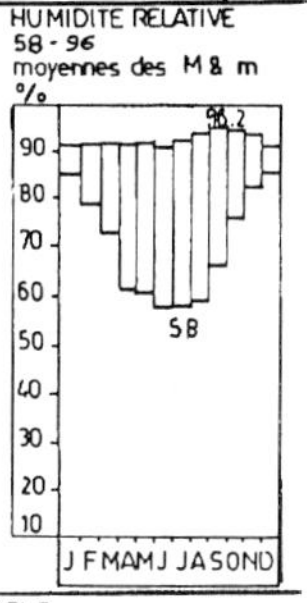

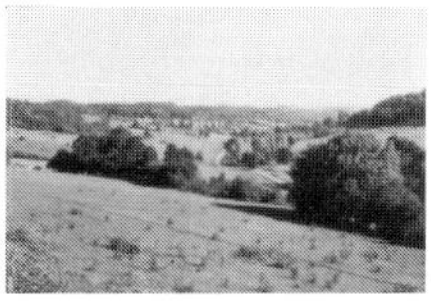

ENVIRONMENT BIOGRAMME

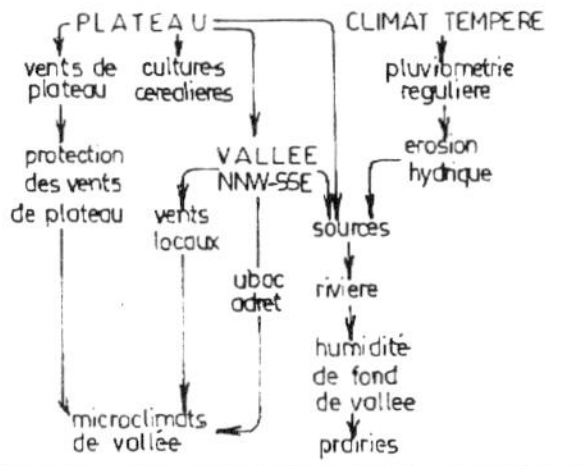

ENVIRONMENT BIOCLIMATIC PRINCIPLE

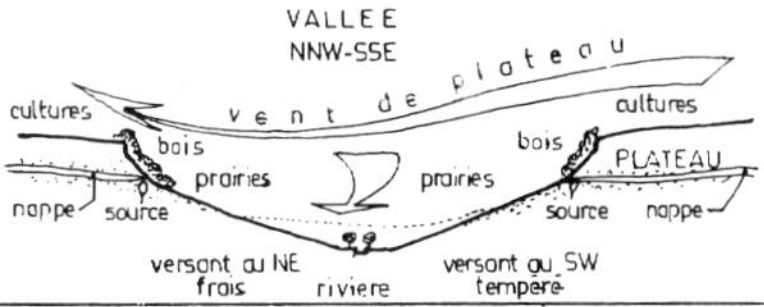

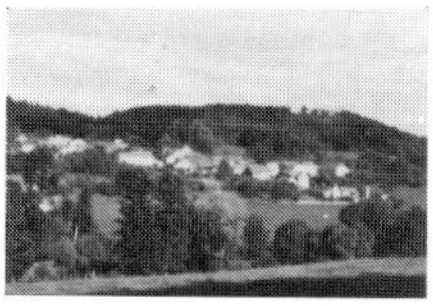

SITE BIOGRAMME

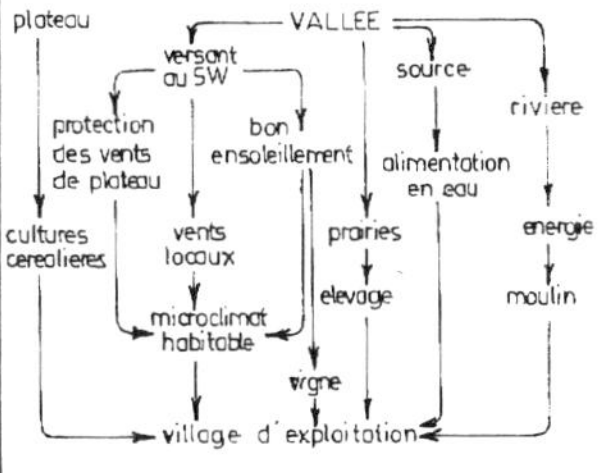

SITE BIOCLIMATIC PRINCIPLE

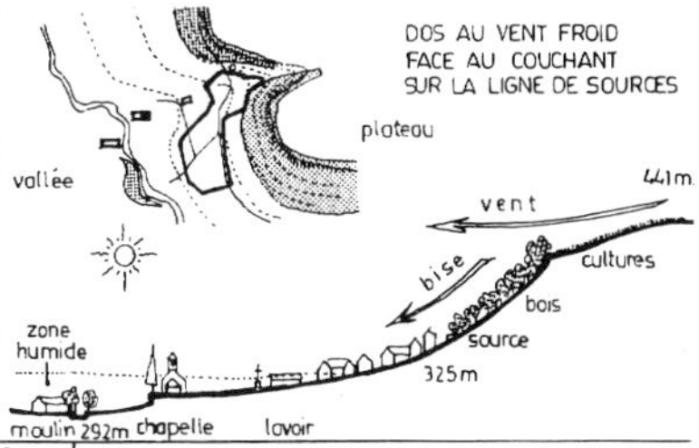

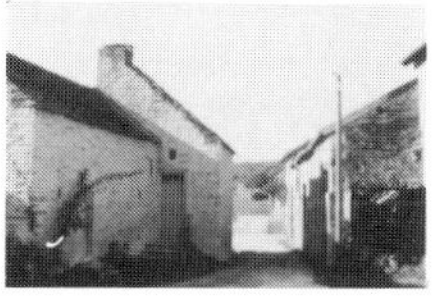

URBAN BIOCLIMATIC CHART

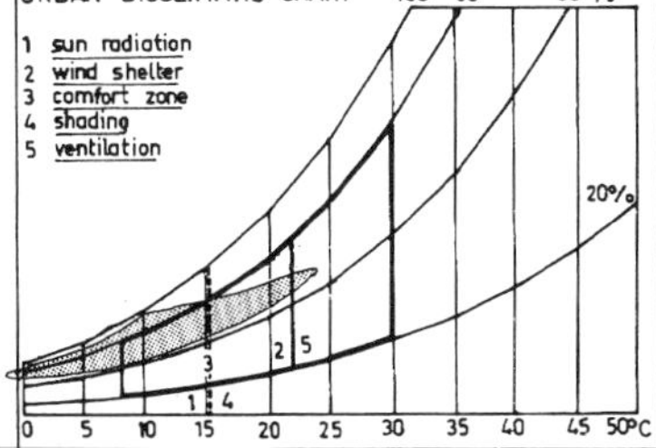

URBAN SPACE BIOCLIMATIC PRINCIPLE

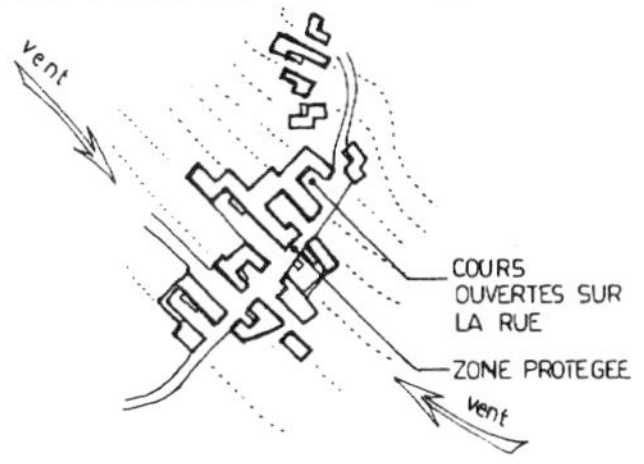

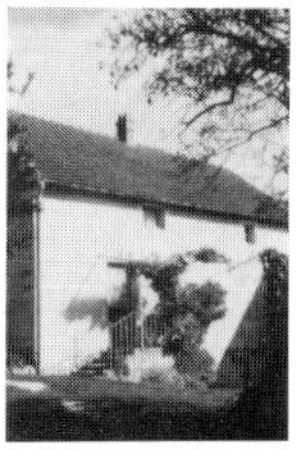

BUILDING BIOCLIMATIC CHART

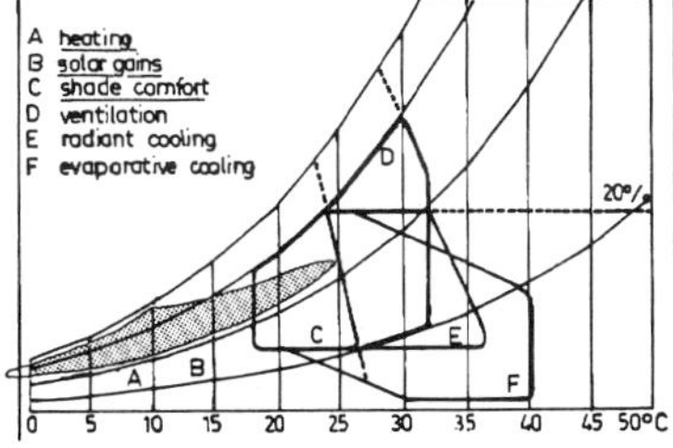

BUILDING BIOCLIMATIC PRINCIPLE

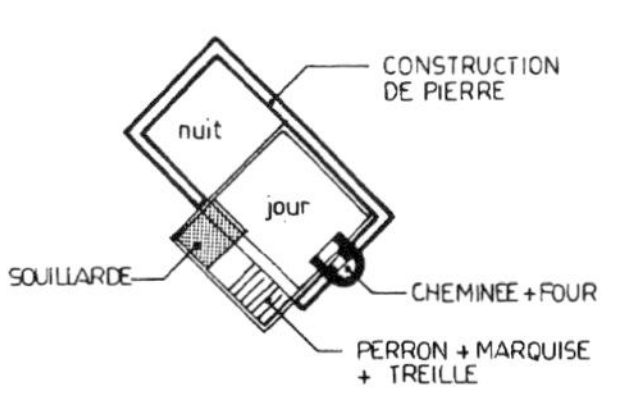

III HOT-ARID Middle East Iran	3 4. CONTINENTAL-MONT.	D. PLATEAU	b RURAL Moorcheh - Khort	III 3-4 D b

CLIMATOLOGICAL DATA OF ESFAHAN (1893 1941)

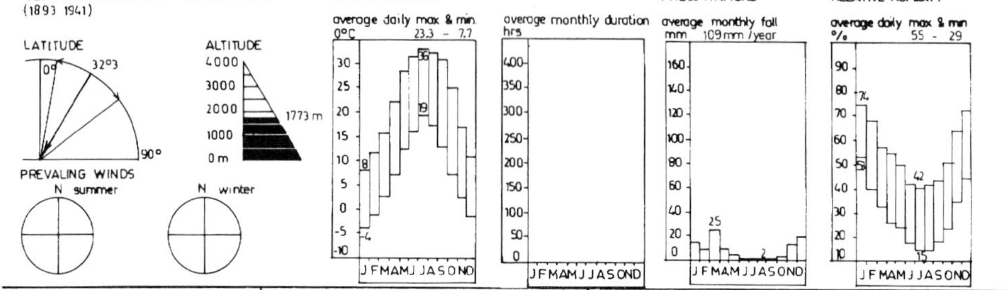

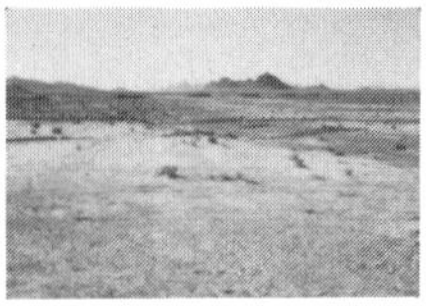

ENVIRONMENT BIOGRAMME

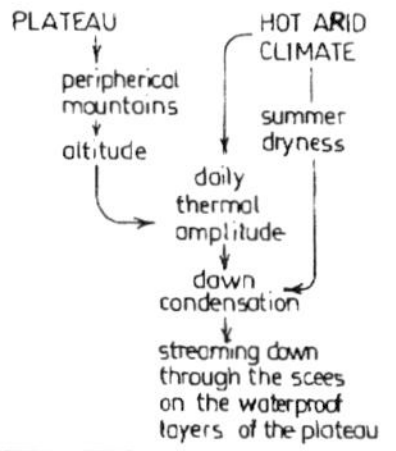

ENVIRONMENT BIOCLIMATIC PRINCIPLE

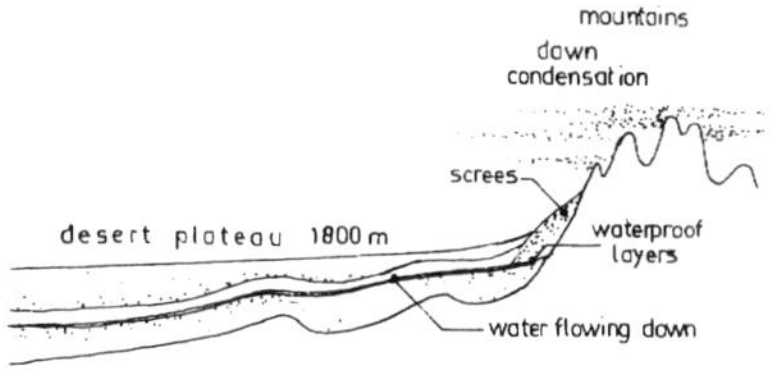

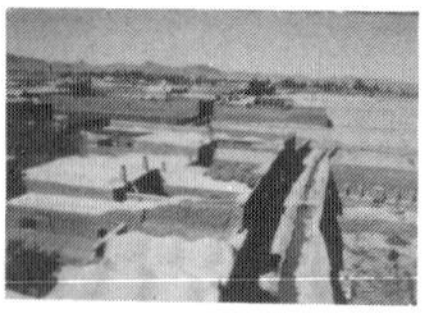

SITE BIOGRAMME

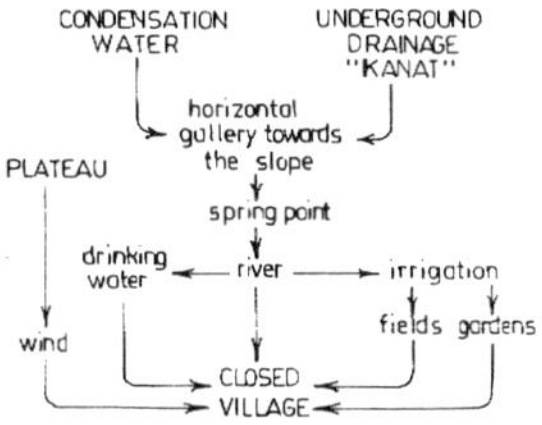

SITE BIOCLIMATIC PRINCIPLE

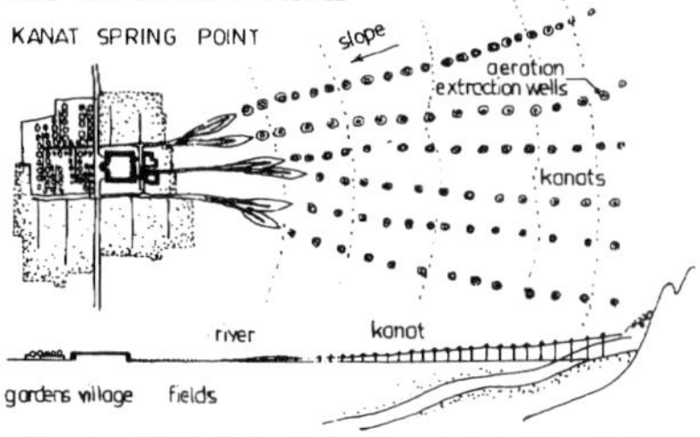

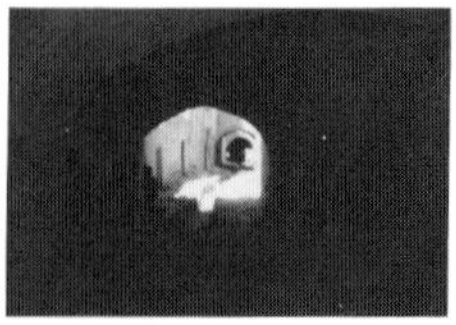

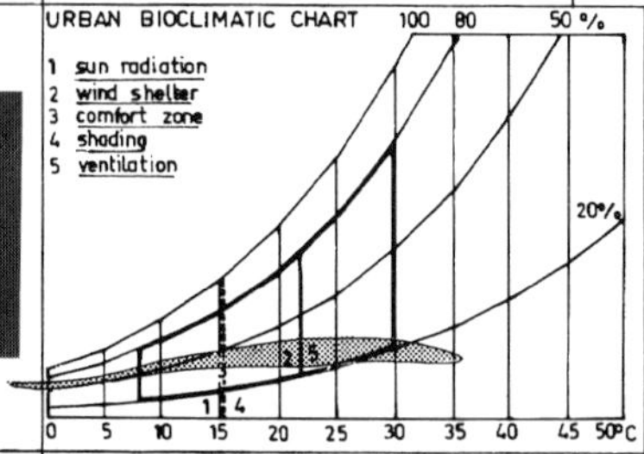

URBAN SPACE BIOCLIMATIC PRINCIPLE

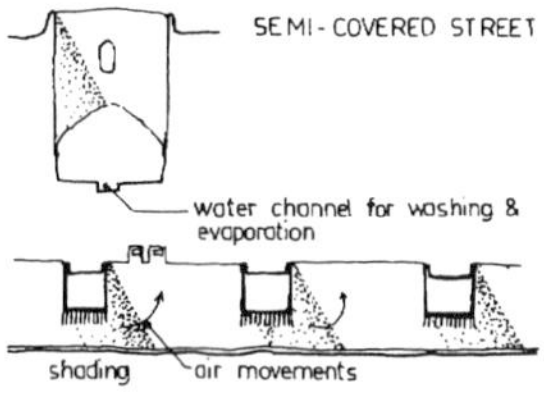

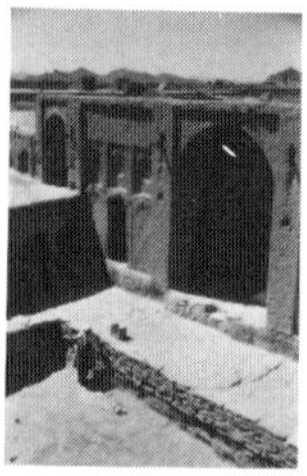

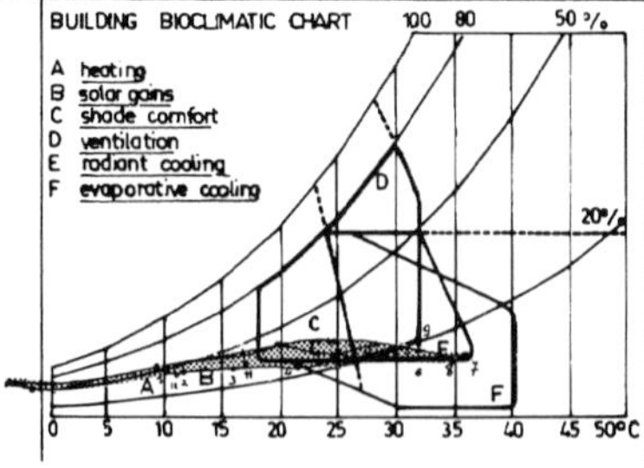

BUILDING BIOCLIMATIC PRINCIPLE

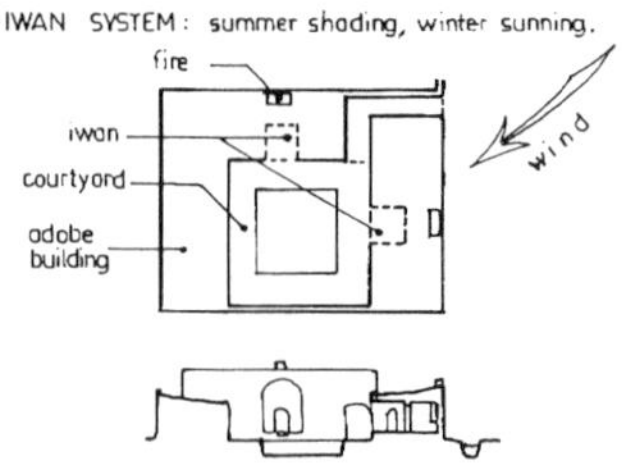

IV HOT HUMID West Africa Togo	3.2. CONTINENTAL LITTO. Litime	A.B. PLAIN-SLOPE Badou	b. RURAL Kpete Bena	IV 3-2 A.B.b.

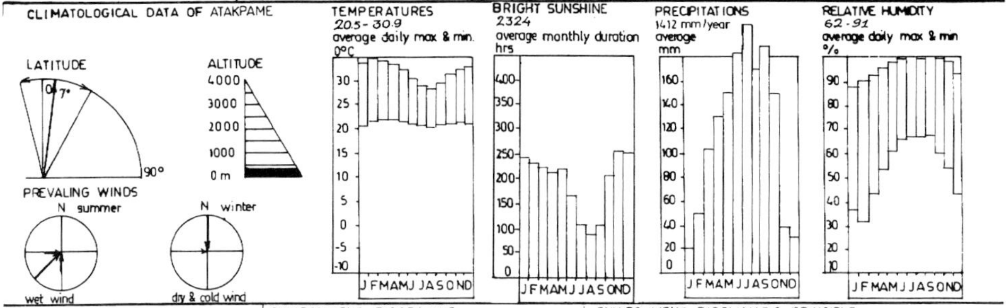

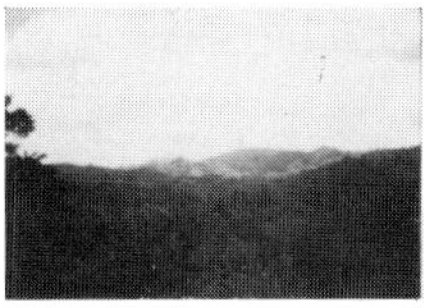

ENVIRONMENT BIOGRAMME

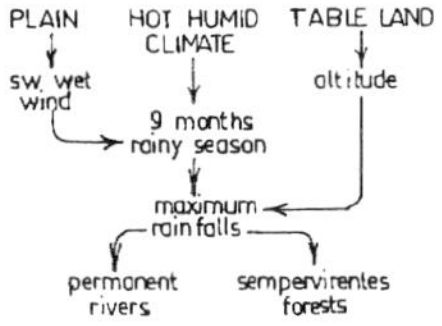

ENVIRONMENT BIOCLIMATIC PRINCIPLE

TROPICAL FOREST

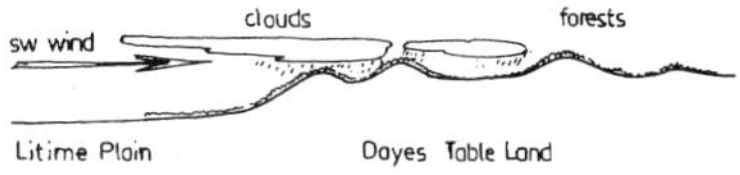

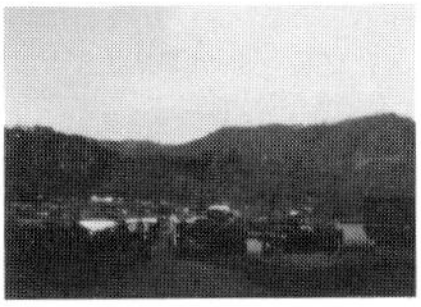

SITE BIOGRAMME

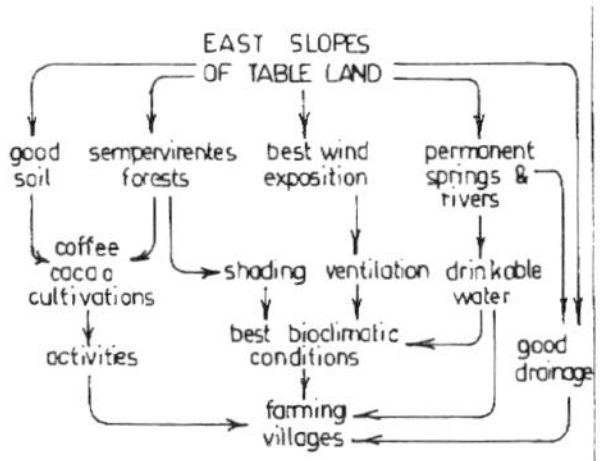

SITE BIOCLIMATIC PRINCIPLE

A WELL VENTILATED GLADE

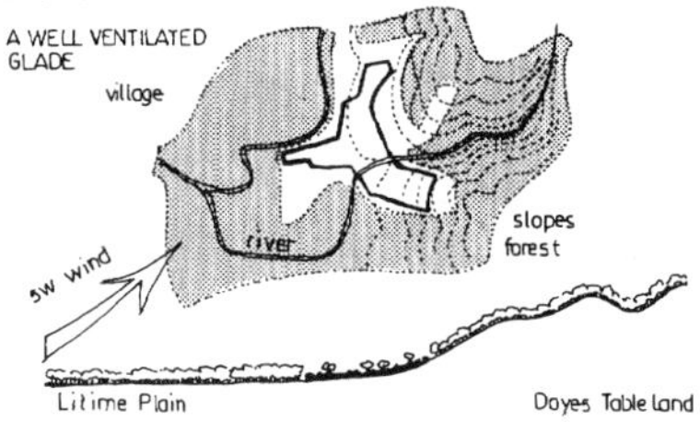

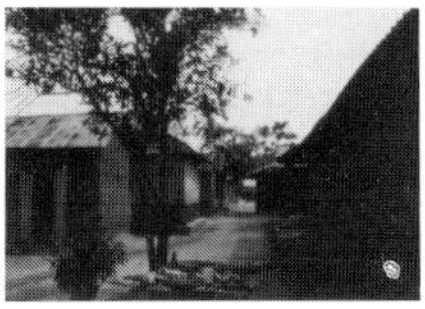

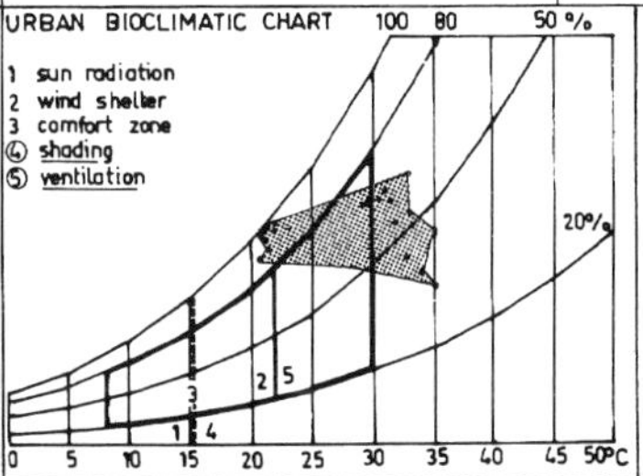

URBAN SPACE BIOCLIMATIC PRINCIPLE

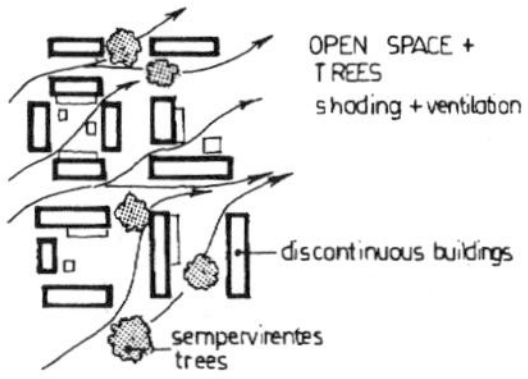

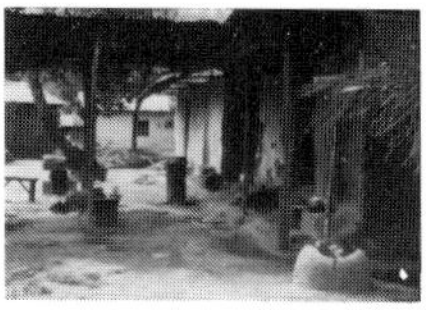

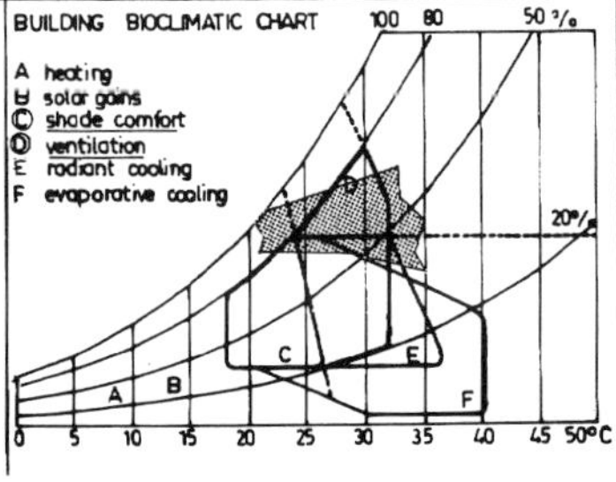

BUILDING BIOCLIMATIC PRINCIPLE

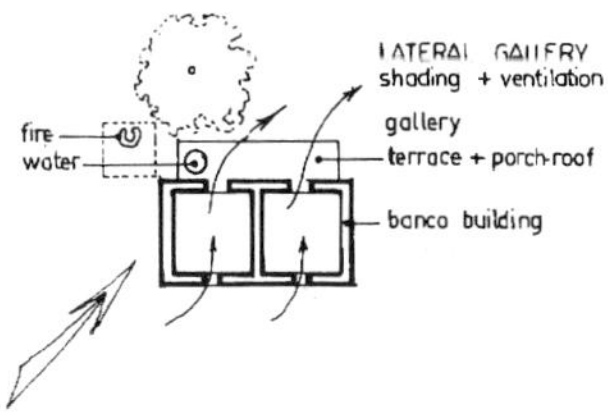

The previous cards illustrate vernacular sites chosen among the four basic climates in the world: nordic, temperate, hot-arid, hot-humid.
If one use these informations facing the topic "Architecture and Urban Space", one can do some observations, vertically card by card and horizontally scale by scale.
Vertically, the topic is introduced by environment and site considerations which allow, at each time, to situate the problem.
Horizontally, at each scale, the comparative approach allows to describe the values specific to each geographical situation.

After a sufficient number of observations, one can express these results:
In Architecture, the light transparent element of the building presents the following distorsions from north to south: the "integrated sunspace" in the nordic becomes an "alternative seasonal space" in the temperate, a "patio" in the hot-arid and a "peripherical gallery" in the hot-humid.

In the Urban Space, the free space presents connex distorsions: the "large solar space" in the nordic becomes a "mixed solar and green space" in the temperate, a "shaded and watered minimal space" in the hot-arid and a "green shaded space" in the hot humid.

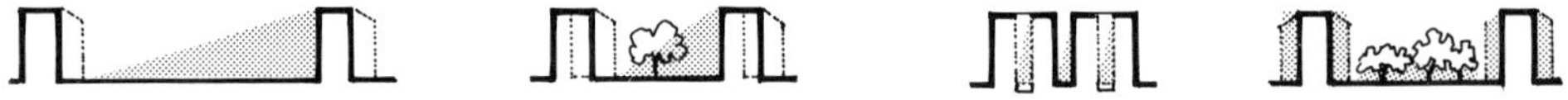

In all the urban space cases, the built and free spaces are in a superimposition generating a shared area which can be identified as the external part of the light-transparent element of the architectural combination, except for the hot-arid-climate.
So, from north to south, the free urban spaces are in progressive penetration within the built spaces, till to be completely absorbed like in hot-arid climate; at the same time, the proportion between the free and built space areas goes down so as to reduce the sun penetration, as much as the latitude is low.

Considering the impact of the environment on the quality of life in the city, the card index can help us to class the questions following the geographical scales input on the top of each card.
For, the environment biogramme contains the climatic and continental informations, while the site one contains the topographical and land-use ones; so analysed, all the chain can reveal the delicate connexions.
So it appears that the water constraints form the first conditions for the human establishment, as well for supply and access as facing problems like pollutions, flood, dryness, frosts, etc,.
It appears also that large scale problems, as acid rains or global warming, find their roots at lowest scales, industrial, urban, individual (smokes, CO_2, aerosols), and would find their solutions at these scales.

SCIENTIFIC KNOWLEDGE AND URBAN DESIGN THE ROLE OF MODELLING AND SIMULATION

Jean Pierre PENEAU
Laboratoire CERMA
Ecole d'Architecture de NANTES
Rue Massenet - 44300 NANTES
FRANCE

ABSTRACT. The author draws attention to the work in energy simulation of global solar lighting and fluid flow simulation carried out by the CERMA Laboratory. The two activities : scientific research and design project are very often, in the field of urban architecture, kept apart. He aims to show how parts of scientific knowledge in climatic adjustment can stimulate creativity in urban project practice.

1 - ENERGY EFFICIENT ARCHITECTURE AND URBAN PROJECT

How can we reconcile the aims of the P.L.E.A. with the constraints and requirements of urban design ?

Is there not a contradiction between the approaches coming from bio-climatism which aim to strike a balance between human housing and nature and the latest trends which seek to ensure the continuity and the coherence with the city of the past ?

We could —in a urban site— improve the shapes of a building to limit its energy consumption : this is an architectural problem ; but the urban project assumes a broader vision. It concerns the building group as well as the open space which it delimits. The empty space of the square, its ground, its fountains, its kiosks, its benches, its plants are as important for the image and the use of the city as the shapes and the layout of the buildings which border it.

On an urban project scale, the problem of energy savings should be treated while keeping in mind the questions of the comfort and amenities requirements of the users of open spaces. We must avoid treating the energy consumption of the buildings at the expense of the quality of urban spaces. On the contrary, not designing the urban spaces leads to excessive energy consumption by surrounding buildings.

More radically, we can put forward the view that the limitation of energy expenses (by using passive cooling, by recovering the solar gain, by natural lighting...) is not enough to guarantee on a city scale a real energy savings. The problems of urban energetics should be set up in general terms. If our building —carefully improved in terms of energy— is located far from work and service areas and with an unsufficient public transport network, the energy consumption of the inhabitants will far exceed the average level of a city with better equipment and a greater density. The problems of urban energy therefore pose

S. Alvarez et al. (eds.), Architecture and Urban Space, 93–98.

formidable questions in terms of density, costs of networks, travelling habits, energy sources, and the design of the urban project alone is quite often completely powerless to intervene at these levels.

In the face of these questions which are not simple, we make a plea for the development of scientific knowledges enabling us to clarify this complexity, to control it and, in addition, to teach something other than empiricism and ideology . This scientific knowledge comes in at two levels. The first one concerns :

- the scientific knowledge of urban energetics,
- the scientific knowledge of urban forms,
- the scientific knowledge of the history of the energetics of urban forms.

It is at this level that the design references of the urban project are made up.

The CERMA Laboratory works in this field on the systematic study of the climatic adjustment of cities with regular layouts.

The second level corresponds to the design techniques of the urban project. It concerns:

- the physical knowledge of energetic phenomena inside the building,
- the physical knowledge of phenomena related to urban ambient environments of open spaces.

It is only by comparing the development and the clear awareness of the necessity of this scientific support and of its duality that we can consider the problem of instrumentation. Among its numerous modes, we will particularly emphasize the modelling and simulation tools, knowing that they are pertinent, both to the development of the scientific knowledge required by the project we have mentioned and to its direct design. (1)

2 - MODELLER, MODELLING, SIMULATION

The work we are carrying out could be summed up by the following tasks :

- to model the fullness (buildings),
- to model the emptiness (open spaces),
- to model the physical phenomena in question.

The modellers which enable us to capture the shapes of the buildings are widely distributed and are, as regards some of them, real standards. This does not mean that everything is known in this field. We must devote effort to simplifying and accelerating some processes which are still slow to implement.

Aside from a few studies on declarative modelling of regular urban structures (2), we did not explore the question much. We use our own simplified tools or standards like ARC +. As for the open spaces : the "urban emptiness", we have been led to use the meshing technique. We have choosen "SIMAIL", developed by the French National Institute of Research in Computer and Automation which is distributed by the SIMULOG company (figure 1).

We use SIMAIL, as a volumic meshing, to discretize the open spaces in order to simulate the aerodynamic flows by using a finite elements method. This tool may also be used for per unit of area mesh division to carry out :

- the simulation of shadows and insolation,
- the simulation of the incidental direct solar energy,
- the simulation of illumination and radiation (by discretization per unit of area).

Among the extensions of the application of this technique, we can consider, without having yet begun the work, defining some cutting up per unit of area to transfer the crossed results of the simulations. Volumic meshing may also be used to simulate, with the help of the finite elements, the sound propagations, but this study is outside the scope of our research program.

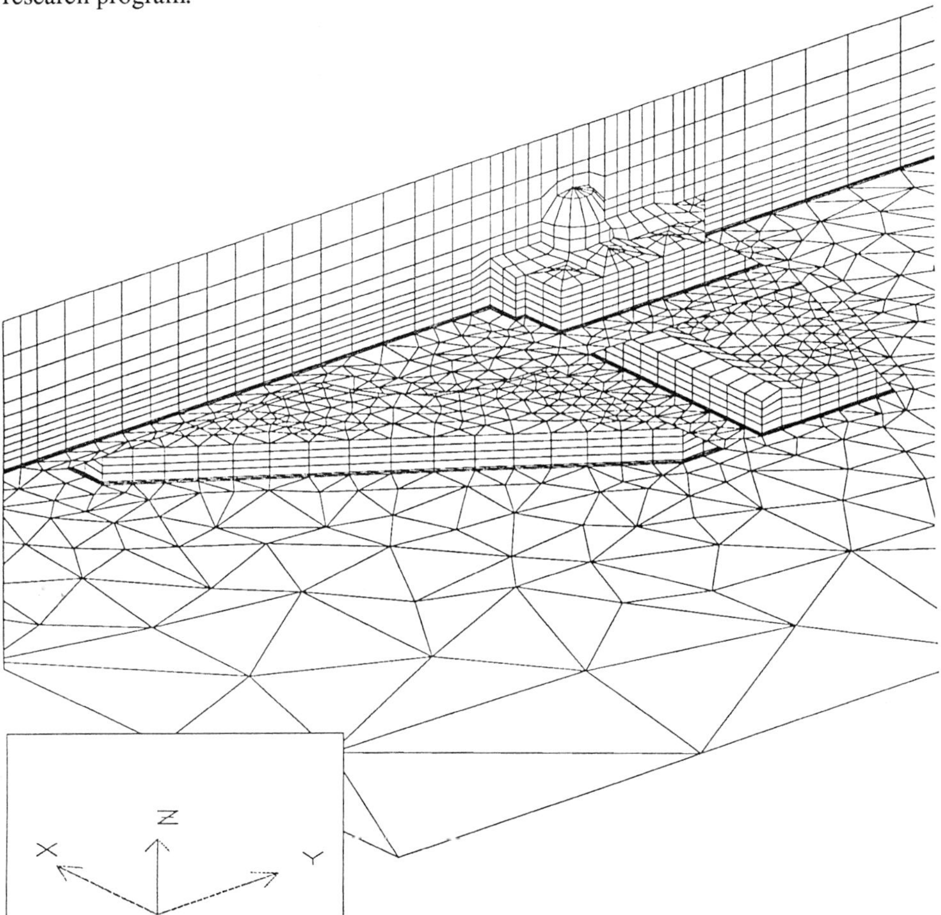

figure 1

3 - THE SIMULATION OF TURBULENT AERODYNAMIC FLOWS

To analyse the comfort of urban open spaces, known as intermediate spaces, it is essential to be able to predict the characteristics of the flows ; this for the sites where cooling is sought as well as in the case where, on the contrary, protection against strong winds is needed. The turbulence at the lower levels of the atmospheric boundary layer corresponding to urban areas is all the stronger as the buildings are discontinuous and high. This instability, added to the complexity of these morphologies, makes the prediction of the flows difficult. Only wind tunnel boundary layer simulations allowed, until now, a good approach to corresponding phenomena. We are trying, for our part, to implement a calculation code using a finite elements method realized by the Laboratoire National d'Hydraulique de l'Electricite de France.

This code, called N3S, solves the NAVIER STOKES equations in three dimensions with turbulence and viscosity by means of a K, ε model (3); this for each point of the mesh which has been kept. We are now testing this code on a square in NANTES from the early nineteenth century whose geometry has been simplified. Simultaneously, we are carrying out validations of the results of the numerical simulation with the analogic simulations made in a wind tunnel on the scale model of the square (this is part of an agreement with the CSTB of NANTES).

The code theoretically makes it possible to obtain the components in the three directions of reference of the speed vector for each point of the mesh. It also gives the value of the turbulence at this point (we know that aerodynamic comfort is a function of the combination of average air speed and fluctuation). We hope to be able, later on , to tackle, by means of this N3S code the modelling of some thermic phenomena which are related to the flows.

4 - SUN LIGHTING SIMULATION IN COMPLEX URBAN SITES

The techniques of sun lighting simulation generally involve implementing the "ray tracing" methods. The "SOLENE" software, adapted by two CERMA research workers: D. GROLEAU and C. MARENNE (4), relies on a geometric method to evaluate the lit up sides or part of sides (that is to say seen by the sun) or shadowed sides (not seen by the sun).

The algorithms have been generalized from a technique involving counting up the pixels of the lit up or shadowed parts. The algorithms have been generalized to be directly used with different post-treatment devices and representation of results by projection of sides. As far as we are concerned, we use a developed version on a "SUN SPARC II" station, with outputs realized with the "IKOLIGHT" image synthesis software . We can also realize some trims of pictures by means of IKO to plate the shadowed or lit up parts. This up-to-date SOLENE version requires a cutting out into plane sides which -as it has already been mentioned - can be realized by means of the mesher "SIMAIL". This technique makes it possible to solve the problem (generally badly treated) of the sun lighting study of irregular or rough sites (figure 2) . Until now, we used to liken the ground to a plane. The relief of the ground is henceforth obtained thanks to the mesher. The buildings and fittings are also broken down into sides ; it is the group of sides which allows the lighting balance to be obtained (5) (figure 3). It is clear that, without any buildings, the lighting balance of undeveloped site is obtained.

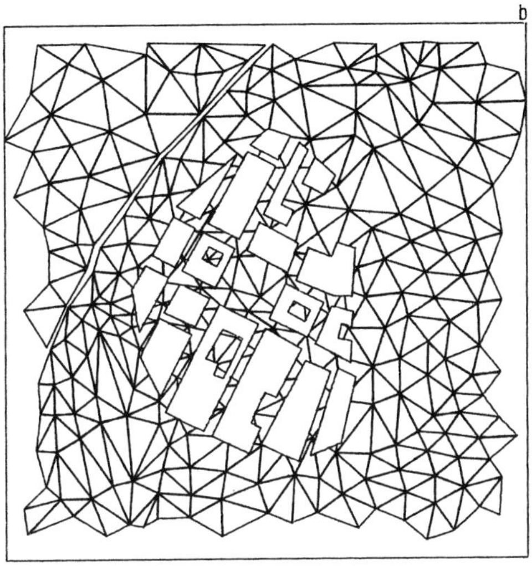

figure 2

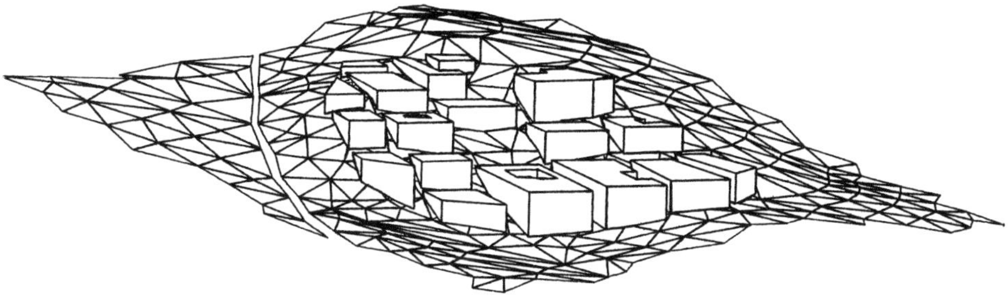

figure 3

5 - PROGRESS IN SIMULATIONS AND FUTURE PROSPECTS

There is still much to be done before the simulation techniques we are starting to implement can become truly useful tools for studying urban projects. Their use is currently subjected to calculation methods and devices which remain quite difficult to implement (particularly for aerodynamic simulation). But, besides this work of instrumentation proper, it is essential to constitute knowledge based system of reference on this subjet. This will be achieved only at the cost of an effort of investigation and systematic analysis on significant architectural and urban examples.

Our work must, consequently, be extended in both directions :

- the improvement of the performance of solar and aerodynamic simulation tools from the above mentioned techniques ; the study of the possibilities of applying these methods to model the propagations of sounds and lighting ;
- the application of these simulations to urban examples of reference to constitute a real knowledge-base on the energetic and climatic adjustments of the project.

Both of these activities should make it possible to bring closer the scientific and urban cultures closer together. They should be implemented at the different levels of teaching and research :

- in initial training,
- in continuing education of practitioners,
- in research and development.

We will conclude this paper by pleading in favour of the effort which has been made by architects and urban designers to gain a better understanding of the physical mechanisms related to energy and urban ambient environments, an effort which should be followed by scientists and technologists to understand the approach to urban project design and acquire some references in this field.

6 - REFERENCES

(1) PENEAU J.P. : "Some problems relating to the numerical simulation of urban ambient environments" In Environment and Planning B : Planning and Design, 1991, Volume 18, pp. 107-117

(2) PIERNOT Ph : "Application des procédures de l'intelligence artificielle à la morpho logie urbaine" Mémoire de D.E.A. de Génie Civil, octobre 1989 CERMA - Ecole d'Architecture de NANTES

(3) CHABART J.P. : "Projet N3S de mécanique des fluides : Manuel théorique, version 2.0" Rapport E.D.F., D.E.R., L.N.H., n° HE 41/48. 14 (1988) p. 150

(4) GROLEAU D., MARENNE C. : "Procédures géométriques de la simulation d'enso leillement" Rapport interne CERMA - Ecole d'Architecture de Nantes (1991)

(5) PENEAU J.P. : "Numerical modelling of ambience factors in urban environment" In "Proceedings of Workshop on Passive Cooling", ISPRA, 2.4. April 1990, Joint Research Center, Commission of the European Communities, pp. 63-71

QUALITY FOR PUBLIC URBAN SPACE AND PEDESTRIAN MOVEMENT

NORMAN PRESSMAN
Professor, Urban Planning & Design
School of Urban & Regional Planning
University of Waterloo
Waterloo, Ontario, Canada, N2L 3G1

ABSTRACT. This exposition examines pedestrian movement systems and urban space in the northern climatic context. Its purpose is to develop a better understanding of how inhabitants adapt both to public and private environments not only on a year-round basis but also during the winter and marginal seasons. Walking, as a form of transit, is viewed as part of an overall strategy of creating high-quality public space in cities. Elements such as sun, wind, rain and snow and how these impact upon urban space are explored. Aspects of socio-cultural vitality crucial to healthy public environments are emphasized together with climatic variables which are assuming greater importance in urban policy interventions. Concern for the gradual erosion of public space in the wake of 'privatization' forces in central city zones is addressed. Examples are drawn from settings indicative of those qualities which assist in creating successful, humane places for all urban dwellers.

1. Importance of Social Space in Urban Culture

Streets as a distinct form of public space have always reflected the spatial diversity and human scale so essential to our physical and psychological well-being. They assist in creating environments which can become agents of social contact, reflection and vitality. Successfully designed public spaces accommodating collective needs almost always bear witness to the primordial demand for social intercourse.

> The street speaks a universal language. Its signals are part of everyday learning; its rules for movement are among the most widely understood of all public codes of conduct; and even its most bizarre variations offer, upon close examination, familiar goings-on in the School of the Street.[1]

The street is a teacher of urban culture. It helps to remind us of the sensory qualities of the city -- those aspects which heighten the experiential realm. After all, the city is much more than a ledger book of statistics on land uses, modal splits, hectares of open space/1000 inhabitants, number of vehicles/available parking spaces, population density and all other data which are part of the urban policy analyst's compendium even if many of these contribute to life quality. Walking for pleasure (<u>spazieren gehen</u> as the

S. Alvarez et al. (eds.), Architecture and Urban Space, 99–108.

Viennese call it) has been an old-world tradition. Taking a "Sunday drive" has been the new world equivalent -- hardly encouraging sociability. Intensity of public life has always been one of the three most critical characteristics that make a city.[2]

Walking for improved health has become one of the great trends toward the end of the 1980s and will undoubtedly continue well into the next century. Urged upon us by the medical profession we take it seriously. We are warned that not-to-walk constitutes a fundamental health hazard. Very recently walking speeds in 36 major American cities were measured as a factor in determining which were "fast-paced" and "slow-paced" cities. On the whole, it was concluded that people living in fast-paced cities were "more prone to coronary heart disease than those in slower cities".[3]

It has been demonstrated through a number of authoritative epidemiological studies that new diseases tend to occur "most frequently among those with few social contacts, lower hierarchical positions, and those disconnected from their biological and cultural heritage".[4]

Studies further suggest that sickness -- often resulting from multiple stress conditions -- which attacks with severe consequences most often seeks out those who are "out of connection" and lack meaningful social contact with others.[5]

In many of the above instances, disease agents are the primary cause of illness. Factors which engender stress -- emotional, physical or otherwise -- can exacerbate the difficulties. And that is precisely where good environmental urban design can play an important role. The physical milieu can be structured in ways which enable social connectivity to occur. It will not always determine such connections but can act as a modifier of human behavioural response since the location and orientation of buildings and spaces, proximity determined by the placement of urban objects (such as benches, chairs, tables, etc., in public space), and micro-climatic improvements can, in harmony, either promote or impede the interactions of humans in three-dimensional space.

2. Climate-Responsive Public Urban Spaces and Streets

In harsh climatic conditions, "protected" public (and private) spaces will be desirable. These should both physically and visually relate and be connected to the ground-level of the city. In doing this, it is probably imperative that the open channels of the streets and public urban spaces retain their dominant roles. Eberhard Zeidler, designer of Canada's largest downtown shopping mall -- the Eaton Centre -- has stated that "within cities the open street is still the best artery, regardless of the climate."[6] Notable exceptions such as the "galleria" (early prototypes are found in Brussels and Milan) and the "passage" (with prototypes situated in 19th century Paris and other major European centres) have managed to co-exist harmoniously with public street life. These have perhaps been the most powerful leitmotifs for newer city centre redevelopments to be experienced in places such as Hamburg, Gothenburg, Stockholm, Stuttgart and Bern to mention a few examples.

Hamburg's impressive "Passagen-viertel" combines seven newly constructed passages (the Hanse-Viertel is the most renowned) with the "fussgangerzone" comprised of open-air pedestrian streets, and the Colonnaden (built in 1877). Together there are roughly 1.2 kms. of interiorized passages running through existing buildings meshed with approximately 1.5 kms. of pedestrian streets resulting in a choice for users of being either indoors or outdoors -- or both, depending on the weather and the season. Hamburg boasts more passages than any other city on the continent -- nine in total dating both from the 19th and 20th centuries. These include hotels, shops, restaurants, coffee-houses, department stores amid historic buildings, the cathedral and public open spaces along the riverbank of the Alster. Already in the 19th century, the Alster Arcades were built under which people, protected from inclement weather (usually rain), could stroll, eat, shop and meet friends and passersby. Today, the city centre illustrates precisely how the combination of enclosed (privately developed) passages can be integrated with the city's network of public pedestrian streets to produce a high-quality environment for inhabitants and visitors alike. In the heart of town, pedestrians appear to have been accorded priority over vehicles without resorting to any vertical separation.

Many German cities have lately taken to an approach which protects pedestrians -- mainly from precipitation. In the Hamburg case, overhead canopies continuously shelter people by projecting over the sidewalks from buildings to which they are attached. Recently, the City of Wilhelmshaven (in the Bremen region) situated on the North Sea, sponsored a competition for a continuous system of galleries (arcades), canopies and passagen to be connected within the "fussgangerzone" and the areas in which parking has been provided. The idea is to create what are called "trockene weg" (dry paths) making walking under weather-protected conditions pleasant and attractive.[7]

The Swiss cities of Zurich, Bern and Geneva have employed underground movement systems in locations connecting the main-line railway stations to major pedestrian thoroughfares beneath the railway station "squares" which generate heavy motorized traffic conditions. An example is the Shopville-Bahnhofpassage linking Zurich's Hauptbahnhof with the Bahnhofstrasse "transit-mall" (for pedestrians, trams, taxis and emergency vehicles only). These underground malls with restaurants and shops are relatively modest in size and are intended not so much to compete with the street as to complement it -- by promoting pedestrian flows in the directions in which they naturally wish to occur.

Bern is an especially worthwhile example to analyze with respect to "climatic-protection" from wind, rain and snow. The "lauben" or linear colonnades/arcades, which run the entire length of this ancient city's major streets (raised slightly above the actual roadway level), are integrated within extremely well preserved 17th and 18th century houses under which they are incorporated (the city itself was founded in 1191). Within this network of "arcaded" streets accommodating cars and pedestrians alike mid-block, covered cross-connections and open-air narrow lanes and alleys link the arcades to parallel streets within the old town. Newer "passagen" were skilfully built within the older network and many are heated with infrared lamps during winter and the marginal seasons creating desirable micro-climatic conditions both in the mid-block connections and within the arcades themselves, making year-round use for cafes, snackbars and even vending stalls possible.

> the medieval arcades and covered cross-connections, modern passageways and shopping concourses amount to a system of 5 miles (about 8 kms) of weather-protected pedestrian ways.[8]

If one were to adopt, as an index of pedestrian comfort, the ratio between the total length of protected pedestrian ways and the number of inhabitants[9] an interesting comparison could be made between Bern and Calgary -- with its +15 skywalk system.

- Bern, with roughly 300,000 regional population supports 8 kms, of protected pedestrian walkways, or 1 km/37,500 inhabitants.
- Calgary, with approximately 600,000 population supports 9 kms of +15 walkways, or 1 km/66,600 inhabitants -- only slightly more than half of that found in Bern.

If one were to use only the actual 'urban' population of the City of Bern which is 135,000 then the ratio of protected ways in Bern would increase to 1 km/16,875 persons making it probably the most completely climate-protected system in the world -- and certainly unsurpassed anywhere in extent or importance north of the Alps. The "lauben" in Bern serve as a living demonstration of how arcades can be truly significant in the organization of town form and the use of urban space.

The design principles embodied in the arcades of Bern's "altstadt" (and in other Zahringer towns such as Murten, Thun, Burgdorf, Fribourg, etc.) are superb examples of how old ideas can, in the present era, structure daily life and have retail behaviour adjust to an historically imposed urban form. The central area of Bern is one of the best shopping environments anywhere. It is full of vitality while offering diverse activities to a broad spectrum of users who drink and eat merrily, make purchases, stroll, and visit art galleries and museums. The centre has direct access to the centrally-situated Bahnhof and makes provision for underground parking garages with electronic panels indicating whether free space is available.

Although historic examples cannot readily provide magic solutions to contemporary dilemmas, they can offer guidelines and valuable insights which serve as a basis for new directions -- even when these are nothing more than the modification of older, time-tested responses. The relationships between density, weather-protection, urbanity, mixed land-use, collective transport, accommodation of the automobile, compactness, and the creation of pedestrian movement systems can all be witnessed in Bern's urban structure. This city performs well -- within its centuries old container -- even when evaluated by the most modern performance standards. It still serves as a potent prototype for inner-city redevelopment designs. Lessons extracted from such historic contexts have applicability to present-day problems because the vocabulary of the past constitutes a repository of ideas and forms which, if successfully adapted, can prove useful.

3. Guidelines for Designing Walking Networks

Although we are witnessing a "renaissance" of the passage and arcade -- examples abound[10] especially in the German cities like Koln, Hamburg, Dusseldorf, Frankfurt am Main, Karlsruhe, etc., as well as in The Netherlands, Switzerland, Sweden, and Denmark -- these "arcades" or glazed-over pedestrian lanes/streets, as commercially vibrant as they are, still are pedestrian islands surrounded by noisy, polluted and sometimes visually unattractive motorized environments.

To achieve a healthy urban organism, walking must be considered as an important movement mode. In fact, walkability can be considered a significant indicator of livability, and should be accorded the same attention given to the various modes of motorized locomotion.[11] Almost all trips commence and terminate on foot and it would be wise to create continuous networks of walk-ways throughout parts of the city -- and even partially, throughout all of the city. Giving equal weight to the pedestrian mode ("slow-mode") as to that which exists for motorized modes would seem a reasonable objective. However, this has not been the normal practice even though, in many European cities, greater efforts are being made to reduce and even eliminate cars from city centres.

If walking -- the soft/slow mode -- could be encouraged (by further restricting car movements and through improvement of collective transport modes) and continuous paths for pedestrians implemented where people live, work and play, then the public and semi-public spaces of the city's fabric which are woven into these paths could be appropriated for a wider range of uses.

Clearly, new attitudes -- especially in North America -- will be required if funding, participatory measures, legislative means and marketing are to evolve with the intent of making serious changes to our behaviour with respect to how we move around. The micro, meso, and macro-scales will have to be tackled simultaneously. Solutions developed for an urban region will significantly affect strategies targeted at specific, localized parts of the city. If quality in urban life is demanded, then walking must ultimately be viewed as part of the overall plan, and as a constituent element of the broader civic space.

If people are to be encouraged to walk, transport research will be required to:[12]

> determine the sensitivity of pedestrians to distance, to show what standards are appropriate for kerb heights, pavement widths and traffic or pedestrian lights phases, and to determine the benefits of such design features as seating or arcades or other forms of cover on the pedestrian network.

Each city shall have to develop a framework -- within its public mechanisms and policies -- for developing criteria capable of answering the question "what is a good city?"

Preservation of historic architectural and urban tissue, the creation of multi-functional public spaces at prestigious locations in the city, connections between these spaces and other imageable 'places' via linked pedestrian pathways including cultural, administrative and leisure 'magnets' must all combine to generate a "livable" central city district.

> The name of the game is captivating people. The methodology is to exploit and enhance every conceivable attraction, to make enjoying these attractions easy, and to spread the magnetism over as wide an area as possible. Success can be measured in terms of how far people can be lured to walk, how long they can be enticed to stay, and how frequently they can be motivated to return.[13]

The important issue is to ensure that public policy envisages a state of equilibrium between collective concerns and private interests.

Current trends indicate a strong concern for climatic protection, among all the other social, economic and political goals operative in urban centres. Combined with functional considerations -- which provide the 'raison d'être' of central zones -- aesthetic qualities, climatic comfort and programmed events (such as parades, carnivals, musicians, street-performers, etc.) are becoming more important in sustaining the urban tempo and rhythm we have come to expect of downtowns. The following principles should be followed in developing areas for walking, standing and sitting especially in uncomfortable conditions:[14]

- provide shelter from the wind.
- create an urban form which avoids increased windspeeds.
- designate sunny and sheltered open spaces for non-motorized activities.
- save existing vegetation (for wind protection) or provide windbreaks.
- provide for comfortable walking conditions (design effective rainwater run-off areas, provide selected snow deposit areas and respective drainage for melting, eliminate or modify kerb-cut details for ease of movement, etc.).
- include ramps (as well as stairs) where changes in elevation occur. Heat them in winter.
- design for ease of access.
- extend the outdoor comfort zone especially in the marginal seasons, providing "transition areas" mediating between indoor and outdoor spaces and activities.
- furnish the public spaces with comfortable, visually pleasing, and well located "urban furniture".
- serve the pedestrian zones and networks with highly visible and accessible collective transportation.

4. Conclusion

The recent wave of underground passages and above-grade skyways linking private "atrium-style" developments and enclosed shopping malls (often connected to railways' terminals and subway systems) has created a pseudo-public environment of substantial size. While offering weather-protection, these projects guarantee social filtering and careful "selection" of patron users. Commercial objectives are first and foremost on the "development agenda". Hence, increasingly, the traditional role of the city centre -- to provide goods, services and amenities to the entire urban population -- no longer can be said to prevail, as the newer projects generate incremental privatization, resulting in traditional users being displaced elsewhere to find locations which meet their needs. This dilemma must be offset by opening up more genuinely public space -- parks, streets, lanes, green-areas, squares -- to counteract the erosion of the public domain. Public space has become one of the most critically studied areas falling within the scope of urban design, architecture, and planning for human well-being. The configuration of this public space, its appropriation, its meaning and content, its connection to the natural elements and the symbolic interpretations given to "genius loci" (sense and spirit of 'place') have strong implications for affecting and shaping the quality of urban life. Participation in public events, the relationship among the various urban functions, the siting of "solid" objects within the spatial "void" of the city -- all contain opportunities for attracting or repelling human life; for making people proud of being citizens of certain towns or alternatively, for causing depression, and even debilitating illness. So-called healthy places must exhibit a powerful combination of factors enhancing sociability in well-designed public space so as to promote connectedness between people when this is desirable; to offer a range of choices between the extremes of solitude, on the one hand, and togetherness, on the other.

Social enjoyment (in the spaces of which the city is composed), visual pleasure, climatic comfort (a relative concept), security, strong identity, contact with nature, and sensory stimulation should be viewed as goals to which the physical organization of the city must aspire. "Genius loci" will occur when ordinary space is elevated, in the minds and hearts of urban dwellers, to a sense of place, with very special attributes, universally recognized and appreciated.

Public life for all must be viewed as a legitimate concern of local and national authorities. Attractive and well managed streets and open spaces -- and the activities, both planned and spontaneous, which occur in them -- form the essence of urbanity. Once out of either the car or public transportation everyone becomes a pedestrian. Therefore, it is essential to direct greater attention to achieving a milieu in which the slow mode is dominant. This will be the only way to enrich the use of the city -- in all seasons -- and to make it both economically and socially viable. The city can, in this way, not only provide the means to life but also answer more fundamental questions pertaining to the meaning of life.

5. References

1 **Clay, Grady** (1987) "Why Don't We Do It in the Road?", **Planning,** May, (American Planning Association), p.19.

2 **Gruen, Victor** (1967) **The Heart of Our Cities,** Simon and Schuster, N.Y., p.28. (Mix of wide-ranging human activities and compactness are the other two characteristics).

3 **Levine, Robert** (1989) "The Pace of Life", in **Psychology Today,** October, p.46.

4 **Lindheim, Roslyn** (1985) "New Design Parameters for Healthy Places" in **Places,** Vol. 2, No. 4, MIT Press, p.17.

5 **Ibid.**, p.17.

6 **Zeidler, Eberhard** (1985) **Multi-Use Architecture in the Urban Context,** Van Nostrand Reinhold, N.Y., p.141.

7 The jury for the above competition met in September 1989.

8 **Galantay, Ervin** (1967) "New Old Towns" **Progressive Architecture,** December, p.92.

9 This ratio is suggested by E. Galantay, op.cit., pp.92-93.

10 Excellent case-studies are found in Heidi Kief-Niederwohrmeier's book **Neue Glaspassagen**, Verlagsanstalt Alexander Koch, 1986.

11 **Garbrecht, Dietrich,** (1981) "Foreword", **Creating Livable Cities** (Contact - Journal of Urban and Environmental Affairs - Norman Pressman, guest editor), Vol. 13, No.2/3, pp.xiv-xv.

12 **Hillman, M. and Whalley, Anne** (1981) "Walking: The Neglected Transport Mode" in **Creating Livable Cities,** op. cit., p. 22.

13 **Wiedenhoeft, R.** (1981) **Cities for People,** Van Nostrand Reinhold, N.Y., p.90.

14 These principles are particularly important in cities with harsh or even mild winters, as well as with marginal seasons which do not comfortably support outdoor activities for extended periods.

HAMBURG'S "PASSAGEN-DISTRICT"

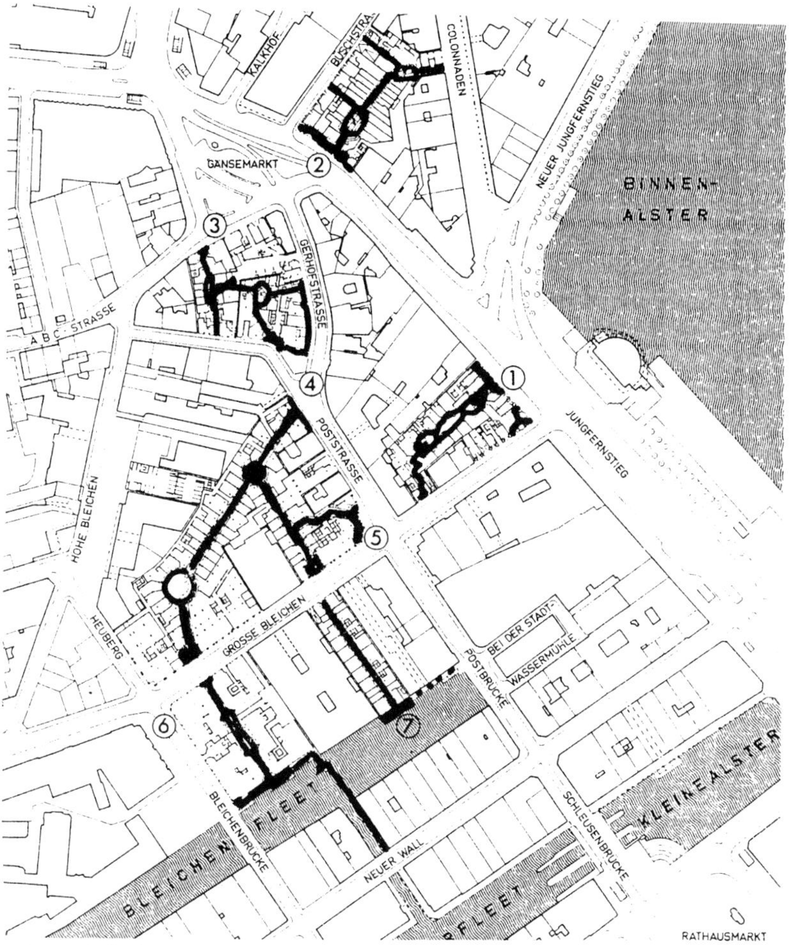

Passagen in Hamburg
1 *Hamburger Hof*
2 *Gänsemarkt-Passage*
3 *Neuer Gänsemarkt*
4 *Gerhof*
5 *Hanse-Viertel*
6 *Kaufmannshaus*
7 *Galleria*

Source: Deutscher Stahlbautag 1986 - Ausgabe 5
Gerd Ascher: Stadterneuerung in Hamburg, p. 33

+15 SYSTEM of BRIDGES, WALKWAYS and OPEN SPACE

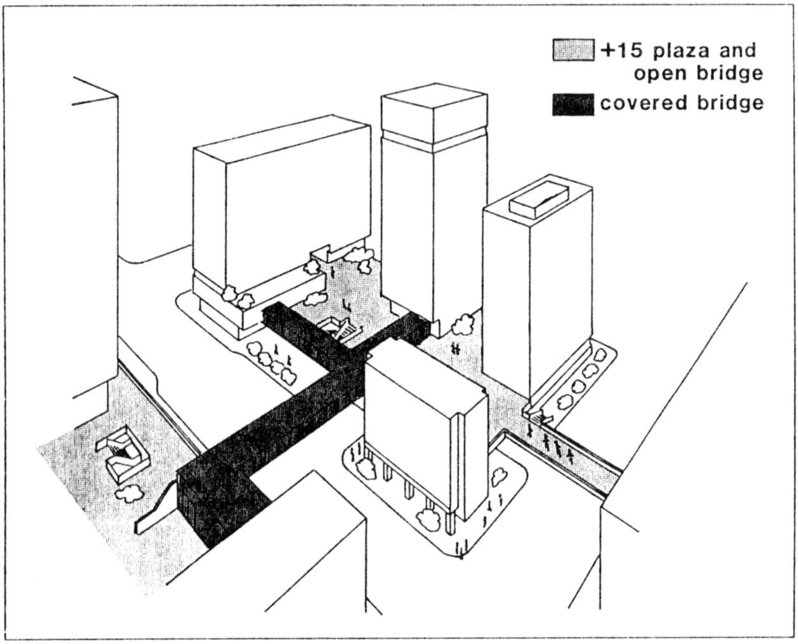

Source: PLAN Canada
January/janvier 1988
p. 273

DESIGN OF OUTDOOR SPACES:
SOCIO-POLITICAL TENDENCIES AND BIOCLIMATIC CONSEQUENCES

S. de SCHILLER and J. M. EVANS.
Research Centre Habitat and Energy,
Faculty of Architecture, Design and Urban Planning,
University of Buenos Aires.
Casilla de Correo 1765, Correo Central,
(1000) Capital Federal, Republica Argentina.

ABSTRACT. This paper aims to detect trends in the bioclimatic quality of outdoor urban spaces. The generic qualities of successful traditional urban spaces are analysed to identify the characteristic bioclimatic ingredients. The paper then considers a series of factors that can lead to a reduction in bioclimatic quality. A scenario of the evolution of bioclimatic urban space is then proposed to describe current trends. This suggests that both supply and demand for bioclimatic urban space may be diminishing. The conclusions summarize these findings and call for a more careful use of the limited bioclimatic potential available.

1. INTRODUCTION

Bioclimatic criteria have helped to create attractive traditional urban spaces that provide shelter from undesirable climatic factors while taking advantage of favourable elements.

Current urban spaces respond to socio-economic and political trends such as maximizing financial benefit, privatization of social housing and communal spaces, vandal proof design and security measures, traffic rationalization and development of enclosed private commercial centres.

All these factors reduce the variety and bioclimatic potential of social outdoor space. Technical innovations and increased vehicular use, impose strict limits on the geometrical form of circulation spaces as well as an increase in atmospheric and acoustic pollution which discourage the social use of urban spaces. Architectural trends also tend to ignore spatial definition and regional bioclimatic requirements of urban spaces leading to a reduction of environmental quality.

Many of these tendencies are magnified in developing countries where foreign influences are "imported" without sufficient consideration for social, administrative, economic and climatic differences.

This paper analizes the characteristics of traditional bioclimatic urban spaces in different regions to identify the ingredients of successful examples. Architectural and technological trends are discussed to identify trends and detect factors that limit the formation of bioclimatic urban spaces and reduce environmental quality.

S. Alvarez et al. (eds.), Architecture and Urban Space, 109–114.

2. TRADITIONAL BIOCLIMATIC URBAN SPACES

Traditional bioclimatic analysis of urban spaces aims to identify spatial and other design characteristics such as vegetation use, surface treatments, etc. Bioclimatic design resources that respond to the specific activities and climatic conditions can be identified.

These bioclimatic design resources will also relate to the social context and life styles that interact with the seasonal and daily variations of comfort in outdoor urban spaces. Table 1 and Figure 1 show some of these design resources and their micro-climatic objectives.

Table 1. Typical Bioclimatic Design Resources for Urban Spaces.

Arcades	Summer shade at mid-day, while allowing winter sun penetration.
Street width	Adequate for winter sun, partial summer shade or wind protection.
Orientation	Catch summer breezes, avoid cold winds, provide summer shade or take advantage of winter sun.
Vegetation	Provide summer shade, wind protection and air humidification in dry climates.
Colour	Avoidance of glare, improvement of day light, reflection or absorbtion of solar radiation.
Materials	Avoidance of dust, suitable for watering to humidify the air or absorption of heavy rain.

However, in addition to the specific bioclimatic design resources, there are certain generic characteristics that are found in most successful urban spaces for outdoor social activities.

Alternative microclimates: The simultaneous provision of different microclimates in one space, for example: shade and sun to allow a choice of spaces for activities according to variations in comfort conditions.

Daily variety: The provision of changing microclimates during the day, for example: orientation to achieve exposure to sun during cool mornings while providing shade in the hotter hours of the afternoon.

Seasonal variety: The use of changing microclimates in different seasons, using low angle winter sun while achieving shade from high angle summer sun or catching summer breezes from favourable orientations while avoiding cold winds from other orientations.

Transitional spaces: Avoidance of a clear limit between public and private to create intermediate spaces that favour social activities and use of outdoor space; restaurant tables that spill out onto the pavement, seats and alcoves adjacent to buildings entrances.

Degree of enclosure: Spatial definition creates appropriate micro-climates and also forms spaces that encourage casual and formal use.

Vegetation: Versatile element of bioclimatic urban design that can be used to define spaces, provide shade or wind protection, humidify air and achieve seasonal variety with deciduous plants.

Multi-use spaces: It is desirable that different activities can co-exist in the same space simultaneously or sequencially.

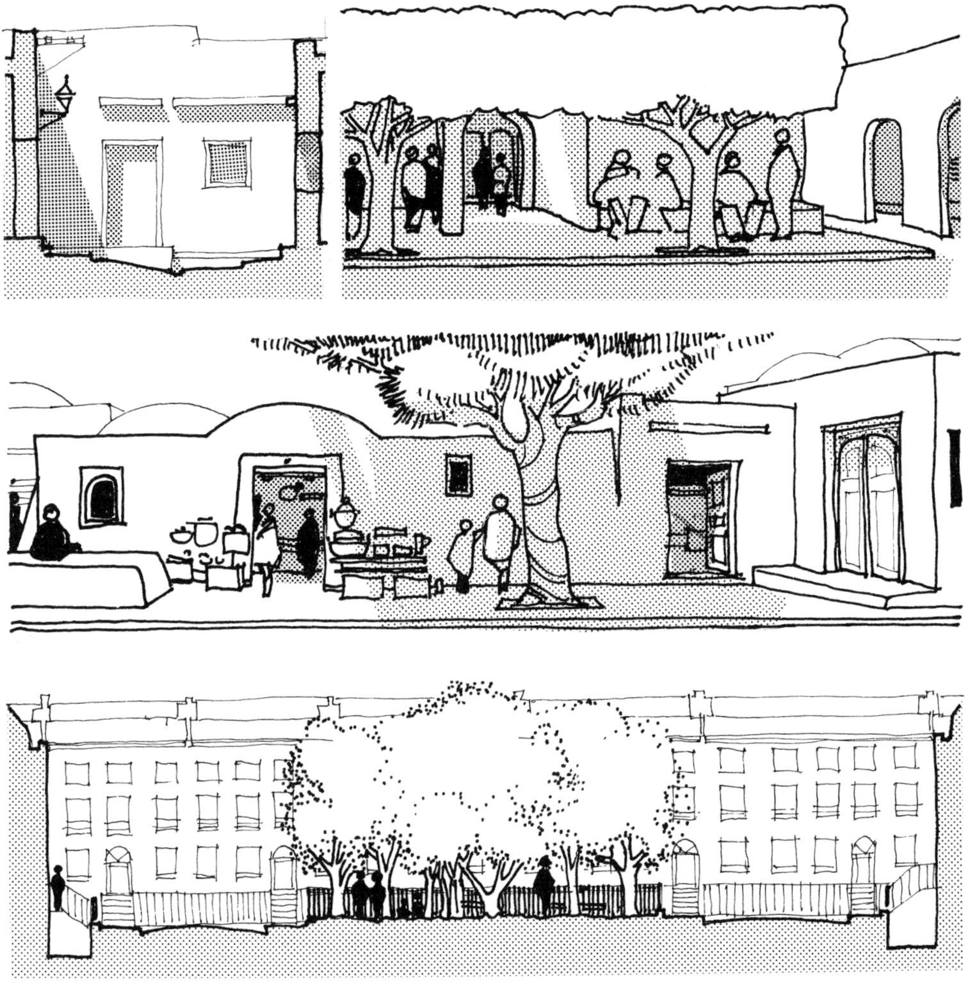

Figure 1. Examples of traditional urban spaces with high bioclimatic quality.

3. FACTORS REDUCING BIOCLIMATIC QUALITY

Many of the generic bioclimatic characteristics decribed in the previous section are becoming more difficult to achieve due to a series of political, legal, social and architectural tendencies.

3.1. Legal Influences

Municipal regulations encourage clear limits between public and private space to avoid gradual invasion of public space for private activities, thus eliminating the desirable transitional zones.

3.2. Functional Separation

Traditional streets combine spaces for circulation and access with areas for casual social contacts. They may also house formal activities such as street markets, craft workshops and eating spaces. Modern planning tends to separate such activities in different private spaces, while traffic discourages social interchange and children's play.

3.3. Privatization

The tendency to privatize social housing projects and certain urban infrastructure such as railways will transfer spaces, open to the public or partially open as transitional areas, to the private sector. To facilitate maintainance, improve security and increase commercial benefits, many such spaces will be fenced off, separated or used for new construction, effectively removing them from the public sphere.

3.4. International Influences

International innovations, fashions and influences, spread by powerful images shown in magazines and T.V., are adopted without considering the specific social, economic and climatic context. Thus, spatial design is not based on first-hand experience or scientific evaluation and project optimization, but on second-hand images which cannot convey the bioclimatic quality of the space.

3.5. Monumental Tendences

Large scale urban design can only be produced when planners have strong political and financial support. Unfortunately this combination is often found in dictatorial regimes where the urban plan symbolises political power; grandiose urban plans that can only be appreciated in aerial views. Examples frequently show wide symetrical and monumental axis flanked by high buildings, tall symbolic blocks placed in the main intersections. Versailles in XVII c. France, Albert Speers plans for Hilter and post-revolution planning in Moscow are examples of this tendency. However, planners may also adopt monumental tendences in democratic regimes as shown by the cases of Canberra and Brasilia.

3.6. Vandalism and Security

Traditional public spaces with good bioclimatic qualities have extensive use and close contact and surveillance from nearby dwellings. On the other hand, excessive urban spaces with poor spatial definition, lack of climatic protection and excessive separation between public and private areas limit the use and social control of such anomimous zones. These spaces may lack security and suffer from vandalism: everyone's space; no-man's land [1]. Defensible space requires limited small scale and well defined public spaces, closely connected to the buildings with which they are related [2].

3.7. Specialization

Specialization of activities and specific use of urban space, mentioned before, may also reduce the absolute demand for social urban space, as more activities are housed in special purpose buildings and private areas. Tecnological innovation such as increased telephone use, urban growth and increased cars use reduce pedestrian activity and opportunities for casual social contact in outdoor spaces.

3.8. Traffic

Traffic requirements limit the geometric form of urban spaces for circulation. Turning circles, sight-lines, parking requirements, horizontal and vertical curves according to the design speed tend to demand ever greater spaces. Fluid traffic movement at reasonable speeds require continuous corridors that are open to the wind and difficult to shade.

Traffic improvements in existing urban areas tend to degrade environmental quality: trees are cut down to widen streets or increase safety; curves are opened and building on corners demolished to improve sight-lines; noise and pollution also degrade urban spaces, making then less suitable for casual conversation; "better roads" mean faster traffic, incompatible with children's play; etc.

3.9. Excessive Space Standards

Another tendency, especially common in many developing countries, though not absent in developed ones, is the excessive increase in the dimensions of urban spaces in the belief that this will achieve higher standards. The increase in overal road widths, combining the generous requirements of transport engineers with wide footpaths and additional public "green space" lead to the prarie planning of suburban areas with little spatial enclosure, lack of identity and poor utilization.

The desire to provide areas for outdoor social activities in housing projects often leads to excessively dimensioned spaces which have the opposite effect. Large unprotected areas lacking bio-climatic quality not only limit outdoor space use but also increase physical separation between dwellings which may also lead to reduced social contact.

3.10. Lack of Maintainance and Social Organization

Increased areas of public spaces with planting, grass, street furniture such as benches, hard surfaces for pedestrian and vehicular use require additional resources and administration for their maintenance. However, resources are not directly proportional to the surface area; they are also related to design complexity and the use and abuse of the space. This is linked to scale, social control, mixed use and design quality. Financial restrictions at the municipal level may also reduce the levels of maintainance and consequently the quality of outdoor spaces.

4. GENERAL TENDENCIES

Four basic stages in urban space development can be identified:
1. Small scale traditional spaces, exemplified by medieval layouts, encourage common outdoor activities and favour high bioclimatic quality.
2. Rational city planning with grid-layouts produce rigid spaces with generally poor bioclimatic quality. Land subdivision and maintanance is simplified though spaces do not encourage social contact.
3. Social city planning introduced flexible planning with wide (and often poorly defined) spaces which favour winter sunlight. Undefined common zones often complicate space delimitation and maintainance which requires strong municipal or cooperative administration.
4. Present tendancies with design for security, privatization and spatial specialization may reduce maintainance and facilitate legal spatial definition. Technological and social developments reduce the demand for outdoor urban spaces for social activities.

Table 2. Urban Planning and Bioclimatic Quality

Spatial character	Traditional	Rational	Social	Tendency
Bioclimatic potential	High	Low	High	Diminishing
Delimitation & maintainance problems	Low	Low	High	Diminishing

5. CONCLUSIONS

Architectural designs should respond to genuine user needs within the changing legal, socio-economic and political context outlined, which does not favour the extensive creation of social outdoor space. Current urban trends in developed and developing countries does not appear to provide a generally improved outdoor bioclimatic environment for social activities. Indeed current technological trends may reduce the need for public open urban space and outdoor facilities.

If architectural approaches follow current economic and social trends, confirming our hypothesis, it is essential that the increasingly limited and restricted social space is optimized to provide better, more usable and comfortable common outdoor spaces. The bioclimatic criteria, vital component of the examples of attractive traditional urban spaces, will also help to create succesful new urban outdoor spaces.

6. REFERENCES

[1] de Schiller, S. and Evans, J. M. (1984) 'Espacio de Todos, Tierra de Nadie' (Everyone's Space, No Man's Land) Summario 80/81, Ediciones Summa, Buenos Aires, Argentina, August-September 1984, pp 20-41.
[2] Newman, O. (1972) 'Defensible Space: People and Design in the Violent City' Architectural Press, London.

Chapter 3

Urban Projects

PROPOSED HISTORIC TRAIL FOR ALMADINAH ALMUNAWARAH

ABDELMOHSEN M. FARAHAT and ASHRAF TURKI
SED, College of Engineering
King Abdulaziz University
P.O.Box 9027,Jeddah-21413
Saudi Arabia.

ABSTRACT. The paper deals with the main Historic Trail proposed to connect the most important historic sites in Almadinah Almunawarah. The trail was planned and designed to be predominantly for pedestrians. Each landscape zone (natural or urban) along the trail was given appropriate treatment emphasizing historic values, environmental values and energy conservation.

1. Introduction

Almadinah Almunawarah is the second most important religious and historic city in Saudi Arabia. It is blessed by many important historic sites beside it's main landmark -- the Prophet's Mosque (P.B.U.H). Inspite of generous projects by the the governmental authorities, these sites have not been utilized up to their full potentials yet, neither at the site specific scale nor at the scale of the whole city where they could have given it a well structured image emphasizing cultural values and environmental values as well. Accordingly, a landscape planning study was done in which the objective was to select and plan the main route connecting most historic sites to the central area where the Prophet's Mosque is located (P.B.U.H)[1]. This "Historic Trail" was developed from analytical studies of the distribution of historic sites, contextual land-uses, traffic and circulation, the natural environmental context, and the spatial pattern (figure 1). The "Historic Trail" was planned and designed to be predominantly for pedestrian circulation, enabling pilgrims and visitors to find their way easily and use appropriate transportation means when necessary. It was also planned and designed to give each landscape zone along it a proper treatment in terms of both the natural and urban environments, and to give the whole city its due image.

In the following analysis is given to the Historic Trail as a whole, followed by both analysis and proposals for each landscape zone along it. Following this a conclusion is given.

S. Alvarez et al. (eds.), Architecture and Urban Space, 117–124.

2. Analysis And Proposals For The Landscape Zones Of The Historic Trail

2.1. THE HISTORIC TRAIL AS A WHOLE

The Historic Trail begins in the South at Quba Mosque and ends in the North at Said Alshuhada Tomb, passing through the central area where the Prophet's Mosque (P.B.U.H) is located. Several landscape zones are in between, ranging from rural to urban character. At least the Southern half of the Trail represents in general the same path the Prophet (P.B.U.H) took when he first came to Almadinah [2]. In addition, this is the same route most visitors to Almadinah use in order to visit it's most important historic religious sites [3]. Accordingly, giving this Historic Trail special care and appropriate treatment (for each of it's landscape zones as explained in the following) would help to establish the backbone of a spatial structure for the whole city out of which other spatial branches could grow, linking to other significant sites. Since indigenous date palms are extensively used in the proposals for various landscape zones, even the densely urbanized ones as will be explained, it is expected that the old image of Almadinah as an "oasis city" could be re-established in contemporary terms, besides the positive microclimatic benefits in such hot-arid climate[4]. Figure 1 shows the Historic Trail, with numbers indicating various landscape zones discussed in the following.

2.2. LANDSCAPE ZONE NO.1

Located in the South, this zone consists basically of date palm orchards. The main vehicular road has limited side-walks that are not adequate, in terms of width and shading, to pedestrian movement especially during peak seasons, and Washingtonia palms are used in the median (figure 2).Proposals for zone 1 include the following: having a wide pedestrian path penetrating palm orchards on the East side, main vehicular road is devoted to public transportation only, private vehicles use a loop connected to the road, date palms are used in the enlarged median (figure 3).

2.3. LANDSCAPE ZONE NO.2

This is mainly a low density residential area with vacant lots assigned for 4 to 5 floors residential buildings, followed by some date palm orchards (figure 4). Proposals here include the following: decrease of the set-back to increase the width of the Eastern side-walk used for the pedestrian Historic Trail, limited height (2 floors) is assigned for future buildings on vacant lots in order to allow views to palm orchards, road median condition and proposal are similar to zone 1 (figure 5)

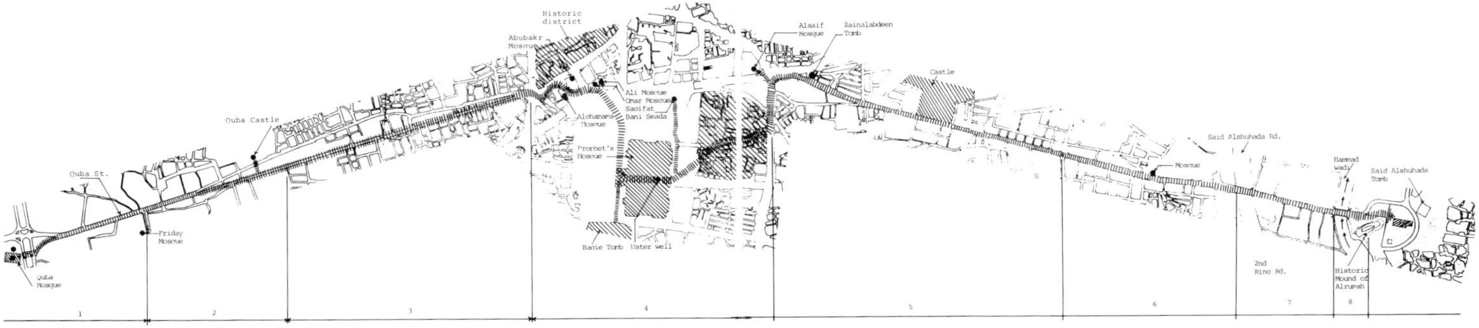

Figure 1. Plan of the proposed Historic Trail of Almadinah Almunawarah passing by most important historic sites. Numbers indicate landscape zones dealt with.

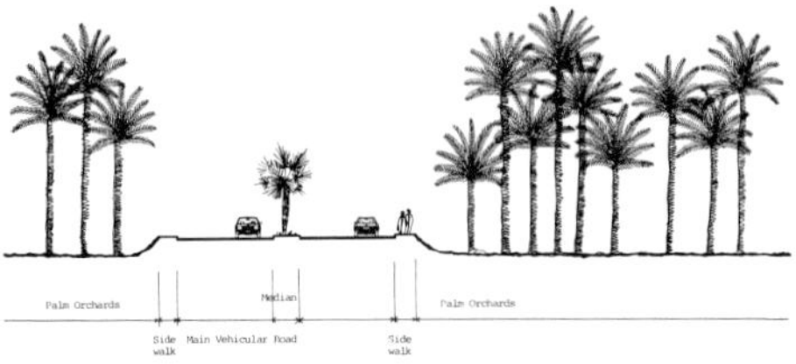

Figure 2. Zone 1 (existing)

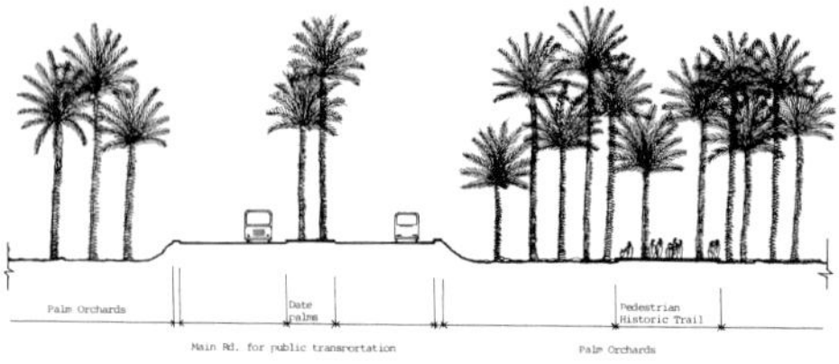

Figure 3. Zone 1 (proposed)

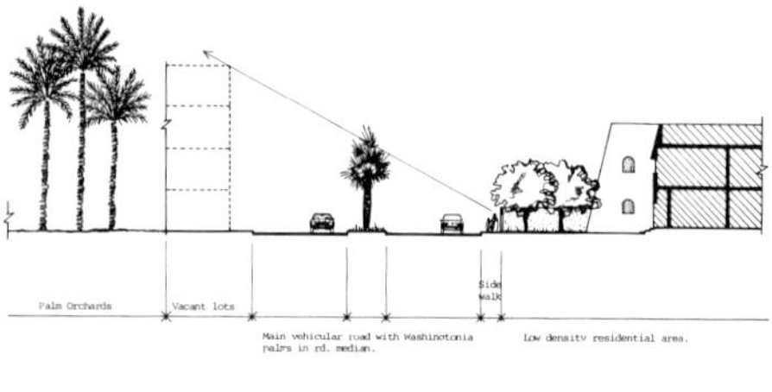

Figure 4. Zone 2 (existing)

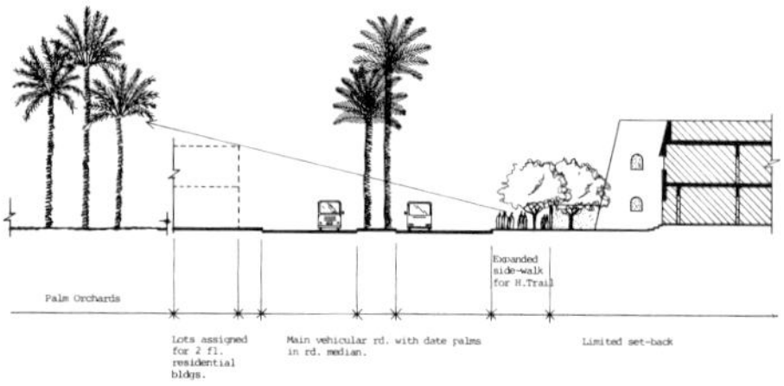

Figure 5. Zone 2 (proposed)

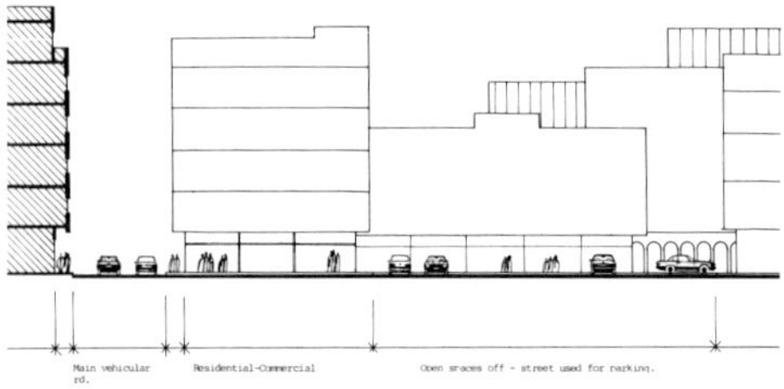

Figure 6. Zone 3 (existing)

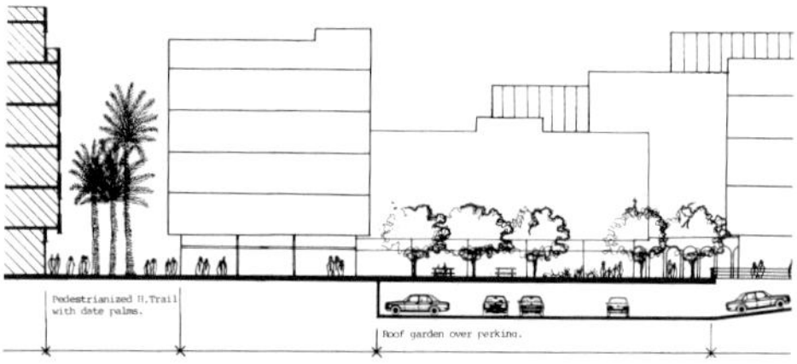

Figure 7. Zone 3 (proposed)

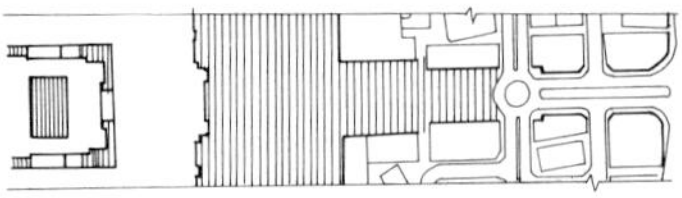

Figure 8. Zone 4 (existing)

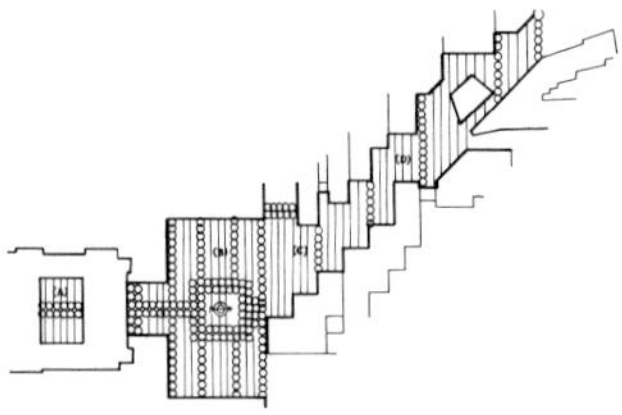

Figure 9. Zone 4 (proposed)

2.4. LANDSCAPE ZONE NO.3

This is a high density residential and commercial zone, overcrowded, almost completely deprived of soft landscape elements, and open spaces off-street used for parking (figure 6). Proposals here include the following: traffic is diverted to a parallel road forming a connected loop, main road is pedestrianized and used for the Historic Trail with date palms, large parking areas are kept under grade with roof gardens for the open spaces above (figure 7).

2.5. LANDSCAPE ZONE NO.4

This is the central area containing the Prophet's Mosque (P.B.U.H). Current plans (which are being implemented) for the area include the following: huge extension of the Mosque building, huge rectilinear U-shape plaza surrounding the Mosque, high-rise buildings surrounding the plaza, and the Historic Trail is left undefined and used as a vehicular road (figure 8).

Alternative proposals in this study for this very important area are based on the basic principle of integrating the Mosque with the surrounding urban fabric in terms of buildings and open spaces that are also shaped in appropriate locations to define the pedestrian Historic Trail. While S.Bianca has done a very successful scheme in Architectural terms following this principle of integration[5], an attempt is made here to follow the same principle with more emphasis on Landscape Architecture and open spaces. Accordingly 4 space categories are proposed for the transition from the Mosque to the city along the Historic Trail which are: formal courtyards inside the mosque unifying existing and extension buildings (A),transitional formal spaces between the mosque and the surrounding community (B), community open spaces (informal and rather organic) serving a mix of different activities especially shopping (C),linear spaces with mix of functions(D) The last three categories of spaces are also used for prayer as extension of the Mosque at peak times. Mentioned categories of spaces and the resulting free overall form would enable the "incremental" growth of the Mosque through several stages as needed (figure 9).

2.6. LANDSCAPE ZONE NO.5

This is a high density residential area. Most buildings on the West side are in bad condition (figure 10). Proposals here include the following: set-back of newly constructed buildings on the West side, expansion of side-walk on this side with arcades to accommodate the Historic Trail, and using date palms on expanded road median (figure 11).

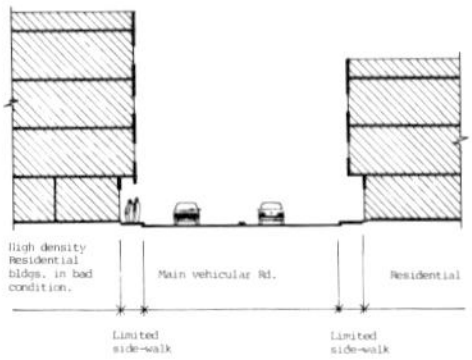

Figure 10. Zone 5 (existing)

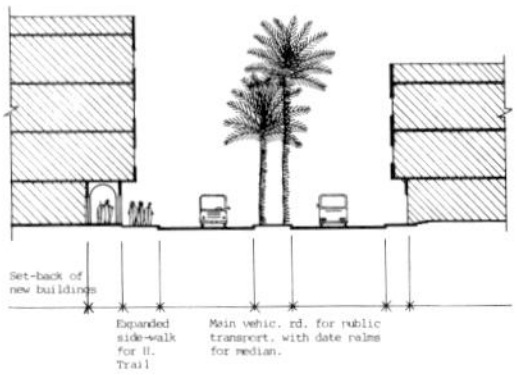

Figure 11. Zone 5 (proposed)

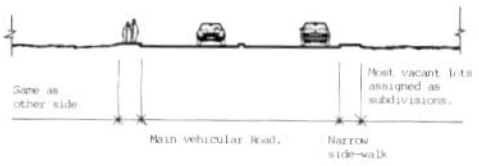

Figure 12. Zone 6 (existing)

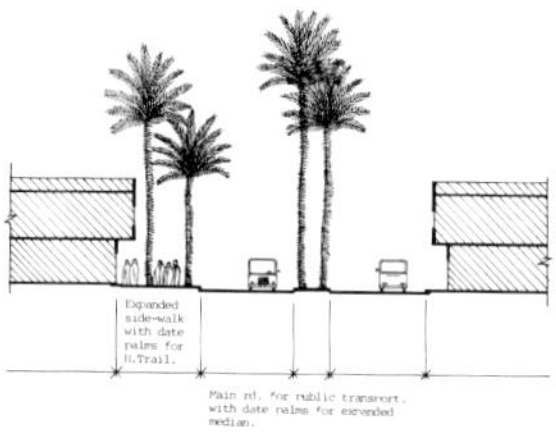

Figure 13. Zone 6 (proposed)

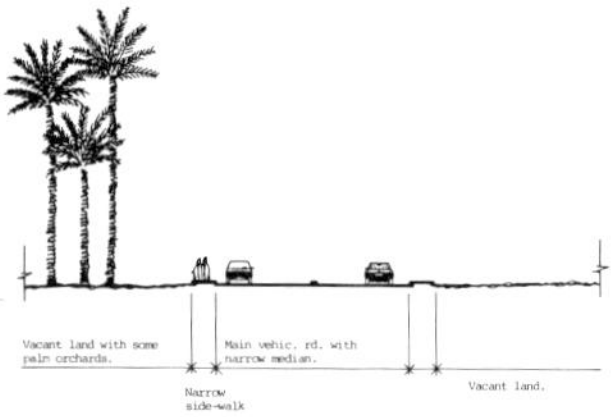

Figure 14. Zone 7 (existing)

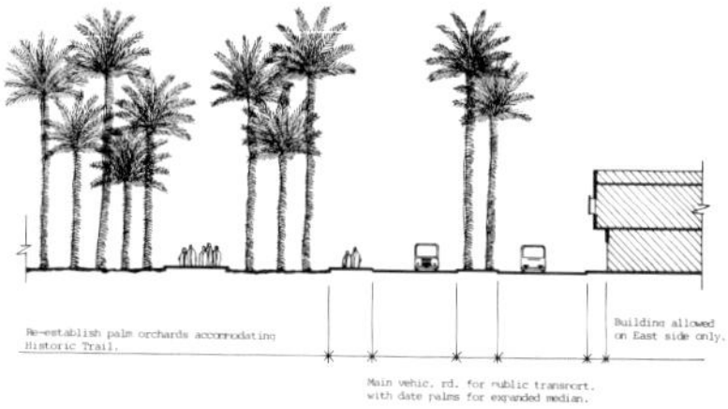

Figure 15. Zone 7 (proposed)

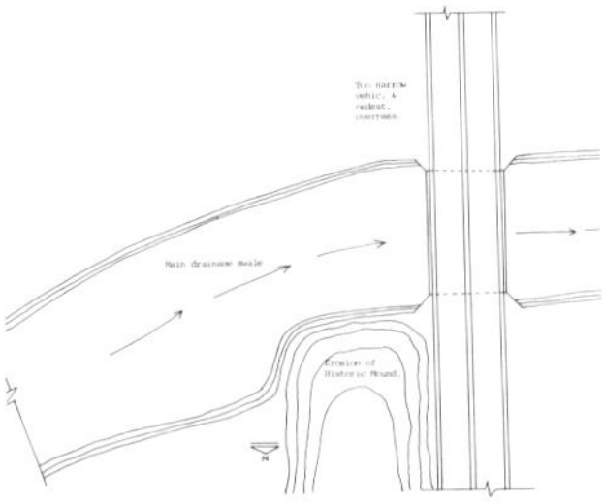

Figure 16. Zone 8 (existing)

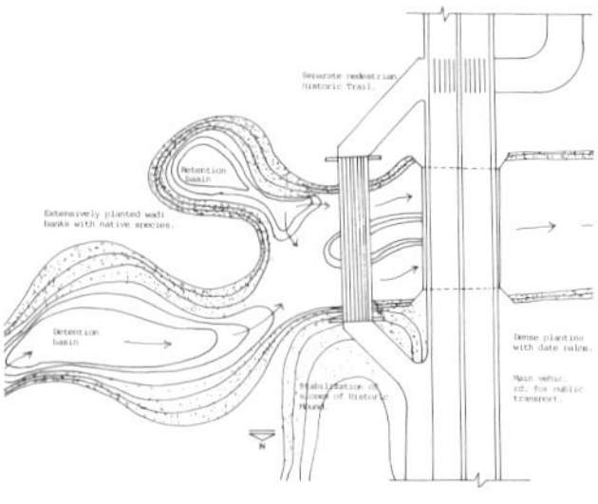

Figure 17. Zone 8 (proposed)

2.7. LANDSCAPE ZONE NO.6

This zone is mostly vacant lots assigned as land subdivisions for future buildings. Side-walks and median are too narrow (figure 12). Proposals here include the following: expansion of side-walk with date palms for Historic Trail, main road devoted to public transportation only, private vehicles use a connected loop, expansion of median for date palms (figure 13)

2.8. LANDSCAPE ZONE NO.7

This zone is mostly vacant land, with some palm orchards more concentrated on the West side. Side-walks and median are too narrow (figure 14). Proposals here include the following: re-establishment of date palm orchards on the West side accommodating the pedestrian Historic Trail, buildings allowed on the East side only, and treatment similar to landscape zone No.6 is proposed for the road (figure 15).

2.9. LANDSCAPE ZONE NO.8

This is the Northern most zone which is mostly a natural landscape. It's main features include the following: main drainage path crossing Historic Trail, erosion of the Historic Mound of Alromah located beside the drainage path, too narrow vehicular and pedestrian overpass (figure 16). Proposals here include the following: planned and designed expansion of the drainage path to include detention and retention basins, path and basins sides are planted with native species with the use of rocks for stabilization, similar stabilization of slopes and appropriate grading of path for the protection of the Historic Mound, and separation of the pedestrian Historic Trail overpass from the vehicular overpass (figure 17).

3. Concluding Notes On Improvements In Terms Of Environmental Quality And Energy Efficiency

- Re-establishment of date palm orchards and rows along streets would re-establish a natural resource native to the city and would reverse the trend of it's diminishing due to urban expansion[6].
- Dates have been a well known and desirable product of Almadinah. Re-establishment of this resource would contribute to self-reliance at both the city and national scales in economic terms and would decrease the need for energy wasteful importation[7].
- Rows and orchards of date palms inside the city would give it an "urban forest" with all accompanying benefits in terms of micro-climate (i.e. energy conservation) and other environmental benefits[8].

- Energy conservation is also expected to result from the proposals in several landscape zones that discourage the use of the private automobile[9], while encouraging public transportation (that can be battery-operated) and pedestrianization[10].

Notes And References

1. This study includes some of the basic ideas taught in a project in Spring 1989 in LA-610 Studio, SED, College of Engineering, King Abdulaziz University. Students were Ashraf Turki and Eid Ali. Teachers were Abdelmohsen Farahat (Studio Coordinator), Judge Khurana and Geoffrey Ricks.

2. Rasch, B.(1980) The Tent Cities of the Haj, Institute for Leichte Flachentragwerke and Bodo Rasch, Stuttgart, pp.225-227

3. Ibid., pp.225-229

4. Farahat, A., et.al. (1988) Evaluation and Development of Building Codes and Land-Use Regulations, final report, KACST, Riyadh (unpublished).

5. Bianca, S. (1984), "Madinah Almunawarah, Planning and Urban Design Concepts for the Central Area", in Mimar, No.12.

6. Rasch, B. (1980) Op Cit, p.226

7. Newland, K. (1980) "City Limits: Food, Energy and Employment Restraints on Urban Growth", in Ekistics, Vol.47, No.284.

8. Spirn, Anne W. (1984), The Granite Garden, Urban Nature and Human Design, Basic Books Inc., New York, pp.204-206.

9. Steadman, P. (1975), Energy, Environment and Building, Cambridge University Press, Cambridge, USA, p.286.

10. Breines, S. & Dean, W. (1974), The Pedestrian Revolution, Streets without cars, Vintage Books, New York, pp.92-95.

COVERED PEDESTRIAN SPACES ALONG REVOLUTION BOULEVARD IN BELGRADE

Dr DARKO RADOVIć
CEP,
11000 Belgrade,
Zahumska 34,
Yugoslavia

ABSTRACT. Deatiled planning of one of the most important areas for future development of the Yugoslav capital, the stretch along Revolution Boulevard, is approaching. Author has proposed the development of diverse continual covered pedestrian spaces along the Boulevard. This paper discusses typological issues, measurable and non-measurable potentials of such structures in urban tissue and their bioclimatic qualities, in general and at the example of the proposal for Revolution Boulevard.

Revolution Boulevard in Belgrade is a linear, fast developing centre of the City whose prosperity is being initiated by strong centripetal force of the basic core of Belgrade. Among variety of its functions, already existing and those whose development, according to the very nature and position of these spaces in wider urban context, is certain, commercial functions and services are dominating; they are present at all levels, from the local one (typical shop at-the-corner) to urban and even regional (big supermarkets, markets, department stores). The briefest survey shows that, just in the most central part of the Boulevard itself, without adjoining urban blocks, we can speak about more than 400 shops.

It is clear that development of this traditional activity is to be encouraged and stimulated. The great stimulus for its development is in enhancing **the level of comfort of users**. That compounds the quality of service, goods, functional and esthetic qualities of interior design, facades, shop-windows, appropriate advertizing, convenient parking facilities and public transport ... but besides those, we should say, usual methods to improve conditions for commerce, one of the best, historically proved ways to attract users is by covering pedestrian communication spaces that are in direct contact with central functions, i.e., **development of the continual colonade**.

Colonade is one of urban archetypes, rooted in the furtherest past. Born from the very need to protect still unglazed interiors, development of covered pedestrian spaces could be traced back to Minoian Haggia Triada, Greek stoas and Helenistic avenues to Roman urban planning enterprises. In the great Roman Empire they become everpresent urban motive - just in August's capital it was possible to walk more tha 20 kilometers without loosing their protective shade.

Outliving the "eternal Rome" colonades are spontaneously developed along streets of cities almost Europewide - let us mention just brilliant

S. Alvarez et al. (eds.), Architecture and Urban Space, 125–128.

examples from Mediterranean countries or those further nord, in England, Chech cities or Insbruck, for example. During Rennaisance arcades and colonades are established as architect's and planner's reliable device to achieve *belezza* and comfort of open urban spaces. Baroque develops this type even further; Rue de Rivoli is just the largest example from the classicist XIX century.

So natural and, in climatic conditions like those of Belgrade, so appropriate form of architectural and urbanistic expression has its historic examples in this part of Europe, too. In places that lack stronger urban tradition (or at least the continuity of urban tradition) it finds its expression at least punctually, as almost inevitable, form of porch, *ajats*, especially as a part of buildings of some central activity (very often pubs and bars, of course).

This wide acceptance of covered pedestrian spaces is rooted in the fact that colonades are but **a spatial expression of one of reasons of architecture itself**, determined by the nature of human beeing: that is architectural type based in the natural need to be protected from hostile natural conditions, as well as in social needs of each human.

Aristotle, in Libanus's words, says that people "in towns without colonades feel separated and are separated. Nominally, they are in the same town, virtually distant as beeing in different towns." Bernard Rudofsky, following the worldwide presence of this motive in "architecture without architects", finds one of the most beautifull definitions of its nature: "The most sophisticated and refined way of covering streets, **a tangible expression of civil solidarity** - we might even say **philanthropy** - are arcades."

Natural basis of colonades is in the closest relation to its social roots. To be concrete: in Belgrade we have 2084 hours of sunshine per year, more than 50% (1055 hours) in period june/september, i.e. during overheated months with pronounced need to have protection from direct sunshine and heat. Need to be protected from prepicitation is everpresent, during all 165 days with rain or snow (28 snowy, 136,7 rainy days).

The simplest calculation shows that, developing continual colonades along street like Revolution Boulevard, possibility to live outdoors, what means to have social contacts, walks, trade, play, is increased at least for **1200 hours per year** (based at average of four hours of rain in rainy days, six hours of unpleasant sun during summer months). That means **150 eight-hour working days!**

Our experience from urban planning excercises of those spaces shows that physical protection of pedestrians given by columns as dividing membrane from vehicles is not to be forgotten, too.

We are sure that there is no need to explain the esthetic component. The change of rhythm, shade and light, big chance for architectural design, could find illustration in any historic survey of European tradition.

Revolution Boulevard has almost the uniform WNW-ESE direction. At its nord side the need for colonades is based on natural conditions and we find them indispensable there. Protection from sun and prepicitations is needed during more than a half of the year.

At the south side sun is not the problem. "Natural" reason for colonades could be find in 165 rainy days, but, we shouldn't forget social reasons, too.

Lesser amount of sun would ask different typology of colonades at the southern side - maybe wider spans, possibly two-storey hights, creative approach to each situation for sure. The "problem" is the source of inspiration.

Besides that level of the problem we would like to illustrate some of the other interesting potentials of colonades at the concrete space of Revolution Boulevard at just one example.

There is a tendency, based in modernistic tradition still present in some important planning institutions, to plan and develop the new width of Boulevard, to establish new, wider regulation lines, in order to accomodate high demands for future vehicular trafic. Usual justification of such proposals is found in numerical data of Parisian, Munich or some other boulevards; the existing nature and unique charm of **our Boulevard** is beeing forgotten.

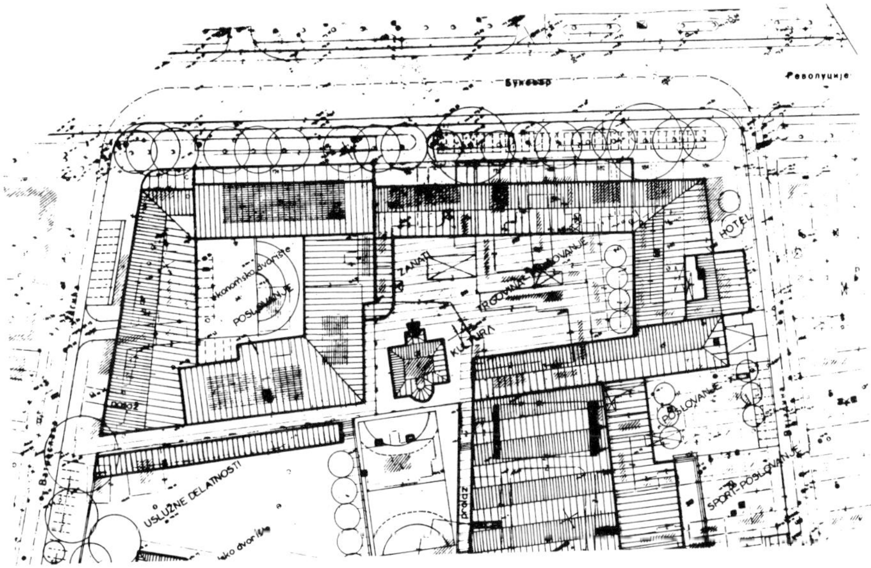

Figure 1. A part of detailed development plan for Block 55 at Revolution Boulevard in Belgrade done in 1985. by CEP team (D. Radović et al.). The first planning application of colonade as bioclimatic element at the Boulevard.

Japanese have a nice proverb which says that genius and philosophy are saved in the shape, not in material (at least in both of them, we believe). Relatively dilapidated state of urban tissue of the right side of Revolution Boulevard shouldn't be the reason to neglect its historic dimension and to look for a new one based at any single functional aspect. Historic city, and this part of Belgrade is its historic part, asks for *stabilitas*.

This space has its soul, one of whose components is the intimacy that can't be found in wide urban prospects of different cultures. Charles Bodleire speaks that in the castle "there is no nook for intimacy". It is easy to design a castle, it is impossible to design intimacy.

Development of the continual colonade enables the virtual widening of the existing regulation by simple coming with the width of those new pedestrian spaces below existing buildings at the narrowest parts of the street, without factical chang of established regulation.

Maybe this way to preserve the width, one of the essences of our Boulevard, is just another, indirect contribution of the **bioclimatic urban element** as **colonade** to **the regional expression of that part of Belgrade.**

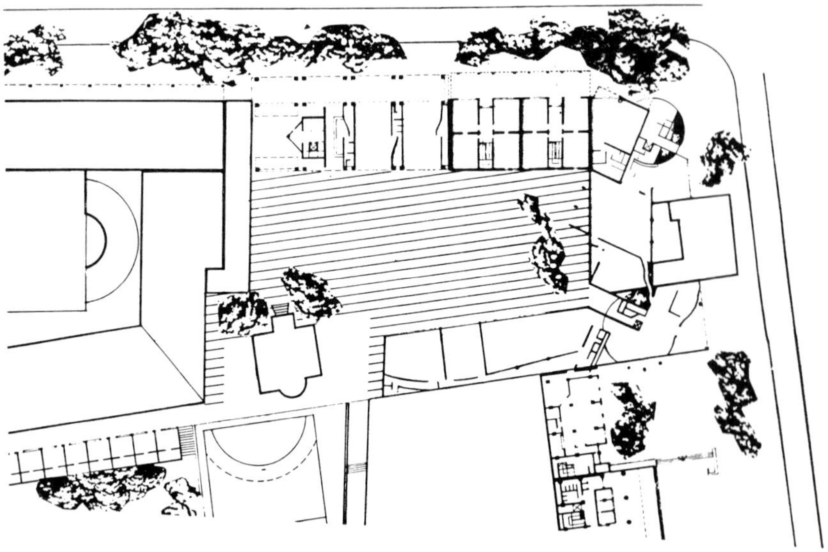

Figure 2. Block 55 at Revolution Boulevard in Belgrade; part of the ground floor level with covered pedestrian spaces

URBAN PLAN FOR A NEW NEIGHBOURGHOOD IN SEVILLE.

JAIME LOPEZ ASIAIN
JOSE MARIA CABEZA LAINEZ
Architects. (Seminar of Bioclimatic Architecture)
School of Architecture of Seville
Av. Reina Mercedes, 2
41012 Seville
SPAIN

ABSTRACT. A new residential area is being developed in South Seville. Such area was conceived and designed in an environmental way to minimize climatic impact and at the same time to allow for passive conditioning of the projected buildings. To ensure these objectives, " environmental " urban regulations were produced. Besides, acoustical problems were taken into consideration.

1. INTRODUCTION.

The new plan was intended to regulate the future buildings of a developing area of Seville in a climate responsive way. This was considered important by the architectural Service of the City Hall, once they became aware of the significant impact of buildings on the urban climate.

The main idea was somehow suggested by early modern architects like the late Walther Gropius or Ludwig Hilberseimer but mainly regarding sun availability and disposal of green areas.

Similar regulations do exist in Northern Europe's countries probably due to sunlight scarcity but in Southern Spain solar radiation is available most of the time and therefore such requirements were not introduced before though later they proved a principle question for passive heating and cooling of the built environment.

The plan provided for residential areas, garden areas and a primary school. Because of this not only thermal but also acoustical problems were taken into account.

2. THERMAL PROBLEMS.

2.1. Orientation.

In this type of climate, orientation of buildings along an East-West axis (i. e. main façades to South and North) is important not only for heating in winter but also for cooling in summer provided that South-oriented vertical planes receive the less possible amount of solar radiation during summer (almost equal to North façades)

S. Alvarez et al. (eds.), Architecture and Urban Space, 129–133.

while maximizing it in winter. Thus all the new buildings should have this possibility and that is the first condition for the projected groups of dwellings, determining the lines of the open spaces and the streets of new creation.

2.2. Solar protection.

Once main orientation is defined, solar protection becomes important for mutual shading should be restricted in winter but encouraged in summer through adequate building distances and shading devices.

An adequate parameter to avoid obstructions in winter seems to be around 30º altitude measured from the foot of each row of dwellings. In this way, maximum height of adjacent buildings is fixed.

On the other hand, shading for overheated conditions has been promoted with overhangs, protected walkways, pergolas with vegetation and to a lesser extent, white colour and canvas.

2.3. Ventilation.

In Southern Spain air movement around buildings is critical for cooling and ventilating the indoor spaces. However, attention must be paid to the outdoor conditions; in fact some types of buildings act as obstructions to the wind and the areas behind them remain unventilated.

As the prevailing wind in summer is South-West, the former undesired effect has been avoided through the use of spread plans and a variety of heights in the same area, thus creating pressure and suction alternately. Obstructions are only encouraged when they restrict the flow of North-East cold winds in winter.

The distances between buildings achieved as explained in the former chapter and the fact that orientation of dwellings (South) is not perpendicular to the cooling breezes (South-West) enhance the availability of air movement.

This ventilation (and specially cross-ventilation) is difficult to obtain in dwellings more than double-banked; besides, apartaments located in the rear have little sunlight and that is the reason why this typology (known as H-type block) is seriously discouraged though it was one of Modern Movement's main features.

2.4. Vegetation.

The favourable effects of vegetation in urban climate have been summarized by many researchers including the authors.

Due to this fact, intensive use of vegetation all over the area has been promoted, reaching a proportion of more that sixty per cent (60%) of total floor plan.

Plants can also be used in façades and roofs according to the proposed regulations.

2. ACOUSTICAL PROBLEMS.

The new area has dense traffic requirements and the consequent noise conditions, therefore the buildings should be designed to balance this negative effect not only with insulation and windows but also with protecting shapes folded to the inside which can provide a better sound condition in the core of the area.

The buffer spaces such as bathrooms and kitchens can also be located in from of noisy streets to reduce sound intensity in resting/living areas.

Also, with the use of such dense zonal vegetation reductions of noise level up to 4 dB can be expected in the neighbourghood, and specially near the school facility.

4. CONCLUSIONS.

A new kind of Urban Development has been established by conscious use of legal instruments already available. Further generalization of these procedures can prove helpful in the field of a new urban " environmental " culture.

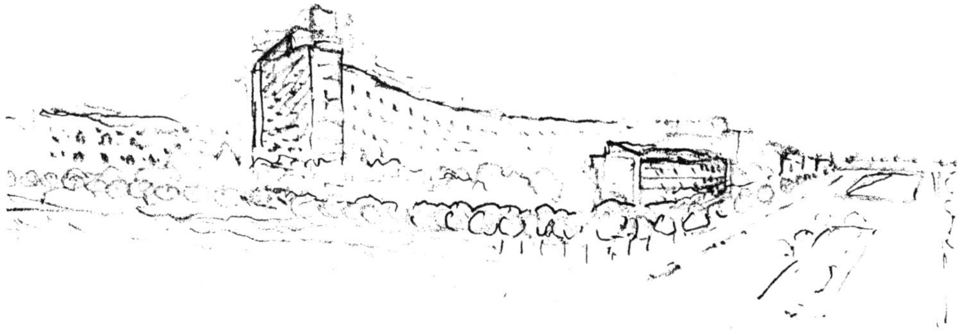

Figure 0. General views.

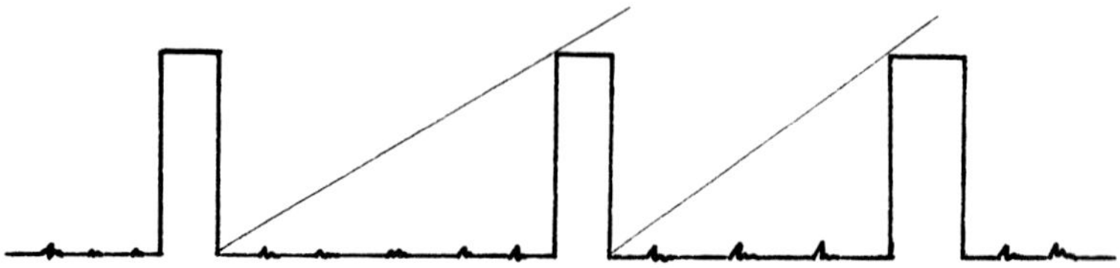

Figure 1. Section through blocks.

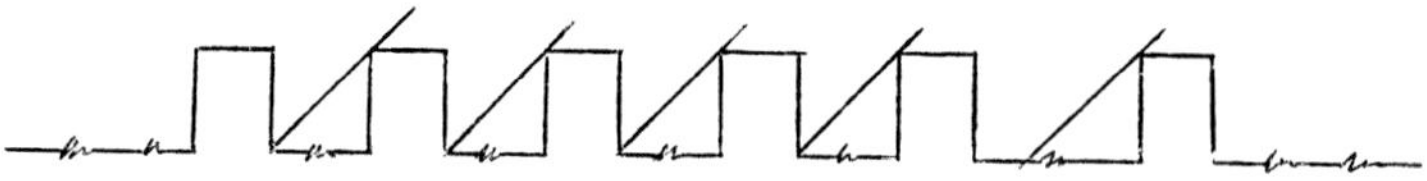

Figure 2. Section through dwellings.

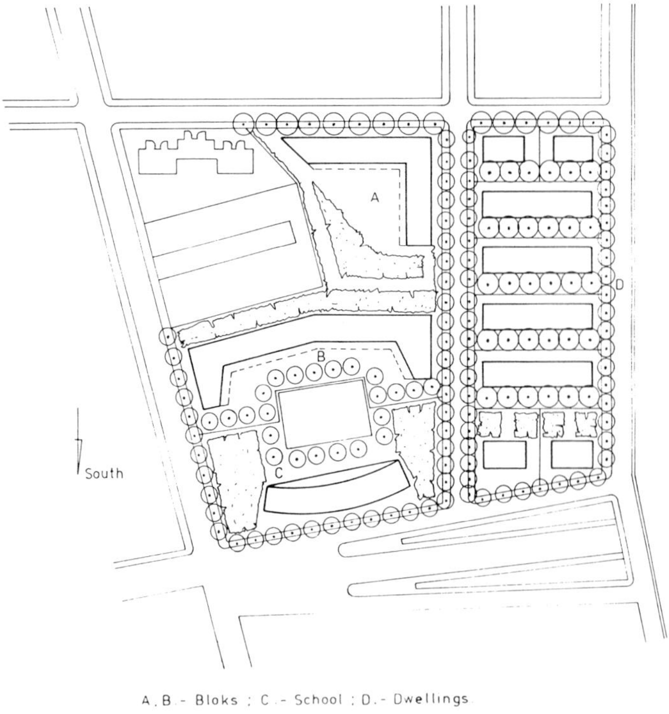

Figure 3. General Plan.

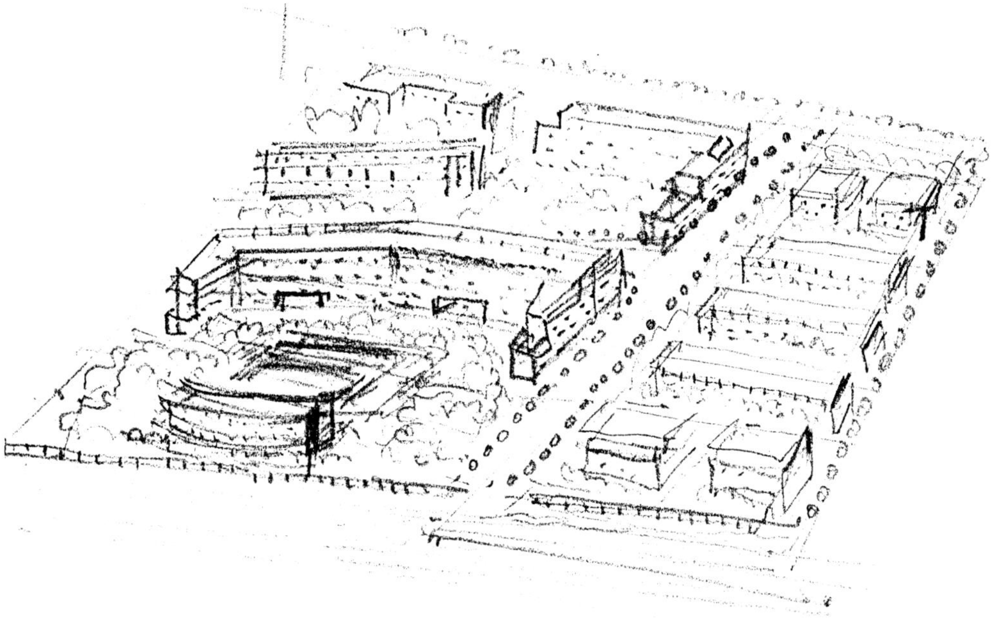

Figure 4. View from the South.

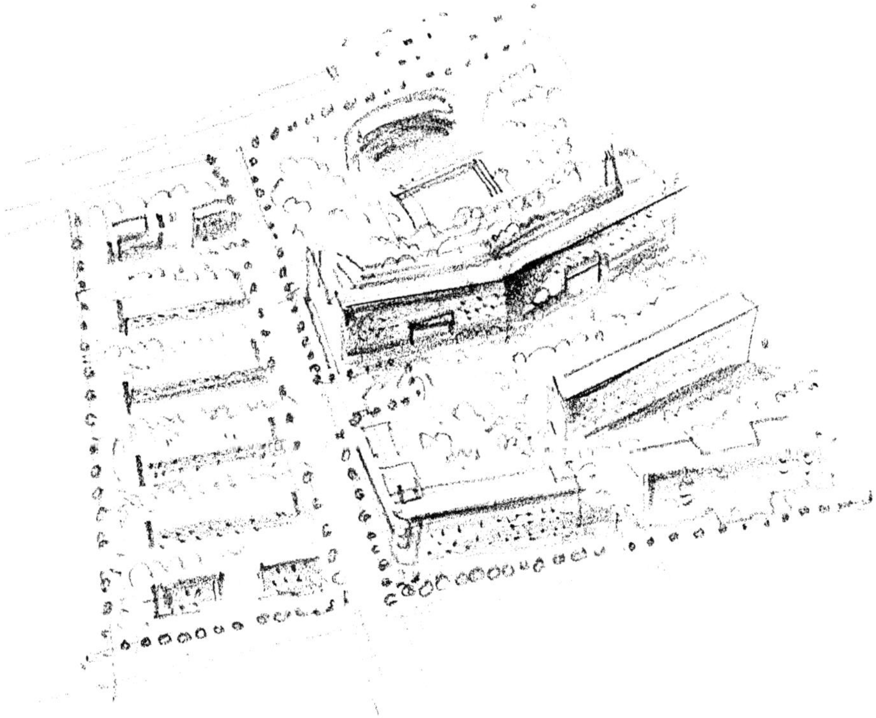

Figure 5. View from the North.

BARKANTINE 2000: A GREEN PLAN FOR LONDON DOCKLANDS

D TURRENT
The ECD Partnership
11-15 Emerald Street
London
WC1N 3QL

ABSTRACT. This paper describes outline proposals for a major low energy urban renewal project in London Docklands. The aim would be to achieve a 50% reduction in CO_2 emissions by the year 2000.

1. BACKGROUND

1.1 UK CONTEXT

During the past year or so we have all become much more aware of the problem of global warming caused by greenhouse gases, particularly carbon dioxide. It is estimate that the UK emits some 600m tonnes of carbon dioxide into the atmosphere every year. Half of this is related to buildings and the power that needs to be generated to keep them warm and to provide lighting and services. The UK government has made a commitment to reduce current CO_2 emissions by 30% by the year 2005. However experts point out that this only results in a stabilisation of the current situation, not an improvement.

Energy efficiency in buildings is the single most effective way of combatting the greenhouse effect. In particular, thermal upgrading of the existing housing stock could play an important role in reducing CO_2 levels. There are some 20m dwellings in the UK, the majority of which are draughty and poorly insulated. Most of these could benefit from the application of energy saving measures which will in turn contribute to significant reductions in CO_2 levels. By contrast the construction of new, energy efficient housing will only have a very small impact, because the rate of replacement is less than 1% per annum.

In June 1990 The ECD Partnership was appointed to carry out a feasibility study on the Barkantine Estate in London Docklands. The study involved a team of 8 consultants

S. Alvarez et al. (eds.), *Architecture and Urban Space*, 135–138.

working in close consultation with the community and was funded through the Social Accord, a joint agreement between London Docklands Development Corporation and the London Borough of Tower Hamlets.

1.2 ESTATE DESCRIPTION

The Barkantine Estate is situated on the west side of the Isle of Dogs and covers approximately 30 acres. It is bounded to the west by Westferry Road and the river, to the north by Olympia and York's Canary Wharf development, to the east by the LDDC Enterprise Zone and to the south by the Daily Telegraph printing works. The estate comprises some 1000 dwellings, together with a primary school, health centre, shops, pub and community centre. It is indeed the size of a large village or a small town. Due partly to its physical isolation, the estate has a strong sense of community and is well liked by the majority of its residents.

During the nineteenth century the site was developed with two storey Victorian terraced housing for dockworkers and their families. In the 1920's one or two blocks of flats were added and in the 1930's, Hammond House was built in the 'modernist' style. Extensive bomb damage during the war created a number of empty sites and a further phase of development took place in the 1940's with the concrete frame Orlit houses. A major redevelopment of the site took place in the 60's when four tower blocks were built, together with a large number of 4 storey deck access maisonettes. A block of sheltered flats for the elderly was added in the 70's and a small estate of Barratt houses was developed for sale in the 1980's.

With the exception of 1950's semi-detached housing, the estate therefore has examples of most forms of low, medium and high rise accommodation. The majority of properties are public sector rented although 10-15% of the households have exercised their right to buy. In addition there are a number of housing association properties, mainly in St Huberts House and in Mellish Street.

Like most public sector estates of this type, the physical fabric of the buildings has deteriorated due to poor maintenance. This gives the estate a run-down appearance which is compounded by bleak windswept spaces between buildings. Virtually every window on the estate is in need of replacement and the existing flat roofs have been a constant maintenance problem. External walls are either solid brickwork or uninsulated cavity brickwork. In the case of the four storey deck access blocks, concrete cladding panels supporting brickwork have slipped and major expenditure is required to fix them back to the structure.

Generally, the tenants are satisfied with the interiors of their flats or houses, although replacement of kitchens and bathrooms is now due, as is re-wiring of the electrical installations. The major complaint from tenants however concerns the heating, which is on a pooled billing system and is only available 6 months of the year. In general, communal gas fire plants supply heating to one tower block and an adjacent low rise block. The output from the system is blown warm air, which operates inefficiently. For this, the tenants pay £6 per week. On top of this they pay for on peak electric water heating and top-up on peak electric space heating. Total annual fuel bills for a two bed flat can amount to £600 a year. In addition to the communal plants, there are a variety of heating systems operating around the estate, mainly gas fired wet systems and electric systems. Generally the systems on the estate could continue to function for 5-10 years, although repair and maintenance will be a growing commitment.

1.3 ENERGY SAVINGS

As part of the overall feasibility study of the estate, ECD is developing a strategy for upgrading the thermal performance of the buildings and replacing the inefficient heating systems. The principle behind this is that everyone should be able to afford to heat the whole of their dwelling to an adequate level and to have a plentiful and affordable supply of hot water. Further to this, the strategy aims to achieve equality of heating costs, independent or flat location, but not size. It is also felt that individuals should wherever possible, have direct control over their heating and it should be paid for on an individual basis.

Replacement heating systems would therefore ideally consist of individual gas fired wet systems in the low rise properties. This would allow the communal systems to be converted to small scale combined heat and power plants, serving each tower block. More efficient year round heating systems would improve comfort standards and could result in lower annual fuel bills if a package of insulation measures is included to bring the fabric up to 1990 Building Regulations standard.

- Insulated roofs to 0.25 W/m^{2}k
- Insulated walls to 0.45 W/m^{2}k
- Double glazed replacement windows
- Ventilation to kitchens and bathrooms
- Low energy light fittings and appliances

Ultimately it would be desirable that all properties on the estate could achieve a National Home Energy Rating of 6.

2. CONCLUSIONS

Our calculations show that an estate the size of the Barkantine generates 10,000 tonnes of CO_2 a year, both on and off site at the power station. A feasible target would be to reduce this by 50% by the year 2000. Our proposal is to promote the estate as a major demonstration project to show the range of measures that are necessary to achieve this objective.

The estimated cost of implementing the demonstration project is £50m and the construction period would be 6-7 years. The medium term objective would be to quantify CO^2 reductions and energy savings achieved by the 2000.

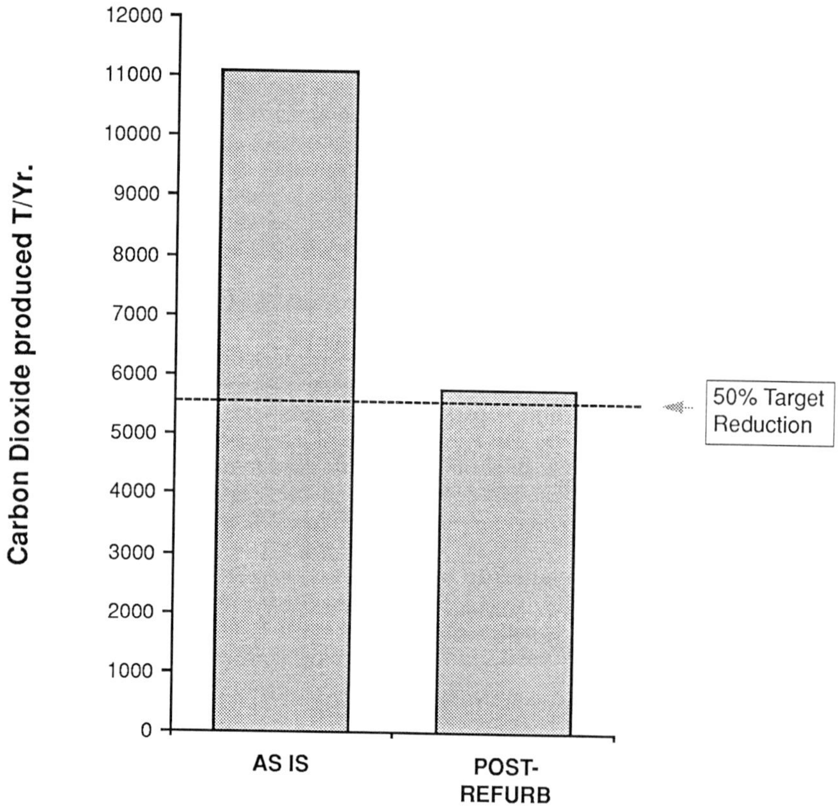

POTENTIAL CARBON DIOXIDE SAVINGS – Barkantine Estate Housing

URBAN CONCEPTION OF THE OLD CITY OF NICE TO PROVIDE SUMMER THERMAL COMFORT.

E.RAUZIER[*] and X.BERGER[**]
* Ingineer, 203 ch. de la Gorgue
06410 Biot (FRANCE)
** Laboratoire de thermodynamique - Université de Nice
C.N.R.S. Groupe Ecothermique
rue A.Einstein, Sophia Antipolis 06560 Valbonne (France)

ABSTRACT. Besides classical elements for summer comfort, the old city constitutes a whole conceived from width of streets to opened courtyards and wells, in the aim of gradients and condensations avoidance. Measurements and a mathematical model analyse these aspects.

1. INTRODUCTION

The origin of the old city of Nice goes back to the fifth century B.C., which testifies the multiple influences come under. In the past, the life was less sedentary, and, in this mediterranean climate, comfort problems were mainly related to the summer period. The city was conceived consequently. But, going further than a survey of the classical wellknown elements found as everywhere, measurements were done, and a mathematical model built to point out unforeseen considerations which would have been taken into account.

2. CLASSIC ELEMENTS FOR A NATURAL COOLING OF THE OLD CITY

Let explain that, for security considerations, the old city was established in a triangular area delimited by the Paillon stream, the coast, and a high steep rock, named castle hill (fig.1).

Along the seaside, a double row of one floor buildings (Ponchette bar) constitutes a breaker of the mistral wind.

The east-west streets are larger than the north-south ones. They are slightly ascending, which increases the air velocity by venturi effect. In the last centuries, the castle hill was without vegetation and waterfall. So, its insulation lead to a draught (acceleration of the air movement by chimney effect).

The building height insures a protection against sunlight for inhabitants perambulating at the bottom of the narrow streets (3.84m width in opposition to 15m for the modern part of Nice). For north-south streets, italian penthouse caps minimize the solar direct incidence at midday. Shutters have down opening panels.

Fountains and watering at midday, after the market, the warm time in the day, insure an evaporative refreshment.

S. Alvarez et al. (eds.), Architecture and Urban Space, 139–144.

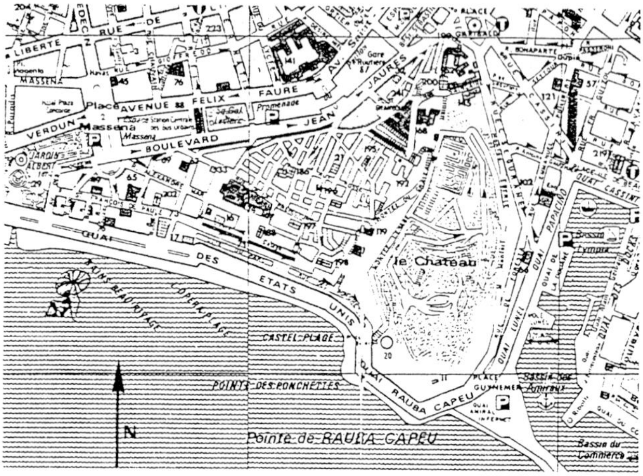

Figure 1.

Map of the old city of Nice.

3. THE MATHEMATICAL MODEL

Let be an elementary unit of city constituted by a parallelepiped, itself divided into meshes (fig.2). The air passages include streets (width e), opened courtyards and wells. A square section represents flats, and include both massonery and motionless air.

Four agents are responsible of the thermal exchanges:

* the air running inside the city, with a different velocity than above.
* the roofs sunlighted.
* the lower part of buildings and streets, sunlighted at some times only. Some heat is transfered to ground.
* massonery partly sunlighted. Its contribution to inertia is majority.

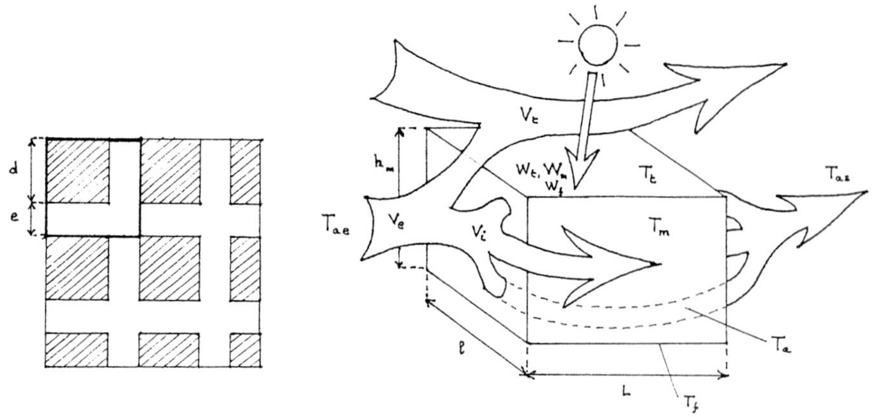

Figure 2. Skeleton diagram of the porous environment that constitutes an urban quarter. Thermal exchanges (W: flux, T: temperature, V: wind, m: massonery, a: air, t: roof, e: entrance...)

It was also considered the exchange area air/material S_m, and the porosity, which is the ratio between the volume of external air and the total volume. The massonery temperature T_m was assumed only depending on the altitude z, and the air temperature T_a on the horizontal abscissa x.

The four equations (Rauzier 1990) establishing the thermal balances for the four agents were solved by integration along a daily cycle (convergence) with an implicit method. Comparisons with measurements permitted to adjust the most uncertain parameters as those linked to ground (conductivity, diffusivity), to convective coefficients (walls h and roofs), and to geometry (dimensions of the lementary unit, part R_m of massonery in the whole volume).

4. RESULTS

Three main families of parameters govern the thermal state of the system:
* those linked to inertia and thermal diffusivity, especially R_m
* those linked to convection, and consequently to dimensions S_m and e
* those linked to solar incidence: height H of buildings, and e (fig.3).

It is not the lower inertia of buildings in the recent part of Nice (hollow bondstones) which is of strong influence. The size e of the streets matters, and the impact is not only on the air heating, but also on the radiant temperature. Indeed, this last affects the felt comfort (sunlight, radiation from warm streets and walls).

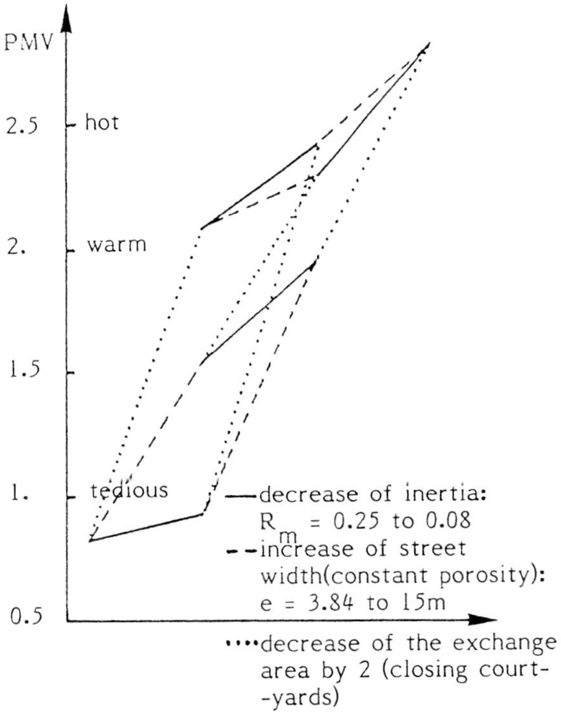

Figure 3

Variations of PMV when going from the old city to a modern quarter.

But the most efficient effect is got by S_m: in fact, it is the product h * S_m which intervenes in the equations: shutting the "clairoirs" (wellknown openings above the main doors in the old city of Nice), or covering wells and courtyards by glassroofs, it's equivalent to reducing the exchange area without modifying the sunlight heating (d, e), and so, h. In the old city, the ramification in lanes and backyards, and the wells opened at bottom and top, have a positive contribution to the comfortable coolness.

Besides this, conceiving buildings with empty central parts opened at two ends, is a way to avoid condensations by chimney's effects: the dynamical system runs at nighttime, and, by air renewal, slowly throws away warmth and humidity of walls piled up at daytime. Indeed, the nocturnal air comes down from mountains; it is drier (lower water content) than the diurnal air which is the sea breeze (fig.4). Maintaining confined the diurnal air leads to condensations when cooling it by conduction through walls (Berger 1989).

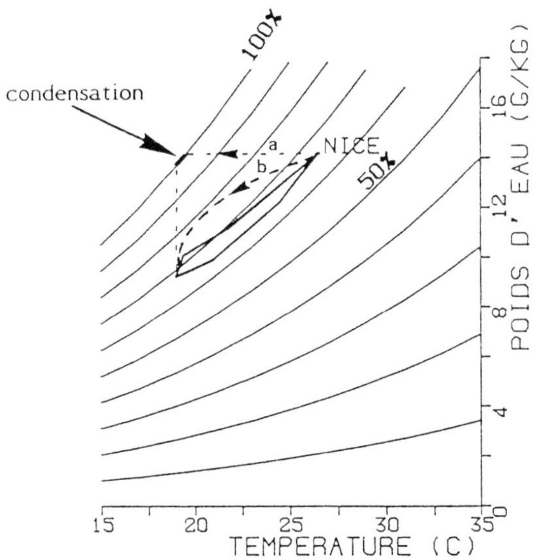

Figure 4

Mean summer daily cycle.
a : cooling a confined air
b : cooling a renewed air.

5. THE FELT SUMMER THERMAL COMFORT

Contrarily to winter comfort linked to a null balance for human body (homeothermy - dry exchanges), summer comfort is connected with sweatiness (secretion of water, salts, greases - humid exchanges), and changes in the body working (internal blood circulation slowed down whereas the external one is speed up...). Wettedness w (ratio between the required skin evaporation to insure a null balance, and the maximum possible evaporation) is only one visible parameter of the felt discomfort (Berger 1990). The other parameters are less quantified, being linked to the fatigue of a body aloof from equilibrium (Kerslake 1972). These parameters are of two kinds:
* thermal ones: gradients in time and place are associated with variations

of the body working, which requires a delay. The old city is conceived to avoid them, two large sunlighted areas, to maintain a same ventilation in all the streets, whatever its orientation, to lower by -3°C the daily amplitude of air temperatures, compared with the modern city.
* Psychological ones: comfort is not only the fact of a thermal state, but also of the five senses and brain states. Fatigue is resulting from all of them by its additive aspect. So, interaction exists, which is felt as a thermal contribution: absence of cars (stress, noise, congestion), odors of plants, pleasantness of the ancient architecture (a thermal system as a clairoir was made beautiful - photo 1), pleasantness of noises of the common life, all this has a positive contribution.

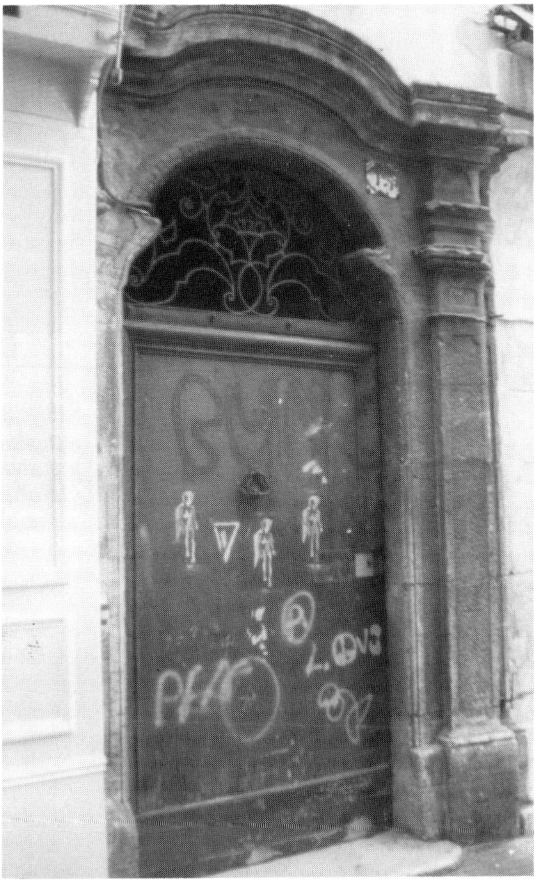

Photography 1

A "clairoir" to ventilate the internal parts of the building.

6. CONCLUSION

Such a study demonstrates that not only an architectural patrimony can exist for the old cities, but also a thermal one. A secular experience has been slowly gained, and technology never made up for aesthetics. Summer comfort was then finely analysed: the impact of gradients was felt more important than skin wettedness itself! Today, the modern way of live should not reject a priori the past. Rehabilitation of ancient buildings or quarters, conception of the present day should result from a continuation through an understanding of the former periods. Committed errors could have been avoided, efficient and aesthetic solutions found.

ACKNOWLEDGMENTS

The authors acknowledge their thanks to the A.F.M.E. which granted the study.

REFERENCES

Berger X. (1989) "Transferts parietodynamiques en climat chaud et humide", Congress J.I.TH Alger

Berger X. (1990) "Thermal comfort" Workshop on passive cooling Ispra (Italy) - Commission of the European Communities.

Kerslake D.McK. (1972) "The stress of hot environments" Cambridge University Press

Rauzier E. (1990) "Etude climatique du vieux Nice" Contrat AFME n°09 04 0004

PREHISTORIC PUEBLOS & PLAZAS OF THE ANASAZI

Jeffrey Cook, Regents Professor
School of Architecture
Arizona State University
Tempe, Arizona 85287-1605
U.S.A.

1. **Introduction**

1.1 THE ANASAZI AND THEIR RUINS

Anasazi is the Navajo word for "the ancient ones." It can also mean "the other ones" or even "the aliens." The prehistoric Anasazi came before and lived upon and with the land differently than today's inhabitants - the Navajo, the Hopi and the Anglos.

There are over 25,000 Anasazi settlement sites in Arizona (AZ), with as many in New Mexico (NW) and lesser numbers in Colorado (CO) and Utah (UT). These are the states of the Four Corners whose straight lined borders meet in the southwest of the USA.

The Anasazi abandoned their stone communities without evidence of violence. Most of the ruined stone architecture we admire today comes from their Pueblo II period (AD 900-1125). Their best known and admired stone city is Pueblo Bonito at Chaco Canyon (NM). However, the cliff communities at Mesa Verde (CO) gave the Anasazi their name as "the cliff dwellers."

1.2 THE LAND AND ITS CLIMATE

The Anasazi territory covered most of the Colorado Plateau, a high arid windswept land nested near the end of the rocky Mountains. It bridges the Great Divide of the continent, where precious head water originate the flow of great rivers in opposite directions to the Atlantic and Pacific Oceans. The arid land is rocky and inhospitable, but spectularly beautiful visually.

The elevation is 5500' to 7500' (1700 to 2300 M). The latitude is 36° N. The air is dry and clear. Winters alternate blizzards and severe storms with brilliant sun and freezing cold. The summers are warm and equally brilliant with sun. July and August bring occasional afternoon showers.

1.3 THE PUEBLO

The Anasazi invention of the pueblo form was the basis of their distinctive architecture. We may call them urban villages or miniature cities, these multifunctional megastructures of stone and mud. Within a single articulate building were accommodations for day to day life, for the storage of wealth, and for the social and spiritual activities that ultimately shelter and succor any society. Thus the pueblo was not only an impressive architectural unit, it was a confirmation of land planning philosophy, of community responsibility and a concretization of culture.

We don't know what the Anasazi called their invention. Our word 'pueblo', is borrowed from the Spanish, meaning both a people and a village or town. Architecturally it is defined as one or more flat roofed multistory structures of adobe or stone arranged in terraces, housing a number of families. As a design for community it is the opposite of a rural farmhouse or a suburban house.

S. Alvarez et al. (eds.), Architecture and Urban Space, 145–150.

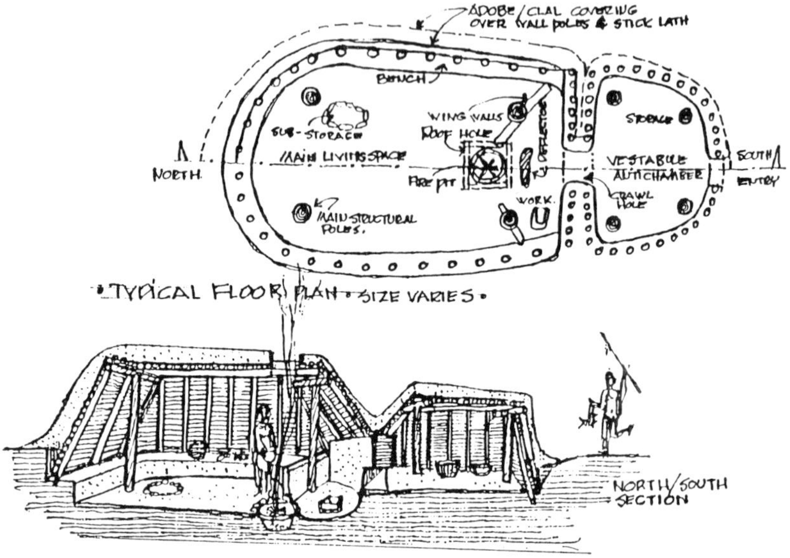

•ANASAZI•B./\. III •PITHOUSE•

1.4 ANASAZI EVOLUTION

The pueblo as a preplanned architectural urban unit emerged as an evolutionary stage of Anasazi culture. The earliest Anasazi, known as Basket Maker I, were wandering hunters and gatherers without fixed addresses. They used spears for hunting and digging sticks to plant corn or squash. They did not use bows and arrows like their neighbors to the south but did use snares.

The earliest Anasazi settlement sites are known as Basket Maker II (AD 450-750) and include caves, rock shelters, and a few small open villages where they built shallow pit houses. Without the pottery already known in the south they stored food in baskets and rock lined pits.

By Basket maker III (AD 450-750) weaving and pottery had been introduced and ceremonial structures or kivas made their appearance. The pithouses were now more substantial, deeper and better built and the villages were larger. See figure 1. Orientation of the pithouse appears to have been important with entrance to the south. But it was the Pueblo period (AD 750-900) when the Anasazi emerged as a distinctive and even revolutionary people. They invented the first pueblos by moving from pithouse clusters to surface structures of rectangular contiguous rooms that formed larger villages. The enlarged social unit became the architectural unit - a composition of rectangular stone cell assembled with circular ceremonial kivas and linked by the plazas and terraces where daily life was conducted outdoors.

1.5 THE CULTURE OF THE ANASAZI

This neolithic people with their egalitarian culture had no wheel, no metal, and no beast of burden. The first three cultural signatures that distinguish the Anasazi from their neighbors and indeed identify them from all other peoples are architectural: the <u>kiva</u>, the round underground clan ceremonial spaces, derived from their ancient dwelling, the pithouse; the <u>unit pueblo</u>, a cluster of two to twelve rooms for both storage and living, together with outdoor plaza or workspace that includes the kiva; and lastly the <u>orientation</u> to the world, the kiva and plaza facing the sun.

Thus the Anasazi are among the few cultures distinguished by their concern for <u>orientation</u>, and for their commitment for <u>public outdoor space</u>.

2. The Daily Settlement

2.1 SAND CANYON

At an elevation of 6,800 feet (2060 m) Sand Canyon, CO close to the Four Corners provides a large "U" shaped land form facing south . A perimeter wall reinforced the protective and sheltering form of nature. Continuous water in the canyon was practical and necessary. But possibly it also had mystical significance, accounting for the unusual number of kivas or circular rooms. The settlement consisted of subset clusters of rectangular rooms associated with kivas and plazas. These blocks or unit pueblos were organically terraced into the terrain. Systematic excavations were begun in 1983, but areas are being refilled after study has been completed.

At Sand Canyon the pueblo design is characteristic of Anasazi standards. The exterior wall defining the settlement site is also part of the housing and is not developed as fortifications. Rooms could be accessed by ladder through a hatchway in the roof, or through a doorway that often had a "T" shape. The 'kiva' was integrated within the architectural agglomeration.

2.2 SECTION THROUGH A CLAN KIVA

The circular rooms known as kivas are associated with clan based religious and social ceremonies. See figure 2. Small kivas served the clan living in adjacent housing. Large ones served the whole community and region. But the basic plan and structure were always similar. Whether dug into the ground, or built above ground with the enclosing walls filled with dirt, it always was a subterranean space.

Access to the kiva was through a central opening in the roof by ladder from the plaza. Descent on the ladder through the smoke of the hearth was an act of purification, aligned on a north-south axis. The recessed hearth, the stone deflector behind the fire pit and the separate ventilator were also on axis. Thus orientation of all the details within a circular room had spiritual, not passive heating or comfort purposes.

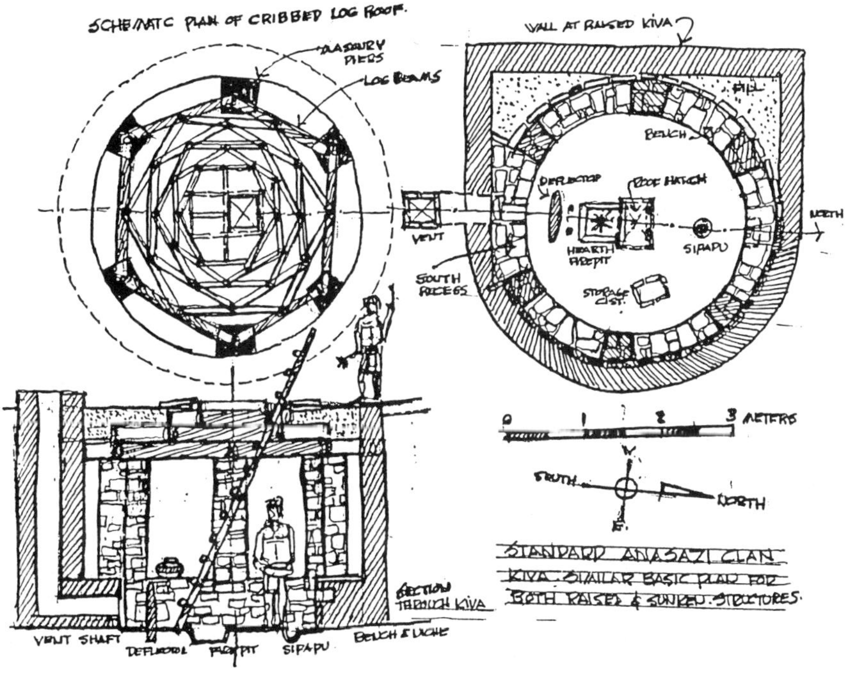

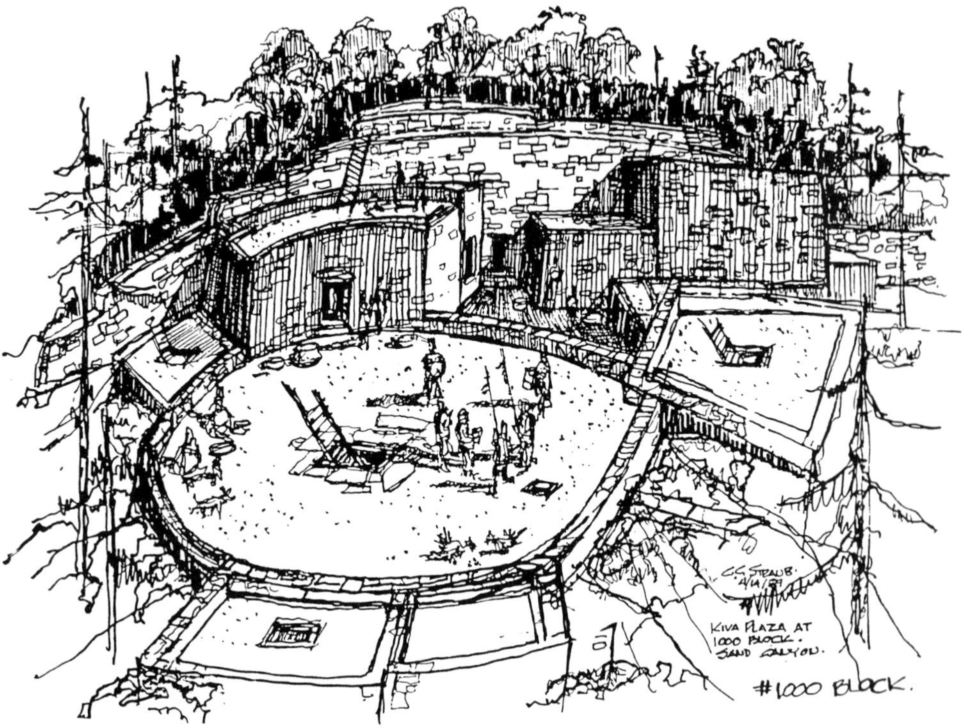

2.3 PUEBLO BLOCK 1000 AT SAND CANYON

The Anasazi were accomplished at geometry. Their straight lined roads could run for 30 miles without deviation, and their kivas could be a perfect cylinder. But this settlement at Sand Canyon was strongly formed by local conditions. See figure 3. However, the relationship of the three parts of the unit pueblo remain constant. The circular subterranean kiva 20 ft.(6.5 meters) in diameter, is the south of the housing and storage building consisting of seven rooms, and forms part of the plaza which is always larger than the kiva roof. The bioclimatic advantages of this community arrangement of outdoor space in a high cool climate are obvious. Yet, the spiritual orientation of the circular roomed kiva is always present.

3. **The Ceremonial Settlement: Chaco Canyon**

3.1 CHACO CANYON

Chaco refers to that dramatic three mile stretch of Chaco Wash that contains the remains of some six large Anasazi pueblos and one large singular kiva, Casa Rinconada. Mostly within less than a half mile of their neighbors, each is of distinct and autonomous design. As impressive is the natural setting. Cliffs especially on the north define this climatic microcosm. They provide wind sheltering and a solar backstop for winter.

The collection of ruins at Chaco Canyon is the largest and most studied prehistoric architectural site in North America, often compared to the great pre European sites of Mezzo America, the previous cultures of present-day Mexico and of Peru. But their vast physical wealth was supported by expansive rich hinterlands. The autonomy of the Anasazi and especially the apparent remoteness of a now desolate Chaco plateau with its scant resources reinforces their importance as an inventive people. They were the original and unique creators of a distinct culture.

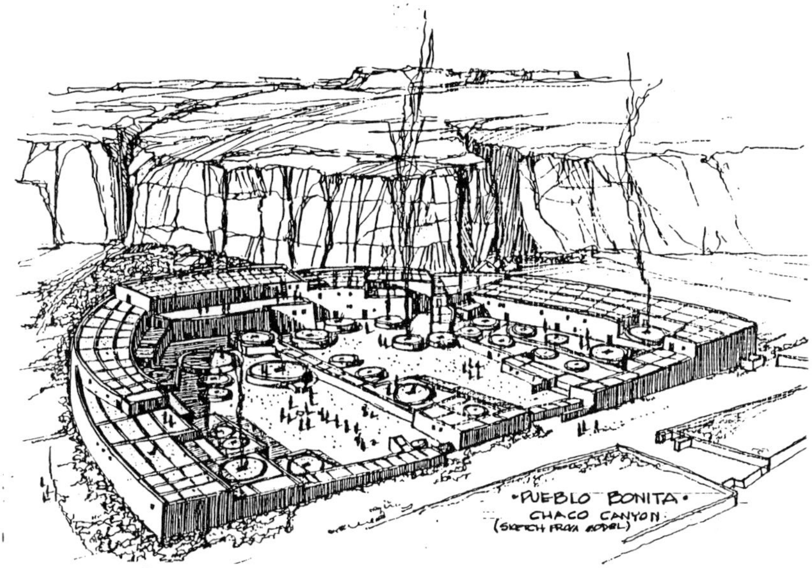

3.2 PUEBLO BONITO

The classic Bonita phase of the Anasazi around 1000 AD climaxed 6000 years of human occupation. It is named after a kind of Anasazi architectural swansong, Pueblo Bonito. This great south facing cup shaped city was hard under the sheltering cliff. See figure 4. Although unchallenged by other people, abandonment began about 1125 AD. Today deliberately unpaved, unpredictable roads to Chaco Canyon impose a small control on the waves of enthusiastic tourism.

A sketch cross section through the outside wall of rooms at Pueblo Bonito illustrates the three dimensional organization of community life. See figure 5. Life was oriented to the sun and addressed the plaza. The roofs of the continuous pueblo rising to 4 and 5 stories also provided open terraces for daily activities and formed arcs like a giant amphitheater framing the plaza.

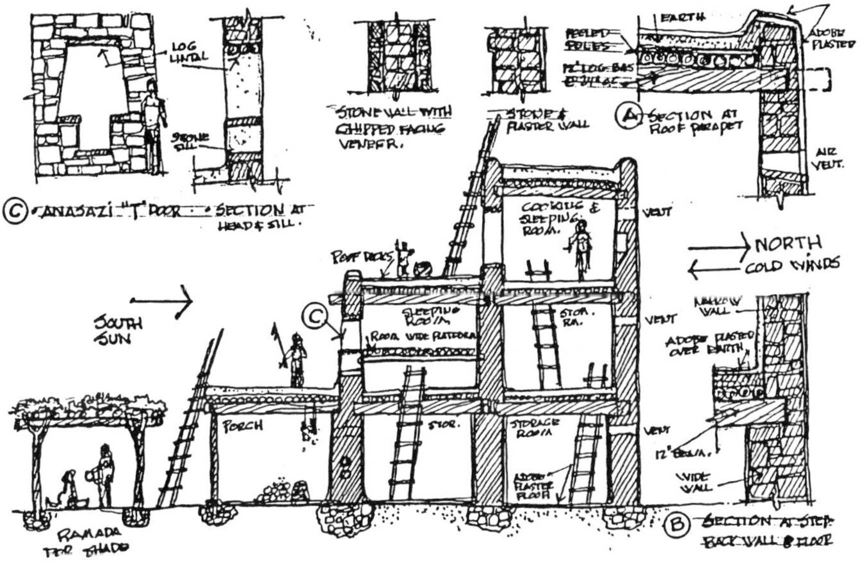

4. The Late Settlements

4.1 CLIFF PALACE MESA VERDE

The well preserved settlements at Mesa Verde, CO, continue the Anasazi record of impressive architectural and planning skills. The Anasazi cultural signature of the three elements and their relationship established by orientation were continued. However, the well known cliff dwellings at Mesa Verde were among their last settlements. In some cases they were constructed and occupied for scarcely more than a generation. See figure 7.

Although the Anasazi had built pueblo complexes of more than 50 rooms and three stories high on the flat mesa tops, around AD 1200 they began to move back to the cliffs that had sheltered their early ancestors. The caves and ledges that became sites for the 600 cliff dwelling we know today were built by the middle of the 13th century but abandoned by AD 1300. These last settlements continued to respect orientation to the sun with their environmentally adaptive designs.

5. Conclusions

5.1 THE ANASAZI LEGACY

The blossoming of the egalitarian Anasazi was of a non-materialistic society of nominal physical possessions. A peaceful non-specialized people they were non-hierarchical. Their most powerful censure was to ostracize, to refuse to talk to or to allow participation by the offender. Each family or kin group made all of their own pottery, tools and clothing. An outdoor people they built magnificent architecture that included only the most modest interior living spaces.

5.2 THE ANASAZI AS SOLAR ARCHITECTS AND AS URBAN DESIGNERS

In his 1978 book Energy and Form, An Ecological Approach to Urban Growth, Ralph Knowles wrote convincingly about the Anasazi: "a large south facing shallow cave in which the Indians have built one of the finest structures for efficient use of the daily and seasonal rhythms of the sun." In the years since then intensive research has confirmed the expertise of the Anasazi in understanding the land and the sun. However, many of the cliff communities do not face south. And the architectural development of their domestic shelter was minimal compared to the refinement of kiva design, and the urban composition of pueblo and plaza.

From the generous dimensions and careful construction of their kivas we know that the Anasazi could have conceived and built a commodious domestic architecture to balance some ideas of comfort and convenience against their demanding natural environment on the high arid plateaus and canyons. They chose instead to live primarily outdoors, in designed public space juxtaposed with nature, and to build for community. Their own private rooms were tiny retreats and store rooms. Thus the Anasazi invented both community architecture and microclimatic design as a unity, as expressions of their love of the outdoors and of their belief in an ordered society.

Although greatly skilled in the naked eye optics of surveying and astronomy they refused rigid buildings or gridded cities. Although sophisticated in their understanding of the geometry of the sun their solar structures were metaphysical instruments of the spirit but limited as thermal systems for the comfort of the flesh. Least understood is why these gentle people chose to live in these desolate wastes of incredible vista. Thus the composition of their communities demonstrating a belief of an orderly universe tied to orientation is still to be understood by peoples considered much more advanced than the prehistoric Anasazi.

6. Credits

This paper is based in part on my book Anasazi Places, to be published in the spring, 1992 by the University of Texas Press. The sketches are by Calvin Straub, FAIA, Professor Emeritus from field studies., including archaelogical excavations at Sand Canyon in 1989.

Chapter 4

Outdoor Spaces

PLANNING FOR VISITOR COMFORT, OUTDOORS, AT THE UNITED STATES PAVILION, EXPO '92, SEVILLE, SPAIN

M. R. Yoklic, N. V. Chalfoun, T. L. Thompson, D. Macloed*, B. Myers*

Environmental Research Lab.
University of Arizona
2601 E. Airport Dr. Tucson,
Arizona 85706-6985 U.S.A.

**Barton Myers Associates*
Paramount Theatre Bldg.
6834 Hollywood Blvd. L.A.
California 90028-6175 U.S.A.

ABSTRACT. A design method for performance improvement and prediction of outdoor comfort at the US Pavilion is presented. Traditional bioclimatic tools and new technologies for enhancing human comfort in the Pavilion's outdoor spaces are employed. Performance prediction using ERL's outdoor climate modelling techniques are described. Strategies and technologies for shading, evaporation and ventilation are developed and evaluated. Building and walkway shade is augmented by adjustable shade sails. Evaporative cooling devices, including passive downdraft chimneys (cool towers) and high pressure fogging systems, are described. Ventilation is enhanced by overhead fans in selected areas. The systematic use of these technologies and strategies are discussed and the bioclimatic performance of the outdoor spaces with climatic adjustments are predicted.

1. INTRODUCTION

Human thermal comfort at the outdoor exhibition plaza for the US Pavilion has been an area of concern and a challenge. Seville, Spain is often too hot for outdoor comfort in the summer. The Environmental Research Laboratory (ERL) at the University of Arizona has reviewed the work of EXPO planners concurrent with its own research and experience on outdoor comfort in arid regions. With the US Pavilion architects, Barton Myers and Associates, outdoor comfort issues have been tactfully addressed in the Pavilion design. Four essential climate control strategies, shading, evaporation, ventilation and radiant heat control, are employed to improve the plaza outdoor climate conditions for human comfort. These strategies are realized in the project through both technology, design and materials selection. The strategies are incorporated into the architectural scheme to amplify the Pavilion's theme while providing comfort. The techniques employed to render strategic effects are "choreographed" to improve the overall bioclimatic conditions in the Pavilion's outdoor spaces.

S. Alvarez et al. (eds.), Architecture and Urban Space, 153–158.

2. DESCRIPTION OF THE US PAVILION

The US Pavilion site measures 260X150 meters (~39000 m²). The project, designed by Barton Myers and Associates consists of a major theatre building on the south and five exhibition and support buildings. The entry plaza separates the exhibit halls from the theatre building. The building layout creates an almost T-shape outdoor space. The plaza areas hold and guide visitors to the exhibits and outdoor performance area. An enclosed walkway provides for second level circulation.

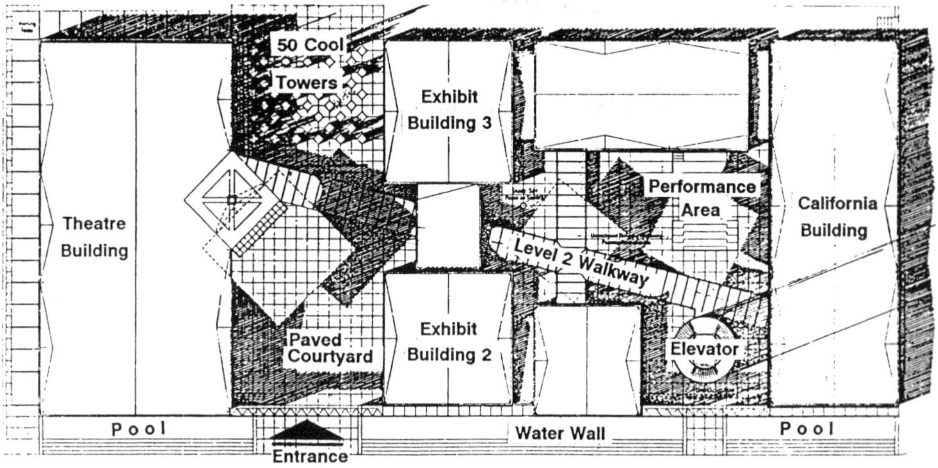

Fig. 1 Site and Exhibit Buildings of the US Pavilion in Seville, Spain.

3. CLIMATIC ANALYSIS

The size of the Iberian peninsula and the influences of the Atlantic and Mediterranean result in three distinct climatic zones; Marine west-coast, Highland humid, and Mediterranean. In Central Spain, inland Andalusia, where Seville, Córdoba, and Granada are located, the Mediterranean climate is classified Semidesert with an average daytime winter temperature of 10°C. Summer temperatures of 32°C and higher are not uncomon. From Seville climate data (EXPO'92, 1990), the average midday temperatures vary between 30°C (45% RH) to 36°C (34% RH) in the four summer months June to September. An analysis of the available climatic data shows that, in general, summer condition falls outside the accepted human comfort standards (see Fig. 2 below).

4. BIOCLIMATIC EVALUATION

Using available Seville weather data, the hourly conditions for the four summer months, June through September, were plotted over the psychrometric chart, which also shows the ASHRAE comfort standard 55-74 (ASHRAE, 1981) noted by the toned area (Fig.2). From the chart it appears that the summer climatic conditions fall outside traditional indoor comfort standards. By transferring the index of thermal stress onto the psychrometric chart (Givoni, 1976), zones of appropriate climatic control strategies for comfort are outlined. The potential effects of ***shading***, ***ventilation***, and ***evaporative cooling*** on outdoor comfort may be estimated. Note that the ASHRAE comfort zone is for indoor conditions in terms of dry bulb temperature (DBT), relative humidity (RH) and still air with no radiant heat exchange (the Mean Radiant Temperature MRT is equal to the DBT). Although it is highly unlikely that in an indoor situation more than approximately 5°C difference could exist between MRT and DBT (Szokolay, 1980); in "outdoor" spaces, with the presense of solar radiation, MRT will be higher than DBT and will have a more profound effect.

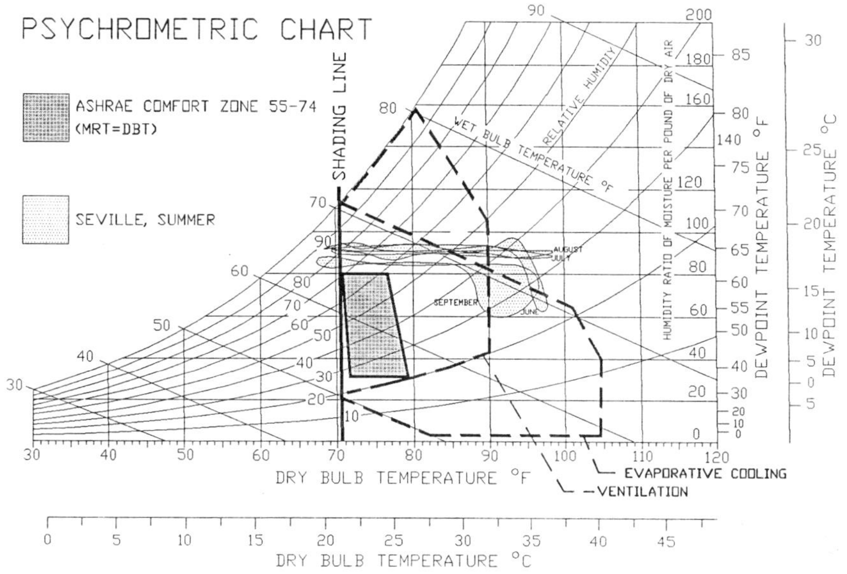

Fig. 2 Comfort/Climate Psychrometric Chart for Seville, Spain.

4.1 Shading

The control of direct solar radiation to the outdoor public areas of the Pavilion was critical. Analytical techniques were applied to identify the areas within the plaza which would not be shaded by the existing layout of the exhibition buildings. Since the terrain of the plaza was uniformly flat, shadow patterns were determined by abstracting buildings corners into a series of poles corresponding to the height of the structure. The shadow length of each is then defined. The composite of the pole shadows gives an approximation of the overall shadow cast by the objects and defines the unshaded areas. These areas were then protected through the use of movable "shade sails". The shade sail devices are designed to provide optimal shade through mobile horizontal and vertical axis. Shade is thereby adjusted through the day and season. The mast and drive were specified from wind power technology (Fig. 3-right).

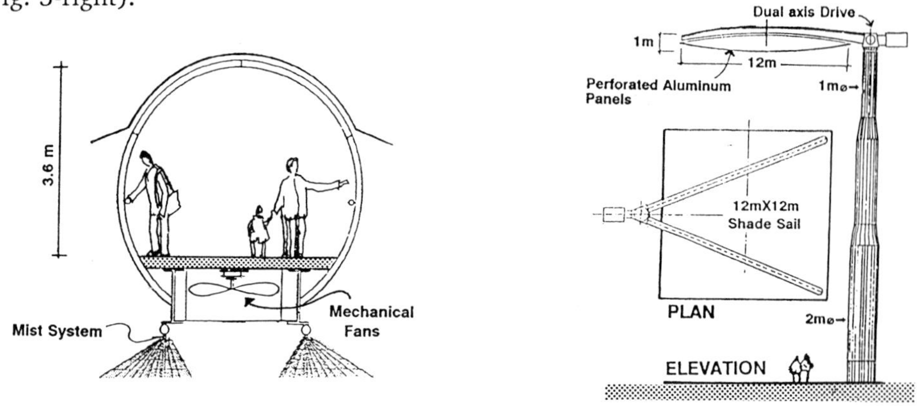

Fig. 3 Ceiling Fans Applied to Walkway (left), and Shade Sails Towers (right)

4.2 Ventilation

The merit of ventilation for enhancing outdoor comfort in Seville is illustrated in the bioclimatic evaluation. Givoni's plot of the thermal stress index extends the comfort with airflows of 1 m/s (Fig. 2). To provide ventilation, cool towers and overhead ceiling fans were used. Cool towers, located on the west center of the main plaza, were estimated to discharge air at between 1.0 to 2.0 m/s. The cool tower location and the prevailing wind direction moves output air across the main plaza toward the pavilion entry. Seville average wind speeds, 1.7 m/s for June and July and 1.4 m/s for August and September, prevail from the southwest (Wallen, 1970). Arens et al, (1980) have shown that active ventilation at 2 m/s speed further extends comfort. Traditional low velocity overhead ceiling fans are used to provide 2 m/s ventilation and are installed into the second level pedestrian walkway system (Fig. 3-left). The higher turbulence and the variable quality of ceiling fan output produces more comfortable effects than uniform air motion (Rohles et al, 1983). This second ventilation system was added to provide active air movement where the effect of the cool towers diminish.

4.3 Evaporative Cooling

The high dry bulb temperature and moderately low relative humidity in Seville are within the ranges favorable for evaporative cooling (Watt, 1986). Under the local climatic conditions, the use of the phase change of water for cooling offered another strategy for improving outdoor comfort. In the pavilion's plazas cool towers and high-pressure mist systems were employed for evaporative cooling. Cool towers are passive natural down draft chimneys which use evaporation to cool the air to near wet bulb temperature (Cunningham and Thompson, 1986). Under average June conditions, 36.4°C and 43% RH, 50 cool towers, each having an estimated output 1.5 m^3/sec, together will produce 75 m^3/sec at 28.9°C. This output volume will provide 2 air changes/minute in the entry plaza (Fig. 4). When warranted, supplemental evaporative cooling is provided by a misting system installed under the second level pedestrian walkway as shown in Figure 3. Evaporative cooling potential may be reduced at some EXPO site locations by humidification from water exhibits.

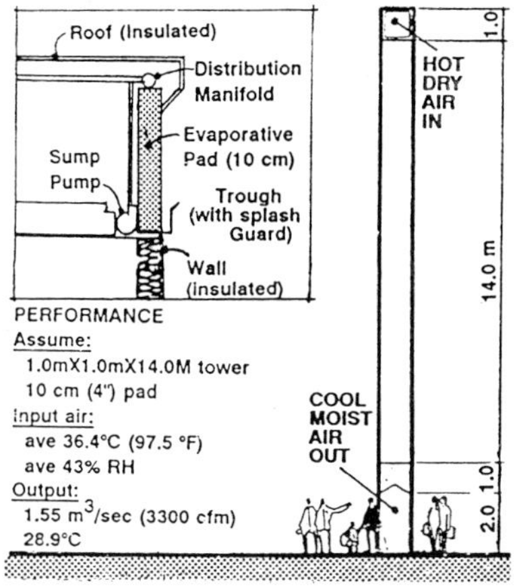

Fig. 4 Cool Tower Performance

4.4 Mean Radiant Temperature and Comfort

Thermal infrared exchange and the effect of MRT on comfort are well understood (Fanger, 1972). The materials at the pavilion outdoor spaces (paving, building surfaces, etc.) absorb solar radiation and become a source of radiant heat to the visitors. Moving air decouples the body from these radiant sources if directed across the person. Shade limits solar gains. The overall effect of the strategic adjustments to the Pavilion's outdoor spaces, combining air temperature, air moisture, air movement and radiation have been assessed using ERL's ***MRT*** program (Yoklic et al, 1990 / Chalfoun et al, 1991). The combined effect is presented by the ASHRAE (1981) Corrected Effective Temperature (CET) for average June 21 conditions. MRT, surface temperatures, CET and DBT are plotted for the ambient conditions (left) and the adjusted conditions (right) in Figure 5. The adjusted conditions assume enhanced ventilation at 2 m/s and fully shaded pedestrian walkways. Controlled operation of ventilation, off in the early morning and afternoon, would extend comfort conditions at these times. Evaporative cooling, although not included in Figure 5, will contribute to outdoor comfort in the afternoon.

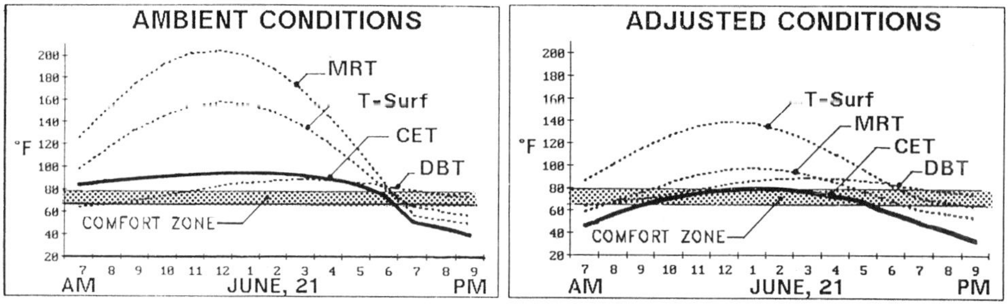

Fig. 5 Adjusted MRT Versus Comfort Conditions

5. CONCLUSION

Outdoor human comfort at the US Pavilion in Seville, Spain should be achievable through the combined and controlled use of shading, ventilation, evaporative cooling and radiant heat control. Shade is optimized through operable shade sails. Ventilation is augmented by the use of cool towers and ceiling fans. Evaporation sources are the cool towers, and high pressure mist systems. The use of building, landscape and paving material aids in radiant heat control.

6. ACKNOWLEDGMENTS

The authors thank Planetary Design Corporation, Tucson, Az. and the US Information Agency (USIA) for their support. For reprints of this paper from the Environmental Research Laboratory, request ERL#91-10.

7. REFERENCES

ASHRAE (1981). Handbook of Fundamentals. The American Society of Heating, Refrigerating, and Air-Conditioning Engineers, Atlanta, GA.

Chalfoun, N. V., T. L. Thompson, and M. R. Yoklic (1991) "*MRT*©" Update: A Study on the Thermal Effect of Dry and Wet Paving and Landscape Materials on Restoring Human Thermal Comfort Conditions at Outdoor Spaces. International Solar Energy Society (ISES) Solar World Congress Conf. Denver Colorado.

Cunningham, W.A. and, and T.L. Thompson (1986). "Passive Cooling with Natural Draft Cooling Towers in Combination with Solar Chimneys", Proc. of PLEA '86, Passive and Low Energy Architecture. Pecs, Hungary. Sept. 1-5 1986.

EXPO'92 (1990) Complementary Rules I.D. 8.05, A.C. and Ventilation Installation.

Fanger, P. O. (1970). Thermal Comfort, Danish Technical Press, Copenhagen.

Givoni,B. (1976) Man, Climate and Architecture, 2nd Edition, Applied Science Publishers Ltd., London.

Rohles et al. (1983) "Ceiling Fans as Extenders of the Summer Comfort Envelope", ASHRAE Transactions, 89(IA):245-263,.

Szokolay,S.V.(1980) Environmental Science Handbook,John Wiley & Sons,New York

Wallen, C.C. (1970) Climate of Northern & Western Europe World Survey of Climatology, v.5, Elsevier Pub. Co. New York.

Watt, J. R. (1986) Evaporative Air Conditioning Handbook, 2nd Ed., Chapman & Hall, New York, pp. 43-89.

Yoklic, M., Chalfoun, N., Thompson, L. and Kent, K. (1990). "MRT: An Original Computer Program for Predicting Thermal Conditions of OutDoor Spaces", Proc. of 1st World Renewable Energy Congress, Pergamon Press, New York., Vol 4.

8. INDEX

Outdoor comfort, Evaporative cooling, MRT, Cool towers, Ceiling fans, Exhibitions

EVAPORATIVE COOLING FOR LARGE EXHIBITION SPACES: A METHODOLOGY FOR SYSTEM DESIGN/ENGINEERING BASED ON HUMAN COMFORT

N. G. HICKS, T. L. THOMPSON, M. R. YOKLIC, N. V. CHALFOUN, and K. J. KENT*
Environmental Research Lab., University of Arizona
*Peroxidation Systems, Tucson, AZ USA

ABSTRACT. The Saudi Arabian Pavilion for EXPO '92 in Seville includes a large exhibition space with a light weight tensile roof effectively surrounded by an air conditioned building. Evaporative cooling was investigated as a more energy efficient cooling strategy for the exhibition space than conventional refrigerative air conditioning. The design techniques and the methods of assuring human comfort are described. Both the PMV (Predicted Mean Vote) and PMV* methods of comfort evaluation were used.

1. Introduction

The Saudi Arabian Pavilion for EXPO '92 includes a space called the "Desert Room," surrounded on three sides by air conditioned exhibit spaces, and on the fourth by a water wall. Exterior to the water wall is an outdoor space which has been modified to improve human comfort. A dominant feature of the pavilion is the roof suspended high over the Desert Room, constructed to resemble a huge carpet draped above the exhibit area. From the outside, the carpet/roof appears as art work. Occupants in the room, who have just entered the building from the outside, stand within desert settings, and observe various displays, including a six minute film. Total time within the space for an observer is roughly 9 minutes, and a typical occupancy is approximately 173 visitors at one time.

1.1 THE EVAPORATIVE COOLING OPTION

A number of features of the building and space design, occupancy profile, and local environmental conditions suggested the possibility of using evaporative cooling as an economical alternative to conventional refrigeration for the Desert Room. First, the high dry bulb temperature and

S. Alvarez et al. (eds.), Architecture and Urban Space, 159–164.

moderately low relative humidity in Seville are within the ranges favorable for evaporative cooling (Watt, 1986). Second, the visitors have just entered the building from an outdoor space that has been modified to increase the human level through such techniques as the use of large natural draft towers (Kent and Thompson, 1990) which utilize evaporative cooling. This makes the transition into the refrigerated spaces from the outside seem more gradual, and reduces the shock. Third, the occupants in the Desert Room will be standing in a sunken space surrounded by a wall of sand nearly one meter high, which creates a low spot within the room for relatively cool air to sink into, possibly providing comfort for the occupants without the need to provide uniformly cool conditions throughout the entire room. This is illustrated in Figure 1.

2. Design Summary

A summary of the design parameters appropriate to describe the evaporative cooling conceptual design is shown in Table 1.

TABLE 1. Design Parameters for Desert Room

Total Air Flow	32 m^3/s
Air Exchange	Once every 2 min
Refrigeration Cooling Load	68 kJ/s
Temp Rise of Evap Cooled Air	1.1 °C

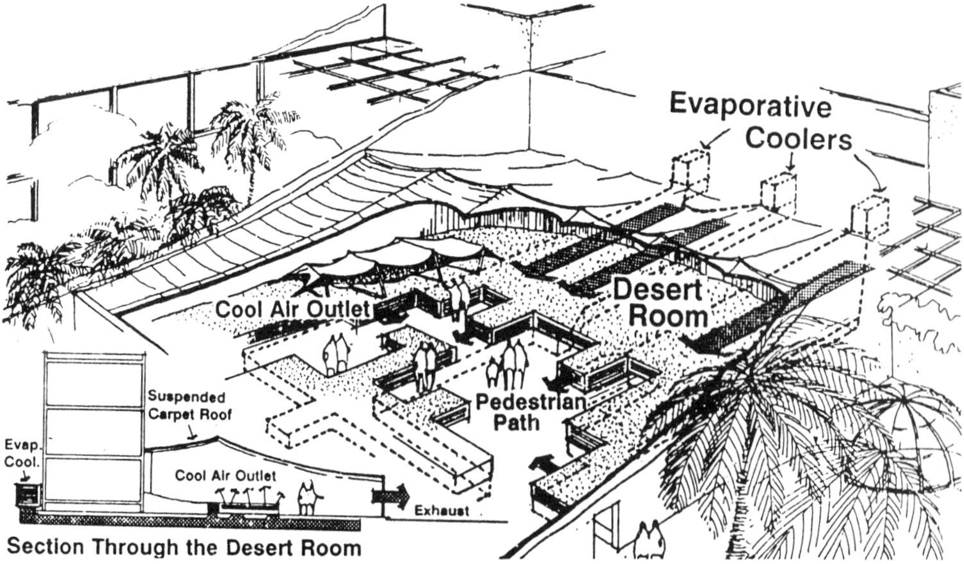

Figure 1. Saudi Arabian Pavilion, EXPO '92.

The evaporative cooling hardware was located outside the building along the back wall, and the entire crawl space underneath the back exhibit area used as a large duct, delivering air to the desert room. The duct was extended into the desert room under the sand and terminated at diffusers discharging cool air directly onto the occupants near or below waist level. Commercially available packaged evaporative cooling units with fan, pads, sump, pump, and controls were used in this application.

Evaporative cooling allows the discharge of the air directly onto the occupants, placing them in a zone of cool, relatively high velocity air. A continuous horizontal vent delivers air at a discharge velocity of roughly 2 meters per second. This combination of cool air and air movement increases the comfort level at a reduced energy cost. Evaporatively cooled air could be delivered to the occupants in the Desert Room for an estimated energy requirement of approximately 11 kW, while refrigerative air conditioning would require roughly 30 kW of electrical energy. Under some evaporative cooling conditions expected at the site, occupants could become too cool, and control would be achieved by reducing the air flow and the diffuser discharge velocity to optimize comfort conditions. With variable speed fan controls, the energy savings are considerable. A 50% reduction in air quantity results in a 87.5% reduction in fan energy under a variable speed control strategy.

Estimates of the water consumed by an evaporative cooling system reach a peak in July in the range 5.5 to 7.8 m^3/day, depending on the thickness of the cooler pad.

To reduce the maintenance required by the system, a relatively permanent, rigid media pad such as corrugated cellulose (Munters, 1990) was selected rather than the lower cost wood fiber materials, such as aspin pads. In fact, the higher efficiency of the corrugated cellulose pads are a required by the design. As will be explained, the design calculations favored a 30 cm thick pad with an air flow velocity of 2.5 meters/second, for a 89% evaporative cooling efficiency.

One of the major design problems associated with evaporatively cooling the Desert Room was the discharge of air from the space. An air curtain was selected for the passageway into the refrigerated section of the building to minimize the flow of evaporative cooled air into that portion of the building, and allow the conventional refrigeration equipment to function independently. Another outlet for the air from the Desert Room was through the entry (Pre-Islamic) corridor, providing a slight breeze to greet the visitors as they entered the building. A third outlet is provided by simple exhaust vents in the ceiling to remove warm air near the roof, while allowing control of the internal pressure.

3. Predicted Comfort Conditions

The feasibility of evaporative cooling is determined by the level of human comfort in the Desert Room. The PMV (Predicted Mean Vote) is the most widely accepted index in use today for comfort evaluation (Fanger, 1970). However, Fanger's comfort equations were derived for air conditioned interior climates, and their extension to the high humidities encountered with evaporative cooling is uncertain. Consequently, we also used the PMV* of Gagge et al. (1986), which more realistically models the evaporation of perspiration from the human skin.

For comfort calculation, a clothing insulation value of CLO = 0.5 was used, corresponding to summer conditions. The permeation efficiency of water vapor through clothing was taken as icl = 0.45, recommended by Gagge et al. (1986) for everyday clothing. A metabolic rate of M = 1.4 met (81 W/m^2), which corresponds to people standing and walking, was selected for the analysis. In addition, the mean radiant temperature for the space was assumed to be equal to the dry-bulb air temperature.

Using the manufacturer's data (Munters, 1990a) for various thickness of corrugated cellulose media, and temperature data for Seville in July (2.5 percentile, EXPO '92, 1990), Figure 2 was prepared to illustrate the effect of pad thickness on dry-bulb temperature, at a constant pad face velocity of 2.5 m/s.

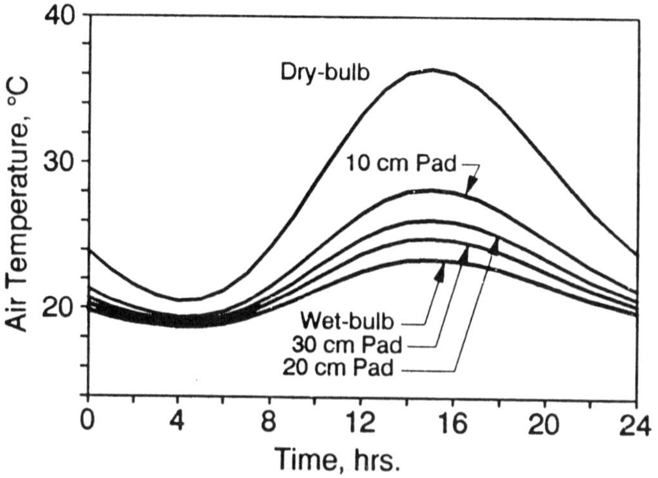

Figure 2. Effect of evaporative cooler pad thickness.

From the air flow rate given in Table 1 of 32 m^3/s, an average velocity for the Desert Room 0.17 m/s was estimated. Note that this value is for the entire room, while the occupants will likely experience higher velocities in the exhibit area. Figures 3 and 4 were prepared from the data of Figure 2 and the comfort parameters previously given. From midnight until 9 A.M. no evaporative cooling is required. The conventional PMV (Figure 3) predicts a greater degree of discomfort during the warmer part of the day when ambient air is used than does PMV* (Figure 4), since the latter term more accurately reflects cooling the skin by perspiration. While the PMV chart suggests that either a 20 or 30 cm evaporative cooler pad would be adequate for comfort; the PMV* chart indicates that a 30 cm pad is required, and the air velocity will need to be increased from 0.17 to 0.4 m/s for the PPD (Predicted Percentage of Dissatisfied) to be reduced to 20% at 3 P.M.

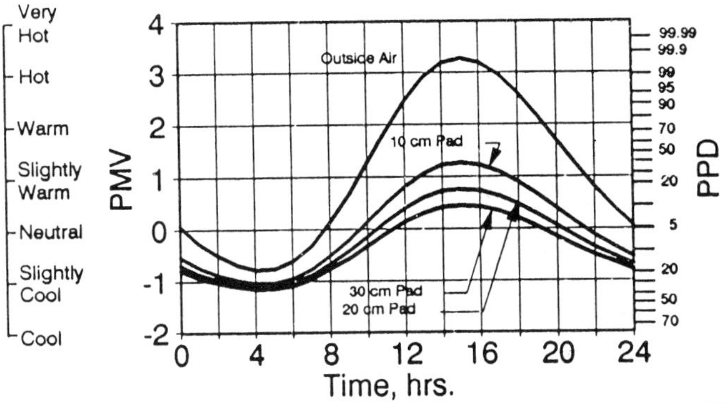

Figure 3. Comfort conditions based on Fanger's PMV.

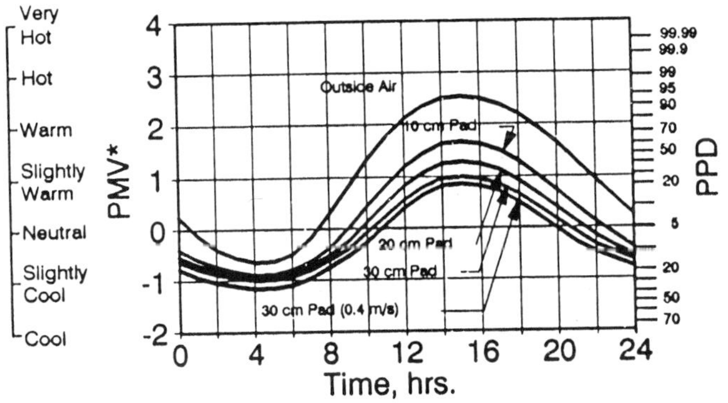

Figure 4. Comfort conditions based on PMV*.

4. Conclusions

Evaporative cooling is an energy efficient method for cooling large exhibition spaces, as illustrated here by the example of the Saudi Arabian Pavilion, EXPO '90, in Seville. The comfort analyses performed to determine the proper evaporative cooler pad thickness using PMV* (Gagge et al., 1986) indicated that a 30 cm pad was required. The results of the PMV (Fanger, 1970) and PMV* analyses differ considerably; because PMV* more realistically models the evaporation of perspiration from the human skin, it is more likely to give satisfactory results with evaporative cooling.

5. Acknowledgments

The authors are indebted to:

H.R.H. Prince Abdullah Bin Faisal Bin Turki Al Saud of the Royal Commission.
Abdulrahman Al-Shathry, Al-Shathry Consulting Engr.
Fitch Benoy, Design Consultants, London.
Planetary Design Corporation, Tucson, AZ.

For reprints of this paper from the Environmental Research Laboratory, request ERL#91-9.

6. References

EXPO '92 (1990) Complementary Rules I.D. 8.05, Air-Conditioning and Ventilation Installation.

Fanger, P. O. (1970) Thermal Comfort, Danish Technical Press, Copenhagen.

Gagge, A. P., A. P. Fobelets, and L. G. Berglund (1986) 'A Standard Predictive Index of Human Response to the Thermal Environment,' ASHRAE Trans., Vol. 92, part 2B, 709-731.

Kent, K. J., and T. L. Thompson (1990) 'Natural Draft Evaporative Cooling,' Proc. of the 15th Nat. Passive Solar Conf., Austin, Texas

Munters (1990) Engineering Bulletin 080-5K. Munters Evaporative Cooling Div., Fort Meyers, FL 33911 USA.

Watt, J. R. (1986) Evaporative Air Conditioning Handbook, 2nd Ed., Chapman & Hall, New York, pp. 43-89.

7. Index

Human comfort, evaporative cooling, Saudi Arabian Pavilion.

SUMMER COMFORT IN URBAN ENVIRONMENT

Rafael Serra, Helena Coch, Antoni Isalgué, Nuria Sierra
Escola d'Arquitectura de Barcelona, U.P.C.
Av. Diagonal, 649 08028 Barcelona

ABSTRACT. The urban environment needs specific consideration in relation to comfort problems. A large amount of the total population lives or works in cities, and therefore some considerations about comfort in the urban environment have to be taken into acount in planning.

Among the problems present in urban situations, it has to be mentioned that acoustics will be very different in urban environments to non-urban environments. Natural light is very often reduced because of obstructions, wich also affect radiation and thermal conditions. Similarly, microclimatic conditions depend largely on artificial environment.

It should be noted that comfort appreciation may be influenced by synaesthesia effects (confluence of stimuli) and by psychological effects -human relationships)-.

Summer problems are usually more prevalent in situations where a lot of people are together in the same space. Temperature is not the only problem: there are others, such as protection to air movement, humidity and chemical effects, wich together play a very important role.

While a realistic approach to comfort must balance accurately all the factors present in normal outdoor situations, this balance becomes critical in summer conditions and in high concentrations such as we find in urban environments.

1. INTRODUCTION

In the evaluation of human environmental comfort, we must first of all take into consideration ENVIRONMENTAL PARAMETERS, which are manifestations of energy of the thermal, acoustic and luminic environment. We can divide thse into "specific parameters", which are those which affect every one of the senses (thermal, acoustic and visual) separately and which in many cases may be measured using physical measurements (ºC, dB, Lux, etc.) and in "general parameters", which are those which act upon various senses at the same time, and which may be measured in terms of dimension or of time or space (dimensions, form, continuity in time, etc.).

S. Alvarez et al. (eds.), Architecture and Urban Space, 165–170.

Secondly, we must consider COMFORT FACTORS OF THE INDIVIDUAL, which are particular to each individual and which have no direct relation with the environment, but which influence the individual's appreciation of the environment. These factors may be classified as follows: firstly, physiological (age, sex, activity, heredity, etc.) secondly a social level (education, fashion, dress, etc.) and thirdly a psychological level (character, personality, phobias, etc.).

2. BASIS OF THE EVALUATION OF GLOBAL COMFORT

2.1. Evaluation of the effect of environmental parameters.

The specific environmental parameters have a double effect on comfort. The first is the direct action over the corresponding sense, such as that of air temperature over thermal perception, that of the level of a sound over the acoustic sense, etc. This area has been extensively researched and there are abundant references to same in the bibliography.

The second of these is the action which a specific parameter, corresponding to a sense, may have over other senses. This effect of associated perceptions, which is commonly known as "synaesthesia", arises in certain occasions. For example red light is normally associated with a greater sensation of warm or hot. The relative importance of this factor is normally lesser than that of the previous case, but nevertheless it may at times exercise quite a significant effect on the sensation of comfort.

The general parameters act in a more indirect manner in the area of comfort, affecting the whole range of senses and thus modifying the sensations produced by the specific parameters. Thus the time of the day, for example, may modify the sensation produced by visual, acoustic or thermal stimuli and in the same way the shape and dimensions of a space may augment the sensation produced by a noise, by a colour or by the temperature.

2.2. Effect of the factors particular to the individual

The factors or conditions particular to the individual depend on the characteristics of each individual. These are what cause the same space and same determined parameters to produce different sensations of comfort in different individuals.

"Biological-physiological" factors, such as age or occupation, have already been the subject of research in so far as the sensation produced by some of the specific parameters, although we believe that we have yet to detect and investigate by complementary experiments a lot of the influences that these exercise. The "social" type factors have also been the subject of research, for example style of clothing and its effect on thermal sensation, but there still remain significant voids in the knowledge. Finally, the "psychological" type factors are perhaps the least well researched and their very nature renders it difficult to devise reliable methods of evaluation.

2.3. Evaluation of global comfort

In order to evaluate the global comfort of a given environmental situation, and also in order to evaluate accurately any of the thermal, luminic or acoustic sensations, we have to consider all the parameters and factors mentioned in the previous paragraphs.

It is a common error in the evaluation of comfort to consider only those aspects which may be measured in terms and units of energy, i.e. those which are related only with those parameters of comfort which we have already described as "specific parameters", ignoring the effects of the "general parameters" and "comfort factors".

Another very common error consists in evaluating the comfort of a given situation taking into account only those parameters traditionally or classically associated with a sensation. Thus thermal sensations are analysed separately, independent of acoustic and luminic sensations, which leads to results which can never be relied upon to give us a global evaluation since the synaesthetic effects, the general parameters and comfort factors, have been overlooked.

Finally, a third type of error consists in evaluating by parallel, separate measurement, the effects of the different parameters, without taking into account that its actions interact and that according to the situation of one of those the effect of the results may be different. As an example of this type of error in practice, sometimes the remedy applied to a situation of thermal discomfort is the application of excess heat, together with lineal reductions on the other specific parameters (air temperature, radiation, humidity and movement of air), in the belief that the final sensation will be one corresponding to the total of the four reductions.

In contrast, our research team is trying to lay down the basis of what should be an integrated system of evaluation of global comfort, which will correlate in one single algorithm the effect of all the parameters and factors already mentioned. In this line of work, we are carrying out various exercises which will help us to arrive at the desired result and we will now describe one of these, in relation to summer comfort in an urban situation.

3. COMFORT IN AN URBAN ENVIRONMENT

3.1. Specifics conditions of the outdoor urban environment.

When we consider the question of comfort in an exterior summer environment, we must take into account the specific features of this type of situation.

Firstly, the characteristics of the environment are much more agressive and variable than those which we encounter in an interiour environment. Since there are no barriers or elements which exercise an effect of inertia and restrict the effects generated by energy, the result is that the environment conditions present intense levels of energy, and at the same time a wide variability of weather conditions.

This very variability, which normally results in a worsening of comfort conditions, also has its positive aspects. We are now beginning to recognize that environments whose conditions are too static produce effects of discomfort to the occupants, due to the lack of sensorial stimuli because of the lack of change. Similarly, a typical urban exterior environment induces a greater stimulus and indirectly an expansion of the conditions which may be regarded as generating comfort.

From the point of view of control and improvement over the conditions of these environments by means of design, there are also various specific actions which we may apply. By their very essence, being spaces which are difficult to control or regulate, the use of means of environmental control, whether these be natural or energetic, are very much limited. For this reason, in approaching the question of global comfort and of systems for the improvement of global comfort in an urban environment, we must determine as accurately as possible what parameters we ought to act upon.

3.2. Parameters for summer comfort

Among the various specific parameters which operate in summer comfort, let us consider briefly those which may be most decisive.

With regard to those related to the thermal sensation, of primary importance are the four classical parameters, **air temperature, radiation, humidity** and **air velocity.** Each one of these as always is important but in contrast to what occurs in interior environments, the role of the latter three, and especially of radiation and humidity, becomes relatively more critical.

In the parameters which relate to the luminic sensation, the most critical are **illuminancy** or level of illuminance, **luminance** of surfaces and **colour**, together with the effects of **glare** which these surfaces or direct radiation may produce. Among these parameters, illuminancy is not generally critical (by day) but luminance and its glare effect may be.

With regard to the parameters which relate to the acoustic sensation, it is the **sound level** with its variation in time which will influence most decisively the sensation of comfort. It is important to bear in mind that in a situation of this kind, **reverberation** is less than in an interior space, although much greater than in a rural situation. Hard surfaces are much more prelevant and the high levels of sound produced by the significant sources of urban noise are further augmented by their echoing against these surfaces.

Together with the individual consideration of each of these specific parameters, we must consider their **interactions**, on a synaesthetic level as well as after the correction by the general parameters of time, space, etc. Of these, the **form and dimensions of the space** is particularly significant, with nearly always an open aperture to the sky and on which the width of the streets and squares, etc., has a very large effect on the sensations generated. The **time of the year** is also similarly important, especially because of effects on thermal comfort.

Finally, <u>individual factors</u> play a more important role in this type of urban situation which we are now considering, among which we should mention **activity type** (in general more energetic than in interior situations), **clothes type**, logically more "formal" and less flexible, and the **psychological attitude**, normally more tense than in an interior space.

4. RELATION BETWEEN VARIATION OF MAGNITUDE AND SENSATION OF COMFORT

4.1. Magnitude-sensation dependence

It has been demonstrated that human sensations do not in general maintain a lineal relation to the stimuli, energetic or otherwise, which generates these sensations. We can affirm from our study that that there are response curves, in the specific parameters as well as in the general parameters and also in the comfort factors which are not lineal at least in the extremes of these curves.

Consistant with this, the variation of any given magnitude, i.e. of the value of any paramater, will influence the sensation (and, consequently, the comfort) in a different way depending on the location, i.e. it will depend on the value of the entry of the parameter considered.

4.2. Strategies for the optimization of comfort

The manner in which the general interdependency between magnitude and sensations produced by same arises, yields results of saturation at extreme points of the magnitudes, which normally coincide with very uncomfortable situations. On the other hand, a slight variation of magnitude, in a situation far from saturation (i.e. in a situation of comfort) may be produced more effectively by physiological resources of the individual. For example, in relation only to thermal comfort, in conditions of relative humidity of 50%, very low air velocity, and radiation temperature equal to air temperature, the effect and sensation of a variation of temperature from 22 C to 23 C, given that we are discussing conditions of thermal comfort, is really hardly appreciable; however, if the temperature changes from 18 C to 19 C, normally the effect is greater, while if the variation is from 35 C to 36 C, the effect is less since both cases (35 C and 36 C) are very uncomfortable thermal situations.

The immediate consequence of what we have analysed is that to correct a situation of discomfort, it is much more beneficial to alter the magnitudes whose effect on the sensation is higher in each particular situation, naturally paying special attention to the secondary effects which variations which are too wide may have.

4.3. Example of Application

Let us consider, for example, the typical summer conditions of Seville. From the Guia Resumida del Tiempo en España (Short Resumé of the climate of Spain, National Meteorological Service, Madrid, 1968), we see the following average values for a typical day: average temperature, 28 C; maximum temperature, 36 C; minimum temperature, 20 C; relative humidity, 50%. The absolute maximum temperature may rise as high as 46 C.

If we plot these values on a psychrometric abacus, when the temperature is near the maximum of 36 C, the relative humidity is only about 35%, which might lead us to think that we could take advantage of this value by evaporating water so as to reduce the air temperature. However, if we evaporate water in order to reduce the air temperature some 8 C, down to 28 C, the humidity would rise to about 65-70%. This situation is not in fact comfortable, and we find ourselves in the situation that the physiological resources to evacuate heat from the human body have been reduced appreciably (evaporation of perspiration is now difficult because of the high relative humidity). If the maximum temperature reaches 40 C, even with the same relative humidity, by evaporating water we may arrive at a temperature of about 32 C and about 60% relative humidity.

Regarding the above observations we ought to note that at the end of the evening, the situation is presumably worse than calculated because of the accumulation of humidity as a result of evaporation, although, in any case, it is also true that the movement of water generates a synaesthetic effect by means of a greater sense of "freshness".

Then, it might be concluded that evaporation of water is not a single good response to summer thermal problems in Seville. Furthermore, if radiation effects are taken into account, it appears that, at least, good air movement conditions and protection against direct solar radiation are needed.

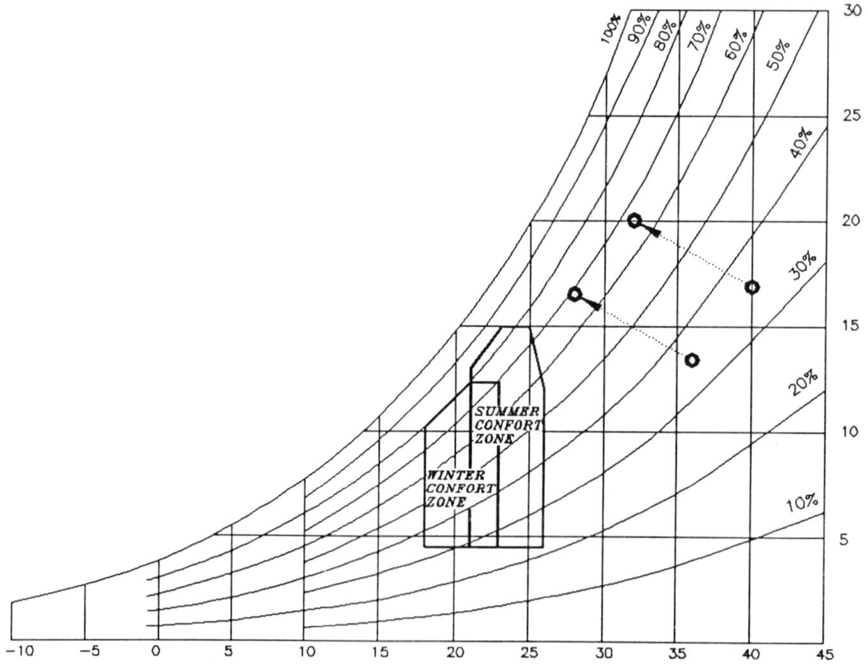

GENERAL ASPECTS OF THE WORK ON CLIMATIC CONTROL OF OUTDOOR SPACES AT EXPO'92

R. Velázquez*, S. Alvarez*, J. Guerra*, E.A. Rodríguez**, J.M. Cejudo*

* *Universidad de Sevilla*
Departamento de Ingeniería Energética
y Mecánica de Fluidos
Av. Reina Mercedes s/n
E-41012 Sevilla. Spain

** *CIEMAT-IER*
División Termosolar
Av. Complutense 22
E-28040 Madrid. Spain

ABSTRACT. This paper describes the general aspects of the research and development project carried-out by the Department of Energy Engineering and Fluidmechanics of the University of Seville, in order to improve the climatic conditions of the outdoor spaces at EXPO'92. Mention is made about the physical, climatic and cultural framework guiding the project. Besides, the general criteria of conditioning based on the analysis of comfort are presented. Finally, the methodology and the different levels of activities developed since 1988 are described.

1. Introduction

Conventional conditioning techniques have concerned themselves almost exclusively with the environmental treatment of the enclosed spaces which make up buildings. Changes in certain social behavioural patterns, the endless search for an improved quality of life and the ever-increasing desire for closer contact with our surroundings, have all contributed to a more important role being played by the outdoor spaces in which urban buildings are located. The ability of these outdoor spaces to satisfy the foregoing requirements are closely linked to achievement of acceptable levels of comfort in such spaces. This requirement must be fulfilled, because of the demands of the conditions to be met, by means of soft techniques which are compatible with the architectural and urban design. It should be acknowledged that the theoretical and technological knowledge at our disposal for solving these questions is still in the early stages and requires conceptual development, a defined methodological corpus and further concrete experience. We describe below the general framework of this problem in EXPO'92 and the stages covered to fill in some of the theoretical and practical gaps encountered at the outset of this research.

2. The Physical Setting of EXPO'92

The Seville Universal Exposition will take place between April 20th and October 12th 1992. The 215 hectares it will occupy are located on what is known as the Cartuja Island, a lowland surrounded by the two branches which form the natural and artificial courses of the Guadalquivir River (Figure 1).

Five large avenues measuring 300×80 meters make up the area reserved for the pavillions of the international participants. Perpendicular to these avenues there is a wide pedestrian way some 2 kilometers in length, along which are strung the various theme pavilions. An 16

S. Alvarez et al. (eds.), Architecture and Urban Space, 171–177.

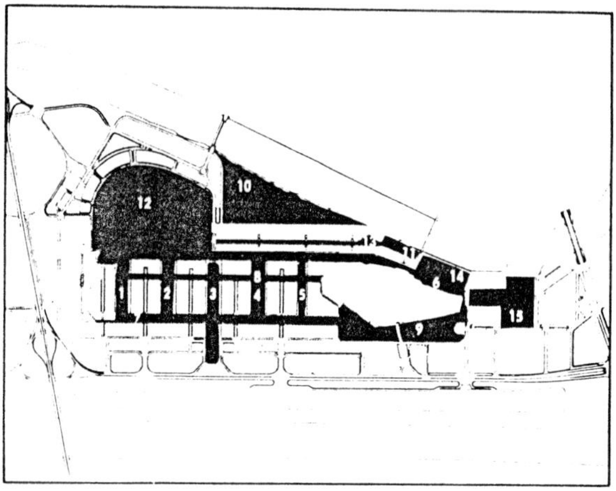

1. *Waterway Avenue.*
2. *Avenue of Europe.*
3. *Palm Tree Avenue.*
4. *Ombu Avenue.*
5. *Maple Avenue.*
6. *Route of Discovery.*
7. *Acacia Avenue.*
8. *Orange Grove Avenue.*
9. *Outer Gardens.*
10. *Guadalquivir Gardens.*
11. *Garden of the Americas.*
12. *Lake of Spain.*
13. *Canal.*
14. *Port of the Indies.*
15. *South Zone.*

Figure 1: General Plan of the EXPO'92 Site

hectares artificial lake, semicircular in shape is bordered by the pavilions which represent the different autonomous regions of Spain.

In all, the built-up area comprises some 600,000 m^2 scattered throughout 170 hectares of open spaces in which restaurants, cafeterias, shops, bars, kiosks and small show areas abound. Over 100,000 trees, 35 kilometres of hedges and 500,000 m^2 of parks and gardens alternate with avenues, promenades and rest areas, forming in a certain sense one immense park.

Forecasts of attendance calculate some 18 million visitors to the Expo, who will generate some 40 million visits to the site.

3. The Climate of Seville

Table 1 shows the most representative meteorological variables of the city's climate for the time of year during which the Exposition will take place.

The hottest period covers the months of June to September, with a maximum absolute temperature of 45 °C recorded. 1% of the hours in this period (29 hours) give temperatures in excess of 37,8 °C ; 2,5% of the hours exceed 36,4 °C ; and in the 5% percentile (144 hours) the levels are around 35 °C .

The foregoing data refer to Seville Airport and no additional overheating of the Expo site will occur due to the following circunstances:

a) The beneficial influence of the two branches of the river and the 16 hectares of water forming the Lake.

b) The abundance of trees and other vegetation.

c) The decision of the Organizers to prohibit thermal emissions into the environment, thus avoiding the formation of heat islands.

Table 1: The Climate of Sevilla

	Maximum Mean Temperature (°C)	Minimum Mean Temperature (°C)	Mean Relative Humidity* (%)	Mean Absolute Humidity (g/kg)	Maximum Mean Radiation** (W/m^2)
April	22,9	9,9	45	8,1	772
May	26,0	12,1	47	9,8	825
June	31,2	15,6	39	11,1	840
July	35,3	17,3	34	12,3	863
August	35,1	17,7	35	12,3	827
September	31,4	16,0	43	12,3	725

* At the peak temperature time ** Upon horizontal surface

4. Cultural Framework

Even with all these beneficial effects, the climatic data clearly show that during the middle hours of the day the thermal conditions do not favour the use of the outdoor spaces unless in parts thereof something is done to attenuate these conditions. This was the origin and goal of the research work discussed in this document.

The research carried out, at its deepest level, is nothing other than a symbiosis between the technologies and knowledge of the 20th century and some of the cultural patterns of our region.

Our streets, squares and most representative historical buildings exhibit and contain numerous techniques for softening the rigors of the Andalusian summer. Awnings, sails, pergolas, fountains, ponds, trees and vegetation are among the expedients which may frequently, even today, be found in the urban areas of our region.

Most of the solutions proposed in this document are based on the same means which, through intuition, our predecessors used. The aim in incorporating them is to favour an urban environment at the Exposition site conducive to human contact and cultural interchange which may be enriching for the participants to the point where one of the achievements of EXPO'92, whose theme is the Age of Discovery, may be the rediscovery of this cultural richness which entails a respectful coming together between people.

The research, a synthesis of which is offered in this document, starts out, then, from a deeply rooted cultural tradition to which a process of innovation has been applied.

At the initial stage we aimed to distinguish in these traditional solutions between what was fundamental and what was secondary, whether for aesthetic reasons or due to the technical limitations of the age. Next, we attempted to re-use and improve the fundamental concepts with the support given by modern technologies.

The physical-mathematical simulation and the subsequent experimentation have this innovatory character and, thanks to these instrumental procedures, so characteristic of our civilization, it is possible to predict and evaluate beforehand the results of the solutions studied.

Table 2: General Criteria of Conditioning

Measures	Generic Actions	Tipical reductions of gains (W)	Specific techniques
Reduction of the solar radiation	Blocking of the direct and diffuse	40 to 70	- Coverings
	Blocking of the reflected radiation	25 to 50	- Confinement - Treatment of adjacent surfaces
Reduction or inversion of the long wave radiant exchange	Reduction of the temperatures of the surrounding surfaces	20 to 50	- Ground: cool pavements water films - Covering: irrigation water films - Vert. surfaces: cascades water curtains
Reduction or inversion of the convective exchange	Reduction of the air temperature	15 to 50	- Confinements - Sensible cooling - Latent cooling
	Movement of cooled air		- Chanelling of breezes - Jets of water

5. General Conditioning Criteria

The comfort model described in [1] provides the starting point for determining the actions to be carried out. In synthesis, in order to improve the level of comfort it is necessary to reduce the contribution of the various unfavourable heat flows (gains), eliminating them whenever possible or even transforming them into favourable flows (losses) if they admit sign inversion.

Table 2 summarizes this strategy and allows a rough evaluation of the degree of effectiveness achievable with each action. It is clear that there are numerous combinations of actions which give rise to the same level of comfort. In each case, the number of variables which may be operated upon and the intensity of manipulation depends on a series of factors including:

- The characteristics of the space to be treated (geometric dimensions, orientation, degree of confinement).
- Activity of the occupants and foreseen length of stay in the area.
- Effectiveness and cost of the possible techniques to be used.
- Possibility of integration of these techniques into the area without distortion of its aesthetic content.

6. Differential Aspects of Conditioning of Indoor and Outdoor Spaces

Although the aim is in both cases the same, there are differences which it is advisable to underline, such as:

a) In indoor conditioning, convective exchange is the most important component of the Thermal Balance. It is for this reason that the temperature of the premises is the parameter upon which we operate most intensely. Outdoors, the convective component is much lower and techniques aimed at lowering the surrounding temperature should only be applied once all other actions have been exhausted.

b) The incident radiation upon the occupant is the greatest gain in situations involving outdoor conditioning. Effective blocking of this radiation is paramount in these cases. On the other hand, indoors this factor is negligible.

c) The residual velocity in enclosed spaces is such as to avoid currents of air. In outdoor spaces, speeds in excess of those suggested in conventional conditioning are desirable. Other implications not deduced directly from the Thermal Balance, but rather from experience acquired during the project are:

d) The inter-relationship between architecture and engineering is very high, due to the fact that the structural, architectural and urban elements which make up the spaces are at the same time the, largely technological, components of the installation.

In the open spaces at EXPO'92 it is not proposed to achieve thermo-hygrometric conditions such that the visitors experience a sensation of well-being comparable to that in the air-conditioned interior spaces.

This is for two reasons: firstly, it is virtually impossible, from a physical point of view, to achieve such an aim by means of passive technology; secondly, such an aim is not economically viable by means of conventional techniques, given the size of the area to be treated. Consequently, when we speak of climatic conditioning of the outdoor spaces at EXPO'92, it is more appropriate to speak in terms of softening, tempering or improving, even though, due to the generalized use of the expression, we frequently use the term "thermal comfort".

7. Methodology

The first stage of the project consisted in modelling the individual conditioning techniques. Subsequently, in the second phase, full-scale experimentation was carried out with a view to validating and adjusting if necessary the models obtained. Once reliable models had been acquired, a process of simulation was used to obtain basic design criteria, cost- benefit ratios, optimum operational conditions for each technique and data for sizing the systems.

Table 3 summarizes the various stages of the project and their duration. The research stage lasted almost three years; the development stage commenced at the end of the second year and lasted approximately one and a half years. Finally, the evaluation stage will begin in June 1991 and will last for at least a year and a half.

Among the models developed, the following are worthy of mention:

- Thermal comfort in outdoor spaces.
- Thermal behaviour of coverings.
- Calculation of shadows cast by coverings.
- Cool pavements.
- Cooling of air by underground ducts.

Table 3: Workschedule

Activities	1988				1989				1990				1991				1992			
RESEARCH																				
- Modelling	■	■	■	■	■	■	■	■	■	■										
- Experiment Design		■			■	■				■										
- Experiment			■				■				■									
- Validation			■	■			■	■			■	■								
- Sensitivity Analysis				■	■			■	■											
APPLICATION																				
- Design's Analysis and Solar Control								■	■	■	■									
- Simulation and Systems Predesign											■	■								
- Engineering Projects													■							
EVALUATION																				
- Overall Testing														■	■	■				
- Actual Operation																		■	■	■

- Ponds and water films with or without sprayers or jets.
- Evaporative cooling with conventional air handling units.
- Droplet evaporative cooling in situ by natural convection (micronizers).
- Droplet evaporative cooling in situ by forced convection (cool towers).
- Droplet evaporative cooling in situ by forced convection (wet barriers).
- Inter-zonal movements of air.

A detailed description of each model can be found in [2]. Since any treatment is made up of the inclusion of various components, it is necessary to have available a tool capable of establishing the couplings between the various subsystems and predicting their joint behaviour. This task was entrusted to the S3PAS computer code [3]. Experimentation was carried out during the months of June to September in the first three years. Tests were carried out in order to characterize materials, individual components and systems (see ref. [4]).

A description of the more important undertakings in the conditioning of outdoor spaces at EXPO'92 may be found in references [5], [6] and [7].

During 1991 and 1992 both the regional and national governments join the project, via the General Directorate for Industry, Energy and Mines of the Andalusian Government and the CIEMAT-IER, respectively. The supplementary finance provided by these bodies is aimed at ensuring the continuation and present and future exploitation of the work developed in order to apply it both to outdoor spaces and to the design and use of natural cooling techniques in buildings.

8. Conclusions

From a conceptual, methodological and experimental point of view, the research has revealed that:

1. It is possible to soften and temper the conditions of habitability of outdoor spaces by means of the use soft technology at low initial and operational cost.
2. It has also been established that such techniques are compatible with the aesthetic and functional aspects required by architecture and urban planning.
3. There are considerable differences between the conventional conditioning of buildings and the treatment of outdoor spaces.
4. A strong link should be established between architecture and engineering from the earliest design stages.
5. To determine in a reliable manner the appropriateness of each specific measure in each specific situation, it is necessary to have a methodology based on the modelling/simulation process.

Acknowledgements

The authors would like to acknowledge the EXPO'92 State Corporation for the financial support of the research and development work on which this paper is based.

References

1. Guerra J., Alvarez S., Rodríguez E.A. and Velázquez R. (1991) 'Comfort Issues in the Outdoor Spaces at EXPO'92' in Architecture and Urban Spaces, PLEA'91, Sevilla
2. Departamento de Ingeniería Energética y Mecánica de Fluidos (1989) 'Acondicionamiento Climático de los Espacios Abiertos en EXPO'92. Modelización' Vol. I and II. Final Report. Sevilla
3. Alvarez, S., Rodríguez E.A., Velázquez R. (1989) 'Thermal Performance of Passive Buildings Modelling and Simulation in the S3PAS Computer Program' in Proceedings of the 2nd European Conference on Architecture. Kluwer Academic Publishers, Paris, pp 416-418
4. Alvarez S., Cejudo J.M., Rodríguez E.A. and Guerra J. (1991) 'Full Scale Experiments in EXPO'92: The Bioclimatic Rotunda' in Architecture and Urban Spaces, PLEA'91, Sevilla
5. Alvarez S., Guerra J. and Velázquez R. (1991) 'Climatic Control Applications in Outdoor Spaces at EXPO'92' in Architecture and Urban Spaces, PLEA'91, Sevilla
6. Velázquez R., Guerra J., Alvarez S. and Cejudo J.M. (1991) 'Case Study of Outdoor Climatic Confort: The Palenque at EXPO'92' in Architecture and Urban Spaces, PLEA'91, Sevilla
7. Alvarez S., Rodríguez E.A. and Molina J.L. (1991) 'The Avenue of Europe at EXPO'92: Application of Cool Towers' in Architecture and Urban Spaces, PLEA'91, Sevilla

COMFORT ISSUES IN THE OUTDOOR SPACES AT EXPO'92

J. Guerra*, S. Alvarez*, E.A. Rodríguez**, R.Velazquez*

* *Universidad de Sevilla*
Departamento de Ingeniería Energética
y Mecánica de Fluidos
Av. Reina Mercedes s/n
E-41012 Sevilla. Spain

** *CIEMAT-IER*
División Termosolar
Av. Complutense 22
E-28040 Madrid. Spain

ABSTRACT. Thermal comfort in outdoor spaces presents significant differences with respect to that indoors. The present paper deals with these differences by a model of the thermal balance on the human body adapted to outdoor spaces. The use of this model permits, among other things, quantify the effect of a specific conditioning action in relation to a reference situation, compare the performance of two or more apparently equal situations (sensation temperature concept), design the best combination of parameters in order to obtain a given level of comfort (isocomfort graphs), or evaluate a measured performance.

1. Introduction

When we compare the conditioning of buildings with the climatic control of outdoor spaces, significant differences arise with respect to comfort issues:

- The number of variables to be manipulated.
- The relative influence of each variable.
- The degree of manipulation admissible for each variable.
- Comfort level required, etc.

These questions are discussed in some detail in reference [1]. However, an important additional difference arises from the methodological viewpoint: whilst in indoor spaces comfort issues usually appear in the final stages, when evaluating the suitability of a given design, in the case of outdoor spaces the question of comfort is used as a permanent guiding factor throughout the entire design process and even in the preliminary stages thereof.

In this paper we describe various applications derived from the use of comfort models. In general terms, we would point out that the aim is not to establish whether a person occupying an open space feels comfortable or uncomfortable in a given situation, but rather to determine, from among a set of possible situations, which one gives rise to a greater level of comfort. In other words, we are not considering absolute terms but relative ones.

2. Thermal Balance on the Human Body

Figure 1 shows the major heat flows over the human body where: H is the direct (H_D), diffuse (H_d) and reflected (H_r) solar radiation absorbed by the subject; ΔR represents the longwave radiation exchange with surrounding surfaces; C is the convection with the air and E represents evaporation.

S. Alvarez et al. (eds.), Architecture and Urban Space, 179–187.

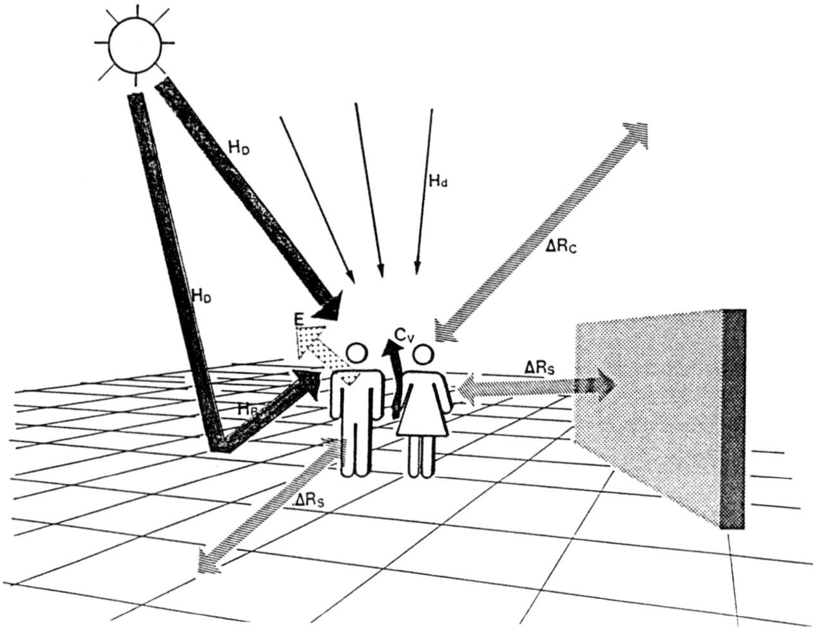

Figure 1: Major Heat Flows Over the Human Body

The thermal balance results in:

$$E_{sw} = M(1-\eta) + (R+C) + H - Cres - Eres - Edif$$

where $M(1-\eta)$ is the net metabolic heat production, $Cres$ and $Eres$ are respectively the sensible heat and latent heat due the respiration. $Edif$ is the skin diffusion heat. The regulation sweating term (E_{sw}) closes the thermal balance and gives a measure of the adaptative effect of the organism and in consequence of the level of comfort.

The expressions used to quantify the different terms assume steady state conditions and derive from widely accepted models. The exception is the effect of solar radiation which is calculated from the incident values using a simultaneous balance over clothing and skin considering their short-wave radiant properties [2].

A reference situation was set up consisting of an average man placed in an uncovered and untreated outdoor space, subjected to climatic conditions representative of those pertaining at 2 p.m. hours a hot Summer's day in Seville. Table 1 shows the resulting values for such a situation and contains in itself the general strategy for successfully undertaking the climatic control of outdoor spaces, namely the priority action (control of solar radiation).

Table 1: The Thermal Balance in Figures

	COMPONENT	VALUE (W)	%	EXTERNAL CONTROLLABILITY
Gains of Subject	1. Net Heat Generation	60	24	Not controllable
	2. Total Radiation (direct + difuse+ reflected)	135	55	Controllable
	3. Long-wave interchange	35	14	Controllable and likely to be negative
	4. Convection	15	7	controllable and likely to be negative
	Total Gains	245	100	
Losses on Subject	5. Exchange with the sky	25		
	6. Sweating	220		Controllable in its actual effectiveness
	Total Losses	245	100	

3. Use of the model

3.1. EVALUATION OF A SPECIFIC ACTION

This is the immediate application of the model of thermal balance on the human body and simply consists in the analysis of the variation occurring in the reference situation when one or several actions are taken with a view to improving the climatic conditions.

By way of example, Figure 2 shows the values for the reference situation and those obtained following the inclusion of a covering of vegetation with zero transmissivity and whose surface temperature is equal to the ambient temperature.

This analysis provides information regarding the degree of conditioning still required in order to obtain a given level of comfort and the relative order in which other complementary actions should be taken.

3.2. COMPARATIVE PERFORMANCE OF COMPONENTS

This is usually done using the concept of sensation temperature and, more precisely, increase in sensation temperature.

The sensation temperature is a readily understandable index for designers, analogous to the effective temperature [3] and can be defined as the temperature of an environment (air and surfaces) at 50% RH, no solar radiation and still air that results in the same total sweating rate (or E_{sw}) as in the actual environment.

The increase in the sensation temperature provides the variation (positive or negative) of said temperature in relation to that obtained with a base component. The reason for using the increase rather than absolute values is because of the need to keep the analysis independent of other factors in the balance which are not affected by the components under study.

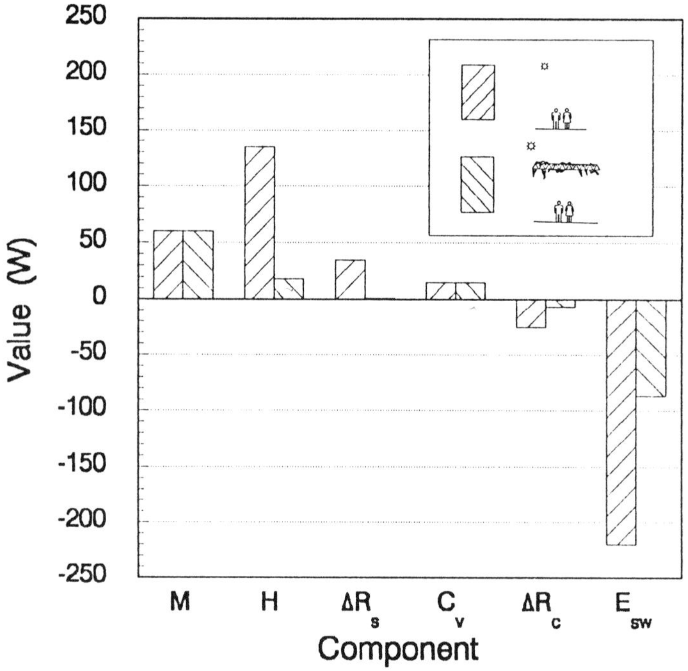

Figure 2: Comparison Between the Heat Flows With and Without Vegetation Cover

When the problem arises of determining the most appropriate component in comfort terms, it frequently happens that the various possible components have advantages and disadvantages which make a decision difficult to adopt. The use of the increase in sensation temperature provides a solution to this problem.

We present below an example of a comparison between various types of covering with different transmissivity and which give rise to a surface temperature which is likewise different. The base case chosen was a covering with zero transmissivity and no overheating (surface temperature = ambient temperature).

Figure 3 shows the results for 8 coverings. These results encouraged us to strongly recommend for the outdoor spaces in EXPO'92 the use of coverings of vegetation (base case), of which more than 50000 m^2 will be installed in pergolas, and as the next alternative white PVC with irrigation. Similar analyses were carried out for other components such as pavements.

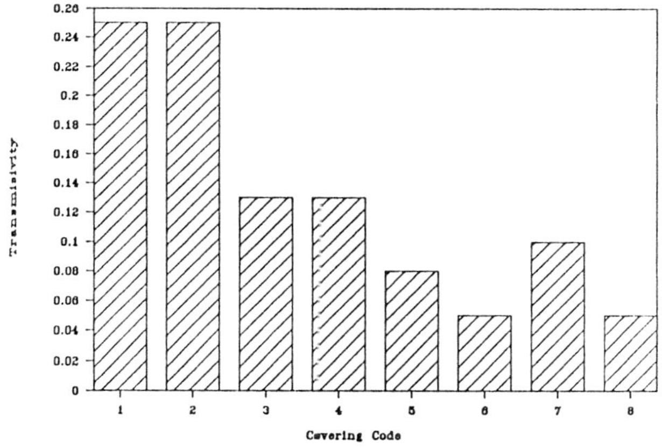

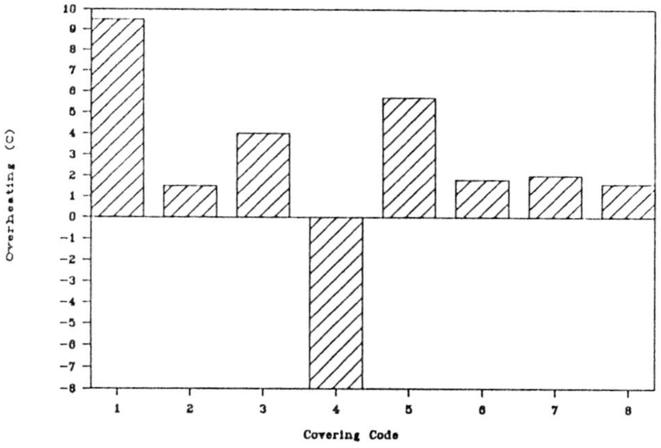

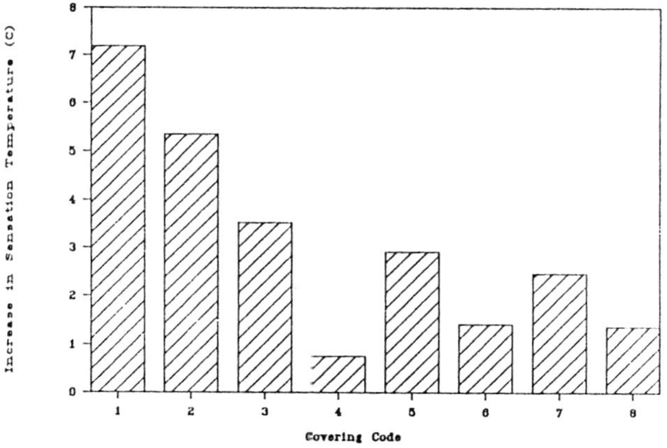

Covering number	Description
1	single, textile, clear, closed shape, dirty
2	single, textile, clear, open shape, clean
3	single, PVC, white, open shape, dirty
4	same as 3 plus irrigation
5	double: upper layer: textile, clear, dirty
	lower layer: textile, clear, clean
6	double: upper layer: PVC, white, dirty
	lower layer: textile, clear, clean
7	multiple: clear, sheets sloped 45°, l/d=2
8	multiple: clear, sheets sloped 30°, l/d=2.5

Figure 3: Comparison Between the Performance of 8 Different Coverings

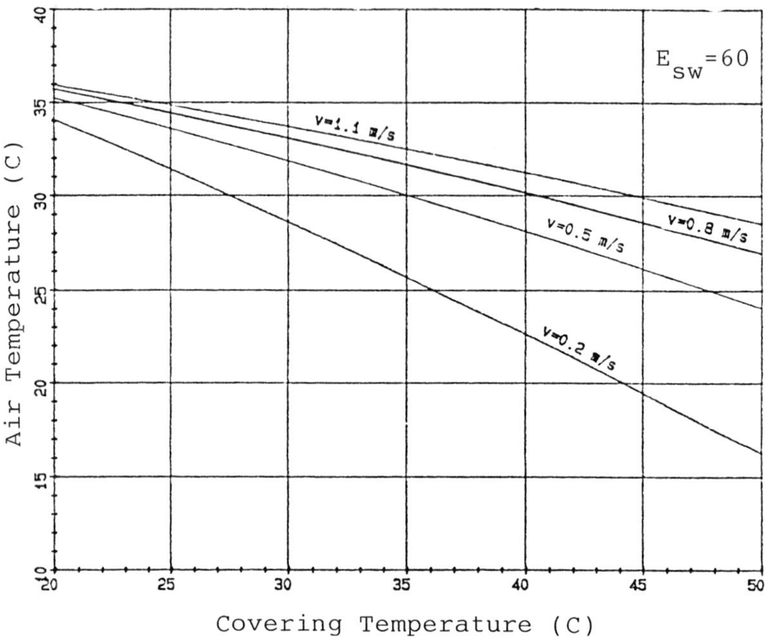

Figure 4: Example of a Isocomfort Graph

3.3. DESIGN (ISOCOMFORT GRAPHS)

In order to achieve the required comfort conditions in a given zone, there are numerous possible combinations relating to the heat flows to be manipulated and the degree of this manipulation.

However, in practice, due to prior design decisions or functional or aesthetic constraints, the number of variables which can be manipulated is at most 3. If such is the case, it is possible to construct an isocomfort graph of the area under study which contains all the possible combinations of the variables which can be manipulated and which lead to the same level of comfort. This permits a decision regarding the most adequate option.

Figure 4 shows a specific case of an isocomfort graph for a sweting rate of 60 g/h obtained for the rotunda (see ref. [4]) with outdoor design conditions. The variables which appear are the surface temperature of the covering (X-axis), the temperature of the air (Y-axis) and the velocity of the air (variable parameter).

It may be seen, for example, that for zero air velocity, the same sensation of comfort is obtained with:

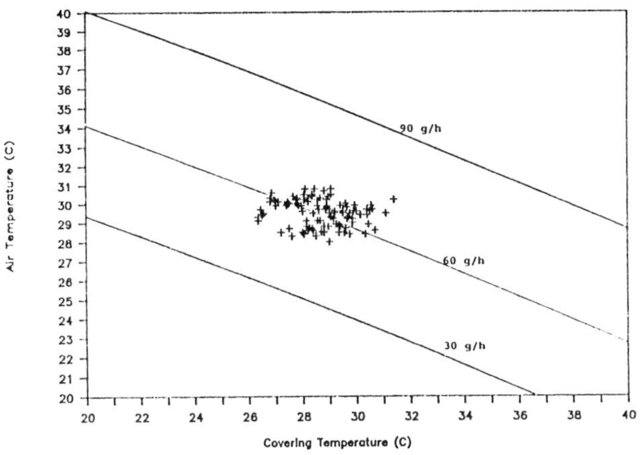

Figure 5: Measured values for the temperature of the covering and the air temperature. (irrigation of the covering and micronizers in trees)

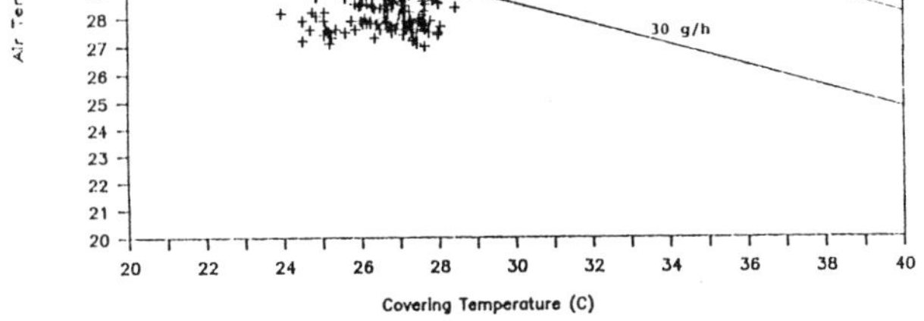

Figure 6: Measured values for the temperature of the covering and the air temperature. (irrigation of the covering micronizers in trees and AHU)

Air temperature	Covering temperature
30	27
26	35
20	44

or that, for a given temperature of the covering (say, 34 °C), the following combinations are equivalent:

Air temperature	Air velocity (m/s)
26	0
30	0.5
32.5	1

It is apparent that irrigation of the covering is essential (in its absence the temperature of the covering would surpass 45 °C) and that there are great beneficial effects to be obtained from provoking movement of the air, however slight.

3.4 EVALUATION

The experimental results derived from the use of a given system are evaluated by means of graphs which are similar to the isocomfort graphs. To continue with the example of the rotunda mentioned in the previous paragraph, we compare two conditioning strategies using the proposed graphs.

In the first one (Figure 5) we show the measured values for the temperature of the covering and the air temperature. The conditioning strategy includes irrigation of the covering and evaporative cooling of the air by means of micronizers located in trees which keep the air still.

In the second one (Figure 6), the same variables are represented using a conditioning strategy which includes irrigation of the covering and evaporative cooling of the air by means of an evaporative air handling unit supplemented by the micronizers in trees. This setup gives rise to an air movement of the order of 0.5 m/s.

In both figures, we include the isocomfort lines for sweating rates of 30, 60 and 90 g/h respectively. The experimental dots plotted refer to times of the day when the outdoor ambient temperature ranged between 34 and 41 °C .

It may be seen that the second strategy ($E_{sw} < 30g/h$) is clearly superior to the first one, although the latter might be valid in zones which require an E_{sw} of the order of 60 g/h. In any case, it should be noted that sweat rates below 90 g/h (Givoni, personal communication) do not produce an unpleasant sensation in occupants.

Acknowledgements

The authors would like to acknowledge the EXPO'92 State Corporation for the financial support of the research and development work on which this paper is based.

References

1. Velázquez R., Alvarez S., Guerra J., Rodríguez E.A. and Cejudo J.M. (1991) 'General Aspects of the Work on Climatic Control of Outdoor Spaces at EXPO'92' in Architecture and Urban Spaces, PLEA'91, Sevilla

2. Departamento de Ingeniería Energética y Mecánica de Fluidos (1989) 'Acondicionamiento Climático de los Espacios Abiertos en EXPO'92. Modelización' Vol. I and II. Informe Final. Sevilla

3. ASHRAE 1989, Handbook of Fundamentals, American Society of Heating, Refrigerating and Air-Conditioning Engineers. Atlanta, USA, 1.989.

4. Alvarez S., Cejudo J.M., Rodríguez E.A. and Guerra J. (1991) 'Full Scale Experiments in EXPO'92: The Bioclimatic Rotunda' in Architecture and Urban Spaces, PLEA'91, Sevilla

CLIMATIC CONTROL APPLICATIONS IN OUTDOOR SPACES AT EXPO'92

S. Alvarez, J. Guerra, R. Velázquez

Universidad de Sevilla
Departamento de Ingenieria Energética
y Mecánica de Fluidos
Av. Reina Mercedes s/n
E-41012 Sevilla, Spain

ABSTRACT. Many applications of the most relevant techniques used for the climatic control of the outdoor spaces at the EXPO'92 site are collected in this paper. These techniques are classified following the main conditioning actions: blocking solar radiation (coverings), reduction of surface temperatures and reduction of air temperature. Some photographs illustrate all these realizations.

1. Introduction

A significant percentage of the EXPO'92 site consists of outdoor spaces which are the place of social activities. Many waiting, resting, shoping and restaurant areas are located in such spaces.

Due to the very harsh climatic conditions of Seville during the summer months, the outdoor spaces of EXPO'92 will be conditioned in order to make the intended activities feasible.

It should be made clear that the aim is not to achieve a complete climatic treatement of all the outdoor spaces because this is neither necessary nor technically possible nor economically feasible. Only the areas of intensive use by visitors will be treated and, within these areas, the conditioning intensity will vary from high in the case of the rest zones, to medium in that of the walkways and low or zero for other spaces.

The design of the outdoor spaces in EXPO'92 has meant a constant interaction process between architecture and engineering from the very early design stages. In all cases, the solutions finally adopted represent a compromise between climatic control, aesthetic and urbanistic positions, preserving at the same time a set of economic and functional constraints.

The current state of the research permits us to declare that there will be implemented a large proportion of the conditioning techniques emerging from the work whose general aspects are described in [1]. This paper summarizes some of the more representative actions taken.

2. Covering

With respect to coverings, we have basically opted for the use of vegetation, due to its ideal properties which allow the obtention of low levels of transmissivity, whilst the radiation absorbed does not entail overheating since this is compensated by the plants' transpiration.

S. Alvarez et al. (eds.), Architecture and Urban Space, 189–193.

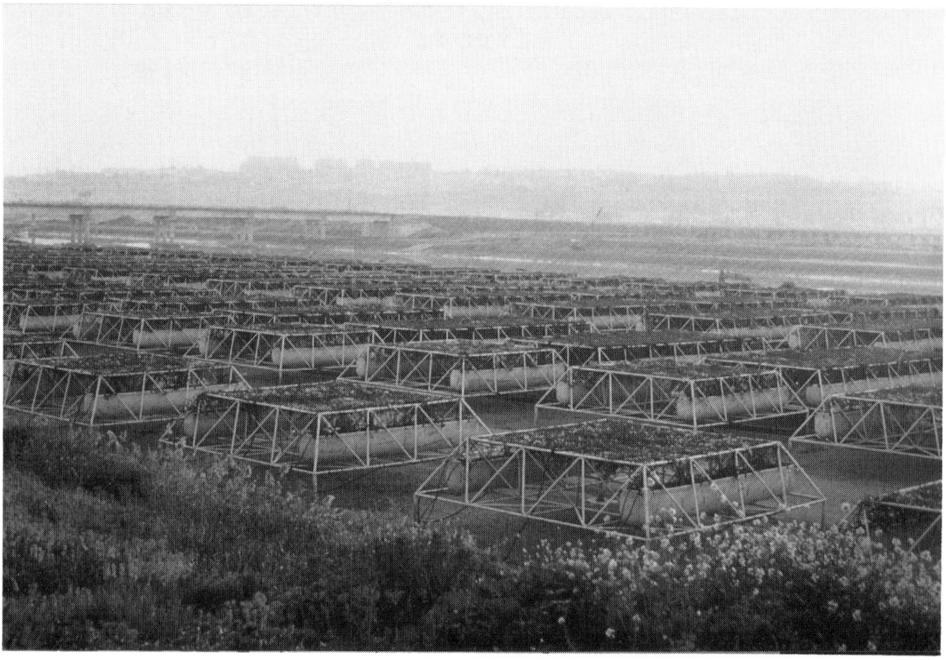

Figure 1: EXPO Pergolas Growing

Nevertheless, in order to ensure this good performance, the vegetation has to produce a continuous and uniform covering. The growth of the plants has been assured by use of the so-called "EXPO pergola" which incorporates plant containers situated in the shade structure itself, thus avoiding the need to have the plants grow from ground level. At the present time, 50000 m^2, in 6×6 and 12×12 m modules, are at an advanced stage of development (Figure 1) in the future parking areas of the Exhibition site. This modules has been transferred to the open spaces and situated at different heights (Figure 2).

Independently of the vegetation, double coverings are used, as in the case of the Railway Terminal (Figure 3), and single PVC as in the Palenque (Figure 2) and the kiosk in Avenue I (Waterway Avenue).

About 40% of the areas of intensive occupation are shaded in order to permanently provide the visitors an alternative to avoid the solar radiation.

3. Reduction of surface temperatures

Water, both in the form of ponds and cascades, is a common feature in all the outdoor spaces. By way of example, we would mention the cascade of over 400 meters in length and 6 in height which runs all the way along Avenue V (Maple Avenue, see Figure 4) and the 15 hectare artificial lake (Spain Lake).

The covering temperature is usually reduced with controlled irrigation, as in the Palenque and the kiosk in Avenue I.

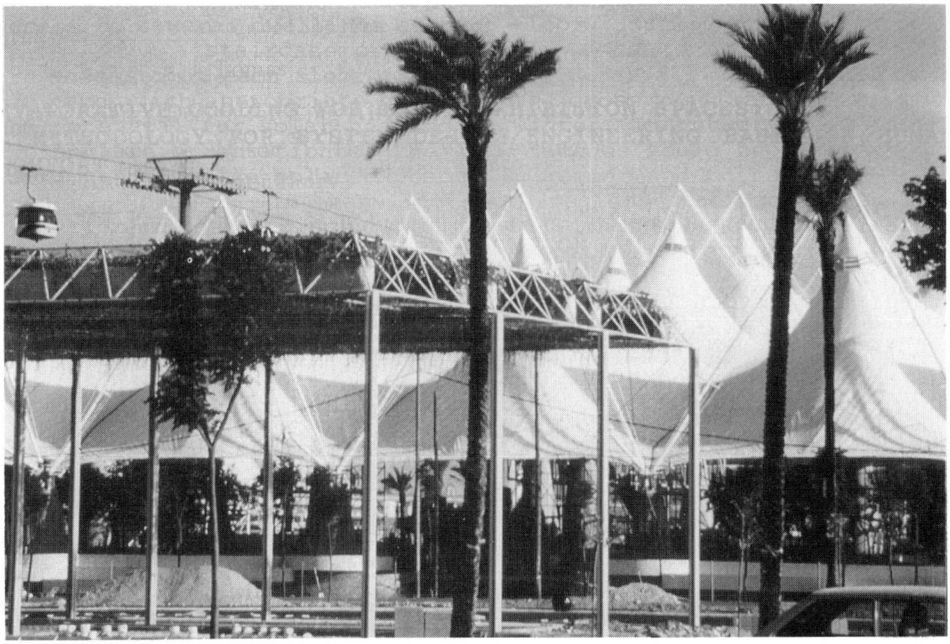

Figure 2: EXPO Pergola placed in site and Palenque Covering

Only the cool pavements have been ruled out due to their low level of efficiency in shaded areas.

4. Reduction of air temperature

Among the air cooling techniques, all the subsystems tested in the Rotunda [2], will be used in the open spaces. It is worthwhile underlining the presence of the 12 cool towers, 30 metres in height, in the Avenue of Europe. These cool towers contain at their upper end wind catchers and micronizers distributed along the towers in different sections. The towers can provide more than 3.5 million frigories/hour (Figure 5).

Practically in all the avenues micronizers placed in the foliage of trees and in the structure of EXPO pergolas produce a continuous descending flow of cool air.

In the Palenque and the Avenue IV (Ombu Avenue) there are micronizers and water curtains in peripheral rings (wet barriers) which cool the air by forced convection with the wind as driving force.

In the outdoor spaces at EXPO'92 there are installed 12 different systems with 7000 micronizers approximately.

Finally, in the Palenque, Avenues I, III and V air handling units are used which have both a sensible cooling and an evaporative cooling sections. The sensible sections uses cold water from the nearerst pond. To prevent this water from heating the ponds are equiped with spray systems.

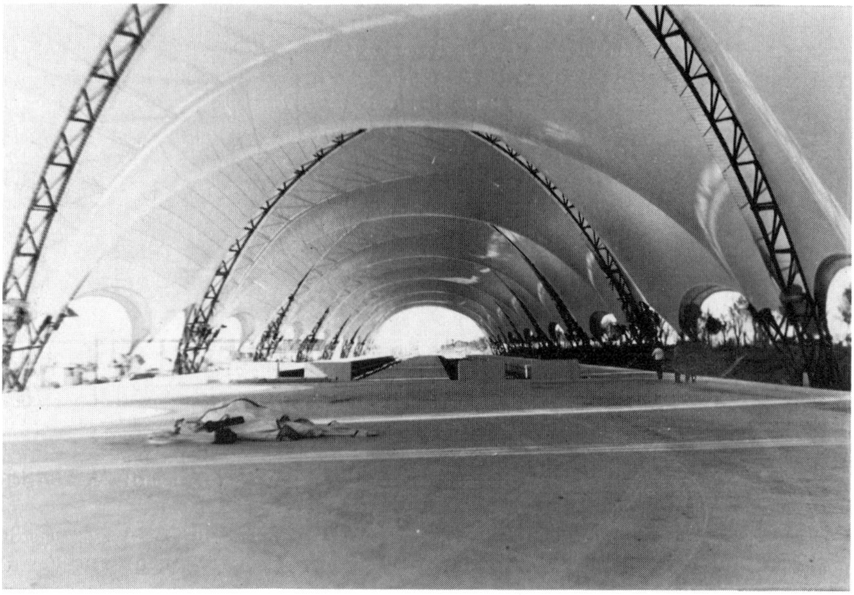

Figure 3: Double Cover of the Railway Terminal

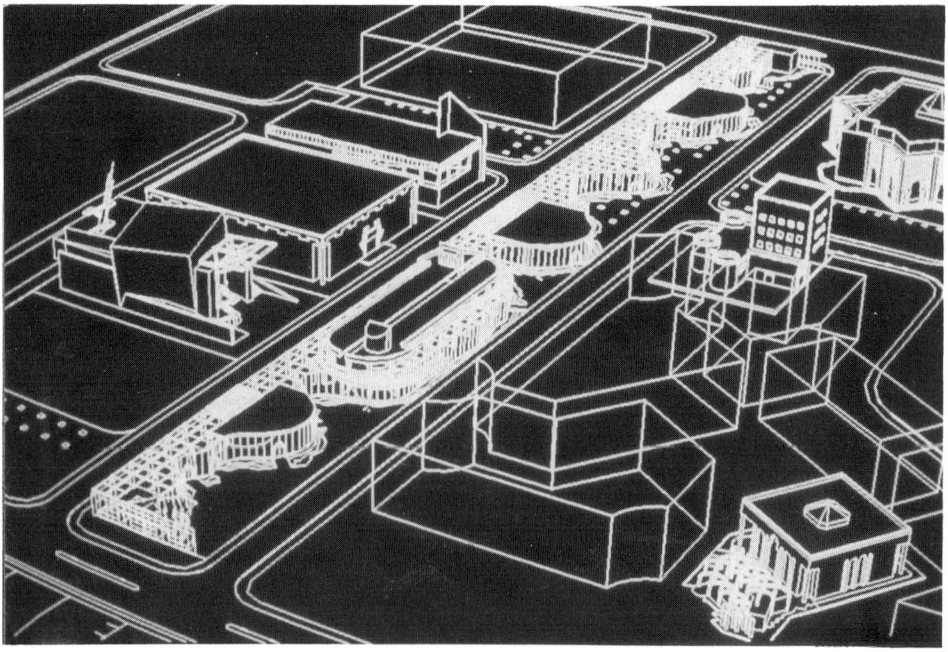

Figure 4: Avenue V (Maple Avenue)

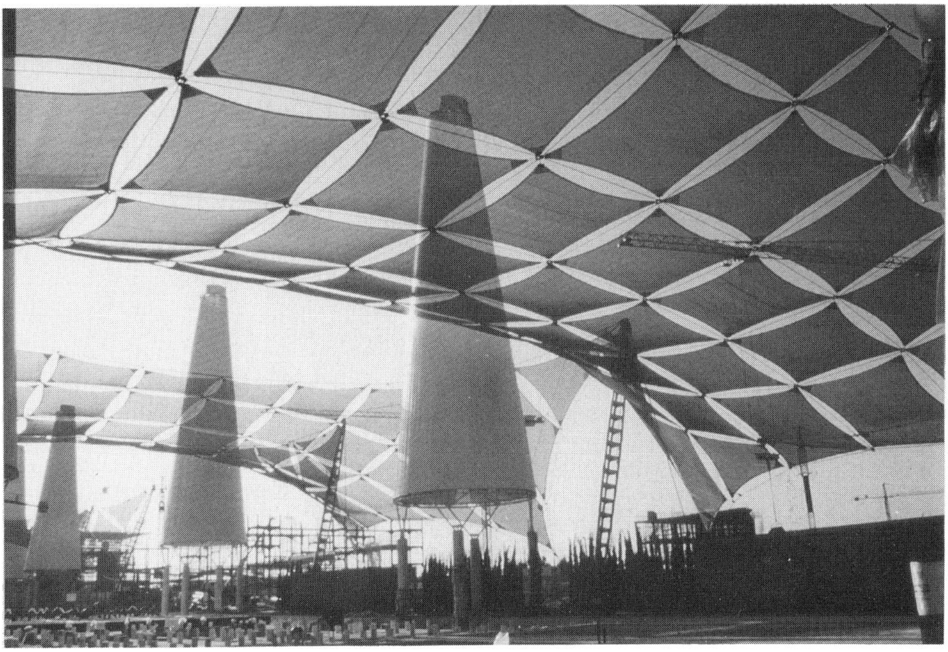

Figure 5: Cool Towers at the Avenue of Europe

Acknowledgements

The authors would like to acknowledge the EXPO'92 State Corporation for the financial support of the research and development work on which this paper is based.

References

1. Velázquez R., Alvarez S., Guerra J., Rodríguez E.A. and Cejudo J.M. (1991) 'General Aspects of the Work on Climatic Control of Outdoor Spaces at EXPO'92' in Architecture and Urban Spaces, PLEA'91, Sevilla
2. Alvarez S., Cejudo J.M., Rodríguez E.A. and Guerra J. (1991) 'Full Scale Experiments in EXPO'92: The Bioclimatic Rotunda' in Architecture and Urban Spaces, PLEA'91, Sevilla

THE AVENUE OF EUROPE AT EXPO'92: APPLICATION OF COOL TOWERS

S. Alvarez*, E.A. Rodríguez**, J.L. Molina*

* *Universidad de Sevilla*
Departamento de Ingeniería Energética
y Mecánica de Fluidos
Av. Reina Mercedes s/n
E-41012 Sevilla. Spain

** *CIEMAT-IER*
División Termosolar
Av. Complutense 22
E-28040 Madrid. Spain

ABSTRACT. The Avenue of Europe, one of the most emblematic open space designed for EXPO'92, constitutes a welcomed and successful colaborative joint work between architecture and engineering. This paper describes the main architectural elements of this avenue and the microclimatic issues related to them. Special emphasis is put on the twelve cool towers, concerning the theoretical and practical studies that maximize their performance.

1. Description of the Avenue

The design of the Avenue of Europe was carried out by the architects Lippsmeier and Hennin-Normier. The Avenue is bordered by the pavilions of the 12 countries of the European Economic Community, with the EEC pavilion occupying a central point within the Avenue.

The Avenue is made up of two lateral walkways from which access is gained to the pavilions and a large central zone which functions as a garden. The central zone alternates between plazas with walkways which connect one plaza to another and to the EEC pavilion (Figure 1). The basic elements of the Avenue are: garden, towers and covering.

a) The garden:

The garden, 300m in length by 40m wide, was conceived as a meeting place and cultural centre. The essential idea is the presence of water in the form of a long river upon which there are four plazas or islands: the white, green, yellow and blue plazas.

Visitors may walk over this river, from island to island, via a framework forming a bridge made of reinforced concrete elements placed at a height of 30cm above the water level.

The Garden of Europe is surrounded by perimeter hedges which provide a certain degree of confinement. Entrance to the garden is via 12 gates in the shape of 12 towers.

b) The towers:

There are a total of twelve towers in the garden, distributed in two parallel rows. Each tower is 30m high in all, 5m of which are the mast and the remaining 25m the trunk covered with a white plastic material. The diameter of the trunk decreases from 8m at the bottom to 3m at the top (Figure 2). The total weight of each tower is approximately 15 tons.

S. Alvarez et al. (eds.), Architecture and Urban Space, 195–201.

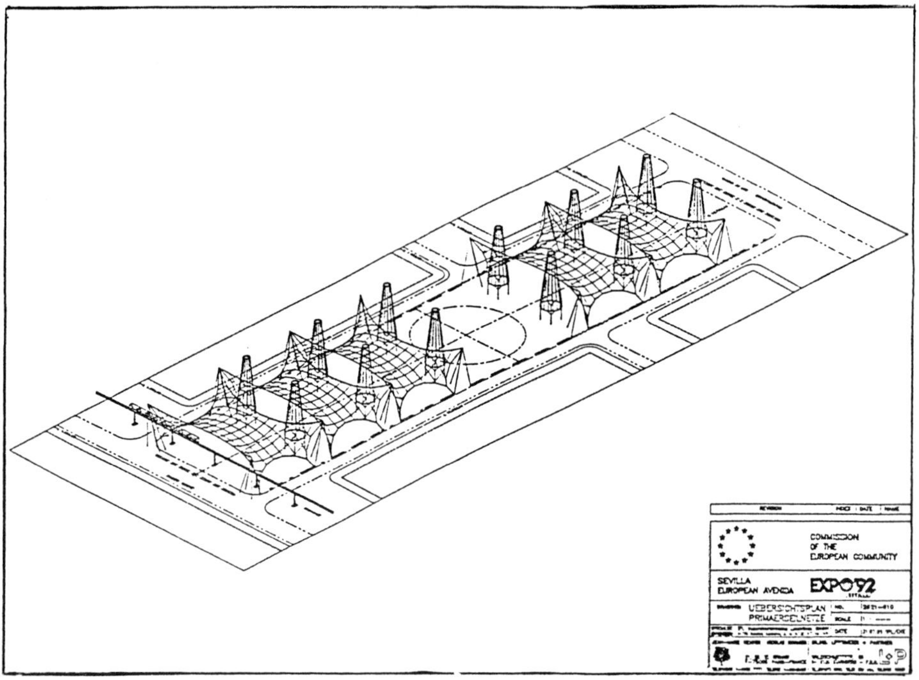

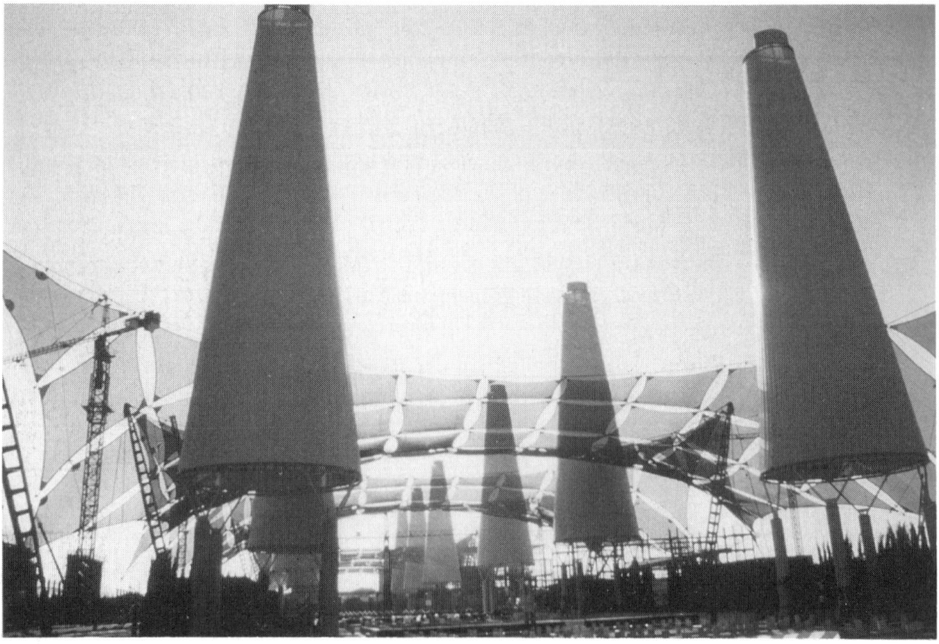

Figure 1: The Avenue of Europe

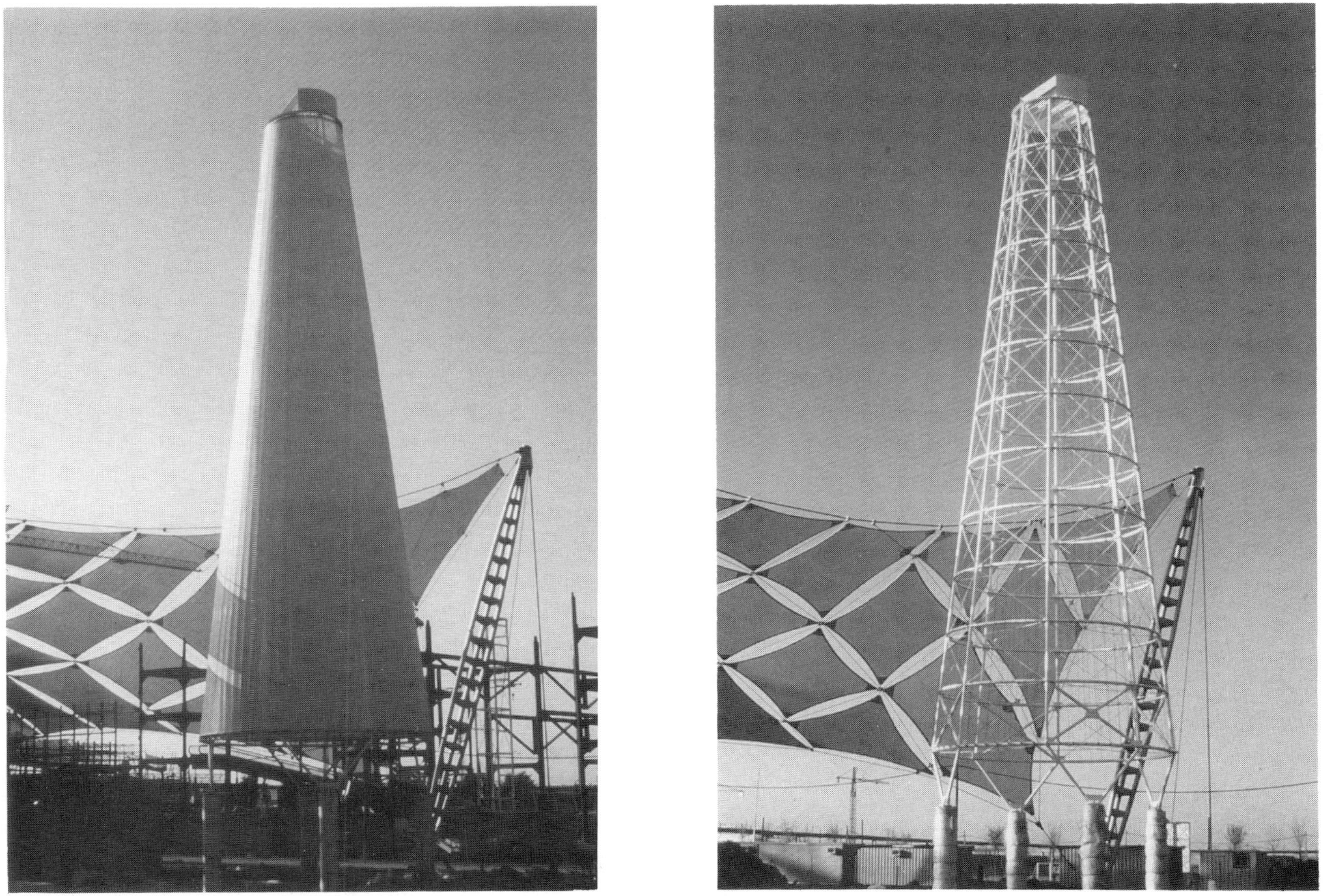

Figure 2: Cool Tower at the Avenue of Europe

The towers symbolize the 12 countries which make up the European Community and are inspired by the ceramic kilns built in the 19th century on the EXPO'92 site itself.

c) The covering:

The covering of plastic material with openings extends over the Garden of Europe between the 12 towers and is divided into two parts, leaving a space between them for the EEC pavilion.

2. Climatic Analysis

Control of solar radiation is achieved by means of the covering which exists in the Avenue. The material and nature of the covering ensures an appropriate surface temperature, although the amount of solar radiation obstructed does not permit a sufficient level of comfort to be reached by itself.

The perimeter hedge which surrounds the Garden of Europe improves its level of confinement, hindering the entrance of untreated air.

The central ground area of the Avenue, with the exception of the plazas, alternates between pavements and ponds. The latter occupy a significant proportion of the total surface area and improve the long-wave radiant exchange, fundamentally in the spaces adjacent to the plazas.

The 12 towers include in their interior a system of micronizers and a device for catching breezes at the top (cool towers). The evaporative cooling of the air inside enables large-scale treatment of air in the Avenue, the purpose of which is to improve the level of comfort therein.

3. Cool Towers

3.1. CONCEPT

The idea of providing cool air by wind or gravity flow within a tower has precedents in the oldern Middle Eastern wind towers as described by Bahadori [1] and especially in the natural downdraft evaporative cool towers cited in [2] or [3].

The major difference between the above application and the design proposed in this paper is that the former operate by air passing through evaporative cooler pads where as the towers of the Avenue of Europe use fogging devices (atomizers) as cooling source.

The use of atomizers is due to two main reasons:

- The need to minimize the air pressure losses in order to obtain a good efficiency under boths still air and wind conditions.
- The need of having a fine regulation capability so that, independently of the outdoor climatic conditions (air temperature and humidity, wind velocity and direction), the cooling capacity can be maximized while preventing people from getting wet because of non-evaporated water droplets.

The height of the tower serves to increase the down draft chimney effect, to use cooler inlet air and to avoid changes in wind direction due to adjacents elements (the wind catcher orientation is fixed, facing the SW prevailing wind direction).

The current design of the cool towers and wind catchers represents a compromise between aesthetic and climatic control criteria.

3.2. MODELLING

In general terms, any atomizing-based cool tower can be modelled by making use of the equations appearing in [4]. The modelling of naturally or wind driven cool tower makes the additional assumptions of steady state and unidirectional movement of the air. With these hypotheses the momentum equation is written as follows:

$$\rho\, v \frac{d\,v}{d\,s} = \frac{-d\,p}{d\,s} - \rho \frac{f\,v^2}{2\,D} + \rho\, g \tag{1}$$

where the viscous force has been modelled through the friction factor f. The boundary conditions of this equation should be one pressure value and one velocity value at a given spatial coordinate. However, we know two pressure conditions at the inlet and outlet sections of the tower:

$$s = 0 \qquad p_0 = p_{atm.} + \Delta p_{dyn} - \Delta p_{in} - \rho_\infty\, g\, H - \tfrac{1}{2}\,\rho_\infty\, v_0^2$$

$$s = L = H \qquad p_L = p_{atm.}$$

This particularity guides the integration process which is:

1. An inlet velocity v_0 is chosen arbitrarily and the corresponding inlet pressure p_0 is calculated. The wind pressure is a function of the wind velocity at the top of the tower through a pressure coefficient C_p:

   $$\Delta\, p_{dyn} = \frac{1}{2}\,\rho_\infty\, C_p\, v_{wind}^2$$

   This term may be positive or negative, depending on the favourable or unfavourable effect of the wind.

   The inlet pressure loss $\Delta\, p_{in}$ is modelled through a pressure loss coefficient C_{in} affecting the inlet kinetic pressure:

   $$\Delta\, p_{in} = \frac{1}{2}\,\rho_\infty\, C_{in}\, v_0^2$$

   Both C_p and C_{in} depend on the inlet configuration of the tower. Values for these coefficients can be found in many engineering handbooks ([5], [6]).

2. The tower is discretized along its length and equation (1) is solved in finite diferences to obtain the static pressure in each section k of the tower:

   $$p_k = p_{k-1} + \rho\, g\, \Delta s - \rho\, \Delta(v^2/2) - f\, \rho \frac{v^2}{2D}\, \Delta s \tag{2}$$

   The solution of both energy and water vapour equations [4] lets us update the value of the air density along the tower's length.

3. At the outlet section the calculated pressure must equal the real outlet pressure. Since the initial inlet velocity has been chosen arbitrarily, this must be corrected succesively until the outlet pressure is adjusted.
4. Once the adjustment of the outlet pressure has been reached, the resulting air velocity is the real induced by both the evaporative cooling and the wind pressure.

3.3. SIZING AND CONTROL

The placement of the towers in the entrances to the avenue impose severe restrictions in their operating condition, because it is always necessary to guarantee a total evaporation of the atomized water in order to avoid drops falling onto the visitors walking at the bottom.

Applying to these towers the model described in the previous section, we can size the water atomizing system in terms of the maximum water flow rate needed for design conditions.

Though a more effective air movement is induced if the air cooling is produced at the higher point, the risk of drop interaction and the wetting of the tower's structure that might cause a continuous rain, advised us to distribute the nozzles in several sections of the upper part of the tower. The final design consists of nine atomizing rings with eight $7.3 l/h$ impaction type pressure nozzles each. This rings are supported by the structural rings of the tower. These nine rings are grouped in four circuits with 3, 1, 2 and 3 rings respectively, that permit a nine-staged partialization.

The control of these atomization rings is achieved by measuring the outdoor conditions and checking the difference $(T_\infty - T_{wb})$. Also, the wind velocity is measured at the inlet section where a wind catcher takes advantage of the dominant winds. A fine adjustment of the number of rings working everty time is made by checking the outlet conditions which must not be excessively humid. Figure 3 plots the number of rings working as a function of the difference $(T_\infty - T_{wb})$ and the wind pressure.

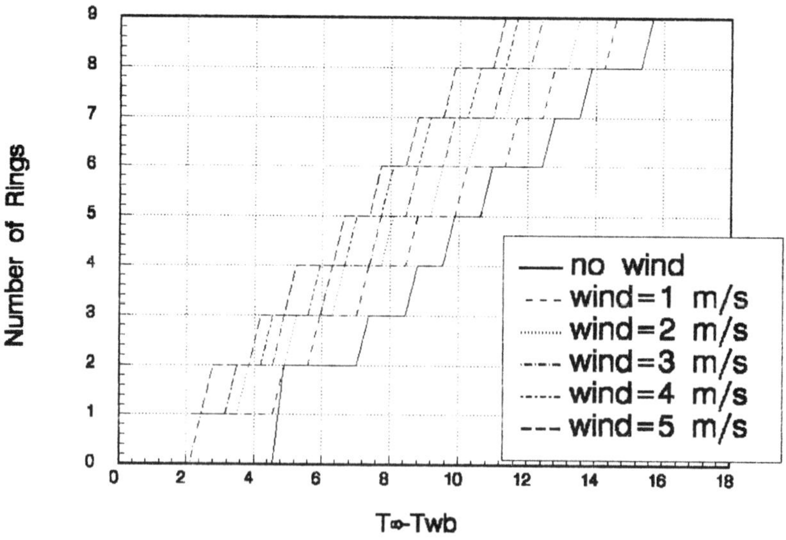

Figure 3: Control Strategy of the Cool Towers

An additional control based on the humidity level of the treated zone, reduces the number of rings working whenever the humidity set point is exceeded.

Acknowledgements

The authors would like to acknowledge the EXPO'92 State Corporation for the financial support of the research and development work on which this paper is based.

List of symbols

D	tower diameter
g	gravity acceleration
H	tower's height
p	static pressure inside the tower
p_0, p_L	inlet and outlet pressures
$p_{atm.}$	atmospheric pressure
s	directional coordinate
T_∞, T_{wb}	ambient dry and wet bulb temperatures
v, v_0	tower inside and inlet velocities
v_{wind}	wind velocity
Δp_{dyn}	wind pressure
Δp_{in}	inlet pressure loss
f	viscous friction factor
ρ, ρ_∞	inside and outside air densities

References

1. Bahadori M.N. (1986) 'Natural Air-Conditioning Systems' in Advances in Solar Energy, Vol 3, Plenum Press, New York, pp 283-356
2. Cook J. (1989) 'Phoenix Solar Oasis: Cooling Design of Urban Spaces' in 14th National Passive Solar Conference, Denver, pp 350-355
3. Kent K. and Lewis T. (1990) 'Natural Draft Evaporative Cooling' in 15th National Passive Solar Conference, Austin, pp 157-160
4. Rodríguez E.A., Alvarez S. and Martín R. (1991) 'Direct Air Cooling from Water Drop Evaporation' in Architecture and Urban Spaces, PLEA'91, Sevilla
5. ASHRAE, 1989 Handbook of Fundamentals, American Society of Heating, Refrigerating and Air-Conditioning Engineers. Atlanta, USA.
6. Idelchik I.E. (1986) Handbook of Hydraulic Resistance. Hemisphere Publishing Corporation. New York.

CASE STUDY OF OUTDOOR CLIMATIC COMFORT: THE PALENQUE AT EXPO'92

R. Velázquez, J. Guerra, S. Alvarez, J.M. Cejudo

Universidad de Sevilla
Departamento de Ingeniería Energética
y Mecánica de Fluidos
Av. Reina Mercedes s/n
E-41012 Sevilla, Spain

ABSTRACT. This paper presents the climatic conditioning strategy designed for an outdoor space located on the site of the Seville Universal Exhibition (EXPO'92). The purpose of the space is to serve as a theatre with capacity for 1500 people, surrounded by lateral staying zones and resting and refreshment areas. The design is based on natural cooling techniques together with a conventional support system to complement the former at peak times. The analysis of specific isocomfort graphs leads to the establishment of the values of the parameters which govern comfort and allow us to decide the basic criteria for the thermal design. We present the features of the irrigation system for the covering and its control strategy. Also described are the air handling units and the combinations thereof which lead to the maximum usage of soft techniques (ponds and humidification) and minimum usage of conventional techniques.

1. Description

The Palenque is located at the geometric centre of the EXPO'92 site and covers an area in excess of 8000 m^2. Functionally, it is divided into two zones (see Figure 1).

The central zone constitutes an area for shows with seating capacity for 1500 people. The stage is separated from the spectators by two ponds situated at different levels and connected by a cascade. The ponds have fountains which are not only decorative but also a key element in the conditioning strategy which has been designed. The North and South lateral zones constitute a rest area and contain kiosks, restaurants and shops.

The central zone is separated from the lateral zones by a barrier of trees and fountains. The entire Palenque is in turn separated from the exterior space by a virtually continuous peripheral barrier of similar characteristics to the former. The entire area is covered with white PVC (see Figure 2), forming a structure made up of 50 cone-shaped elements.

In addition, the Palenque contains other functional spaces (offices and subsidiary services), situated in the basement and conditioned by traditional methods. For this reason, they will not be dealt with in this paper.

The architectural design is the work of Jose Miguel de Prada and the climatic conditioning was carried out by the authors of this paper.

2. Cooling strategies

- Control of solar radiation by means of a continuous white PVC covering.Trees at the

S. Alvarez et al. (eds.), Architecture and Urban Space, 203–208.

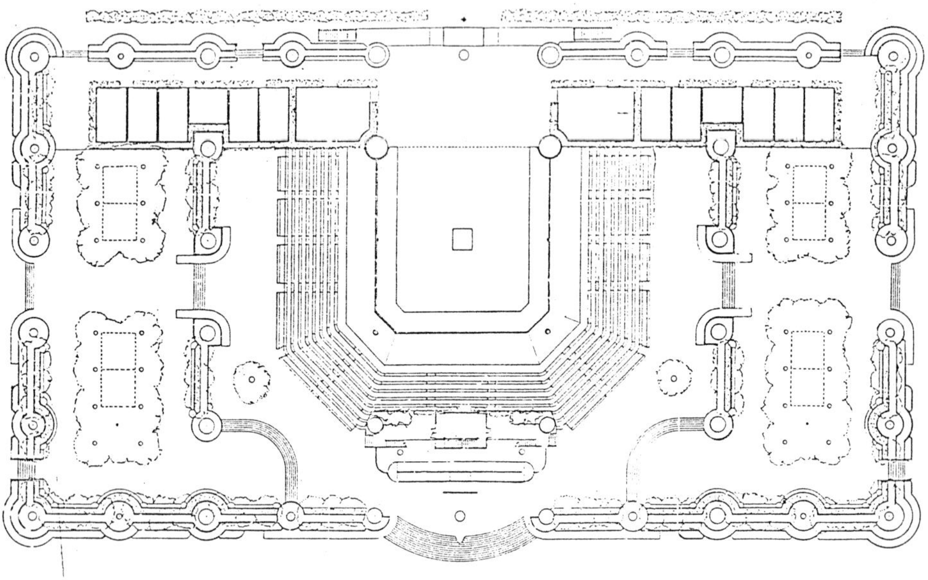

Figure 1: Floor plant

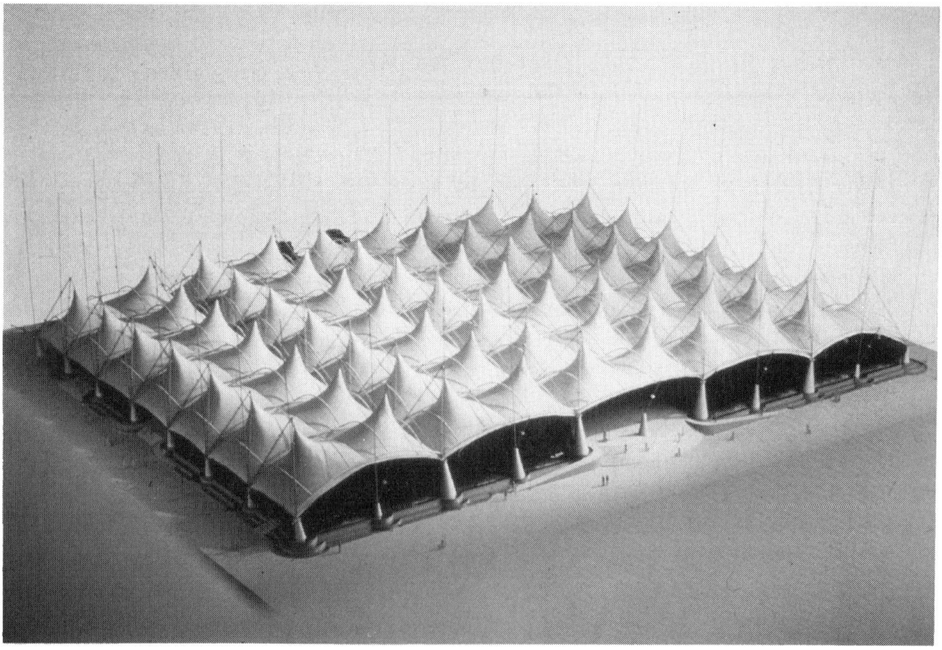

Figure 2: View of the scale model

peripheral barrier act as blocking elements for the reflected component and serve as a complement to the covering during periods of low solar elevation.

- Reduction of the temperature of surrounding surfaces, achieved fundamentally by means of controlled irrigation of the covering.

  The presence of peripheral and intermediate rows of fountains provides cool surfaces with a favourable local effect, negligible in any case on an overall scale.

  Other possibilities, such as cooling the ground, were discarded due to its low efficiency and difficulties of implementation and control.

- Reduction of the air temperature, achieved by the use of 5 air handling units(AHU), each with three different cooling sections. The efficiency of this action depends to a great extent upon the confinement of the zones to be treated which is accomplished by sinking the space in the seating area and by means of barriers of vegetation and fountains for the remaining zones.

  Additionally, the installation of micronizers incorporated into the peripheral fountains prevents the wind from neutralizing the reduction of the air temperature in the lateral rest zones.

The practical implementation of the above actions requires the determination of the best combinations of values for the different variables which lead to a given level of comfort. This has been done with the aid of isocomfort graphs, the basis of which is described in [1], assuming a required level of comfort (E_{sw}) of 30 g/h in the central zone and 60 g/h in the lateral rest areas. Figure 3 shows for instance the isocomfort graph for the central zone, in which the covering temperature, air temperature and air velocity appear as the variables to be manipulated.

The following sections describe the major technical issues for the cooling strategies.

3. Covering

Theoretical models revealed, and experimentation confirmed, that the surface temperature of the covering would surpass 45 °C even if openings were left at the top to favour elimination of the hot air.

Under such conditions, it is clear from the isocomfort graphs that, in order to obtain the desired comfort levels, the air temperature required would have to be very low and the subsequent implementation cost would be too high. For this reason, several methods of cooling the covering were examined, including different possibilities with water films and irrigation.

The solution selected consists of controlled irrigation that maintains an average surface temperature below 30 °C during peak outdoor conditions. The maximum water flow rate required for such a reduction is 50 m^3/h. It should be clarified that cooling is not due to the water temperature (convection) but to its evaporation. The monitoring system guarantees that the covering is permanently wet and ensures that the maximum surface temperature (there are three differently orientated sensors) never exceeds a value 7 °C below the outdoor air temperature. An additional advantage of the irrigation is the fact that the covering is kept clean and a lower absorption of solar radiation is obtained.

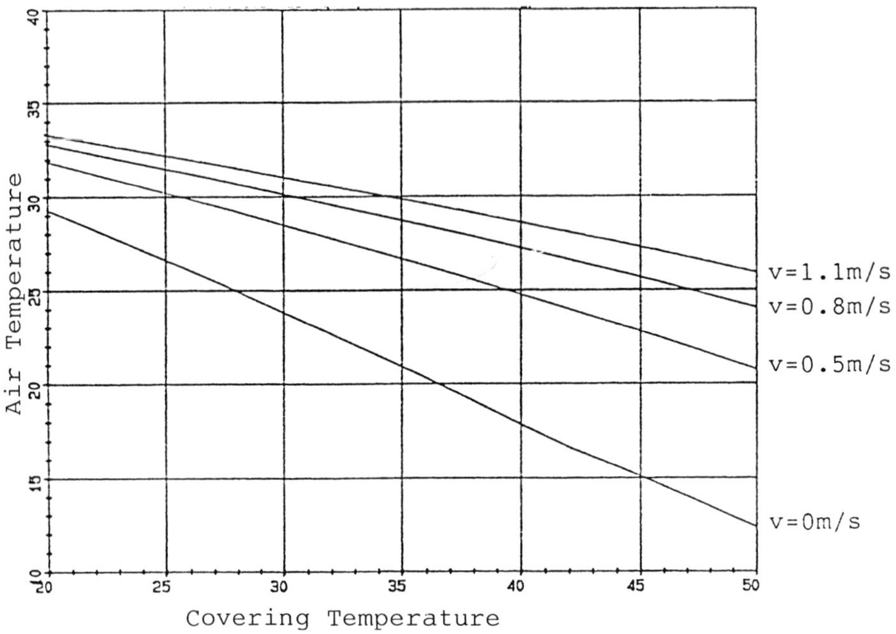

Figure 3: Isocomfort Graph of the Palenque; Central zone; $E_{sw} = 30g/h$

4. Cooling of air. Air Handling Units

In order to obtain the required comfort levels, and once the covering temperature has been controlled, the air temperature of every zone is a function of the estimated air velocity. The diffusion of air to the central zone gives rise to an air velocity of 0.5 m/s which implies a temperature of 27 °C . For lateral zones (less critical from comfort viewpoint) no air movement is to be expected and their maximum temperature will be of 30 °C .

Cooling of the air is achieved by means of 5 AHU's which are distributed as follows:

- Seating area: 1 AHU with a capacity of 70000 m^3/h of air;
- South lateral zone: 2 AHU each with a capacity of 20000 m^3/h of air;
- North lateral zone: 2 AHU each with a capacity of 15000 m^3/h of air.

The output conditions of the treated air are 19 °C and 90% relative humidity.

Each AHU is composed of three cooling sections, as shown schematically in Figure 4.

The first section precools the outside air by means of a coil which is supplied with water from the pond which surrounds the stage in the case of the AHU in the seating area and from the intermediate ponds in the remaining 4 AHU's.

If the dry-bulb temperature of the air on exit from the precooling coil is above 19 °C the evaporative section comes into operation. For peak load periods (maximum occupancy and extreme exterior conditions), the temperature on exit from the evaporative section may be above 19 °C . Under these conditions, the evaporative section pump is halted and the cooling coil, supplied with cool water from the auxiliary chilling plant, comes into operation.

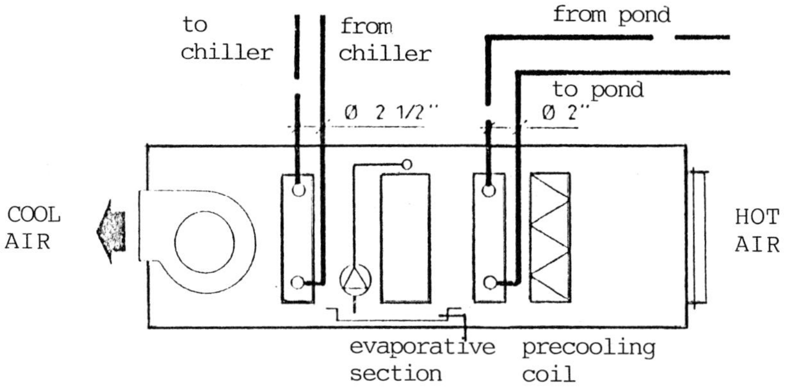

Figure 4: Scheme of air handling units

Under design conditions, natural air-cooling techniques provide 60% of the cooling load. During the time the Exposition is to be held (April to October) the auxiliary chiller plant will provide only 25% of the total cooling requirements of the Palenque.

Finally, the currently instaled capacity is only a 43% of the requiered capacity if only conventional chiller plants would be instaled.

5. Water in ponds as a Cooling Source

One of the fundamental aspects of the air-cooling strategy described in the preceding section is the design and sizing of the ponds and fountains which provide the cool water to the first section of the AHU's.

In the stage pond, which has a capacity of 200 m^3, cooling is accomplished by having the fountains operate during the night and during the intervals between shows. In the intermediate ponds, the water is cooled by sprayers which operate permanently the whole time the Palenque is occupied (10 a.m. to 2 a.m.)

Selection of the different design variables, such as the capacity of the ponds, the water flow rate to be sprayed by the fountains, the droplet size (type of nozzle), etc., was effected by means of the model described in [2].

Figure 5 shows the predicted results for the stage pond with the final design selected. We represent the outdoor ambient temperature, the wet-bulb temperature and the water temperature under design climatic and operating conditions. As may be seen, the water temperature varied between 18 and 24 °C despite the significant heating deliver to the pond due to the return of hot water from the precooling coil section.

Substantially lower values are obtained for average days and/or for fewer than 1500 occupants, which is of course foreseeable in most cases.

Similar performance is obtained for the intermediate ponds which provide water to the other AHU's.

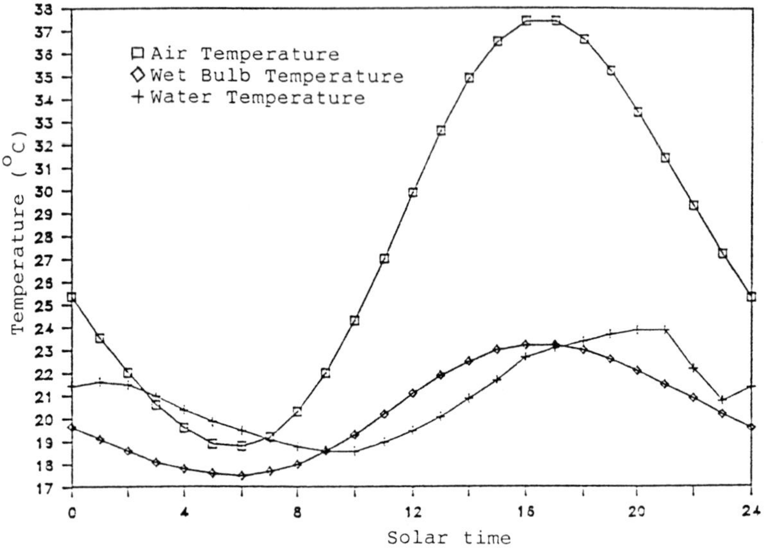

Figure 5: Daily evolution of pond temperature (design conditions)

Acknowledgements

The authors would like to acknowledge the EXPO'92 State Corporation for the financial support of the research and development work on which this paper is based.

References

1. Guerra J., Alvarez S., Rodríguez E.A. and Velázquez R. (1991) 'Comfort Issues in the Outdoor Spaces at EXPO'92' in Architecture and Urban Spaces, PLEA'91, Sevilla
2. Guerra J., Molina J.L. and Cejudo J.M. (1991) 'Thermal Performance of Water Ponds: Modelling and Cooling Applications' in Architecture and Urban Spaces, PLEA'91, Sevilla

FULL SCALE EXPERIMENTS IN EXPO'92. THE BIOCLIMATIC ROTUNDA

S. Alvarez*, J.M. Cejudo*, E.A. Rodríguez**, J. Guerra*

* *Universidad de Sevilla*
Departamento de Ingeniería Energética
y Mecánica de Fluidos
Av. Reina Mercedes s/n
E-41012 Sevilla. Spain

** *CIEMAT-IER*
División Termosolar
Av. Complutense 22
E-28040 Madrid. Spain

ABSTRACT. A pilot plant was set as a full scale experimental support for testing and evaluating different systems designed for the climatic treatment of outdoor spaces at EXPO'92. This paper describes the characteristics of this pilot plant and of the major cooling techniques included in it. Mention is made of the objectives and workschedule at the experiments. Finally, we include some comments about the actual value of the experimentation process and the need to include it within a wider research programme.

1. Introduction

The methodology used during the work on climatic control of outdoor spaces at EXPO'92 [1] includes a wide stage devoted to full scale experimentation. The aim of the experiments is threefold:

- To provide data to characterize the thermal performance of every individual component.
- To provide data to characterize the higrothermal conditions resulting at the different zones.
- To provide data to validate the theoretical models.

A full scale experimental setup, containing the most significant elements, was designed taking into account the size, functionality and aesthetic appearance of the areas that will be distributed throughout the outdoor spaces at EXPO'92.

The architectural design was made by the Seminar of Bioclimatic Architecture from the University of Seville. The Energy Engeneering and Fluid Mechanics Department of the same University was in charge of the engineering design, monitoring and evaluation.

2. Site description

The experimental setup includes basically two walkways with different sizes and coverings-type, a central pond with jets and sprayers, and essentially, a rest zone known as the bioclimatic rotunda in which the natural air-cooling techniques have been tested.

Geometrically the rotunda (Figure 1) has three levels linked by stairways and forms a square with sides of 31 metres. The two lower levels constitute the area intended to be used by occupants. They form two concentric circles with diameters of 24 and 16 metres

S. Alvarez et al. (eds.), Architecture and Urban Space, 209–216.

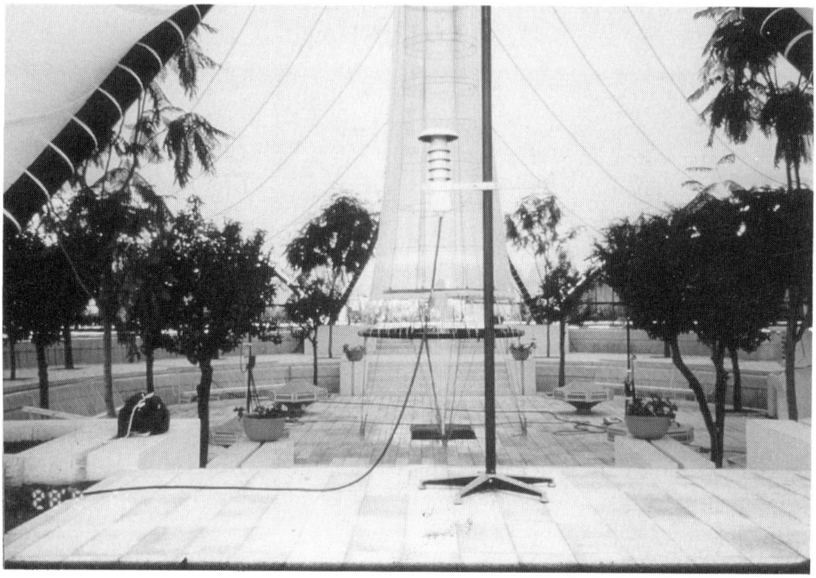

Figure 1: The Rotunda

respectively. The difference in height between the various levels is 80 cm. and cascades fall from one level to another. The rotunda is covered with a white PVC pyramidal covering open at the top.

As previously mentioned, the purpose of the rotunda is to serve as a full scale pilot experiment. As a consequence, it houses a considerable number of techniques and strategies aimed at the thermal conditioning of outdoor spaces.

As a matter of example, we will mention the natural air-cooling techniques implemented, perhaps the more innovative elements of the project. All of them are based on the evaporative cooling which takes place when air and water are put into contact:

- Air handling unit (Figure 2)
  Concept: mechanical evaporative cooling.
- Micronizers (high pressure nozzles) in trees (Figure 3)
  Concept: evaporation of droplets (natural convection).
- Micronizers in tower (Figure 4)
  Concept: evaporation of droplets (forced convection; driving force: fan)
- Micronizers in peripheral rings. Wet barriers (Figure 5)
  Concept: evaporation of droplets (froced convection; driving force: wind).

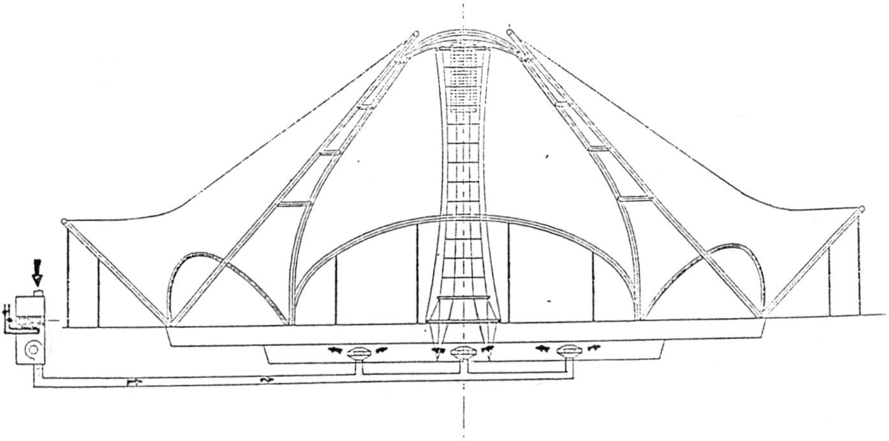

Figure 2: Air Handling Unit

3. Experimental workplan

Experiments took place during summer of the years 1988, 1989 and 1990. Three different stages can be identified.

During the first year, the aim was to evaluate and validate the performance of individuals components (pavements, coverings, ponds) as well as to characterize the influence of jets and sprayers on both the water temperature and the surrounding air temperature.

The second year was mainly devoted to evaluate the air-cooling techniques installed at the rotunda.

The third year, after a redesign and resize of the different techniques and operational strategies, the objective was the evaluation and validation of the rotunda as a whole. For that purpose, an adaptive-predictive control system was in charge of automatically decide which subsystem should be operated and at what intensity, in order to assure the prescribed comfort levels.

The monitoring includes a total of 96 sensors, 41 of which are located at the rotunda, recording air temperature and humidity surface temperatures (coverings and pavements) and water temperatures.

4. Some experimental results

Figure 6 shows the cooling capacity of the cool tower (12000 m^3/h of air) on the 16th and 17th of August 1989. The outdoor ambient temperature and the temperature of the air on exit from the tower are shown.

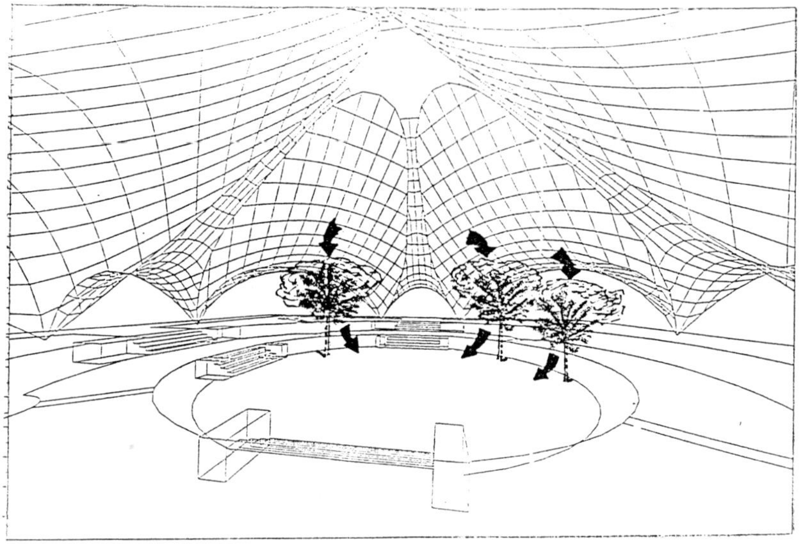

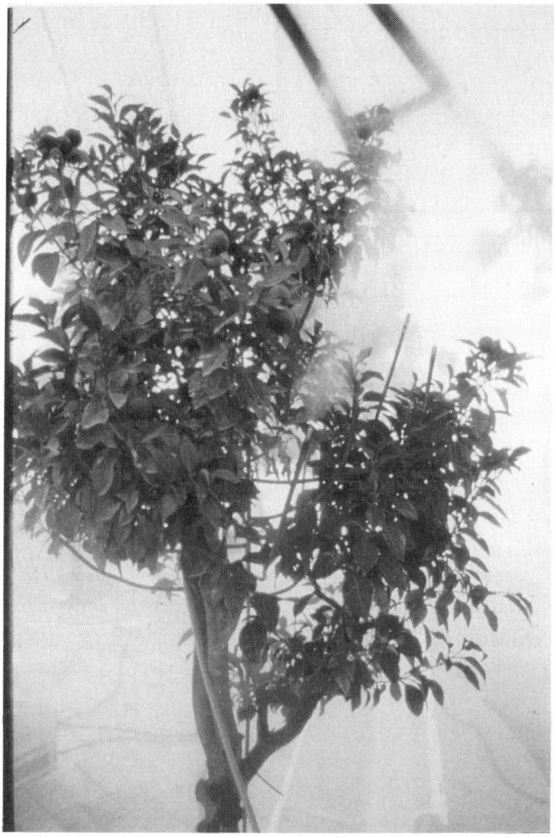

Figure 3: Micronizers in Trees

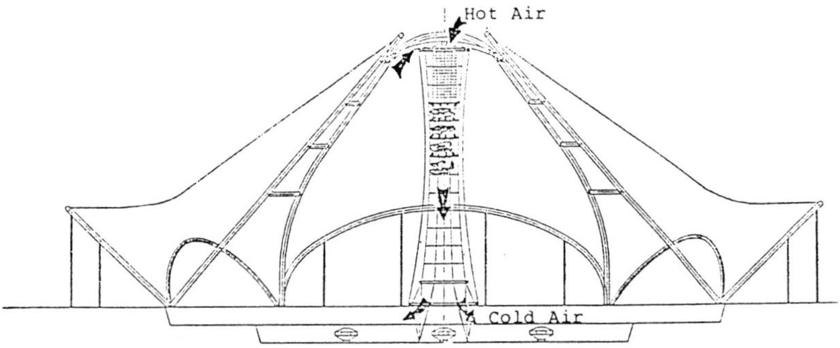

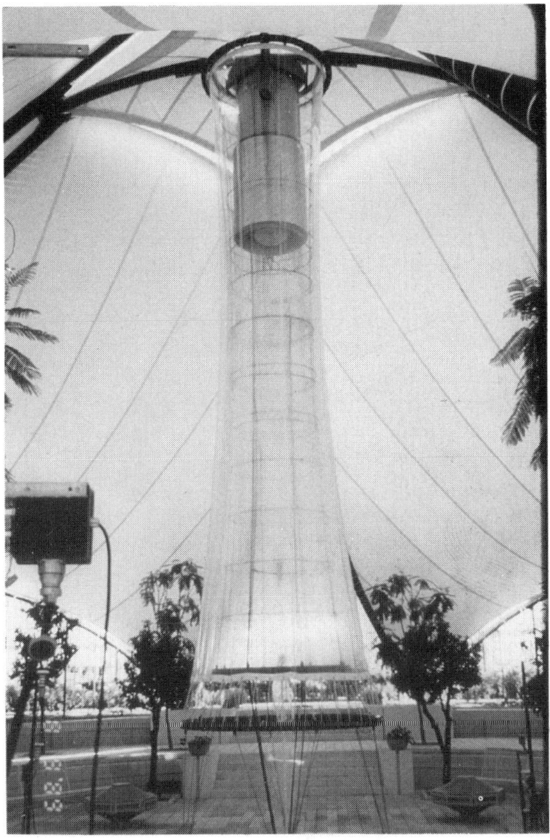

Figure 4: Micronizers in Tower

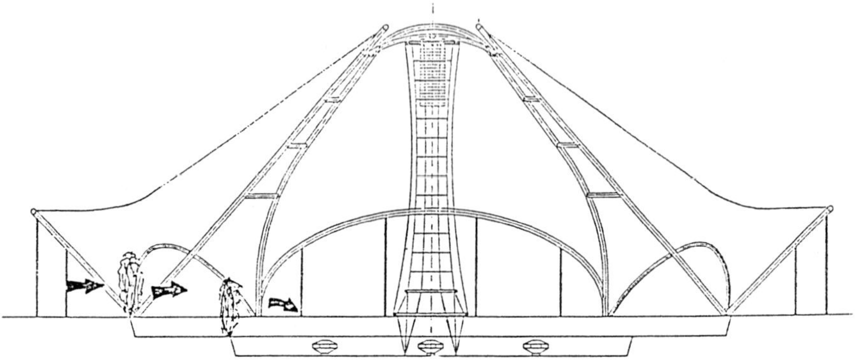

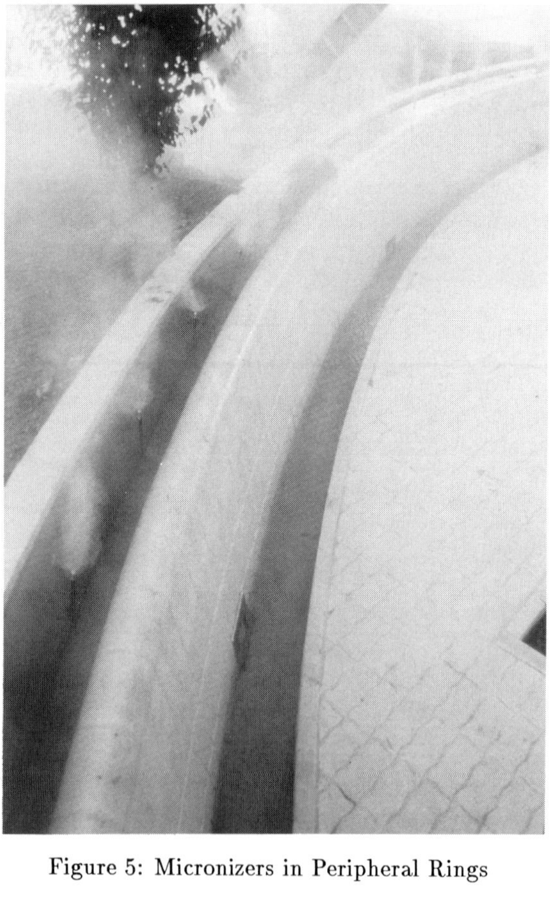

Figure 5: Micronizers in Peripheral Rings

Figure 7 shows the covering temperature at SW facing plane (the least favourable) during August 16th (with irrigation) and 17th (without irrigation) and the outdoor air temperature. A reduction of the surface covering temperature in the region of 16 °C due to the irrigation can be observed .

Figure 8 and Figure 9 show respectively the air temperature and the relative humidity at different heights of the rotunda during August 14th, 1990, using micronizers in trees. The control set-point was established to maintain 65% average relative humidity between heights 0.5 and 1 m.

5. Conclusions

The research works carried out within the frame of the presented project revealed the necessity of having a systematic and comprehensive experimentation programme which allows to understand, to evaluate and to validate the performance of the different techniques and components.

At the same time, it was (once more) demostrated that experimentation alone is a clearly insufficient tool, unable to guide by itself future designs. In fact, the analysis of an experiment tell us what is happening but not:

- Why is it happening?
- Which subsystems worked correctly and which didn't?
- Why did this happen?
- Which design variables are relevant in each subsystem and which are secondary?
- How may the behaviour of each subsystem be improved and what is its point of optimum performance? etc.

In short, experimentation does not allow us to obtain a solid basis for systematically extrapolating the results nor for laying down the subsequent design guidelines. This basis may only be obtained via a process which includes:

MODELLING → VALIDATION ← EXPERIMENTS
↓
SIMULATION (Sensitivity analysis)

Acknowledgements

The authors would like to acknowledge the EXPO'92 State Corporation for the financial support of the research and development work on which this paper is based.

References

1. Velázquez R., Alvarez S., Guerra J., Rodríguez E.A. and Cejudo J.M. (1991) 'General Aspects of the Work on Climatic Control of Outdoor Spaces at EXPO'92' in Architecture and Urban Spaces, PLEA'91, Sevilla

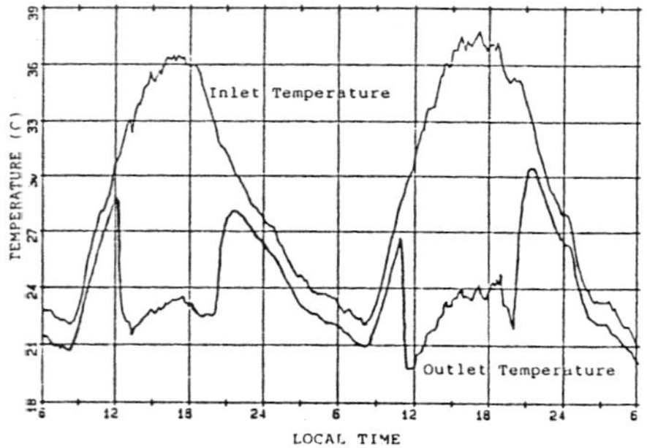

Figure 6: Cooling Capacity of the Cool Tower

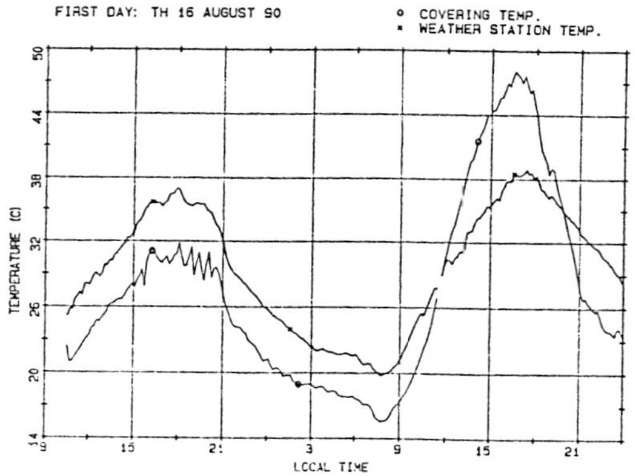

Figure 7: Covering Temperature at SW Facing Plane

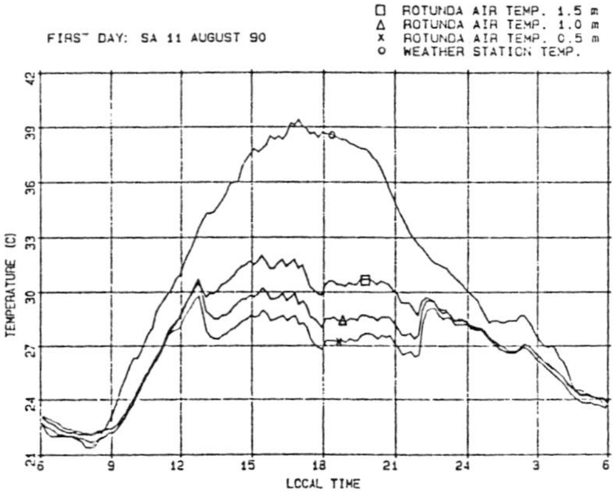

Figure 8: Air Temperature in Rotunda at Differents Heights

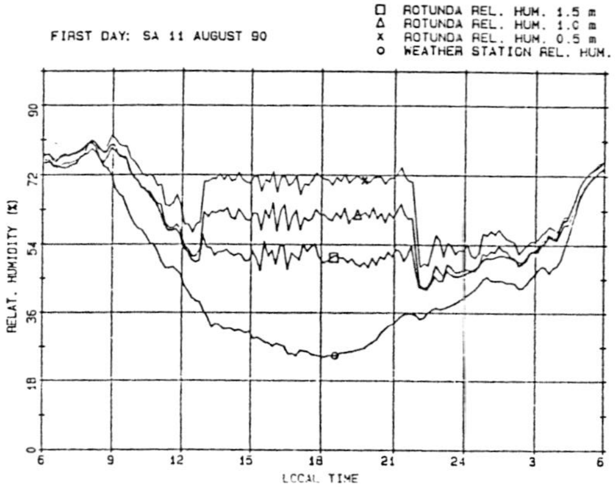

Figure 9: Relative Humidity in Rotunda at Differents Heights

THE OPEN SPACES OF EXPO 92

Jaime López de Asiain
José Pérez de Lama
José Mª Cabeza Laínez
Alberto L. Ballesteros
Arquitectos

Seminario de Arquitectura Bioclimática. Reina Mercedes, 2. 41012 Sevilla. Spain. Ph. 34.5.4556555. FAX 34.5.4231993

After the selection of Seville as the site for the Universal Exhibition 1992, it was necessary to decide upon suitable dates for a six months' event. A condition of paramount importance was to provide for the somehow extreme climatic conditions of the twon, under the starting point of 12th October, anniversary of the discovery of the New World by the Spanish vessels.

From the 15th July to the 15th September the climatic conditions in Seville remain well above the comfort limits and can be just intolerable. Not to mention that even in May, June and October temperatures can reach very high levels and become a nuisance specially for non-acclimatized persons.

These notions may have leaded to the conclusion that the appropriate period to celebrate the World Fair was from the 12th October on, trough autumn and winter but this could have been against the trend of international tourism and summer holidays; besides, the meeting with the Olympic Games at Bacelona was an unmistakable attractive for potential visitors.

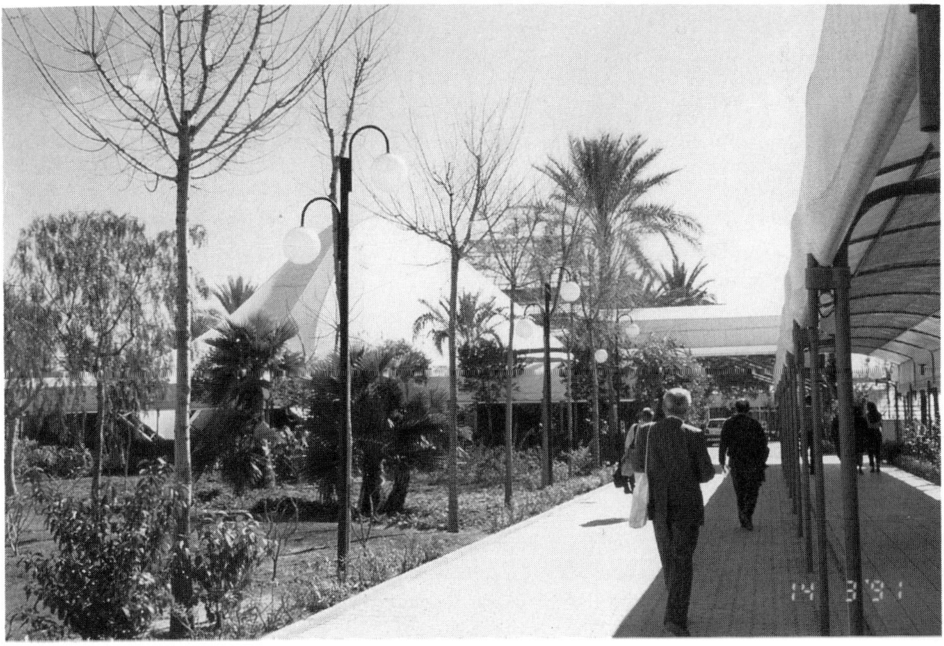

S. Alvarez et al. (eds.), Architecture and Urban Space, 217–222.

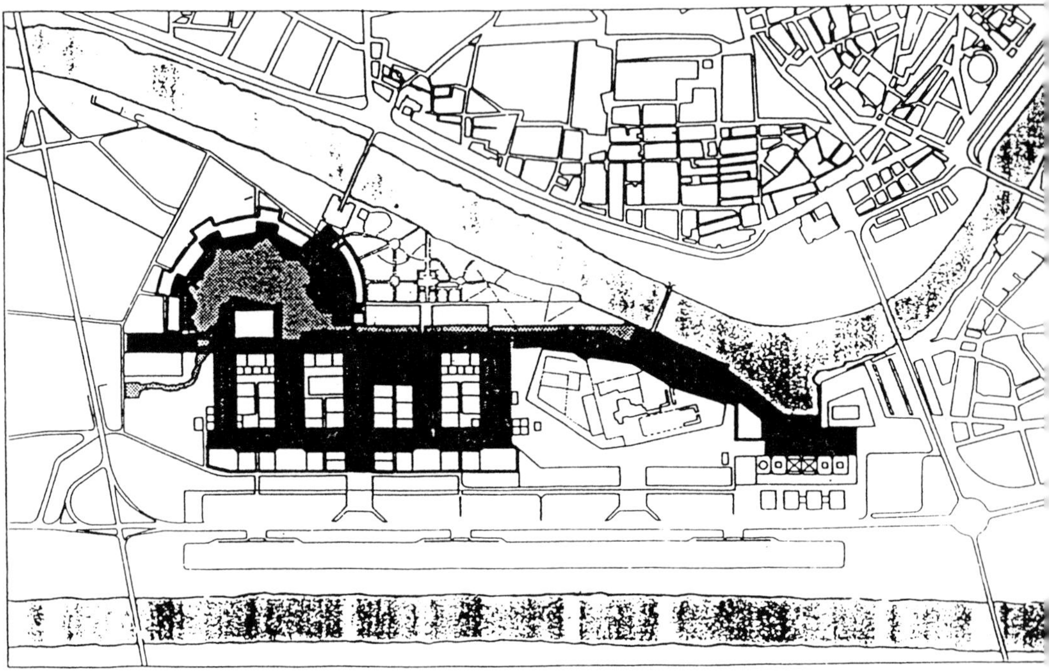

3.2 Open Space

The scheme utilizes groupings of pavilion parcels to organize larger open spaces in specific areas designated as high-traffic public zones. In these special areas EXPO will have more complete control of the environment and will concentrate resources to provide a unique ambience, special climate controls and many rest and leisure facilities. Major international pavilions will Look onto these special areas.

A team from the Bioclimactic Architecture Department of the Seville School of Architecture has studied ways to improve climate conditions during the summer months via the use of shade structures , special uses of water, concentrated vegetation and devices to promote air movement. These will be concentrated in areas where visitor density is expected to be high.

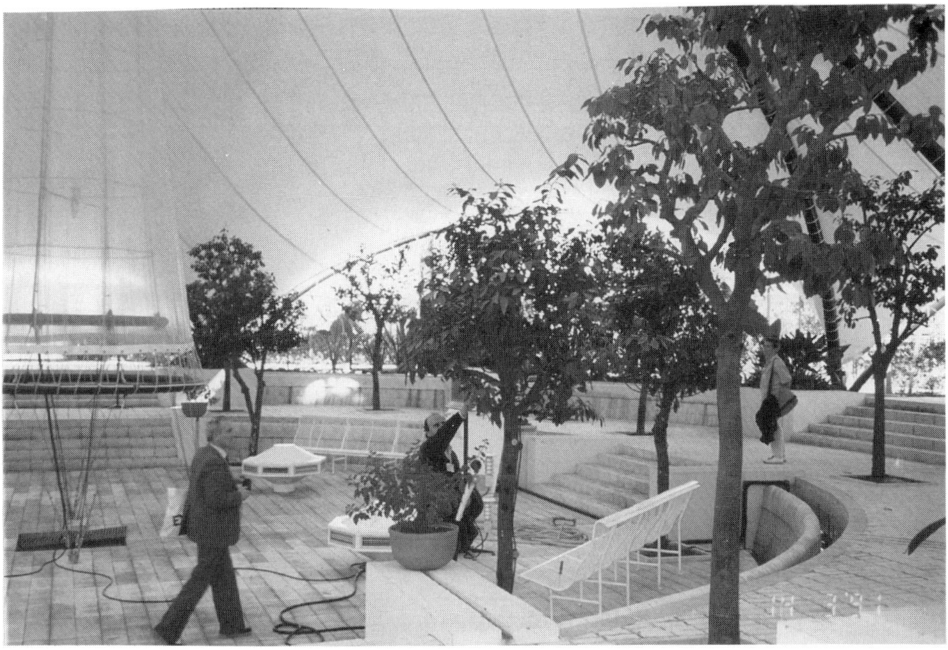

Therefore, it was decide that Expo'92 was to be held between the 20th April and the 12th October. Thus, the need for passive cooling of Expo spaces arose anew as a principle objective. From the precedent discussion the Seminar of Bioclimatic Architecture (School of Architecture of Seville) took an important part, and suggested through several issues the possibilities of passive cooling of the, by then, desert plains of the Island of La Cartuja and the necessity of massive forestation, an item that was rapidly taken into account by the team of consultants of Expo Commissionate, creating several greenhouses with more than 250.000 plants and trees.

However, from the moment that a steady decision han been taken, the question was growing in urgence together with the other aspects of organization, etc..., and so the design team in charge of the Director Plan whose head was, at that time, the renowned Canadian architect Richard Blackborney integrated the Seminar of Bioclimatic Architecture within the work.

At that time, the item of passive cooling of the outdoor spaces was developed in deep and was distinctly included in the Director Plan, issued in the Working Document number 5 of the State Society for the World Fair of Seville 1992.

As a parallel with this notion of outdoor cooling introcuded by SAB, the overall energy problem had been studied with that of the echological impact due to the World Fair. The Departments of Thermotechniques and Thermodynamics of the School of Engineering of Seville and the Department of Ecology of the Faculty of Biology collaborated with the SAB in the production of a series of issues regarding important aspects such as the convenience of developing a deep energy study, or the necessity to create a cooling network to avoid emissions of hot-air into the outdoor spaces of Expo, also the necesity to implement a meteorological station and monitor the climatic data around the island of la Cartuja.

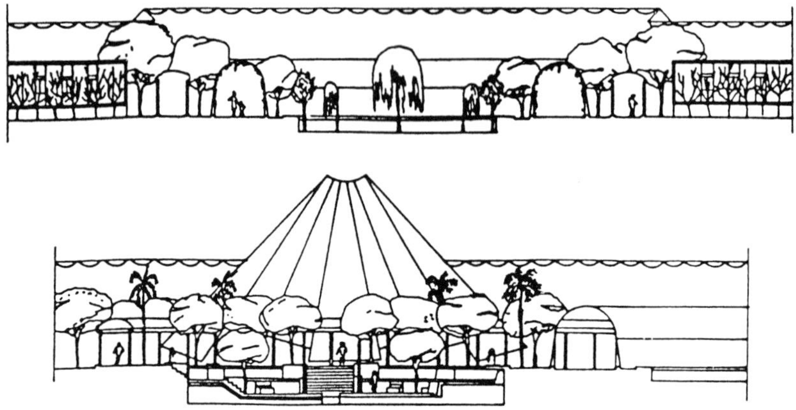

BIOCLIMATIC ROTUNDA. Sections.

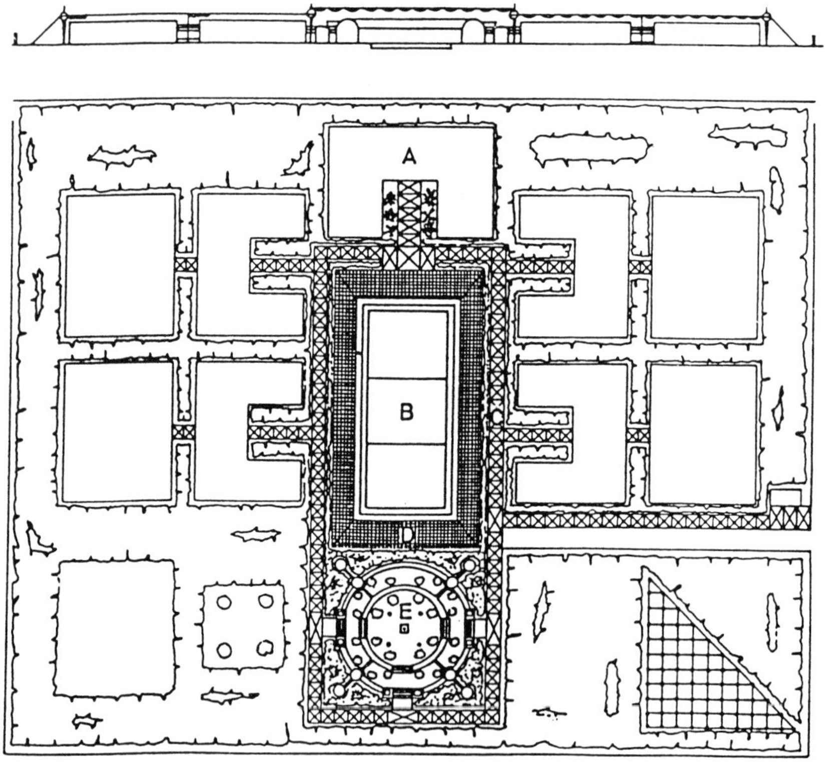

PILOT EXPERIMENT. A. OFFICES.B. POND. C. PERGOLAS. E. BIOCLIMATIC ROTUNDA.

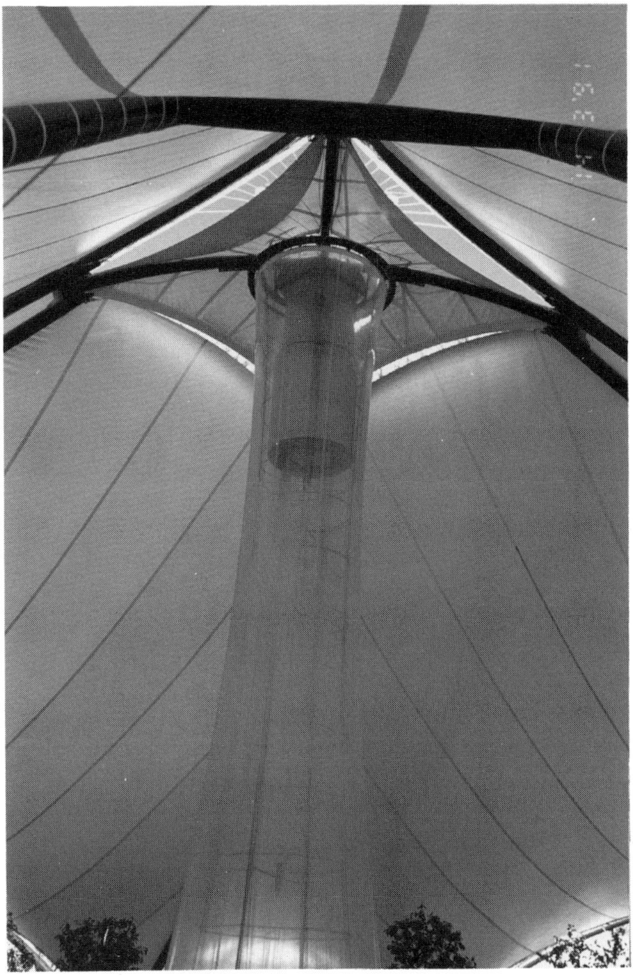

Everything was started at once, establishing the appropiate measurements. Nothwithstanding the true preoccupation of the Commissionate as Professor Olivencia revealed in many statements, was to settle down the required guarantees to ensure that the Avenues, pedestrian walkways, promenades, resting and waiting areas, gardens and so on...in Expo could be livable spaces at the hardest moments of Sevillian summer; whilst achieving this objective with entirely passive or natural methods.

These methods, on the other hand were related to the historical record of Andalusian architectural productions like the Alhambra in Granada, Medina Azahara in Cordoba or the Alcazar in Seville and directly entwined with the new ecological paradigma demanded by the culture of the 21st century, as defined by the Seminar of Bioclimatic Architecture in the guidelines released through the Director Plan.

In this aim, the SAB had a pilot experiment built, at full scale over a surface of more than 15.000 m2 containing the temporary offices of the diverse working teams of Expo. All this under the coordination of its Delegate Councillor.

This project commissioned to SAB and built during the years 1987 to 1988 has been the field for some of the most important experiments of bioclimatic treatment of open spaces in warm climates and the biggest in size ever realized in the world.

The bioclimatic spirit was well within the minds of the Expo responsibles, the public opinion approved enthusiastically though a little bit skeptically this intent and the architects of SAB faced one of the most interesting challenges that the Architecture of this century has presented at an institutional level.

Then, with a hectic timing, a whole group of interdisciplinar teams including architects, engineers, technicians, biologists and landscape experts started to work in coordination to finish a task that extends over a surface of more than 250.000 m2, creating the ideal conditions to receive 18 million visitors in 102 national pavillions, 60 enterprise pavillions, 17 regional pavillions, and showing what the "Age of Discoveries" and the past two millenium have been able to bring about.

We can proudly state that we first presented, impulsed and promoted this bioclimatic and ecological intelligence that today is sheer reality in Expo'92. Even when the final projects were not produced under the guidelines of the SAB and many of them seriously contradicted the former, next 20th of April the stage will be set as a field of discussion for the coming years.

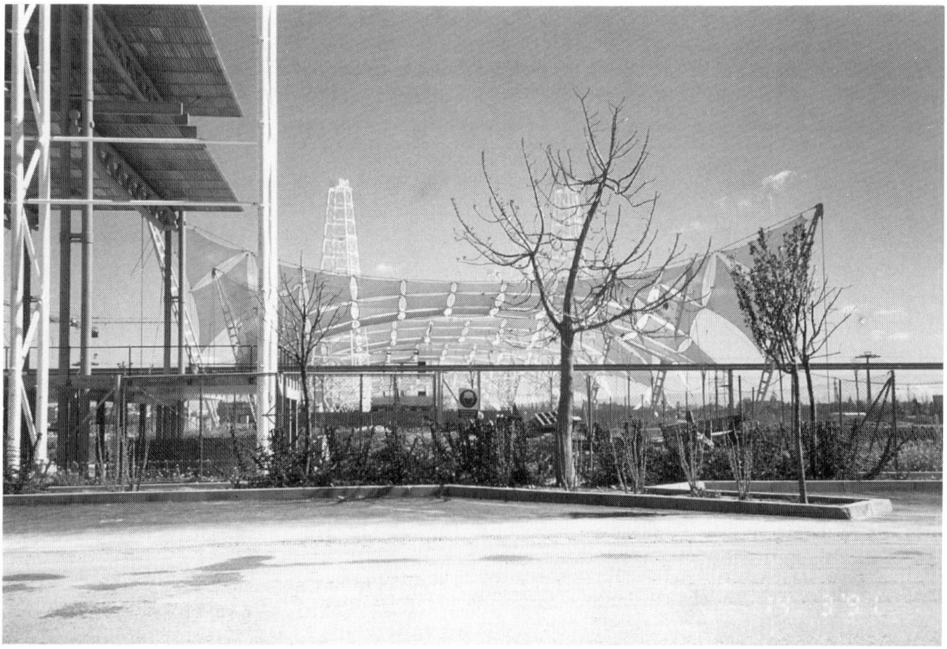

Part 3

BUILDING DESIGN

Chapter 1

Building Design Concepts

SMART BIOCLIMATIC HOUSE

ALEŠ KRAINER
University in Ljubljana
Department of Architecture & Civil Engineering
61000 Ljubljana, Jamova 2
Slovenia, YU

ABSTRACT. Design of living environment will demand in the near future the orientation on high-tech high-touch principle. His basic tasks are improvement of life quality, protection of the environment and rational use of the resources and energy. The pattern of the use and effective dynamic "passive" elements and systems, connected with the information technology, will probably create the new living environment.

1. BACKGROUND

The design of living and working environment ought to be the result of the interactive influences of site conditions. Living conditions of a certain site are formed by: 1. geomorphological conditions, being more or less unchangeable, 2. climatic conditions, being changeable and unpredictable, but in a certain sense harmonic and 3. men's behaviour patterns and living habits, being the consequence of culture and creativity, in this way changeable, often nonharmonic and unpredictable.

Connections of mutual influences and theirs diversity are the starting-points for the design strategy of bioclimaticaly oriented design on different levels: from the site and the building to the element and material.

If the demands and needs of users and geomorphological and climatic conditions in the natural environment are different and changeable, the response of the building on the identical conditions and needs must be the same: dynamic, changeable and adaptable.

A man used to consider this condition in the design of living environment during all his history: sometimes more, sometimes less successfully, but in many cases in a really genious way.

The basic condition for good operation (rational and har-

S. Alvarez et al. (eds.), Architecture and Urban Space, 227–232.

monized) is the consideration of conditions of system's elements interactivity and the consideration of the principle of space and time continuity.

2. CONCEPT

The recognition of this principle in buildings and settlements is one of the keys in the conceptualisation of new bioclimatic architecture. It can be achieved by moving of living and working zones between the opened and closed spaces in certain intermittent temporary periods, guiding light, heat and air flows, doing this thermal, light and air flows and atavistic conections with nature.

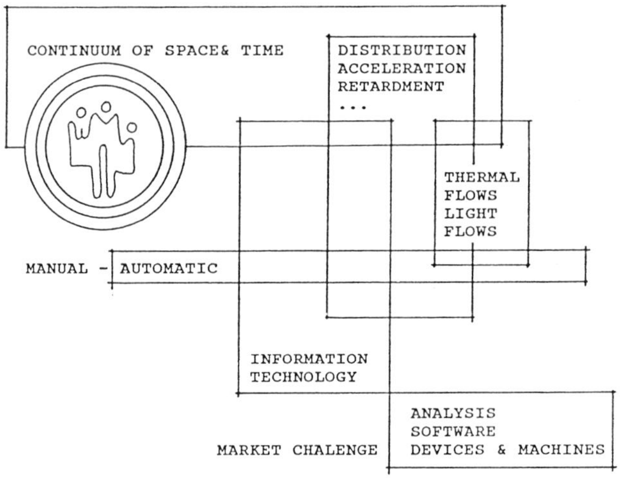

Figure 1. Interaction of the principle of continuum of space and time with the information technologies

The interval of tolerance between the natural and the artificial environment diminished more and more with the development of mankind.

Manual and trained guidance was sufficient to a certain level of the development. The development of living patterns, demands and technologies, parallel with the reducing of tolerances, can be no more efficient.

By definition unfortunately, represent the technolo-gies whose objective is the design of human friendly living conditions in physical, physiologic and cultural sense unfavorable charge to the natural ecological system.

The quality of the environment is the goal of the design of living and working space. The condition for such environment is, first of all, the smallest charge with with negative, harmful influences. Ideal thermal, light, sound, olfactory, and also cultural conditions are worthless in degraded and polluted air, water and food.

Between the living environment on one side and natural environment on the other continuous interface appears being formed by various materials, components, elements and entire buildings. Their different and changeable combinations represent new generations of building technologies. With dynamic control devices the adaptability of the built environment to the conditions of the outer environment is improving and necessary amount of supplementary energy is diminishing.

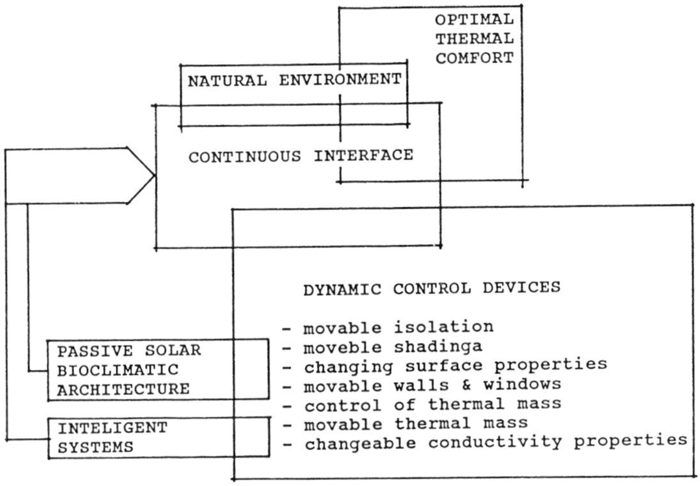

Figure 2. Model of the thermal part of kinetic architecture

To achieve this in as large extend as possible, it is smart to use besides good examples from history, also the possibilities of available new technologies and management systems.

The new interface appears in a contradictive conse-quence. One pole represents the requirements of comfort living and working environment, being assured by permanent constant energy flow through the system. The amount of energy is increasing together with the enlargement of the preformance conditions accurcy.

The second pole represents loads on the natural environment. The loads are increasing together with the rise of living conditions quality.

The goal of the interface is the regulation of differences between the natural and artificial environment. From the technological point of view we are in the first place dealing with thermal energy, natural light and air stratifi-cation. The level of energy flows needed for maintenance of desired conditions is in inverse proportion with the level of adaptability of the interface.

Its structure is formed by dynamic control mechanisms. From the material point of view these are different constructional complexes and their combinations, from movable insulation and shading devices, thermal mass, movable transparent and opaque parts of the envelope, to the control of the thermal mass and changeable properties of thermal conductivity and light transparency of building materials.

The system is formed by flows of information, energy and matter and by functional elements. Thermal and luminous levels and relative humidity together with the air exchange and contacts with natural environment, depending of psychological, physiological and cultural needs and/or demands of men, represent information source that stimulate corresponding responses of functional elements. Functional elements integrate and combine into the building. The building should finaly express comprehensive and a rounded off socio-ecological system.

The basic ecological systems activity criterion is the rationality and the equilibrium in the operation of its parts and complexes. The same criterion passes for socioecological systems whose parts are build environment and men.

The correct action of information system leading energy and matter flows and movements of men is a prerequisite for socio-ecological cycle to approach the natural ecological systems. The application of automation enables recent develop-ment of information technology and computers.

These facts shape the system of the smart bioclimatic house.

A new market was formed and the demand for research and development in the field of the analysis, model design and software, a new market for the application of new contexts in the field of existing and new technologies and a demand for new products.

The market of microelektronic devices, the market of appliances, of software, the market of redesigned classic building elements: windows, shading devices, ventillations, are tied up into the market of new living environment.

The development of the information technologies en-ables practically unlimited possibilities of management of different flows in an users friendly, balanced and rational way.

To create optimal thermal, visual and communication conditions in the environemnt where these conditions are not taken for granted we must design corresponding functional systems and supply necessary thermal energy, light and healthy environment for their operation.

These systems are not classical building systems of past centuries. They represent dynamic control mechanisms being able to utilise renewable energy sources and natural light. The consideration of these parameters in the design of build living environment is in accordance with principles of men's positive communication with his natural environment.

The technological solutions of the transport and distribution of energy and matter, connected with the natural and artificial environment by intelligent information systems are nearing the model of kinetic, responsive architecture.

The development of new technologies, where one of the goals is energy efficient design, is not an restriction, it is a challenge for architectural design.

The introduction of passive technologies into biocli-matic architecture is going on on the level of living space like a comprehensive system where performance problems, heating and cooling principles, daylightning, ventilation and physical shifting are solved. On the other side, the response of different materials, products and systems is treated from the point of view of special demands being represented by the reduction of thermal loses using new materials from aerogels and transparent insulations to PCM and cromogenics. The conductability of performance is very important if we want to achieve harmonic activity mutually with functional demands and needs.

Smart or intelligent handling organizes two sets of in-fluences. The first set is represented by the users demands. They are general and individual but both changeable, depending on live patterns of different groups on different levels (cultural, national, religious etc.) and on particular ways of use. Various kinds of use should not represent a disturbance in the system but an enrichment although they look irrational at first sight. On the other side cyclic and unpredictable changes in the natural environment take place. They are usually not in accordance with momentary demands and needs of the individuals and groups.

The increase of energy prices (direct and indirect ones naturally) is followed by the increase of the level of precision and complexity between the two sets of influences.

The application of information technologies and intelligent systems enables more efficient system performance.

In spite of the fact that in case of the smart house we are dealing with new and boundary technologies, where radical changes are forcasted until the year 2000, some significant changes in the basic concept of houses are not expected. Because of the fact that dramatic changes in the organiz-ation of the living space are not going to happen, important changes will appear in the first place in the building envelope structure and in the field of material properties.

For dynamic demands and for dynamic changes a dynamic building and its elements respectively are needed.

Different levels of design appear from the smart element to the smart house and settlement. The problem can be solved through parts functioning discrete, in the framework of systems where some parts unite and finally, global where the entire living space act as an interconnected and dependent organism on the principle of the biggest rationality.

The efficiency is the goal of performance of the three groups of influential parameters represented by comfort living conditions, higher life quality and intelligent systems. They are lying in the space, which is defined by cultural and physiological characteristics and demands on the one side and by technology level together with necessary energy with all positive and negative impacts on the natural environment on the other side.

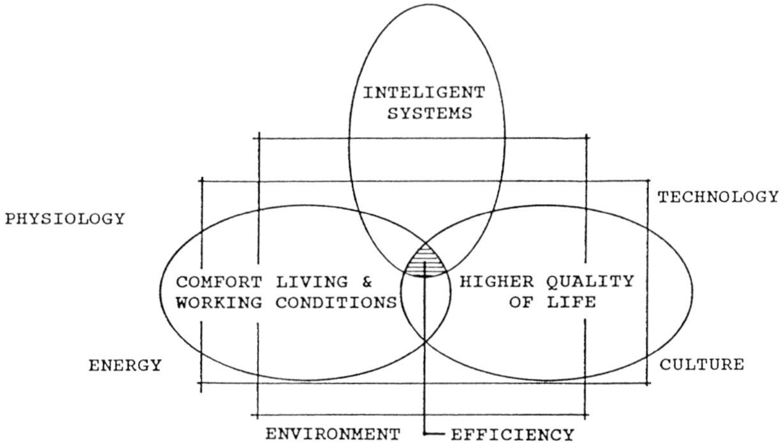

Figure 3. The cross-section of all participating domains is the efficiency

The basic characteristic of the system is its duality. It is acting in both directions. It takes care for higher life quality and for the lowest impact on the environment.

The building as a part of an ecosystem is the objective of efforts for bioclimatic design of smart living environment. High technology can help to harmonise the diversity of influences and conditions.

The flexibility of use being in the past a function of place and time becomes also the function of technical capability of appliances.

The development of new technologies and their appli-cation is a challenge, not a limitation, for architectural design. Energy efficient design is oppening new formal and functional possibilities.

The old romantic are old models, the new romantic are new contents.

THE HIDDEN DIMENSION (1)

An eye-opener for the architect

A.N.TOMBAZIS
85, Marathonodromou str.
GR-154 52 Psychico
Greece

ABSTRACT. How a building responds from the bioclimatic point of view is not as easy to perceive as other issues of architectural design. However this "hidden dimension", apart from being so basic, can add a further inner beauty to architectural creation.

One of the most basic concerns of architecture is that of the organization of space. It is so basic that one could even say that space is architecture itself. But what do we define as space? Space is the universe beyond the reaches of mankind, space is nature - the vastness of the ocean or the Himalayas - and the architect has nothing to do with these. Space is also our immediate surroundings at this very moment and this is intimately related to architecture.

The role of the architect is, amongst others, to create space, which really means to define its shape, proportions and size, which means its scale in relation to something else, usually the human being.

But what are the conventional considerations of an architect when organizing space? What usually influences its shape and size?

The first dimension is that of the purpose, or function for which it is being created.

The second dimension is the population density it is going to serve.

The third dimension is that of movement of people and/or goods and that means both direction and flow. From this point of view, space can be dynamic and/or static.

The fourth dimension is its sculptural quality as perceived from both the outside and/or the inside. Texture and colour come here into consideration. Hierarchy, the ordering or randomness of space could also be mentioned. In a general term we could call this group of considerations the aesthetics of space.

The fifth dimension is that of structure, materials and construction.

The sixth is that of the servicing of the space.

The seventh dimension are the psychological effects that the space will have on its users. Will it be calm and relaxed, dynamic, warm or cold?

I am sure that one could mention a few more dimensions that usually come into the consideration of the architect. But this would not really matter for the argument that I wish to develop.

Space can be completely enclosed with opaque surfaces as enclosures. In this

S. Alvarez et al. (eds.), Architecture and Urban Space, 233–238.

case, space will be completely artificially dependant. To the other extreme, space could be completely open with no enclosures at all. In this case, space would be exterior and completely dependant on nature. In between, there exist a whole range of degrees of enclosure and dependance on natural or artificial means of existence. The elements that define the degree of enclosure are the different kinds of apertures such as windows, doors, skylights, etc. glazed or unglazed. (Fig. 1)

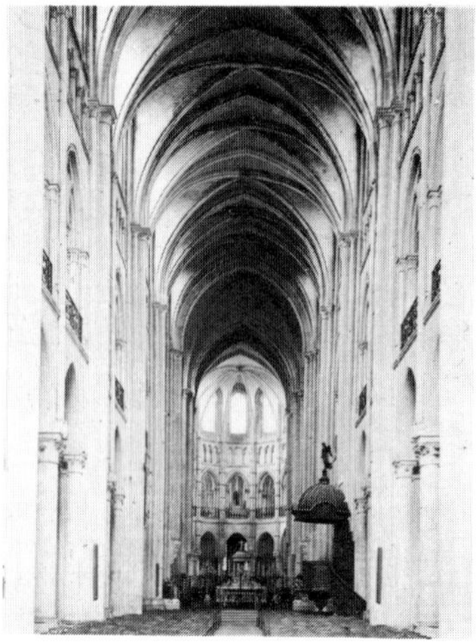

Fig. 1 One of the most basic concerns of architecture is to develop space. Space has many dimensions.
(Notre-Dame, Noyon 1145-1150)

But let us now consider the ingredients of space. In naming the main ones I will list:
- mass, the different materials that make up the different surfaces that define space
- gravity, which is the condition within which every element of space performs and lastly,
- void, which is space itself. One could say that it is what is left over, just vacuum, the negative part defined by the positive materials.

If one assumes that void is only a vacuum, then there would be no further reason for debate but reality is very different. As long as we are on this planet, void means AIR and it is then that a whole series of further considerations come into action. Considerations that are usually overlooked by the architect (but not the engineers). A whole series of issues that, if considered by him, usually enter the picture only in the form of technical considerations that influence architectural design but not on the same basis as the other ingredients already mentioned. It is for this reason that I wish to name air as the HIDDEN DIMENSION of architectural design. Air in the sense of this paper is really no more than the expression of the existence of the laws of physics and, in a broader sense, the relation between architecture, nature and the universe.
More specifically, the presence of air immediately means air movement due to temperature differences on a microscale or macroscale basis which in a building could mean within a room or throughout a building and its immediate surroundings. From there one is lead into the issue of when movement of air is wanted or not and what measures can be taken to induce or prohibit it. And this relates to the shape and proportions of a space, to the size and position of apertures, to their control devices, insolation and shading. And all these issues intimately relate to the actual form and appearance of the

building, if one limits oneself to the visual aspects only of the whole issue. (Fig. 2)

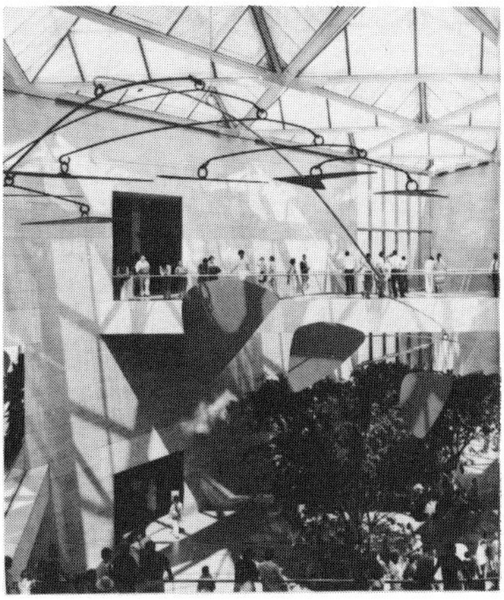

Fig. 2 Within every space there is movement of air that depends on many not so apparent factors one could call "the hidden dimension" (East Wing National Gallery, Washington, I.M.Pei 1968-78)

So this issue of sensitive bioclimatic design although in first appearance is "hidden", very much exists in reality and influences the organization of space just as much as many of the other issues that are more obvious to the architect.

There are a number of reasons why this hidden dimension tends to be overlooked by the architect during the process of architectural creation.

To start with, there often is the lack of knowledge or the necessary information. The basic training and education of the architect is much at fault on this point. Conventional training either ignored such subjects or treated them as purely technical which in my opinion is the wrong approach.

A second reason is the natural fear that the common architect has for everything technical, mathematical and calculable or quantifiable. These aspects he very often believes he cannot understand and leaves to the specialists and then usually it is too late!

A third and very important reason is that the architect often feels that these issues when related to the process of architectural design, will inhibit his architectural creativity and freedom of expression. He here takes a completely negative approach: "If something is going to be a nuisance, and what's more, restrict my creativity let's just pretend it doesn't exist". It is exactly the opposite that is true and could be the basis for a different approach: "Let's see how this new and different parameter can influence my design, what new ideas it can generate".

A fourth reason is a very practical one. This issue the architect can ignore while others even if he wanted to he cannot. To give you an example: the architect cannot ignore structural issues even though he does not make the calculations himself. His building would just not stand up and even if it was so simple that it would, this is not what he usually is striving for. The norm is that he is demanding from the structural engineer more that what can be done. In contrast, issues of sound bioclimatic design can, to a certain extent, be ignored by substituting their existence by increased energy consumption, as in the case of extra artificial lighting or air-conditioning. This not only can be done but for many years has been the main philosophy of our consumer society. So on this point the architect is not the only culprit.

Let us now look briefly at the process of architectural design in order to see when the bioclimatic issues should enter the process. Architectural design is an evolving continuous process. However it can be divided into two broad phases, that of preliminary design (including the initial conception) and that of design development. Bioclimatic issues should enter at both phases in a different way.

During the first phase, when the architect is (at least at the very beginning) on his own, it is his basic knowledge, interest and prior experience that count. The architect, although many would on different occasions disagree, is an animal of logic. Architectural design is like a crossword puzzle, a stepped process of action and recreation and inspiration too. The architect has been trained to reason, to solve problems and even more, when he has talent, to be inspired by other phenomena. It is for this reason that, at this point, when the basic seeds of creation are being sown which at a later stage will develop into full architectural creation, the architect, who is a generalist, posseses both the ground knowledge but is also motivated to include these issues in his vocabulary.

During the second phase (and of course even during the later part of preliminary design) the architect is no longer on his own but relates to the other consultants and engineers as a member of a team. However, as far as buildings are concerned, as the architect is usually the initiator and leader of the team, it is important that there develops a common language and understanding between him and the other consultants. This means that he understands enough of their problems and ways of thinking not to follow them, but, more important, to be able to foresee and guide them within the overall framework of which he must keep control.

There are many justifications and reasons for the promotion of a sensitive to the environment architecture. One is the enhanced comfort that this provides to its user[(2)]. Another is energy conservation. These two alone would suffice if not make it the architect's duty to make these issues part of his approach. But if we are to succeed, it is often better, further to reason, to address ourselves to emotion. When one is told all the time what is reasonable and what is not and even more so what is one's duty one tends to rebel, to leave duty to others who are more dutiful.
It is for this reason that I suggest that it is preferable and will be more rewarding to show the architect the inner, hidden beauty that can be added when bioclimatic issues take their proper place within the process of architectural design. My choice of the words "proper place" is intentional because I believe just as strongly that exaggeration is just as destructive. In fact I believe that one of the mistakes made up to now is that, when dealing with these issues, we have tended to isolate them from the overall aspect of architectural creation. Many so-called solar designs behave as if nothing else mattered, the mere reference to solar buildings is a misnomer, as if there are any buildings that are not solar! Environmental issues are soft and sensitive, they are part of a very complex and broad system and they must be treated within architecture as such, as part of a new set of issues.

I suggest two thoughts which could bring these issues close to the heart of the architect.

The first thought is that of harmony and harmony with nature and its laws and the laws of physics. Harmony is, after all, one of the basic concerns of architecture. In recent times architecture has become more sensitive to historical precedence. The man-made environment of the past, having been so brutally mutilated, due to our accelerated rates of growth and development has once again become a matrix of reference. The natural environment, climate and geography should become just as

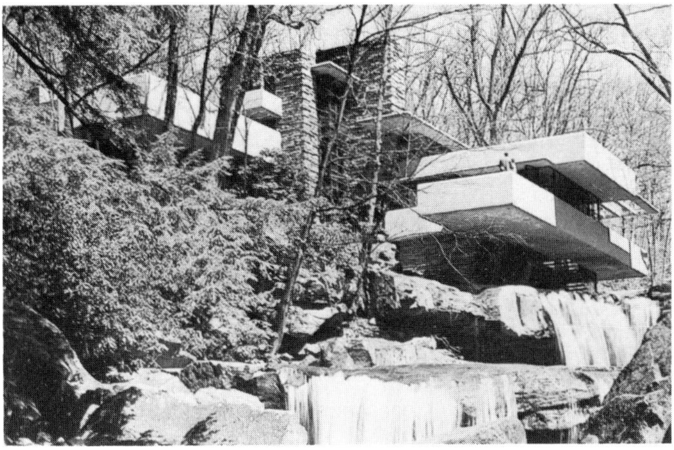

Fig. 3 Architecture in harmony with nature
(Falling Water, Bear Run - F.L.Wright)

much part of one overall matrix. This should not be approached in a sentimental or romantic way but in a practical but poetic manner. After all, no child has ever been convinced to eat up its food because there are other children starving somewhere else in the world. By this I mean that the beauty of nature that one should relate to is that of simplicity and the economy of appropriate means. (Fig. 3)

The second thought is that of natural lighting. Heating and cooling and all the other issues are of course just as important but in the end are taken care of to a great extent when proper daylighting is incorporated in a design. The reason for addressing ourselves to the issues of daylighting is that light is architecture itself. Architecture cannot exist but with light. It has always done so and great architecture has always existed within the vast range of sensitive lighting. It is dismal to find how much the issue of sensitive daylighting is ignored in present-day lay architecture. At the same time, since it is an issue that every architect understands even if he ignores it, I suggest that it could become the vehicle to both motivate the architect and achieve more sensitive architecture from the environmental point of view. (Fig. 4)

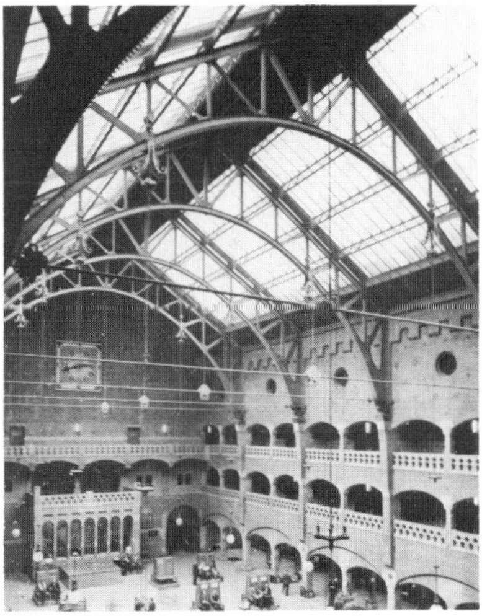

Fig. 4 Daylighting and Architecture are inseparable
(Stock Exchange, Amsterdam - H.Berlage - 1897-1909)

In conclusion, my point is: yes, in order to awaken awareness, at the beginning one may have to overkill, to give labels and to exaggerate. However this, concerning architects, has put many of them off just as much as it has made some of them more sensitive. My suggestion is to promote a more natural approach. The hidden dimension of bioclimatic design, however important an issue, is but one of many issues within the realm of architectural creation. One only, but one with a hidden inner beauty.

References

1. "The Hidden Dimension" Man's Use of Space in Public and Private by E.T.Hall, published by: The Bodley Head Ltd, London 1966.
   The use of the term "the hidden dimension" has been influenced by this wonderful book although E.T.Hall uses the term in a different meaning.

2. E.T.Hall (op.cit. page 60) states: "Man's sense of space is closely related to his sense of self, which is in an intimate transaction with his environment. Man can be viewed as having visual, kinesthetic, tactile, and thermal aspects of his self which may be either inhibited or encouraged to develop by his environment".

PASSIVE SOLAR FOR SMALL COMPACT BUILDINGS: A NEW DESIGN PARADIGM

DIMITRI PROCOS
Faculty of Architecture
Technical University of Nova Scotia
P.O. Box 1000
Halifax, Nova Scotia
B3J 2X4, Canada

Abstract

The Research compared according to solar orientation the space heating costs of 119 developer housing Sample groupings of about 100 residents each, located in the Ottawa Census Metropolitan Area. Heating cost information was obtained from the 1986 Canadian Census of Population.

The findings from Univariate analysis and Analysis of Variance were:

1) The older large house is the only beneficiary of location on East-West running streets.
2) Smaller houses of 6.5 to 8 rooms respond haphazardly to solar orientation.
3) Newer, relatively large and well-insulated housing built on narrow lots is not affected by street and house orientation.

These findings show that there is no appreciable financial benefit in persuading or requiring builders of ordinary market housing to orient their subdivision layouts for theoretical solar benefit. Instead, it is the appropriate design of the housing unit which should be emphasized. Three new design principles are advanced to suggest that, particularly for the compact house, passive solar advantages can be had from selectively opening up its well insulated building envelope to limited insolation. This new architectural approach is seen as part of a gradual progression of passive solar design from its early preoccupation with full southern exposure to more subtle forms involving internal daylighting and spatial design.

Implications for Architectural Design

The emphasis of the Research on housing produced without benefit of Architect does not strip it of design interest. On the contrary, its results point to important design initiatives. These begin with the acceptance that, for a number of reasons -by no means all of its own fault- the Architecture profession has not made inroads into a mass housing market which is not as indifferent to energy-efficiency as we like to believe. Insulation standards have improved over the years and many a builder brochure has included them in its sales pitch. The mass housing market supports solar accessories and cannot be said to be insensitive to the general passive solar ambience. The fact that its solariums may be net heat losers or that they at times end up on the wrong side of the house should, however, be seen by architects as a hopeful sign for an expansion of their passive solar role in the mass housing market. **The issue is for Architects to base their promotion of passive solar on existing design interest and performance in precisely the building type that has been found in this Research to stymie solar orientation: the Compact Building.** Before the means of achieving that are presented here, a brief overview of passive solar design practice in the recent past will serve to provide both background and contrast.

The earliest era of contemporary solar design, was characterized by two major tendencies: Firstly, the tendency to exploit the full southern exposure of the building with a combination of diurnal thermal storage, sunshading, air movement and insulation. The second -and related- tendency was toward outward reaching and expansive buildings, often for sunshading or equipment installation reasons.

S. Alvarez et al. (eds.), Architecture and Urban Space, 239–244.

In Figure 1 we begin to witness an early reversal of these tendencies. With the Crowther House (1b) we see the rays of the sun admitted through openings of limited size, mainly from the top of the building, then reflected, focused and made to heat and light the interior. Less introverted, Watson's Gross House (1a), with its lantern-like loft, shows the possibilities of passive solar design that is not wedded to the southern exposure. The third example of Figure 1, the Guest Quarters of the University of Berlin's International Meeting Center by Steidle (1c), with its rooftop conservatory and its stacked floor solariums, litterally adds another dimension to passive solar design, the vertical one.

Inherent in the examples of Figure 1 is the admitting of the sun from different points in the building envelope. Implicit in them are the possibilities of letting it play its role -whether thermal or just daylighting- deep in the interior of a building which, because of its compactness, affords limited peripheral solar penetration. Figure 2 embodies these possibilities in two examples, Herzberger's Central Beheer building in Appeldorn, the Netherlands (2c) and the Thresher Building in Minneapolis by BRW (2a). In both these buildings scale and land use posed a particular daylighting challenge, one which had been largely ignored in the early years of energy-conscious design. For, squeezed between the priority of reducing heat losses on the one hand and the need to moderate the glare that sometimes came with passive solar on the other, daylighting was once suppressed in buildings with large floor plates. But what of the building scale and type with which this Research has been concerned, that of the compact residential building?

The image at the center of Figure 2 shows the skylight of the recently completed "Hoagie" House by the Design/Build firm of Jersey Devil (2b). It represents a distillation of some of the design possibilities of small scale solar installations in compact domestic buildings. There is a rotating system which alternately concentrates sunlight in the interior of the house or repels it; there is a closure system to retain the heat at night; the skylight assembly imparts a definite form which is followed up by other elements of the house interior and exterior; and the daylighting effect is unmistakeable.

The "Hoagie" House provides a convenient quasi-technical, quasi-esthetic introduction to Figures 3 and 4, both dedicated to the building type that, according to this Research, has brought into question accepted theory about solar orientation: the Compact House. These Figures are an array of views of recent compact small buildings -not all domestic- which display a daylighting approach that could easily have been married to an explicit passive solar function. The fact that the readily recognizable ones among them are best known for their form suggests two things: **<u>One</u>, that the slippage of architectural design away from passive solar opportunities has nothing to do with real formal barriers; and <u>Two</u>, that by its very concentration of internal space, the compact house or small building intensifies the formal and visual impact of a passive solar design centered on daylighting.**

Nowhere is this more apparent than in an architectural transition which has seen a traditionally introverted but expansive dwelling archetype become more and more compact under conditions of land scarcity and cost: the Japanese house. Figures 3 and 4 show four examples of this new introversion: Hasegawa's design for a house in Nerima (3a) gives the basic outlines of the so-called "Three generation House", a three story, two unit building on a site as small as 150 sq. meters, with an inner atrium. This atrium re-appears in indoor version in Hexa Architects' exuberant "Chrysler Building" House, of which only the facade appears here (4c). A section-perspective of the Solar Chimney House by Yoshio Kato Atelier (3b) gives us an explicit example of how easily a passive solar function can be accomodated into this building type: a centrally placed solar chimney, "fed" by the sun primarily from the upper reaches of the house, culminates in a daylighting lantern whose form, while directly reminiscent of the Nerima House above, also takes us back to the Gross and Crowther Houses discussed here earlier. But it is Riken Yamamoto's Gully House that really transports us into the passive solar design paradigm that is being presented here.

The Gully House (3a) was given some prominence on Figure 3 not as a solar house, which it is not, but because it addresses solar penetration and lighting. It does so by incorporating into the building transparently roofed outdoor space that could readily have taken on the characteristics of a ground and upper level conservatory; by drawing that space into the interior of the house and by turning it into a form of outdoor -in this case- "den"; and by extensively using sky- and clerestory lighting to open up that interior. Turning such compacteness into an architectural virtue is not limited to Japan with its traditional land constraints. Recent architecture elsewhere has also sought compactness as its own reward in terms of interior space but also exterior form. Figures 3 and 4 focus on three western buildings to highlight that point. First, two houses by Gehry: his own, in Santa Monica (3b), and the Spiller House (3a), the latter a direct analog of the Japanese compact house, particularly of the "Chrysler Building" House. The third is is a somewhat unusual example.

The capping of a 18th century building with an aggressively transparent capsule housing company meeting facilities may at first seem un-related to this discussion. No thermal storage, nor for that matter solar or daylight penetration of the floors below, takes place and the design motivation is clearly formal rather than environmental. Then why is the Penthouse addition to the offices of a Vienna law firm by Coop Himmelblau (4c) being treated here as exemplary ?

In dealing with the transfer of research into the realm of design, one can be better served by examples of what could be than by a search for typicality. The Vienna Penthouse addition

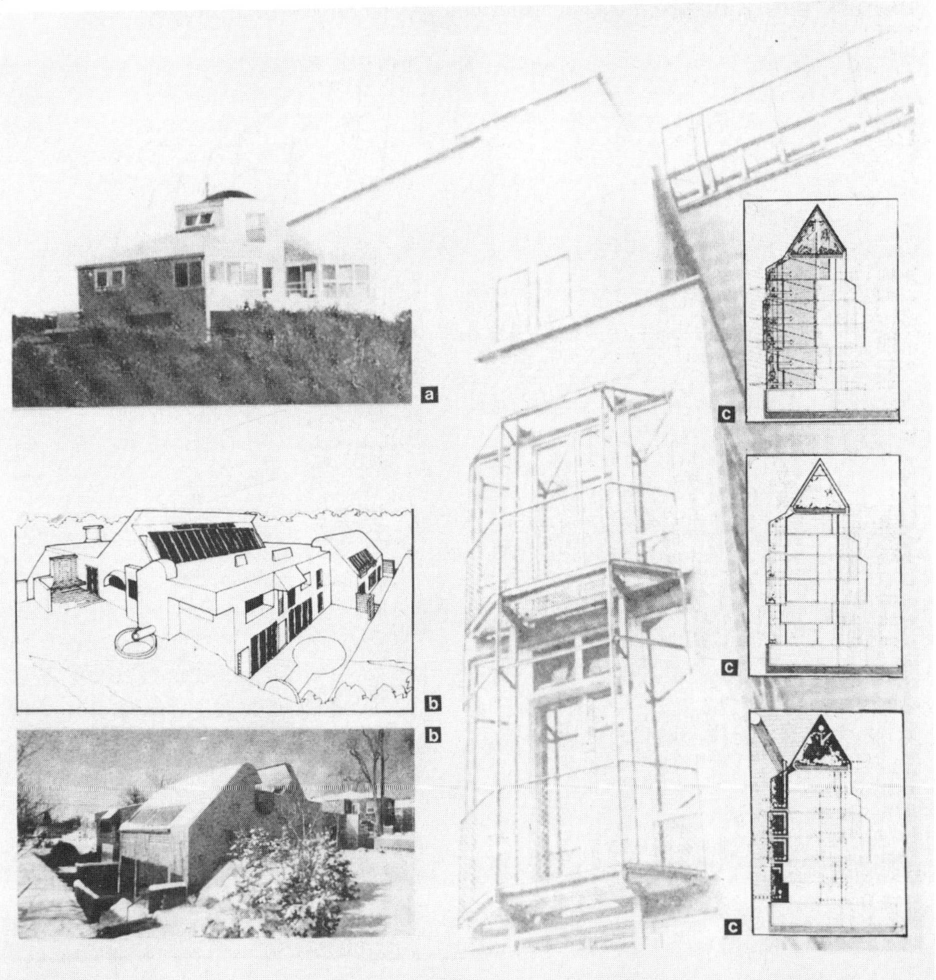

FIGURE 1

does that well, both by commission and omission. On the former count it strongly projects the two basic design principles advocated here as corollaries of the Research:

4) Solar admission should be independent of the street orientation constraints of the building. This is clearly represented by the Penthouse's irregularly disposed configuration.

5) Solar admission can take place through openings of varying size and location over those parts of the building envelope affording adequate solar access. Both the rooftop location of the Penthouse and the restraint in its scale support the point.

The omissions -from the point of view that animates this discussion- of the Penthouse are equally instructive. They lie primarily in the sharp contrast between an excess of glazing on the one hand and a total exclusion of the heat and light of the sun from the interior of the older building on the other. This observation, transported through many contextual and climatic layers to the scope of this Research, yields a third basic design principle:

6) Solar admission should be intricate to allow for a combination of the thermal, daylighting and visual functions of the sun with little heat loss.

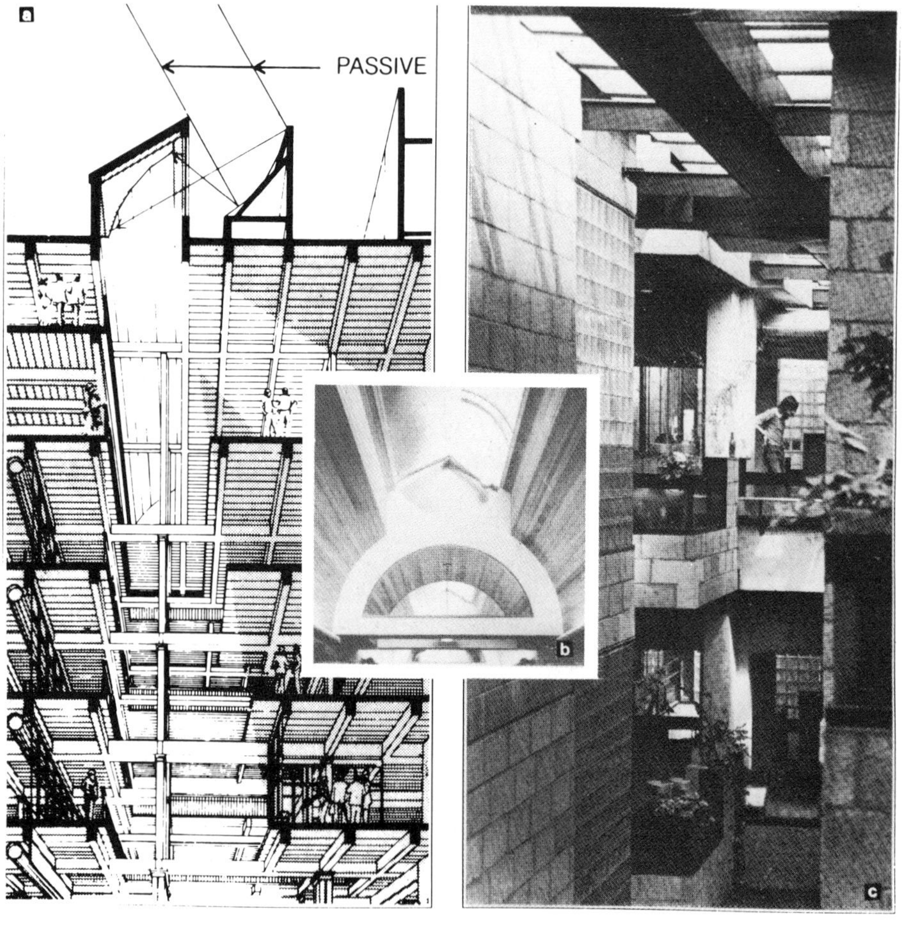

FIGURE 2

Conclusion

This Research Report has come a long way, from refuting some deeply held convictions about the importance of solar street orientation in the mass housing market, to advocating architectural design approaches to replace what may have been lost in the process. These approaches do not just reintroduce the sun into a building type that would appear from the initial Research results to be destined to exclude it; nor do they simply borrow fashionable building form to make passive solar design more palatable. What they do is exploit a coincidence: Recent design -driven by considerations as disparate as architectural form or land scarcity, but as yet not by solar ones- has nevertheless created the preconditions for the reintroduction of passive solar design in all its ramifications: daylighting, passive heating and cooling, heat loss control, all cast in desireable forms and spaces. And nowhere is this coincidence more apparent than in the small compact building, which already represents a major portion of the housing market. Architects who are seeking a larger share of that market would do well to base their efforts on the conjunction of solar technique with the spatial design possibilities inherent in this building type.

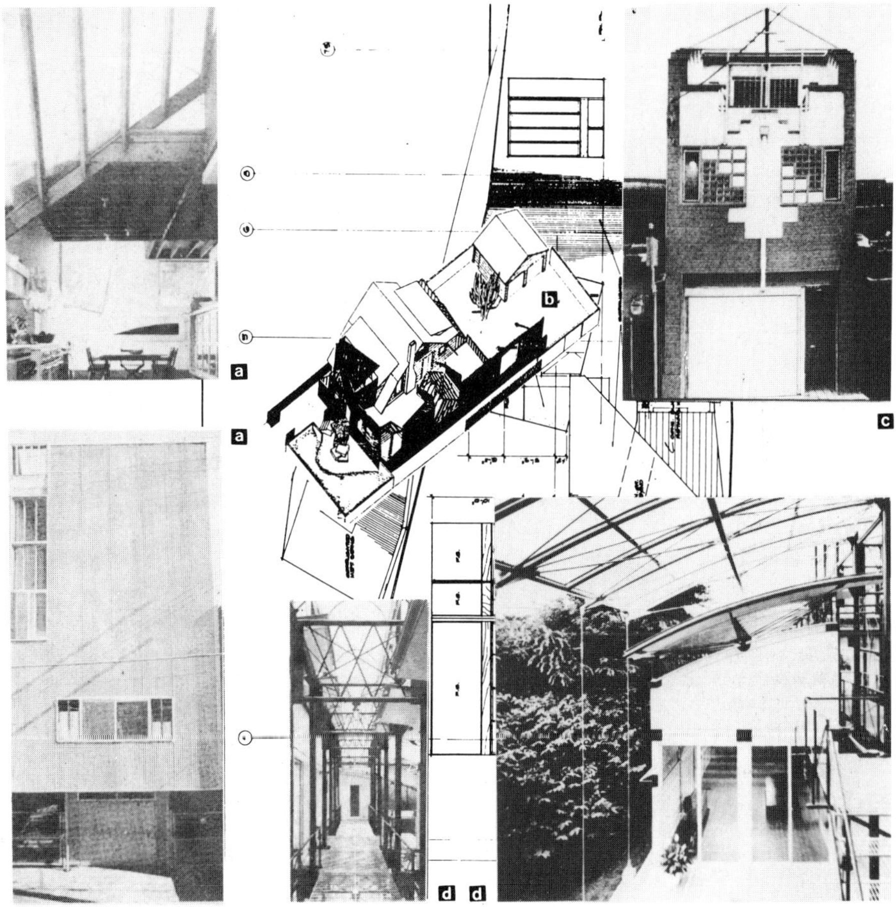

FIGURE 3

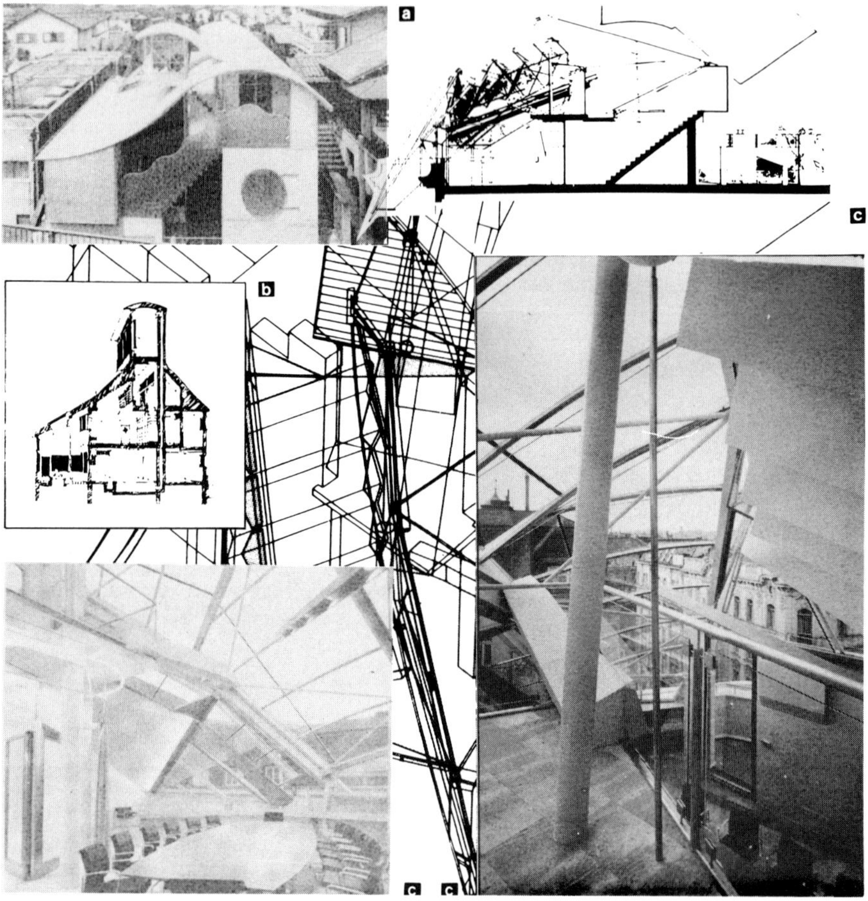

FIGURE 4

Illustration Credits

Fig 1. (a) Gross house, Block Island, Rhode Island. Architect: Donald Watson. PA, 9/85, pp.15-5
(b) Crowther House, Colorado. Architect: Richard Crowther. PA, 4/80, p.150
(c) International Meeting Center, Berlin.Architect: Otto Steidle. AdA, 9/84, p.42

Fig 2. (a) "Hoagie" House, near Washington, D.C.Architects: Jersey Devil. A,7/87. p.55
(b) Thresher Building, Minneapolis. Architects: BRW Incorporated. PA, 4/85, p.83
(c) Centraal Beheer, Apeldoorn. Architect: Herman Hertzberger. A+U, 12/83, p.39-74

Fig 3. (a) Spiller House, San Francisco. Architect: Frank Gehry. Gehry, Buildings and Projects,p.124
(b) Gehry House, Santa Monica, California. Architect: Frank Gehry. Ibid, p.143
(c) "Chrysler Building" House, Japan. Architects: Hexa Architects, Planners. JA, 4/86, p.101
(d) Gully House, Japan. Architect: Riken Yamamoto. JA, 8/89, p.62

Fig 4. (a) House in Nerida, Japan. Architect: Itsuko Hasegawa. JA, 8/89, p.62
(b) House with Solar Chimney, Japan. Architects: Yoshio Kato Atelier. JA, 2/86, p.50
(c) Remodelled of Office Building, Vienna. Architects: Coop Himmelblau.AR, 8/89, pp.82-89

BIOCLIMATIC CHARACTERISTICS OF SOME TRADITIONAL URBAN AND ARCHITECTURAL FORMS IN MONTENEGRO

D. VUKSANOVIC
Univ. of Titograd, Dept. of Civil Eng.
81000 Titograd, Cetinjski put bb
Yugoslavia

ABSTRACT. The purpose of research in traditional urban and architectural forms is to rediscover those forgotten attitudes and main elements of an anonimous builder's method. This method had enabled, in spite of poor technology, the architecture which was satisfying the needs and was adapted to climate and soil in a fascinating way. Analysis, presented in this paper, deals with development and spatial organisation of settlements and typical buildings, technological solutions and building materials, outdoor spaces and ambiental values. Chosen settlements are mostly of rural type, where environmental factors have had stronger influence on building design, than in towns.

1. Introduction

Teritory of Montenegro lays in the south part of Yugoslavia, between 41°52' and 43°32' of north latitude. This is mountainous terrain, which includes a part of the Adriatic sea coast.

Only the coastal part has been considered, where the sea moderates the climate, especially the winter temperature extremes. Mean annual temperature is 15.5°C with the monthly minimum in January 7.6°C, and the maximum in July 24.6°C. Mean annual duration of sun is 2700 hours and mean global solar radiation is 4200 Wh/m^2 per day.

2. Review of the Traditional Bioclimatic Forms

2.1. PASTROVICI (ADRIATIC COAST)

Compact groups of houses with the standardised shape, and the streets with little squares are the main features of Pastrovici villages appearence. These physical characteristics are inherent to the urban structures, what allows the conclusion that these villages represent the initial urban forms (Figure 1.). Their compact morphology was conditioned by the reasons of economical land usage.

The basic form of Pastrovici house originates from the primitive shelter, a single-pitch roof leaned on a rock. Influences from Venice

S. Alvarez et al. (eds.), Architecture and Urban Space, 245–250.

caused the development of the house in vertical direction up to the final solution with the ground floor and the two floors (Figure 2.).

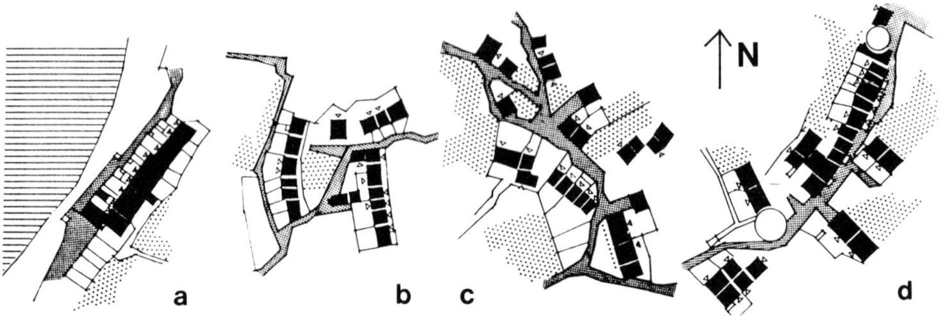

Figure 1. Urban dispositions of Pastrovici villages: a - Przno, b - Blizikuce, c - Celobrdo, d - Rafailovici

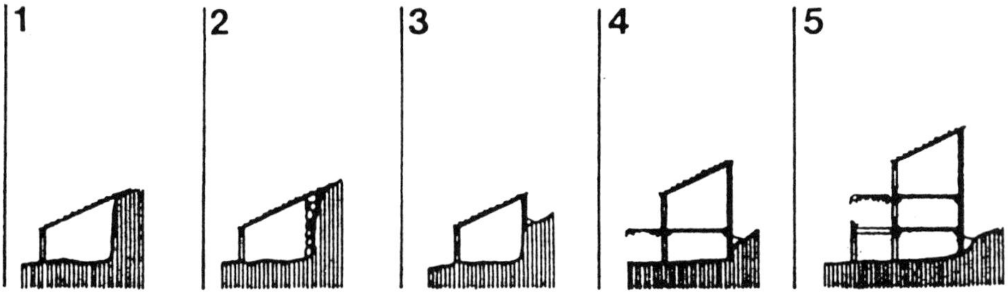

Figure 2. Development of the house in Pastrovici

The concept of a house is predominatly summer oriented, having its best interpretation in the large terrace ("taraca") with pergola ("odrina") in front of the house, on the first floor level (Figure 3.). Local climatic conditions allow that terrace can take place as a center of the social life during the half of a year - "a summer living room".

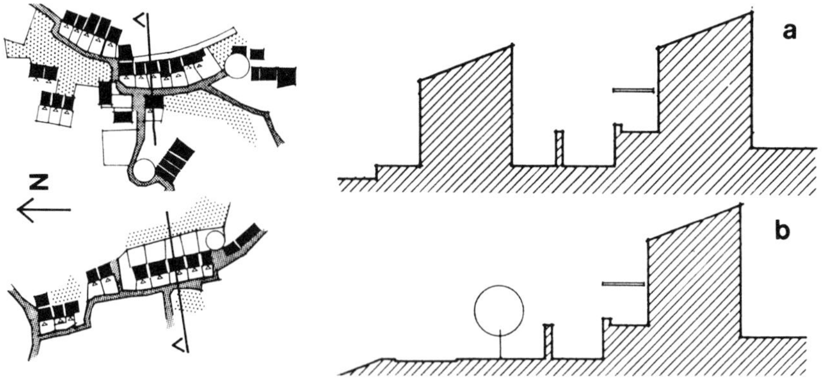

Figure 3. Position of the terrace in concept of the village street (a - Vrba, b - Tudorovici)

Very obvious traditional builder's bioclimatic solution is the single -pitch roof, which organicly follows the lines of relief by its form and slope. At the same time it is adapted to resist better the north wind ("bura") influences and the rain, carried by wind.

One of the examples which illustrates the authentic building sensibility for the landscape is the groupe of houses in Srzentici. It makes a kind of extension of the surrounding terrain by its bioclimaticly incorporated forms and contour, so it is almost difficult to distinguish the built structure from the natural rocks (Figure 4.).

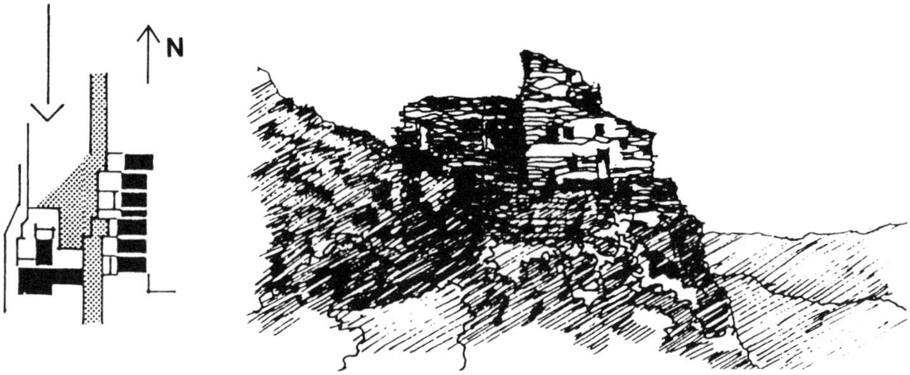

Figure 4. Bioclimatic relation of house grouping in Srzentici to the surrounding terrain

3.2. KRTOLI (ADRIATIC COAST)

Krtoli settlements, similarly to Pastrovici, have the dense spatial organisation. They are formed above the fields with tendention of radial expanding. Special variation represent the "shore magasines", arised from the needs of traditional trading, which was naturally oriented to the sea (Figure 5.).

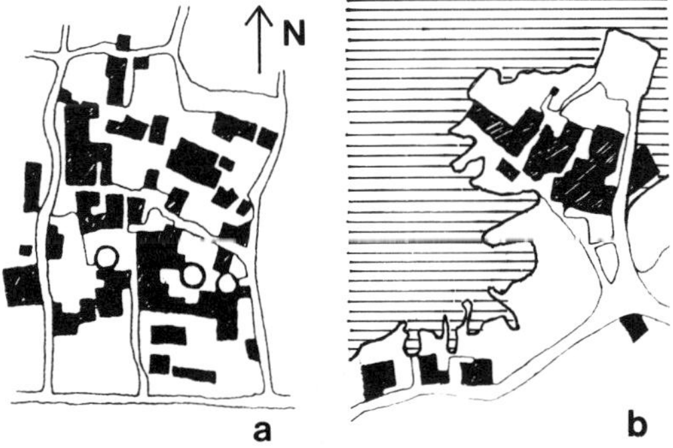

Figure 5. Urban dispositions of Krtoli villages: a - typical village (Djurasevici), b - "shore magasines" (Bjelila)

Houses in Krtoli are placed both longitudinal and transversal to the contour lines, depending on the hillside slope: on the steep slopes they have transversal position and longitudinal position on the mild slopes. The orientation is always to the South, Southeast or Southwest.

In the building's shape and in its relation to the terrain, especially in the case of transversal position to the hillside, the influence of Helenistic megaron is obvious. The stone wall fenced yard, as a part of the house, and the principles of house grouping originate from the Roman courtyard house (villa rustica) (Figure 6.).

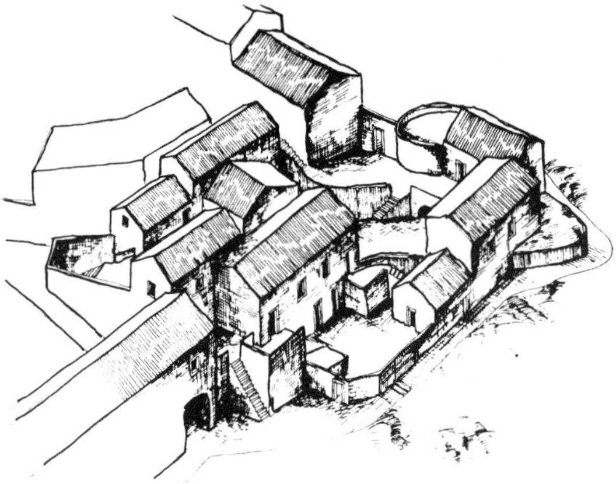

Figure 6. Example of housing units grouping (Bogisici)

These two story houses are made from stone, they have external stairway and the double-pitch roof. All building solutions are the result of adaptation to the local environmental conditions (Figure 7.).

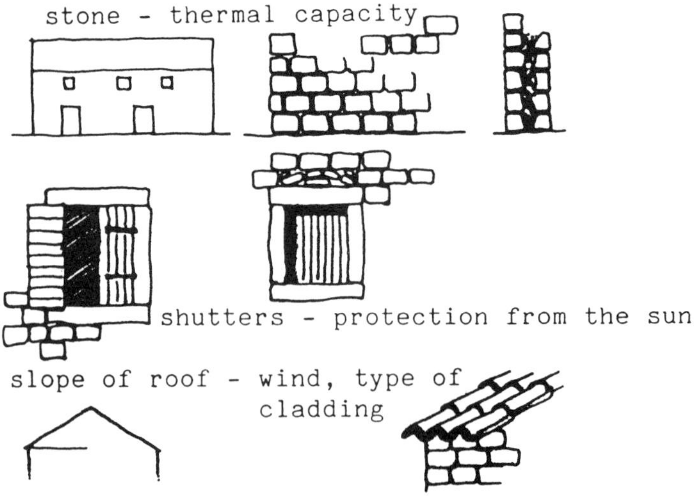

Figure 7. Bioclimatic character of the architectural elements

3.3. RIJEKA CRNOJEVICA (SKADAR LAKE BASIN)

The worth example of adaptation to conditions on the lake shore (seasonal flooding) and to summer heat conditions is accomplished in Rijeka Crnojevica. This little town was the main port and the trading center ("pazar") of Montenegrian principality in the XIX century (Figure 8.).

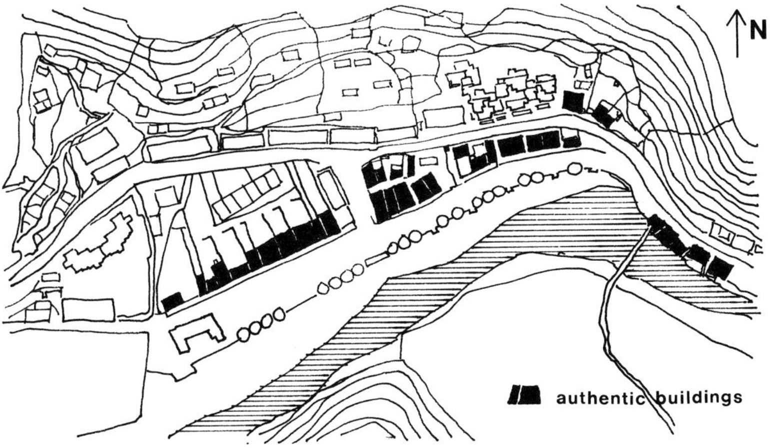

Figure 8. Urban disposition of Rijeka Crnojevica

The house in Rijeka Crnojevica represents the specific type of Mediteranien-oriental house. It has some features of the pile dwelling: it is placed near the water, so its first floor is flooded during the high lake level. As in usually functional organisation of the traditional house, first floor has economic or both economic and trading character and the flat is on the second floor (Figure 9.).

Large terrace under the roof, with position immediatly above the water, enables pleasant and atrractive stay during the hot summer days. As a open living room terrace represents a succesfuly modulated form on the interior - exterior relation.

4. Common Features

The reasons for going in for design concept of old traditional buildings are not in simple taking on their architectural forms. These well oriented, simple houses, built from local materials, located on the firm soil represent the worth models of energy and environment concious architecture.

Amazing reach in adaptation to the natural environment of the houses, their groupings and whole settlements originate mostly from their organic principle of growth during the time. All of these facts enabled the arising of architecture like "the crystallised piece of landscape" (Figure 10.).

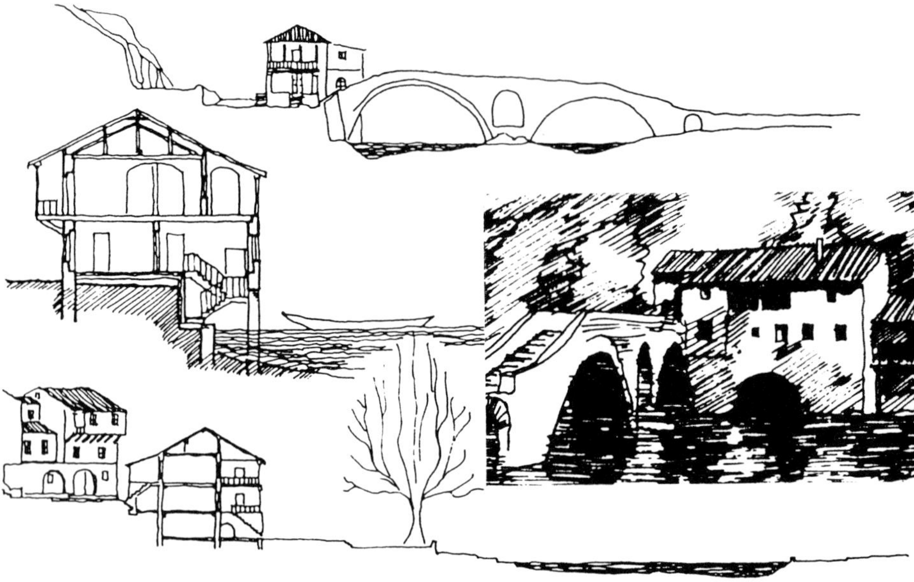

Figure 9. Characteristic architecture of Rijeka Crnojevica

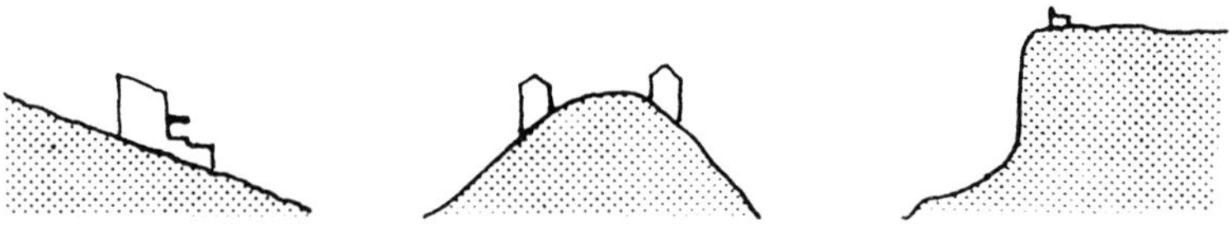

Figure 10. "Crystallisation of the landscape" (Pastrovici)

REFERENCES

Olgyay, V. (1962) Design with Climate, Princeton University Press, Princeton, N.J.

Krainer, A. (1986) "Equality in Variety, A Rewiev of Bioclimatic Growth of Buildings on Yugoslav Territory", Proceedings PLEA'86, Pecs, Hungary, pp. 168-219.

Gakovic, S. (1979) The House of Pastrovici, Center for Urban Development Planing, Belgrade.

Djurovic, G. (1981) Revitalisation Project for Thirtheen Ambient Locations in Tivat Community, Center for Urban Development Planing, Belgrade.

Vuksanovic, D. (1990) "Autochthonous Bioclimatic Building Approach", Proc. International Congress Energy and Environment, Opatija, Yugoslavia, pp. 259-266.

ATRIA-Temperate Climate Responses.

Graeme Robertson
Department of Architecture
University of Auckland
Private Bag
Auckland, New Zealand.

ABSTRACT. This paper reports on the continuing work in Auckland, New Zealand, to formulate design guidelines for architects operating in temperate climate zones to produce atria spaces associated with commercial buildings- particularly offices- which are energy efficient and architecturally plausible.

The apparent lack of recognition of the inherent characteristics of the atrium form has lead to the production of too many examples which , if not air-conditioned, are uncomfortably hot or cold; and if air conditioned are particularly large energy users.

It is argued that in temperate climate conditions the advantages of a non-air conditioned atrium are particularly attractive- and vital if adequate responses to Global Climate Change phenomena are to be adopted by the architectural community in the decades ahead. Work is progressing to finalise design nomographs for use by architects at the preliminary design stage, which is seen as the critical time in the process. The 'art' of the architect must prevail at this stage with fine tuning by computer analysis later in the design process. The complexity and often contradictory nature of the thermal design components in office space design is recognised. If energy efficient buildings are to result, then these components must be considered at the earliest stage or just energy efficient mechanical plant systems will result. Designer checklists are provided.

1 INTRODUCTION.

Climatic severity and the actual use of the building are important to determine the typical thermal responses of the proposed building- the basic question being asked is if the building is internal load dominated or external load dominated and for what period of time does each scenario

S. Alvarez et al. (eds.), Architecture and Urban Space, 251–256.

apply. In temperate climates where the cut-off point between the two basic systems occurs, careful consideration of this issue is necessary. Does the office space typically need heating or cooling and for what proportion of the day/year does this happen.
To many, the term 'commercial building' includes all buildings of an overall commercial nature ranging from offices to retail spaces to hotels. All are 'commercial' but exhibit wide variations in energy performance. We must be more specific in temperate climate zones and forever come back to basic definitions-

INTERNAL LOAD DOMINATED;
components; people, lighting, equipment;
strategy-reduce internal loads, lighting, cooling;
EXTERNAL LOAD DOMINATED;
components; conduction gains, conduction losses, infiltration, ventilation, solar gain;
strategy-reduce envelope losses, use solar gains, possibly use thermal mass.

In the New Zealand context this type difference is well exhibited by the NZERDC Report studying the Annual Energy Use for a range of building uses in two contrasting climate zones -
Auckland 37S° Hotels 550 MJ/s /annum, Offices 850 MJ/sqm /annum,
Christchurch 44S° Hotels930MJ/sq m /annum, Offices 370MJ/sqm /annum.
This shows that even allowing for possible variations in comfort standards the problem in Auckland offices, which are internally load dominated, is essentially a cooling one- quite at variance to other commercial building energy use profiles which tend to be externally load dominated. Throughout this paper, the term 'commercial building' is taken to mean 'office building'.

2 TYPE/FUNCTION.

In temperate climates all five well defined basic type of atria have an application-

Figure 1.

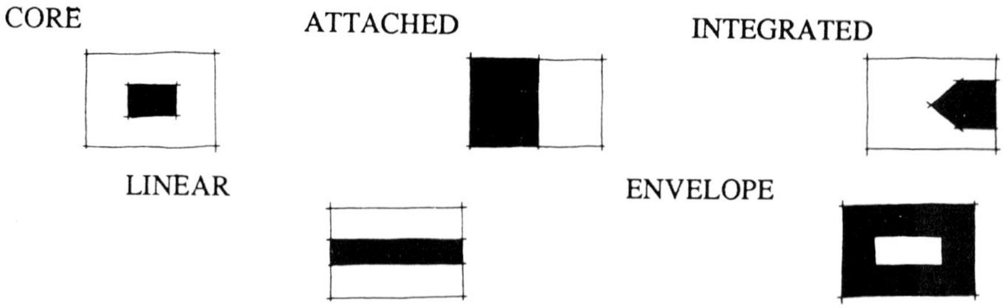

Each of the above offer a range of possible connection between building interior and atrium space with the degree of buffering with each being constrained by aspects of economics, atrium use, site and climatic severity. The acceptable comfort range within the atrium has a major bearing on type, economics and overall thermal efficiency. If the temperature in the atrium is allowed to run free then the auxilary heating/cooling demands will be zero. Whereas if the atrium temperatures are to be kept at precise levels equating to those in air conditioned spaces, then the auxilary loads and associated costs increase dramatically. In temperate climates the possibility of zero auxilary energy load and letting the temperature in the atrium run free is a legitimate design aim and is necessary if the full energy saving potential of the atrium is to be realised. Recent monitoring of a variety of un-conditioned spaces indicate the following for the two climate zones of Auckland and Christchurch Fig2/3.

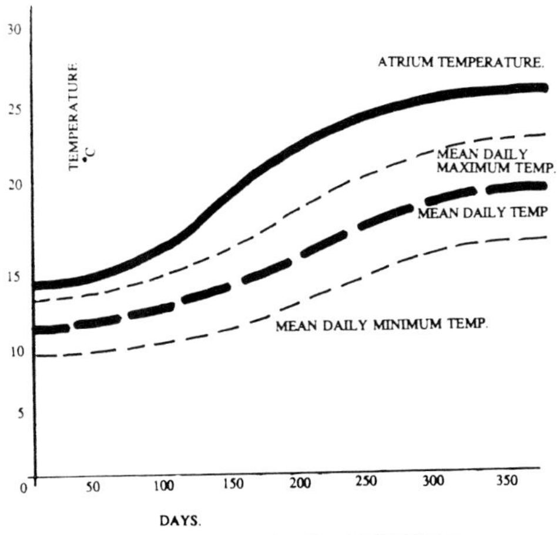

FIGURE 2 AUCKLAND ATRIUM.

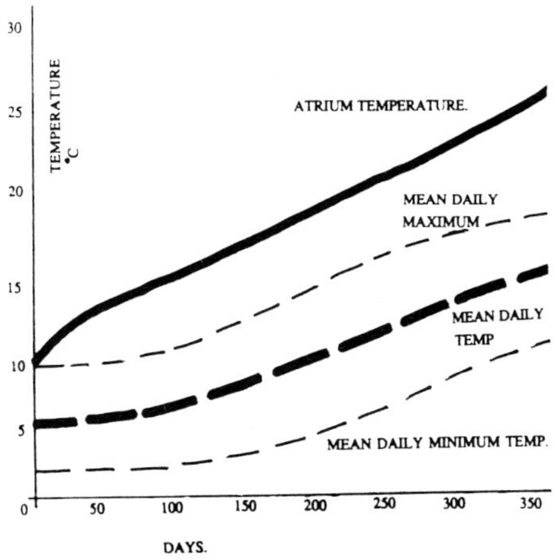

FIGURE 3 CHRISTCHURCH ATRIUM.

Considering the internal climate of the atrium, the above two graphs show -assuming reasonable design control- two things; the relative latitude available to the designer in the milder climate of Auckland and the relatively higher importance of atrium cooling techniques in the milder climate.

Also when the connection back to the surrounding office space is considered- Fig 4/5- the results are similar with need for heat loss reduction techniques- insulation, double glazing- to the Christchurch office space.

The allowable thermal climate conditions in both the atrium and the office space is a critical design decision. A wider range of thermal conditions is generally accepted in spaces that have a good connection to nature and avail the occupant of better control of their environment.

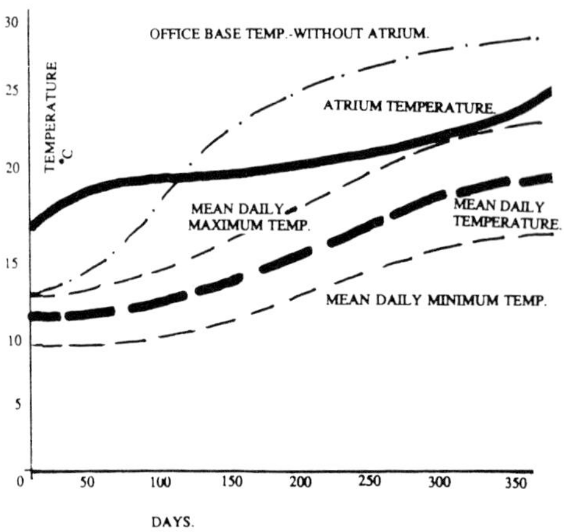

FIGURE 4 AUCKLAND OFFICES.

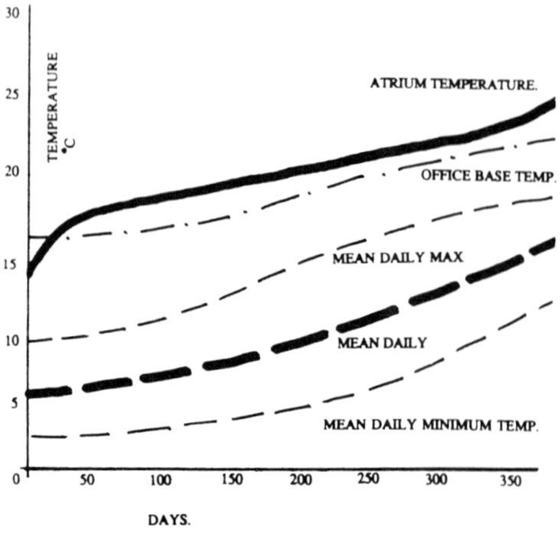

FIGURE 5 CHRISTCHURCH OFFICES.

3 TEMPERATE CLIMATE CRITERIA.

Without the extremes of climate common in other regions, architects operating in temperate climate conditions such as in New Zealand must exhibit care by not adapting the atrium form without carefully assessing the inherent advantages and disadvantages. Generally as the climate becomes less extreme the energy saving justifications of atria become less significant with many of the over heating potentials becoming more dominant if the atria is poorly considered.

The main architectural criteria for the design of atrium- connected office spaces not requiring auxilary energy to achieve acceptable comfort conditions, include-

DESIGN FACTOR	TYPE	NOTES
Building use-internal loads	High-people, computers,etc.	Internal loads may result in cooling needs.
	Low	External climate may become dominant.
Climate zone	Extreme temperate	Better need for atria.
	Mild temperate	Internal loads dominate with stable comfort needs easily met.
Atrium comfort	Small variations	Precise control- particularly solar and ventilation.
	Wide range	Truly passive atrium form is possible.
Office comfort	Small variations	Precise individual controls for fine tuning
	Wide range	Passive non-involvement will be possible.

Atrium configuration	Core	Mild climates with careful design of external wall.
	Attached	Minimal impact on office interior.
	Integrated	Better integration possible.
	Linear	More severe climates good connections.
	Envelope	Ultimate buffering form.
Nature of top skin to sky	Filter	Glare,solar gain and control possible-and needed in high-load applications.
	Clear	Over-heating/over-cooling potential high, therefore mass and ventilation important.
Orientation	Dominant external wall orientation is important to allow optimum solar control and ventilation inducement.	
Atrium width, height,volume	Large	Close to ambient, with a range of thermal conditions.
	Large shallow	Roof skin important with good daylight potential to office spaces.
	Deep	Reduced potential.
	Small shallow	Minimal impact.
Wall surface	Large solid area	Greater buffering potential.
	Small solid area	Less daylight reflectance.
	High reflectivity	Daylight to deep atrium areas possible.
	Low reflectivity	Reduced daylight potential.
Connection to offices, thermal and daylight	Direct	Maximum potential.
	Dilute	May be desirable if atrium thermal conditions fluctuate widely.
Natural ventilation	High	Always desirable
	Low	Dangerous but appropriate in high humidity,low load situations.
Insulation	High R value	Extreme climate locations.
	Low R value	Warm temperate high internal load situations.
Mass	High	Extreme climate
	Low	Most temperate climate atrium uses.
Atrium glazed area	High	High glare, solar gain and heat loss potential.
	Low	Passive solution which may be adequate.

4 CONCLUSIONS.

Whilst the above are presented as available design options, it must be realised that many are inherently contradictory in their impact. For instance, shading- increases lighting loads, diminishes cooling loads and increases heating loads. There must always be an analytical back up to test intuitive decisions taken at the earlier stages. However with so many atrium thermal disasters appearing in particularly temperate climates, an awareness of the design factors involved must produce a better more sustainable architecture. The well-conceived atrium has a place on the architectural palette of temperate climate designers. A truly 'regional' architecture will develop appropriate forms of atria associated with varying building uses.

5 REFERENCES.

Baker N.1988 .The Atrium Environment, Building Technical File No 21,U.K.

Bednar M J.1983. The New Atrium, McGraw-Hill.

Breuer D.1988. Energy Efficient Passive Solar Commercial Building Design. NZERDC 173, Wellington.

BOMA 1989. Optimising Energy Conservation, Seminar, Auckland.

Bryn I. Atria IEA Monitoring and Evaluation of 16 Buildings SINTES, Technical University of Norway, Trondhein.

Gentilli J.1971. Climates of Australia and New Zealand, World Survey of Climatology. Volume 13.

IPCC 1990 Working group 1 Report, Scientific Assessment of Climate Change.

NZERDC 1985. Commercial Sector Energy Use. Report 115. Wellington.

Robertson G.1990. Energy Efficient Commercial Buildings-A Realistic Market Objective. Proceedings ANZScA/ADTRA Conference, Sydney.

Robertson G. 1990. The Marketing of Energy Efficient Commercial Building Design. Proceedings C I B Symposium, Singapore.

Robertson G.1990. Survey of 150 Auckland CBD Buildings. University of Auckland.

U.S.A. D O E. 1983. Passive Solar Commercial Building Programme Case Studies. DOE/CE 0042.

LE CORBUSIER AND THE "MUR NEUTRALISANT:" AN EARLY EXPERIMENT IN DOUBLE ENVELOPE CONSTRUCTION

HARVEY J. BRYAN, Ph.D.
Graduate School of Design
Harvard University
48 Quincy Street
Cambridge, MA. 02178, USA

ABSTRACT. This paper presents a detailed discussion of Le Corbusier's concept of the "mur neutralisant" (neutralizing wall). It tracks the concept's development, from its first application in 1916 to its abandonment after it failed to gain acceptance for use in the United Nations Building in 1947. The paper concludes by discussing two completed and operating double-envelope buildings that have a number of similarities to Le Corbusier's concept.

1. THE EARLY EXPERIMENT

In 1916, Le Corbusier designed for the Swiss industrialist Anatole Schwob, a villa in La Chaux-de-Fonds. While the Villa Schwob is widely acknowledged for marking the transition between Le Corbusier's vernacular and modernist periods, it is also the first building to be completed since the development of the Domino concept (1914). The notion that Le Corbusier was most concerned with during this period was the use of reinforced concrete frame construction, which according to the Domino concept allows the wall to be liberated from its traditional structural role and thus becomes free to perform new functions. Le Corbusier began just such an exploration in the Villa Schwob by incorporating a thermal distribution function into the exterior walls. Le Corbusier would later term this strategy the "mur neutralisant" (neutralizing wall), which can be simply described as double wall construction with conditioned air being circulated in the cavity between the two walls.

This early version of the "mur neutralisant" incorporated heating pipes laid at the bottom of the cavity created by the double wall. Convection would heat the air within the cavity, raising the temperature of the surrounding wall surfaces. The temperature of the inner wall surface would be raised sufficiently to radiatively heat the adjoining space. Thus, by controlling the radiant fields, Le Corbusier introduced a level of comfort into this design that was unattainable by then current heating technology. While the Villa Schwob is thermally quite comfortable, this condition cannot be attributed only to the action of the "mur neutralisant" because there are other thermal phenomena at work.

S. Alvarez et al. (eds.), Architecture and Urban Space, 257–262.

Le Corbusier properly oriented the Villa Schwob to take maximum advantage of the sunny Alpine climate by placing its two-story living room and its large 5m x 6m window facing south; this, coupled with the thermal storage capacity of its building materials and the improved thermal resistance of the double wall construction, provides an excellent application of what we now term passive solar design (see Fig. 1).

2. THE EVOLUTION OF THE EXPERIMENT

Upon the completion of the Villa Schwob, Le Corbusier moved to Paris where he could engage in more challenging commissions. Over the next ten years his activities focused on theoretical writing, competitions and the design of several residences. The environmental concerns that he had begun to explore in the Villa Schwob, particularly the use of a south facing two-story glazed living room were present in a number of his designs of this period. In 1927, Le Corbusier brought together many of the ideas that he had been developing into the competition design for the League of Nations building. In this project the "mur neutralisant" evolved into an elaborate affair, with large mechanical ventilators blowing heated air into the cavity between the double glazed walls of the eight-story assembly hall. Le Corbusier spent considerable time on the technical aspects, working out many of the details of its design in consultation with physicist Gustave Lyon. From a historical standpoint this project is quite noteworthy because it is the first use of the hermetic glass curtain wall, which allows in light but not air, thus beginning the functional redefinition of the window.

Although Le Corbusier's design for the League of Nations building was never built, the notoriety he received began to attract larger commissions. In 1928 he was invited to Moscow to develop a design for the Centrosoyus building. In this design Le Corbusier re-used many of the features from the League of Nations design (see Fig. 2). However, the technical features that were proposed suggested an even more elaborate scheme than the one presented in the League of Nations design. Working again with Gustave Lyon, Le Corbusier introduced Lyon's scheme for mechanical ventilating the interior and the "mur neutralisant" for the perimeter of the building. These elaborate concepts caught the concern of the client, who requested additional information on its practicality. Hard pressed to find support, Le Corbusier turned to engineers at the American Blower Corporation, thinking that they would relish the idea of providing mechanical equipment for this innovative design. However, the American Blower Corporation could not find much support for his scheme. In a letter to Le Corbusier they state that ". . . your proposal will demand approximately four times more steam and more than twice the fan energy to heat and ventilate the building than that which would be necessary with current methods employed in our country to heat and ventilate a building exposed to the same atmospheric conditions." (American Blower Corporation (1930). Lacking credible technical support and reliable cost data, the client chose to employ only a double glazed wall.

Undeterred by the second failure to employ the "mur neutralisant,"

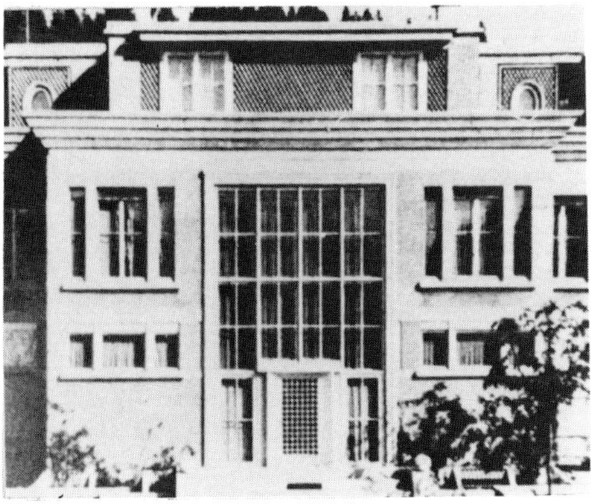

Fig 1. Villa Schwob - South Facade

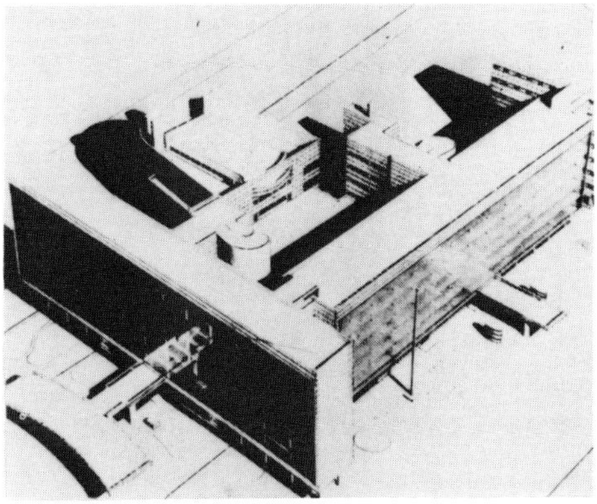

Fig 2. Centrosoyus - Perspective

Le Corbusier went on to propose the concept of "respiration exacte" (exact breathing), which was a closed fan driven heating, ventilating and air-conditioning system that was largely based on Gustave Lyon's ventilation work (Le Corbusier (1930). Like the "mur neutralisant" the "respiration exacte" is an important concept because it is the forerunner of today's hermetically sealed, air-conditioned buildings.

Le Corbusier lost no time applying these concepts to projects like the Cite de Refuge in Paris (1930) and the Palace of the Soviets in Moscow (1931). The Cite de Refuge, a multistory hostel for the Salvation Army, is the more interesting because it was actually built (see Fig. 3). Like in the previous building, the client was concerned about the concepts that Le Corbusier was proposing and wanted to see technical support. This time Le Corbusier approached Saint Gobain, the French glass company, to see if they would test the "mur neutralisant." Saint Gobain, intrigued by the idea, not only agreed to undertake testing but also to involve Gustave Lyon in establishing the testing procedure. An insulated 4m x 2m x 3m(high) test-chamber was built to house a 1.5m x 2.5m double glazed window that allowed hot air to circulate between the panes (see Fig. 4). This window divided the test-chamber into two small rooms; in the first outdoor temperature was simulated, and in the second the resulting indoor temperature. Tests were performed by placing dry ice in the first room and then measuring the change in temperature of the second as well as the energy used by the heater and fan. Performance was determined by comparing the above tests to a window that used a non "mur neutralisant" design. The resulting report was not very flattering to Le Corbusier's concept. While it did acknowledge that the warmer glass surfaces would improve comfort, it did so at a tremendous expense in terms of thermal and fan energy (Saint Gobain Glass Company (1932), thus confirming what the American Blower Corporation had told Le Corbusier two years earlier. This report and the high cost estimates that the client received for the "mur neutralisant" necessitated its rejection. What the client did agree to was a 1,100m^2 sealed single pane curtain wall on the south facade and the heating and ventilation but not the air-conditioning component of the "respiration exacte."

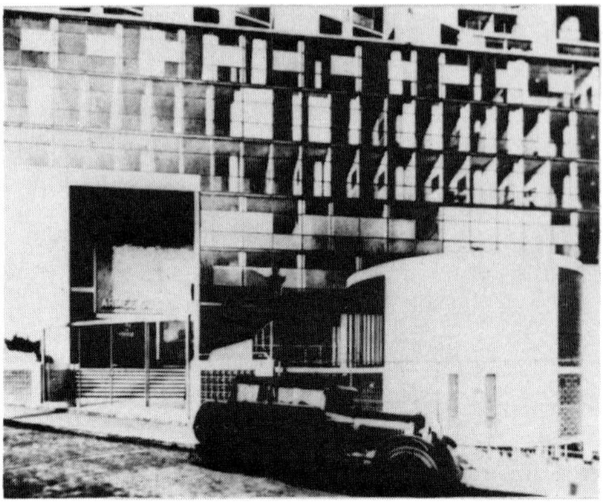

Fig 3. Cite de Refuge - South Facade

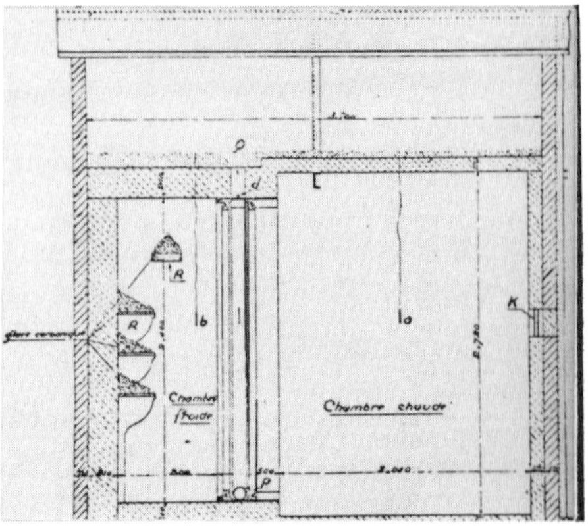

Fig 4. Saint Gobain's Test Chamber

On December 7, 1933, the Cite de Refuge opened with considerable attention being paid to its architect. However, within a few months the client was complaining about severe winter underheating and summer overheating, and proposed the installation of operable windows in the south facade to correct the overheating problems. Le Corbusier countered by strongly arguing against such a change, suggesting instead that the money would be better spent purchasing the air-conditioning equipment in order to complete the "respiration exacte" concept. The controversy continued until the summer of 1935 when Parisian officials threatened to condemn the building unless operable windows were installed (Taylor (1987). While the operable windows remedied much of the overheating, problems persisted until 1952, when a major alteration (suggested and partially executed by Le Corbusier) introduced sunscreens, opaque panels and additional operable windows to the south facade.

It was not until 1947, during the design of the United Nations Building, that Le Corbusier again proposed the "mur neutralisant." As the dominant member of a 10 member Board of Design, Le Corbusier found his design ideas being enthusiastically received by his colleagues on the Board (see Fig. 5), while his technical suggestions received little attention. This was particularly the case with his insistance that for this building to work from a comfort standpoint that it had to incorporate the "mur neutralisant." Late in 1947, as the "mur neutralisant" began to lose out to a single pane curtain wall and a state-of-the-art air-conditioning solution, Le Corbusier became bitter and wrote to U.S. Senator Warren Austin, President of the U.N. Building Commission, complaining that his ideas were not being given proper consideration (Le Corbusier (1947). His letter did not change matters and soon after he disassociated himself from the building (see Fig. 6).

What is interesting about Le Corbusier's proposal for the U.N. Building is that it marked an important transition in his thinking about the design of the building envelope. He never again would attempt to use the "mur neutralisant" and his subsequent designs started to rely less on mechanical devices and more on passive strategies such as the "brise-soleil" and natural ventilation through the use of clever

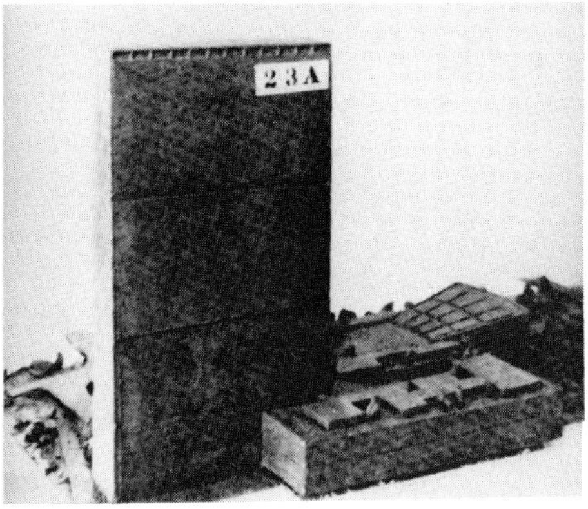

Fig 5. Le Corbusier's Model for the United Nations Building

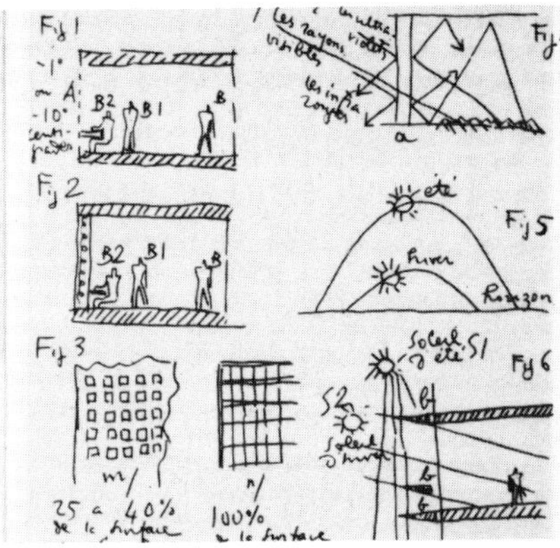

Fig 6. Drawings from Le Corbusier's Letter to Senator Austin

building cross-sections and operable windows.

3. THE VALUE OF THE EXPERIMENT

While historians such as Reyer Banham have dismissed the "mur neutralisant" as an "unreal and unworkable" concept (Banham (1969), I do find merit in the concept. I say this even in light of the negative evaluations done by both the American Blower Corporation and Saint Gobain because both of these studies neglected to take into consideration the solar contribution, just as Le Corbusier failed to understand how it was working for him in the Villa Schwob. If Le Corbusier had taken the sun into account he would have discovered that the issue was not how to blow hot air into the double glazed cavity but rather how to use the solar energy that became trapped within the cavity. Such a version of the "mur neutralisant" would utilize a double glazed wall with operable louvers at the top and bottom and a horizontal grate at each floor which also work as a sun shading device as well as allow air to move vertically. In the winter, this wall would work by reducing the building's heat loss by having both vents closed, thus trapping within the cavity the solar radiation comming through the outer pane. In the summer, both vents would be open and the shading devices would prevent solar radiation from striking the inner pane. The radiation coming through the outer pane would cause expansion of the air and rapid exiting out the upper vent which would result in cooler outside air being drawn into the cavity.

Additional evidence as to the feasibility of a modified version of the "mur neutralisant" is the fact that there have been several window prototypes as well as buildings completed over the last decade which utilize such a strategy. The window prototypes range from the "air-flow window" which is beginning to see application in Scandinavia to the "return-air window" recently used in The Lloyd's of London Building. The buildings can all be classified as having double envelopes and range anywhere in size from small institutional to large commercial office buildings. While there are a number of these buildings in operation, space limits the mention of only two.

The first is the Equitable Life Assurance Society Building in Charlotte, N.C. (1977). The envelope of this building utilizes several horizontal glass plenums called "solar belts" that surround the perimeter and act as the neutralizing cavity as well as form the route by which solar heated air moves around the building in winter and from which interior exhaust is vented in summer. While the building performs well, considerable energy is needed to move the air horizontally through the solar belts.

The other is the Occidential Chemical Building in Niagara Falls, N.Y. (1981). The entire envelope that surrounds this building is comprised of two glass curtain walls which create a 1.5m x 9-story neutralizing cavity on each of its four orientations. Open louvers at the top and bottom of the 9-story cavity and grates at each floor allow solar heated air to vent via the chimney effect in summer, while closing the louvers traps the warm air in winter. Although the building is performing better than design estimates had predicted, the building's cube-shape and identical treatment of each facade suggests a building that is insensitive to orientation.

4. CONCLUSION

While Le Corbusier misunderstood the physical principles that were at work in the "mur neutralisant," he at least was a protagonist in the technological debate. He was not content, like so many present-day architects, to let the consultants or manufacturers solve the technical problems for him. The two recently completed double-envelope buildings are cited not as a validation of the "mur neutralisant," but rather to suggest that the questions concerning double-envelope wall construction with which Le Corbusier was so concerned are still with us and need to be investigated further.

5. REFERENCES

American Blower Corporation, (1930) Letter to Le Corbusier dated January 24, Foundation Le Corbusier, Paris.

Banham, R. (1969) The Architecture of the Well-Tempered Environment, University of Chicago Press, pp. 159.

Le Corbusier, (1930) Precision, Paris, pp. 64.

Le Corbusier, (1947) Letter to Senator Warren Autin dated December 4, reprinted in Architectural Record, July 1950, pp. 69-71.

Saint Gobain Glass Company, (1932) Verres et Glaces, Aug/Sept.

Taylor, B. (1987) Le Corbusier, The City of Refuge, Paris 1929-1933, University of Chicago Press, pp. 115.

RESIDENTIAL DENSITIES AND APPLICATION OF PASSIVE SOLAR SYSTEMS

JOVANOVIĆ POPOVIĆ MILICA
University of Belgrade
School of Architecture
Bulevar revolucije 73
11000 Belgrade, Yugoslavia

ABSTRACT. The aim of investigation was to determine the possible residential density of multi-storey residential buildings with applied passive solar systems and to make a comparison with the conventional approach. The dwelling density for the given location is determined mathematically for different building size on the basis of the accepted standard apartment area.

1. PROBLEM DEFINITION

The residential densities and the corresponding "density index" are specified for cities in Yugoslavia by Master and Detailed Urban Planning Documents.

The residential density is defined as the ratio of the number of residents to that part of an urban area reserved for residential buildings and local driveways. The density index, as a density indicator represents the ratio of the total floor area of residential buildings to the location area. This index thus directly depends on the number of storeys. The accepted standard in the city of Belgrade is a net apartment area of 21 m2 per resident.

There is a general opinion in practice that such a high residential density, especially for buildings over 5 storeys, cannot be provided with passive solar buildings due to significant distances between buildings imposed by the specific geographic latitudes, as is the one of Belgrade. For this reason the passive solar concept is presently excluded from development of urban design projects, defining the exact location and orientation of buildings, their shapes and sizes and their mutual relationship. Such a practice makes it virtually impossible to apply passive solar design principles later on, at the stage of architectural building design.

The purpose of this paper is to prove that it is possible to build multi storey residential settlements at the specific geographic latitude of Belgrade, applying the passive solar design principles

S. Alvarez et al. (eds.), Architecture and Urban Space, 263–267.

while at the same time complying with the given residential density standards.

2. ANALYTICAL METHOD

The dwelling densities have been calculated under the following assumptions:

- the geographic latitude for the considered case study is $44^{o}48'$ (city of Belgrade);
- the topography is essentially a flat one; for the South-oriented slopes, the residential densities can be expected to be higher, and the opposite is true for the North-oriented slopes;
- all buildings are daily sun exposed from 9 - 15 hours (winter solstice);
- the accepted resident standard is net apartment area of 21 m2 per resident; considering the legally prescribed ratio of the net floor area to the total floor area of 1 : 1.3, the total floor area per resident is about 27 m2.

The building size has been varied in the following way:

- the building width A = 10, 15, and 20 m;
- the building length B = 15, 20 30, 45, 60, and 90 m (high values stand for rows of buildings);
- the number of storeys n = 1 - 9, with the storey height of 3 m.

The residential density can be calculated by the formula:

$$g = \frac{A\ B\ n}{27\ P} \times 10^{4} \quad \text{(residents / ha)}$$

where P-sum of areas of a building and its shadow (Fig. 1). No other building is to be constructed on this restricted area.

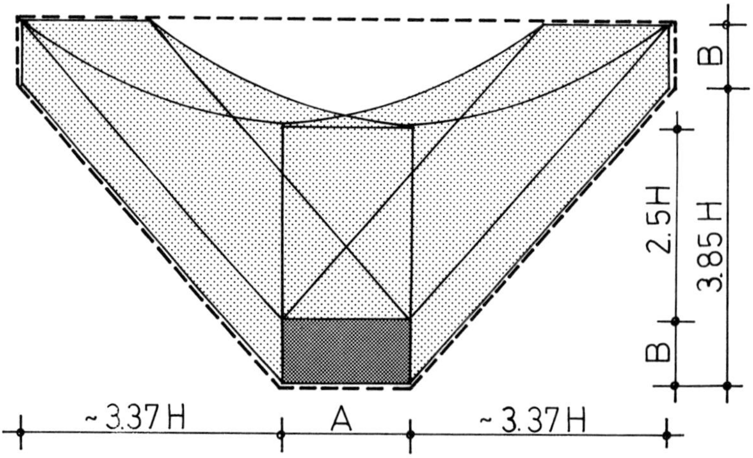

Figure 1. Restricted area for a single building (geogr. lat. $44^{0}48'$)

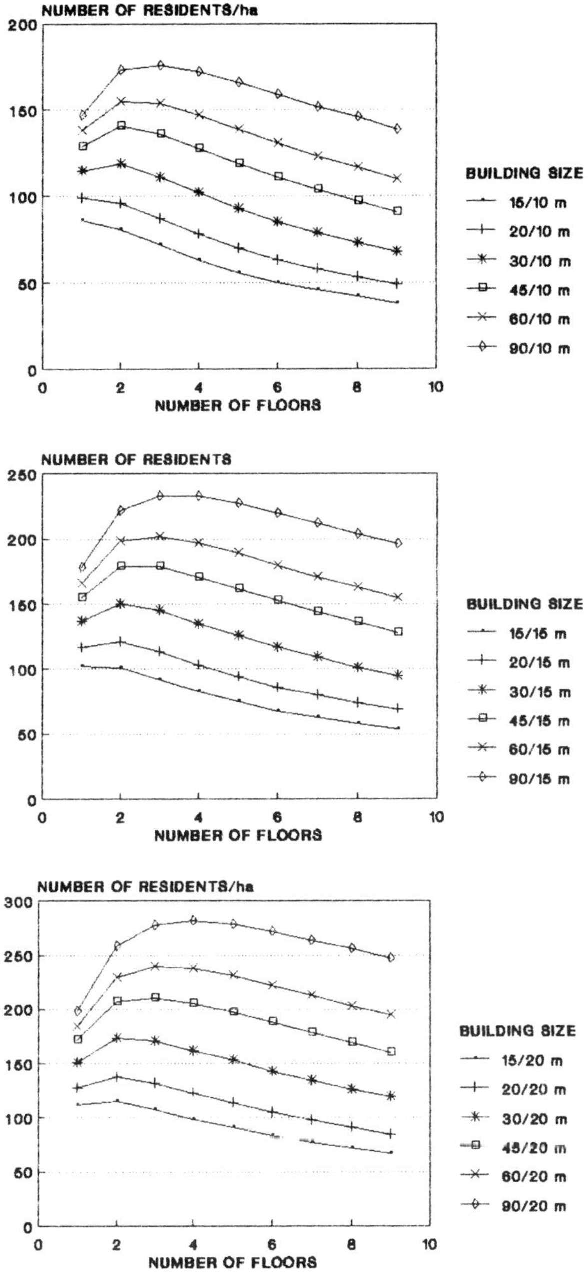

Figure 2. Residential densities as a function of the building size

3. RESULTS

The obtained results are presented in the graphic and tabular form. Fig.2. depicts relationship between the residential density and the building size while the density index and the residential densities specified by the Master Urban Plan (1) and calculated (2) are compared in Table 1.

Table 1. Comparison of calculated residential densities and density indexes with densities and indexes specified by the Master Urban Plan

degree of resid. density (1)	density index (1)	density index (2)	residential density (1)	residential density (2)	number of storeys
small	0.1-0.3	0.2-0.7	35-90	80-258	1-2
medium	0.4-0.7	0.1-0.8	105-140	56-282	3-5
high	0.8-1.0	0.1-0.7	155-180	56-272	5-9
very high	1.1-1.5		200-350 (central town area)		

Considering the given results, several observations can be made:

-the highest residential densities for small houses correspond to houses 1-2 storeys high;

-as the size of the building increases, the highest residential densities and the corresponding values of density index are attained for the number of storeys 3-5; increasing the number of storeys results in a decrease of the residential density;

-residential density obviously increase with the increase of the building width for a given (constant) height and the increase of the building length, for a given (constant) width;

-Extremely high densities of over 280 residents/ha are not possible under given conditions.

4. CONCLUSIONS

1. All degrees of residential densities as specified by the Master Urban Planning Document for Belgrade may be obtained by building of various sizes (number of floors) providing architects with a freedom of urban planning design.

2. Residential densities specified by the Master Urban Plan can be ensured by multi storey buildings with applied passive solar systems, (as is shown by the case study described in this paper), and the current urban planning and urban design practice in Yugoslavia need to be changed accordingly.

REFERENCES

Jovanović Popović, M. (1991.)'Possibilities of passive solar systems application for multi-storey residential buildings in Belgrade', Ph thesis, in preparation.

Djordjević, A. et all (1976) 'Master urban plan of Belgrade' Beogradski izdavačko-grafički zavod, Belgrade

Szokoley, S.V. (1980) 'Environmental science handbook', The Construction Press, Lancaster, England.

THE PATIO; A CONSTANT IN MEDITERRANEAN ARCHITECTURE; CLIMATE AND SIMBOLISM

JOSE PEREZ DE LAMA HALCON
& JOSE MARIA CABEZA LAINEZ, architects
Seminario de Arquitectura Bioclimatica
ETS Arquitectura / Avda. Reina Mercedes s/n
41013 Sevilla / SPAIN

ABSTRACT. The patio characterizes Mediterranean architectures throughout history. Besides its daylighting, ventilation and climatic functions patios are of simbolic importance. The same spatial element sincronically assumes different functional and simbolic roles. The profane and the sacred, simbolisms of the center, the mother, recreation of cosmos and of paradise, appear to be significant to patios. Domestic structures incorporating patios as its center show to be extremely permanent throughout history and geography around the Mediterranean. Patio architecture deals with all senses, within, but beyond, the visual/intellectual tradition of clasicisms. Historical study shows, as pointed by Aldo Rossi, patios to be the only fixed thing in domestic architecture, and therefore, the "empty space" where to find the meaning of the building . (1) Eventually, conditioned by social evolution and bougeoise culture patio spaces become public.

1.1. Presence of patio architecture in the Mediterranean

The patio, i.e. an open space enclosed inside a building, is one the most characteristic constants to architecture of the Mediterranean culture. Patio can be found in the early archeological urban sites of the Neolithic, in the Middle East, as for example in Ur, (2) and stay since then in most Mediterranean temples, palaces and houses. It is difficult to think about the architecture of Egypt, Minoan and Mycenaean cultures, Greece and Rome, Islam and the Middle and Modern Ages in Southern Eurpoe without imagining patios, usually surrounded by columns, very often containing water and plants.

S. Alvarez et al. (eds.), Architecture and Urban Space, 269–274.

1.2. Function and climate

First of all, the existence of courtyards is determined by functional needs. Within a dense urban structure, as the first cities had, -density determined by multiple reasons as: social structure, function and economy of streets, limited space, climate...-, courtyards are needed in order to illuminate and ventilate lot interiors.

The benign Mediterranean climate allows open space to be used as livable space during long periods of time. The patio becomes, as well, an environmental control device for the covered spaces attached to it. Its objectives are multiple:

- provide daylighting
- provide ventilation
- collect rain water
- provide climatic control

Climatic control in the geographically homogeneous Mediterranean climate is two folded: winter heating and summer cooling. A well designed patio should milder extreme exterior conditions, becoming an intermediate space between outdoor and indoor conditions.

In wintertime a well oriented patio captures solar radiation and stores the heat, distributing it to indoor space. At nightime indoor spaces should be insulated towards the patio.

In summer time a well designed patio is fresher than the outdoor space. Living rooms relate to the courtyard. Patios cool down at night by long wave radiation and ventilation -crossed with other patios, helped by wind towers, underground devices and evaporation, storing the coolth of the night in its massive walls and floor. During daytime protected from solar radiation by verandas, textiles or vegetation, cooled by evaporation with fountains and floor watering, with its walls covered by climbers, patios keep the coolth of the night and distribute it to indoor space as long as it lasts. (3)

1.3. Origin of patio

In our opinion, functional and simbolic circumstances contribute to the origin of patios. In his text, Adam´s House in Paradise, J. Rykwert, (4) points out how the invention of agriculture is necessarily related to that of the house. The tribal familiy cultivates a piece of land, which they surround by a wall, administrates and celebrates the harvest ..."these imply containers and vases, and these again pieces of furniture to keep them, consequently rooms ..., i.e.: a house. An orchard without a house is like a cart without a mule."

Organization of agricultural production makes possible the appearance of cities. Families associate to cultivate fields in a larger scale, take advantage of production excedents and benefit from social organization. At this stage, the tribal family relates to urban society in a way symmetrical to the one the house relates to the early city. The house is an introverted microcosmos, that reproduces the structure of the **original orchard** . The patio, placed in the center of the house, is the simbolic representation of nature. Palaces and temples, the houses of the powerful and the gods, reproduce analogous concepts in complexer contexts.

The **sacred-profane** complex, so important to the archaic man mentality characterizes the patio. In this sense, patios are not indifferent, profane, open spaces; but sacred ones. This can be appreciated, for example in Roman Villas, which even though located in the middle of the countryside, incorporate numerous patios within, as can be seen for example in the Villa of the Mysteriae, in Pompei (2). To the antique man every action with a precise sense, participates to some extent of the sacred. (5 y 6) The house, the place where family life is possible, -eating, sleeping, sexuality-, is a definite sacred place. Sacred to the archaic mind is that corresponding to an archetypal model, as wil be developed later on, that gives to it its reality and validity. The patio is one of the elements in the house where its sacredness is realized and showed.

1.4. Simbolism of the center

Every sacred place is considered to be the **center of cosmos** . As M. Eliade explains, man can only live in a sacred space, in the center. The center is the place where original creation took place, the *axis mundi* , that unites earth center with heaven, and as well, man with the divinity. Paradoxically the center is in every sacred place, in every city, in every temple, in every house. The courtyard, as it is clearly expressed by the Roman atrium with its impluvium (Fig 1), is the materialization of the *axis mundi* inside the house. This explains to some extent the paradox of the empty center.

A third simbolic aspect, that will only be pointed out in this brief paper, is that of the patio as the representation of the maternal womb, that can also be intuited from the Roman atrium with its impluvium-compluvium system, in fig 1.

1.5. Cosmogonia repetition

Repetition of the archetypal model implyes in construction the repetition of cosmogonia or **cosmos creation** . Construction rituals and sacred space reproduce cosmos creation and cosmos itself, respectively. In the patio, besides the center, the four original elements of creation are represented, -fire,

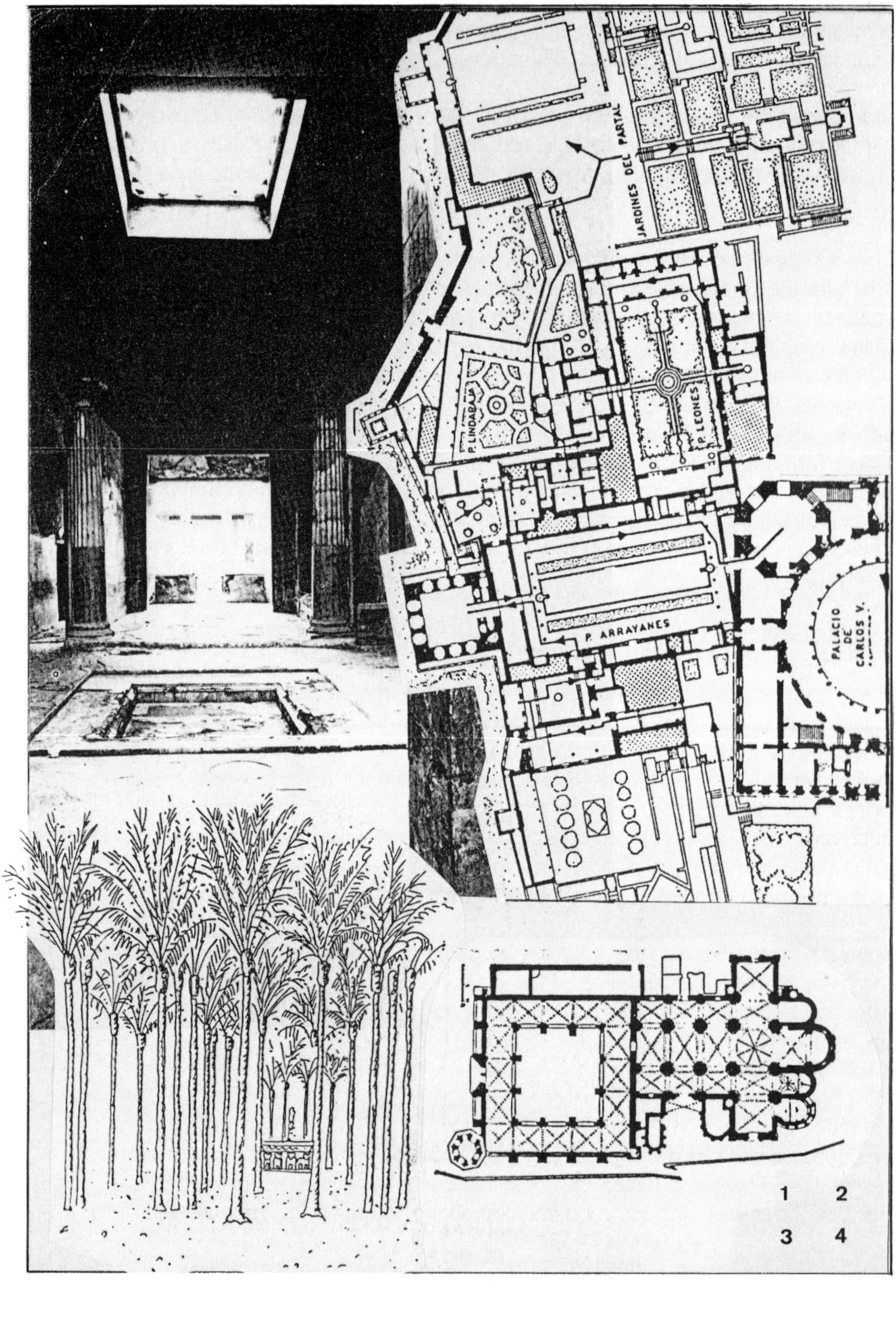
JARDINES DEL PARTAL
P. LINDARAJA
P. LEONES
P. ARRAYANES
PALACIO DE CARLOS V.
1 2
3 4

earth, water and air-, as can be observed in Islamic tradition, and sometimes, **lost paradise** , the original garden, -often in a metaphoric way, through the use of columns. (Fig 2 & 3)

This theme and the one of the center, already introduced, relate very precisely, patios to the architecture of domes. All the simbology attached to domes is applicable to patios. "The sky itself is the patio dome." A sensitive observation of the Pantheon in Rome demonstrates the hypothesis.

The second prototype, derived from cosmos, to all Mediterranean architecture in respect to patios and domes, should be found in Solomon´s temple and palace, as explained by Oleg Grabar in his book about Alhambra. (7) (Fig 2)

Again in Islamic architecture, the relationship between patio and the original orchard or garden, the place where happiness was possble, can be observed.

1.6. Paradise nostalgie

In the Middle Ages patio and cloisters can be compared to bonsai, an artistic recreation of the whole nature contained within a built environment. In monastic cloisters and gardens of sufic inspiration we recognize a space directed to all five senses, that make possible acces to religious or metaphisical experience. The quote from El Quijote illustrtes the statement in a humorous and minor way: "Apenas vio el ama que Sancho Panza se encerraba con su señor ... tomando su manto toda llena de congoja y pesadumbre se fue a buscar al bachiller Sansón Carrasco... Hallóle paseándose por el patio de su casa, y viéndole se dejó caer ante sus pies, trasudada y congojosa."

Buildings like Alhambra or Lerida cathedral (fig 4) show the importance of patio, transcending, no hesitation, its functional mission.

1.7. Permanence of patio typology in modern urban structure

The very permanence of the Roman house typology can still be observed in Seville´s modern urban structure. This is a large house with little fassade and a succesion of three courtyards. The first one, the Roman atrium and modern *apeadero* surrounded by those rooms related to arrival and the street. The second one, surrounded by columns, is the space of representation, the actual center of the house. Eventually, the third is a garden the place of family life and intimacy

"It is said that years ago in Seville whoever had a house built instructed the architect, or simply the mason, what the dimensions of the patio should be, and then added that he should draw around it a possible configuration of rooms." writes A. Rossi (1) and goes on like this: "this seems to me connected with the problem of freedom and of imagination: since there are few things fixed, they cannot be miscalculated; they become **the meaning of the building** ." Our initial hypothesis is confirmed. The meaning of the building is in the patio.

The bourgeoise house in the XIX and XX centuries, in a process explained by several authors, reduces the size of its spaces and the number of patios, which nevertheless retain their simbolic importance; patios sometime covered by the XX century passion: glass structures, that deteriorate their summer performance.

Social evolution, makes possible an interesting acculturation phenomena: the *plaza salón,* the salon square. Figuratively, the outter patio of the three patio house, lost by the smaller XIX century house, becomes public, shared by the neighbours; it is a bourgeoise space, closer to the *corral* and the patio than to the *agora*. Its aspect, furnished with trees, fountains and banks, is closer to domestic patio than to public space dedicated to politics and commerce.

1.8. Conclusions; permanence and change of patios

The same scenery, the patio, is the place that makes possible history of Mediterranean culture happen, strictly following Rossi´s theories, loading itself, as time passes, with meaning, evocation and life. Let´s finnish with another quote from Aldo Rossi: "These forms of dwelling are stored in the history of man to such a degree, that they belong as much to anthropology as to architecture. **It is difficult to imagine other forms, other geometric representations** ."

2. Bibliography

1. Rossi, Aldo, A Scientific Autobiography, Oppositions Books, Cambridge 1985
2. Kostof, Spiro, Historia de la Arquitectura, Alianza, Madrid 1988
3 Pérez de Lama Halcón, Climatic Design for a Holyday Resort in the South of Spain, directed by B. Givoni, UCLA 1988
4. Rykwert, Joseph, La casa de Adán en el Paraiso, GG, Barcelona
5. Eliade, Mircea, Imágenes y símbolos, Taurus, Madrid 1989
6. Eliade, Mircea, El mito del eterno retorno, Alianza, Madrid 1985
7. Grabar, Oleg, La Alhambra, iconografía, formas y valores, Alianza, Madrid 1988
8. Vázquez Consuegra, G., 100 edificios, JA, Sevilla 1987
9. Prieto Moreno, Francisco, Los jardines de Granada, MEC, Madrid 1983

LOW ENERGY DESIGN IN AN URBAN CONTEXT

K. A. STEEMERS
The Martin Centre for Architectural and Urban Studies
University of Cambridge Department of Architecture
6 Chaucer Road, Cambridge CB2 2EB, England
Tel: (0223) 332981 Fax: (0223) 332983

ABSTRACT. The general theme of this paper is urban climatic contextualism. The specific objectives are to investigate the influence of the urban context on the energy balance of non-domestic buildings. The impact of the main parameters related to urban design are analysed using an integrated energy computer model. Results indicate the significance of lighting as a main design determinant for overall energy performance when considering urban geometry, building form and facade design. The urban context significantly alters the simplified lessons that have been learned from energy efficient design on open sites.

1. Introduction

Much has been written about the nature of cities from a general urban design and architectural theory aspect in work by people such as Rossi, Krier or Rowe. Alongside this debate, but divorced from it, researchers such as Landsberg, Miess and Chandler have produced a great amount of work on the urban environment and microclimate. However, between the two extremes of urban academic theory and urban geography, lies a body of urban research work that needs to be addressed. The author's interest and research work, some of which is described in this paper, attempts to bridge the gap.

The environmental impact of buildings on the urban microclimate has been investigated by others, but little has been done to study the effects of the urban context on buildings. Energy efficient design of individual buildings on 'green field' sites have received much attention in research and practice. Domestic energy efficient architecture, in particular, is known to rely largely on rules-of-thumb about insulation, efficient systems and maximising winter solar gains whilst encouraging shading and ventilation in the summer. However, in non-domestic buildings the energy balance is more complex, and includes lighting and cooling considerations. It is in this respect that the complexities of the urban environmental interactions have not been fully explored. For example, how significant is orientation in a built up environment ? What are the integrated energy effects of obstructions to daylight, heat loads and solar gains through windows ? Perhaps the most significant question is, how does the architect respond to the environmental conditions of an urban context ?

Unfortunately, the answers to these questions are not simple because they depend on conditions and assumptions made about the nature of a building's internal and external arrangement, and how this interacts with the conditions of the urban context. Thus, an extra layer of complexity intervenes between the systems that make up the building (energy flows, occupant behavioural patterns, etc.) and climate.

S. Alvarez et al. (eds.), Architecture and Urban Space, 275–280.

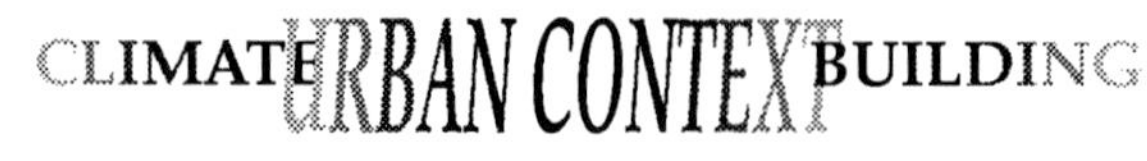

Figure 1. The urban context defracts environmental interactions between building and climate.

2. Cities and Climate

The nature of existing, well-established cities in extreme climates is generally influenced by highly environmentally responsive structures which have been fine-tuned over time to deal with adverse conditions. Whether it be narrow streets to provide shade from the intense sun or protection from wind and dust in the hot arid regions, or dispersed open forms to encourage cool breeze in hot humid climates, or smooth ground-hugging forms to reduce exposure to cold wind and snow drifts in arctic regions, there are many examples of climate responsive urban planning. These are however examples which reflect strongly asymmetric climates with only a few factors dominating the synoptic conditions, such as solar radiation, air temperature or wind. In many instances life could be at risk if protection was not provided.

In the relatively moderate climates of Europe and North America the seasonal and diurnal variations are more complex and require a more sophisticated response. Typically, diurnal conditions and internal use patterns of office buildings may call for heating in the morning and shading, ventilation or even cooling in the afternoon. Many office buildings even require simultaneous heating and cooling: Perimeter heating to balance the fabric losses, and cooling in the centre of a building to balance the internal gains from office equipment; or heating in the north zone and cooling for south facing accommodation. Seasonal climatic variations present further challenges to building designers, and require a response in terms of seasonal adjustability to address changing solar intensity, ambient temperature and daylight availability. These aspects represent the complexities of the interaction between building and climate. The interaction of the urban context on bioclimatic architecture is more complex and requires further discussion. This is particularly so if we are interested in environmental issues ranging from minimising global impact and a building's energy use, to improving internal health and comfort conditions of users.

The effects of urbanisation on climatic factors in towns and cities are well understood. Extensive research of a wide variety of cities around the world (Landsberg 1981) seems to indicate that although the urban microclimate is closely influenced by macroclimatic factors; it consistently alters certain environmental factors.

> *"Basically,... the city modifies the natural energy balance and air circulation through its multiple reflection and absorption patterns, its rough, uneven surface, its lack of water and vegetation over many tracts and the additional sources of heat and dust provided by human activity."* (Douglas 1983)

The extent to which environmental factors are affected in cities as compared to rural environs is indicated in the following table (Landsberg 1981):

Factor	*Compared to rural environs*
Temperature (annual mean)	0.5 - 3.0 ^{o}C more
Radiation (total horizontal)	0 - 20 % less
Wind speed (annual mean)	20 - 30 % less
Relative humidity (annual mean)	6 % less

Table 1. Urbanisation effects on the micro-climate in cities compared as with rural environs.

These general, overall changes are relatively small in terms of the effect on building energy use when compared to effects of the specific urban context. Furthermore, although the variations have an impact on the overall energy use of cities, they do not affect the relative energy savings due to a building design response to site conditions.

3. Energy Modelling of Urban Design Factors

The immediate urban context presents a range of conditions which will significantly affect the energy balance of buildings. Perhaps most important is the degree of obstruction from surrounding buildings, which mainly has repercussions on daylight availability and solar gains, and thus lighting, heating and cooling loads. The extent of the influence of context on energy use is in turn determined by the amount of glazing. Thus one of the first tasks is to study the interrelationship between glazing ratio and obstruction angle (or Urban Horizon Angle, UHA) and effects on heating, cooling and lighting energy use.

A series of parametric studies were carried out using a computer programme specifically developed and adapted to address issues related to the urban context (Baker 1989, Steemers *et al* 1989). Some of the results and conclusions are presented in this paper. The graphs (Figure 2) show results from a series of parametric studies on office accommodation in the UK. Many assumptions have been made about the building and systems, as well as occupant behavioural patterns. Basically, these assumptions reflect current UK building regulation standards and sensible energy measures such as the use of solar blinds and controlled lighting regimes.

3.1 HEATING

The heating curves are relatively smooth and show some interesting characteristics. For a north facing facade, obstruction has little effect on heating energy. This is because north elevations in the UK receive limited solar gains to displace heating and so heating energy is unaffected by the angle of obstruction. However, increasing the glazing ratio will unilaterally increase the heating requirements due to increased losses through the glass, irrespective of the angle of obstruction. For south facing rooms there is a relationship between glazing ratio and obstruction. When there is little or no obstruction, increasing the window area will increase useful solar gains and increase heat losses. These tend to balance each other out to a large extent and result in no significant change of heating energy. However, any obstructions will quickly reduce solar gains without affecting heat loss, so that increasing the glazing will result in increased heating requirements. Eventually, with high obstruction angles the south facade will have similar heating loads as the north facade.

3.2 COOLING

The curves labelled cooling energy do not take into account fan or pump power and could thus be referred to as the refrigeration energy only. In many cases fan and pump energy use can typically be twice as much as the refrigeration energy, which accounts to a large extent for the fact that air-conditioned buildings consume significantly more primary energy than naturally ventilated buildings (Figure 3). The results of the parametric study indicate, as expected, that increased glazing on south facades results in higher cooling loads due to solar gains. There is also some increase in cooling loads with increasing obstruction with small windows, probably due to the lack of natural light and therefore increased casual gains from artificial lighting. On north elevations, increasing window openings generally reduces cooling loads due to a combination of increased heat losses and reduced internal gains from lighting. However, the cooling curves are not smooth and thus indicate the more complex and less predictable nature of the results.

3.3 LIGHTING

The first impressions of the results reinforces the importance of lighting energy use in offices, as the graphs indicate, largely irrespective of orientation. As expected, both increasing the glazing ratio and reducing the obstruction angle will significantly displace lighting loads by the increase of natural lighting. The particularly steep area of the curves at around 30 % glazing ratio and 40 degrees obstruction, typical for urban buildings, highlights the sensitivity of lighting energy to urban design. A small increase in obstruction, or a reduction in the window size will result in a substantial increase in lighting energy.

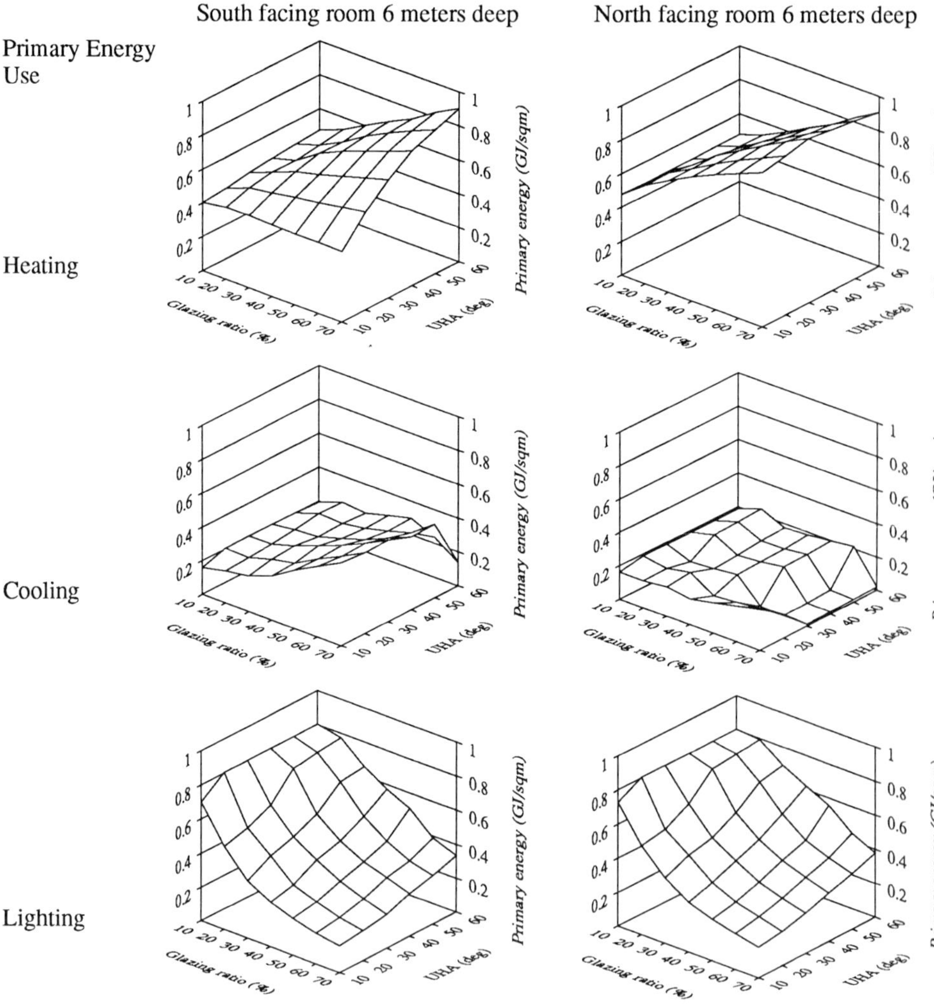

Figure 2. Example of results from a parametric study comparing the effects of glazing ratio and obstruction angle to primary energy use for heating, cooling and lighting for a south facing and north facing office space 6 meters deep.

3.4 REAL DATA

Some of the theoretical results are confirmed by monitored data from real office buildings. Figure 3 indicates the significance of lighting energy and highlights the difference between naturally ventilated offices and air conditioned buildings. It is furthermore interesting to note that air conditioned buildings generally have higher lighting loads due to the fact that they tend to be deep plan buildings, as opposed to shallow plans to provide natural cross ventilation. This is also the reason for reduced heating loads in air conditioned offices, which have reduced external surface to floor area ratios.

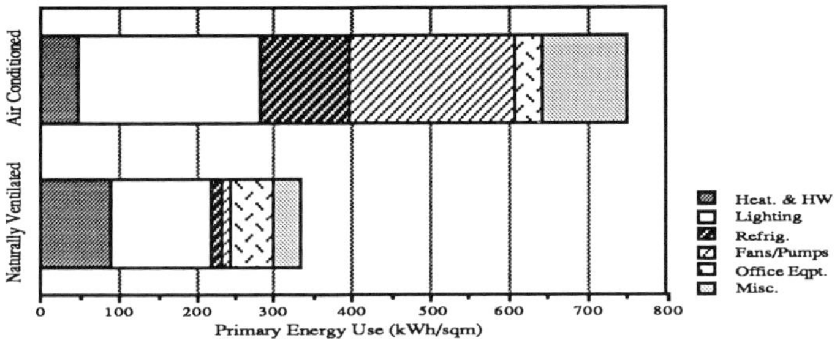

Figure 3. Measured breakdown of average primary energy use in UK office buildings (Source: BRECSU Internal Report, W. Bordass, 1989).

4 Built Form and the Urban Context

Questions of built form in relation to energy use can be reduced to the analysis of orientation and building depth, in relation to surrounding obstructions. The following graphs (Figure 4) indicate the effects of room depth, comparing an unobstructed facade to one with a limited obstruction angle of 30 degrees from the centre of the window. They are a sample of results from a broader study, and relate to the UK climate and the office building type.

The graphs indicate the beneficial influence of increased glazing ratios irrespective of room depth, for unobstructed south facades. Energy use is reduced, as expected, due to the increased useful solar gains and access to daylight. However, this effect is significantly reduced when there is an obstruction, and room depth starts to play a slightly more significant role. The results of an example indicate that in the obstructed case, a room depth of about five meters and a glazing ratio of 50 % would optimise energy efficiency.

For unobstructed north facing rooms, the energy balance is similar to an obstructed south facing room. The effect of obstructions to the north is to increase the energy use for lighting and generally reduce the influence of glazing ratio and room depth on energy use.

Generally, obstructions reduce the difference between the energy balance for north and south facing spaces. Furthermore, the presence of obstructions for any orientation result in a direct relationship between optimum room depth and glazing ratio, which is not the case for an unobstructed south facade where maximising useful solar gains leads to larger glazing ratios irrespective of room depth (assuming effective shading and ventilation will maintain comfort conditions). In the above graphs, showing particular results for obstructed spaces, deep plans require large glazing areas to maximise daylighting, for all orientations. Shallow plans require less glazing, with the optima only slightly more dependent on orientation because even small solar gains in shallow rooms can have a significant impact. The optima for the particular example shown indicate glazing ratios of about 35 % for south, 30 % for west and east, and 15 % for north facing three meter deep spaces. It is important to note that these ratios will be different for each climate and building specification.

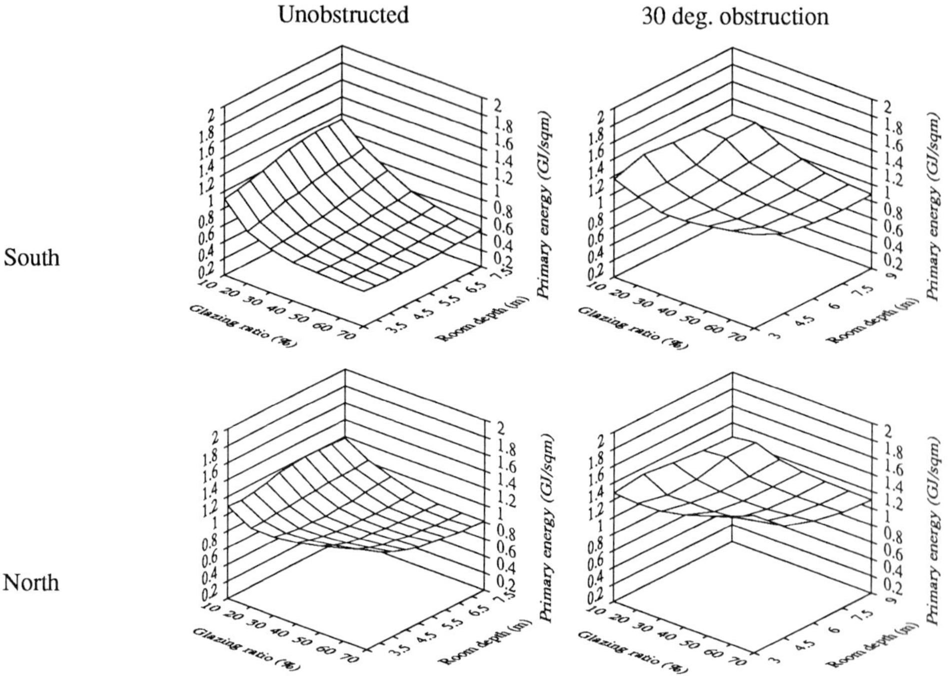

Figure 4. Heating plus lighting primary energy use in unobstructed and medium obstructed (30 degrees UHA) offices facing north and south.

5 Conclusions

In an urban context, the traditional low energy rules-of-thumb are no longer robust enough to ensure satisfactory results. Orientation plays a minor role in energy efficiency, having an impact only on very shallow plans or the top floor of buildings. Much more significant are the implications of natural light in relation to obstructions (urban geometry), plan depth (building form) and glazing ratio (facade design). Shallow plan building forms benefit from a greater interface with beneficial climatic factors such as daylight and solar gain. Covered courtyards or atrium designs, adopting appropriate glazing strategies, increase this potential whilst reducing heat losses. This paper has highlighted some of the lessons learned from research, and begins to suggest the most important considerations to be taken into account when designing in an urban context.

Baker, N. and Hoch, D. (1989) 'Experience with a design tool which combines the energy value of daylight and the thermal value of solar gain', UK-ISES Conf. Proc., London.

Douglas, I. (1983) The Urban Environment, Butler and Tanner Ltd., Somerset.

Landsberg, H. E. (1981) The Urban Climate, Acad. Press, London.

Steemers, K. *et al.* (1989) 'Energy use and urban development' and 'Parametric energy modelling in urban situations', Second European Conference on Architecture, Kluwer Academic Publ., Dordrecht.

ADVISORY PROJECT RULES FOR THE RE-USE OF RURAL SETTLEMENTS IN LOMBARDY

M. BOTTERO, N. BRAULIN, E. COSTA, C. FONTANA
Dipartimento di Programmazione Progettazione Produzione Edilizia
Politecnico di Milano
Via Bonardi 3
Milan, Italy

ABSTRACT. The object of this research is to develop a body of advisory project rules, in order to supply methodological indications for an energy conscious rehabilitation of the built heritage; these indications are mainly concerned with the bio-climatic and environmental characters both of the buildings and of the context.
This body of rules holds the features of a technical guide to the evaluation of all the steps that must be undertaken in the development of a project on existent buildings, with the purpose of improving their bio-physic comfort, by means of controlling the climatic (sunlight and temperature) and electro-magnetic parameters.
The methodological indications shall be conceived as a series of advisory rules: that is to say, as a series of steps proceeding towards a better knowledge of the excisting building, for an appropriate project. Advisory rules should be implemented by local authorities as a community service to architects, tenants and owners.
The case-study buildings have been analyzed in order to:
1) evaluate the lag between the present thermal-light behaviour and the comfort conditions which have been stated as a project aim; 2) evaluate a number of technological methods, means and devices that may be employed to improve comfort levels; all of these means must be adequated to the local environment conditions and to the existing buildings structure.

1. THEME AND OBJECTIVES

Rules concerning the built environment enclose a number of regulations in order to: 1) regulate space dimensions and functions (quantitative rules); 2) guide the project towards stated quality standards (qualitative rules).

Generally speaking, rules are effective as long as they are proper and can be actually applied; there are either rules which are given general consent, or compulsory rules which need to be imposed by strong authorities, in order to reach a general interest.

Rules may regulate different behaviours either by means of orders

S. Alvarez et al. (eds.), Architecture and Urban Space, 281–285.

and prohibitions (compulsory rules) or by means of advice and directions (advisory rules). In the latter case, rules do not impose a stated behaviour; they rather suggest one or more ways to attain a stated purpose.

To disobey an order causes the transgressor to be punished; on the contrary, not to follow proper advice causes a failure in the pursuit of a stated aim: therefore, advisory rules call upon the citizens' responsibility and culture.

In building rehabilitation and re-use, advisory rules can take the features of technical instructions or guide to rehabilitation project. They should outline an "information path" to recognize and analyse the specific and local characters and qualities of the existent buildings and of their environmental context.

Advisory rules also tend to promote a cultural improvement in the operators: actually, these ones are expected to work out responsible and well-grounded project outputs, rather than to carry out an order passively.

Such a body of technical advice should become the core of a public information system, that ought to be run and implemented by local authorities.

In the outlining of the "information path", the advisory rules this research is developing should point out the main steps towards an energy conscious rehabilitation project: these steps concern bioclimatic conditions and comfort levels performed by a group of Lombard rural dwellings.

Advisory rules will direct the analysis of buildings, materials and conditions, and lead to the understanding of the relationship between the elements of buildings and their environmental context; all project advice will be based on the understanding and knowledge of building materials, shape and constitution.

Such a proceeding has another purpose, too: that is, to preserve the rich heritage of information and technical knowledge which are embodied in traditional architecture; this aim can only be obtained by a deep comprehension of the technical features and material characters of the old buildings.

In fact, traditional buildings often display a very appropriate relationship between construction materials and techniques and climate and environment in general: therefore, one of the main purposes of the rehabilitation project must be improved leaking situations without damaging or destroying the existent elements.

It is well known, in fact, that a large portion of the built heritage seems to have suffered from unadequate repair as much as from neglect; the point is, to provide a guide to avoid all kinds of unrespectful or technically abusive methods.

2. THE CASE STUDY

The Lombard dairy-farm is a residential-agricultural typology dating back to the 18th century, a distinctive feature of the Lombard Plain landscape and a rich testimony of a harmonious and functional relationship between the built and the natural environment, between bulding and climate. Usually, dairy-farms have a good orientation, with the main side (occupied by dwellings) running east-west; the southern front always has a deep porch, roofed in fired clay, which both protects the dwellings and cowsheds from the summer sun (whilst allowing the sun's rays to penetrate in winter) and provides a sheltered outdoor space to the homes.

Dairy farms have thick, solid brick walls; floor and roof frames are in timber; the two-pitched roof has fired clay curved tiles roofing; the windows are few and small, those facing north have thick wooden shutters.

3. METHODOLOGY

3.1 The thermal-light parameter control

Dairy-farms deserve to be restored and preserved; the main problem is how to evaluate any possible deed, in order to achieve the thermal-light comfort without destroying the character of the dairy-farm.

Therefore, the advisory rule system should provide the user with information about a range of different optional technologies to perform on the building and its environment.

The evaluation criteria of these technologies will be on one side, the improvement rate of the building thermal-light seasonal behaviour, and on the other side, the degree of fitness to the existent building of the technical and architectural output.

To implement such a method, we need an operative tool in order to evaluate:

a) the thermal-light behaviour of the existent buildings;
b) the actual critical parameters influencing the thermal-light behaviour of the existent buikldings;
c) the efficienty of the improvement techniques;
d) the susceptibility of the building components to be transformed.

Raphael Serra has developed a model, CLA, that we think is the better answer to such requirements. It provides a computer simulation of the thermal-light conditions inside the buildings related to summer and winter comfort by quantitative/qualitative analysis of the relationship between climate, site and building parameters.

The climate parameters are:

a) average temperature in extreme months; b) average thermal interval (excursions?) in extreme months; c) average wind speed and prevailing directions in extreme months; d) average hourl radiation on south vertical surface in extreme months; f) average absolute humidity in extreme months.

The site and building parameters are referred to:

a) site features; b) nearest environemnt; c) building shape; e) building interiors; f) solar gain devices; g) thermal storage systems; h) ventilation and air treatment systems; i) sunlight system and shading devices.

This model enables:

1) evaluation of the specific role of the descriptive parameters in controlling the indoor climate and the threshold indicating significant change in this control;
2) proposals about the transformation of the descriptive parameters in order to reach a steady state of comfort, either by modifying the outdoor micro-climatic condition or by designing a different control system of the energy flows through the building.

A range of technologies which will improve the thermal-light behaviour of existent buildings will be identified. Each of these technologies will be evaluated, by weighing the degree of their impact on the characters of the dairy-farm typomorphology.

3.2 The electro-magnetic parameter control

The case-study buildings have solid brick walls and timber framed flooring: these structures look sound and steady, and likely to be re-used; nevertheless, they badly need to be improved as far as comfort conditions are concerned; in addition to the improvement of the thermal-light behaviour, important comfort conditions can be achieved by employing ecologic materials that must be suitable to the existing structure.

Ecologic materials available at present can be employed for acoustic-thermal insulation, to strenghten bearing structures, to obtain good inner finishing results.

Besides, scientific research by physicians, geologists and biologists demonstrated the importance of analysis and of measuring the geologic and electromagnetic features of sites. Actually, radioactivity, electro-magnetic fields, electric pollution - as well as water and air pollution - may severely affect the health and physic comfort of people living in sites; information about these features would provide advice to improvement works as well as to proper techniques and materials.

The collection of data on these features, which would be of the greatest importance at the territory scale, may start from the analysis of the ground immediately around the buildings, then proceed to the interiors and to the building materials.

Data collection will concern:
- measure of overall radioactivity (ß and) and radon-caused radioactivity (a). Improvements could be obtained by providing ground level ventilation, or by removing radioactive materials;
- location of faults and geologic cracks and of underground water streams; location of Hartmann network; evaluation of the natural electro-magnetic field and of other geo-pathogenous anomalies. This survey will be useful to provide suggestions about the distribution of functions for living and working: this means to enable the users to avoid geo-pathogenous places, and to use them to locate only service functions, such as corridors, cupboards, wardrobes, etc.;
- evaluation of electro-magnetic pollution, divided into three frequent groups: under 50 Hz, from 1 to 1000 MHz, from 1 to 30 GHz. Special attention will be paid to controlling the leakage of electro-

magnetic artificial fields, by employing special techniques in plant equipment: such are open patterns in electric installations for every single room, shielded wires, bio-switchers, grounding systems easily maintainable in time.

4. REFERENCES

Kelsen, Hans (1979) Allgemeine Theorie der Normen, Manzsche Verlags- und Universitätsbuchhandlung, Wien.

Bobbio, Norberto (1980), "Norma", in Enciclopedia Einaudi, Vol. IX, Torino, pp.

Bianchi, A. and Bottero, M. (1988) "Rural Architecture in Lombardy", Proceedings of the Sixth International PLEA Conference, Porto.

Bottero, M., Rossi, G., Scudo, G., Silvestrini, G. (1984) Architettura Solare, Clup, Milano.

Serra, R., Clima, Lugar, Arquitectura, ed. CIEMAT, Barcelona.

Schmid, Peter (1988), Bio-logische Architektur, R. Müller GmbH, Köln.

NOT EVERY COURTYARD IS NECESSARILY A GOOD ONE CLIMATICALLY
Physical Characteristics
design of domestic courtyards in the hot-dry climates of the sub-tropics

Subhi Al-Azzawi, AA dipl, PhD (London), RIBA
Architect, Climatic & Environmental Design Consultant
18 Farm Holt, New Ash Green, LONGFIELD, Kent DA3 8QA

ABSTRACT
One of the aims of this paper is to identify the physical characteristics of the courtyard which have direct influence on its performance micro-climatically and thermal environmentally. Here, the objective is to take these into account at the conceptual design stage so that these physical characteristics of the three-dimensional form of the courtyard work in harmony with other aspects of it; the goal is to ameliorate the severe micro-climatic conditions prevailing in summer, and ultimately to produce acceptable, if not desirable, thermal environmental conditions within the courtyard and inside the habitable rooms and spaces grouped around it and looking inwards towards it; this is specially so in the regions of the hot-dry climates of the sub-tropics where buildings should be designed to be energy efficient structures by incorporating concepts and principles of passive solar design due to acute shortage of energy for continuous cooling in summer and intermittent heating in winter.

KEYWORDS
courtyard; physical characteristics; three-dimensional form; traditional courtyard houses; hot-dry climates; sub-tropics; thermal environment; vernacular architecture; developing countries.

1. INTRODUCTION
The incorporation of a courtyard in the design and execution of a house or a building does not necessarily mean that such a structure is specifically designed to control the severe micro-climatic conditions prevailing in summer in the regions of the hot-dry climates in the sub-tropics. This is because not every courtyard is necessarily a good one micro-climatically. In other words, the concept of incorporating a courtyard is right, but its detail design (in terms of size, shape, form, etc.) may not be correct thermal environmentally.

There are many factors which determine the quality of the performance standards and performance specifications of an internal courtyard in the hot-dry climates of the sub-tropics. A detailed discussion of them is outside the scope of the present paper; however, the identification, analysis and appraisal of the broad principles are not; some of these are presently discussed in detail.

Here, some of the factors which make a courtyard climatically desirable or otherwise are identified, explained and appraised, particularly those relating to its physical characteristics. Therefore, when designing new buildings incorporating one or more courtyards (whether within traditional houses, residential buildings, or any other type), these factors should be taken into account; hence, such courtyards are consciously designed to be climatically desirable in the hot-dry climates of the sub-tropics. Besides, their thermal environmental performance in winter should also be considered so that they do not become a source of discomfort even during this short and cool-to-cold

S. Alvarez et al. (eds.), Architecture and Urban Space, 287–298.

season. These conditions directly influence those prevailing in the habitable rooms and spaces surrounding it and looking inwards towards it. Here, the courtyard also acts as an in-between place or a transitional space physically, visually and sensually between the external micro-climate and the internal thermal environment.

Internal courtyards are a characteristic design feature of vernacular domestic architecture as well as of indigenous non-domestic architecture in the hot-dry climates of the sub-tropics, particularly in the Developing Countries of the Third World; these have been developed through the ages by a process of trial and error; here, the accumulated design experience of the master-builders and the technical know-how of the craftsmen were passed on from one generation to another to be understood, assimilated and improved upon before they were in turn passed on to the succeeding generation. The proto-type of such indigenous courtyard houses date back to antiquity; in the case of Baghdad, for example, this is found in the milieu of the Sumerian, Akkadian, Babylonian and Assyrian civilisations. (1), (2) and (3).

As used in this paper, the term courtyard refers to an internal courtyard which is the focal space of the house and the hub of the family domain; it is surrounded by habitable rooms and spaces (as well as ancillary areas) along two, three or four sides of it; these look out onto it visually and communicate with it physically either directly or indirectly; additionally, they receive their daylight and natural ventilation through it; at the same time, the courtyard provides them with contact with the external environment and the weather outside as well as with a view of the sky above. (4), (5) and (6).

It should be pointed out that from the point of view of micro-climates and internal thermal environments, what is considered to be a disadvantage in summer may well be an advantage in winter in terms of sun penetration, duration of sunlight, solar heat gain, high surface temperature and high air temperature. It is a case of striking a balance between the requirements needed to deal with the duration and severity (or otherwise) of micro-climatic conditions prevailing in one season with those of another; here, a set of compromise design criteria related to micro-climates and internal thermal environments for all seasons are rationally arrived and deliberately carried out.

2. SHAPE OF COURTYARD IN PLAN

The plan of the courtyard is usually rectangular in shape, but it can be of a perfect square; there are also a few rare examples where it is of a trapezium shape (7). The influencing factors in all cases are the shape and size of the building plot and the size, number and use of the habitable rooms and spaces grouped around it and looking inwards towards it at each floor level. The function and use of the courtyard is also an important factor in determining its size. All these are naturally related to the financial means of the owner-occupier as well as to his social standing in the community. It would seem that for the same floor area, a rectangular-shaped courtyard in plan with its axes oriented in the correct directions vis-à-vis the sun, would perform better micro-climatically than a square one of the same orientation.

3. SIZE OF COURTYARD IN PLAN

The size of a courtyard in terms of its floor area and plan dimensions varies considerably. It can be very small (with a floor area of 12 to

15 sq m (8), (9) and (10), etc.) to small, medium, large and very large (with a floor area of 80 to 90 sq m (11), (12) and (13), etc.). (Fig. 2, 3, & 6-8). There are also exceptionally small courtyards (with a floor area of up to 10.0 sq m (14) and (15), etc.) and exceptionally large courtyards with a floor area of 140 sq m (16), of 220 sq m (17), and even 330 sq m (18), etc.); but, these are rare cases representing the exception rather than the rule. It should be emphasised that from a micro-climatic point of view, the smaller the courtyard, the better its performance is.

4. NUMBER OF FLOORS

The great majority of houses incorporate courtyards which are usually two storeys high, consisting of ground floor and first floor. However, there are also courtyards which are only one storey high. There are also a few rare cases where the height of the courtyard can be of partially single storey and partially two storeys; alternatively, it can be of partially two storeys and partially three storeys. From micro-climatic point of view, two-storey courtyards perform better than single-storey ones having the same orientation, and three-storey courtyards are more effective than two-storey ones. However, this is also related to the size of its floor area, and to its plan dimensions. Naturally, the number of floors is related to the number of habitable rooms and spaces (as well as ancillary areas) grouped around it and looking inwards towards it. These are also related to the size of the building plot, the financial means of the owner-occupier, as well as to his status and social standing in the community.

5. FLOOR-TO-CEILING HEIGHT

At ground floor, the floor-to-ceiling height is relatively low and varies between a low of 2.70m (19) and 2.80m (20) and (21), a medium of 3.10m (22), and a high of 3.50m (23), or occasionally exceptionally high of 4.00m (24), with the great majority of houses having a floor to ceiling height of about 3.0m; this limits the extent of sunlight penetration into the interior in summer, and therefore reduces heat gain from direct solar radiation which is undesirable during this season. (Fig. 10 & 11).

At first floor, the floor-to-ceiling height is comparatively high and varies between a low of 3.00m (25) and (26); a medium of 4.00m (27) and (28), and a high 4.40m (29), or occasionally exceptionally high of 5.00m (30), with the great majority of houses having a floor-to-ceiling height of about 4.00m; this increases the extent of sunlight penetration into the interior in winter, and hence increases heat gain from direct solar radiation which is desirable during this season. (Fig. 10 & 11).

It should be pointed out here that the ground floor is designated mainly for summer habitation (and therefore designed to limit the extent and duration of sunlight penetration into the interior) while the first floor is designated for winter habitation (and hence designed to increase the extent and duration of sunlight penetration into the interior). It is also important to emphasise that all habitable rooms and spaces at first floor are not used in summer; they are completely deserted. Therefore, this floor acts as a barrier against the sun in summer by shielding the ground floor from its scorching heat; in a way, the first floor functions like a well-ventilated attic of a pitched roof in protecting the space immediately underneath it from direct solar radiation as well as from high surface temperature and air temperature.

6. HEIGHT OF COURTYARD

The actual height of a two-storey courtyard varies between a low of 6.50m (31) and (32), a medium of 7.50m (33), and a high of 8.50m (34), or exceptionally high of 9.0m (35), depending on the size of the courtyard in terms of its floor area and its plan dimensions; in turn, these are related to the size and number of habitable rooms and spaces (as well as ancillary areas) grouped around it and looking inwards towards it at each floor level; they are also related to the size of the building plot, to the financial means of the owner-occupier, and his status and social standing in the community. From a micro-climatic point of view, the higher the courtyard, the better its performance in summer is. (Fig. 10 & 11).

7. PROPORTIONS OF COURTYARD DIMENSIONS

From a micro-climatic point of view, the performance of a courtyard depends on, inter alia, the orientation of its axes, its three dimensional form, and the optimum proportions between its physical dimensions of length, width and height. These help to keep part of the floor area of the courtyard as well as its walls in the shade even around mid-day when the sun is nearest to the zenith. The extent of the shaded area of the floor and walls of the courtyard depends on the cut-off angles to the rays of the sun provided by the walls surrounding it and defining its physical size, and on the orientation of its axes in plan. The cut-off angles are determined by the three-dimensional form of the courtyard (especially the proportions of its length, width and height or that of its length plus width to its height), and by the altitude of the sun above the horizon throughout the year, especially in summer in the case of the hot-dry climates. The combination of all these factors keeps part of the floor area of the courtyard and of its walls in the shade; they also help to avoid high surface temperatures which in turn alleviate high air temperatures in summer.

The proportions of plan dimensions of width:length in the courtyards of existing traditional courtyard houses tend to approximate towards 2:3. However, their height is related to the size and height of their habitable rooms and spaces. In general, those at ground floor level are about 2.70m in small courtyard houses, 3.00m in medium-size ones, and 3.50m (or exceptionally 4.00m) in larger ones. On the whole those at first floor level vary between 3.00m in small courtyard houses, 3.50m in medium-size ones and 4.00m (or even exceptionally 4.50m) in larger ones. (Fig. 10 & 11).

8. OPEN AREA OF COURTYARD AT ROOF LEVEL

The micro-climatic performance of the courtyard also depends on the projected or elevated area of the courtyard at roof level open to the sky, namely, the size of the opening (or hole) in the roof immediately above the courtyard.(Fig. 4 & 9). This area may be smaller or larger than the floor area of the courtyard itself depending on the latter projected at first floor level and at roof level; these in turn depend on the four profiles of the courtyard in section, on the projection (towards the centre of the courtyard) of the access gallery (**Mamsha**), and on the projection of the roof (towards the centre of courtyard) at eaves level (**Sachagh**). (Fig. 10 & 11).

The intensity of direct solar radiation and the duration of sunlight in the courtyard depends on the size of this opening at roof level as well as on the proportions of its three-dimensional form. Therefore, the smaller this area, the less sunlight is admitted into the courtyard, and the less direct solar energy penetrates the interior; hence, the better thermal environmental conditions prevailing in the courtyard in summer as well as in the surrounding habitable rooms and spaces grouped around it and looking inwards towards it.

9. SECTIONAL PROFILES OF THE COURTYARD

The sectional profile of the side of the courtyard can take any of the following forms: first, a vertical shear profile characterised by the lack of sizeable protrusions or recesses; secondly, an articulated, recessed profile; and thirdly, an articulated protruding profile. All of these may or may not incorporate an inward projection (towards the centre of the courtyard) of the roof at eaves level (**Sachagh**) beyond the face of the walls overlooking the courtyard at first floor.

In theory, of course, each side of the courtyard could have a different profile; however, aesthetically and practically, the profile of all sides could be the same, every two opposite sides could be identical, the two main sides could be the same while the remaining two sides could be either similar or different. In other words, there could be one or two common profiles and three or four different profiles.

A shear vertical profile would not contain appreciable recesses or protrusions; it can be either of a blank wall or may contain windows in the same plane (36). An articulated recessed profile may contain a recessed access gallery (**Mamsha**) (Fig. 11) or a recessed colonnaded gallery (**Tarma**) (37). An articulated protruding profile may contain a protruding access gallery (**Mamsha**) (38), (Fig. 10), a protruding colonnaded gallery (**Tarma**) (39) or bay-windows (known as the **Shanasheel** in Baghdad, **Mashrabiya** in Cairo and **Rawshan** (40) in Jeddah).

An inward protruding eaves would not only reduce the projected area of the courtyard opening at roof level, but would also shade the walls underneath it at first floor from the sun in summer and the rain in winter. (Fig. 10 & 11). Similarly, an inward protruding access gallery would not only reduce the projected area of the courtyard opening at first floor, but would also shade the walls underneath it at ground floor from the sun in summer and the rain in winter. (Fig. 10). In both cases, there would be less sunlight and direct solar radiation reaching the floor of the courtyard; shade means reduced surface temperature and consequently reduced air temperature. The combination of all these factors result in better thermal environmental conditions prevailing in summer in the courtyard and its surrounding habitable rooms and spaces.

10. DESIGNATION, GROUPING AND ORIENTATION

It is possible to group habitable rooms and spaces in such a way so that: first, those designated for use in summer are oriented away from the sun to reduce direct solar radiation, but at the same time they are oriented towards the prevailing wind to encourage convective cooling, especially overnight when the air temperature outside is lower than that inside; and secondly, those designated for use in winter are oriented towards the sun to encourage direct solar radiation to warm up the interior, but at the same time, they are oriented away from the prevailing wind to avoid heat loss by convection. Such concepts have actually been used in the past by the master-builders of Baghdad.

It is equally possible to group all habitable rooms and spaces along two or three sides of the courtyard, leaving the remaining two sides or fourth side respectively facing any undesirable orientation and incorporating all the ancillary areas or leaving it blank. Even these can be shaded using an access balcony (**Mamsha**) at first floor and inward-projecting eaves (**Sachagh**) at roof terrace level (**Satih**). Again, such concepts have also been used in the past by the master builders of Baghdad.

11. NUMBER OF COURTYARDS

In the great majority of houses belonging to the low and average income groups, there is usually only one courtyard. However, in the houses of well-to-do families, there may be one, two, three or four courtyards, each with special designation and with its own habitable rooms and spaces as well as ancillary areas grouped around it and looking inwards towards it. The first may be the family courtyard (**Housh al-Haram** or simply the **Haram**), the second may be the business and guests' courtyard (**Housh al-Diwan-Khana** or just the **Diwan-Khana**), the third may be the kitchen complex courtyard (**Housh al-Matbakh** or simply the **Matbakh**), and the fourth may be the stable courtyard (**Housh al-Tola** or just the **Tola**) where animals are kept. The use of more than one courtyard was adopted for family, social, business, and hygienic reasons. However, their use has helped to reduce the size of the various courtyards considerably which led to better micro-climatic and thermal environmental conditions prevailing in them in summer.

12. INFLUENCING FACTORS

There are many factors which determine the physical form of the courtyard and which ultimately influence its micro-climatic and thermal environmental performance. These include the following:

- the use and function of the habitable rooms and spaces grouped around the courtyard and looking inwards towards it;
- their size and number at each floor level;
- their plan dimensions;
- their floor-to-ceiling height;
- the proportions of their three-dimensional form;
- the use and function of the courtyard itself (ground floor level to roof level;
- its size in plans, sections, elevations and three-dimensional form;
- the size of the house as a whole;
- the size of the building plot;
- the floor-to-ceiling height at each floor level;
- the total height of the courtyard itself;
- the number of floors the courtyard and the house are composed of;
- the financial means of the owner-occupier;
- the status and social standing of the family in the community;

All these factors are inter-related and have a direct influence on each other; hence, they should be considered in total and not in isolation.

There are other factors which affect the courtyard performance; these are related to prevailing micro-climatic conditions (and ways of ameliorating them), building materials, landscaping, and procedures followed by the inhabitants themselves. These are the subject of a separate paper by the author.

13. OTHER ASPECTS OF THE COURTYARD

There are other aspects of the domestic courtyard which are equally important and which have a direct influence on its performance; however, these are not discussed here because they are outside the scope of the present paper. They include the following topics:

- functional aspects;
- the courtyard as the hub of family life and of daily living activities;
- the courtyard as the gathering place for social festivities;
- the courtyard as the centre for religious ceremonies;
- the courtyard as a mitigator of micro-climates and attenuator of internal thermal environment;
- the courtyard in winter;
- what makes a courtyard climatically desirable;
- visual characteristics;
- non-functional aspects; and
- the courtyard in a historical and archaeological context;

Some of these aspects have already been discussed in separate publications while others will be dealt with in future papers.

It is not always possible to satisfy the requirements of all these aspects of the courtyard; a compromise solution would have to be adopted which would satisfy most of these requirements according to a weighting system of priorities adopted for each case. Therefore, this paper is only one in a series which the author has been writing; these deal in detail with the micro-climatic and thermal environmental performance of courtyards and of their houses in the hot-dry climates of the sub-tropics and cover wide-ranging topics.

14. CONCLUSION

The physical characteristics of the courtyard, especially of its three-dimensional form, affect the duration (and intensity) of sunlight in it throughout the year. This duration has a direct impact on its performance micro-climatically and thermal environmentally.

In summer, the design objective is to shorten this duration to a minimum in order to reduce heat gain from direct solar radiation to avoid overheating the interior because of the hot-dry micro-climatic conditions prevailing during this season.

In winter, the design aim is to lengthen this duration to a maximum in order to increase the heat gain from direct solar radiation to warm up the interior because of the cool-to-cold conditions prevailing during this season.

The designer's goal is to reach a compromise between the contradictory requirements of the design criteria for these two different seasons. The balance depends on the severity and duration of the particular season in question. Here, a weighting system may be used to tilt the balance between the requirements of the design objectives of the two seasons and to prioritise them. Therefore, the resulting design would produce environmental conditions which would not be too hot in summer nor too cold in winter.

It is perhaps pertinent and to the point to mention here an Arabic saying used by the people of Baghdad which states that "not every nut is rounded, and not every rounded object is a nut". Therefore, not every courtyard is necessarily a good one micro-climatically and thermal environmentally; in other words, there are climatically good courtyards, but there are also climatically bad courtyards; that is, the performance of each courtyard depends on its own merits which are directly related to all the influencing factors discussed here. This is especially so when there may be contradictory requirements between the functional aspects of the courtyard (which may dictate its physical characteristics in terms of size in plans, sections, elevations and three dimensions) and its micro-climatic and thermal environmental performance.

Hence, a check list should be used to identify, analyse, and assess the performance of the courtyard from all aspects: functional, physical, social, religious, cultural, micro-climatic, thermal environmental, planning, architectural, visual, and psychological.

When these are arranged in a hierarchical order of priorities and a weighting system is applied, the resulting design of any particular courtyard would be rational, appropriate and fulfilling the necessary requirements in question. Because in most cases the priorities are different, the resulting courtyards would also be different, reflecting the priorities themselves; however, the concept, design and detailing of the courtyard itself would be correct for the circumstances under discussion.

BIBLIOGRAPHICAL NOTES

1. AL-AZZAWI, Subhi. (1985a). Oriental Houses in Baghdad: Concepts, Types and Categories. Part I: Concepts and Types. UR: The International Magazine of Arab Culture. No.1; pp. 2-14 (incl. 14 fig.). London.
2. AL_AZZAWI, Subhi. (1985b). Oriental Houses in Baghdad: Concepts, Types and Categories. Part II: Categories. UR. No. 2; pp. 30-41 (incl. 23 fig.). London.
3. AL-AZZAWI, Subhi. (1985c). Oriental Houses in Baghdad: Concepts, Types and Categories. Part III: Social Factors, Analogy with Ancient Houses, Conservation and Conclusions. UR. No. 3; pp. 2-21 (incl. 29 fig.). London.
4. AL-AZZAWI, Subhi. (1969). Oriental Houses in Iraq. In: Shelter and Society (P. Oliver, ed.), pp. 91-102 (incl 26 fig.). Barrie and Jenkins. London.
5. AL-AZZAWI, Subhi. (1986). The Courtyards of Oriental Houses in Baghdad: non-functional aspects. In: The Arab House: proceedings of a colloquium (A.D.C. Hyland and A. Al-Shahi, ed.), pp. 53-60 (incl. 2 fig.). CARDO, University of Newcastle-upon-Tyne, UK.
6. AL-AZZAWI, Subhi. (1989). The Courtyards of Indigenous Oriental Houses in Baghdad: functional aspects and visual characteristics. In: The Courtyard As Dwelling (N. Alsayyad and J.P. Bourdier, ed.), pp. 59-84 (incl. 16 fig.). CEDR, University of Berkeley, California.
7. REUTHER, Oscar. (1910). Das Wohnhaus in Bagdad und Anderen Städten des Iraq. Verlag von Ernst Wasmuth, A.-G. Berlin. Fig. 73, p. 34 and fig. 74, p. 35, **Bayt** Sha'shoor, Baghdad.
8. And plan dimensions of 4.0m x 3.0m.
9. And plan dimensions Of 3.65m x 3.65m. **Bayt** Ibin Khadh-dhoori, house no. 249/206; **Mahallat** Siraj Al-Deen, Baghdad; measured drawing by the author (1971-72).

10. And plan dimensions Of 3.90m x 3.90m. **Bayt** Ibin Chireem; house no. 245/206; **Mahallat** Siraj Al-Deen, Baghdad; measured drawing by the author (1971-72).
11. And plan dimensions of 11.55m x 7.25m. **Bayt** Al-Haj Ali Al-Kaw-waz; house no. 259/206, **Mahallat** Siraj Al-Deen, Baghdad; measured by the author (1971-72).
12. And plan dimensions of 10.0m x 8.65m. **Bayt** Al-Haj Abdul-Ameer Al-Khateeb; house no. 112/14, **Mahallat** Al-Qattana, **Bab** Al-Murad, Kadhimiya, Baghdad; partial measured drawing by the author (1971-72).
13. And plan dimensions of 10.40m x 8.45m. **Bayt** Falih Al-Qassab; house no. 24/1/19, **Mahallat** Karradat Maryam, Baghdad; measured drawing by the author (1971-72).
14. And plan dimensions of about 4.0m x 2.5m. **Bayt** Usta Hasan Faraj (formerly, **Bayt** Abu Shawqi); house no. 275/206, **Mahallat** Siraj Al-Deen, Baghdad.
15. And plan dimensions of about 3.0m x 3.0m.
16. And plan dimensions of about 14.0m x 10.0m. The **Haram** courtyard of Bayt Hussein Pasha, Baghdad; measured drawing by Oscar Reuther (1908-10); op. cit.; fig. 63, p. 29.
17. And plan dimensions of about 14.5m x 14.5m. The **Haram** courtyard of **Bayt** Menaheem, Baghdad; measured drawing by Oscar Reuther (1908-10); op. cit.; fig. 48, p. 20.
18. And plan dimensions of about 22.0m x 15.0m. The **Diwan-Khana** courtyard of **Bayt** Hussein Pasha, Baghdad; measured drawing by Oscar Reuther (1908-10); op. cit.; fig. 63, p. 29.
19. In the **Tarma** of **Diwan-Khanat** Al-Estarabadi; op. cit.
20. In the **Tarma** of **Bayt** Ibin Khadh-dhoori; op. cit.
21. In the **Tarma** of **Bayt** Ibin Chireem; op. cit.
22. Measured by others, but not by the author.
23. In the **Tarma** of **Bayt** Falih Al-Qassab; op. cit.
24. In the **Tarma** and **Talar** of **Bayt** Al-Haj Ali Al-Kaw-waz; op. cit.
25. **Bayt** Ibin Khadh-dhoori; op. cit.
26. **Bayt** Ibin Chireem; op. cit.
27. **Diwan-Khanat** Al-Estarabadi; op. cit.
28. **Bayt** Al-Haj Ali Al-Kaw-waz; op. cit.
29. **Bayt** Sha'shoo'; house no. 3/8/1. **Mahallat** Al-Sinak, Baghdad; partial measured drawing by the author (1971-72).
30. Measured by others, but not by the author.
31. 6.575m in **Bayt** Ibin Khadh-dhoori; op. cit.
32. 6.50m-6.60m in **Bayt** Ibin Chireem; op. cit.
33. 7.38m in **Diwan-Khanat** Al-Estarabadi; op. cit.
34. Measured by others, but not by the author.
35. 9.05m in **Bayt** Al-Haj Ali Al-Kaw-waz; op. cit.
36. **Menzil (Bayt)** Ali Efendi Labib; house no. 4 **Harat** (alleyway) Al-Lebbana, Al-Qal'a, Cairo; occupied by the late Egyptian architect Hassan Fathy.
37. **Diwan-Khanat** Al-Estarabadi; op. cit.
38. **Bayt** Ibin Khadh-dhoori; op. cit.
    Bayt Ibin Chireem; op. cit.
    Bayt Falih Al-Qassab; op. cit.
39. **Bayt** Al-Haj Ali Al-Kaw-waz; op. cit.
40. **Menzil (Bayt)** Ali Efendi Lebibb; op. cit.
    Menzil (Bayt) Jamal Al-Deen Al-Dhahabi; Harat Hosh Qadam, Cairo.
    Menzil (Bayt) Al-Suhaymi, **Harat** (alleyway) Al-Darb Al-Asfar, Jammaliya, Cairo.

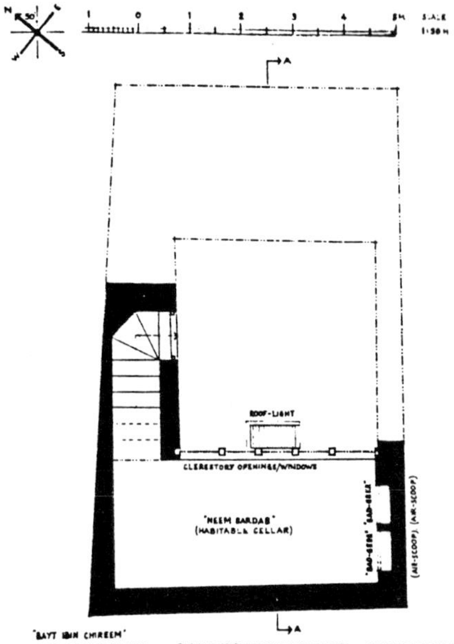

"BAYT IBN CHIREEM" MAHALLAT SIRAJ AL-DEEN HOUSE NO. 245/206, BAGHDAD

PLAN AT SEMI-SUBTERRANEAN LEVEL ("NEEM SARDAB")

SURVEY & DRAWING BY SUBHI AL-AZZAWI.

FIG. 1

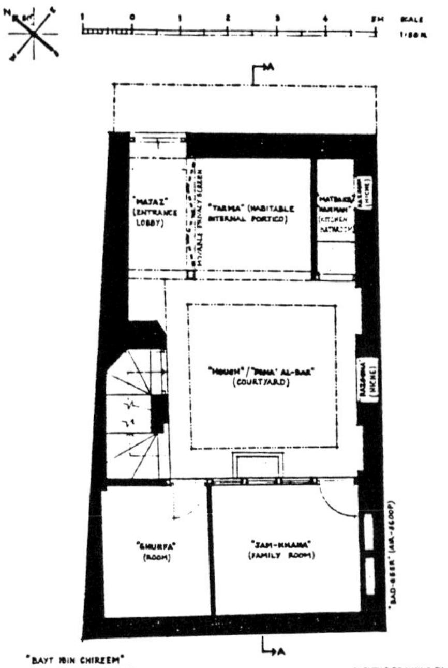

"BAYT IBN CHIREEM" MAHALLAT SIRAJ AL-DIN HOUSE NO. 245/206, BAGHDAD.

PLAN AT GROUND FLOOR LEVEL.

SURVEY & DRAWING BY: SUBHI AL-AZZAWI.

FIG. 2

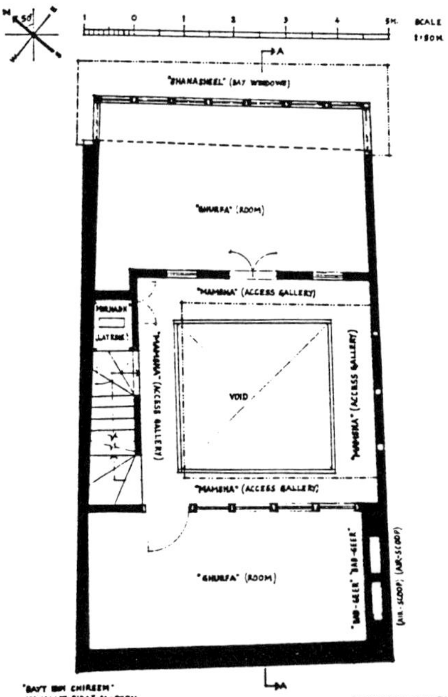

"BAYT IBN CHIREEM" MAHALLAT SIRAJ AL-DEEN HOUSE NO. 245/206, BAGHDAD.

PLAN AT FIRST FLOOR LEVEL.

SURVEY & DRAWING BY SUBHI AL-AZZAWI.

FIG. 3

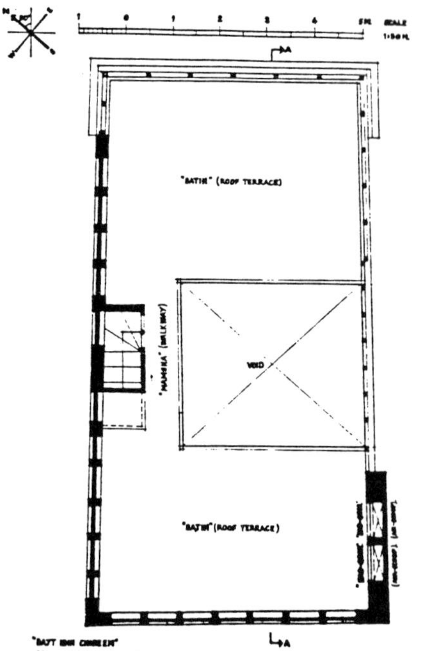

"BAYT IBN CHIREEM" MAHALLAT SIRAJ AL-DEEN HOUSE NO. 245/206, BAGHDAD.

PLAN AT ROOF TERRACE LEVEL ("SATIH")

SURVEY & DRAWING BY SUBHI AL-AZZAWI.

FIG. 4

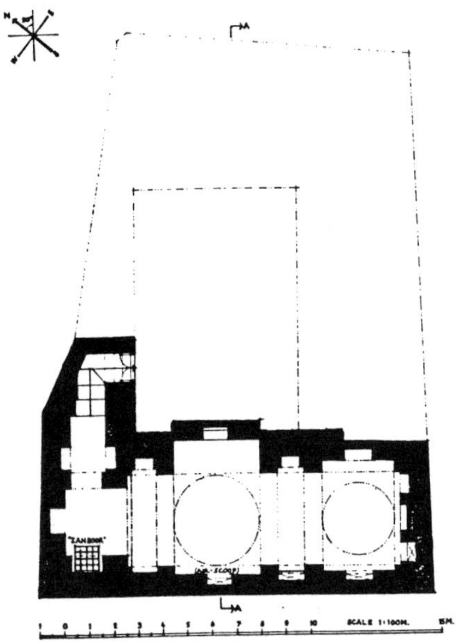

HOUSE AT KADHIMIYA, BAGHDAD. (DIWAN-KHANAT AL-ESTARABADI). PLAN AT SEMI-SUBTERRANEAN LEVEL ("NEEM SARDAB") SURVEY & DRAWING BY: SUBHI AL-AZZAWI.

FIG. 5

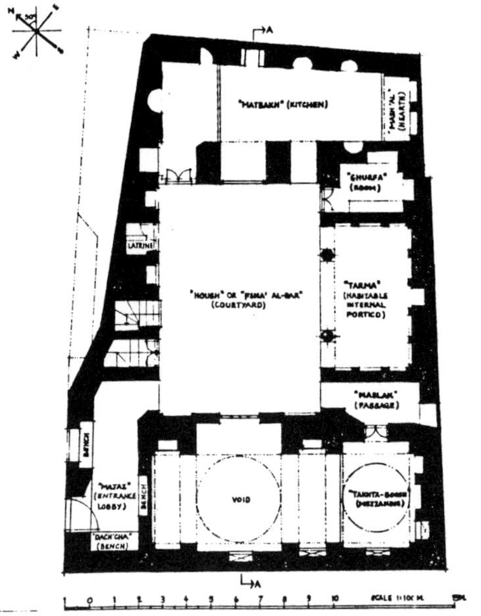

HOUSE AT KADHIMIYA, BAGHDAD. (DIWAN-KHANAT AL-ESTARABADI). PLAN AT GROUND FLOOR LEVEL SURVEY & DRAWING BY: SUBHI AL-AZZAWI.

FIG 6

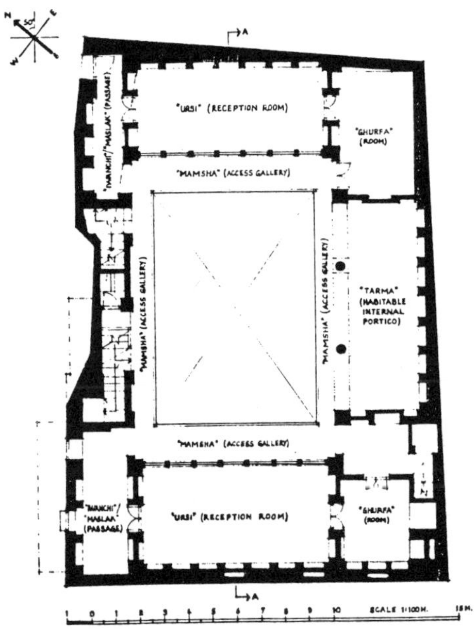

HOUSE AT KADHIMIYA, BAGHDAD. (DIWAN-KHANAT AL-ESTARABADI). PLAN AT FIRST FLOOR LEVEL SURVEY & DRAWING BY: SUBHI AL-AZZAWI.

FIG. 7

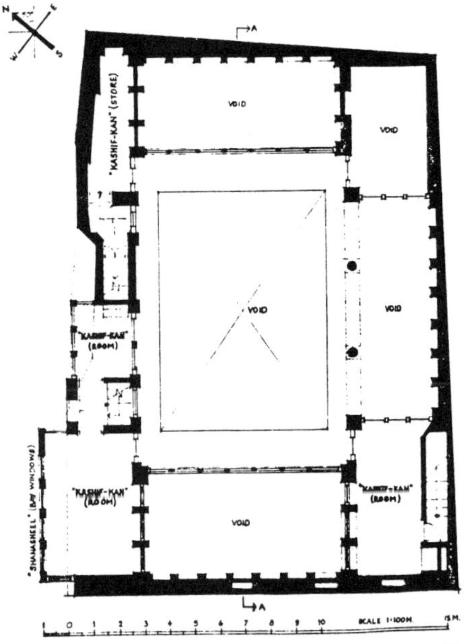

HOUSE AT KADHIMIYA, BAGHDAD. (DIWAN-KHANAT AL-ESTARABADI). PLAN AT "KASHIF-KAN" LEVEL (MEZZANINE AT FIRST FLOOR) SURVEY & DRAWING BY: SUBHI AL-AZZAWI.

FIG. 8

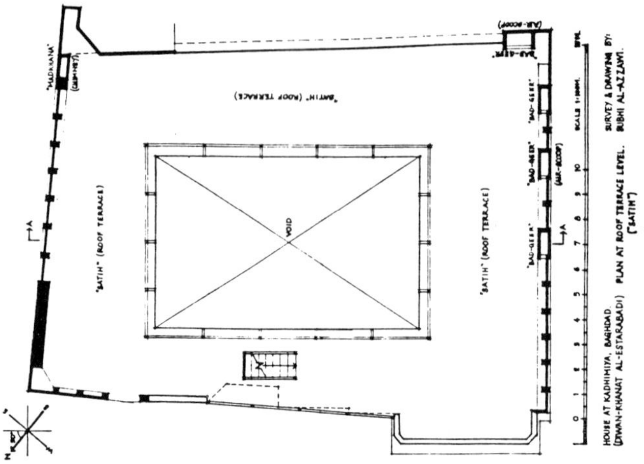
HOUSE AT KADHIMIYA, BAGHDAD.
(DIWAN-KHANAT AL-ESTARABADI)
PLAN AT ROOF TERRACE LEVEL.
("SATIH")
SURVEY & DRAWING BY:
SUBHI AL-AZZAWI.
"SATIH" (ROOF TERRACE)
VOID
FIG. 9

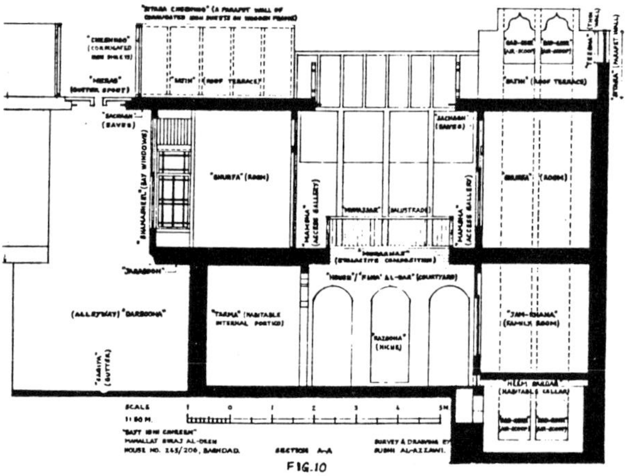
"GHURFA" (ROOM)
"TARMA" (HABITABLE INTERNAL PORTICO)
"HOUSH"/"FANA' AL-DAR" (COURTYARD)
"SATIH" (ROOF TERRACE)
SECTION A-A
SURVEY & DRAWING BY:
SUBHI AL-AZZAWI.

FIG.10

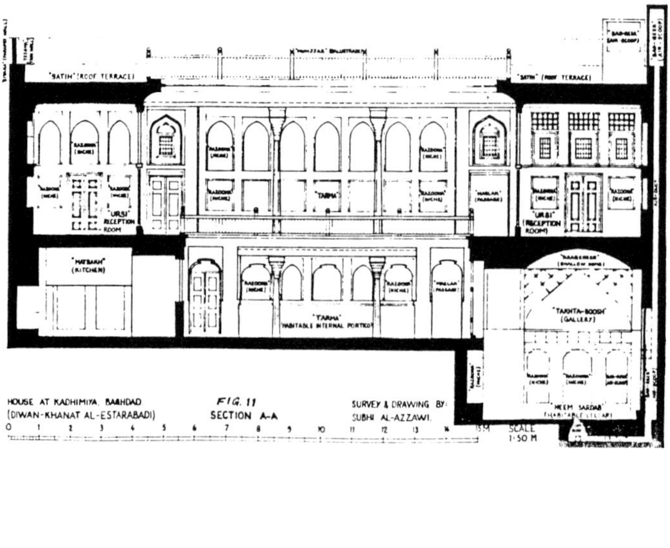
"SATIH" (ROOF TERRACE)
"MATBAKH" (KITCHEN)
"TARMA" HABITABLE INTERNAL PORTICO
"TAKHTA-BOOSH" (GALLERY)
"URSI" RECEPTION ROOM
HOUSE AT KADHIMIYA, BAGHDAD
(DIWAN-KHANAT AL-ESTARABADI)
FIG. 11
SECTION A-A
SURVEY & DRAWING BY:
SUBHI AL-AZZAWI.
SCALE 1:50 M

THE THERMAL GESTALT

S.V.SZOKOLAY
Director, Architectural Science Unit
The University of Queensland
Brisbane, 4072, Australia

ABSTRACT. 'The whole is more than the sum of its parts' - a statement as true for the thermal behaviour of buildings as in perception psychology. The validity of this tenet is demonstrated by a number of case studies and the term 'thermal Gestalt' is introduced to describe it. A proposed working method for approaching the complex design problem is outlined.

1. INTRODUCTION

The word `Gestalt' came into general use from the German, denoting a branch, or school of psychology. The literal translation of the word would be `shape, form or figure', but in the context of perception psychology it is better translated as `configuration'. It means the functional whole of the experience, together with the relevance, weighting or relative importance of its component parts. Emphasis is laid on the structure of the phenomenon or thing perceived, or on the structure imposed by the perceptual process, rather than on its parts or elements.

The word now appears in the Oxford English dictionary (1978), where the following definition is given: "perceived organised whole, (in the 1951 edition the following words were also inserted: 'in which each individual part affects every other') that is more than the sum of its parts, eg. a melody, as distinct from the separate notes of it."

(Subsequently Gestalt psychology, particulary its unsubstantiated hypotheses have been severly criticised, but its principal contribution survives, eg. in the study of pattern recognition.)

2. THESIS AND EVIDENCE

Analogia non probat, - but analogy can be a useful illustration, even if it does not prove anything. It is the thesis of the present paper that the thermal behaviour of a building depends on its `thermal Gestalt'; not just the thermal properties of its components, but on

S. Alvarez et al. (eds.), Architecture and Urban Space, 299–304.

the thermal interaction of all these components, their interaction with any mechanical systems installed, with the outdoor environment and with the user.

2.1. The incident that triggered these thoughts occurred a few years ago, when a research student was doing a parametric study on the roof insulation of a standard house in the Brisbane summer. He reported that something was wrong with the simulation program used (HARMON), as the more insulation he puts in, the hotter the house will be. He expected (not unreasonably) a reduction of solar heat gain by an increased amount of insulation.

There was nothing wrong with the program. What happened was that the solar gain through windows plus the internal heat gain caused the indoor temperature to rise well above the outdoors, so there was an outward heat flow. The added insulation reduced this heat dissipation, hence the indoor temperature remained higher.

Since that time dozens of similar examples came to the writer's notice. Some are obvious, some were surprising. A few will be discussed.

2.2. An article in a glass industry journal ('Glasstime',1977) was promoting glass (large glass windows) as the solution to our energy problems. Even a cursory examination will show that there is nothing inherently good or bad about glass. The U-value (or thermal transmittance) of a single glass window is at least ten times greater than that of any reasonable wall, so it will lose ten times more heat. In some situations it may admit a solar gain greater than the heat loss through it over a season. In other cases it may become a net loser of energy.

2.3. A question was asked what would be the optimum size for a north-facing window (southern hemisphere!) to provide direct gain in the Canberra winter. There is no simple answer. It depends on a whole host of factors. Given the temperature and radiation regime, a larger window is possible if we use some form of night insulation. However, if the building has insufficient thermal mass, severe overheating can occur, even in winter and there will not be enough heat stored to keep the house warm overnight.

2.4. There is a battle raging over the question of the courtyard house type in a hot-dry climate. The conventional wisdom advocating such a courtyard is now seriously attacked (eg. Roaf, 1990). Several research workers measured the courtyard temperature when the high altitude summer sun caused it to increase well above that of the free air. There is no doubt of the results in the case of an unshaded, paved courtyard, devoid of any vegetation and water. However, with at least partial shading, a pond or a spray and ample vegetation the courtyard can be a very pleasant place and a source of cool air for supply to the house.

2.5. A prominent British professor of building science categorically stated that lightweight buildings have a lesser heating requirement than otherwise identical heavyweight ones (Burberry, 1970). This was demonstrated by calculating the heating demand of both, when the heating was intermittent for the 9 to 5 working day. The heavy building would have had a longer heating up period and - if the heating was switched off at 5 pm. - the significant amount of stored heat would have been lost overnight.

The statement is not true for a continuously heated building or for one which is continuously occupied but intermittently heated. The latter would be possible with the storage mass inside of a resistive insulation layer, as this would ensure a very stable indoor thermal regime even with intermittent heating. A direct gain passive solar system just would not work without adequate thermal storage mass.

So, the categorical statement should be avoided and perhaps replaced by a precisely qualified conditional one.

2.6. In some texts the use of skylights is stongly advocated as an effective energy conservation measure (eg. Mazria, 1979). Unfortunately the skylight will admit more solar radiation in the summer than in the winter. It is true that it can provide daylighting to an internal (windowless) space, thus save the electricity which would otherwise be consumed, but - in most cases the skylight would be a net loser of energy: heat loss in winter, solar gain in the summer. At low latitudes (where the sun is near the zenith) it can become a serious liability. If a shading system is fitted to it, this would defeat the daylighting purpose.

If the skylight is used to achieve a certain quality of lighting, as a mood element, we cannot argue with it, but the energy saving claims would rarely be true and are no more than a pretext,

2.7. In a parametric study (Szokolay, 1991), using a small standard house design, it was found that in Townsville (latitude 19°S) a concrete slab-on-ground floor has an advantage (in terms of indoor overheating degree-hours) only with very light wall and roof constructions. With brick or masonry walls (even with lightweight roofs) there is already sufficient mass available, the additional concrete floor would make it too much. The peak indoor temperature would be reduced by the concrete floor, but the cumulative overheating would be increased.

Consequently, if the designer asks the question: what is thermally preferable in Townsville, a concrete or a timber floor, the answer must be "It depends ... "

3. DISCUSSION

Some may conclude from a multitude of such examples that science is useless in aiding design, it is unable to answer the simplest questions. Such a conclusion may be popular and may give support to the increasing anti-science trend in architecture, providing ammunition for 'the scientophobic backlash' (Templer, 1990).

Clearly, this is either a hostile or a defeatist attitude. Science in general and architectural science in particular may have their shortcomings, but they provide the only reliable set of measures and the only technique better than sorcery for the prediction of behaviour (in this case: thermal behaviour) of a future building. Science provides the only method which has predictive value.

The dilemma is not unique to architecture, it occurs in the design of any complex system. It is a question of design methodology. There is no single method that could be prescribed (eg. Heath, 1984). The choice depends on the type of building to be designed and many other objective and subjective factors, but some general points can be made.

Gestalt psychology itself may point towards a working method. Vernon (1962) suggests that "we tend to perceive only as much as will enable us to identify what we see .... The observer tends to perceive only as much detail as he thinks necessary to perform the particular task and ignores the rest," but makes the point that for complex shapes more detail must be observed to perceive, to identify, classify and describe them. It is this ability to select the important elements and relationships in any perception which is uniquely human: the power of abstraction, the ability to recognise a pattern behind the surface phenomena. In the creative process of design we may also start with the broad outline, the overall pattern and gradually add the details, increasing the complexity.

In architectural design the essential skill lies in identifying the constraints, rank-order them and satisfy first the dominant ones. From the thermal (solar) point of view the dominant constraint will probably be orientation. If this conflicts with high-ranking constraints in other fields (eg. a view or existing access), then probably only subjective value-judgements can provide a resolution.

Having arrived at a resolution, a design decision would be made, this particular characteristic of the design (in this case: orientation) would be fixed. The next highest ranking constraint would then be considered, and so on until an overall pattern of the design would emerge. Subsequently the process would be continued until all the details are resolved.

This cannot be a linear process, much toing-and-froing, back-tracking and side-stepping may be involved, but the principle remains: a gradual method is followed, meaning that certain characteristics are fixed (or provisionally fixed) and the next characteristic is then considered and possibly optimised within the context of those already fixed. It is important at any step in this process not to foreclose future options without a conscious decision. Indeed in some cases the result of a sub-optimisation study may explicitly allow later choices between alternatives in the form of if... then... type statements.

Often it may be possible to consider several building characteristics simultaneously and introduce some interim abstractions to assist in reaching concrete decisions. Eg. we may side-step the lightweight versus heavyweight wall problem and consider it as part of the thermal mass question. Do we need any thermal storage mass? If so how much? What is the best location for such a mass: the floor, the walls, the roof? (Again a sequence going gradually from the general towards the particular.)

In some instances a sub-system within the complex thermal system of the building may be identified, where the interaction among its elements is closer, greater or more important than its 'external' relationships with the rest of the building. If so, than this sub-system can be designed with some degree of confidence, largely independently and only later to be integrated with the rest of the building (with the other design considerations). The thought processes involved can be illustrated by the following examples:

a) the climate is strongly underheated (cold winter), - we want winter solar heat input; this is so important that due north orientation is decided regardless of any conflicting demands; the window + mass sub-system will be essential: large north-facing windows will be used; this demands a large thermal mass; the disposition of this mass can then be decided in conjunction with mass surface qualities: if an unclutterd dark massive floor is possible, other mass may not be required, but if the floor is reflective then ceiling and wall mass will be necessary;

b) the climate is overheated, with large diurnal variations, but even the day's mean is too high: we must have a night ventilation + mass system: internal surfaces with large admittance (and small transmittance) and provision for ample night ventilation; this thermal sub-system is decided on in an abstract form: it will take a concrete form when other design requirements are integrated with this dominant principle.

4. CONCLUSION

In the thermal/climatic/solar design of buildings we must get used to dealing with the total building, with the configuration of all its parts and design features, for which the new term "thermal Gestalt" is proposed. The design process, the creation of a desirable thermal Gestalt will be very similar to the design of any complex system.

REFERENCES

anon. (1977): Glasstime, No.20

Burberry, P. (1970): Environment and services. Batsford, London

Heath,Tom (1984): Method in architecture. Wiley, New York.

Koffka, K. (1935): Principles of Gestalt psychology. Kegan Paul, London.

Mazria,E. (1979): The passive solar energy book. Rodale Press, Emmaus.

Roaf, Susan (1990): The traditional technology gap: stereotypes of middle-eastern traditional building types and technologies. Trialog 25

Szokolay,S.V.(1991): Climatic design of houses in Queensland, - Supplementary report. University of Queensland, Architectural Science Unit.

Templer,J. (1990): Architectural research. J.of Architectural Education, 44(1):3, November

Vernon,M.D. (1962): The psychology of perception. Penguin.

Chapter 2

Projects

BIOCLIMATIC PROJECTS IN FLORENCE

A.Gioli, P.Puccetti, M.Sala, A.Toti
in cooperation with G.Festa, G.Musa, L.Ceccherini Nelli
Department of Architectural Design
University of Florence
via Cavour 82
50129 Firenze
Italy

The research has been developed by the Architectural Design Department of the University for the Municipality of Florence in 1989/1990 .

The purpose was the project the new center of a large suburbian expansion and in the same project , to improve the passive solar architecture in urban and building variables and to give design rules for planning the further expansion .

The research to define the parameters to choice the orientation of the buildings and their morphological aspects, was carried out with two principal goals :
the respect of the rules of passive solar building as well the quality of the architecture and innovative urban planning .

Specific points of the work were :
a) urban design of the area and studies of the variables (orientation, horizon angle, street with, high and morphology of the blocks) .
b) building design of the various possibilities according to the conventional variables (glazing ratio, room depth, plant efficency, apartment disposal) .
c) Passive solar devices, technological elements and architectural details for the different solution of the blocks.

This project also introduces a different approach to the passive solar building design , based on a new concept of urban planning and not only on the solution of a single building .

The complexity of the urban and the bioclimatic approach create the various solutions that improve the quality of the town as show in following project :
1) General plan of the area. The north/south axis cross the existing main road and connect the old village with the new centre (we call it "rambla") with a pedestrian route that board all the buildings and finish on a final plaza with the auditorium and a passage to Arno river. Parking and car circolation are outside the centre.
2) This building is orientated on the eliothermic axis (19 degree east) that is also the old roman grid "centuriatio" in sourranding areas. The building has an internal gallery, enlighted from above, that give access to the duplex apartment and to commercial basement.
3) The "tower" building is based on a east/west porch that connect the residential center with sport areas. Passages to north front can be closed in winter with light structures, while the south facade presents direct gain captation throgh a curved glace frame that reduce the possible overheating in middle seasons.
4) This building is north/south oriented and board the "rambla" with a long porch for commercial facility. The duplex apartment have bow-window on both facades, while the simplex apartment on the last floor are separated and take the sun from a green-house facing south . The north edge is an offices building.
5) Computer aided design of the "rambla" in day and night view .

S. Alvarez et al. (eds.), Architecture and Urban Space, 307–312.

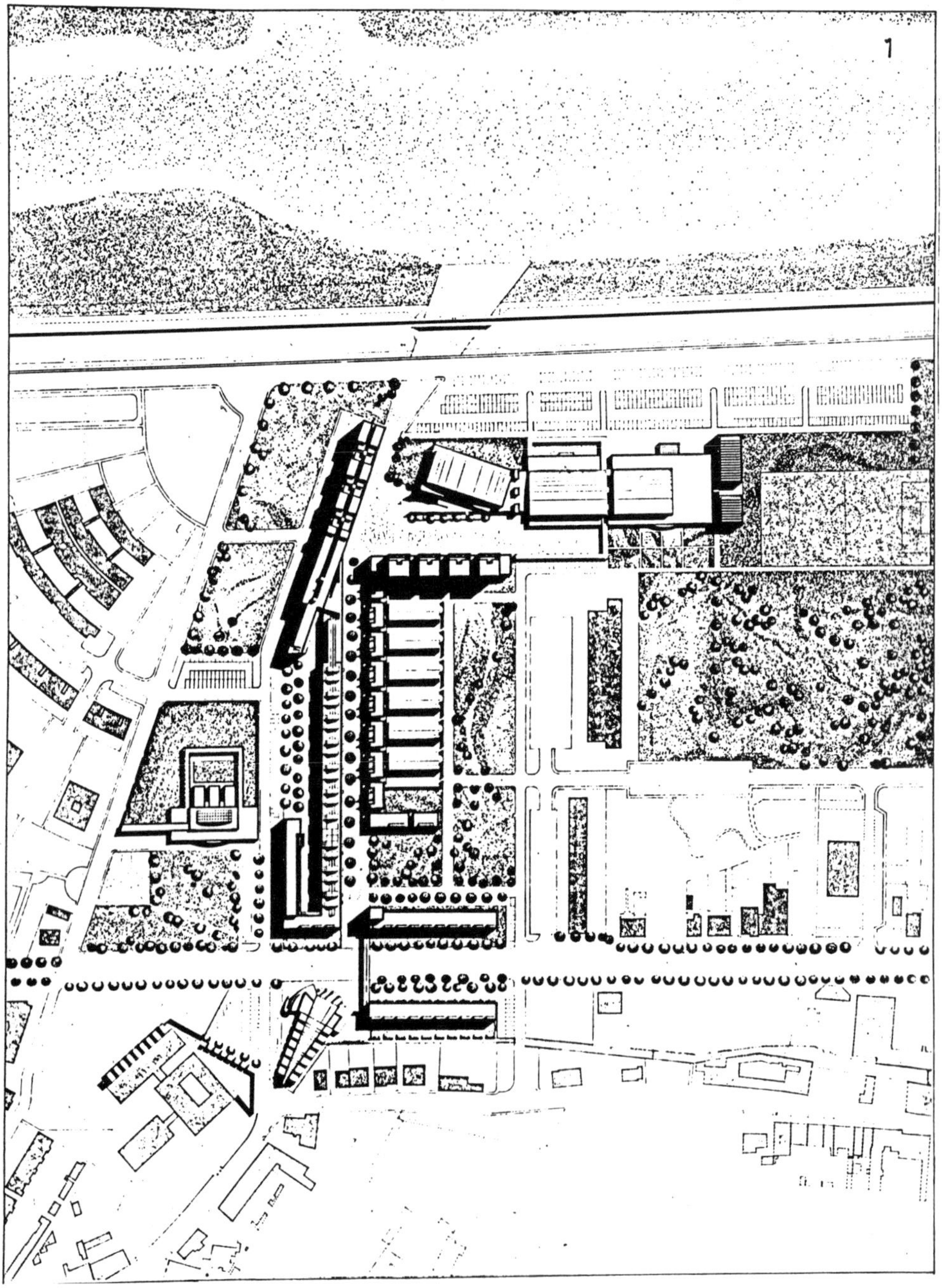
1

COMPARTO 8: EDIFICIO NODALE PER RESIDENZE, COMMERCIO E UFFICI CONCLUDENTE IL SISTEMA URBANO VERSO L'ARNO

2

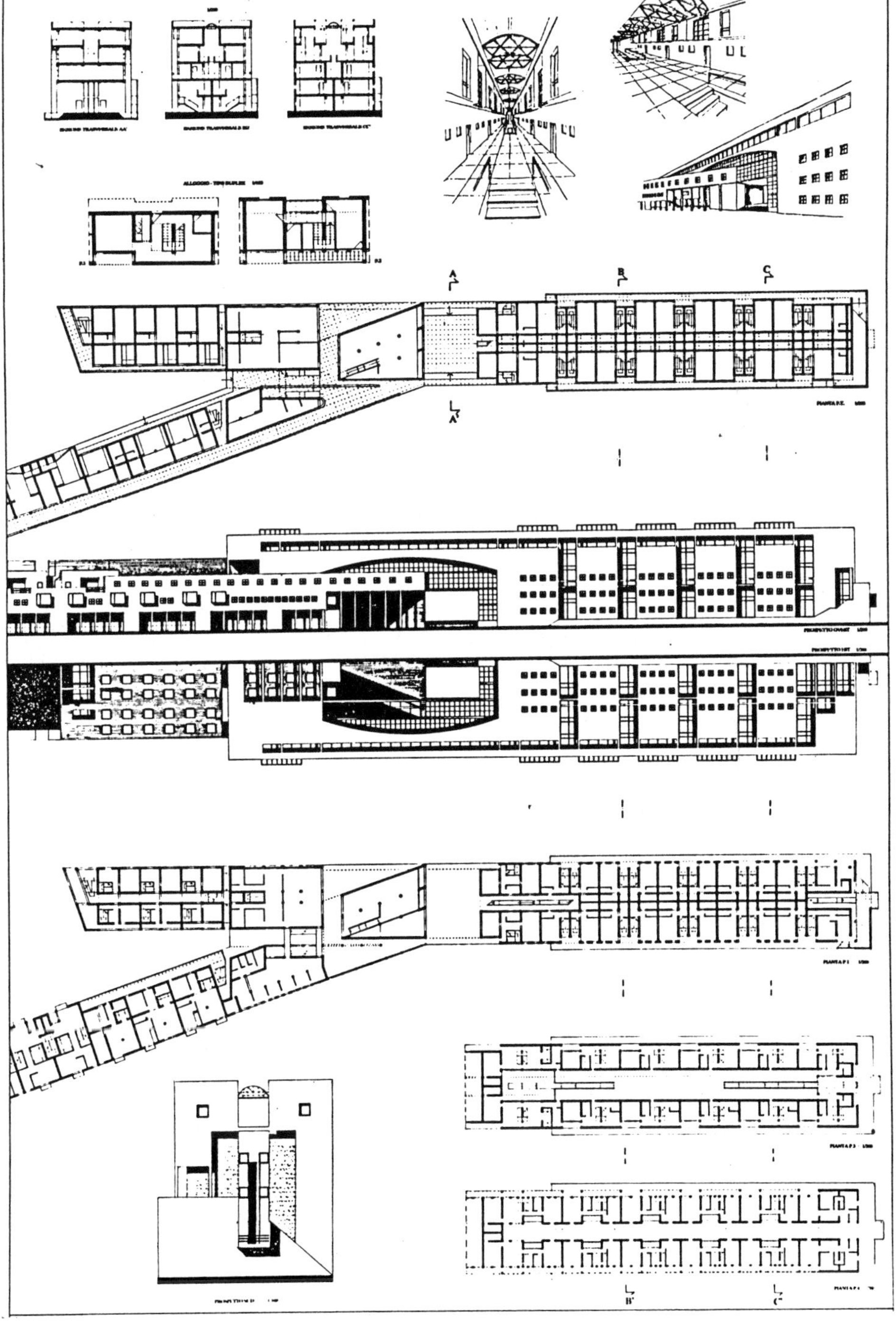

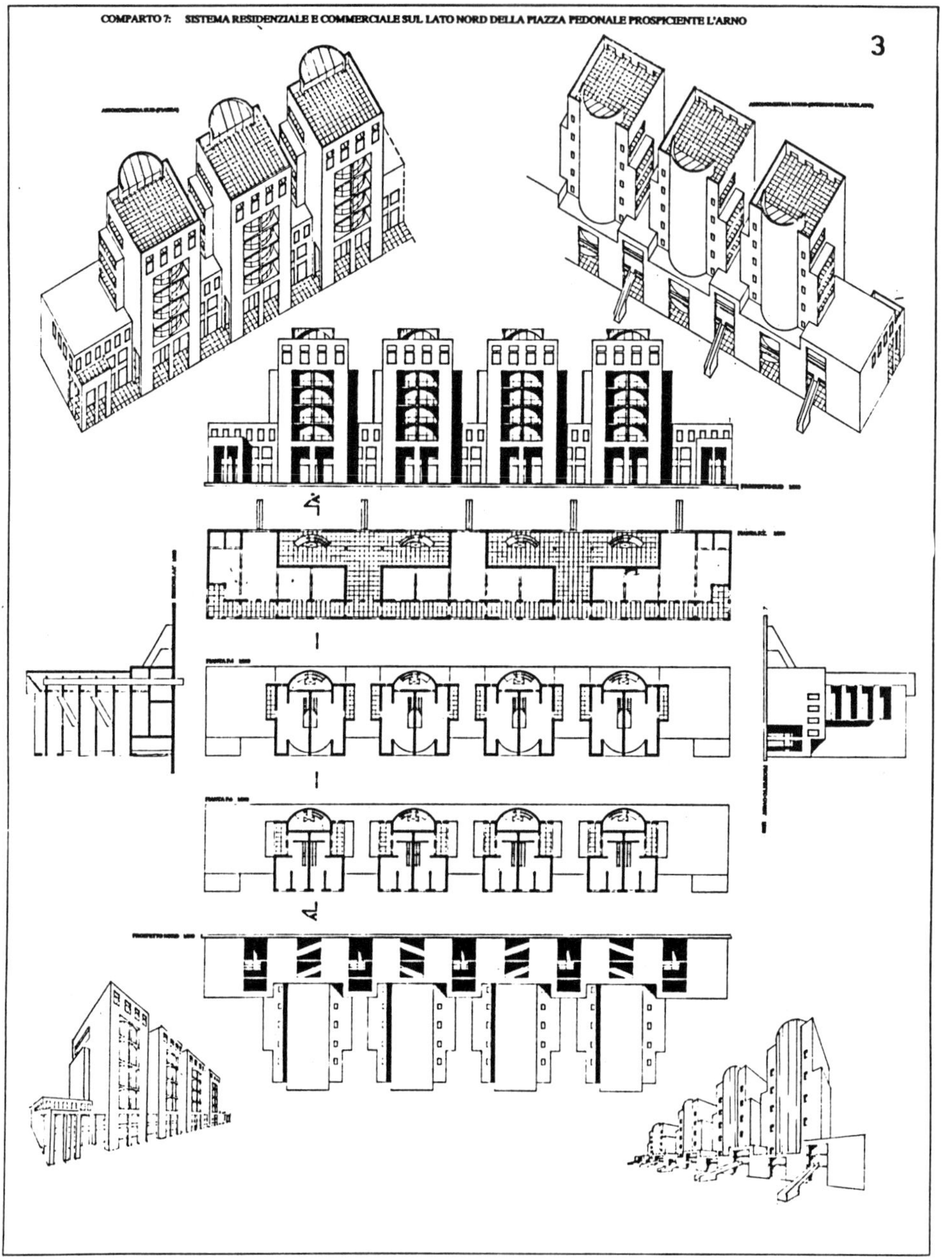
COMPARTO 7: SISTEMA RESIDENZIALE E COMMERCIALE SUL LATO NORD DELLA PIAZZA PEDONALE PROSPICIENTE L'ARNO
3

COMPARTO 6: SISTEMA RESIDENZIALE E COMMERCIALE SUL PERCORSO PEDONALE CENTRALE

4

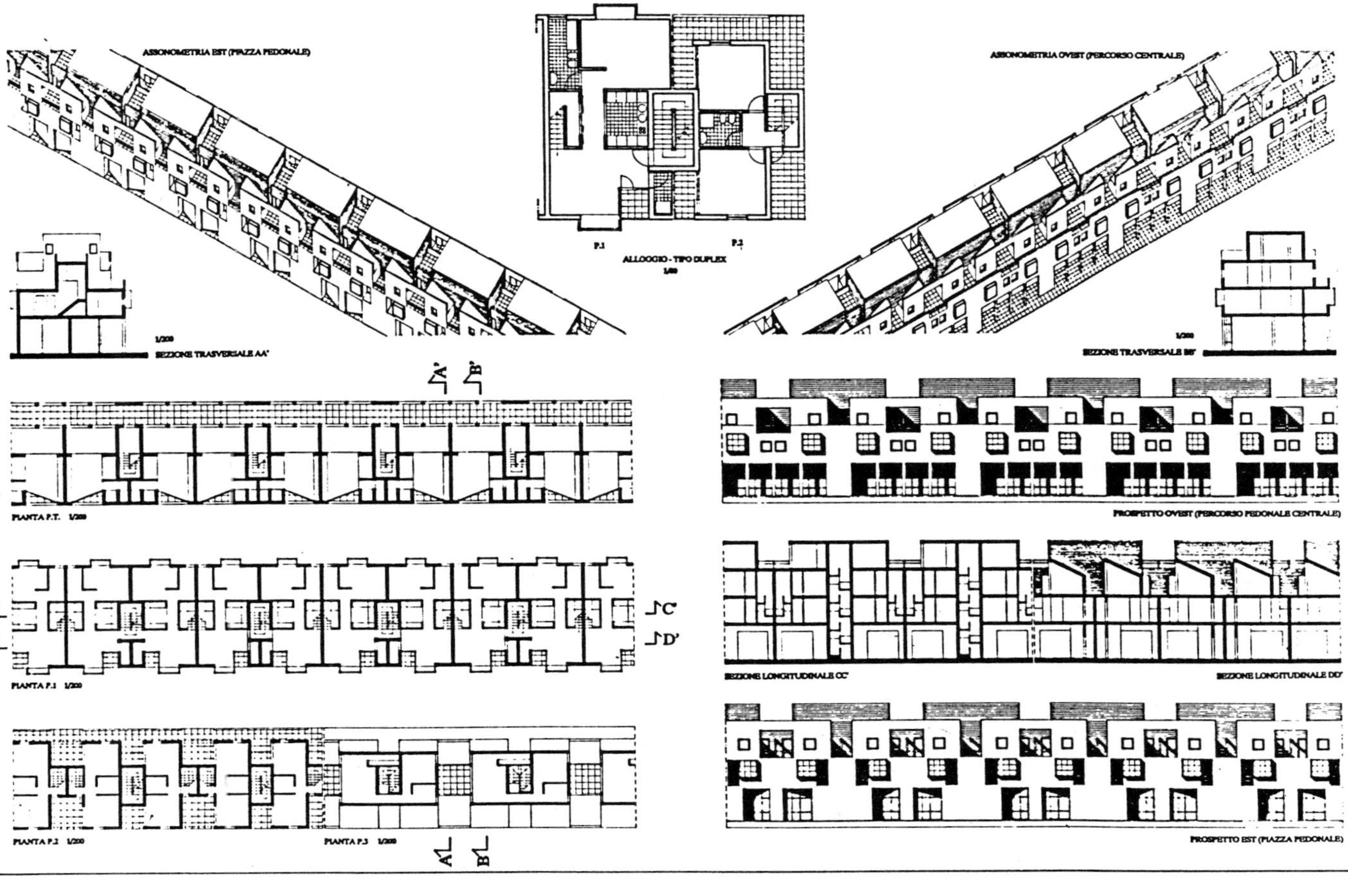

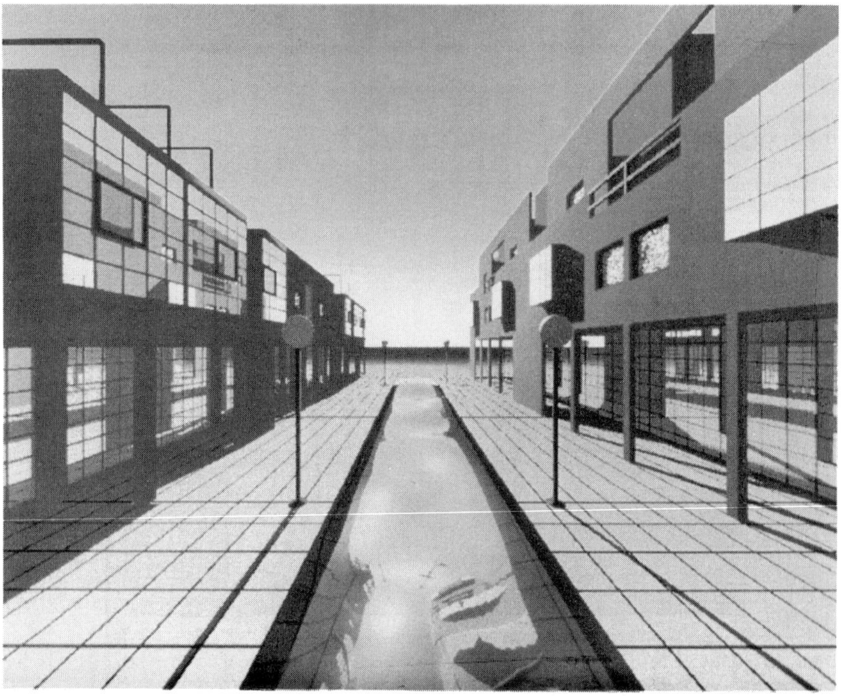

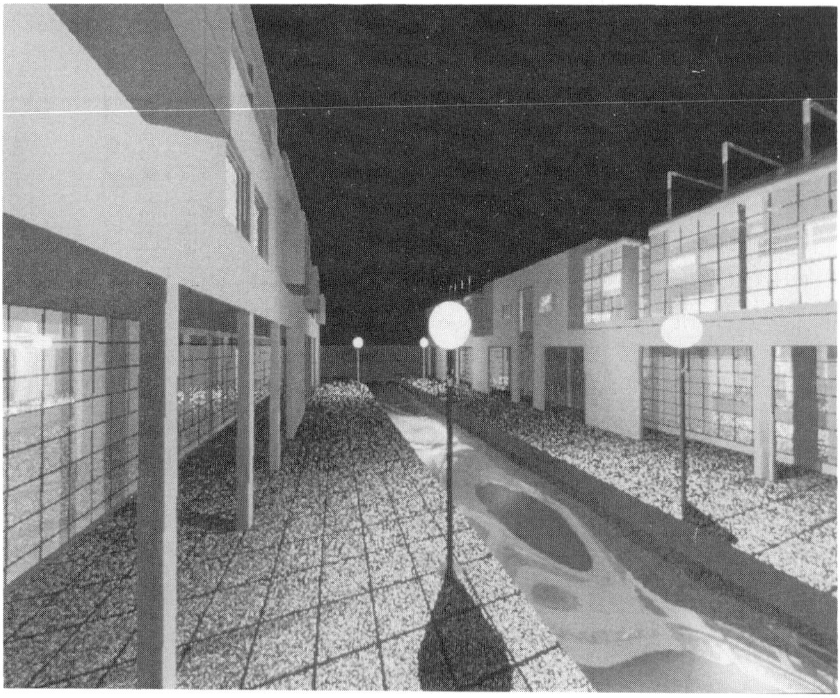

"THE ARCHITECTURE OF A.N.TOMBAZIS AND ASSOCIATES"

A.N.TOMBAZIS
85, Marathonodromou str.
GR-154 52 Psychico
Greece

ABSTRACT A number of office projects illustrate the issue that bioclimatic design is an important, though one of all the considerations of architectural design.

Architectural design should involve a many faceted approach. One of the most important design issues is that of a sensitive approach to climate and environment, man-made or natural. This aspect, contrary to others, exists in every case for every project in every part of the world. Bioclimatic considerations should not be exaggerated but should be incorporated from the very first steps of design in collaboration with the other members of a multidisciplinary design team. A number of office projects illustrate this design philosophy. In some, heating issues were of primary importance, in others, heating and cooling were of equal importance while in the last examples from Dubai the important issue was that of cooling. In all projects, apart from those of housing, daylighting was a governing factor of design.

1. Solar House Helios I - Trapeza Aigialias 1977-79

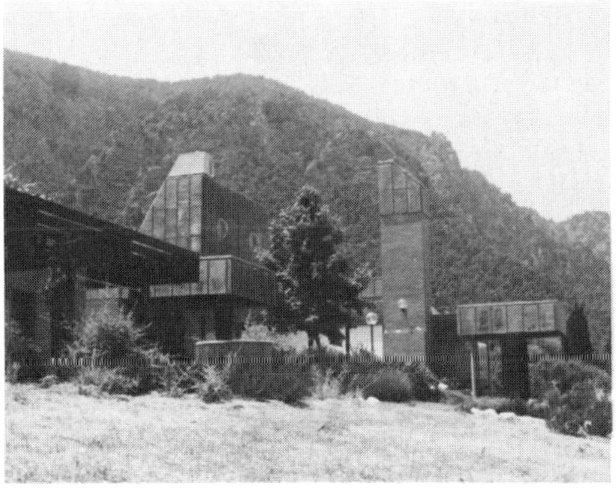

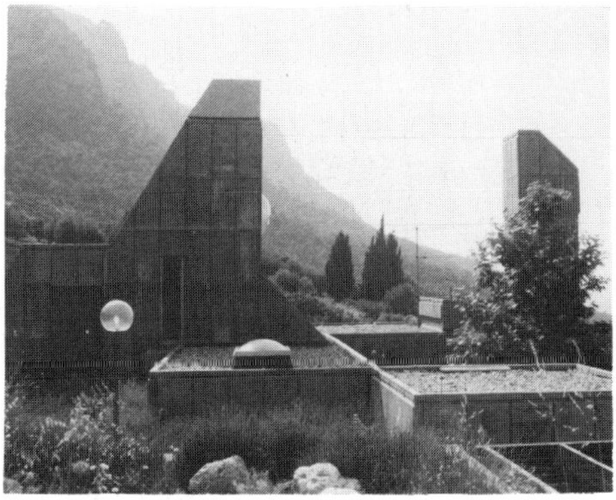

This weekend home of 80m2 is heated by an active solar system with 25m2 of flat plate solar collectors and a 8m3 water storage tank built into the building fabric on a steep, north-facing slope. The house is built in an L shape around an open courtyard which is

S. Alvarez et al. (eds.), Architecture and Urban Space, 313–318.

dominated by a strategically planted plane tree which provides shading both to the house and the open space which functions as an extension for open-air living. The house has performed well but 14 years later the architect wonders why one would have chosen such a complicated system instead of a straightforward passive solution.

2. Solar House Helios II - Ekali, Athens 1980-83

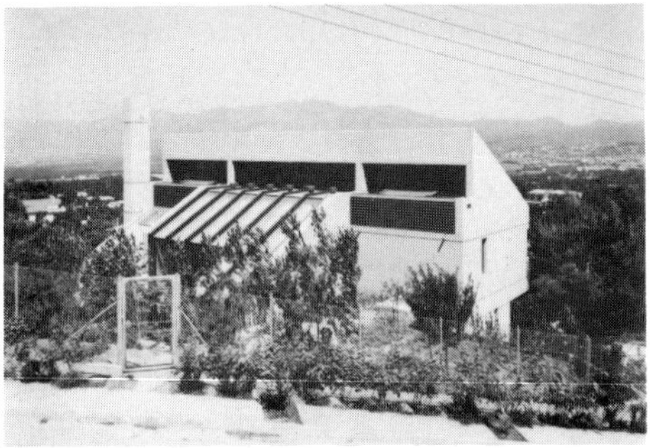

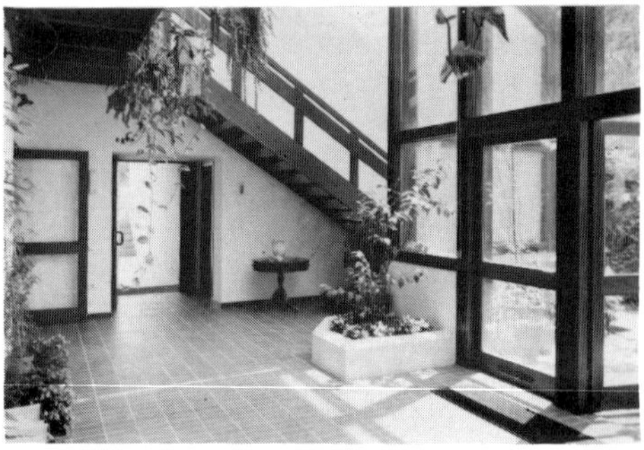

This 300m2 house on a north-facing steeply inclined slope surrounds a double height greenhouse and an open patio dug into the mountainside towards the south. The greenhouse heats the massive walls which radiate to the ground floor living areas. The upper floor bedrooms are heated from the greenhouse via their doors and clerestories that are provided. Shading is provided by interior awnings in the greenhouse. Cross-ventilation is easily achieved for all spaces.

3. Solar House Helios III Perdika - Aegina 1980-83

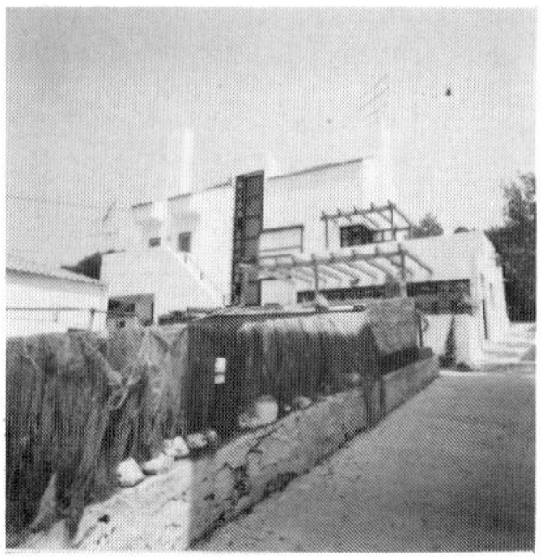

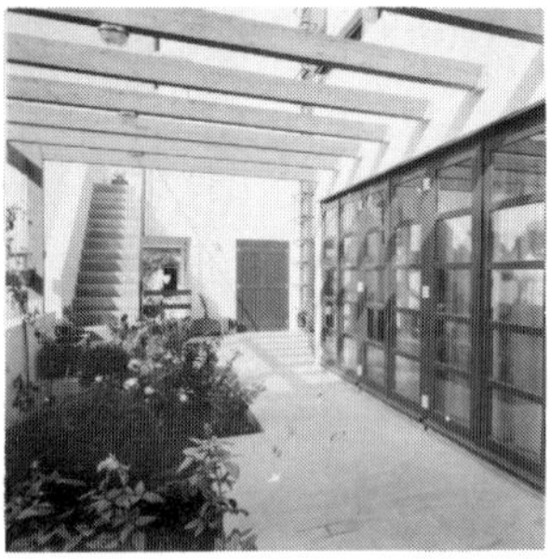

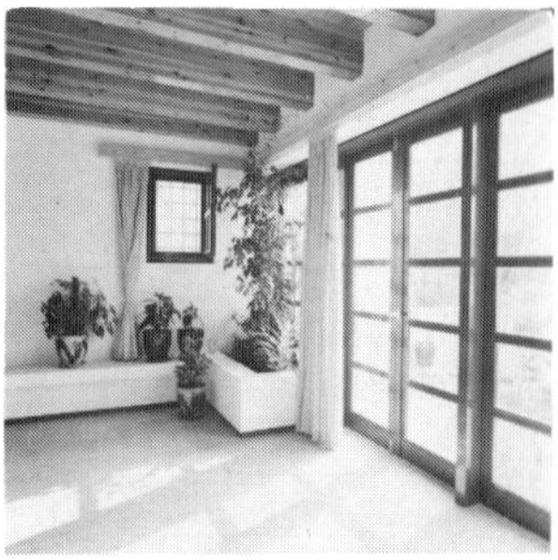

This house for two families is built in a small community on the island of Aegina. It has been provided with a greenhouse, TPU adjacent to the stairs that connect all levels, a mass wall, direct gains and clerestories. The greenhouse is shaded in summer by a vine, while fixed projections and awnings are used to shade the other south-facing apertures.

4. Solar Village, Athens 1978-1989

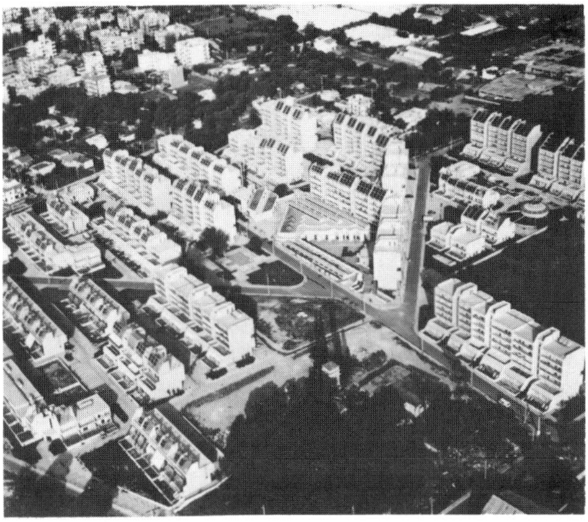

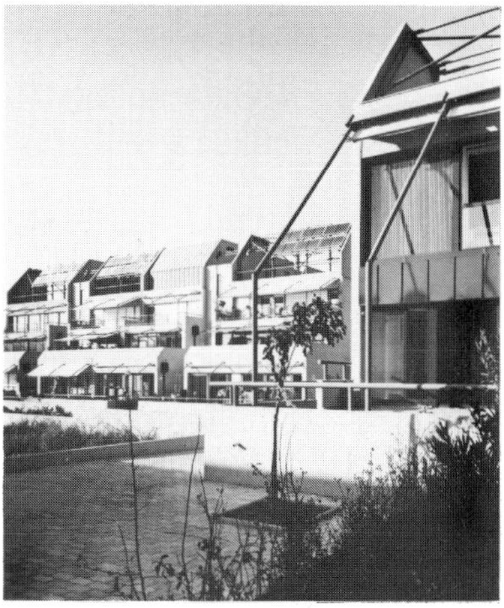

The Solar Village built in the suburbs of Athens has 435 dwelling units of subsidized housing incorporating five different active systems ranging from district heating to inter-seasonal heating and individual building flat plate collector heating. 32 units have been provided with different passive solar systems under similar lived-in conditions. The project is at present being highly monitored.

5. Vila do Conde housing project, Portugal 1988

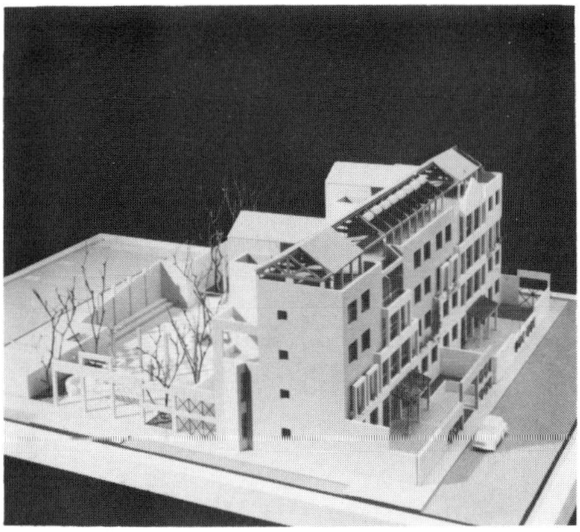

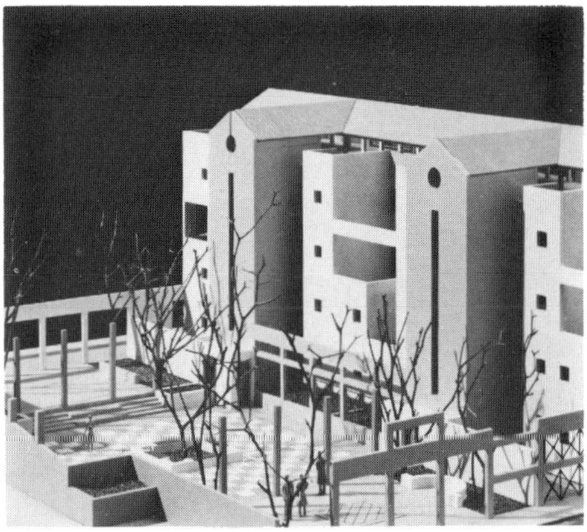

Many of the principles concerning the passive systems of the Solar Village project have been applied to this 16 apartment building of social housing in Vila do Conde, Portugal, which was the object of an international competition. A double south-facing facade was proposed both as a means of shading and for aesthetic reasons.

6. Solar Village - Kalamata 1989-91

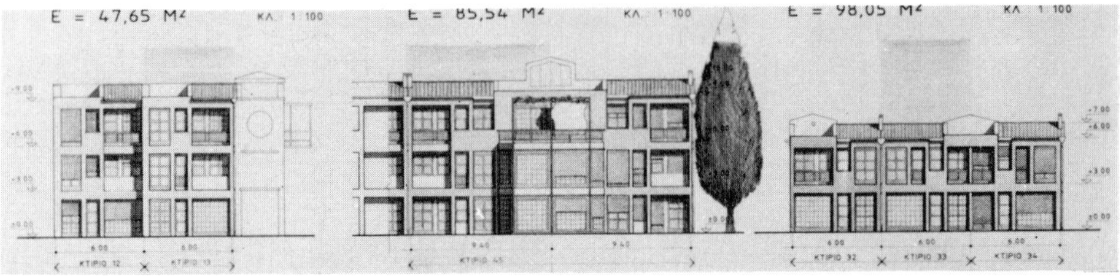

The design of this project for 120 passive solar row-houses in South Greece has just been completed. It provides for one and two level units in buildings of two and three levels. The main passive systems employed in order to keep the project very simple are direct gain and mass walls. Ground floor apertures are provided with safety grills to enable for night ventilation. The landscaping has been designed to complement the solar features.

7. Kindergarten - Athens 1980-84

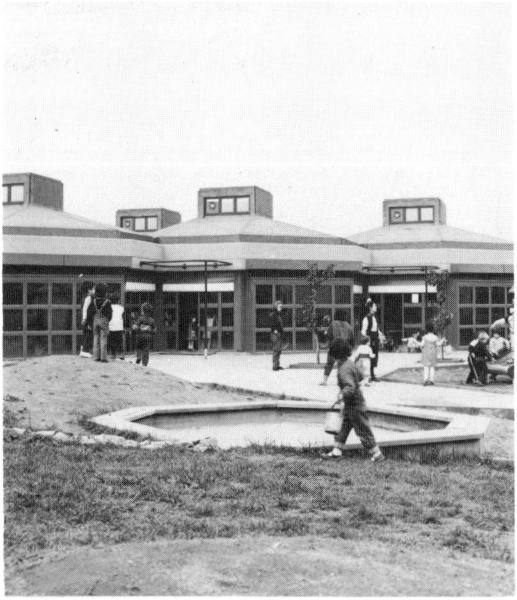

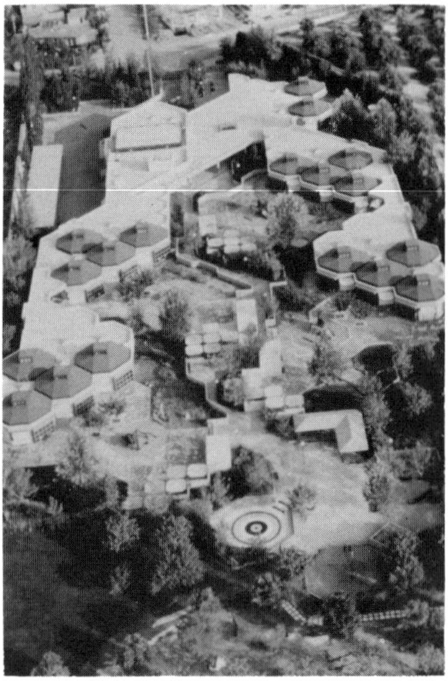

This kindergarten has been built in five semi-independant units for 300 children. The main passive features are direct gain and clerestory windows, mass walls and thermo-siphoning units. Daylighting plays an important role in the project.

8. HHH Tourist Complex, Ag.Nikolaos Crete 1984-89

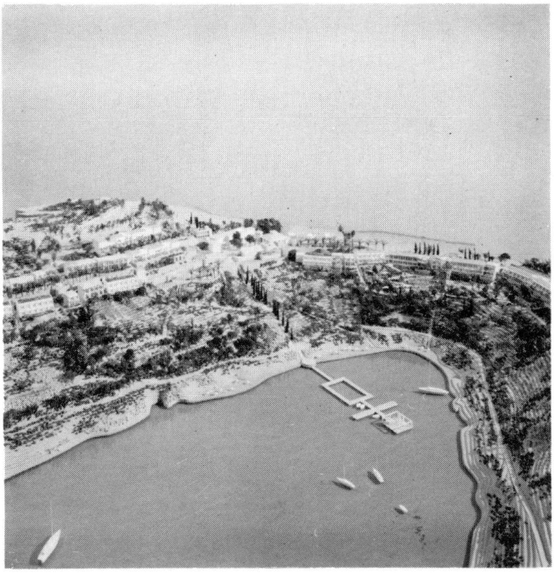

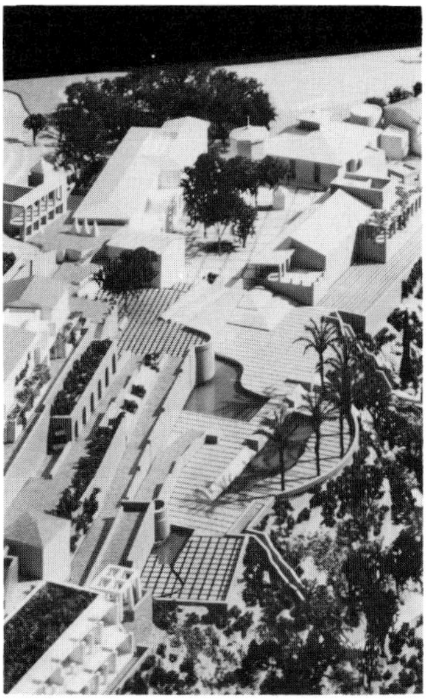

The linear layout of the buildings in the form of a traditional village follows the form of the bay around which the complex develops. A number of buildings are to be dug out of the natural slope and will be top lit. The main concern in this project was that of cooling due to the seasonal use of the project. This will be achieved by cross and night ventilation and the use of buried tubes, ground to air heat exchangers for the precooling of air of the A/C system of the public areas.

9. Extension to the Delfi Museum 1986-88

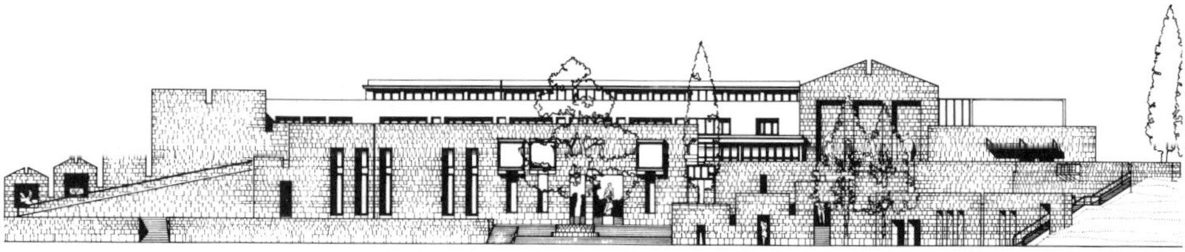

The main feature of this project is that of daylighting. A scale model was constructed to study the daylighting of the main hall of the famous statue of the Charioteer. Different roof apertures and shades of wall finishes were investigated.

10. Mosque, Dubai 1982-1990

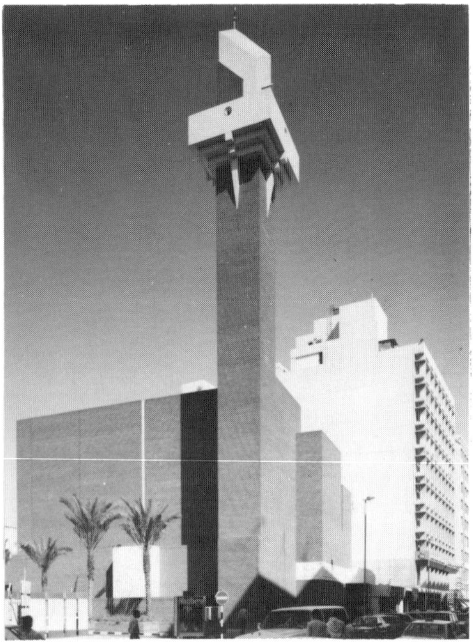

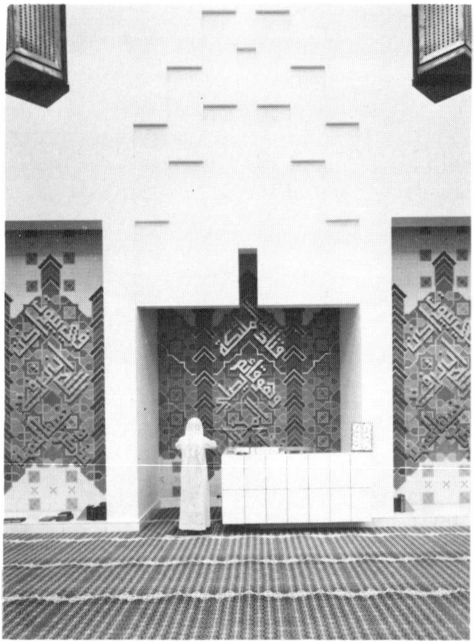

This building has a stepped roof that provides natural lighting to the main praying hall. After dark, artificial lighting, concealed along the clerestories, provides lighting with the same geometry as daylighting. Light shafts introduce light to important parts of the building.

11. Office building - Dubai 1990

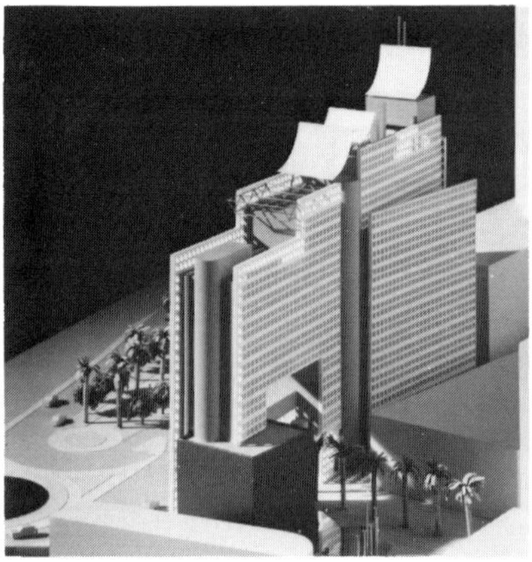

This project for central Dubai has its two long facades created by huge sunscreens moulded according to orientation. In this manner shading is provided and natural light is filtered into the shallow office space behind.

DESIGN OF A NATURALLY VENTILATED LABORATORY BUILDING IN LEICESTER, ENGLAND

BRIAN FORD AND ALAN SHORT
Peake Short & Partners
15 Prescott place
London SW4 6BS
United Kingdom

ABSTRACT

This paper provides an overview of the design of the new School of Engineering and Manufacture for Leicester Polytechnic, incorporating new laboratories for electrical and mechanical engineering students plus general teaching spaces and two auditoria. The site for the new building is at the heart of the Polytechnic's city centre campus. Its urban location has placed particular demands on the design. Nearly all laboratory, teaching and offices are naturally ventilated and daylit. Detailed design is nearing completion and construction on site is due to start in June 1991.

1. Building Brief

The new School of Engineering and Manufacture at Leicester Polytechnic has been designed to combine the two Departments of Electrical and Electronic Engineering and the Department of Mechanical and Production Engineering (Fig. 1). The School has become a focus for research and innovation in the field of 'Computer Integrated manufacture'. The activities of both departments overlap in many areas and computer workstations (rather than experimental rigs) occupy much of the laboratory space. There are a range of activities which have fundamentally different environmental requirements. For example, the main Mechanical Laboratory includes a two-storey volume to accommodate large machines for experiments in hydraulics, thermodynamics, etc. The activities and machinery here generate high periodic internal heat gains (> 100 w/sqM), intermittant high noise generation (110dBA +), and require high light levels (1000 lux) for safe operation.

S. Alvarez et al. (eds.), Architecture and Urban Space, 319–324.

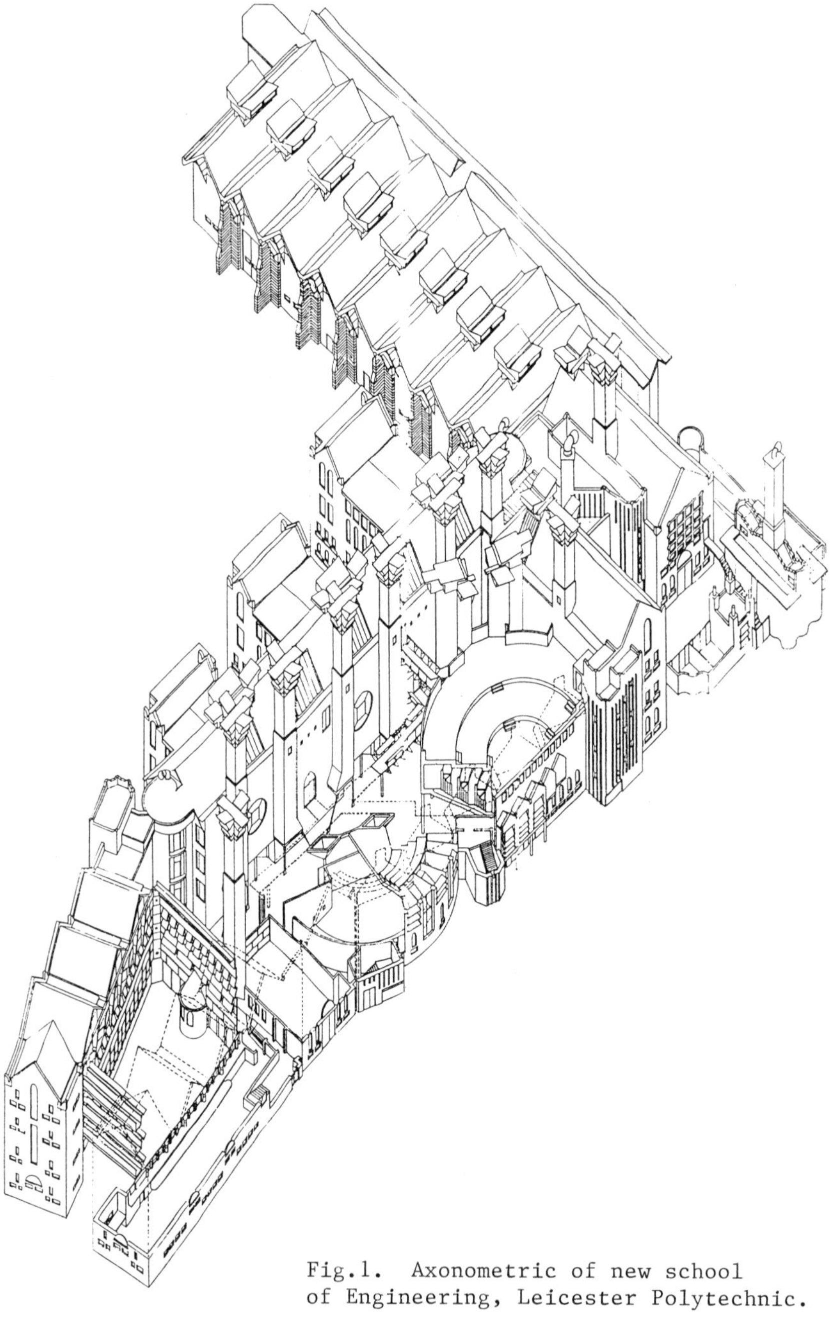

Fig.1. Axonometric of new school of Engineering, Leicester Polytechnic.

This contrasts with the Electrical and Electronic teaching laboratories which generate almost constant high internal heat gains (85 W/sqM), require generally low-level controlled lighting (min 300 lux) and are generally quiet apart from the chatter of keyboards.

2. Environmental Strategy

From an analysis of the nature of the different activities which are to be catered for it became apparent that most laboratory, teaching and office areas could be naturally ventilated and daylit. This avoids the provision of mechanical ventilation and permanent artificial lighting to most areas, with the advantage of reduced capital and running costs and a more pleasant working environment for students and staff. This strategy has had a profound influence on the form of the building. The extended perimeter allows good daylighting and cross-ventilation to many parts of the building, while shaded courtyards provide reservoirs of cool air which is pulled into the deepest parts of the plan in summer.

Internal heat gains from people, computers and other machinery have such a significant influence on the internal thermal environment, that the avoidance of over-heating in summer is a major design consideration. The potential for stack ventilation of each area has been assessed by a series of specialist studies using both physical and computer based models.

Preliminary calculations and simple performance modelling was undertaken by the architects as part of the development of the initial design. Assessment of the detailed design was undertaken after funding approval had been obtained, and included both physical and mathematical modelling. A description of this detailed modelling, and the contribution it made to design development is described elsewhere in papers by Dr Paul Linden of Cambridge University, and Dr Kevin Lomas of Leicester Polytechnic. This paper describes the initial design and how this developed into the final solution.

3. Electrical Laboratories.

There are many precedents which achieve the broad intention of providing daylit, naturally ventilated interiors by means of a narrow section which facilitates cross ventilation and well distributed daylighting. Two narrow section 'wings' on four storeys, with a narrow courtyard between, were designed to accommodate the Electrical Laboratories (Fig.2). Low and high level openings are sized to provide sufficient ventilation to remove high internal heat gains from computers and other small machines. Daylight is controlled partly by 'light shelves' which protect occupants from direct sunlight and reflect light onto the ceiling, providing an even destribution of light across the lab.

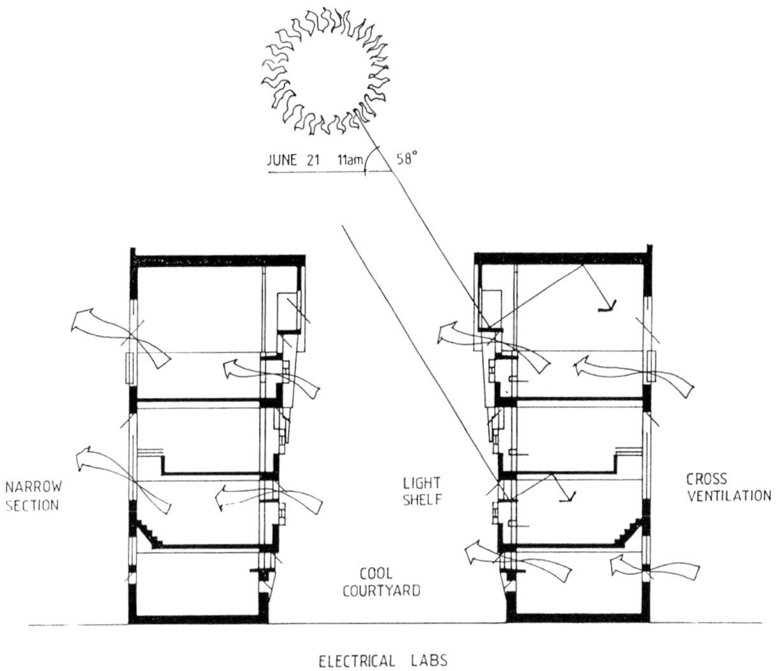

Fig.2. Diagramatic Section through Electrical Laboratories.

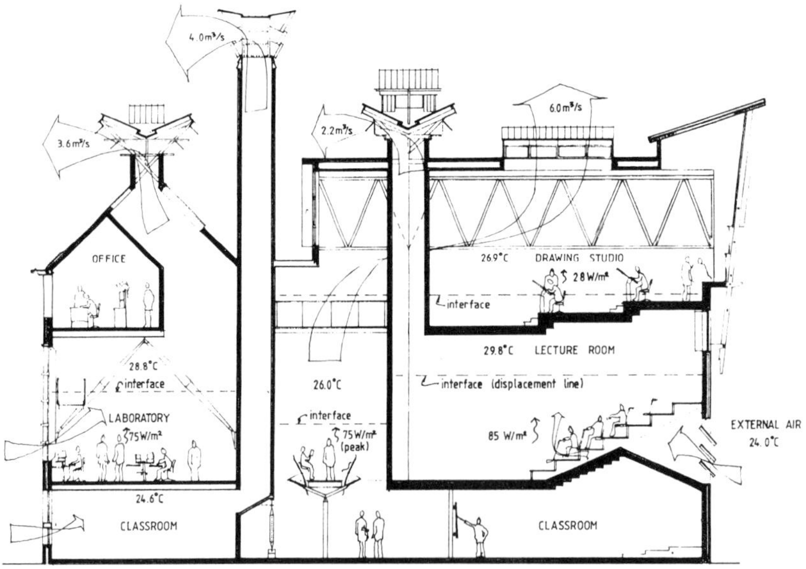

Fig.3. Diagramatic section through central building.

The narrow shady courtyard between the 'wings' provides in summer a source of cool air for the central building and in winter protection from the wind in approaching the entrance.

4. Central Building

The provision of all the accommodation in narrow section buildings would result in a highly extended perimeter, increase capital cost and involve a considerable energy penalty in winter. The main part of the building has therefore been folded back on itself, giving rise to the problems of a deep plan. The provision of natural ventilation and daylight to every part of such a deep plan, can generate a range of formal solutions. In this case, each activity has been treated separately, promoting stack ventilation through each area from ground to third floor. Ventilation shafts from the general lab. also allow light down into the deepest part of the plan.

The main body of the building was originally conceived as five zones, each of which was to be ventilated by separate shafts which rise above the general roof level: 1. Concourse; 2. Classrooms; 3. General Laboratory; 4. Offices; 5. Drawing Studios (Fig.3). A preliminary evaluation of heat gains and summertime temperatures in zones 2, 3 and 4 indicate the importance of avoiding solar heat gains in summer and the importance of high air change rates in maintaining comfortable internal temperatures. The concourse and drawing studios are now one zone, ventilated via ridge vents. This general arrangement can be seen in Fig.3, which is a section through the main part of the building.

5. Mechanical Laboratories

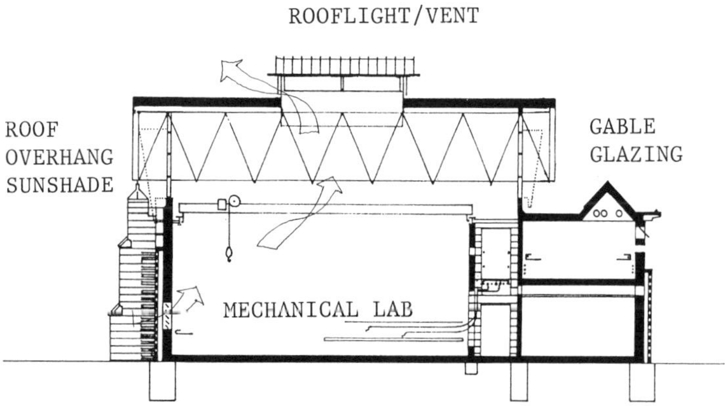

FIG. 4. Section through Mechanical Lab.

The main hall of the Mechanical Laboratory has high periodic internal heat gains from large machines. Ventilation is introduced at low level via 'perforated' buttresses, which incorporate acoustic quilt to reduce noise transfer. The buttresses' primary structural function is to absorb the thrust from a travelling crane gantry. High light levels are achieved by gable glazing and rooflights, which also act as ridge ventilators. Roof overhangs and deep reveals prevent penetration of direct sunlight onto the floor of the laboratory.

Conclusion

The design of the new School of Engineering for Leicester Polytechnic demonstrates that highly sophisticated mechanical and electrical engineering laboratories and associated teaching spaces can be designed to be naturally ventilated and daylit. This is achieved in an urban environment and using traditional building construction providing a building which is economic in both capital and running costs, and which will hopefully be a pleasure to its occupants.

Design Team

Architects:	Peake Short & Partners London, UK Alan Short, Brian Ford, Anne Goldrick and Catherine Hoggard, with Mark Hewitt, Mike Betts, Joseph Ki, Garry Stewart, Bruce Graham, Peter Sharratt, Oliver Chapman.
Structural Engineers:	YRM Anthony Hunt Associates Sheffield, UK.
M & E Engineers:	Max Fordham Associates London, UK.
Quantity Surveyors:	Dearle & Henderson London, UK.
Landscape Architect:	Livingston McIntosh London, UK.

LIBRARY AND CULTURAL CENTER HERTEN CITY

DIETER SCHEMPP
Design: F. Möllring, Construction Management: W. Klimesch,
Cooperation: W. Gans, E. Sedelmaier
LOG ID Group
Sindelfinger Str. 85
D-7400 Tübingen
Federal Republic of Germany

1. Introduction

LOG ID is an interdisciplinary team working towards the goal of employing solar energy in the heating of buildings. Our projects involve the planning and construction of buildings that function for the most part passively, that is without much technology, and which at the same time offer a high living quality.

2. Architectural Concept

The building was planned as a cultural centre, to be combined with a library and a small park. We solved the problem of application of solar architecture in the inner-city with its often very shady sites and estates by spreading a glass-roof horizontally over the building. The space between roof and building has the function of an air-collector.

Via a system of pipes, the surplus warm air is actively utilized for heating. In the air-collector there are absorbers, supplying a water storeage tank and collectors, generating warm water for domestic use. Silicon cells in the glass roof produce electricity for the ventilation equipment and lighting system. The glasshouse is passively utilized. The library building consists of massive brick walls, capable of storing heat energy. The glasshouse contains subtropical plants. This type of building presents new applications of solar architecture in the city.

LOG ID develops architectural concepts that strive for optimal living quality and the employment of solar energy. A building was conceived, consisting of a well-insulated massive structure and a glasshouse, capable of absorbing energy, due to its generous glass surfaces. Subtropic plants make possible a symbiosis between man and nature.

Prior experience in the construction of private homes, office buildings and laboratories led us to draw upon this experience in the consideration of the particular demands that a combination library and cultural center might place.

The massive building is a corner structure displaying a compact outer surface area and generous floor space. The glasshouse is a rotunda that enables the optimal integration of library and indoor/outdoor "park". It houses subtropical plants and serves to transmit solar

S. Alvarez et al. (eds.), Architecture and Urban Space, 325–329.

energy into heat enery. The glass roof protects the outdoor area from rain, snow and wind. It also allows the Jacob Street and Square to be utilized in ways that would not otherwise be possible. At the same time solar energy collected in the area between the massive building and the glass roof can be utilized (see ENERGY CONCEPT).

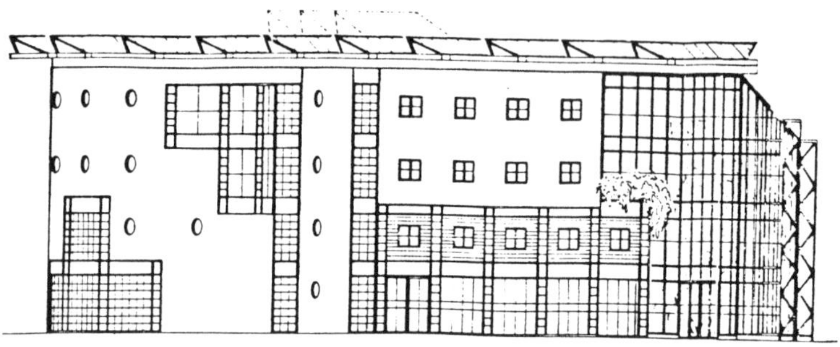

Illustration 1. View from the southwest/Jacob Street

2.1. GREEN CONCEPT

The ground floor of the glass rotunda is to a large extent covered with plants. The free surfaces serve as exhibition space or space for lectures and other public or private events. A generous opening to the east and west connects the rotunda to the adjoining Jacob Square. Outside the rotunda two rows of trees form a transparent green partition to the surrounding building. To the north a similar green area provides seating and reading possibilities and at the same time defines the entrance to the adjacent residential neighborhood in the Hermann Street.

2.2. ENERGY CONCEPT

We have constructed a large all-encompassing glass roof, through which 600,000 kWh of solar energy passes each year. The glass roof is positioned at a distance to the building beneath and to the glasshouse. The massive building is covered by a black roof, thereby creating a kind of buffer zone or air-collector from which solar energy can be retrieved.

The energy thus collected is used
--to heat the building
--to heat the tap water in the building
--to provide electricity.

2.2.1. *Photovoltaics* Silicon cells are integrated into the glass roof. These cells provide the driving energy for the ventilation system and the electricity required by the absorber and collector systems. The electricity also supports the lighting of the premises.

2.2.2. *The Glass House* The glass house is made of highly insulated glass (1 - 1,3 W/m^2 *K). Additonal insulation is provided by the upper glass roof and by its lateral distance to the lower glass roof of the glass house.

The glass house is heated directly by the sun on sunny days in winter and in the transitional periods before and after summer. The upper warmer layer of air is distributed via two air channels. If necessary, the glasshouse is heated conventionally on winter days without sunshine. When no events take place, the temperature is reduced and the doors to the library and cafeteria are closed. In summer surplus warm air is released through the glass roof.

Energy that cannot be gained through solar energy is supplied conventionally by the communal heating supply.

3. Utilization of the Building

The utility rooms of the library, the glass rotunda and the cafeteria are located in the basement. An additonal room in which readings and lectures can take place is located in the round tower in the corner of the building.

The ground floor houses the facilities that inform and invite users and visitors to make use of the library: information, newspapers and magazines, exhibitions, etc. can be found here.

The second floor contains the main library facilities including "reading balconies" that look onto the glasshouse. Work areas for individual and group study are located in the tower.

Offices and administrative facilities are located in the third floor. These rooms are almost exclusively facing the facade, thereby profiting from natural light and fresh air.

Children's literature can be found on the gallery, from which the ground floor floor and the glass rotunda can be seen. The round tower structure houses several rooms that can be used for a variety of events.

4. Materials

The entire supporting frame is made of reinforced concrete. The outer walls of the library consist of reinforced concrete and glass. Insulating glass elements alternating with a galvanized, colored steel frame make visible the geometric structure of the inner building.

The facade of the glass rotunda is supported by a framework. The glass is double-glazed insulating glass. The glass roof integrates the diagonally located glass rotunda and library. It consists of tempered safety glass.

The stone floor of the ground level underlines the integration of street, library and glass rotunda. All other floors are carpeted. The supporting inner walls and ceilings are plastered and painted white. The offices are equipped with flexible partitions. Some of the library rooms have lowered ceilings.

The inner doors are constructed of wood and glass; the outer doors are of steel and glass.

TABLE 1. Volume and area

Massive building	9,000 m^3
Glasshouse	3,800 m^3
Basement	205 m^2
Ground floor	350 m^2
Gallerie	465 m^2
3rd floor	580 m^2
4th floor	601 m^2
Total area	2,201 m^2

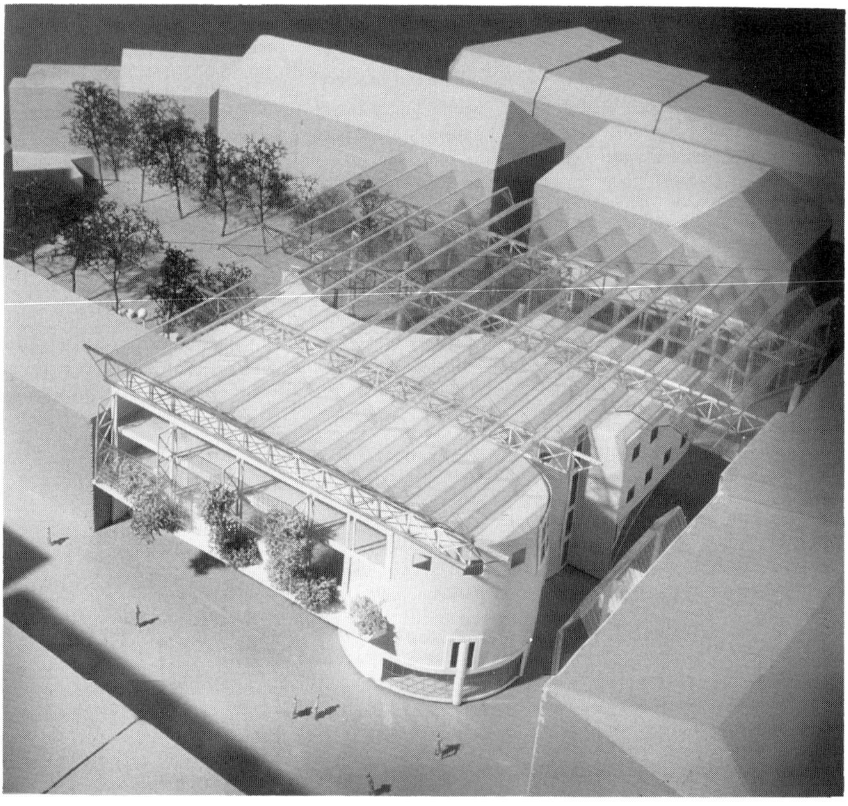

Illustration 2. View from the west onto the glass roof and the round tower

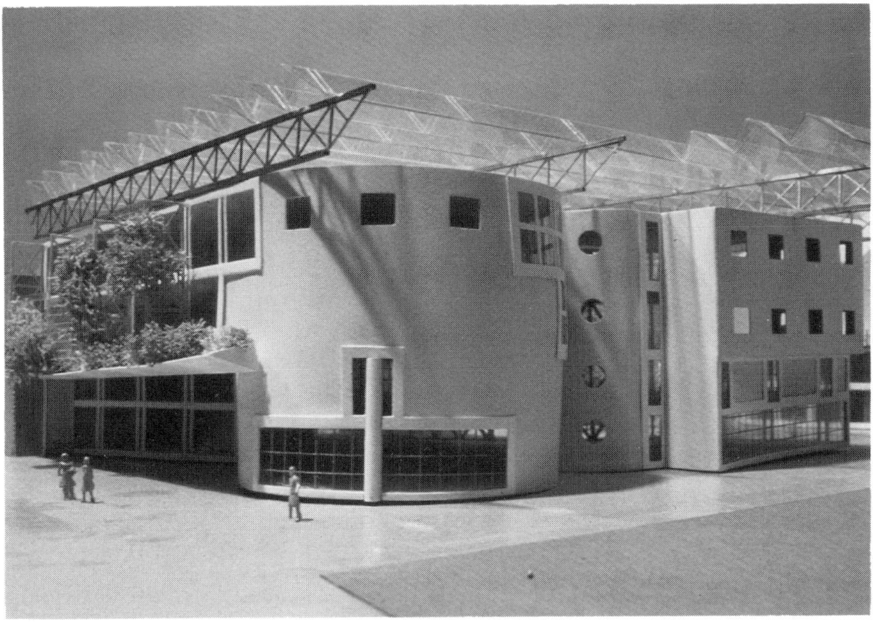

Illustration 3. Facade facing the pedestrian zone

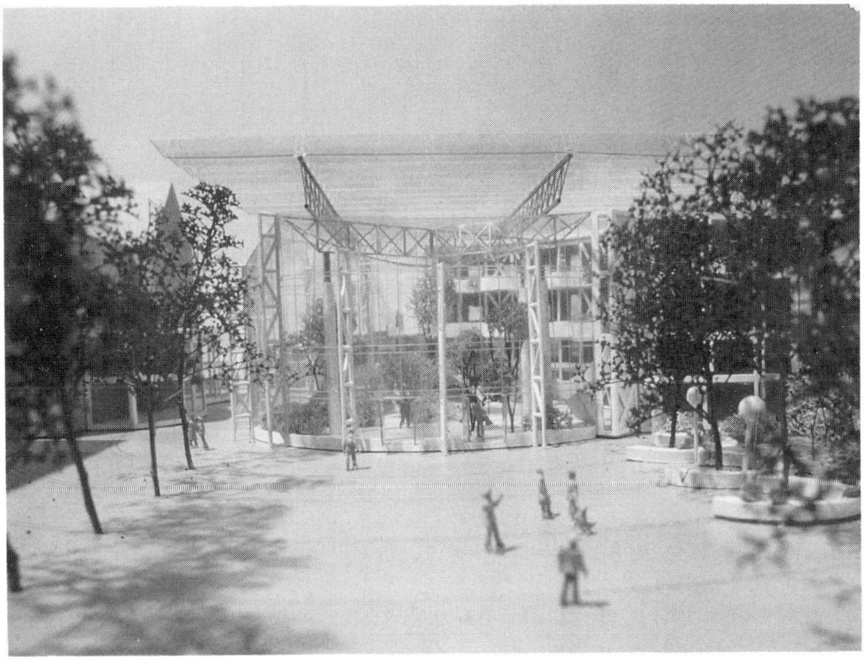

Illustration 4. View of the round glasshouse as seen from the courtyard.

URBAN HOUSING TENDER IN DORTMUND-WELLINGHOFEN, GERMANY

DIETER SCHEMPP
Design: Winfried Klimesch,
Cooperation: Kathrin Miebach, Andrea Schmid
LOG ID GROUP
Sindelfinger Str. 85
D-7400 Tübingen
Federal Republic of Germany

1. Introduction

LOG ID is an interdisciplinary team working towards the goal of employing solar energy in the heating of buildings. Our projects involve the planning and construction of buildings that function for the most part passively, that is without much technology, and which at the same time offer a high living quality.

2. Urban Housing Concept

The purpose of the urban housing tender was to construct 180 apartments for rent in Dortmund-Wellinghofen. Particular attention was paid to the conservative use of land in costructing the apartment buildings, the design of the communal and recreational space, and ecological considerations. The development of the area was aggrevated by the following difficulties:
--a high noise level due to a heavily trafficked highway to the north
--a high voltage cable running across the grounds
--the northern orientation of the area.

Our solution was to construct four types of buildings containing various types of apartments. All four types were designed with the wish to conserve energy. Some of the buildings include glasshouses and are conceived according to the principles of "green solar architecture". The rich plantlife inside these structures is an essential part of their natural system of air-conditioning and improvement.

A long, massive structure facing the busy street to the north protects the residents from the traffic noise. The individual buildings are arranged around a large courtyard or inner zone. Three buildings contain glasshouses that project into the courtyard, thereby optically setting off this communal area from the high voltage cable located to the south.

Parking possibilities exist around the outside of the complex, leaving the courtyard a quiet, green area with pedestrian and bicycle paths. The use of rain water and the conservation of surface area are further goals that were pursued by our planning group.

S. Alvarez et al. (eds.), Architecture and Urban Space, 331–335.

A long, winding structure, to be built along a heavily trafficked highway, was to block off traffic noise and thereby create a quiet courtyard for the inhabitants of this community. A series of individual, smaller houses borders on this courtyard to the east. These houses are not positioned parallel to the street, but face the south. Their position and the space between them thus provides transparency of structure. The southern boundary of the courtyard is formed by three long buildings with glasshouses that project into the coutyard, thereby optically setting off these buildings from the high voltage cable to the south. The rooms in the upper levels are oriented to the courtyard.

The position of the buildings to each other conveys the impression of transparency of construction, in spite of the fact that the main building is a long, winding structure. Glass staircases contribute to this transparency while still offering noise protection. The passage ways in between the buildings are framed. The shape of the structure outlines and immitates the course of the adjacent street.

3. Communal Development

The roads and parking lots were planned along the street and near the high voltage cable, i.e. in areas that could not be designated as residential or recreational sectors. In this manner a large, atomobile-free inner zone was conceived. Delivery vans and ambulances have sole access to this zone.

An important goal was to establish a network of pedestrian and bicycle paths that would not cross streets. This network connects the community with other suburbs of the city. The future industrial area to the north was to be connected to this network as well.

4. Recreational Zones

The inner zone offers playgrounds, gardens, and various other recreational and communicative possibilities. The planned vegetation includes trees and plantlife native to the region. Further noise protection measures include a wicker willow fence along the main road.

5. Energy Concept

The buildings were planned according to the principles of "green solar architecture" with the goal to conserve energy. Some of the buildings have integrated glasshouses and make passive use of solar energy. The glasshouses are "air-conditioned" by their subtropical plantlife and are conventionally heated to a maximum temperature of 5° C. The apartments themselves are thermally separate from the glasshouses, but can be joined with them by opening folding windows, should the temperature permit. The massive building is extremely well-insulated. Its heat storeage capacity is relative to the glasshouse.

The smaller, individual houses represent another type of energy-saving building. In this case the compactness of the buildings prevents loss of heat energy.

The active use of solar energy can be seen in the solar panels that are attached to metall constructions in the parking garages. These photvoltaic modules face the south and convert sunlight directly into electricity. This electricity contributes to the outdoor lighting system.

5.1. WATER

Rainwater is utilized as much as possible and is allowed to percolate into the ground. Rainwater is stored decentrally and used for the flushing of toilets and for watering the plants in the glasshouses. These measures will become more and more important with the growing shortage of water. Other rain water is routed through a system of ditches and gutters to a basin which lies at a lower altitude than the community. In the event of a flooding of this basin, the flood water automatically flows into the communal sewage system.

5.2. GROUND

Wherever possible, paths and sidewalks have a porous surface, therby allowing rainwater to sicker into the ground.

5.3. RECYCLING

A central recycling station is to be established at the main entrance to the community. Biodegradable wastes can be disposed of on an individual basis or centrally at a designated place in the inner zone.

6. Types of Buildings

The many types and sizes of apartments and buildings make them suitable to various needs. The smaller apartments on the ground level may, for example, provide suitable housing for senior citizens or handicapped persons. The glasshouses are used collectively and can aid in preventing the isolation of older persons. The larger apartments or the individual houses, on the other hand, may be better suited to families or collectives.

Education, care of sick or elderly and office work are activities that will become more and more a part of our residential environment in the future. For this reason, an area designated "offices" has been integrated into the architectural concept (see illustrations).

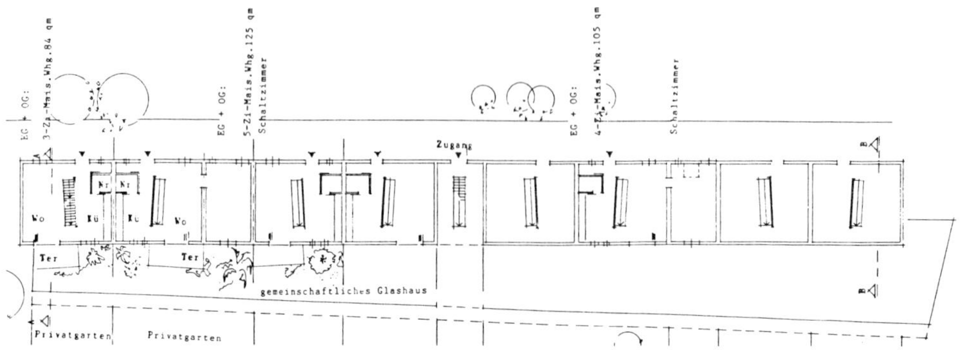

Illustration 1. Floor plan ground level

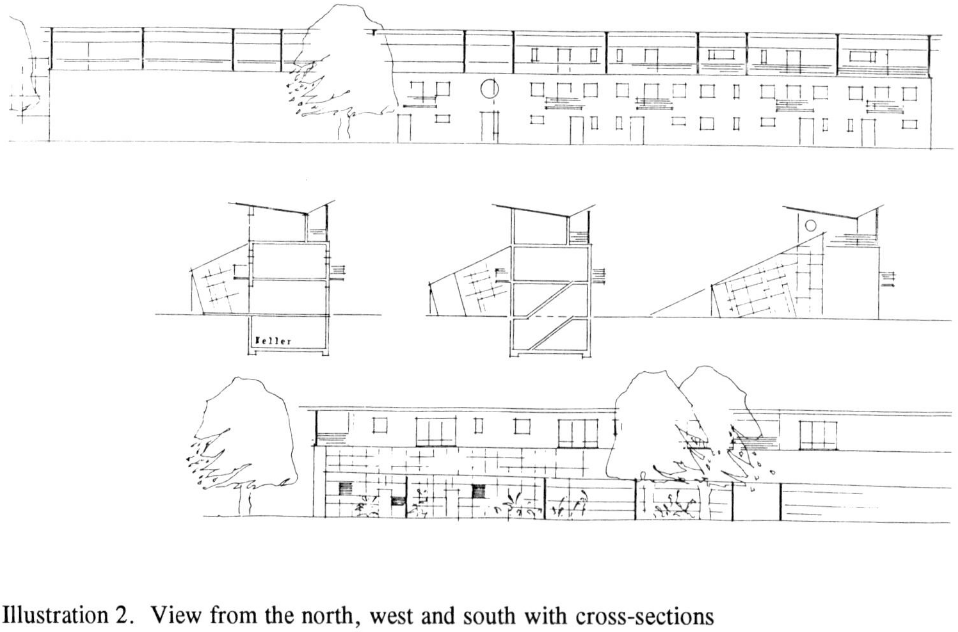

Illustration 2. View from the north, west and south with cross-sections

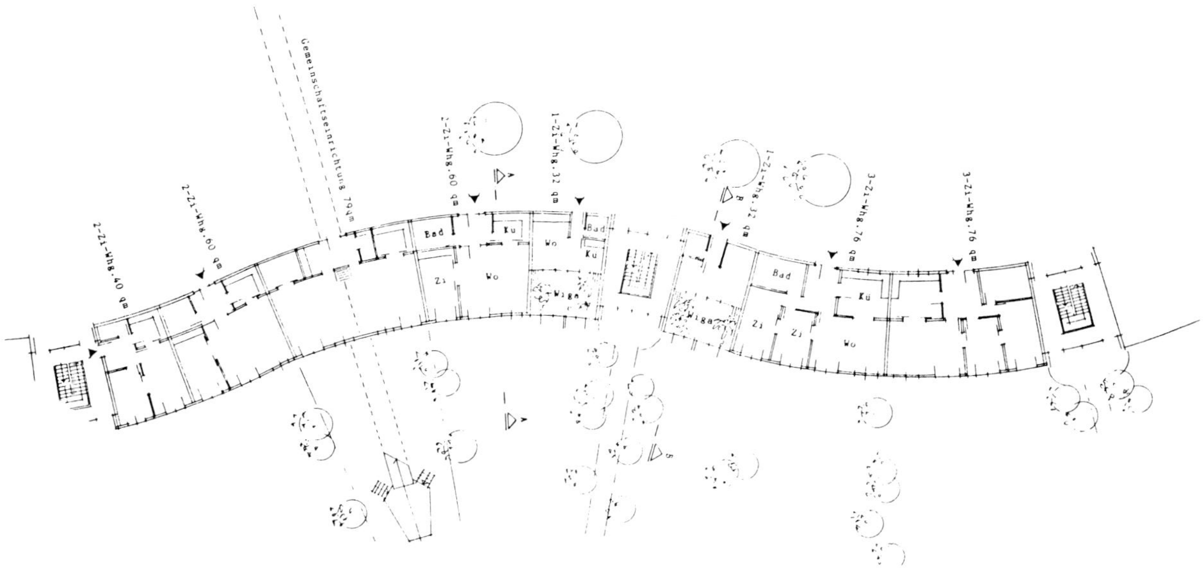

Illustration 3. Floor plan ground level

RESEARCH LABORATORY FOR EXPERIMENTAL TRAUMATOLOGY (EX-TRA)

DIETER SCHEMPP
Design: F. Möllring, Construction Management: G. Steiner,
Cooperation: W. Gans, E. Sedelmaier, Botanical Planning: J. Frantz,
Photography: R. Blunck
LOG ID GROUP
Sindelfinger Str. 85
D-7400 Tübingen
Federal Republic of Germany

1. Introduction

LOG ID is an interdisciplinary team working towards the goal of employing solar energy in the heating of buildings. Our projects involve the planning and construction of buildings that function for the most part passively, that is without much technology, and which at the same time offer a high living quality.

We were commissioned by Professor Burri to plan a research laboratory for experimental traumatology. The lab was to serve as a research center for surgery at the university clinic in Ulm, Germany and to include offices for the directors, secretaries, graduate students at the university, biomechanics labs, histology labs, microscopy labs, an x-ray lab, a library, offices for technical assistants and an auditorium with a seating capacity for 60 persons.

All in all we feel that we were quite successful in constructing an interesting building in which modern technological facilities are integrated into an ecological environment. Due to this and the artwork which is displayed in the massive building, we are sure to have created pleasant working conditions. We further believe that "green solar architecture" can offer high standard working conditions at reasonable construction costs, low energy costs and extensive utilization of sun- and daylight.

2. Utilization of the Building

The building was designed for an interdisciplinary research team of surgeons, engineers, biologists and medical-technical assistants. Their research is directed towards the improvement of the operative treatment of injuries to the motoric apparatus, i.e. injured joints, bones, tendons and ligaments. Implantations have proven to be quite successful in such cases. Further progress has been made through the use of innovative implantation materials and the development of new procedures. This progress is only possible through extensive research and investigation of wound treatment and healing, the biomechanics of the joints, the healing of bone fractures, the compatibility of implant material and human tissue and the technical features of implants and artificial limbs. The Department of Surgical Research and Biomechanics was therefore equipped with histology, morphology, biomechanics, x-ray and biology laboratories.

S. Alvarez et al. (eds.), Architecture and Urban Space, 337–342.

In addition various offices, a small library and conference rooms for up to 50 persons are housed in the new institute.

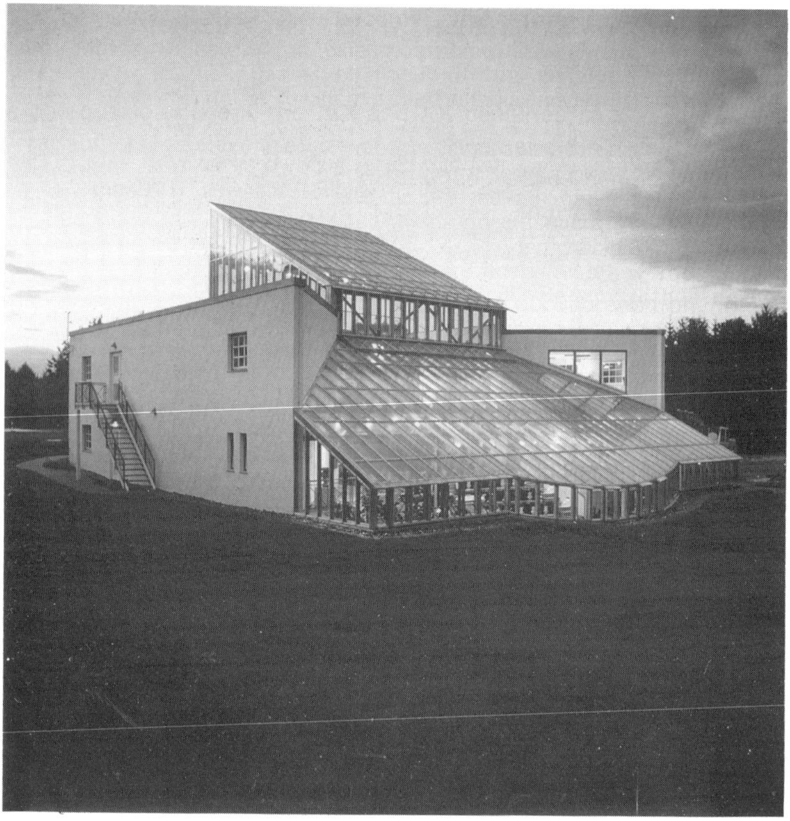

Illustration 1. View from the southwest

3. Archtectural Concept

The building was planned according to our principles of "green solar architecture". The core structure is a well-insulated massive building capable of storing heat energy. It is combined with a glasshouse facing the south. The glasshouse contains plantlife which serves to air-condition the building by providing air that is rich in oxygen and free of dust.

The building was conceived so that the massive building forms a "U" around the glasshouse. The massive building contains all offices and labs. The kitchen, auditorium and conference rooms are located in the glasshouse. All rooms are oriented towards the glasshouse and have large, wall-to-wall folding windows. We decided not to construct an enclosed staircase. The floors are connected to one another by means of an open, spiral staircase in the glasshouse. Access to the individual rooms is through the glasshouse. The building has no cellar.

3.1. GREEN CONCEPT

The plants are subtropical and planted directly in earth substrate. They are watered automatically by a watering system that is installed in the plant-bed. The care of the plants is minimal and involves the collecting of fallen leaves and the occassional cutting of branches.

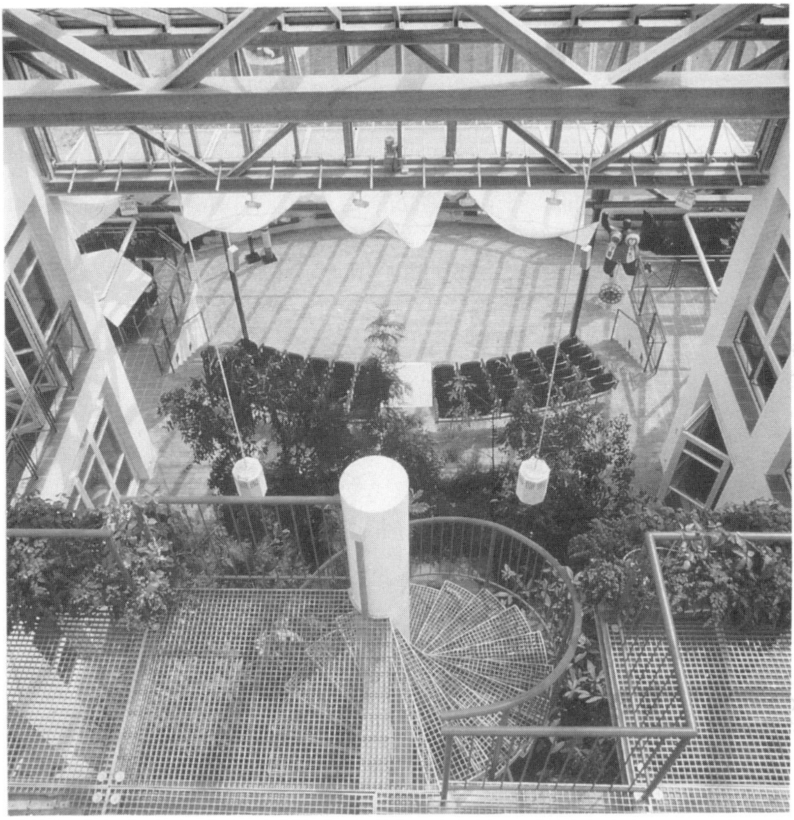

Illustration 2. View from the upper glasshouse platform onto the lecture hall.

3.2. ENERGY CONCEPT

Consequent planning and the use of solar architecture can mean an energy savings of 10-40%.

3.2.1 *Solar Energy* Solar energy is utilized in such a way that the warm air of the glasshouse flows into the massive building when the folding glass doors are open. In addition ventilators which are thermally regulated, are installed in the folding doors. These ventilators draw the warm air into the offices, in which case the heating system automatically shuts down.

3.2.2. Heating Located under the roof of the glasshouse are two large industrial ventilators, automatically regulated, which provide for a turnover of air in the glasshouse. The building also has an additional conventional heating system which can be employed when solar energy is not available. When there is no sunshine in the colder months of the year, the folding glass walls are closed and the massive building is heated conventionally. The glasshouse is then kept at a temperature of 15°C during the day and is reduced to 5° C at night. The heating and air-conditioning systems are optimized and adjusted to the passive solar system. This is achieved by means of an "alert" heating and intelligent control system.

3.2.3. Ventilation Large, automatically regulated ventilator shafts located on the south and north walls and under the roof provide for adequate air circulation in the summer. The air shafts alone would not suffice to achieve an acceptable climate in the glasshouse on very hot days. Shade fixtures and plants therefore play an important role. Cooling occurs through the transpiration that takes place on the leaf surface. Of course the leaves also provide shade. In this way we are able to reduce the temperature in the glasshouse to 2° or 3° C below the outside temperature.

Some plants, for example figs, lose their leaves in winter. This allows the warm sunlight to shine through the vegetation and heat the glasshouse.

4. Materials

The massive building has walls of 36 cm poroton and a roughcasted surface. The ceilings are of concrete. The floor of the glasshouse consists of a brick surface capable of storing heat. The floor of the entrance hall was constructed of smooth travertine, so that very sensitive instruments could be easily transported into the labs. The glasshouse is a galvanized steel construction with thermally separated aluminium profiles and containing well-insulated glass (K = 1,3 Climaplus-N). All windows in the building are made of equally well-insulated glass.

TABLE 1. Area in m^2

total area excluding glasshouse	690 m^2
total volume of offices and labs	1,834 m^3
total volume of glasshouse	1,318 m^3
total cost of the building	2,000,000.00 DM
architects	LOG ID, Tübingen, Germany
owner	Prof. Burri Foundation, Ulm, Germany
date of completion	spring 1988

5. Reaction of the Users

The workers reacted positively to the changing temperature and to the scent, form and variety of plantlife in their new environment. The climate in the glasshouse proved to be quite stimulating. The workers soon learned to use the glasshouse for relaxation and recovery from stress or nervousness. They also reported that they did not miss a conventional air conditioning unit. In fact, the temperature in the building was felt to be pleasant, even on warm days. The relative humidity in the office lies between 40 and 65%.

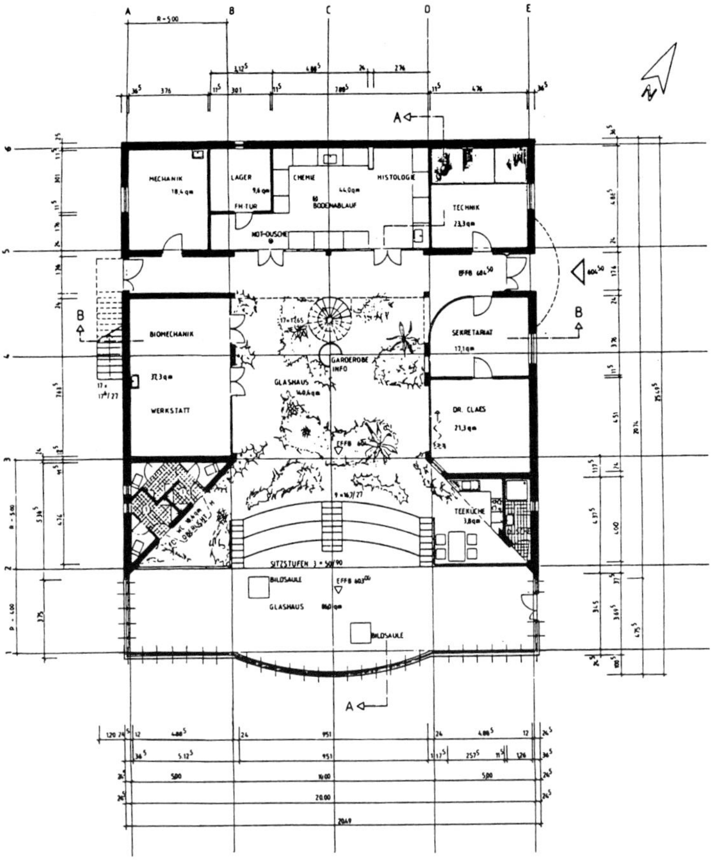

Illustration 3. Floor plan ground level

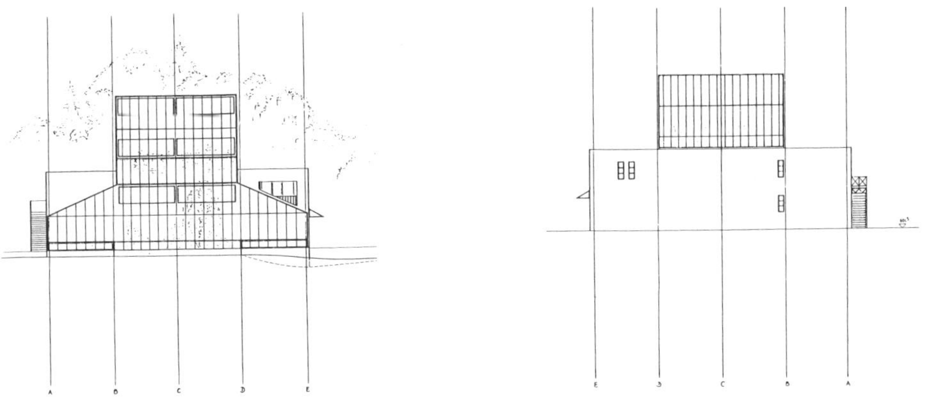

llustration 4. View from the south and north

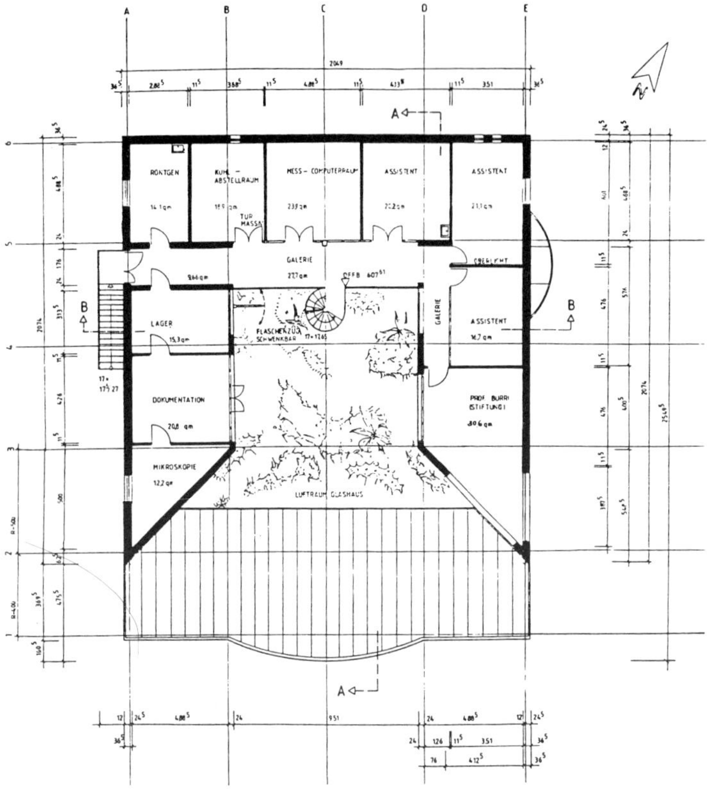

Illustration 5. Floor plan of 2nd level

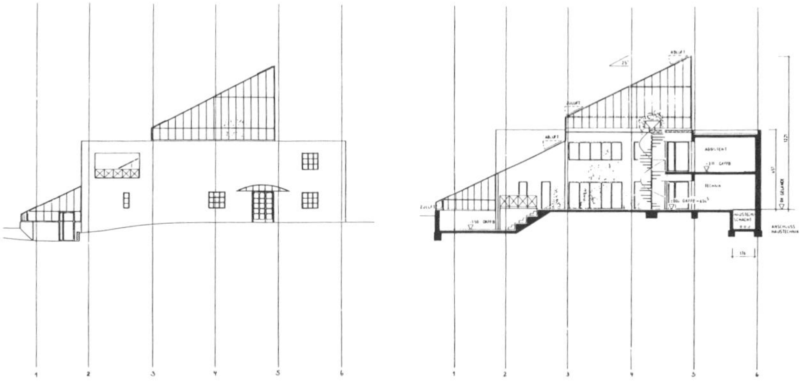

Illustration 6. View from the east and cross-section A-A

HOTEL, 42 THE CALLS, LEEDS URBAN REGENERATION

DAVID CLARKE
David Clarke Associates
1 Castle Row
Horticultural Place
London W4 4JQ

ABSTRACT. David Clarke Associates latest project, the conversion of four warehouses into a 39-bed hotel overlooking the River Aire was completed in March 1991. The £1.5m conversion is a focal point of the regeneration of central Leeds, the major city of Northern England. The project is an important example of the way derelict inner city areas can be revitalised with a combination of private initiative and public assistance. The design juxtaposes original features and new, old features are retained and repaired, new elements are deliberately modern. The layout maximises the southern aspect over the river which has extensive views. Existing walls and roofs are insulated to high standards, with 'low E' double glazing. The building has high efficiency boilers with a central energy management system. The project received a £400,000 grant from the Leeds Development Corporation.

1. Context

In common with many major European cities the centre of Leeds is very run down as a result of changes in industry. However this area of the city has several valuable qualities: many of the remaining warehouses and factories are attractive buildings of historical interest, the area had been made a conservation area, preventing further demolition, the riverside setting has considerable potential, this part of the river is navigable by large boats. The area is separated from the city centre by a railway viaduct, opening up of the railway arches bring the heart of the city within a 2 minute walk. 42 The Calls occupies a pivot position on the pedestrian route between the old city centre and the new developments to the South of the river which will be reached by a new footbridge to be built alongside the building.

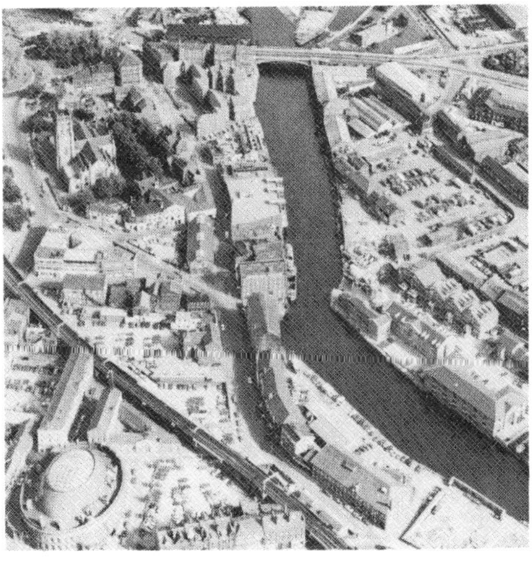

Figure 1 The Calls and the Riverside

S. Alvarez et al. (eds.), Architecture and Urban Space, 343–348.

2. Existing Structures

The existing structures on the site comprised four buildings. On the river side there are three gabled warehouses between 100 and 200 years old, all constructed of brick and timber. On the eastern boundary there was a large building less than 50 years old with a steel and brick superstructure.

The condition of the existing buildings varied greatly according to the age and type of structure. In the newer buildings there were no signs of substantial deterioration, in the older buildings there had been movement in the past and there were many tie-rods where remedial measures had been undertaken.

Extensive repairs were necessary including stabilizing of existing walls, but there were not any fundamental structural problems which affected the viability of the scheme.

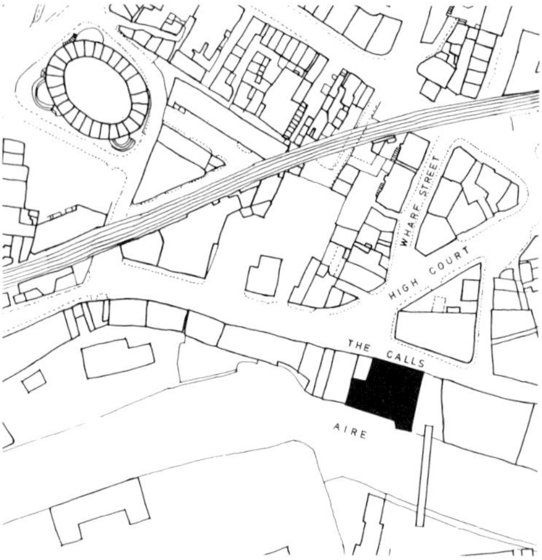

Figure 2 Site Plan

The older warehouse buildings contained many interesting features which have now been restored. The newer structures were of less intrinsic merit but were extremely solid and robustly constructed buildings which contributed to the character of the Calls conservation area and were suitable for retention as part of the proposed scheme.

3. New Uses

The brief was for a 39 bedroom hotel, a large restaurant, and a small shop. The hotel is intended to be at the top level of the market catering for people who want all the facilities of a four star hotel but are tired of the uniformity of the large hotel chains. For this type of hotel the quality of bedrooms and bathrooms have to be high, good space standards are required to allow sitting areas in each bedroom.

3.1 LAYOUT

The existing ground floor levels are all devoted to public uses. The restaurant occupies three of the buildings fronting the river, the western most building on its river frontage is used as a breakfast room for the hotel; this has its own separate kitchen. On the street frontage are separate entrances for the restaurant and hotel.

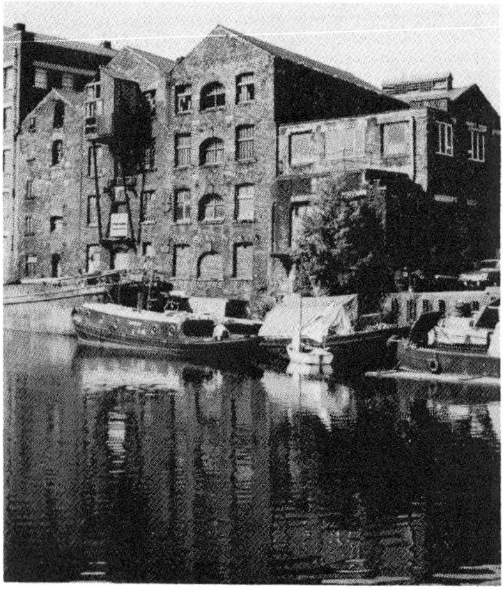

Figure 3 Original Buildings

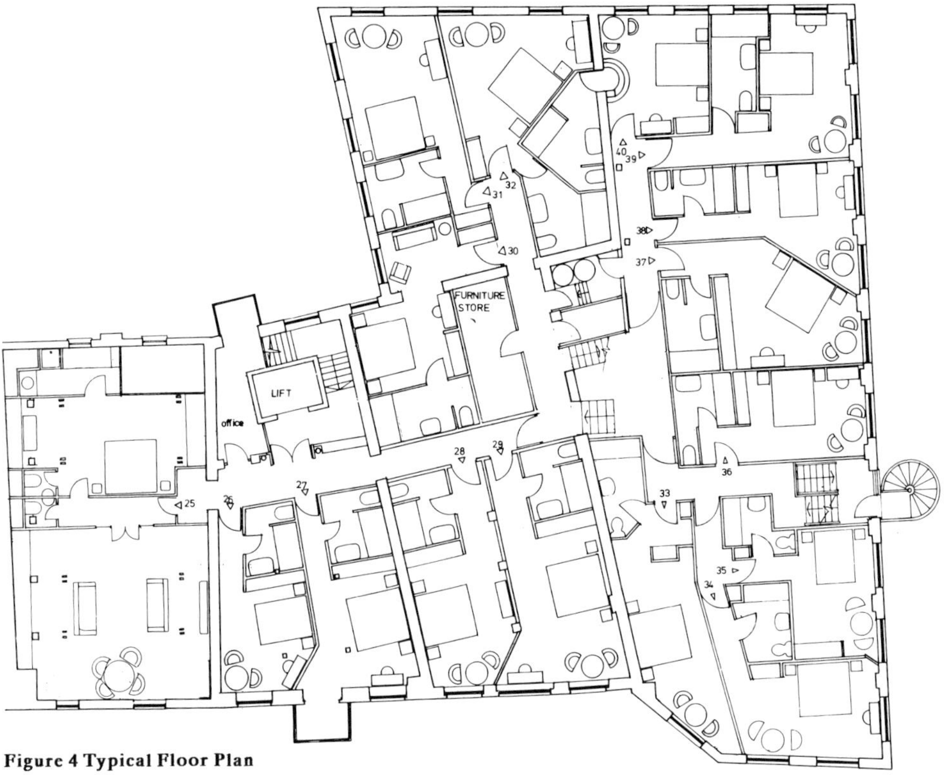

Figure 4 Typical Floor Plan

The upper levels are all occupied by the hotel bedrooms with the associated service rooms. All the bedrooms are unique, taking advantage of the existing features of the building; some have 5m high ceilings with existing roof trusses exposed, others have a sitting area overhanging the river. At fourth floor level a new roof profile provides accommodation for plant and a meeting room which opens onto a balcony overlooking the river Aire.

The layout of bedrooms is dictated by the original internal structure and the fenestration. In the older parts of the building new partitions closely follow column lines and then deviate to take account of the existing window positions. Every opportunity is taken to create interesting spaces and volumes in the bedrooms.

3.2 DESIGN TREATMENT

The interior of the building was planned to retain as much of the atmosphere of the old warehouses as can practically be done within the new uses. Existing walls were cleaned down to the original brickwork which was sealed and painted as a finish. Existing beams and columns were cleaned and retained using fire retardant coatings.

There are several interesting pieces of machinery which remain at roof level and in the winding houses; these were cleaned and retained.

The layout of new partitions was designed so that where there are interesting columns these stand clear of the partitions rather than being enveloped within them.

3.3 EXTERNAL TREATMENT

Externally the elevations of the three older gabled warehouses were fully restored including brickwork cleaning, repointing and re-instatement of the winding houses. Existing features once disguised by grime are now revealed. The existing windows were all in irreparable condition and were replaced with matching timber windows with small panes.

In the treatment of the more recent section it was decided not to attempt to imitate the original warehouses, but to take the opportunity to use some uncompromising modern structures to signal the regeneration of these buildings. A balcony dining area overhangs the river, suspended on stainless steel wires; on the east elevation a spiral escape stair is also suspended from the side of the building, a dramatic sight from the street and the river.

On the street frontage the elevation of the centre building is raised to improve the headroom of the top storey. The loading bay openings are used as the entrances for the restaurant and the shop using frameless glass assemblies.

The hotel entrance is recessed in the existing courtyard and has a fan shaped glass canopy overhanging the street.

Wrought iron planters were attached below first floor windows on the street elevations.

3.4 REPAIRS AND RENOVATIONS

The external elevations were built of a great variety of bricks and had numerous defects. Because of the general deterioration all the brickwork and stonework was cleaned and repointed. The window sills were in very poor condition, and were renewed in cast stone.

Near to the waterline extensive repair work was necessary including renewal of the outer skin of bricks.

The street elevations were of much newer construction but had suffered badly from neglect, patching and over-painting. It was possible to clean these areas without damaging the face of the brickwork. There

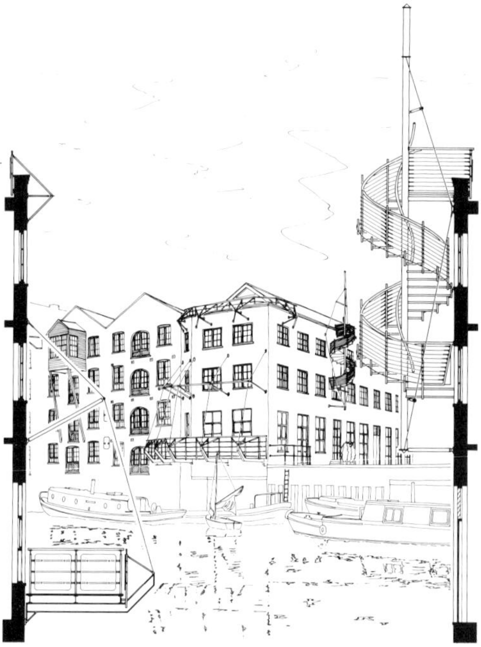

Figure 5 The design proposals for the river elevation

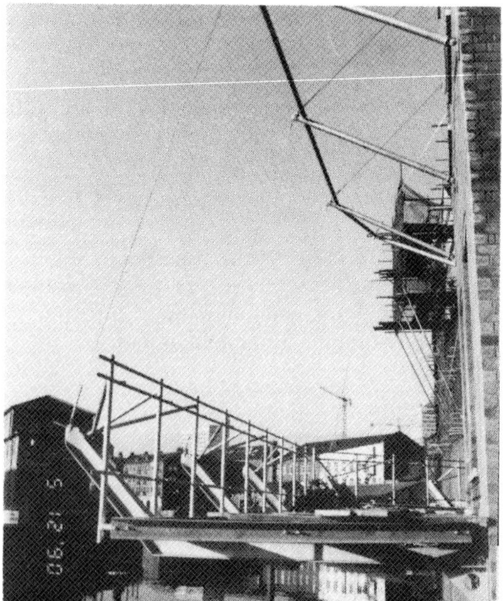

Figure 6 New River Balcony

was an area of serious settlement on one corner which needed underpinning.

On the North and the South elevation there were original winding houses which were used to load goods from barges and from carts. these were in extremely poor condition, and were completely rebuilt to the same profile as the original, but with windows (which were not there before) and a high level of insulation.

4. Energy

4.1 ORIENTATION

The building has an extensive southern elevation which is not overshadowed because of the clear space provide by the river. All the accommodation on this side benefits to the maximum from passive solar gains. All the dining areas for the hotel and the restaurant are sited on the south elevation overlooking the river. The suspended balcony provides outside dining space in summer.

4.2 IMPROVEMENTS TO FABRIC

Some of the walls on the southern elevation had attractive brickwork which was retained as a feature inside and out; because of the aspect the annual energy balance on this wall is not seriously affected. On all other elevations the solid outside walls have been lined internally, improving the insulation value to above new build standards. The very large roof area has been insulated with rigid foam insulation attached to the underside of the existing rafters, thus maintaining a ventilation space between the timbers. The replacement windows had to have narrow glazing bars to match the originals, this made double glazing with sealed units impossible. Instead a lightweight inner frame was used, let into a rebate in the opening timber sashes. In this way the window opens as one pane but the inner pane can be easily opened for cleaning. The inner pane has a low emissivity coating which can be handled and cleaned without damage.

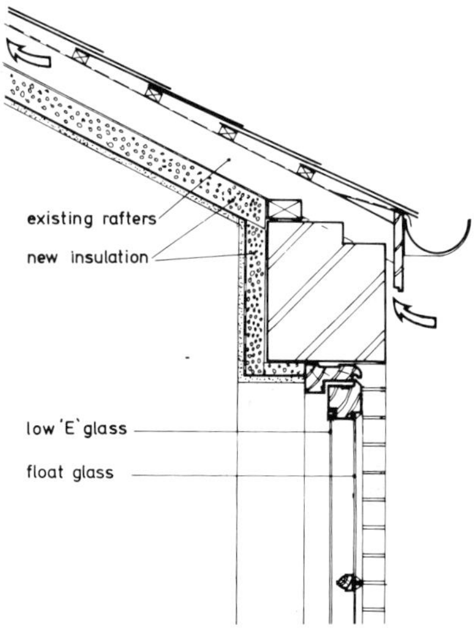

Figure 7 added insulation to walls and roof

4.3 HEATING AND CONTROLS

The building is heated by modular condensing gas boilers mounted in the roof level plant room. The restaurant area has a mechanical ventilation system with heating and cooling capability. There is an energy management system which enables each zone to be separately programmed for time and temperature, so that bedrooms can be set back to lower temperatures when unoccupied and restaurants can be tempered only when required.

5 Conclusion

The project is receiving a great deal of interest as an example of innovative inner city regeneration. It is planned to proceed with similar projects in other urban centres on derelict inner city sites using buildings of historical interest. The project was made possible by the combination of an adventurous hotelier and the support of the Leeds Development Corporation. This was set up by the government to 'make things happen' in inner city areas and is one of several set up in UK cities. Planning processes are speeded up and the corporation is able to provide grants to suitable projects such as this one which would not have been viable on their own.

6 Acknowledgements

Developer/Operator	A Way of Life Limited
Architects	David Clarke Associates
Project Architect	Elspeth Beard
Engineers	Terence E Dudley Associates
Quantity Surveyors	Leonard Stace Partnership
Contractors	Robert R Roberts Ltd

THE AXIS MUNDI AND SOLAR ARCHITECTURE
Some thoughts on the integration of sacred geometry and solar architecture

DAVID J. OPPENHEIM
Taylor Oppenheim Architects
Unit 3 / 743 Glenferrie Road
Hawthorn Victoria 3122
Australia

ABSTRACT This paper sets down some preliminary thoughts on the design of a building in Victoria, Australia, that brings together aspects of sacred geometry and solar architecture. This co-existence is not new. What is new is the re-emergence of sacred geometry as an important design element for modern day architects. The use of solar design principles is really a subset of sacred geometry, the belief that man, god and the whole universe is structured by the same geometric order, all operating in harmony. This building design is begun with the placing of an omphalis in the centre of the site, then developed within a series of Golden Mean logarithmic spirals. The elevations are developed on the same shapes. Traditional solar principles applicable to this Australian climate are naturally integrated.

1. Sacred Geometry

The preciseness involved in solar architecture is evident with sacred geometry. It is not intended here to enter into a full discussion of each of the shapes and volumes listed below, but to note them and to show the rich variety available to the building designer.

1.1 TWO DIMENSIONAL FORMS

The two dimensional forms involved include the following:
circle, square, hexagon, vesica piscis, triangles, pentagon, the golden rectangle, root rectangles and golden mean logarithmic spirals.

1.2 THREE DIMENSIONAL FORMS

The three dimensional forms involved include the following:

the platonic solids - a series of solids which firstly fit precisely inside a sphere and secondly are regular polyhedra: ie have the same shape on each face (either a triangle or a pentagon). There are only five such solids that satisfy both these conditions. They are the tetrahedron, the octahedron, the cube, the icosahedron and the dodecahedron.

the remarkable rectalinear solids - a series of right-angled parallelepipeds (hereafter referred to as R.A.Ps.). The RAP is defined by the three perpendicular dimensions that define the volume (ie length, breadth and height). The golden ratio, 1.618.. , is represented by the symbol ϕ .

RAP 1,1,1 the cube

S. Alvarez et al. (eds.), Architecture and Urban Space, 349–352.

RAP $1,\phi,\phi$	often found in ancient Egypt such as the "Golden Chamber" of Rameses IV
RAP $1,1,\phi$	the vertical ends are squares and all the other surfaces are golden rectangles
RAP 1,1,2	the double cube - often used as the overall volume for Egyptian and Greek temples and Romanesque and Gothic cathedrals
RAP $1,\phi,\phi^2$	the "golden solid" of S. Colman - the height is one and the smaller of the two vertical faces are golden sections
RAP $1,\phi^2,\phi^3$	a vertical volume generally with the plan being a 1, o^2 rectangle and the height being o^3. Often the controlling volume in furniture
RAP $1,\sqrt{5}/2,2$	the "Kings Chamber" solid as found in the kings chamber in the Great Pyramid of Cheops, the base being a double square, the height being equal to half the diagonal of this double square

2 Other considerations in sacred geometry

2.1 THE SAME BUT DIFFERENT

Nature exhibits the quality of diversity and unity. There is a similarity in the design of daisies, yet each daisy is different. Humans can be classified as being the same, yet each human being is a variation upon the theme. No two people are alike. Also in architecture, there can be a overall pattern to a scheme, yet no two parts of the scheme should be the same. A row of columns can be similar, yet no two should be the same.

2.2 FROM MULTIPLICITY TO UNITY

The design should move in the vertical axis from multiplicity and complexity (the human condition) to unity (the godlike condition). This is perhaps best exemplified in Islamic architecture where in plan we move from the square (representing the earth and man) to an octagon, to a circle, and finally to a point, at the the top of the dome (representing the essential unity of god).

2.3 THE UNOCCUPIED CENTRE

So powerful is the presence of the *axis mundi* , the axis of the earth that connects man to god, that in Hindu architecture it is left unoccupied. The Buddha is placed off-centre by one grain of rice.

2.4 FILL THE GEOMETRY WITH LIFE

The geometrical structure that sets up the three dimensional grid should not be slavishly followed. It should be filled with the quirks and humaness of life.

3 The site for this design

The site is set in the gold mining area north-west of Melbourne. It is rural land. The soft rolling hills that are part of, and surround the site, indicate the age of the mountains. The site area is 9.25 hectares and the house will be located in the south-east corner of the site where a previous timber building stood. It was moved to another part of the site to make way for the new building, and will be occupied by paying visitors on a host farm basis. Lovely views spanning from the north to the west can be had well into the far distance.

4 The design and the geometry

A soft geometry should match the rolling hills. The hills that have been formed by violent action to produce sharp shapes have now weathered to soft, feminine shapes. The house will grow out of the site, an extension of the earth, round and smooth, a weathered shape made from local materials.

The *axis mundi* marks the centre of the design. The cardinal axes define the points of entry and exit to the north and the south, and fireplaces mark the rising and setting of the sun to the east and west.

The *golden mean spiral* , growing from the east to the west shows the growth from birth (in the east) to death (in the west). The rooms progress from birth (bedrooms) to life (cooking and eating) to death (the end of the spiral shape)

The spiralling roof reflects the dome of the heavens above, and the skylight over the central stair the opening of man to the gods, the last built evidence of the *axis mundi.*

5 The materials

Local trees will be forested and used to form the main structure. Recycled bluestone will form the walls. Lightweight galvanised iron will form the canopy of the roof, punctuated with windows to let filtered light enter into the building. The building is designed to be built from sustainable materials, and local produced ones at that.

6 Solar design

High internal thermal mass will be provided by the concrete floor and internal bluestone walls. Double glazing will provide year round user comfort. Insulation to the walls and ceiling will provide an overcoat for the cold winter nights (it can snow) and warm summer days (it can be over 40°C). Eaves will shade the windows, and external sliding shutters will protect the west facing windows in summer. The bedrooms are zoned to be apart from the living spaces, since it is Australian practice not to heat bedrooms. The reception area will provide an airlock.

Diagram 1 shows the plan. Diagram 2 shows a notional section

At this stage the building design is being finalised.

7 References

Doczi, Gyogy. (1981) The Power of Limits, Shambhala Publications, Massachusetts

Ghyka, Matila. (1977) The Geometry of Art and Life, Dover Publications, New York

Lawlor, Robert (1982) Sacred Geometry, Thames and Hudson, London

Pennick, Nigel (1980) Sacred Geometry, Harper and Row, San Francisco

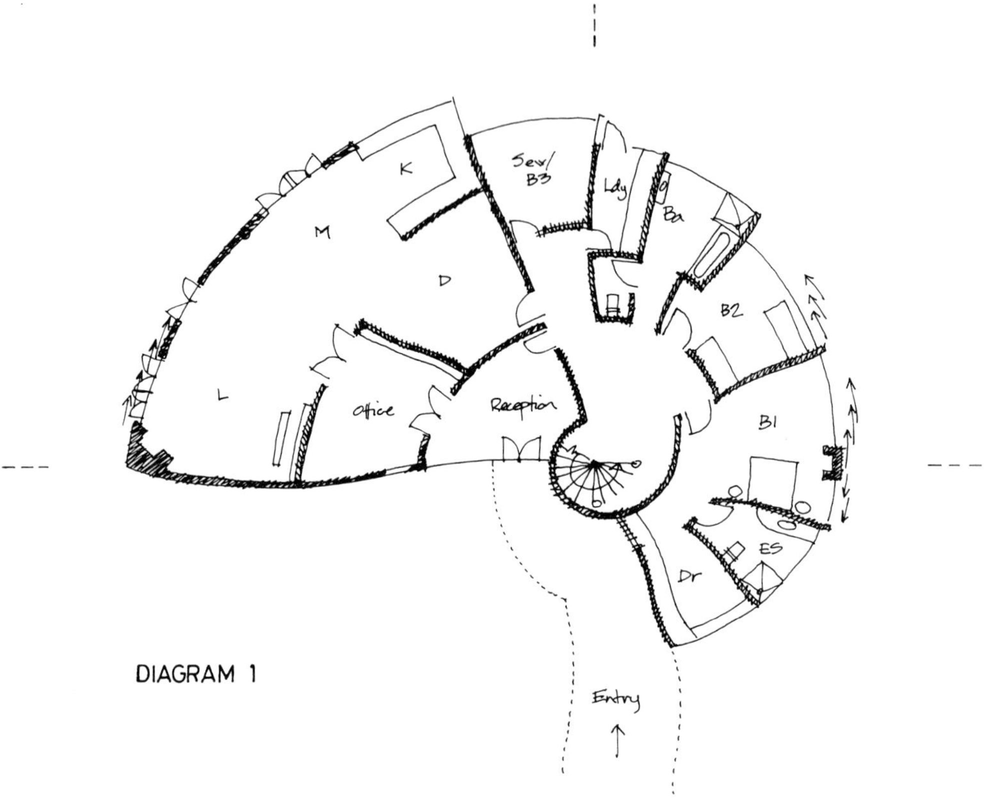

DIAGRAM 1

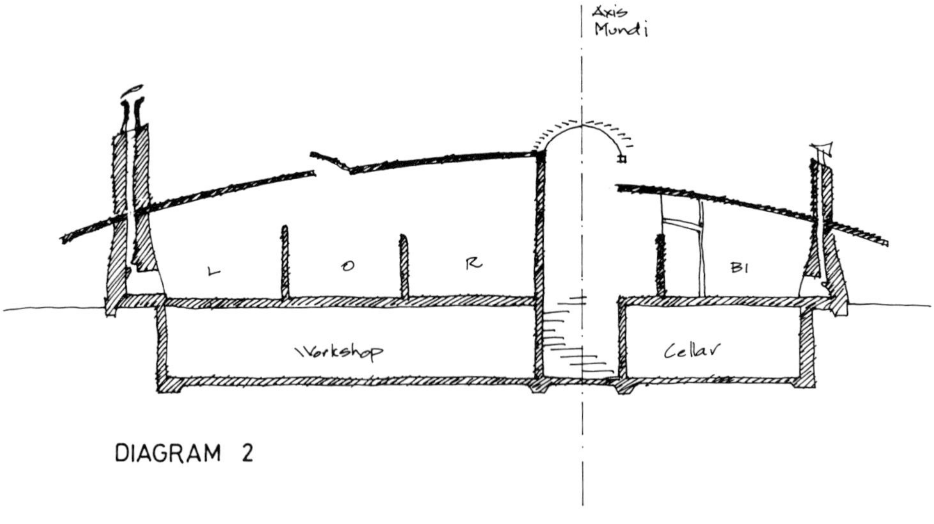

DIAGRAM 2

A SOLAR HOUSE IN THE CENTRE OF ATHENS EVALUATION OF THE THERMAL PERFORMANCE

E. ANDREADAKI*, C. SPYROPOULOU**

*Lecturer, Aristoteles University of Thessaloniki
54006 Thessaloniki.

**Architect, Lykias 18, 11527 Athens, Greece.

ABSTRACT

The main concern of this presentation lies in the fact that a house with passive solar components has been built in the high density, high rise centre of the city of Athens. The project of the house consists of an adaptation of the solar air heating methods to a town-house design. Many problems arising from site restrictions, lot orientation, building regulations and safety requirements had to be solved. The critical point of the design is the "vertical" approach - five floors - resulting from the small available lot and the need of creating the necessary solar energy collecting surfaces. The paper also discusses the thermal performance of the house summarizing some results of the measurements carried out during a year period. The Greek climatic conditions require a suitable adaptation of the solar buildings, in order to satisfy both the heating in winter and the cooling in summer.

1. INTRODUCTION

This project has already been published during the construction stage {1}. The purpose of this new presentation is to follow up the realization of the project and evaluate the degree of the thermal comfort ensured in the building. The monitoring system, by which the measurements were carried out during one year period, consists of forty (40) sensors connected with a data logger. The thermal behaviour of the house based on the above measurements is also discussed.

The passive solar design of buildings becomes more and more acceptable in Greece. In our opinion, the main reasons for this situation are the following:

- the considerable energy saving.
- the building's adaptation to the environment, the local climate and eventually the regional architecture.
- the contribution to the clarification of the atmosphere in our towns, as the air pollution is a rather serious problem.

S. Alvarez et al. (eds.), Architecture and Urban Space, 353–358.

2. HOUSE'S DESCRIPTION - DESIGN PROCESS

The need of space for a large family and the easy access to working area in Athens, as well as a restricted budget, led us to choose the solution of building a town-house in the centre of a high density area of the city (fig. 1). Problems arising from site restrictions, lot orientation, uninteresting surroundings, building regulations and safety requirements had to be solved . The small lot (fig. 1) - only 99,00 m^2 of which 84,00 m^2 to be built on- led us to a "vertical" five-storey solution and to multi-storey passive solar components (fig. 2). The axis of the building has been sligtly turned to provide a better North-South orientation and increase the optical illusion of space (fig. 3). In order to solve the problem of very narrow streets (6,00 m. width) and the proximity of multi storied buildings, the whole concept of design has been focused on the inside area taking advantage of a compulsory inner court and the passive solar components - as the three storey sunspace - which create an artificial view for the sitting areas. Floor spaces, overlooking each other, are connected vertically by an interior opening (fig. 4, 5, 6), thus giving continuity to the space and providing the whole use with good air - flow and ventilation - which proved to be very valuable during the hot summer months.

The glass panels used in both the Trombe-walls and sunspace systems are of the flat wired type made in Greece. Energy loss from the use of that type of glass versus an ordinary one is less than 10% and safety reasons made the residents to accept it, as all described surfaces are located over streets and courts and not easily maintained. From an architectural point of view, that type of glass gives a semi-transparency, needed for privacy, as in the case of the external wall of the sunspace, which is facing a building across the street.

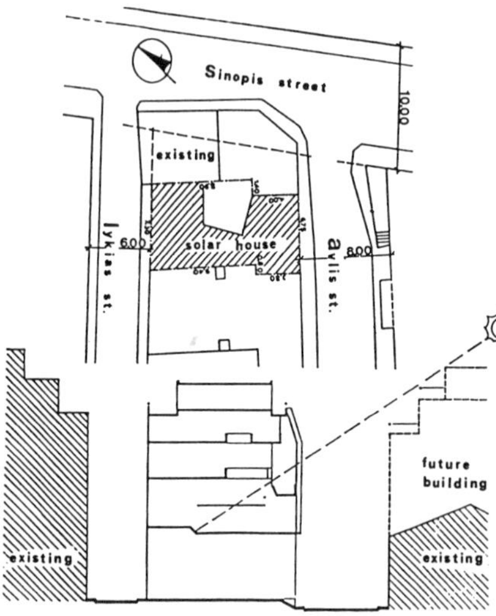

Fig. 1. The high density area.

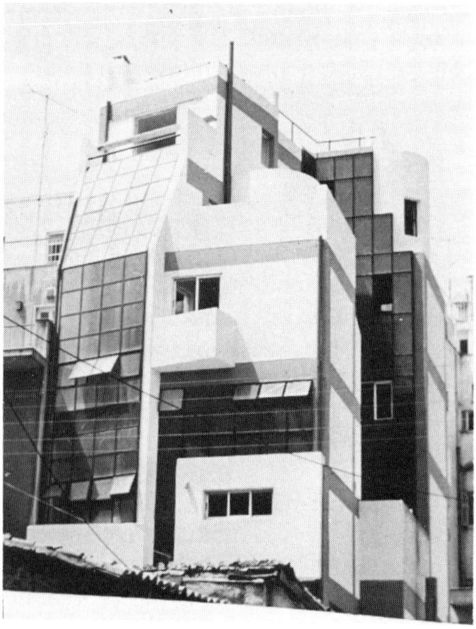

Fig. 2. The solar House from south.

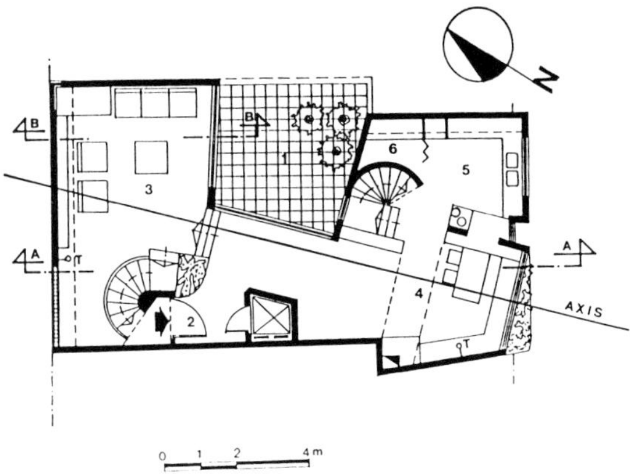

Fig. 3. First floor plan

1. court
2. entrance
3. living room
4. dining room
5. kitchen
6. Store room

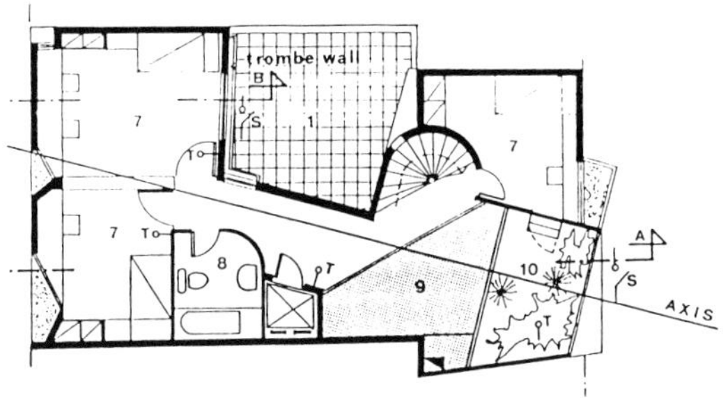

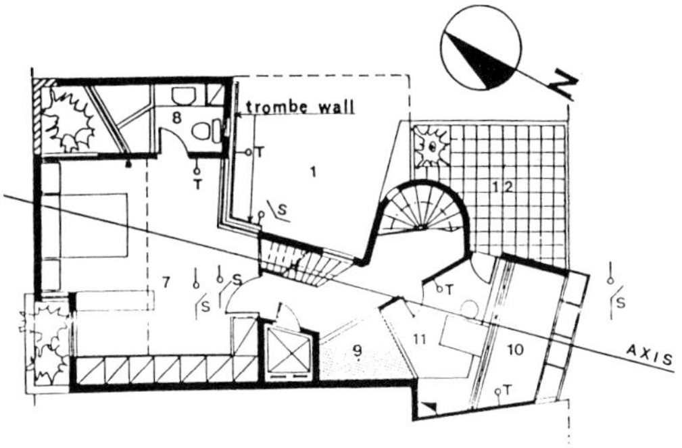

7. bedroom
8. bathroom
9. void

Fig. 4. Second floor plan.

10. sunspace
11. study
12. terrace

Fig. 5. Third floor plan.

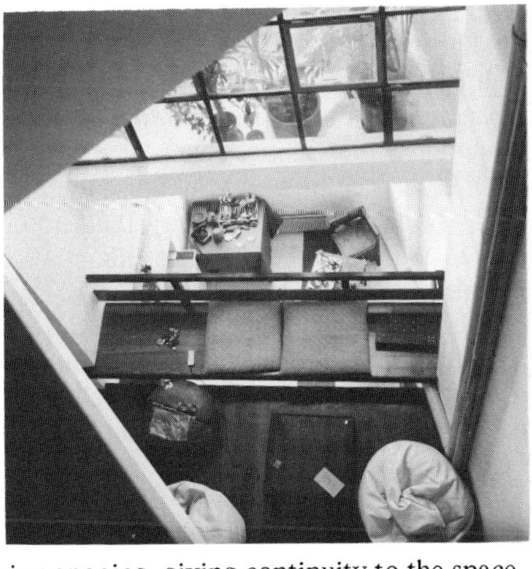

Fig. 6. A view of the interior opening, giving continuity to the space.

3. PASSIVE SOLAR COMPONENTS

a. A sunspace stretches from a height of 9,00 metres (mezzanine floor) up to the bedroom floors. Its total area is 18,00 m^2 and the upper part is tilted by an angle of 20° degrees to increase sun exposure. Warm air flows directly to the bedroom floors. while two small fans force the circulation of hot air to the living room area (fig. 7). Openings are situated opposite to the sunspace giving the opportunity of ventilation during summer (fig. 8).

b. Three-storey Trombe-walls are situated in the courtyard south facade. There are dampers to control air flow in each room. The walls include windows, which open during the summer months increasing the cross ventilation. A small Trombe-wall is also situated on the main south facade.

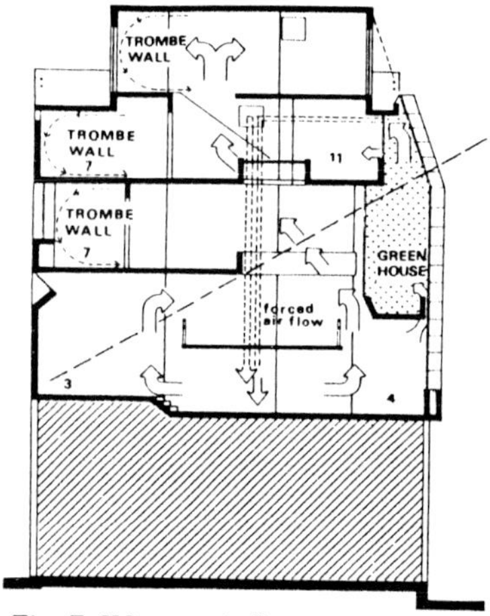

Fig. 7. Winter air flow.

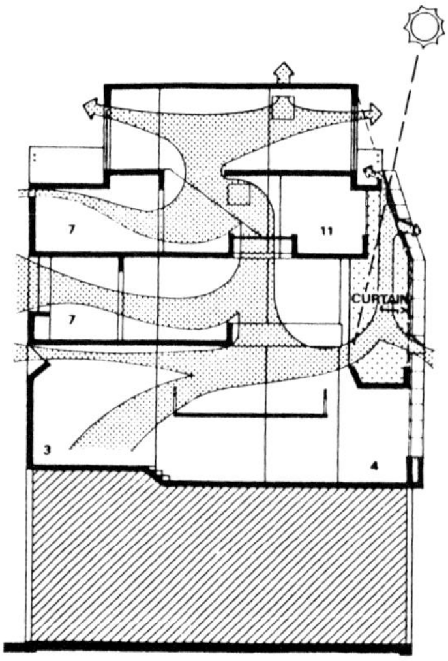

Fig. 8. Summer ventilation.

4. MONITORING SYSTEM AND MEASUREMENTS OF TEMPERATURE FLUCTUATION

The monitoring system consists of fourty (40) sensors connected to a data logger. The physical quantities measured are: the ambient air temperature, the total solar radiation, the diffuse solar radiation, the ambient air relative humidity, the wind velocity and direction. In the internal spaces are measured: the air temperature in all the building´s floors and in the sunspace, the air relative humidity, the electricity consumption, the gas consumption, the space heating system's volumetric flow rate and the domestic hot water's volumetric flow rate {2}. We give by choice some diagrams showing the temperature fluctuation in the different spaces (fig. 9, 10, 11). A general evaluation about thermal performance of the house is that the temperature fluctuation is situated near the comfort zone, however it shows some deviations above the upper limit of the zone during summer and below the lower limit during winter, especially the temperature's mean minimum values.
The measurements started on the 20th April 1989 and lasted a year period, till April 1990. During the summer period increased temperatures have been observed, when the residents of the house were absent and the natural ventilation did not take place.

In the sunspace, at times, the temperature shows considerable fluctuations, especially of the mean maximum temperature, with very large amplitudes during winter, sometimes even exceeding the upper limit of the thermal comfort zone. On the contrary, the mean minimum temperature during winter shows smaller fluctuation, lying close to the thermal comfort's lower limit.
There should be stressed that the comfort zone's limits have been assessed from 19°C up to 28°C, including temperatures that are bearable during winter and summer.

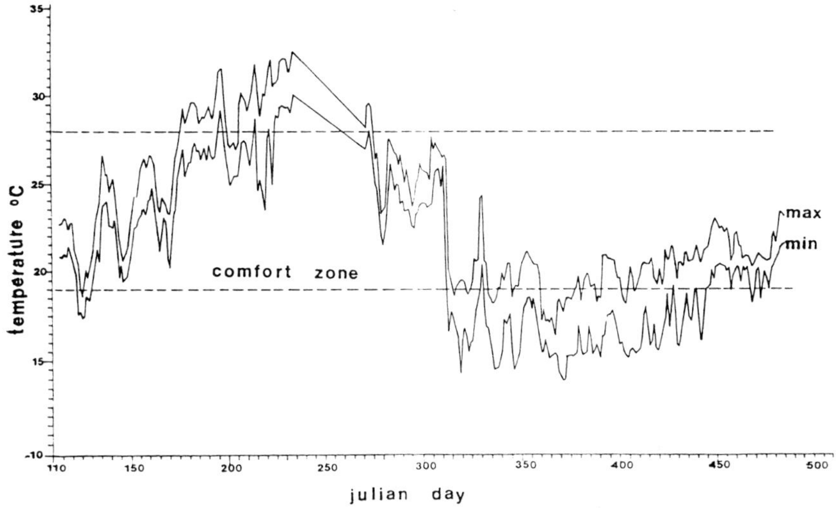

Fig. 9. House air temperature 1st floor, living room. daily Max and Min values, measured during one year.

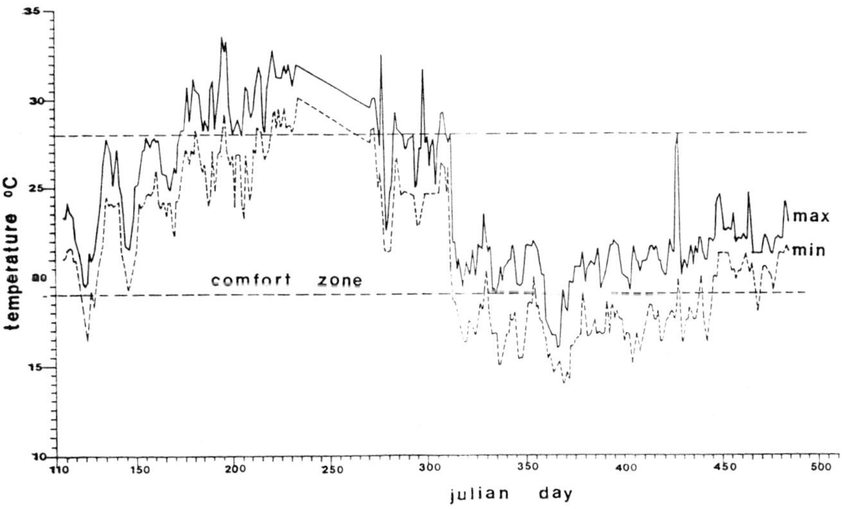

Fig. 10. House air temperature 4th floor bedroom, daily Max and Min values.

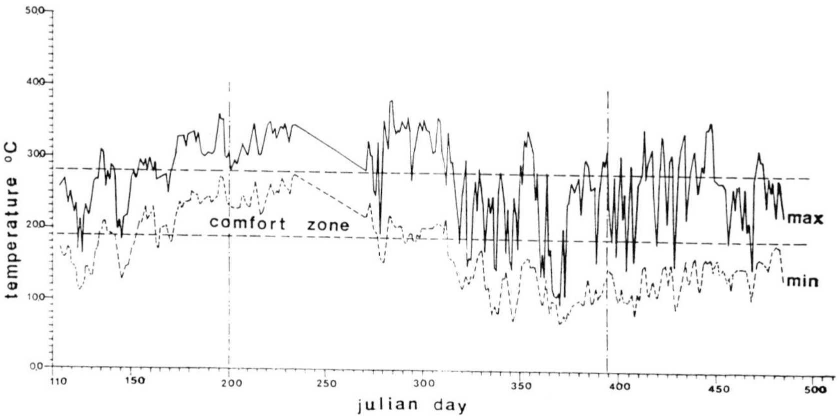

Fig. 11. Sunspace air temperature, daily Max and Min values.

5. CONCLUSIONS

During the four-year period the main remark of the inhabitants is the feeling of comfort and good daylighting of the interior spaces. In summer, good ventilation and continous air-flow creates a pleasant environment, even during the hotest month. Furthermore the carried out measurements of daily temperature's fluctuation affirm the above opinion. This conclusion shows that passive solar components can be effectively applied even under unfavourable town-planning conditions. The solar energy contribution to the building's heating in winter is now developing by a "regression anlysis" method {3}.

AKNOWLEDGEMENT

The house was designed by *C. Spyropoulou*, for solar data and energy calculations, during the planning, was *Prof. G. Veis*. Civil engineers were *C. Telionis* and *G. Theodossiou* and mechan. engineers *C. Tomaras* and *G. Kontoroupis*. *E. Andreadaki* and *G. Paparsenos* were responsible for the monitoring of the house, a project sponsored by the C.E.C.

REFERENCES

1. Spyropoulou, C., Passive Solar House in Athens, *Proceedings of the 1st International Conference, PLEA* 1983, Creta, Greece.
2. Andreadaki, E., Monitoring of two Passive Solar Houses, *Proceedings of the EC Contractors' Meeting*, Brussels, Belgium, October 1987, 133-139.
3. Andreadaki, E., and Paparsenos, G.F., Monitoring of two Passive Solar Houses, *Proceedings of The International Congress*, Saarbrücken, F.R. Germany, October 1988, 379-381.

AN ENVIRONMENTAL HOUSING DESIGN FOR ECOLONIA, ALPHEN AAN DE RIJN, THE NETHERLANDS

F.E. BAKKER and A.P. THIJSSEN
FAGO, Eindhoven University of Technology
P.O. Box 513
5600 MB Eindhoven
The Netherlands

ABSTRACT. This paper deals with ten low cost single family dwellings, designed by the Department of fysical aspects of the built environment, Faculty of Architecture, Building and Planning of the University of Technology, Eindhoven, The Netherlands. They are a part of an urban plan of 101 dwellings, called Ecolonia. Both the urban plan and the dwellings should be a demonstration of how to build with the environment in mind. The plan is located in Alphen aan de Rijn, a fast growing town near Leiden. In May this year the first pile was driven into the ground and for the moment the dwellings are still under construction. The urban layout is designed by the Belgian architect-town planner Lucien Kroll. The 101 dwellings are designed by 9 different Dutch architects. The paper discusses in general the coming into existence and the goals of the project as a whole. Furthermore the design of the Department of FAGO will be discussed in detail.

1. Selection of the Architects

In june 1989, 17 Dutch architects were invited by NOVEM and Bouwfonds Nederlandse Woningbouw NV to produce ideas about housing designs effecting environmental quality both indoors and outdoors and the impact on our natural environment. NOVEM is a governmental institution dealing with stimulating saving energy and our environment, Bouwfonds is the biggest housing developer of The Netherlands. What does it mean to design and to build in an environment conscious way? Well, a lot! The brief for the 17 architects contains:

1. Diminishing the use of energy, especially for the benefit of space heating and hot water;
2. Use of building materials and installations wich has little impact on our natural environment;
3. Adding to the quality of the indoor environment both the fysiological and the psychical aspects including things like safety, limiting chemical and biological pollution, sound insulation and the perception of materials and spaceforms.

Out of the 17 architects 9 have been selected on the basis of their ideas. The 9 were invited to design a number of dwellings within an urban layout designed by Lucien Kroll.

2. The Urban Layout

The site rests on top of a 14 metre thick layer of peat. So the houses have to be build on piles. To prevent the pavement, the sewerage etc. from sinking in the years to come, a 3 metre thick layer of sand has

S. Alvarez et al. (eds.), Architecture and Urban Space, 359–364.

been put on top of the peat. By doing so the water has been sqeezed out of the peat. Now the sinking will be a lot less. Regrettably the layer of sand wiped out all traces of the past that ought to be the starting point of ecological town planning. The only thing that remained was an old small canal. Another handicap was the fact that the future residents were not known when the urban layout was designed (which was by the way also the case when the dwellings where designed). So unfortunaly they could not play their part in the design proces. With the plan Kroll aimed at:

- no abstract, rational arrangement of roads and dwelling but an organic one with coincidences and a slight degree of chaos;
- a great variety of independent elements, like a square, a narrow street, a crossraods, an orchyard, an artificial lake, etc., not submitted to a masterplan;
- not an environment that is all ready but with possibilities for further development. For instance houses that can be extended and empty places for possible amenities;
- a blend of working and living so a lively environment can come into existence;
- an environment for the pedestrian where the motor-car is just a guest. Parking dispersed, not concentrated.
- possibilities for contact with natural elements like there are: sand, stone, water, mud, wood, etc.. Keeping rainwater as long as possible on the surface;
- at least a tree for every car, not in rows but in small clusters, preferably fruittrees. Irregular plantation;
- position and form of secondary footpaths can be decided by the residents and paved later on.

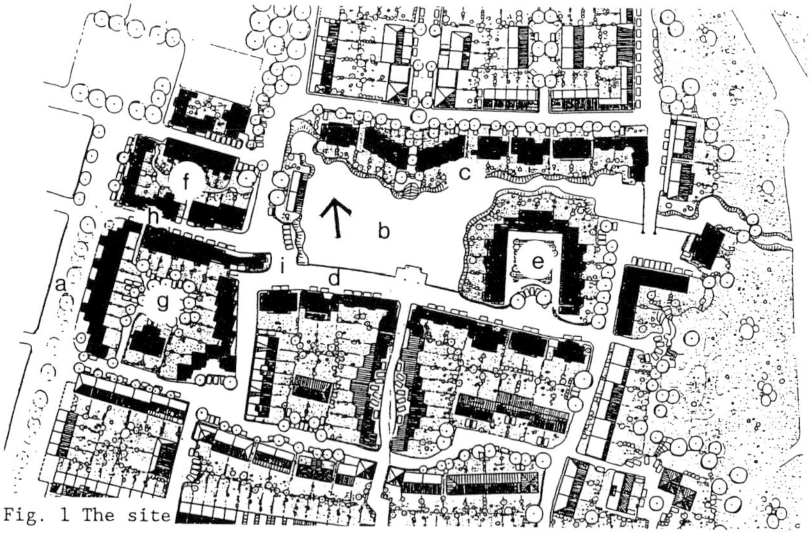

Fig. 1 The site

Ecolonia is a section of Kerk en Zanen, a development of Alphen aan de Rijn. The above mentioned objects make clear that a concise description of the urban layout is hardly possible. Here are some characteristic features (see fig. 1) : The north-south axis of Kerk en Zanen is a route for pedestrians and cyclists only (a). It connects the heart of Alphen aan de Rijn with a future theme park of national significance (600.000 visiters yearly). This axis forms the west border of Ecolonia. In the heart of Ecolonia an artificial lake is situated (b). It has an organic form. On the north bank are dwellings with private gardens oriented to the south (c). On the south side of the lake is a streat with a quay wall and dwellings with a view over the lake (d). On this side is a square enclosed by three rows of dwellings (e). The dwellings adjacent to the north-south axis are partly part of a cluster around a common kitchen garden (f) and partly of a cluster around a common parking place (g). A narrow streat perpendicular to the north-south axis (h) leads to an open space with a towerlike dwelling adjacent to the lake (i).

3. Criteria for the Housing Design

The nine selected architects were commissioned to make a design for about 10 houses in this urban plan. Apart from attention for "ordinairy" aspects each architect had to give attention to:

energy-efficiency: diminishing energy from finite energy sources by diminishing the energy need, use of infinite energy sources and energy-efficient installations.
ecology: diminishing wast of building materials and restemissions. The chain from raw material - production proces - product - wast and emission ought to be considered integrally.
advancement of quality: advancement of environmental quality by diminishing effects and risks caused by materials, products, maintenance, wast and emissions.

On top of this each architect was asked to pay extra attention to one of the nine themes mentioned below:

energy-efficiency:
- extra attention for diminishing heat losses (super insulation);
- extra attention for the use of solar energy;
- extra attention for energy used in the building and living process.

ecology:
- extra attention for diminishing the use of water and recycling of building materials;
- extra attention for durability and maintenance of building materials;
- extra attention for flexibility.

advancement of quality:
- extra attention for sound insulation;
- extra attention for health and safety;
- extra attention for building bio-ecologically.

4. The FAGO Housing Design

4.1 DESIGN CRITERIA ON THE ASPECT OF SOUND INSULATION

The FAGO design consists in one block of ten one family dwellings situated on the north-south axis. So the orientation of the dwellings is east-west. Behind the back garden is a common parking place. The extra attention concerned sound insulation, especially between dwellings and between rooms inside the dwelling. With regard to this specific theme the next design criteria should be taken into consideration:

- sound insulation facade 25 dB(A);
- average airborne sound insulation between adjacent dwellings and between rooms of one dwelling should be 15 dB better than minimum;
- sound level from household appliances should be lower than 25 dB(A)

between adjacent dwellings.

Combined with criteria imposed by the urban plan, these design criteria resulted in the housing design discussed in this paper later on.

4.2 THE URBAN ASPECTS

The dwellings are situated on the axis for pedestrians and cyclists. Although they are in fact simple one family houses their shape is adapted to the importance and measure of the axis and accentuates the ends of the block vertically. On the other hand the shape is tuned to the pedestrian (fig. 2, 3 and 10). Parking of cars and bikes is at the backside of the houses. Consequently residents will enter their houses mainly from that side. Therefore the dwellings are provided with a second adequate entrance at the backside (fig. 4). The sheds are situated in a way to serve as a separation between the private gardens and the common parking space.

The overall shape is simple

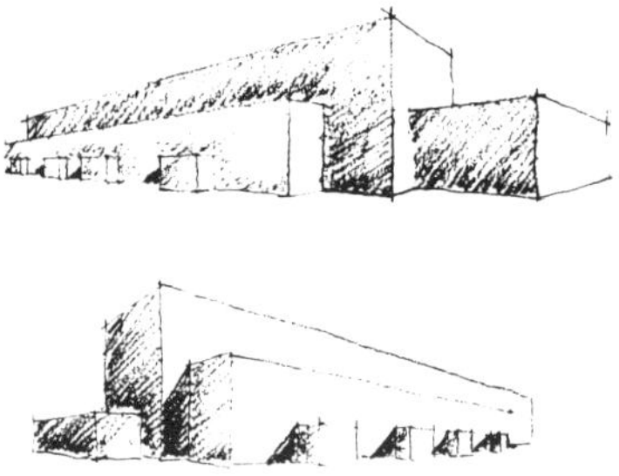

fig. 2

The shape is tuned to the pedestrian

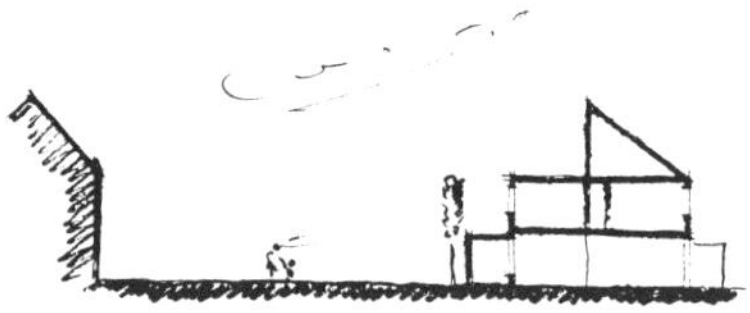

fig. 3

4.3 FLEXIBILITY

The two entrances make it easy for the first floor to give room to another function. It is for instance not hard to imagine that a resident will start an office, accessible at the front side. The second floor, accessible at the back side, can be changed to be a living room and kitchen. If their remains to little room in the attic for sleeping it is possible to add a timber construction on top of the flat roof creating two more bedrooms (fig. 5). Interior walls made out of a wooden frame make it possible to alter the enclose pattern (fig. 6).

4.4 SOUND INSULATION AND THE DESIGN

The dwellings have a so called "sound core", giving accomodation to sound producing provisions like toilets, washing machine, installations and staircases (fig. 7). This core is surrounded by a wall of lime-sandstone with a thickness of 15 centimetres and has sound insulating doors. Futhermore the houses have a "silent room" (fig. 8). This room is situated at the east side of the second floor and makes it possible to play an instrument without annoying inmates and neighbours or to sleep quietly during the daytime (for instance in case of working in a nightshift). The silent room is divided from other bedrooms by a corridor and is provided with a sound insulating door, ventilation opening and dubble glazing.

The dwelling is provided with two adequate entrances

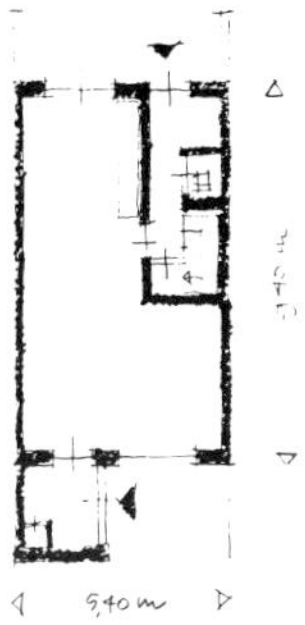

fig. 4

Extention in a vertical direction is possible

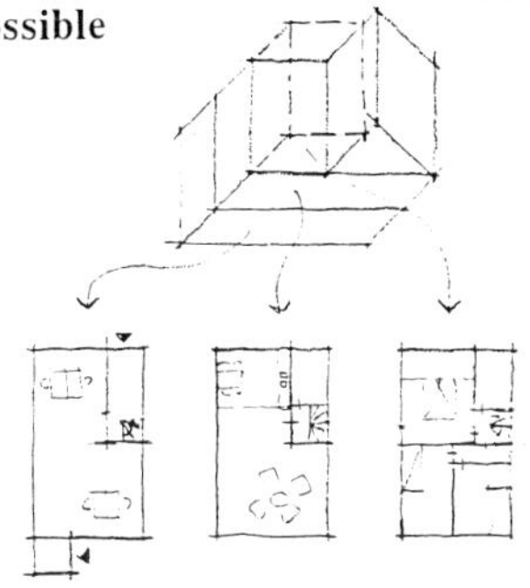

fig. 5 gr.fl. f.fl. s.fl.

4.5 SOUND INSULATION AND THE CONSTRUCTION

The extra attention for sound insulation resulted in a choise for stony materials for party walls and floors. The dwellings are completely dilatated: A separate pile foundation, party walls made out of two separate walls of lime-sandstone, thickness 150 mm, dilatated perimeter walls and roofs (fig. 9). Floors are made of concrete, thickness 200 mm. The interior walls are made of a softwood frame covered with gypsum wallboard and filled with fibrous glass.

The enclose pattern can be altered

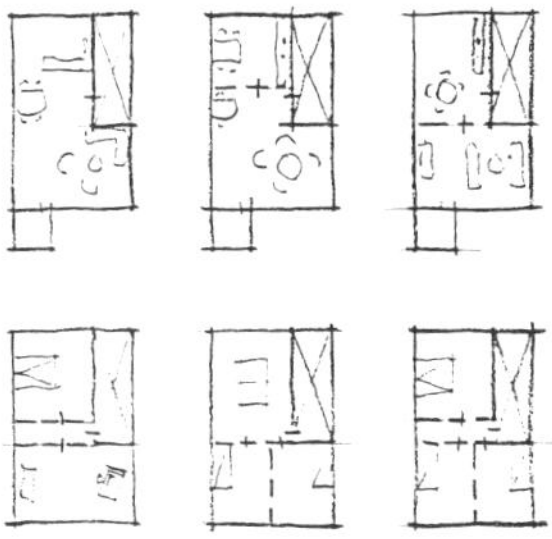

fig. 6

The dwelling is provided with a 'sound core'

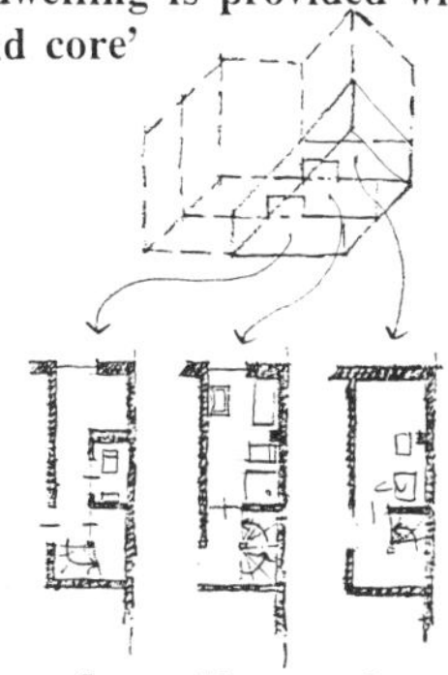

fig. 7 gr.fl. f.fl. s.fl.

4.5 ENERGY-EFFICIENCY

The U-value of walls and roofs is 0,25 W/m2K and of the ground floor 0,33 W/m2K. The windows have double glazing with a coating with a low emissivity factor. The U-value is about 1,8 W/m2K. The cavaty of the party walls is filled with 50 mm of fibrous glass. A thermal bridge between the ground beams and the rest of the house is avoided by a layer of foamglass. As a consequence of the east-west orientation of the houses active solar energy is not applied. The heating appliance has a high coefficient of performance and produces hot water as well.

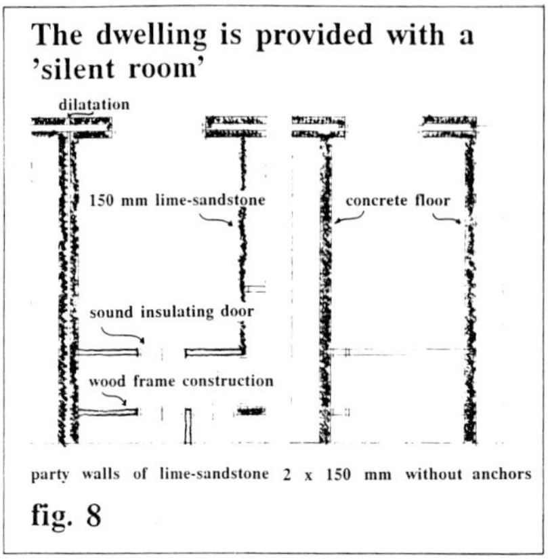

fig. 8

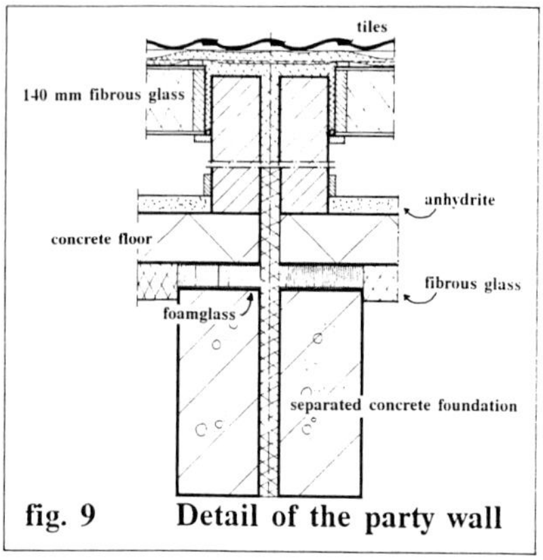

fig. 9 Detail of the party wall

4.6 ECOLOGY

The houses are mainly made of lime-sandstone, brick, concrete and soft wood. The concrete of the piles, ground beams and floors contain as much rubble (coming from demolished buildings etc.) as possible. The pitched roof, the wooden walls and the window frames are made of north European soft wood. The pitched roof is covered with clay tiles, the flat roofs with rubber (EPDM). The woodwork of the facade is of western red cedar, a kind of wood that does not need to be preserved. The insulation material is of fibrous glass, the wallboard is made of gypsum, a wast product of electricity plants. Instead of PVC PE is used where possible. Likewise PUR-foam is not used to stop chinks, instead fibrous glass combined with rubber strips are used. The paint used is water diluted.

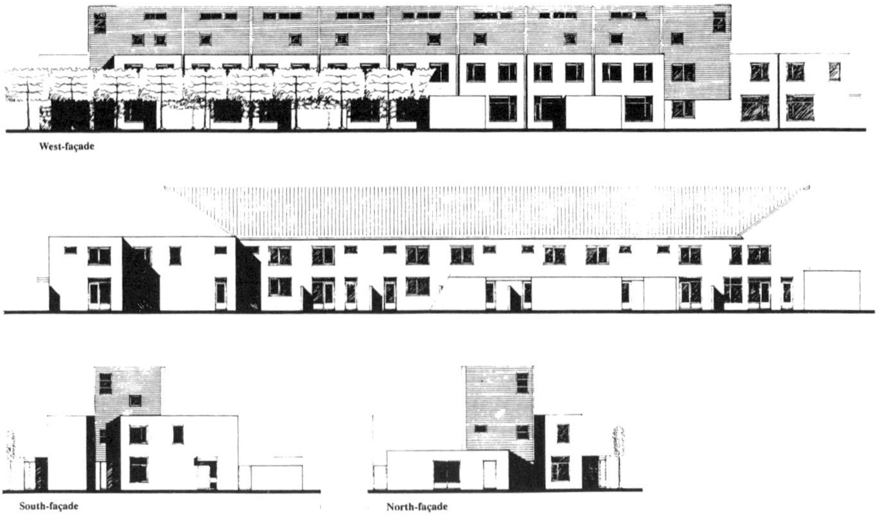

Fig. 10 Facades

EL CENTRO DE INFORMACION E INTERPRETACION DE LA NATURALEZA EN EL PARQUE NATURAL DEL PENYAL D'IFAC. CALPE. (ALICANTE). SPAIN.

IGNACIO BLANCO LOPEZ. Architect.
Architecture Study.
Paseo Alameda 4
46010 Valencia
Spain.

ABSTRACT.

The 19th of January, 1987, The autonomous Goverment of the Valencian Community, Through the Consell of the Generalitat of Valencia, declares the Penyal d'Ifac a Natural Park. This is one of the most beautiful landscape enclaves of the Valencian Community, located in the municipality of Calpe, province of Alicante, wich extends through around 45,5 hectares and constitutes a calcareous escarpment with a maximun height of 332 metres, surrounded by the Mediterranean

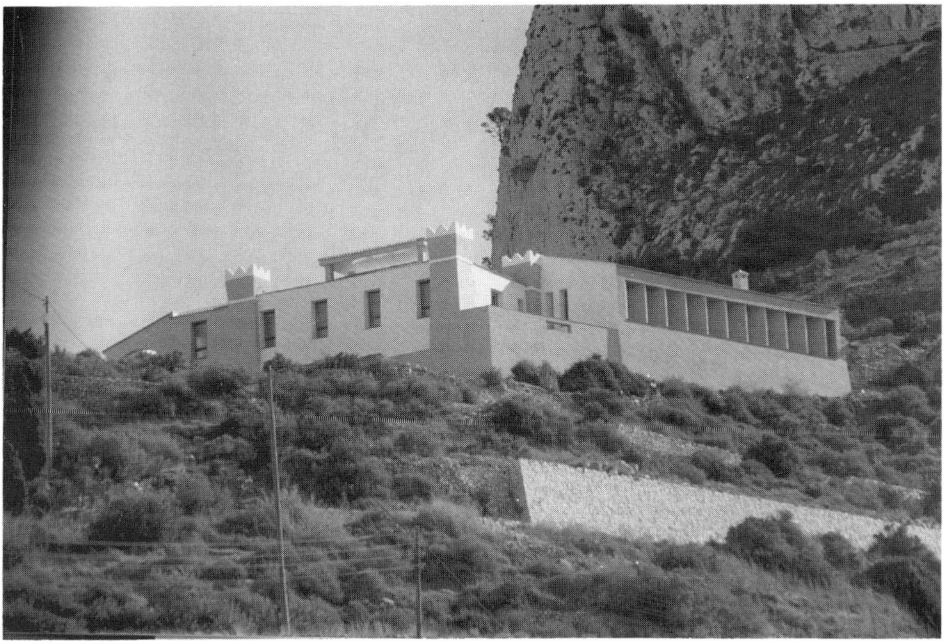

FIG. I

S. Alvarez et al. (eds.), Architecture and Urban Space, 365–369.

Sea and linked to land by an isthmus having big steap slopes. Two buildings an annexes dating from the second half of the 19th century were found there very damaged, wich have been architectonically and functionally rahabilitated as a "Nature Information and Interpretation Center and Hall", according to the requirements program of the Environment Agency (AMA) of the Generalitat of Valencia. Both buildings are located in the eastern slope of the Penyal.

1. SITUATION AND CLIMATOLOGY.

The buildings are found on the north slope of the Natural Park of Penyal d'Ifac (latitude 38237-38241'N and length 3°40' - 3°46' E) (fig. I), which receives direct radiation except at midday in winter: the Mediterranean climate with a high insolation rate, reaches an average temperature in August of 26°C and 12°C in January. The rainfall is scarse, irregular and intense, not excesding 600 mm. a year. The distribution of the winds is irregular depending on the side of the Park, producing irregular air currente caused by the topographic condition itself and due to the winds coming from the sea, which at times can be considered as strong (higher than 90 km/h.). The different microclimatic conditions have conditioned the different and varied vegetation with great scenic interest, which is complemented with interesting species of birds. All the protection and use od this natural region, one of the most beautiful, is protec-

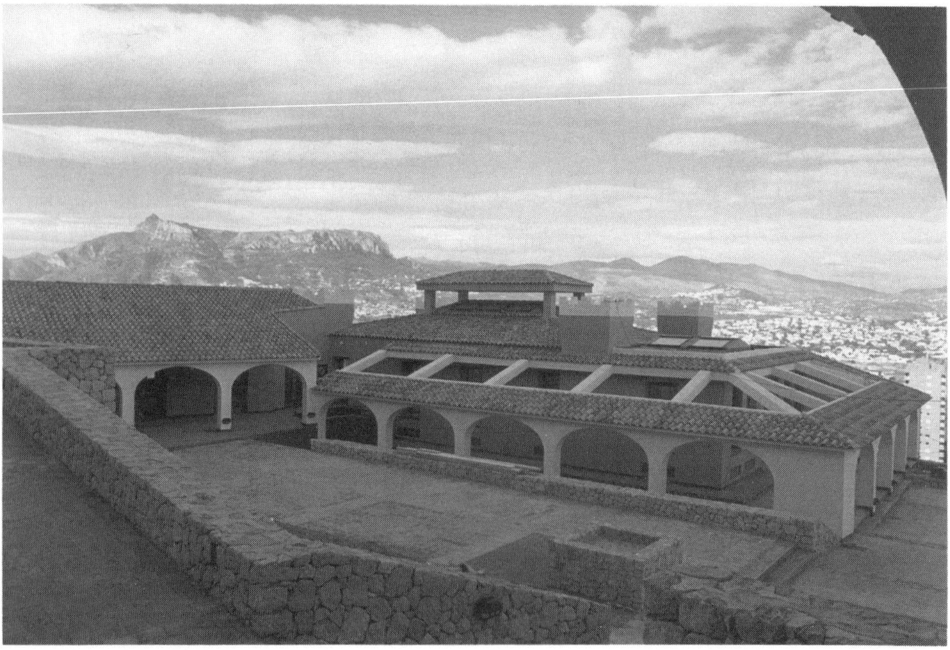

FIG. II

ted by the urbanistic normative of the especial plan which has been drafted with this purpose.

2. ORIGIN OF THE BUILDINGS.

The two existing buildings were built during the second half of the XIX century and beginning of the XX century: one of then, considered as the main building with an attached servios building and the other secondary or complementary one with an inner patio: they are at different levels and separated by 20 m. creating between them a small square. The architectonic composition responds to the Mediterranean typology, both due to the layout and rectangular shape of the windows and due to its open terraces. The closure walls are also the structural elements, with thickness greater than 40 cms., the roof being of Arab type tiles. Both buildings constituted a point of reference in the landscape in spite of their bad state, which was taken into consideration for their present rehabilitation.

3. NEW FUNCTIONAL ADAPTATION.

The new functional adaptation has permitted the recuperation of both buildings as Nature Classrooms. The main building (fig. II)

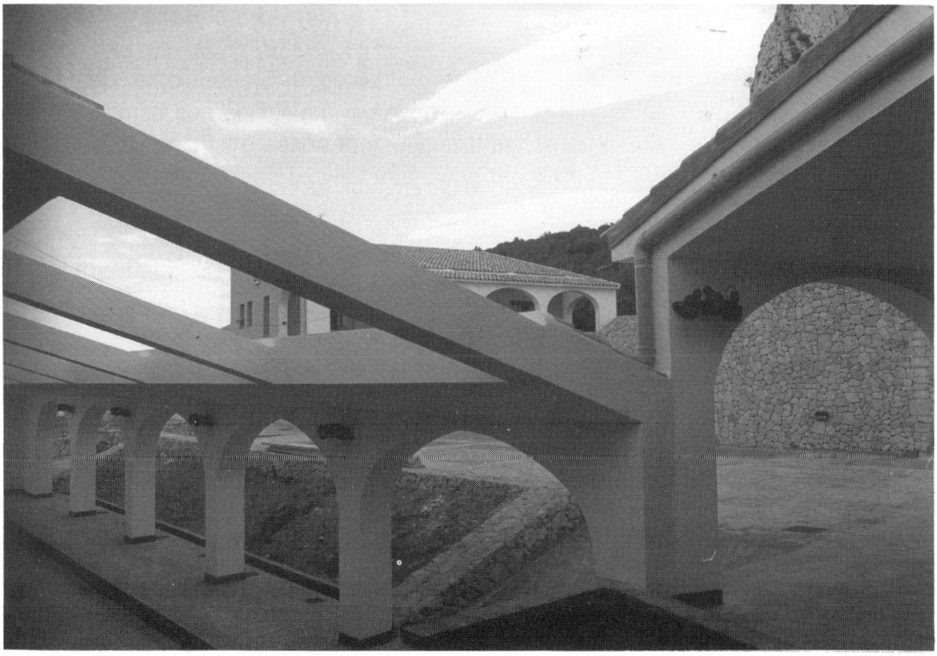

FIG. III

with a surface area of 220 m2. and services of 165 m2. with only one floor, has a maximun height of 7'5 m. and acts as an information-exhibition centre. The secondary building with surface area of 632 m2. (fig. III) and two floors, contains the areas corresponding to the clasrooms themselves, assembly hall and management. Both buildings, due to their unique situation, have views on to the sea (East an West) and to the enormous rock of Penyal (South), wich predetermined that any of the window openings should also be considered as the frame of a beautiful landscape painting. The preexisting layout of both buildings on the Northern slope has meant that different spaces have had to be left on the terrace roofs, to permit the incidence of direct radiation on the windows in the winter time; part of this radiation will be reflected on the water sheet which is situated below the semicovered terrace in the main building. The importance of the transversal ventilation has been considered with the conditioning of the existant microclimate, where the direction of the wind is very varied producing incontrolable effects. Due to this reason, on the main building there is a raised roof wich acts as a chimney, also acting as direct solar protection on the zenithal sloping windows. Part of the air which penetrates the building, does so through small, grills situated on the lowest parts of the walls of 40 cms. having been previously humidified in the fountains and water sheets.

4. BIOCLIMATIC DESIGN.

The bioclimatic design possibilities have been considered based on the conditioning of an already existant old building. The roofs have been transformed, amintaining the window opening to the south; these due to the agressive atmosphare of the sea wind have been made with aluminium on the outside and wood on the inside, with double glazing and air chamber, with hermetic seal.

Different blinsd and inner curtain systems regulate the light intensity which is complemented with the layout of cornices and protective roofs. The thickness of the walls is considered due to its thermal inertia, having been insulated in some cases on the outside of the northern walls. (fig. IV).

Direct radiation is used as the main heating system and indirectly in the natural lighting. An integral water circuit is used in the existing fountains and in the different channels which constitus the natural comfort of the open space, complemented with the use of auctochthonic arboreal species, both deciduous and evergreen.

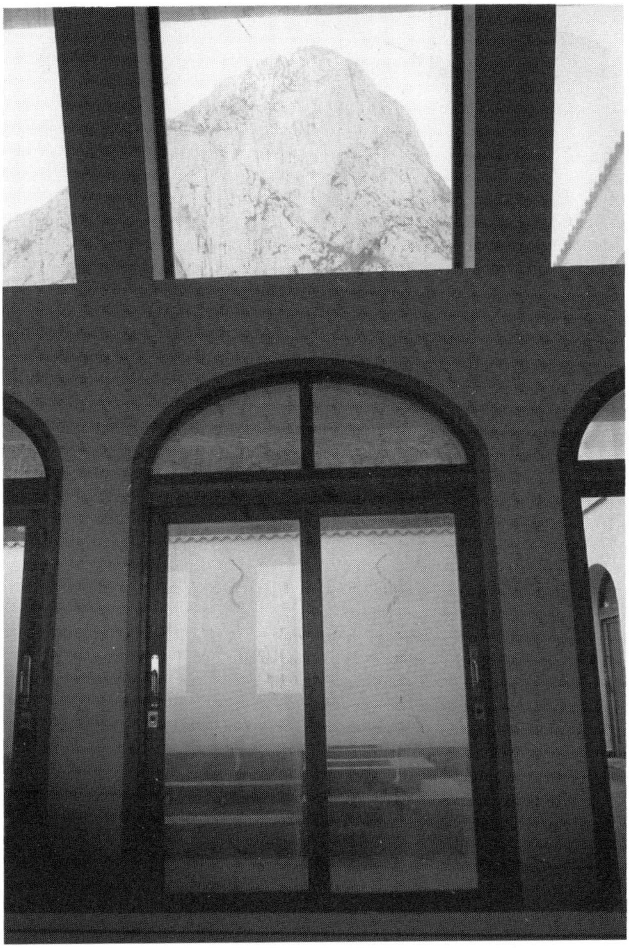

FIG. IV

5. CONCLUSIONS.

The rehabilitation of the two unique buildings permits their inmediate use as Nature Classrooms, where the bioclimatic design contributes to the solar gain in winter and air-conditioning in summer. The existing specific microclimate is considered in the improvement of the transversal ventilation and it is expected, once the results are analyzed, that the buildings will be 80% self-sufficient.

THE HACIENDA PORZUNA ENVIRONMENTAL EDUCATIONAL CENTRE

I. Blanco, Architect, Alameda 4, 46010 Valencia, Spain
A. J. Casanovas, Dept. Termodinámica, Univ. València, 46100 Burjassot, Valencia, Spain

ABSTRACT

The Hacienda Porzuna situated in the township of Mairena del Aljarafe, Seville, Spain is an group of buildings to be retrofited to conform an Environmental Educational Centre. This completely new experience in the Andalusian region has as its challenging objectives the retrieval of historical buildings, the use of bioclimatic strategies of heating, daylighting and natural cooling and the conditioning of the open spaces defined by the inner patios, the gardens and the circulation zones.

INTRODUCTION

From historical data and archaeological rests we can conclude that the Hacienda Porzuna was erected during the Moslem occupation of the south of Iberian peninsula. The building organization around *patios* follows the *villae-romana* structure. (fig.I)

Buildings' description

The 3246 m^2 builded area comprises a main building, the *Caserío*, and several outbuildings, being the *Casa del Guarda* and a turret (an old mill) the most characteristic elements of the annexed buildings. The whole is surrounded by a wall. The external section is typical of the Andalusian *haciendas* with Mauresque tiled roofs.

Most of the buildings are now damaged and all of them will be retrofited to convert them in an 12 ha. environmental sciences educational centre. The most fundamental features of the previous architecture will be preserved.

FUNCTIONAL ADAPTATION OF THE *HACIENDA*

The main building will act as a social and administrative centre, coordinating the activities of the whole *hacienda*, and its inner patio as a circulation zone and generator of a specific microclimate. The old mill, adapted to be a polyvalent meeting zone and dining room, will be provided with zenithal windows, for daylighting and cross-ventilation of the zone. (fig.II y fig.III)

A new building substitutes the former naves. This unit will include the areas of conference hall, ecological museum, exhibition hall, lodge and house of the watchman or caretaker. It is provided, all over its zenithal cover, with vertical glazings and a vertical movable *brisse-soleil* system, allowing for the control of ventilation and daylighting. This formal and structural unit allows, at the same time, a double pedestrian circulation to the patios in the ground floor and to the lodge and assembly hall in the first floor. An inner

S. Alvarez et al. (eds.), Architecture and Urban Space, 371–376.

catwalk and a ramp increases the circulation paths and, in addition to this, act as a exhibition zones. (fig.IV y fig. V)

On the right of the main building, the classroom and laboratories building form with the workshop area a formal unit. On it, a Blanco wall (BW) passive system (see 1988 and 1989 references) adapted to Seville's climate will be built. This highly reliable system supposes very little additional overcost relative to a conventional wall. In the first floor, several thermal mass water tanks are joined together in closed circuit with a large radiation surface situated on the north side. The insulating-reflecting inner shelters and the solar chimneys completes the Porzuna Blanco wall configuration, allowing for passive heating in winter periods, and improved cross-ventilation. (fig.VI)

MICROCLIMATE OF OPEN SPACES

The handling of the common spaces is defined by the buildings and the outer wall. Water is used as an unifying element in the form of waterfall, fountain and water pond; we can found this element in any of the patios organized in recipients of different sizes in the principal axes of the delimited patio spaces. A pedestrian circuit, protected by a pergola, joints all the patios and open spaces, defining a continuous covered space differentiated from the other landscaped zones. An adequate use of vegetation avoids the possibility of overheating due to direct solar radiation.

MAIN ASPECTS OF THE DESIGN OF THE BLANCO WALL IN THE HACIENDA PORZUNA

The Blanco wall (BW) passive solar system were not optimized in Los Molinos for heating due to the very mild climate and high solar irradiation standard, and most of the design recommendations follow from simulation studies performed (see, e.g., A. J. Casanovas (1991) references).

In spite of the fact that the BW is a passive system adequate for European climates, buildings subject to Mediterranean climates will achieve the best relative energetic savings. Some of the points to be considered in the Hacienda Porzuna design of the Blanco wall are as follows:

• Performance will be increased by increasing BW window area up to its geometrical limits and using double o triple glazings in the BW window.

• The water tank thermal mass version of BW performs slightly better than the massive concrete wall version. The heat accumulated in the thermal mass water tanks is distributed by radiators in the north side of the building.

• An insulation thermal resistance of 2.5 m^2K/W or more is sufficient for the Seville climate.

• The optimum tilt of BW windows is about 60º for the Seville latitude.

• A water tank thickness of 20 cm may be sufficient for Seville climate.

REFERENCES

Casanovas, A.J. and Blanco, I. (1988) "Analysis of the Blanco wall passive system in Los Molinos design", PLEA, Energy and buildings for temperate climates, Porto, July 28-31 Paper Number 2.2.172. Edited by Pergamon.

Casanovas, A.J. and Martinez-Lozano, J.A. (1988) "First evaluation of the Blanco wall passive system" PLEA (Passive and Low Energy Architecture), Energy and buildings for temperate climates, Porto, July 28-31, Paper Number 2.1.108. Edited by Pergamon.

Casanovas, A.J. and Martinez-Lozano, J.A. (1988) "Los Molinos, Crevillente, Spain", Project Monitor, Issue 24, Directorate General XII of the Commission of the European Communities, July 1988. Available through J. Oven Lewis, School of Architecture, University College Dublin, Richview, Clonskeagh, Dublin 14

Casanovas, A.J. and Martinez-Lozano, J.A.(1989) "Los Molinos School", IEA, Solar Heating and Cooling Task XI: Passive and Hybrid Solar Commercial Buildings,Basic Case Studies, BCS n.25

Casanovas, A.J. (1991) in "Los Molinos School", IEA, Solar Heating and Cooling Task XI: Passive and Hybrid Solar Commercial Buildings, Advanced Case Studies, (to be published)

Casanovas, A.J. (1991) in IEA, Solar Heating and Cooling Task XI: Passive and Hybrid Solar Commercial Buildings, Sourcebook, (to be published)

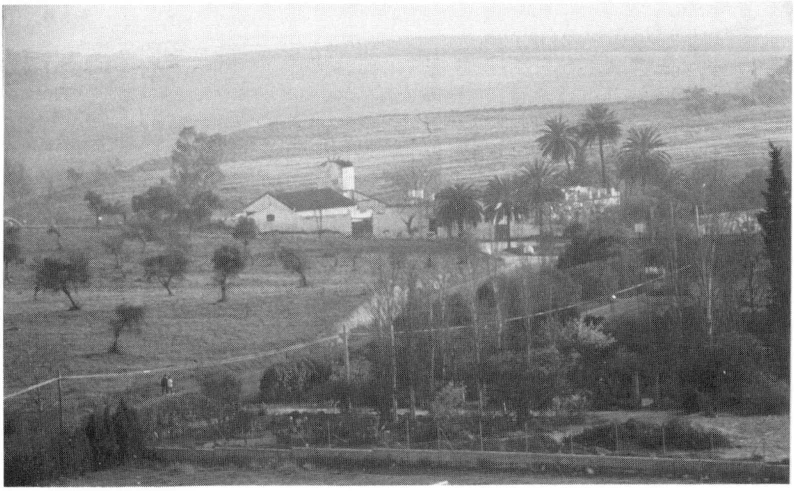

FIG. I

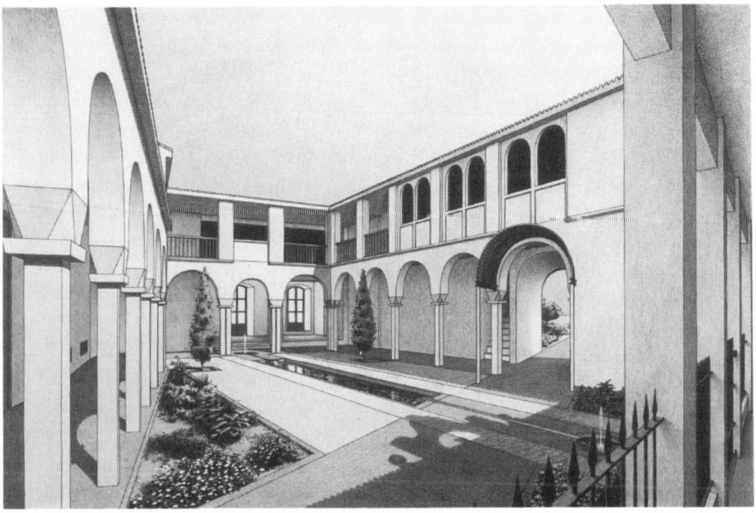

FIG. II

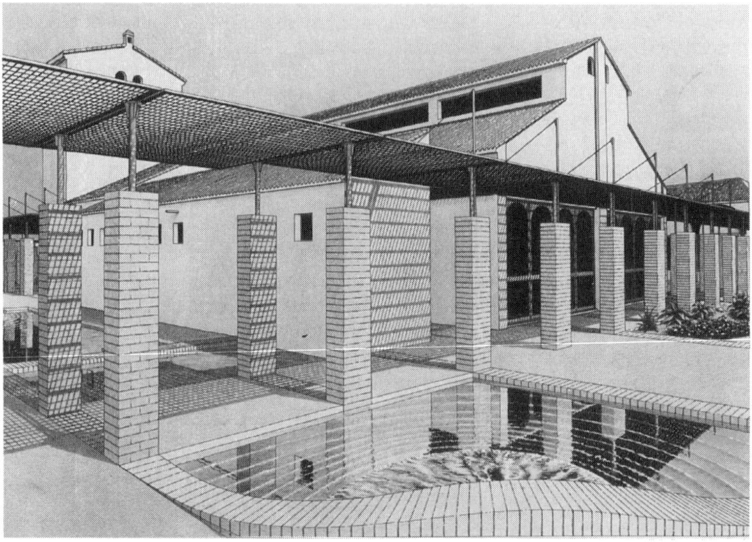

FIG. III

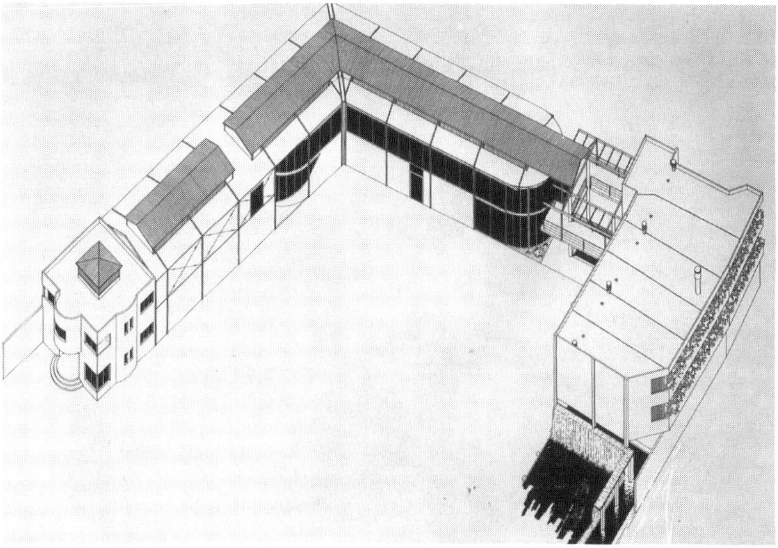

FIG. IV

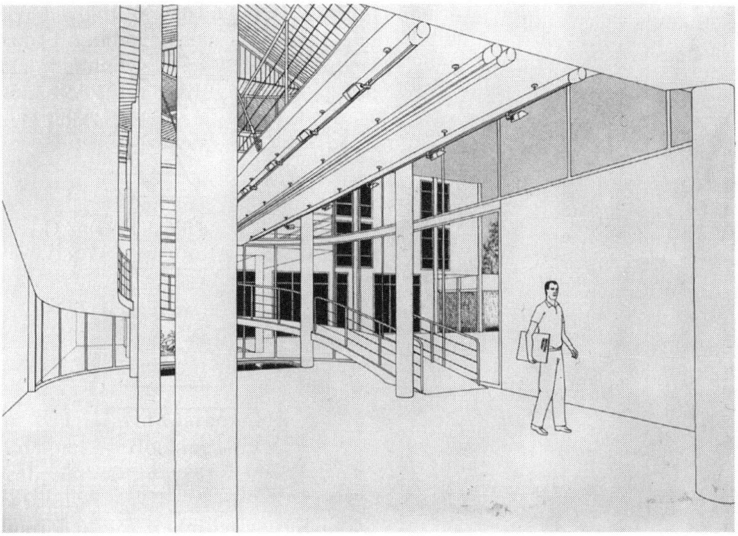

FIG. V

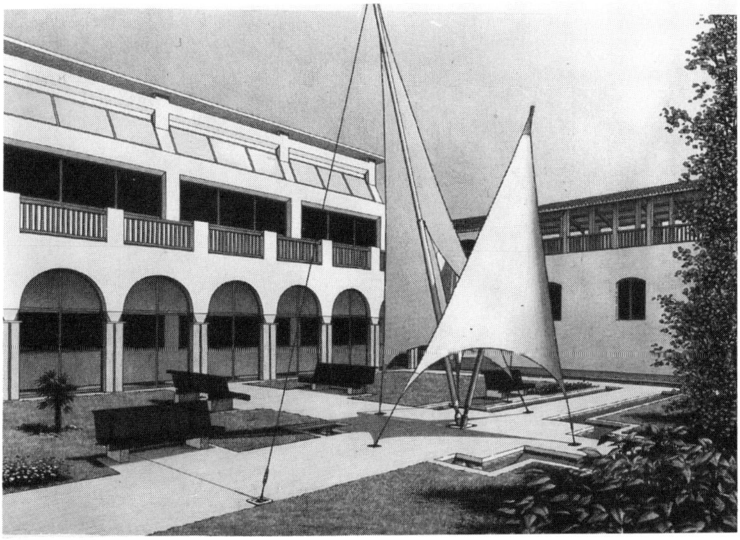

FIG. VI

EDUCATIONAL FACILITY FOR THE DISABLED.

JAIME LOPEZ ASIAIN
JOSE MARIA CABEZA LAINEZ
Architects. (Seminar of Bioclimatic Architecture).
School of Architecture of Seville
Av. Reina Mercedes, 2
41012 Seville
SPAIN

ABSTRACT. An environmental building with classrooms and workshops for the disabled has been designed and built in Mairena del Aljarafe (Seville). Main features are friendly integration with nature, passive cooling and heating, daylighting and use of vegetation. Current activities in the complex are going on with a high degree of acceptance and minimum energy use.

1. INTRODUCTION.

The Spanish Telephone Company had planned for a school and workshop in the vicinity of Seville. Former facilities of this kind had the result of nonenvironmental, strongly disproportionate buildings with a hospital resemblance. Consequently, the idea of integration with nature as friendly users was well received and understood.

A matrix module of South oriented blocks has been displayed and interconnected with two axial pergolas, thus creating symmetrical courtyards, entrance and exit corridors and a central lobby. The roof of the workshops allowed for top-lighting systems to make an efficient use of daylighting although, later, these systems emerged as a strong design element which characterized the building profile.

The courtyards have been designed according to vernacular traditions, with central fountains, loggias and removable canvas for the cycles day/night winter/summer.

Due to security reasons a metal egg-crate device has been proposed to cover the patios, filtering light and supporting visually the canvas.

As gardening is also considered an educational activity for the users, terraces for growing plants have been spread over the site and a North-oriented greenhouse is available.

S. Alvarez et al. (eds.), Architecture and Urban Space, 377–384.

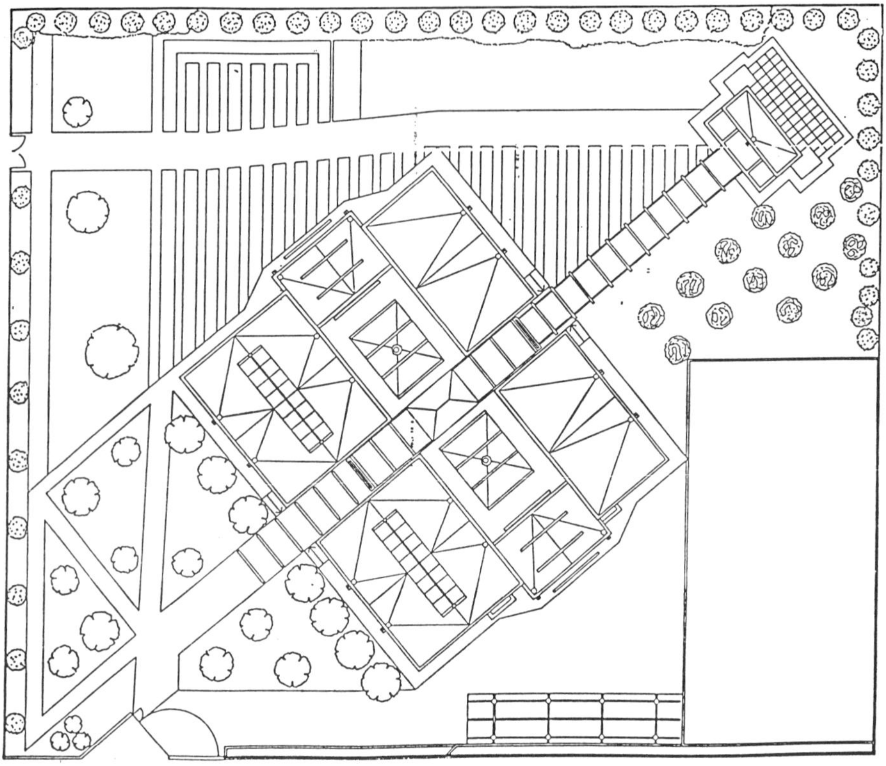

Figure 1. General overview.

0M 5M 10M 15M 20M

2. PASSIVE CONDITIONING STRATEGIES.

In Andalusian climate, South-oriented vertical planes receive maximum solar radiation in winter and minimum in summer. Thus the main strategy both for heating and cooling is to have most glazed apertures facing south. Besides this is an adequate procedure for side-light daylighting concepts provided that beam-daylighting is used.

Then the colour of external surfaces and specially the roofs has been kept as clear as possible while the walls are heavily insulated with 5cm. fiberglass and the structure of the building posses a thick thermal mass of more than 10cm. brick or concrete alternately.

Due to the rooflight position, workshops can be ventilated with cross ventilation through the patios or with stack effect if there is no wind available. The same principle of air buoyancy is used in the lobby.

The protected patios with vegetation and water can help to reduce summer temperatures to a certain extent.

Finally, all glazed apertures are protected against sun with screens, venetian blinds or louvres.

3. DAYLIGHTING STRATEGIES

The main daylighting feature is the rooflight. This rooflight has been designed with unutilizability method, that is, comparing the maximum heat gain produced with conventional lighting for a given illuminance level with the solar gains through the rooflight and leaving them well under this figure once the illuminance level is reached.

Sunlight's use is intended via light-shelves and external louvres though some direct sunlight is admitted in winter months.

The light coming from the roof will be balanced with that of the windows (either from the outdoor or the patios) to avoid excessive contrast and glare.

4. OTHER STRATEGIES.

Intensive use of vegetation has been considered a contributor to passive conditioning of the building and its surroundings. Also some protection from noise, rain and cold winds can be obtained with these plants. Running water in the courtyards improves the physical and psychological comfort conditions.

Finally domestic hot-water is supplied with solar collectors.

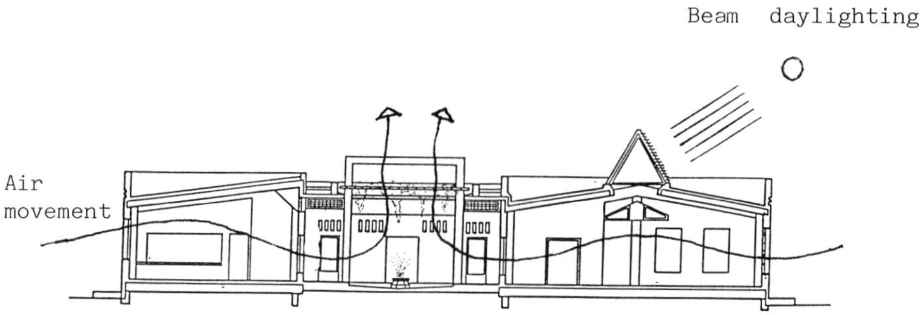

Figure 2. Cross section (South-North).

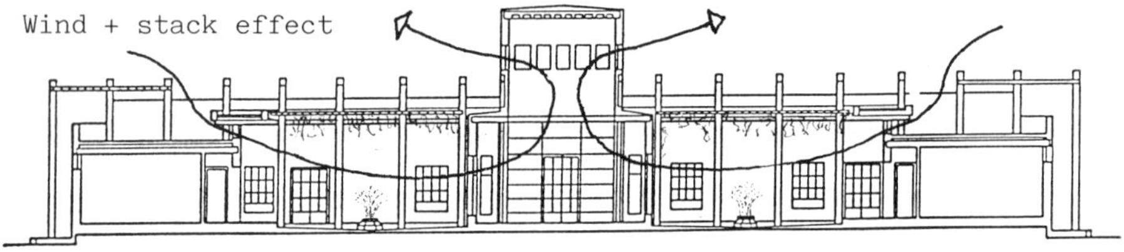

Figure 3. Cross section (West-East). E 1:300

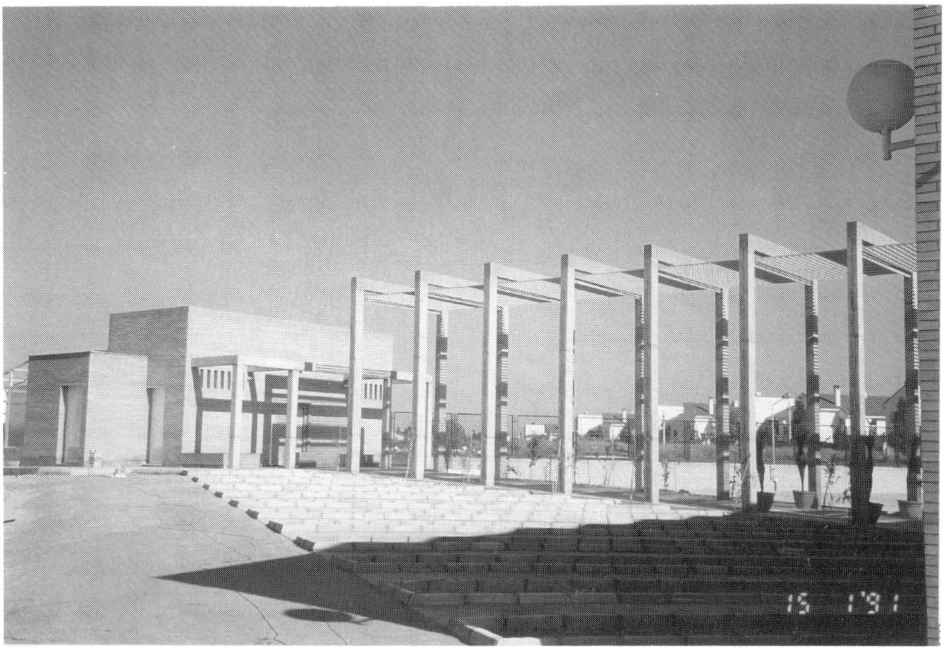
15 1'91

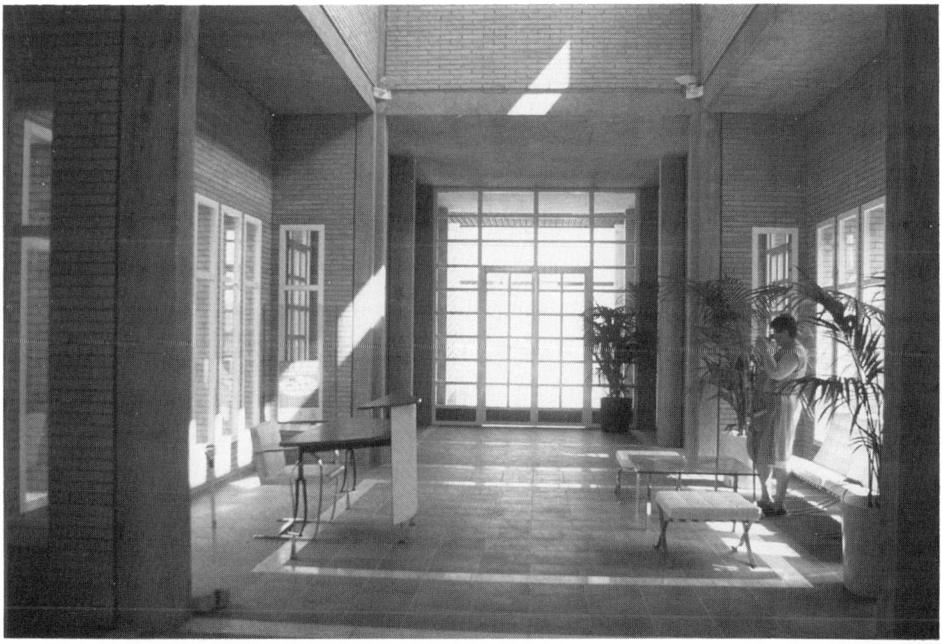

1. LOBBY.
2. PATIO.
3. WORKSHOP.
4. STORAGE.
5. OFFICE.
6. MEETING ROOM.
7. RESTAURANT.
8. MULTI-PURPOSE ROOM.
9. KITCHEN.
10. TOILETS.

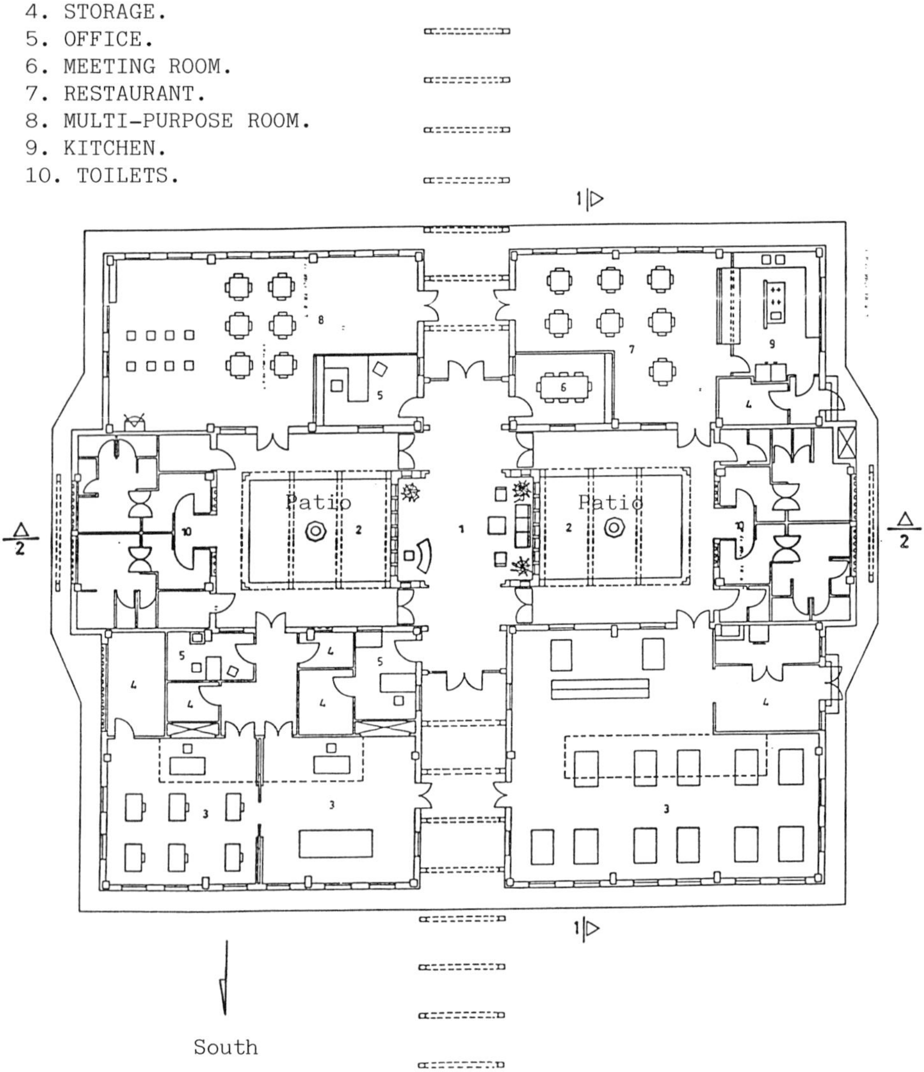

Figure 4. Ground Floor Plan. E 1:300

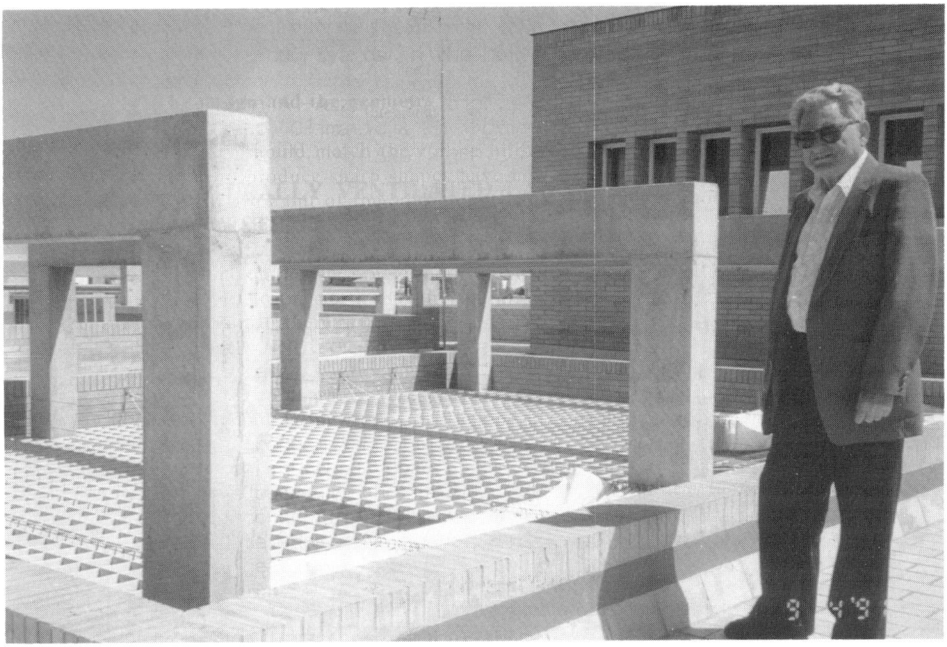

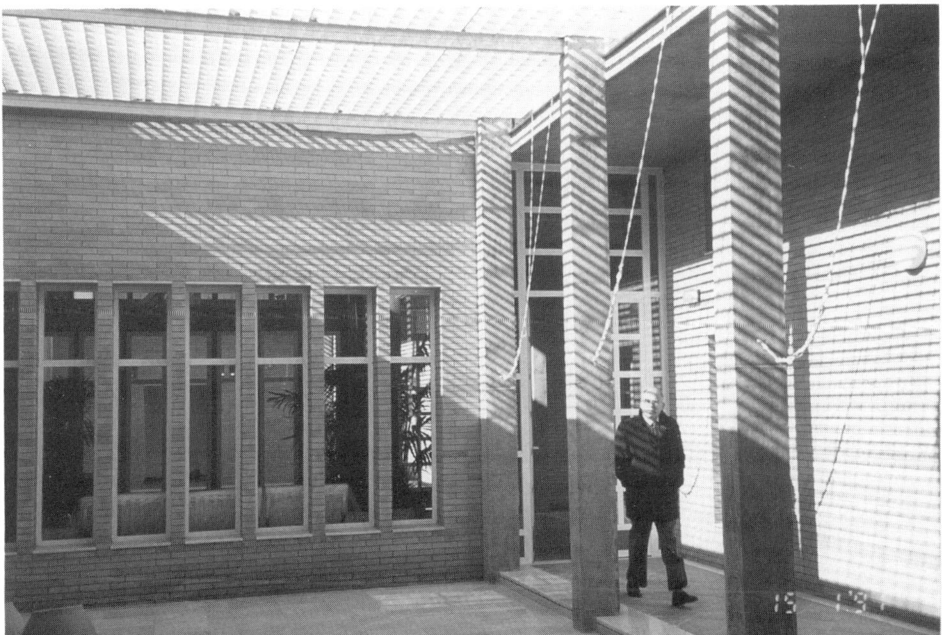

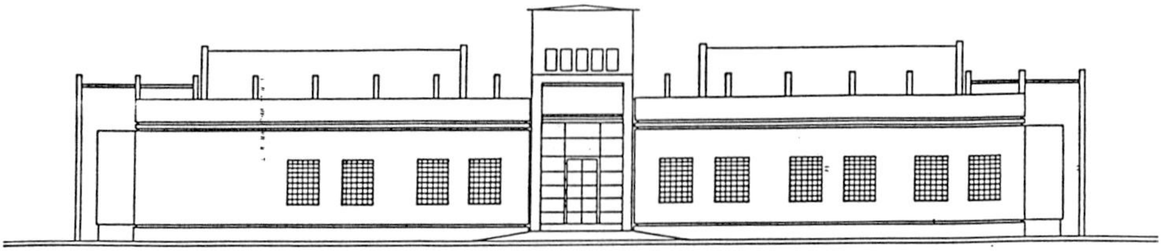

Figure 5. North façade.

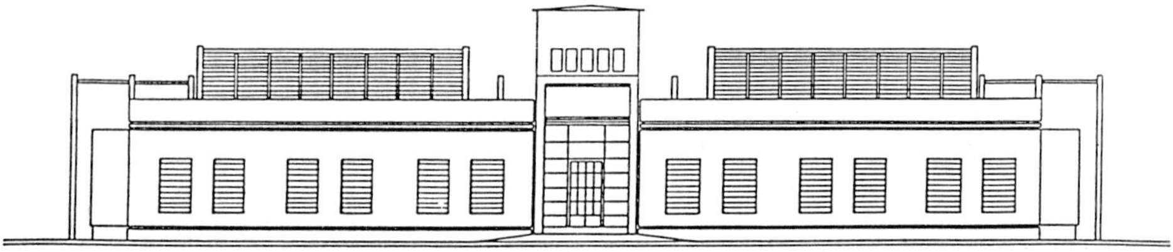

Figure 6. South façade. E 1:300

BIOCLIMATIC DESIGN AS AN ALTERNATIVE SOLUTION TO THE ENVIRONMENTAL AND ENERGY PROBLEMS RELATED TO BUILDINGS IN URBAN AND RURAL SETTLEMENTS OF MEXICO

José Roberto García Chávez

UNIVERSIDAD AUTONOMA METROPOLITANA
DEPARTAMENTO DE MEDIO AMBIENTE
Juan Sebastian Bach 99. Colonia Peralvillo
Código Postal 06220
MEXICO, DISTRITO FEDERAL
Tel. 5 97-02-98. **Fax 781-11-71**

ABSTRACT

Most human activities through history have been characterized by an inappropriate utilization of energy in all its forms. This has provoked a severe damage in our ecosystems. The result of the industrial activities during the last fifty years, based on an intensive exploitation and burning of fossil fuels, and the explosive growth of population, has been a critical damage in our natural habitat. As to the built environment, a significant proportion of the damage has been caused by predatory human activities in the way energy (mainly fossil fuels) is used in buildings, where man live, work, study, and socialize. Nowadays, nearly half of the commercial energy available in the world is consumed in buildings. The objective of this work is to establish alternatives based on a more conscious use of energy in buildings and to integrate them to the environment in a sustainable manner so that buildings are designed and built taking into account the real psycho-physiological needs of the occupants in time and space (**Bioclimatic Design**). The particular objective of this work is to present the necessary mechanisms, using bioclimatic design principles in buildings, to propitiate an integration to the environment whilst conserving energy. This is based upon the utilization of the natural renewable energy resources, as an alternative to help resolving some of the environmental and energy problems of the country and to establish at the same time, the appropriate conditions to achieve an authentic quality of life. Under this approach, some recent bioclimatic projects, used as case studies, are presented and the results examined in this work.

1. Introduction

The present environmental deterioration of our planet is the result of a number of predatory activities by man, and manifested in phenomena such as the increment of the "greenhouse effect", the reduction of the ozone layer, the destruction of the rain forests, the formation of acid rain, and so on. Most human activities through history have been characterized by an inappropriate utilization of energy in all its forms. This has provoked a severe damage of our ecosystems [1 to 4]. At the beginning of the 20th century, neither human beings nor human technology had the power to significantly alter the global ecosystem. Today, as the century draws to a close, human population in an accelerated growing pattern, have the power and so the technology, to severely modify our natural habitat.

Certainly, the combined effect of the industrial activities, particularly those occurring during the last fifty years (based on an intensive exploitation and burning of fossil fuels), and the explosive growth of population, has provoked a critical damage in our planet.

The damage caused to the environment in Mexico is severe: 95 percent of its rain forests have been destroyed, and more than 50 per cent of its low forests and 65 percent of its mixed forests have disappeared [5]. More than one- third of the country land has become useless desert region. The great majority of its rivers, lakes and lagoons are highly polluted. The atmosphere, water and soil of both urban and rural areas are also considerably contaminated. This situation affects people's health and quality of life as well as productivity, which in turn inhibits the development of the country.

S. Alvarez et al. (eds.), Architecture and Urban Space, 385–392.

As to the built environment, a significant proportion of the damage has been caused by predatory human activities in the way energy (mainly fossil fuels) is used in buildings, where man live, work, study, and socialize. Nowadays, nearly half of the commercial energy available in the world is consumed in buildings.

Therefore, it is clear that since the Industrial Revolution, and particularly after the Second World War, the absurd and inappropriate utilization of the energy (mainly in the form of fossil fuels), has led to a number of problems, such as that provoked in the environment. In the built environment, the situation is the same. In most locations of the world, it is common to see buildings designed and constructed without any consideration for the climatic conditions and culture of a region. As a result, the energy is wasted, the environment damaged and the economy and people's health, affected. Most modern buildings built after the 50s', consume extensive amounts of energy to heat, cool, ventilate and light the interiors, squandering energy, becoming greatly dependent on the availability of fossil fuels and polluting the environment. Those buildings can then labelled as excessive energy consumers and highly dependent on fossil fuels.

Therefore, it is urgent and necessary a change of attitude in the design and construction of buildings, by using energy in a more conscious and efficient way and particularly to consider the application and integration of natural renewable energy resources, such as solar, wind, etc. This approach can lead to energy savings, whilst extending the conventional sources, and improving the environment and people's quality of life. One way to achieve this is by means of Bioclimatic Passive Design Systems, which implies the utilization of the building itself to select those components of the prevailing outdoor environment which can contribute to the users' natural comfort in the indoor environment. Unlike most conventional buildings designed with energy consumptive air-conditioning equipment and artificial lighting, most of the time still uncomfortable. Passive Design will enhance users' ambient comfort conditions, providing a more healthy environment, in which people can rest, work and play, at lower capital and running costs. Therefore, the premise of Passive Design can be summarized as follows: maximum human comfort conditions at minimum energy costs and maximum preservation of the environment. In Mexico, the implementation of this approach, can lead to increase productivity and to propitiate a sustainable development, meeting the needs of the present without compromising the ability of future generations to meet their own needs.

2. Background. Present situation of buildings in Mexico related to energy use and the environment

Most buildings of Mexico consume large quantities of energy, in the form of fossil fuels, to provide heating, cooling, ventilation, lighting and to heat water. This situation has provoked a number of health, environmental, economic and sociocultural problems in the country [6,7]. All of those requirements can be satisfied to some extent from the diverse natural renewable energy resources widely available throughout the country, thereby reducing the dependence on energy consumptive services and the emission of pollutants to the atmosphere.

In the housing sector, the majority of the existing buildings do not respond to the local traditional, cultural and climatic conditions. Instead, they seem to follow "international pattern styles" produced for unjustified reasons such as "social status" and "fashion in vogue".

Electricity, natural gas and oil are the main energy sources consumed in urban dwellings [8,9], whereas firewood is the main energy source used in the rural regions [10]. This excessive consumption of firewood is causing an accelerated deforestation, which in turn affects the environment and their living organisms, as the absorption rate of carbon dioxide is reduced and the production of "greenhouse gases", increased.

2.1. BUILDING TYPOLOGY AND ITS RELATIONSHIP WITH CLIMATE

Traditional building construction in Mexico was the result of the interaction of historical, sociocultural, technological, economic and climatic factors. A number of buildings were built under this approach, and are still found in the countryside. Light well-ventilated buildings in hot/humid and high massive constructions in hot/arid regions can be found in traditional designs. Unfortunately, this practice is becoming less common, particularly in the congested urban centres, where buildings are characterized by the high dependence on conventional fossil fuels and their excessive consumption. Certainly, in a big number of buildings constructed in large urban settlements, the

"International Style" and which can be its last manifestation, the so-called "postmodernism" (pure cosmetics?), predominate, without any concern for the local climatic and sociocultural factors of the site where they are built, nor for the necessary integration with the surrounded natural environment.

3. Availability of energy resources in the country

Mexico can be stated as a privileged country due to its great diversity and availability of energy. Conventional fossil fuels such as oil, coal and gas are abundant in the country, however, there is a high dependency on these sources. For instance, 90 percent of the energy consumed in the country comes from hydrocarbons [9] and there is no clear indication of interest to support investigation and application of other energy alternatives. It has been historically demonstrated that the exclusive exploitation and use of a sole conventional energy source can be a costly risk for a country.

In terms of non conventional energy sources, such as solar, wind and so on, Mexico is located in a geographical region, which is plentiful of them. For example, the country itself is situated from 14 to 32 degrees North latitude, where high solar radiation intensities occur. The wind potential of "La Ventosa" in the state of Oaxaca is one of the most important in the world. The 11,600 kilometres of coastal areas surrounding the country makes it potentially important in terms of ocean energy (e.g. tidal, thermal gradient and waves). Mexico has also a considerable hydroelectric potential [11], which at present has only been exploited in less than 20 percent. Hydroelectric power is an indirect renewable resource from the sun. Biomass, another energy source produced from organic waste, in the form of firewood, represents 13 percent of the final energy consumption [9]. In other words, around 15 million people consume more than 20 million tons of firewood. These figures illustrates the importance of biomass in the energy consumption panorama of Mexico. However, the burning of firewood provokes environmental problems such as the increase of "greenhouse gases" and deforestation. As a matter of fact, biomass energy should be obtained from other sources, which do not cause a damage to the environment. Under this approach, recycling is essential.

A suitable utilization of alternative energies and a more efficient use of the conventional ones, taking into account the country circumstances, can lead to energy saving, whilst preserving the environment and extending the time horizon of conventional sources. Furthermore, and perhaps most importantly, people's quality of life can be significantly improved.

4. Application and integration of alternative energy sources in domestic buildings. Case studies

The consciously oriented use of the natural renewable energies in buildings, is the basis of Bioclimatic Design, which implies the application of "Passive Systems", that is, the utilization of the building itself to select those components of the outdoor environment which can contribute to integral ambient natural comfort conditions of the users' indoor environment. This approach includes the satisfaction of the hygrothermal, luminous/visual, acoustic and olfactory requirements of occupants.

4.1. FIRST CASE STUDY: ARAGON BIOCLIMATIC/ECOLOGICAL HOUSE

This house [fig. 1], built in 1985 and located Northeast Mexico City (a temperate climate), was designed to integrate natural ambient climatization with energy production and water saving systems, as well as food production and classification and recycling of domestic waste. From the Bioclimatic point of view, and for the achievement of integral ambient comfort conditions, passive design measures such as solar geometry, texture, colours, thermal mass, ventilation, vegetation, water bodies, acoustics and daylighting, were implemented in this project. Solar collectors for water heating, solar kitchen for food cooking and a passive refrigerator for food conservation were installed in this house. Water systems included grey water recycling, rainwater storage, saving devices and low consumption sanitary equipment. Hydroponics and kitchen garden were also integrated in the house to help domestic economy of the occupants.

4.2. SECOND CASE STUDY: AZCAPOTZALCO BIOCLIMATIC/ECOLOGICAL HOUSE

The design and construction of this house [fig. 2], built in 1989 and located Northwest Mexico City (a temperate climate), were oriented to integrate natural climatization with energy production and water saving systems, as well as food production and classification and recycling of domestic waste.

From the Bioclimatic point of view, and for the achievement of integral ambient comfort conditions, passive design measures consisted of solar geometry, texture, colours, thermal mass, ventilation, vegetation, water bodies, acoustics and daylighting. Solar collectors for water heating, solar kitchen for food cooking, a passive refrigerator for food conservation and low energy consumption lighting fixtures, were installed in this house. Water systems included grey water recycling, rainwater storage, saving devices and low consumption sanitary equipment. Hydroponics, a kitchen garden and compost apparatus, were also integrated in this project.

4.3. THIRD CASE STUDY: ECATEPEC BIOCLIMATIC/ECOLOGICAL HOUSE

The project is located Northeast Mexico City a temperate/dry region [fig. 3]. The design of this house was focussed on the integration of natural climatization with energy production and water saving and potabilization systems (there is a problem of water supply, highly contaminated with heavy metals in this location), as well as food production and classification and recycling of domestic waste. From the Bioclimatic point of view and for the achievement of integral ambient comfort conditions, similar systems to those above mentioned, including olfactory control, by means of the combination of free ventilation and plant materials, were implemented in this project. Solar collectors for water heating, solar kitchen for food cooking, a passive refrigerator for food conservation, a solar still for pure water production, and low energy consumption fluorescent lighting fixtures, were designed for this house. Water systems included grey water recycling, black water treatment, rainwater storage, saving hydrosanitary devices and low consumption sanitary equipment. Hydroponics, a kitchen garden, a compost apparatus, a greenhouse, pisciculture, apiculture, and a fruit/vegetable solar dehydrator were also integrated in this project.

4.4. FOURTH CASE STUDY: BIOCLIMATIC CATHEDRAL AT LAZARO CARDENAS, MICHOACAN

This building is located close to the sea, in the West coast of Mexico, in a typical hot/humid region, with predominant clear skies [fig. 6]. The main concern in this project was the achievement of comfort for the congregation. Maximum solar protection by means of building mass, optimum volume to surface ratio (maximizing the internal space, while minimizing the exposed surface area), building thermal mass, shading devices and the provision of effective ventilation were the main objectives. Ventilation was designed using properly located wind outlet and inlet apertures in the building, oriented 30 degrees from the prevailing wind direction, which was experimentally proven to be the optimum orientation. Daylighting design considerations as well as the use of lighting translucent systems were designed for both thermal and luminous control. The special lighting effects needed for religious purposes in the cathedral were considered in combination with the thermal/luminous design. Vegetation was designed to prevent the buildings from uncomfortable noise and smells from the big iron and steel industry located near the sea, in the direction of the prevailing winds. Hot water production for domestic and semi-industrial applications (using solar collectors) and water treatment/saving systems for grey, black and rainwater were considered in this project, as well as food production and classification and recycling of domestic waste. This had the doble objective to achieve energy and water savings and to serve for training/educational purposes for the inhabitants of the community.

5. Analysis of the results and conclusions

Comparisons of the first case study house (Aragon) with a similar one nearby, show that after a running of six years, water and energy savings are as follows. Water: 75 percent; gas: 70 percent; electricity 28 percent. Moreover, occupants have expressed satisfaction with the internal ambient conditions and said to be happy with the economic results obtained so far. Environmentally speaking, this house contributes favourably, as it uses less water and fossil fuels than other conventional ones, which can be comparable. Problems found in this project were the performance of the greywater recycling system and of the storage tank of the solar hot water collector. It is worth mentioning that occupants took around six months to get used to the handling and maintenance of the systems installed. The second case study presented the opportunity to correct some of the problems found in its predecessor. Similar savings have been obtained so far: Water: 70 percent, gas 64 percent and electricity 39 percent. Occupants of this house were more committed to handle and maintain the systems and learned faster. The third case study in now under construction and completion is expected by the end of 1991. The construction of the fourth case study is about to be finished. On-site monitoring carried out during the most disadvantageous period (summer afternoon,

1990) revealed that the dry bulb temperature in the space being occupied can be as much as 7⁰C below the existing outdoor dry bulb temperature (i.e. 36⁰C/29⁰C). The wet bulb temperature registered inside was 27⁰C. These figures indicate that the conditions inside are just in the upper limit of comfort for most people. Interviews with occupants confirmed this. Twelve percent of individuals manifested discomfort with the conditions though. Mechanical ventilation has been suggested to improve human comfort conditions in this building.

The construction of bioclimatic/ecological buildings in Mexico offers important benefits for the occupants, directly and indirectly, and is also a suitable alternative for help resolving some of the environmental and economic problems that the country is facing. Energy savings, preservation of the environment and improvement of users' economy and quality of life are the main benefits, which can be obtain with the implementation of such approach. However, massive actions are necessary if significant results are expected. All sectors of the society need to collaborate in this objective. A change in our predatory behaviour to the natural environment and a real ecological culture should be of the highest priority in this respect . It is indispensable that we stop thinking that our planet is infinite in resources and that we are the only owners to dominate it and make use of it to our convenience. Nature has laws and we must respect them and maintain a permanent alliance. As designers of habitats, we should be committed ourselves to offer our best endeavours in the search for solutions to the authentic human needs. Bioclimatic design is a promising alternative in this respect, and it should be always oriented to a permanent and sustainable evolution, according to the particular requirements and conditions of a given project.

6. References

1 Clark, W. "Managing Planet Earth", en Scientific American (Special Issue), Vol. 261, Num. 3, pp. 46-54, September, 1989.

2 Graedel, T., Crutzen, P. "The Changing Atmosphere", en Scientific American (Special Issue), Vol. 261, Num. 3, pp. 58-68, September , 1989.

3 Shneider, S. "The Changing Climate", en Scientific American (Special Issue), Vol. 261, Num. 3, pp. 70-79, September, 1989.

4 Ruckelhaus, W. "Towards a Sustainable World", en Scientific American (Special Issue), Vol. 261, Num. 3, pp. 166-174, September, 1989.

5 Liverman D. M., O'Brien, K. "Climate change in México: likely effects and impacts". Conference Proceedings, AAAS Annual Meeting, New Orleans, EE.UU., February, 1990.

6 Moncada, G. "D.F. 2000: Los Años Que Viviremos en Peligro", en Nuestro Ambiente, Revista Mexicana de Ecología, Vol. 1, Num. 1. Respuesta Ecológica, México, D.F. June, 1990.

7 Tijerena, E. " El Problema Ecológico de México" Año 2, Número 11, , pp. 33-37. Universidad Autónoma Metropolitana, Iztapalapa, DCSH, México, D.F. June, 1990

8 Menchaca, L., Conde, C., Gay, C. "El Caso México, un Problema Poco Estudiado". Edición Mundial. Año XIII, Num. 160. Suplemento ecológico, pp. 7-8, México Desconocido, México, D.F. June, 1990.

9 Balance Nacional de Energía 1989. Secretaría de Energía, Minas e Industria Paraestatal (SEMIP). México, 1990.

10 Energía Rural en México. Secretaría de Energía, Minas e Industria Paraestatal (SEMIP) y Comisión de las Comunidades Europeas (CEE). México, 1988.

11 Viqueira, J. "Proyecciones del Consumo Eléctrico en la Cuenca del Valle de México al año 2000". Proceedings: Agua y Energía en la Ciudad de México. Fundación Friedrich Ebert, México, D.F. July, 1988.

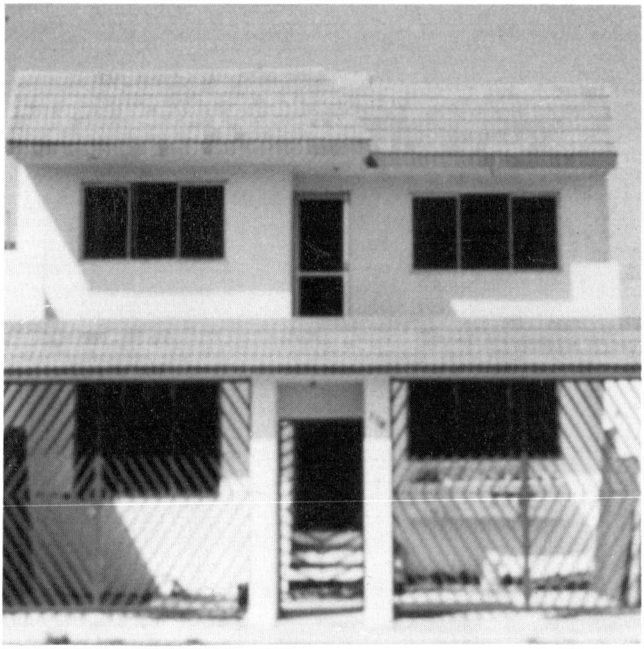

Fig. 1 Aragon Bioclimatic/Ecological House

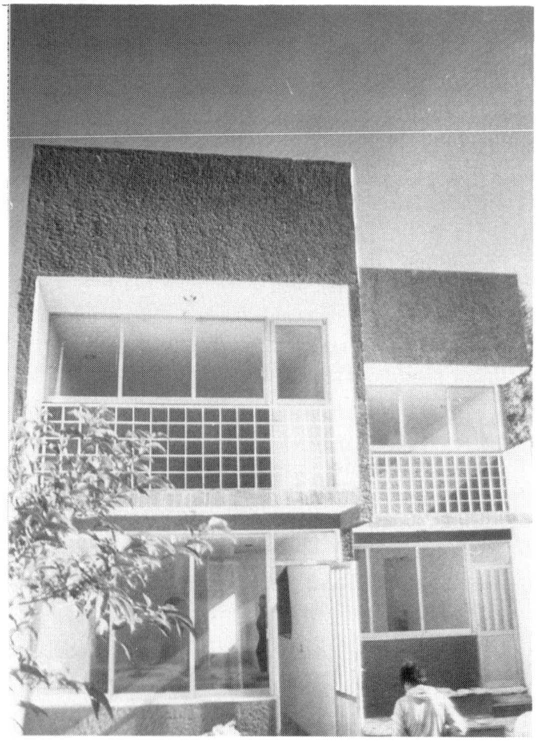

Fig.2 Azcapotzalco Bioclimatic/Ecological House

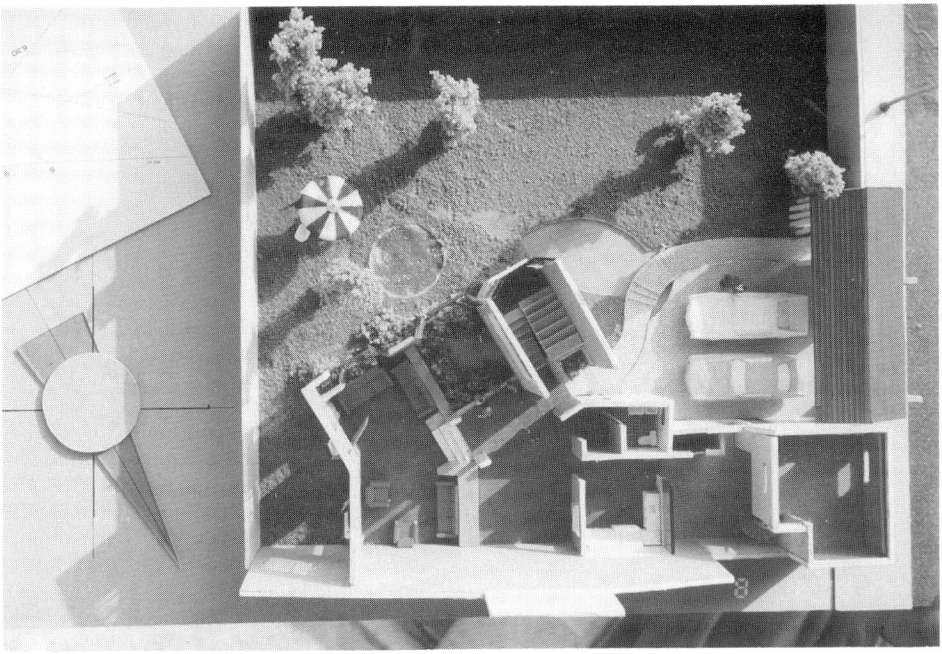

Fig. 3 Ecatepec Bioclimatic/Ecological House

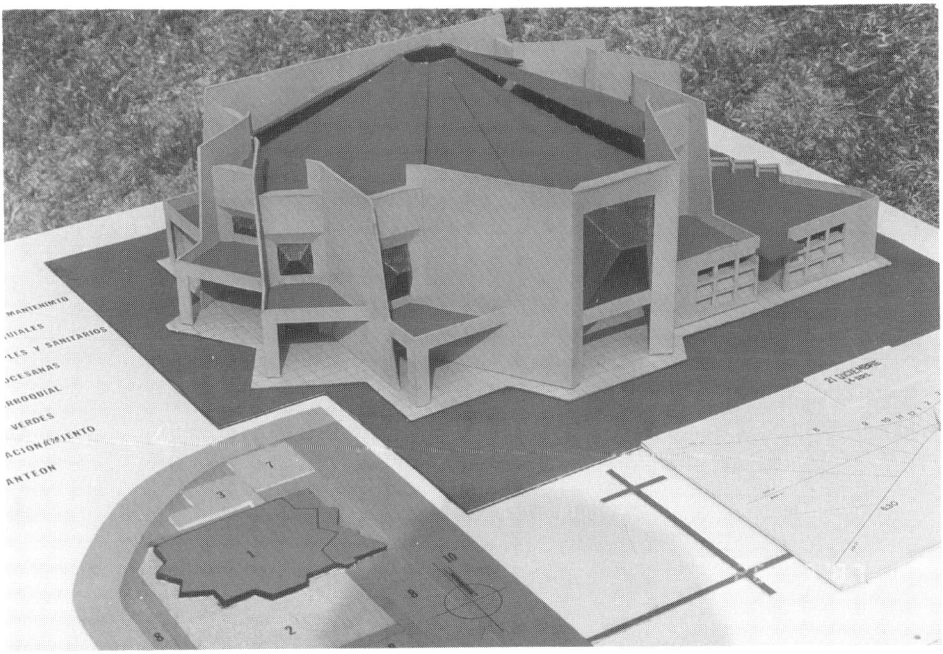

Fig. 4 Bioclimatic Cathedral at Lazaro Cardenas

SOLAR INDOOR SPORTS CENTRE IN THESSALONIKI, GREECE

K. ANTONIOU — Dept of architecture
prof M. PAPADOPOULOS — Lab of building physics
prof K. PATTAS — Lab of thermodynamics
prof B. SOTIROPOULOS — Lab.of process engineering
School of Technology, Aristoteles University of Thessaloniki, Greece
54006 Thessaloniki, Greece.

ABSTRACT. The Solar Indoor Sports Centre is a demonstration project employing energy saving techniques and a variety of passive and active solar systems. The energy balance calculations for the building show that there is no need for conventional fuel for space heating and hot water.

1. INTRODUCTION - PRESENTATION OF THE BUILDING

The Solar Indoor Sports Centre [S.I.S.C] is a demonstration project in Thessaloniki [north Greece, 40oN.Lat.], designed on the principles of energy conscious and energy saving design, and exploiting solar energy, so that conventional fuel for space heating and hot water is not necessary. The knowledge and information gained from this practical application of solar technology will facilitate the future construction of such sports centres.
The building is a general purpose indoor and outdoor Sports centre to be used for a variety of sports activities, as well social and cultural events for the local community.
The construction of the S.I.S.C. has already started; completion is planned for summer 1991.

PROJECT DATA

Site area:	14.190 M^2
Total built area:.	4.500 M^2
Ground floor area:	3.350 M^2
Main Sports Hall area:	2.000 M^2
Seats: Fixed:	900
Mobil:	500

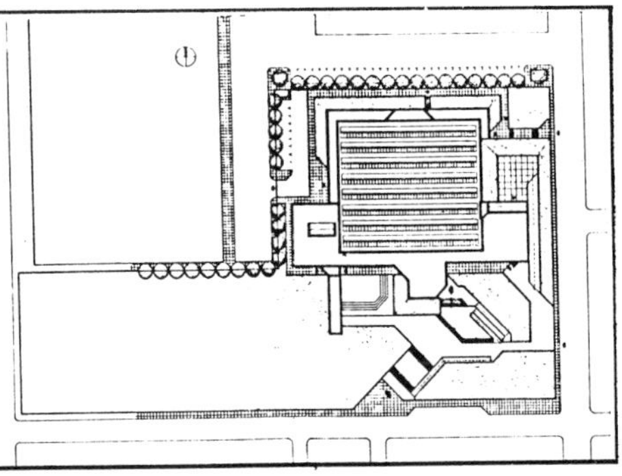

Fig. 1
SITE PLAN

The building consists of:

* a central sports hall with a sports area 48.00 X33.00X9.70 m. With the use of two movable partitions the central hall can be divided into three independant sports areas, each 16.00X 30.00 m, which can be used simultaneously.
  There is also a training Hall, next to the main Sports Hall, and six changing rooms together with the necessary service and storage rooms. [fig. 3]
* Several halls for cultural and recreation activities. These spaces are either fixed, e.g. lounges, library, seminar rooms and offices for the sport clubs of the area, or multipurpose halls. Foyer, entrance hall and bar, a total of 350.00 M^2, can be used by the local community during weekdays and when there are no sports sessions. [fig. 5]

S. Alvarez et al. (eds.), Architecture and Urban Space, 393–398.

2. PRINCIPLES OF ENERGY SAVING DESIGN

2.1. ORIENTATION

The main Sports Hall is located at the northerh end of the site, while the staff offices, the foyer and all the other rooms for social activities, which will be used by the community, are grouped together, forming an elongated, low building on the south of the Sports Hall, so as to receive the maximum solar radiation.
The Centre is protected from the cold northern wind by trees planted along the northern side of the site. [fig. 1]

2.2. MINIMISING THERMAL LOSSES

The main energy conservation measures are the high level of thermal insulation [10 cm], double glazing in all windows, the small openings to the north, and a wind buffer space in front of the entrance hall.
The spatial organisation of the building, which is organised in three climatic zones; closed,semi-closed and open-air zones, and the formation of the facades; glazed south facade and almost windowless north, east and west facades, also contribute to the minimising of thermal losses. [fig. 7, 8, 9,.10]

2.3. EXPLOITING SOLAR ENERGY

The architecture of the building helps in obtaining the maximum benefit from solar energy. The main features of the design are

* the extended south facade which permits the use of passive solar elements; large openings for direct gain and mass walls,
* the shed design of the roof which creates the necessary space for the installation of the active solar collectors, incorporated architecturally into the design and
* the existance of shading devices, to avoid overheating of the building from Spring to Autumn. [fig. 6]

2.3.1. Passive Solar Systems

Direct gain double glazed openings:

* South facing windows, for direct gain are located in the south elevation and on the roof of the lower parts of the building. A row of openings covering the south side of the last shed-roof admits solar energy into the main Sports Hall [total area: 213.00M^2]. [fig. 7]

Mass walls:

* Solar Mass Walls without vents are built into the south facade of the tall building from the level of +6.90m up to the level +9.00m, and six units in the south facade of the ground floor [total area: 135.00M^2].

2.3.2. Active Solar Systems

Auxiliary space heating and hot water are supplied from a water storage tank, which is heated by a 570.00 M^2 array of flat plate solar collectors mounted on the roof of the main Sports Hall.
The heat store is a well insulated concrete tank, located in the basement of the building, where all the mechanical installation for heating and ventilation are located.
The stored water is used for the underfloor heating system, for preheating the fresh air and for heating the hot water. [fig. 5]

2.4. NATURAL DAYLIGHTING

The main Sports Hall is naturally lighted by the roof openings [seven north and one south]. [fig. 4, 6]

2.5. NATURAL VENTILATION AND COOLING

Natural cross ventilation, during the warm periods of the year, is achieved by opening northern windows of the sheds and windows in the south and north facades of the main Sports Hall. Also, cross ventilation on the East-West axis is created at the highest level of the building, through the 16 openings in the 8 sheds.
At ground floor level there are opposite exit doors, which, if open, help the natural ventilation of this level, together with the openings at the level of +9.00m and +11.00m.
In winter, mechanical ventilation is mainly used. [fig. 4, 6.]

3. EXPECTED ENERGY SAVINGS

The energy balance calculations are made separately for a) the main Sports Hall [heated area 1440.00M^2 and volume 14000.00M^3] and b) the rest of the building.

*The main Sports Hall is heated by an underfloor heating system and preheated fresh air [15.000 M^3/h].

*The other spaces are heated only by the underfloor heating system, exept the service areas [showers and changing rooms] and the training hall which also receive preheated fresh air [3000 M^3/h]].

The total thermal load of the building is 311 MWH [4W/M^3].

The expected thermal gains are:

*internal gains from spectators and athletes

*internal gains from lighting

*passive solar gains

*recuperation from the ventilation air

*active solar gains

In fig. [2.a.] the calculated energy balance of the main Sports Hall is presented:

The estimated thermal gains cover the space heating requirement during December and January. In the rest of the heating season there is a surplus. In November the heating demand is supplied only by the incidental gains [from people and lighting]. In March a part of the passive solar gains is required.

The auxiliary heating provided by the active solar systems is required in December and January and partly in February.

The energy balance for the whole building is presented in fig. [2.b.].Only in December and January is the space heating demand not covered by the expected thermal gains. However, if the rate of fresh air coming in is halved, the thermal losses of the building are met by the estimated thermal gains for the above mentioned months as well.

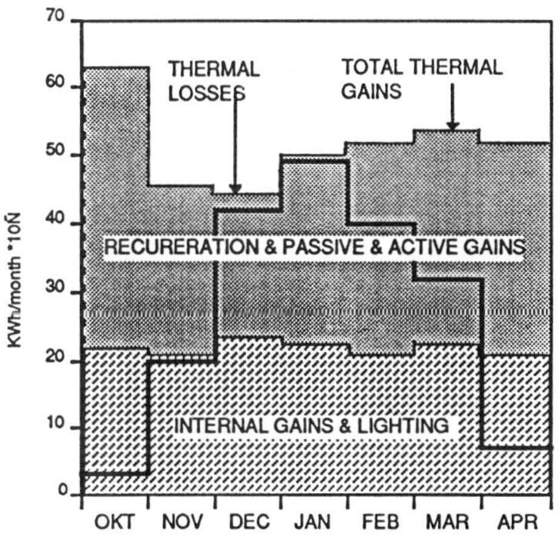

Fig. 2.a. **THE MAIN SPORTS HALL**

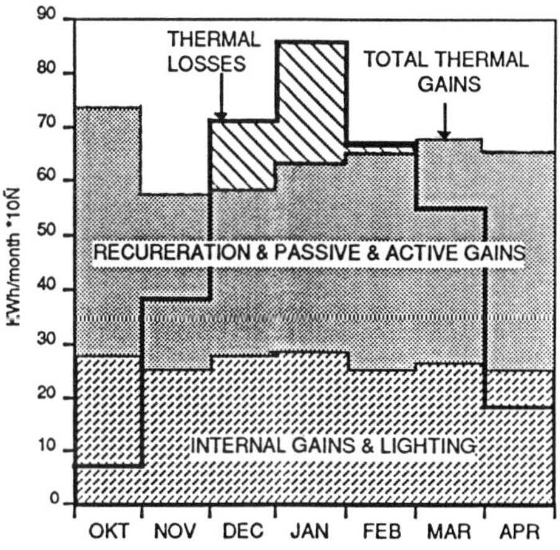

Fig. 2.b. **THE WHOLE SPORTS CENTRE**

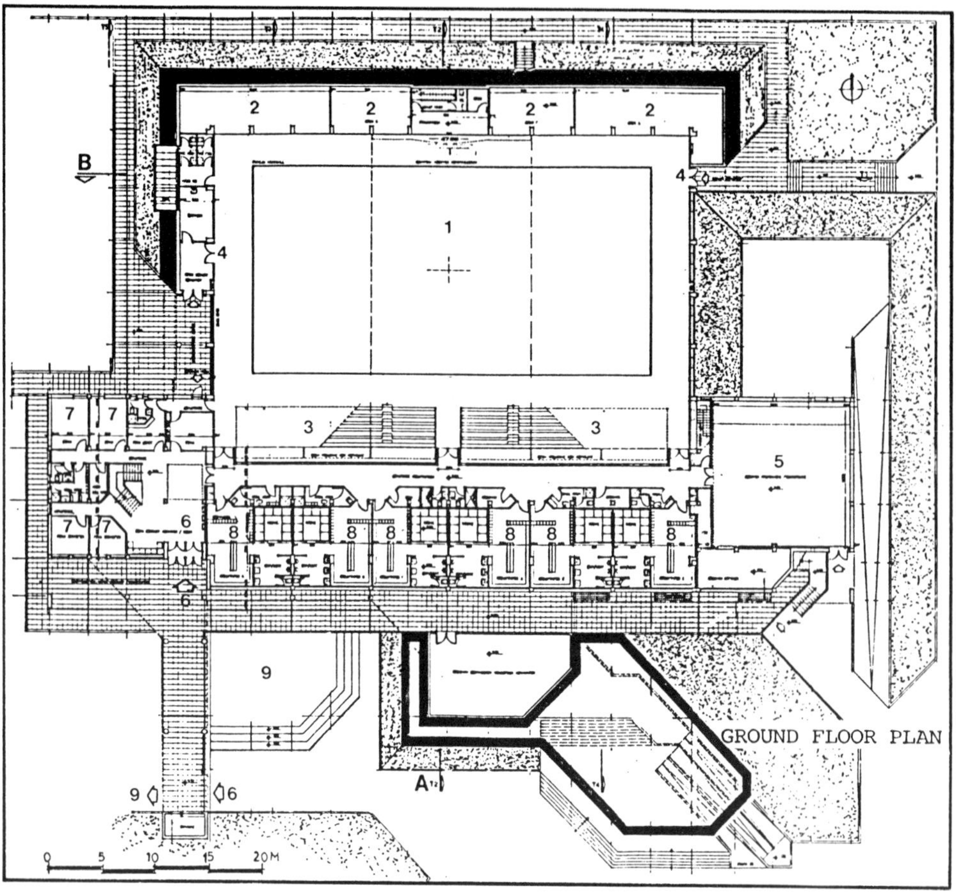

Fig. 3

GROUND FLOOR PLAN ±0,00

1 MAIN SPORTS HALL
2 STORE ROOMS
3 MOBIL SEATS
4 EXIT
5 TRAINING HALL
6 ATHLETES' ENTRANCE
7 SPORT CLUBS OFFICE
8 CHANGING ROOMS
9 OUTDOOR ACTIVITIES

Fig. 4 SECTION B-B

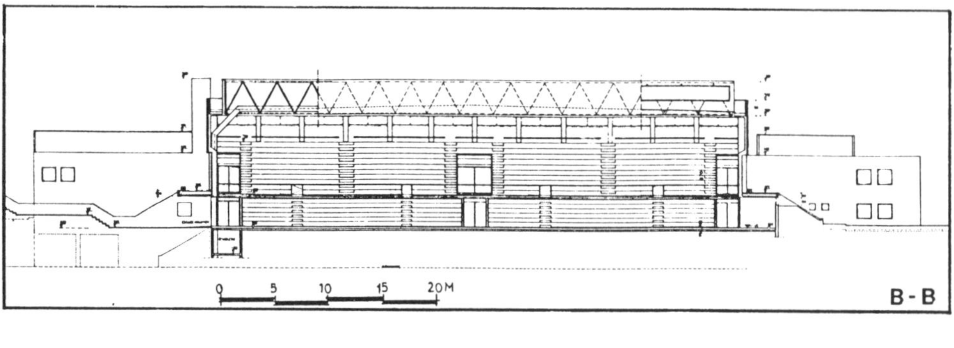

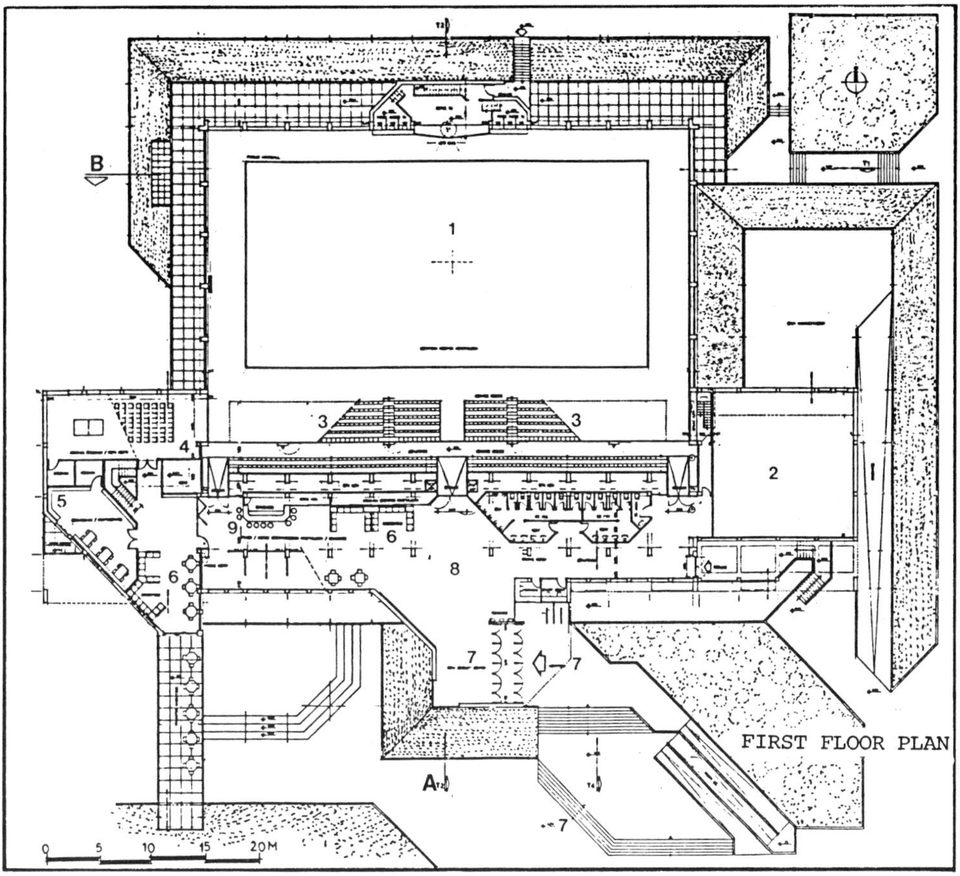

Fig. 5 FIRST FLOOR PLAN +3,30

1 MAIN SPORTS HALL
2 TRAINING HALL
3 MOBIL SEATS
4 SEMINAR ROOM
5 LIBRARY
6 SITTING ROOM
7 SPECTATORS´ENTRANCE
8 FOYER
9 BAR

Fig. 6

SECTION A-A

1 ARRAY OF SOLAR COLLECTORS
2 ROOF OPENINGS FOR DAYLIGHTING
3 DIRECT GAIN WINDOW
4 MASS WALL
5 SPORTS HALL
6 FOYER - SPECTATORS ENTRANCE
7 HOT WATER STORAGE TANK
8 GROUND FLOOR-CHANGING ROOMS
9 BASEMENT-PLANT

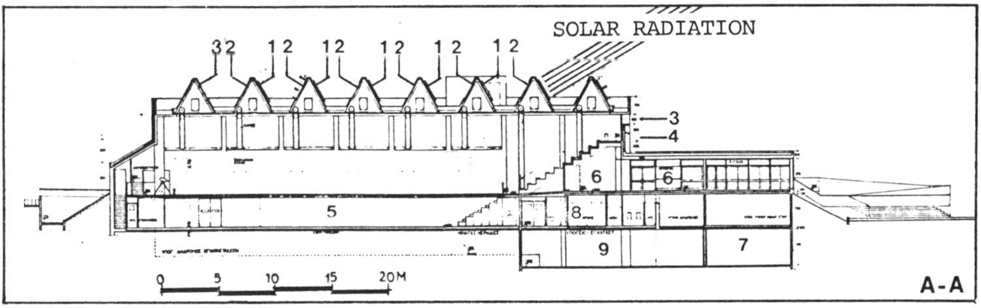

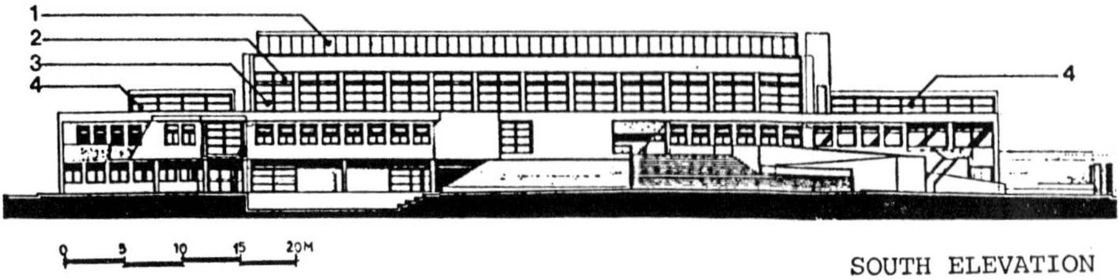

SOUTH ELEVATION

Fig. 7

SOLAR FLAT-PLATE COLLECTORS — 1
DIRECT GAIN WINDOWS — 2
MASS WALLS — 3
ROOFLIGHT — 4

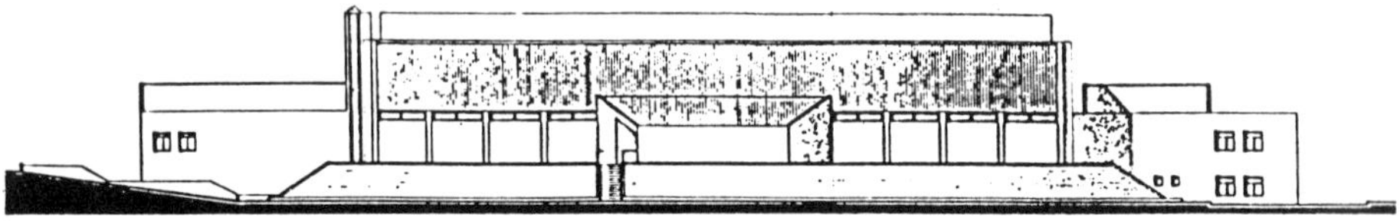

Fig. 8

NORTH ELEVATION

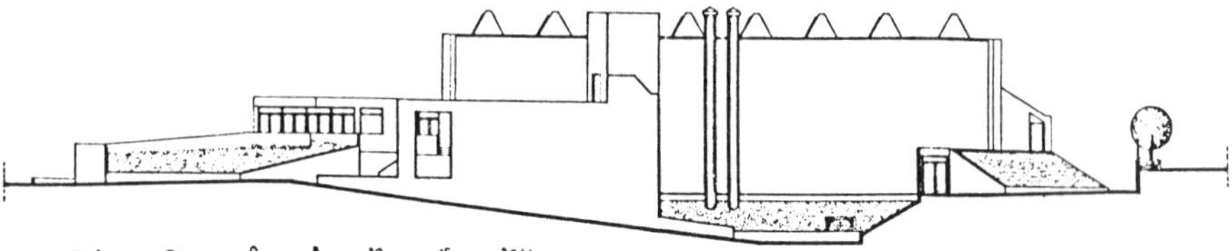

Fig. 9

EAST ELEVATION

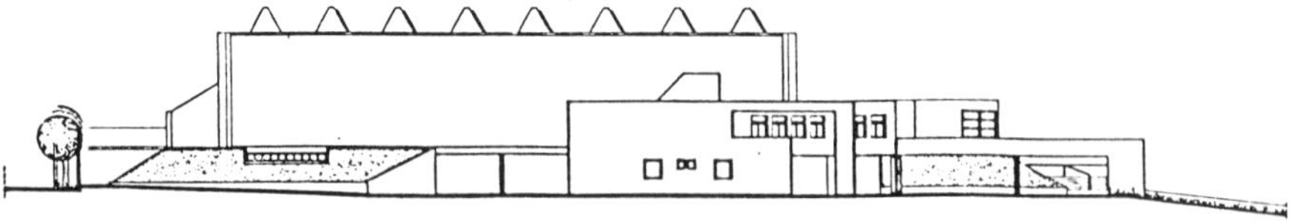

Fig. 10

WEST ELEVATION

A LOW ENERGY HOUSE IN WARM HUMID REGION OF BRAZIL

L. S. BITTENCOURT
Environmnet and Energy Studies Programme
Architectural Association Graduate School
34-36 Bedford Square
London WC1B 3ES, UK

ABSTRACT. The paper presents an environmental analysis of a single storey house located in Maceio', Brazil. It comments on the house design and environmental performance, based on measurements done on site and on computer simulations. The interest of the study lies on the fact that the building has been used for four years with the inhabitants making no use of mechanical equipments for cooling (heating is not used at all in the region). The aim of this case study is to show that there is scope to design adequate buildings, in warm and humid areas, where the use of air conditioning would be drastically reduced, and even eliminated for many of the buildings purposes.

1. Introduction

The paper refers to a house designed to reach comfortable spaces in a warm humid region. This section gives a rough idea of the area's climatic characteristics. Section 2 describes the building design principles and its components and analyses its environmental performance. Section 3 presents the simulations and measurement results obtained for a summer period while section 4 suggests some steps to improve the building performance.

The town of Maceio' is located at latitude 10deg South in the Brazilian coastal area. It has a warm and humid weather almost all year round. The annual mean temperature is 25 degC. The maximum mean DBT is 29 degC and the minimum mean is 21 degC, which gives an annual amplitude of about 8 degC. In the summer the mean air temperature is 27 degC while in the so called 'winter' is about 24 degC (1).

The mean relative humidity of the air is 81%. In summer it reaches 70% as average and in 'winter' it may go up to 99% during some nights.

Sky nebulosity varies from 6.9 (1/10) in June to 4.5 (1/10) in November. Solar radiation is strong and varies from 12 MJ/day on winter, to 27 MJ/day on summer (1).

There is a constant breeze coming from the ocean and predominant wind directions are placed within the East quadrant. In the wintertime the mean wind speed is around 2 m/s while in the summertime time the mean wind speed goes up to 4 m/s. The calm periods represent 8.2% of the day duration as an average, and occur mainly between 2:00 and 6:00 h (2).

It is also important to state that local population always wear light clothes and that there is no heating appliances at all. Cooling is the target.

S. Alvarez et al. (eds.), Architecture and Urban Space, 399–404.

2. The Building Design Principle and Environmental Performance

The house is located on an urban plot measuring 12.00 m on the frontage and 23.00 m depth. It is located some 200.00 m from the ocean in a low density area where most of the buildings are single storey constructions.

The design strategy was to divide the plot into three zones (service, social and private/study) in order to assure that every zone has an inlet facing the predominant winds, and to reach an integration between inside and outside spaces. To guarantee an adequate air speed, the building was located back in the plot, where it could also benefit from the large tree's shading. See Figure 1 and 2.

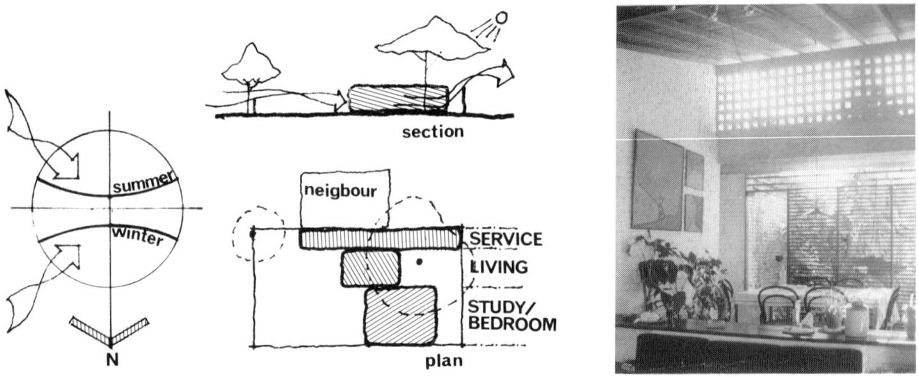

Figure 1. Site Layout (left) and interior view looking from the living room toward the courtyard (right).

KEY:
1- verandah
2- living/dining room
3- bedroom
4- studio
5- bathroom
6- gardening dep.
7- deposit
8- kitchen
9- washing area
10- car port
11- courtyard
12- roof garden
13- water tank

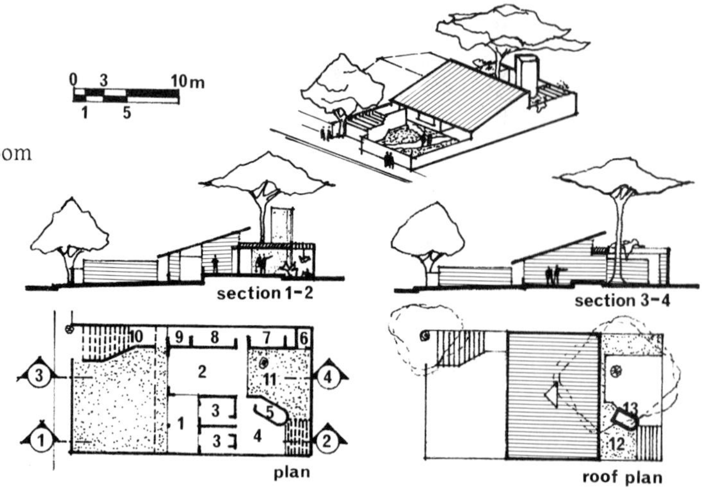

Figure 2. Plans, sections and isometric view.

The building design shows some problems related to the internal layout. The bedrooms, for example, may present some discomfort during hot days peak hours (13:00 to 16:00 h), due to the roof bad insulation. Here, daylighting appears to be insufficient during in cloudy days.
The studio does not present the ventilation rate expected by the architect. Part of the airstream, coming through the bedrooms' windows, goes out through the perforated block located between the roof-garden and the pitched roof. Daylight is insufficient in the back of the room and artificial light is necessary most of the time as shown in Figure 3.
In the living room and kitchen the roof radiation effect is reduced because of the much larger openings (See photo in Figure 1) and also because they are partially protected, by the existing tree, from solar radiation.

Though the kitchen performs well in thermal aspects, it needs artificial lights to be turned on especially in cloudy days. The other rooms present no environmental problems.
The verandah plays an important role to the living and bedrooms comfort, especially during raining days, when it allows the louvres of these two rooms to remain open. It also protect part of walls and openings from the East insolation, though reducing daylighting inside bedrooms. The verandah is located leeward working as a windcachter, improving ventilation in living and bedrooms due to Venturi effect.
Some components design and location play important roles in the house performance, namely roof, windows, perforated blocks and pergolas.
There are two kinds of roofs. The pitched one and the roof-garden. The first consist of blue tiles over a wooden ceiling. Due to its colour it absorbs a lot of solar radiation (68%). The later, consisting of a flat concrete slab and 0.30 m of earth, covers the studio and service area.

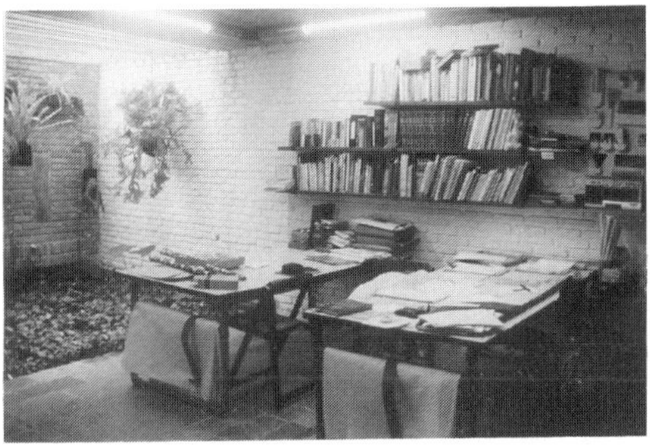

Figure 3. View from the studio, showing the lights on.

The gap between the two roofs are filled with perforated blocks.(See Figure 4). The time-lag produced by the roof-garden has a buffer effect during peak hours but it rises internal temperature during summer nights. Figure 5 shows the time-lag and decrement in external air temperature produced by the roof-garden.

Due to its movable louvres, the windows allow a good solar control and permanent air flow. The sill height (0.50 m) brings ventilation to pass along occupants body. Since the fanlight is open, there is always air movement close to the ceiling. See Figure 4.

The perforated blocks forming a screen wall are located between the pitched roof and the flat one, along the whole building width, facing West. It is made of precast concrete and is coloured grey, absorbing part of the afternoon solar radiation. Due to its shape it doesn't obstruct the sun beans after 16:00 h. The daylight it could provide is reduced by its dark grey colour.

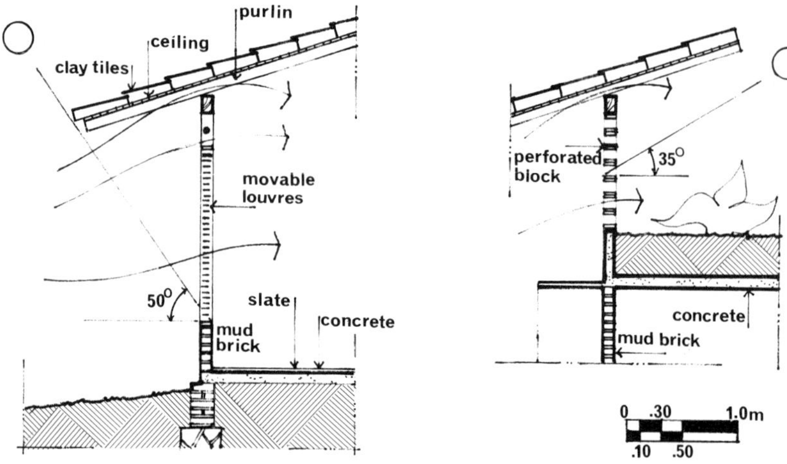

Figure 4. Detail of movable louvres, perforated blocks and roofs.

The pergolas covering the studio's internal garden works as ventilation outlet and avoids completely the sun penetration. However, it turns the room illuminance level lower than the necessary for its function.

3. Simulations and Measurements

Simulations have clarified how effective is the shading produced by the large tree and the consequent reduction on the radiation received by the roofs. Figure 5 introduce the results of software modelling of the house showing shading produced by the trees in summer and winter solstice, considering clear sky. It has shown that considering the tree opaque it would reduce the insolation by 67% and 89% for the tilted roof and the roof-garden, respectively, in the summer solstice. In winter solstice these figures change to 47% and 46% respectively. Internal and external air temperature near the two kinds of roofs above referred, were measured for summer typical days during 1991 summer. Results are presented in Figure 6 below.

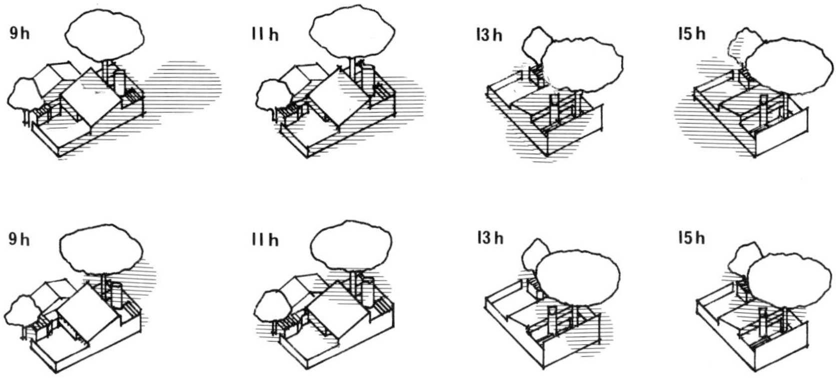

Figure 5. Shading produced by the tree.

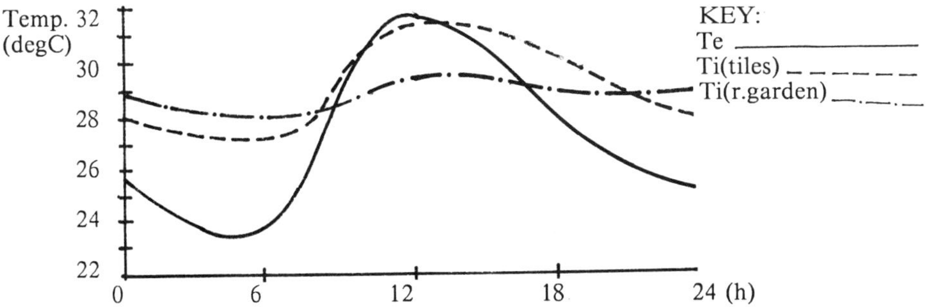

Figure 6. Internal and external air temperature for typical summer days.

Daylight levels were measured for typical summer sky at 9:00 h, 12:00 h and 15:00 h in all rooms having perforated blocks (Figure 4) on it, in the center of each room at 0.80 m height. First considering its concrete grey colour. Then the blocks were painted white and new measurements were carried on in order to examine the effect of this very simple step. Results are presented in Table 1, where room number are the same as in Figure 2. One can note though significant improvements may occur, daylight levels are below those required for these rooms' functions, introducing daylight is the main environmental problem encountered.

The same procedure was performed with the pergolas, where was found an increment in daylight levels of the order of 155% in the center of the studio and 105% in the back of it. Even so, illuminance levels encountered were not enough for reading and drawing tasks.

TABLE 1. Influence of block colour on illuminance levels (lux)

	concrete grey			white paint			increment (%)		
time	9:00	12:00	15:00	9:00	12:00	15:00	9:00	12:00	15:00
room									
2	47	70	41	68	99	59	45%	41%	44%
3	21	23	17	41	33	33	95%	43%	94%
8	26	19	31	41	30	33	58%	58%	6%

4. Conclusions

From the data presented one may conclude that though the building is not using any cooling device, it can perform in a better way if some improvements were to be introduced.

a) The roof tiles can be given a higher reflectance in order to avoid the strong solar radiation. If the ceiling is placed below the purlins with an aluminium foil above it, part of the heat transmitted by the tiles will be reflected and an air gap between the tiles and the ceiling will be created.

b) The bedrooms windows can be better located, in order to propitiate a better ventilation distribution, and maybe a bigger size to improve the air speed (reduced by the necessary use of mosquito screens). The present position do not allow air flow to reach the corners of the rooms where the beds are located.

c) The implementation of white painted perforated blocks and pergolas shows significant improvements. However, the introduction of other perforated block types would reach the required illuminance levels without compromising thermal performance. Also, if the louvres colour (now dark brown) is changed to lighter colours one certainly will obtain much better results regarding daylighting.

Part of the modifications above suggested would change some of the house aesthetical characteristics, especially the roof colour. But most of them would neither affect the architectural composition nor the construction budget and would save on mechanical equipments and future electricity bills.

References

(1) Lima, B. et Bittencourt, L. (1983) 'Analise do Clima de Maceio', Report for Departamento de Arquitetura da UFAL, Caderno de Arquitetura e Urbanismo no.1.

(2) Carneiro, C. et al. (1986) Sistematizacao de Dados Climatologicos do Recife, SUDENE/UFPE, Recife.

PASSIVE COMPONENTS IN RESIDENTIAL BUILDINGS

M.Sala, L.Ceccherini Nelli
Department of Architectural Design
University of Florence
via Cavour 82
50129 Firenze
Italy

The research presented here, is an analysis of solar passive components and technical elements developed in projects and studies carried on by the bioclimatic design group (the authors and A.Gioli, A.Toti, A.Corsi, P.Puccetti) and aimed to investigate and to test a system of rules for projects of bioclimatic buildings in urban areas.

We catalogued design solutions according to the different residential models shape and orientation varying with regards to the different performances. Gools of our research were:

a) to propose a catalogue of architectural solutions capable to integrate energy saving devices, such as by the capture of solar radiation in winter or by protecting from it in the other seasons, in the building scale .

b) to improve thermal and lighting performance of buildings with use of passive solar components, technical and architectural elements for different situations of use.

The project presented here is multistorey apartment house in Greve in Chianti, close to Florence, at the altitude of 250 m., latitude 43.35 N, with daily degree 15.60. The orientation facing south of the hill suggested to test in this building several passive components, integrated in the same architecture, to study different performances in the same condition of use.

We gave special attention in ventilation solutions, for the hot clima in Greve during the summer, and we choice to use the cool passages that separate the building from the hill as ventilation device. The falling green from the upper terraces and the architectural form of terrace's edge protect windows from the direct radiation.

Greenhouses are designed as light structures to be removed in summer and also the windows frames can be removed forward or backward in the thikness of the wall . Moble solar panels for warm water can be used as shadow device.

S. Alvarez et al. (eds.), Architecture and Urban Space, 405–410.

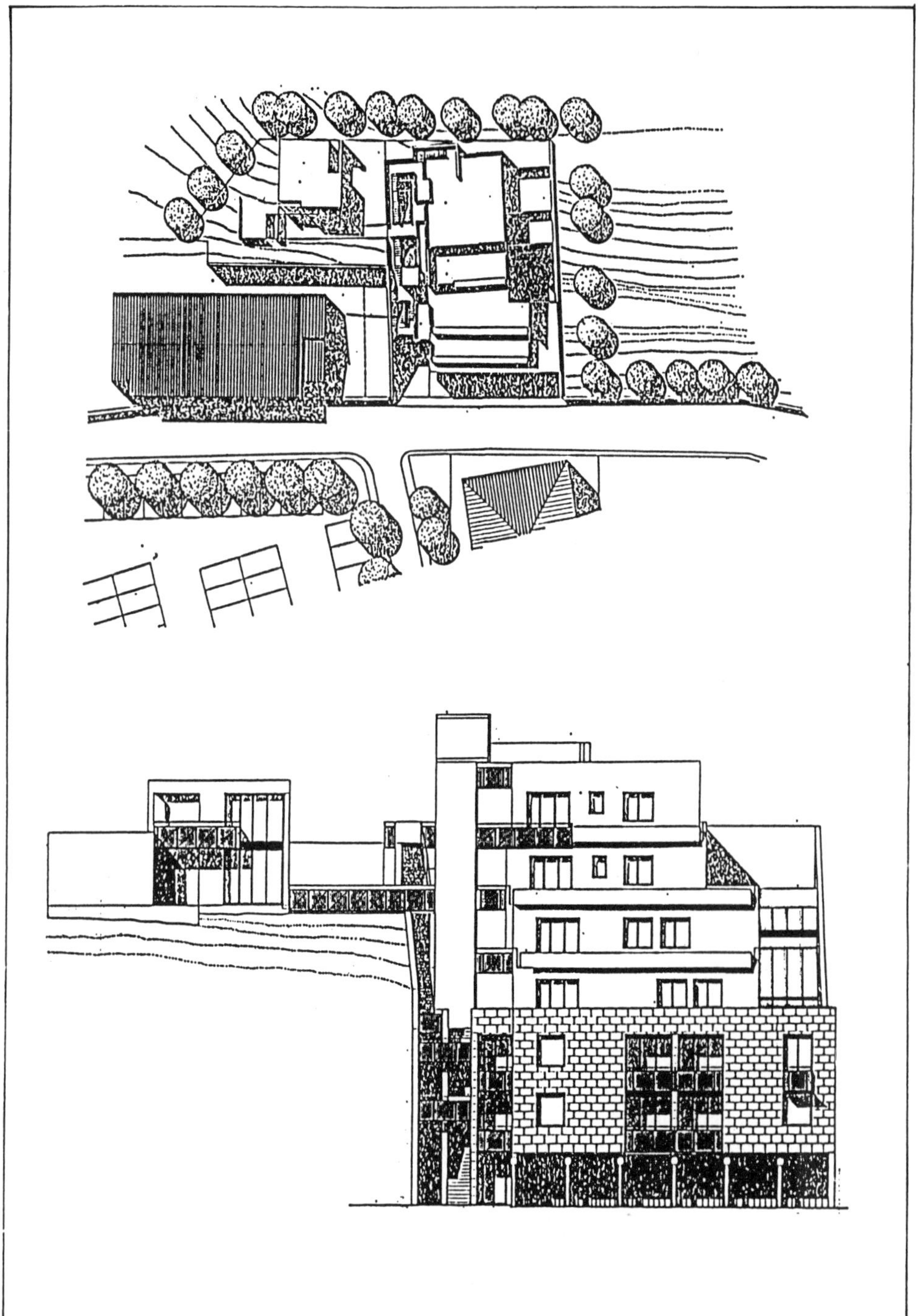

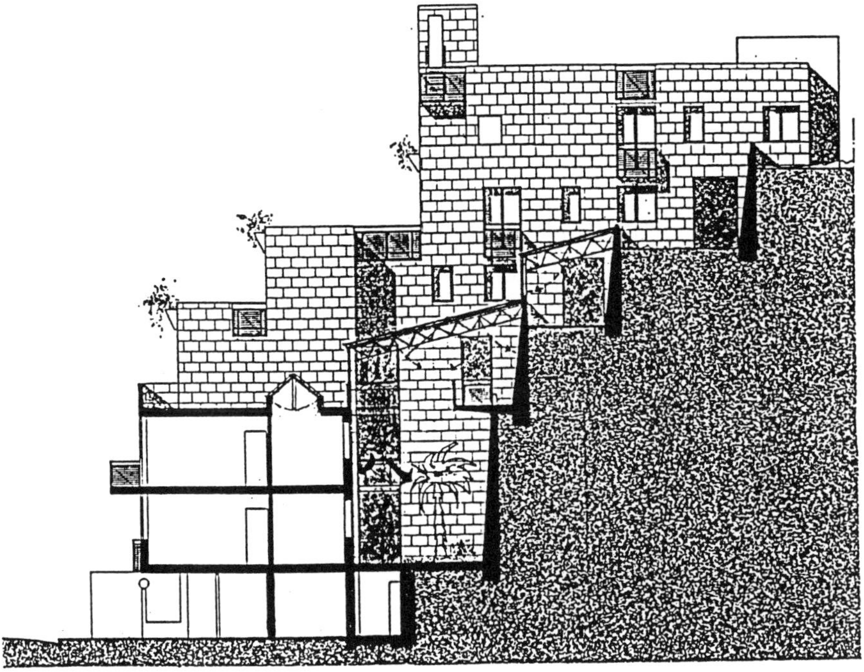

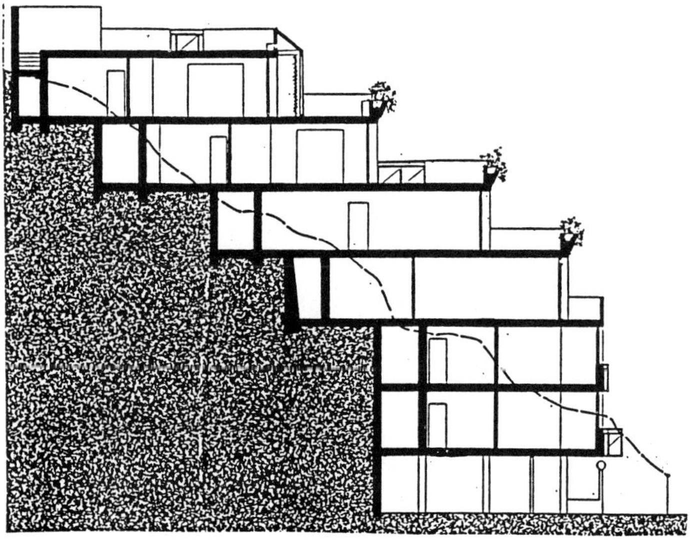

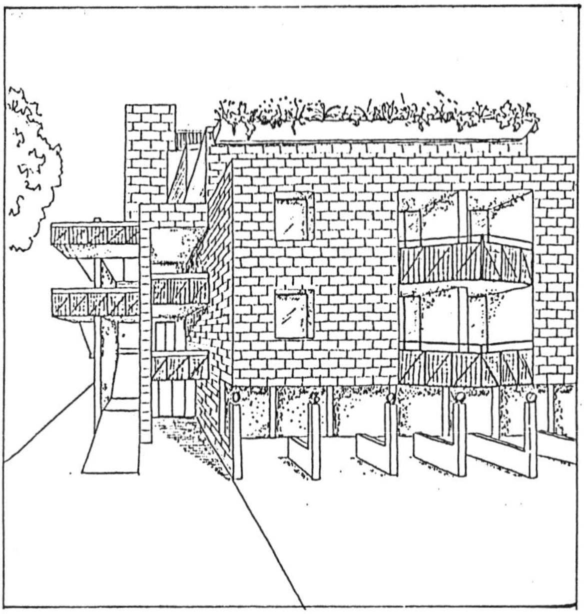

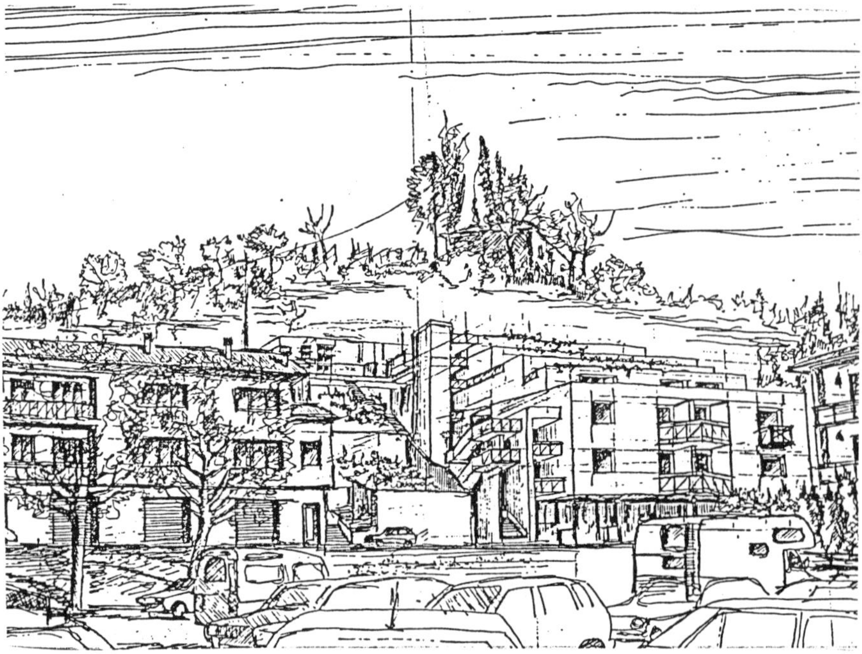

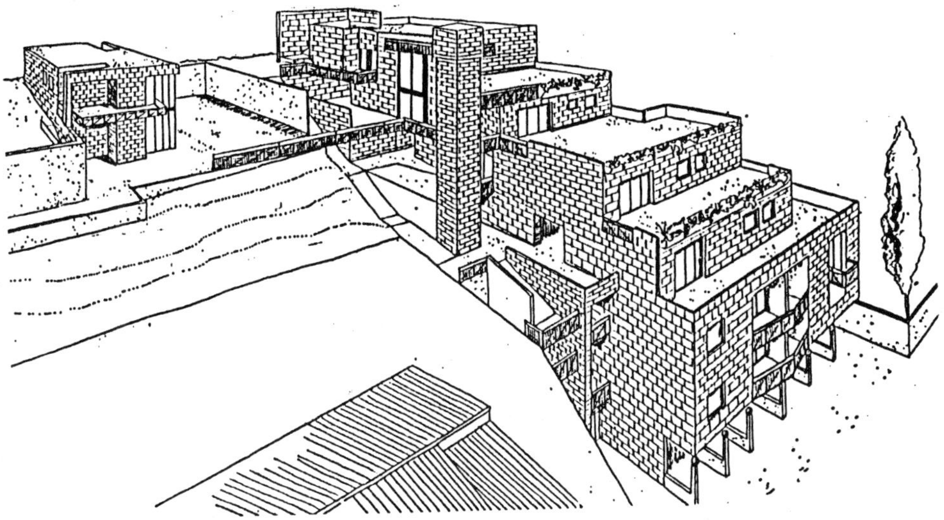

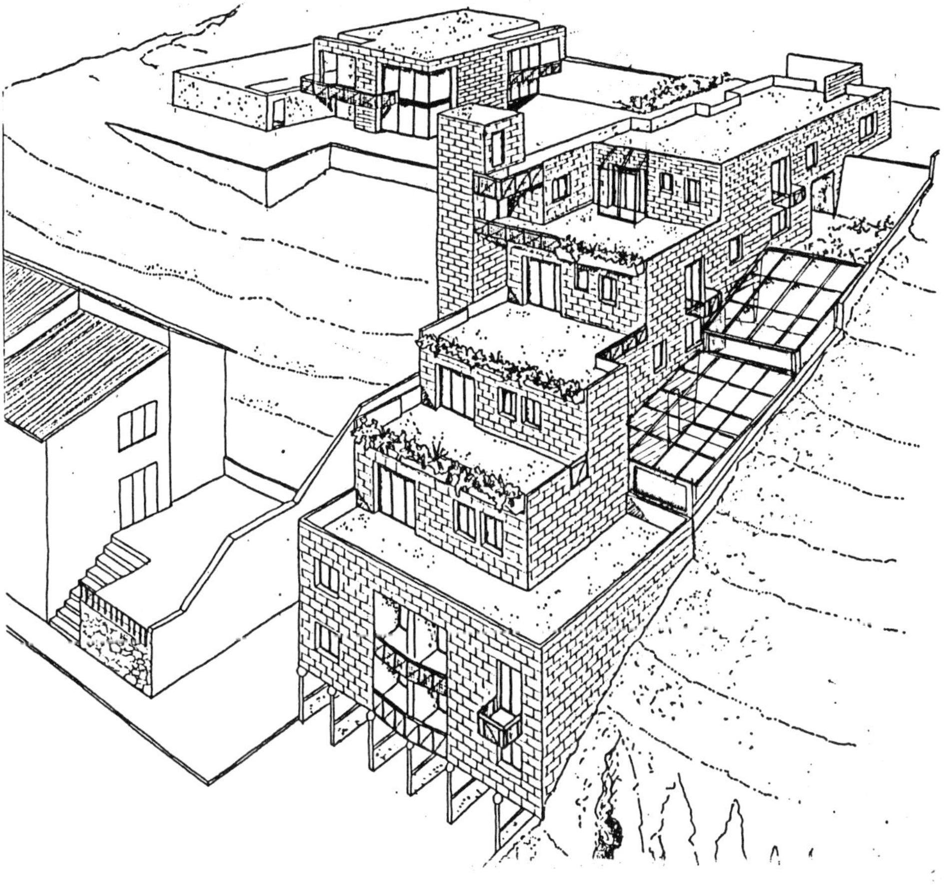

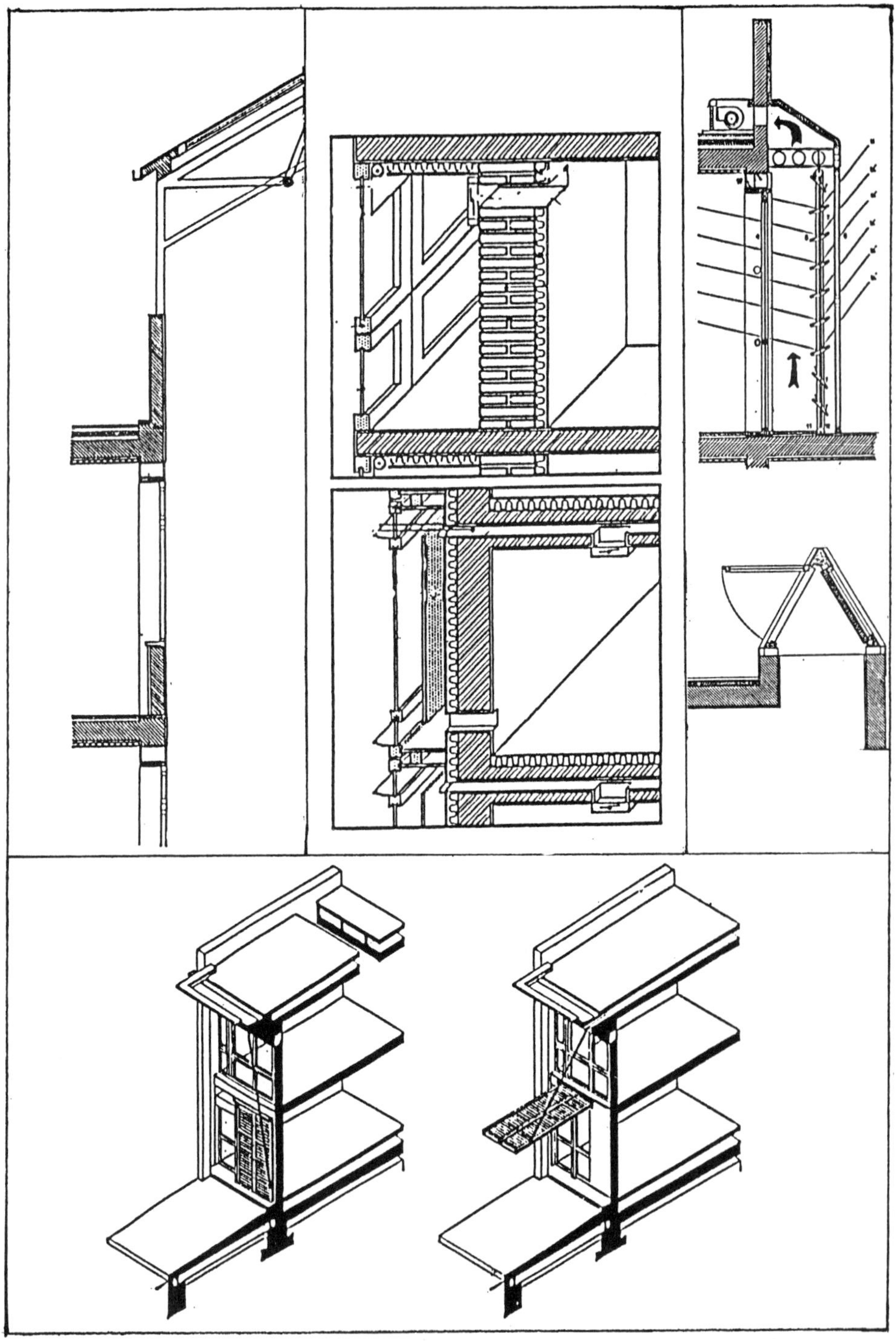

CULTURAL CENTRE - YENITSA GREECE
Restoration -reuse of a neoclassical building and redevelopment of the surrounding area.

Michael E. Nomikos
Lecturer at the Department of Architecture
Faculty of Technologie
Aristotles University of Thessaloniki
540 06 Thessaloniki-Greece

ABSTRACT. Yenitsa is a small town in Macedonia-Greece, founded by the Ottomans in the XIV century on the ruins of two Byzantine castles.
From XV to XVII century the town has been developed to a very important military, commercial and religius Islamic centre, especially because of the existence in the town of Gazi Evrenos Bey tomb, the ottoman conqueror of Macedonia. In the beginning of our century and especially during the Balkan War (1912-14) a big part of the town has been destroyed.
The neoclassical building on which our project is focused is build at the end of XIX cent. as the residence of the town's governor and constitutes with a small number of other important buildings the historic heritage of the town.
The project includes the restoration of the typological and morphological characteristics of the building, the integration of new functions and the rearrangement of the surrounding area which is the commercial and recreational centre of the town.

The town of Yenitsa

The foundation of the town was made in 1383, by the family of Gazi Evrenos, leader of the Ottoman troops (Gazi).[1]
According to Evliya Tselebi, traveller of the XVII cent. the new town has been founded on the ruins of two demolished Byzantine Castles [2]. The name of the town "Yeni-ce" which in Turchish means "new-place" confirms this description.
The town is located on the road which was connecting Thessaloniki with the western Macedonia, near the banks of an old lake. Yenitsa has been developed by the ottomans into an important military centre and used as a base for the conquest of Thessaloniki and other Macedonian towns.
During the next two centuries (XV, XVI) the town had its flourishing state: 17 neighbourhoods 1500 stone-build houses, 17 moslem temples, 740 shops in the market, numerous schools and baths and a big Karavan-serai which could accept 500-600 travellers as well as a water supply system is mentioned.[3]
Towards the end of the XVIII cent. the population of the town reaches the 6.000 inhabitants.[4] In the end of XIX cent. two distinct parts are mentioned: the Greek and the Turkish one (800 Turkish and 400 Christian families).[5]. In the Balkan War (1912-14) and especially during the Yenitsa battle a big part of the town has been destroyed.

S. Alvarez et al. (eds.), Architecture and Urban Space, 411–416.

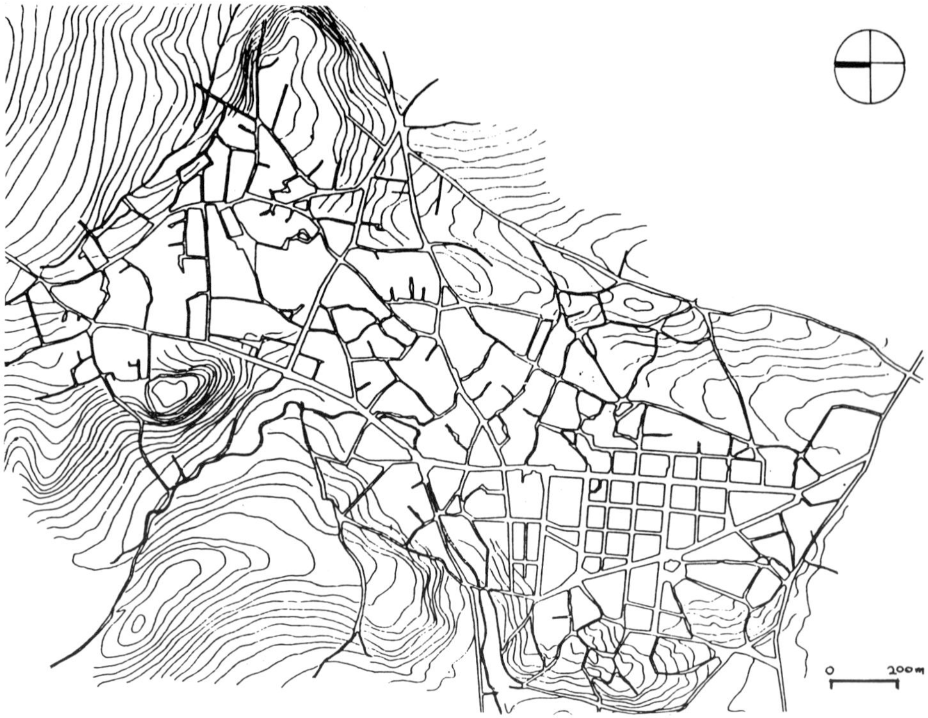

Figure 1. Yenitsa in the beginning of XX century

The modern-rationalized plans for the town of Yenitsa placed into the general frame of rebuilding policy for the Macedonian towns were the beginning for the gradual demolition of the preexisting building stock and the alteration of the old urban tissue.
The few survived religius and civil constructions, which were organic parts of the old destroyed tissue, stand today without any relation to the rationalized axis of the new-realized urban plans.[6]
The old tissue characterized by the curvilenear building ilots and the penetrating blind-alleys, by the differentiation of the neighbourhoods and the Market area and by the organic articulation of public buildings, has been replaced by a rationalized orthogonal urban system mainly influenced from the European models.[7] The old limits of the town have been extended with the creation of new neighbourhoods and there are no more limits between the Greek, Turkish and market area.

The restoration of the neoclassical building and the rearrangement of the town's centre

The historic building is situated into the most important part of the town's centre, on the bigest commercial axis.

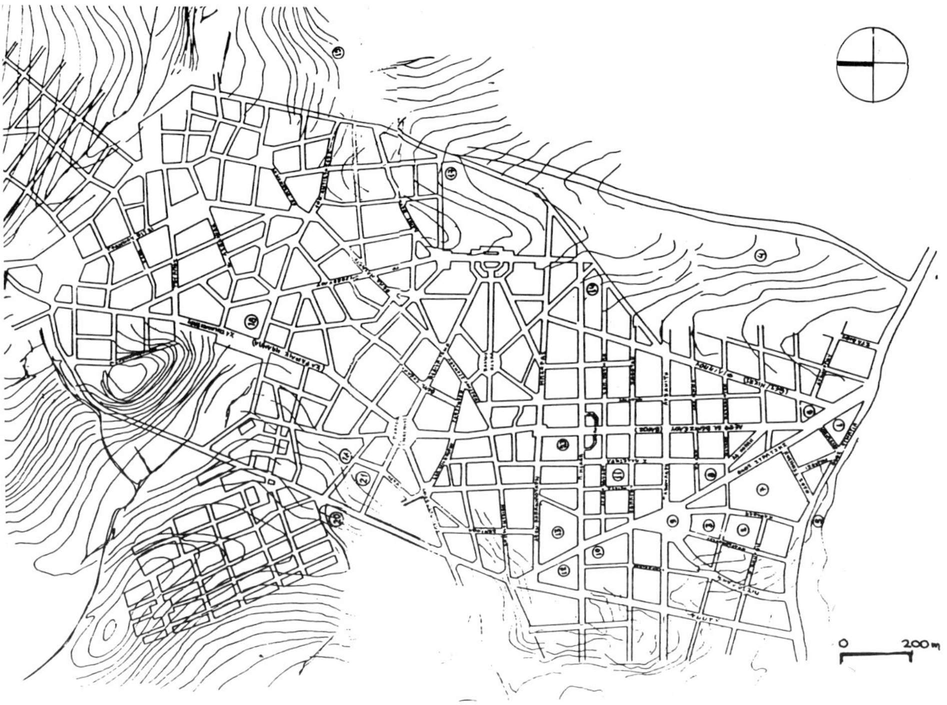

Figure 2. The new town plan of Yenitsa which was greatly transformed when applied, to be adapted to the completely different pre-existing urban contex.

It has been built towards the end of XIX centr. to be a residence of the town's governor. After the town's deliberation from the Turks in 1912, it has been used in different ways: dispensary, orphanage and finally as the town's hospital from 1930-1980.
To accept these different functions many alterations to its typological structure have been made.
In 1950 the main wooden structure has been replaced by concrete interventions (floors, beams etc) and the neoclassical roof endings have been demolished.
In 1988 the Municipality's programme included not only the restoration-reuse of the existing historic building but also the creation of a new building unit next to the old one to be used as a small auditorium and the rearrangement-reevaluation of the surrounding area. The restauration of the historic building attempts to mark out its typological and morphological characteristics based on existing traces and sources using traditional building methods and materials while for the new constructions modern materials and methods are used.
The building is oriented according to a previous urban tissue which is different from the modern one. The main directions of the new urban arrangements are organized in relation with the building axis. This differentition from the modern urban plan emphasizes the previous spatial character of the building.

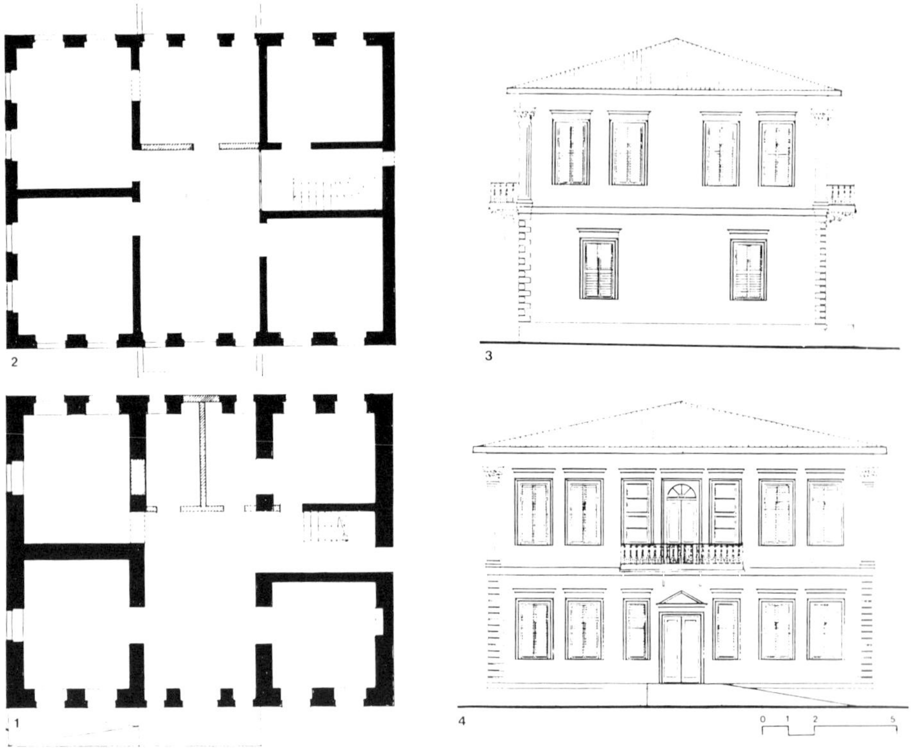

Figure 3. The building before the restauration works 1,2 ground and first floor plan 3,4 south and west elevations.

The differentiation of the two urban tissue's axis (the old and the new one) is indicated with the addition of an urban arcade which penetrates through the light iron-made new construction and the introduction of a diagonal grid which in a way compromises the different geometrical laying out.
The proposed plantations follow and underline the two axis: the new town plan axis and those of the historic building.

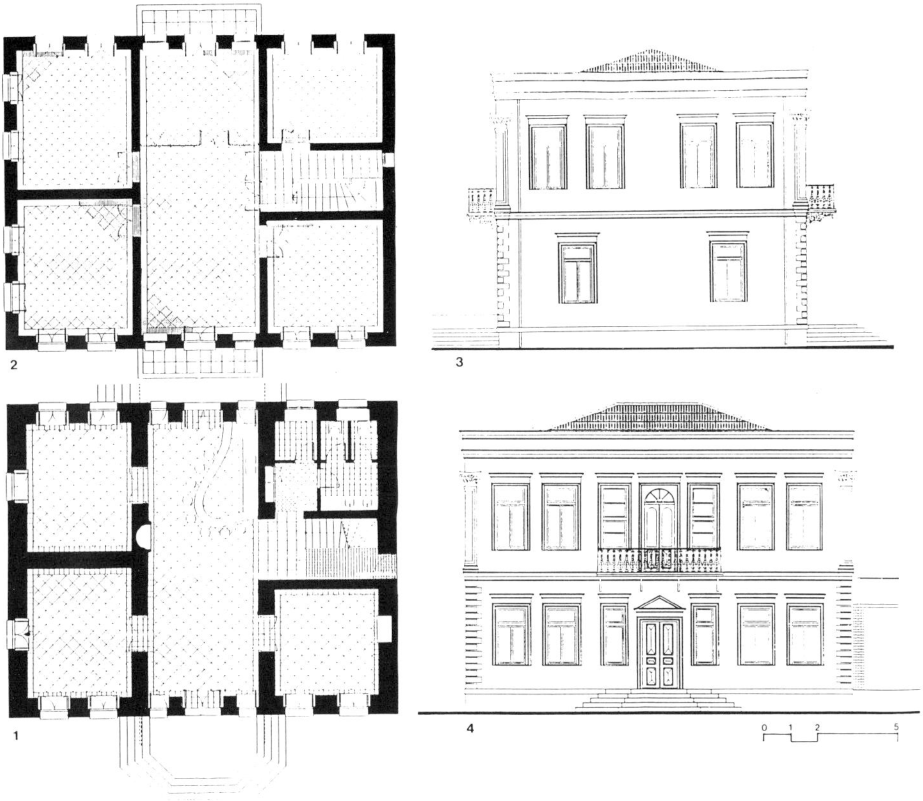

Figure 4. The building after the restauration works 1,2 ground and first floor play 3,4 south and west elevations.

REFERENCES

1. Kill Machiel, "Venice-Vardar (Vardar-Venicesi-Giannitsa) a forgotten Turkish Cultural centre in Macedonia of the 15th and 16th centuries" in Studia Bizantina et Neoellenica n. I, 1972, pp. 300-329.

2. Demetriadis V. I Kentriki ke Ditiki Makedonia kata ton Evliya Tselempi, Thessaloniki 1973, p. 23.

3. Demetriadis, V., op.cit. p. 22.

4. Bequjour, Tableau de Commerce de la Grece Paris 1800, vol. I. p.p. 111,112.

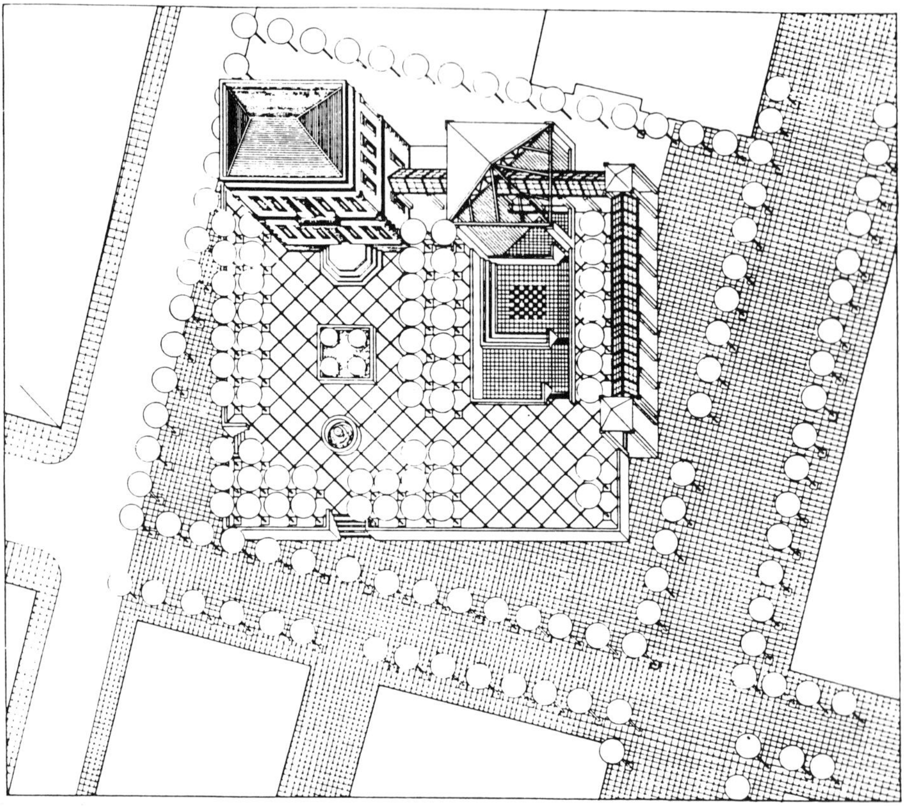

Figure 5. Assonometric view of the building and the new intervantions: new constructions and rearrangement of the surrounding area.

5. For more detailed informations about the social and physical structure of the town see also:
   - Cousinery E.M. Voyage dans la Macedoine, Paris 1831, vol. I. p. 86
   - Leake M.W., Travels in Northern Greece, London 1835, vol. III pp. 266,267.
   - Schinas N., Odiporikes simiosis Makedonias, Ipirou, Athens 1886, vol. I. pp. 193-196.

6. For the preexisting civil and religius buildings see: Ekrem hakki Ayverdi, Osmanli, minari, eserleri, Istambul 1982, pp. 303-307.

7. Further informations about the character of the Macedonian towns see also: Eleni Andreadaki, Michael Nomikos, Regional Architecture of Macedonia: The case of Vernacular Settlements as an Example of Passive Low Energy Design, PLEA '88 pp. 193-198.

PASSIVE HEATING AND COOLING IN DWELLINGS. CASTILLA-LA MANCHA.

HERNANDEZ MARCULETA. SANTIAGO
CONSEJERIA DE POLITICA TERRITORIAL
C/ FERMIN CABALLERO, 24
16071 CUENCA
SPAIN

0. INTRODUCTION

The spanish self-governing Administration in Castilla-La Mancha initiated four years ago a scheme in order to build dwellings which performance were energetecally efficient. The houses must be of low cost of maintenance. So it has been constructed 16 solar passive family houses in three terraces on Montalbo (Cuenca), and at the present it has been finished 118 apartments in Cuenca, with the greenhouse effect.

Before this, the Goverment built 42 dwelling-houses in Alpera (Albacete) with hibrid solar passive system. All these are building with public budget.

1. LA MANCHA CLIMATE

The "La Mancha" plateaw raises 800 metres above sea level and endures the bigger sudden transition between nights and day temperatures in Spain, as it can see in the next relation.

1.1. Table 0. Relation spanish cities with a larger thermic jump

SPANISH CITIES	ANNUAL				SUMMER= MAXIM. JULY			WINTER= MINIM. DECEMBER		
	AVARAGE			THER	AVARAGE		THER	AVARAGE		THER
	ANNU.	MAX.	MIN.	JUMP.	MAX.	MIN.	JUMP.	MAX.	MIN.	JUMP.
MOLINA DE ARAGON 1	10'2	17'2	3'1	14'1	28'9	10'1	18'8	7'9	1'9	9'8
CUENCA, LABORATORIO 2	11'7	18'7	4'8	13'9	35'2	18'6	18'6	9'5	2'4	11'9
ALBACETE 3	13'4	20'1	6'7	13'4	32'6	15'5	17'1	10'0	0'1	9'9
CIUDAD REAL 4	14'5	21'2	7'8	13'4	34'4	16'5	17'9	10'8	1'7	9'1
TERUEL 5	11'6	18'2	4'9	13'3	29'7	12'9	16'8	9'0	1'0	10'0

S. Alvarez et al. (eds.), Architecture and Urban Space, 417–422.

This climate that have a huge termic variation within a same day, needs materials with large thermic mass and with hight specific heat in order to:

- Absorb and mitigate the extreme temperatures.
- To storage heat and cold.
- To delay indoors radiation from walls.

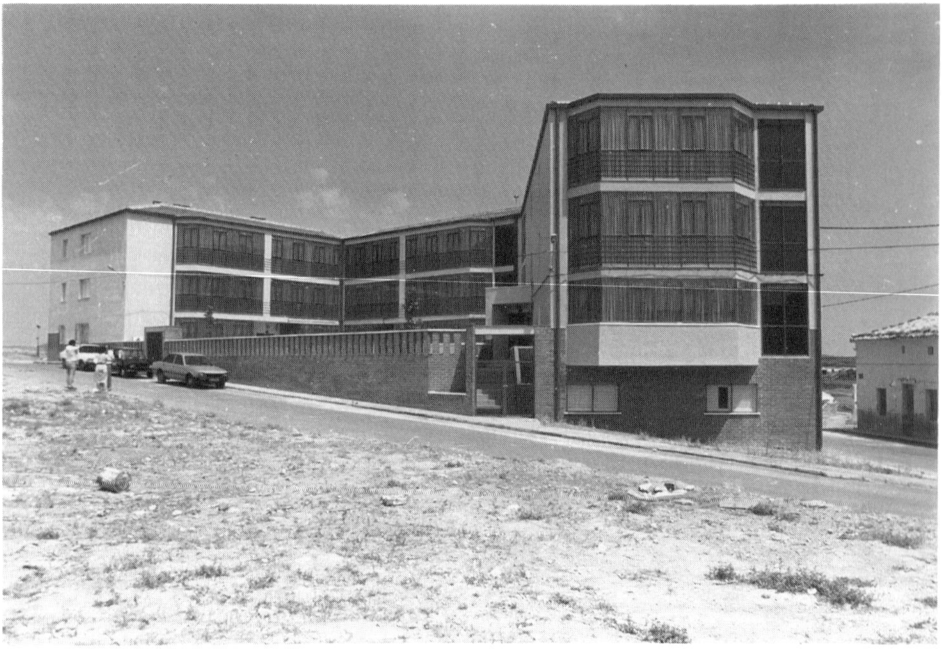

Fig. 1. Alpera. On the right corner we can see the active air collector.

2. DESIGN OBJETIVES

It's necessary a natural device that can keep constant thermal indoors conditions in the widely fluctuating temperatures of the boiling days and freezing nigths.

The most proper systems, for these hard climate, is to put in a control room between the exterior and the living room.

Within these buffer room, it must locate the due thermal storage wall, in order to accumulate the heat and thereinafter transfering it to the ambiente air.

It's essential the contribution of the people who live in the house. They must put in due place the insulations and protectors devices during the winter nights and also during summer mornings, so that they can avoid the direct gain and so the uncomfortable overheat.

These social dwelling-houses intend to prove that energy efficient schemes are both: low cost effective and low cost of maintenance. This purpose isn`t so top priority for private promoter as for the Goverment who must look after the people with low attainment power. Therefore the research and study of the natural solar system passive must be develop by the Public Administration.

The design and method for build solar-passive dwelling-houses belong to soft technologies and their applications don`t develop by itself the industrial increase and the non-renovable energy consumption and also don`t augment the construction cost; instead it cheapens maintenance cost along the building life and moreover, it provides a better integration in the natural environment.

3. PROMOTION AND CONSTRUCTION FEATURES

3.1. 42 DWELLING-HOUSES IN ALPERA (ALBACETE). L.N. 38º 55'

Finished in 1985
Passive and active Solar Systems:

- Greenhouse with reflecting courtins
- Masonry wall heat storage
- Movable isolation
- Active air collector and fan
- Convective hot air loop is carried away by duct throught the walls as far as the north bedroom.

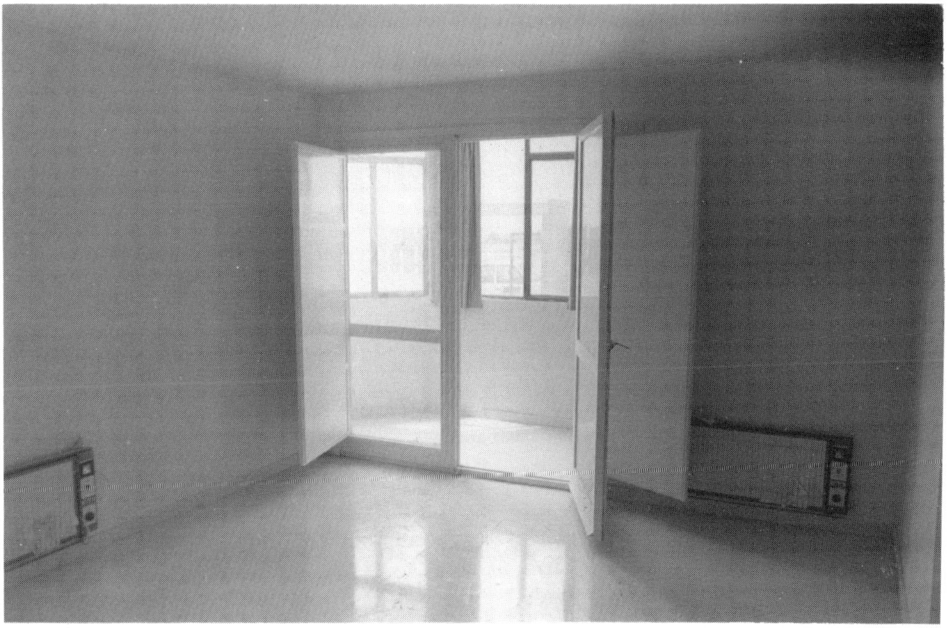

Fig. 2. Alpera. The indoors windows have movable shutter with isolation.

3.2. 16 SEMI-DETACHED HOUSES IN MONTALBO (CUENCA). L.N. 39º 53'

Finished in 1989

Natural passive solar systems:

- Accumulator walls for cooling
- Indoors shutters with high value as insulation
- Natural air movement
- Exterior sunblind and direct gain approach
- Flooring with black flagstone to improve heat acumulation
- All the windows have double glass
- In the first floor has been arranged a sheet of polystyrene under the flagstone.

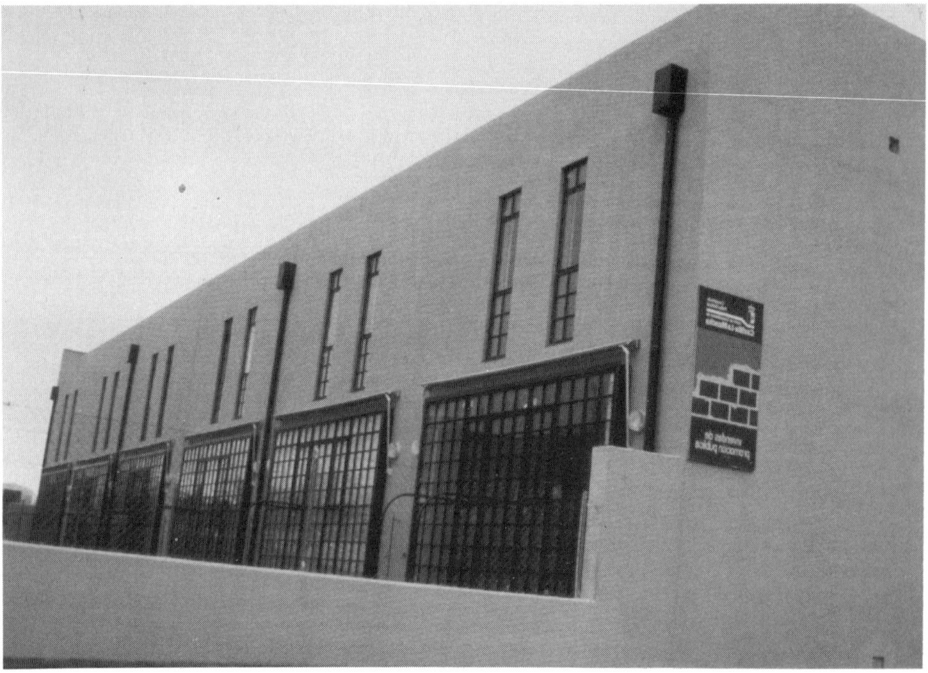

Fig. 3. Montalbo. South elevation in a terrace of 6 dwelling-houses.

3.3. 118 APARTAMENTS IN CUENCA. L.N. 40º 08'

Finished in 1990

Passive solar systems:

- Sun space and direct gain
- Double frame and glacing
- Vegetation and microclimate in the courtyard
- High level of insulation

4. PASSIVE COOLING

Montalbo (Cuenca) as another places in La Mancha are at 890 m. above sea level and has very sunny weather. (103 days without clouds).

The summer shows the widely fluctuanting temperatures. One must prevent over-heating but also must benefit from low mornings temperatures. It is possible thanks massive wall and the inner structural brick-walls wich improve thermal inertia.

The outer layer of the walls have insulation with external cladding. This tipology avoid interstitial dampness and moistness.

There is a vent at the top skylight which enable the air elevation in the central staircase and nexts rooms.

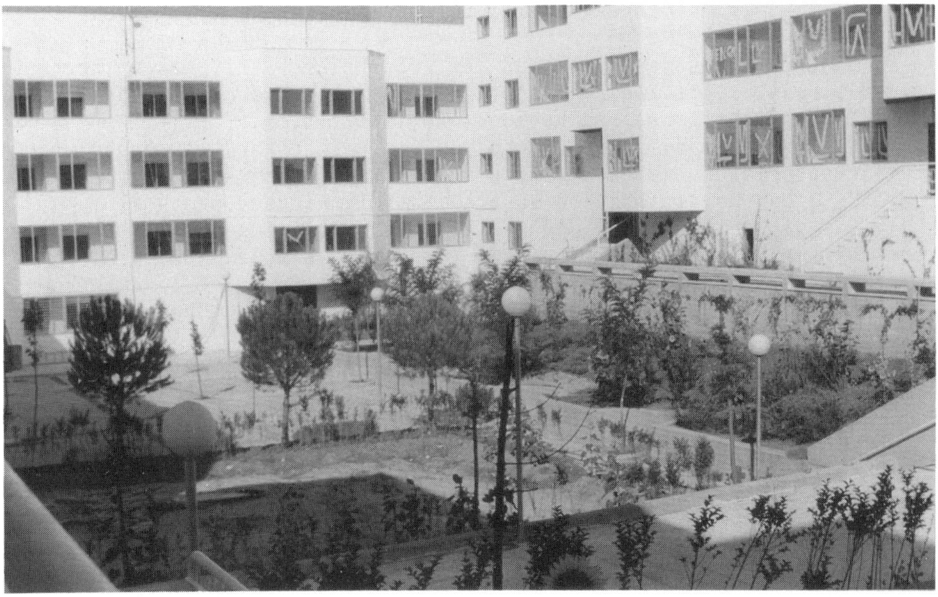

Fig. 4. Cuenca. 118 Apartments. Microclimate in the courtyard.

5. DESIGN OF OUTDOOR SPACES

The aims to reach in Castilla La Mancha with the new enlargement of urban planning are, besides to fall in prices the parcel, the planning for solar access.

Nowdays there is a huge economic pressure over the construction development, so planning for solar access can be seen as another boundary to the hope of a large profit.

The main responsibility of looking after the solar-access covenants belongs to:

1.- The Municipal and Regional Goverment in their own buildings and urban planning.

2.- The community, in the form of zoning-ordinance reform.

3.- Promoters and developers, in the form of covenant protection.

4.- Design professionals; with solar access another advantages are warranted as the quality of our life, our light and vision, our warmth and our energy.

6. CONCLUSIONS

The Consejería of Política Territorial don`t forget that if the solar passive houses are of low cost of maintenance, all social dwellings must be natural designed or solar-passive just as all the families who go into these social houses have very little income to keep up them.

The scheme of dwellings and new urban plans that takes account of solar access, efficiency energetic and low cost of maintenance, must be susidized by the Local and Regional Administration.

When a new land is going to be urbanized and arranged for new settlements it is necessary to protect building owners. So solar-rights covenants can be established to provide solar-access protection.

Example Solar-Access Covenant: All the dwelling-houses must be free of any shading whatsoever, and must be exposed to direct solar radiation. The duration of solar-access protection must be 4 hours a day as mimimum, even on December 21, the shortest day in the year.

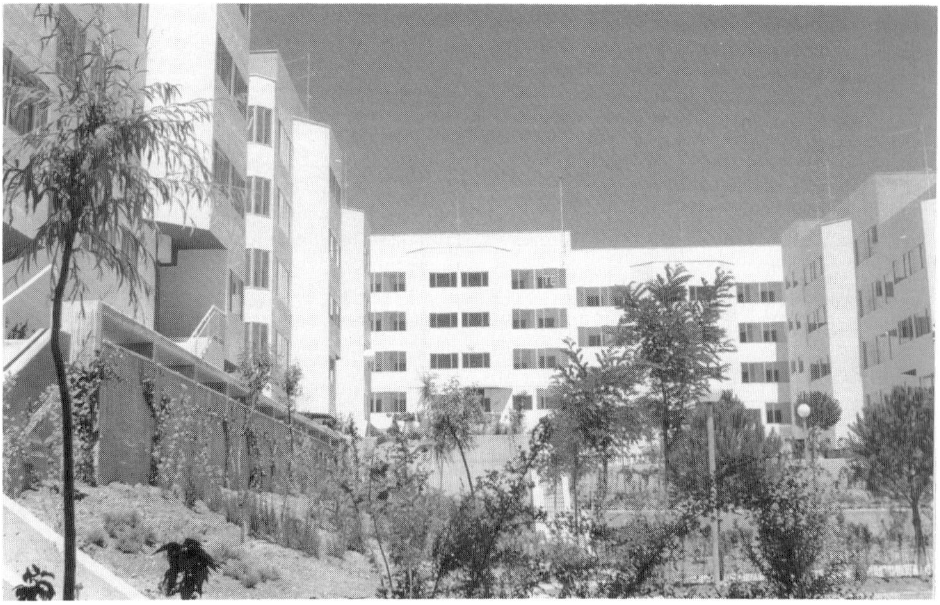

Fig. 5. Cuenca. 118 Apartments. Vegetation in the courtyard.

RETROFIT - FEASIBILITY STUDY

S. KOVIČ, A. KRAINER
Department of Architecture & Civil Engineering
University in Ljubljana
61000 Ljubljana, Jamova 2, Slovenia

ABSTRACT. A study of passive solar elements of architecture has been done with emphasize on a greenhouse, its form and microlocation, conditioned by an inconvenient orientation of the longitudinal axis of the building in the direction N - S. Within the existing shape the plan (three propositions) was rearranged, more quality in a living space was provided and greenspaces of different shapes were incorporated in the existing building shape.

1. Foreword

The multistoreyed dwelling Djakovičeva 9, 11, 13 in Ljubljana consists of three stairway units. Each of them links two flats together. The building is made of bricks and has no thermal insulation whatever. To improve the quality of the residential environment and energy balance, adequate thermal insulation, passive solar elements for direct gain of solar energy, among them greenhouse, treated first of all from the architectural view, were comprehended in the renovation. The inconvenient orientation of the longitudinal axis of the building in the direction N - S conditioned the analysis of form, function and microlocation of a greenhouse and also the preparation of the comparative matrix for possible incorporation of a greenhouse in the existing building shape. Within the existing building shape the plan (three propositions) was rearranged. For the proposal for a renovation the variant 1 has been chosen. By projected openness toward east and west new quality is reached - transparency of the residential environment and sufficient sunshine from morning till evening.

2. Analysis of forms and microlocations of greenhouse

A matrix of different forms and microlocations of a greenhouse has been made on the basis of an analysis of different forms and functions of a greenhouse in certain space and time. The matrix enables us to select a greenhouse which is adapted to various situations and positions on the facade of an optionally orientated building.

S. Alvarez et al. (eds.), Architecture and Urban Space, 423–428.

2.1. FORM

In this case a greenhouse represents a plastic element which is integrated in the ground-plane.

Transformation from circle over triangle to square and modification of basic forms of a greenhouse

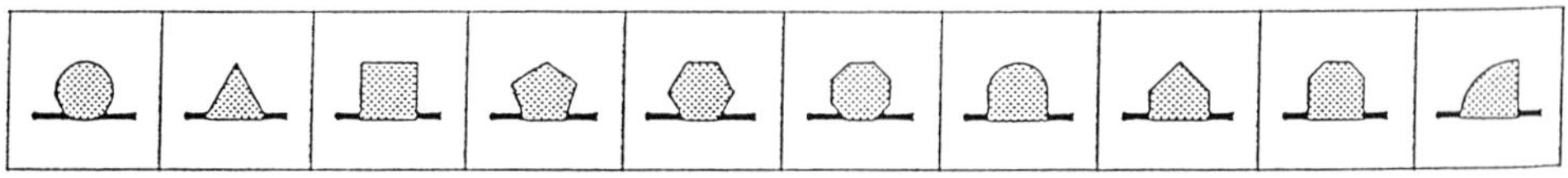

2.2. POSITION ON THE WALL

Different positions of a greenhouse on the wall and in the corner are defined. The area of residential environment is enlarged the most by a complete exposure of a greenhouse and its positioning on the facade. On the other hand, there is risk of overheating present during summer in case of an insufficient ventilation and inadequate shading. In winter very low night-time temperatures can be expected. Therefore this kind of a greenhouse is not treated as a part of a living space during winter. It functions primarily as a buffer-space. Such an exposed location changes the appearance of a facade the most. A greenhouse which is withdrawn into the inside is partly protected against unfavorable oscillations in temperature. A completely integrated greenhouse does not enlarge the residential environment. There are only minor problems present regarding the installation of shading devices. The outlook of a facade is changed the least.

Outside - inside

Transition from an exposed location (on the facade) over halfsheltered (in the wall axis) to completely sheltered location (in the inside).

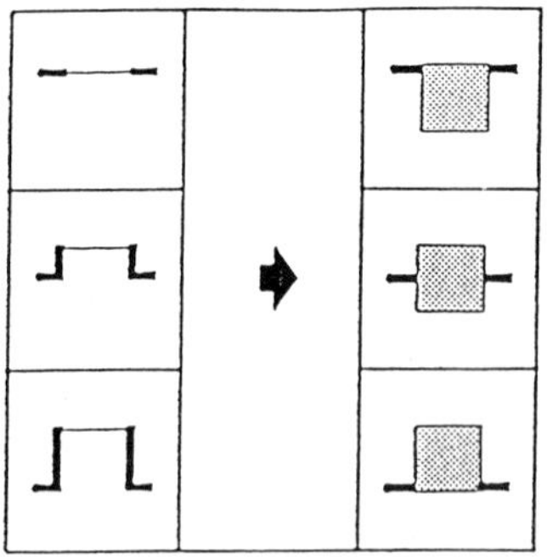

Corner

Position of a greenhouse in the corner and different levels of enclosure regarding the site and orientation of a building.

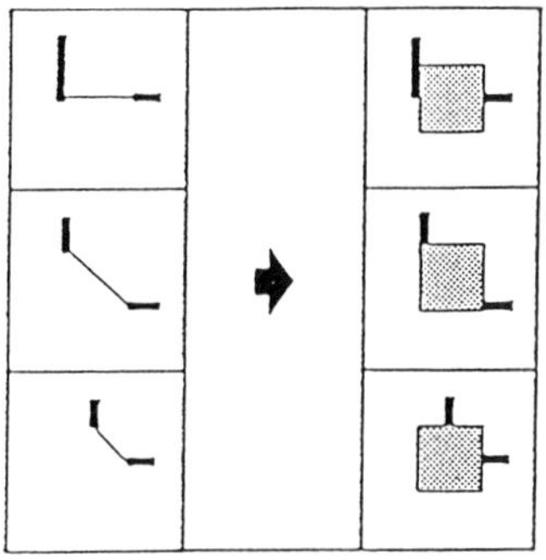

2.3. SHELTERING TOWARD NORTH

The primary function of a greenhouse is conditioned by its orientation and its thermostat - set temperature. A greenhouse can represent a living space usable during the whole year, a temporary living space, a conservatory - winter garden, or a buffer-space. Although the advantages of a strict southern orientation of glazed architectural elements and of consecutive possibility of greater intensity of solar radiation are often emphasized, other orientations should not be neglected. A greenhouse with an orientation shifted from the south functions during the heating period as a buffer-space - a space in which the temperature of the inside and outside ambience is balanced and thus the transmission heat losses are reduced. Because of an inconvenient orientation of the longitudinal axis of Djakovičeva 9, 11, 13 in the direction N - S greenhouses are oriented toward south-east and south-west respectively and partly toward north. A glazed surface exposed toward north is replaced by an insulated thermal storage wall or shielded by shifting the outer wall to cover a suitable part of the northern area of the greenhouse. A certain level of intimacy of the residential environment can so be ensured in multistoreyed dwellings.

Enclosure

East or west - orientated greenhouse sheltered toward north by a wall to desired level (intimacy of space).

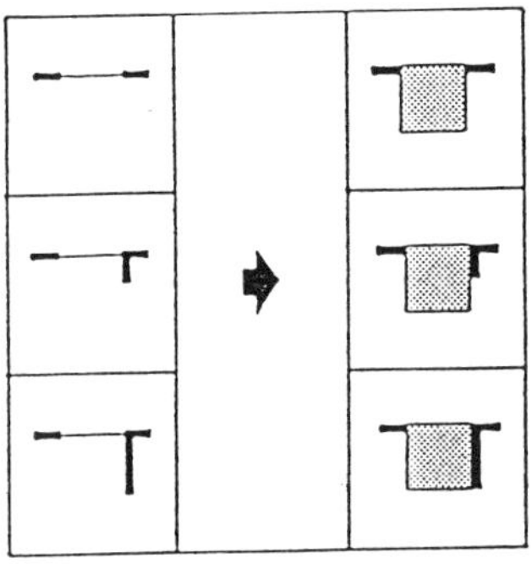

Motion of wall

Greenhouse enclosed toward north by partial or complete motion of wall and combination of possible variants.

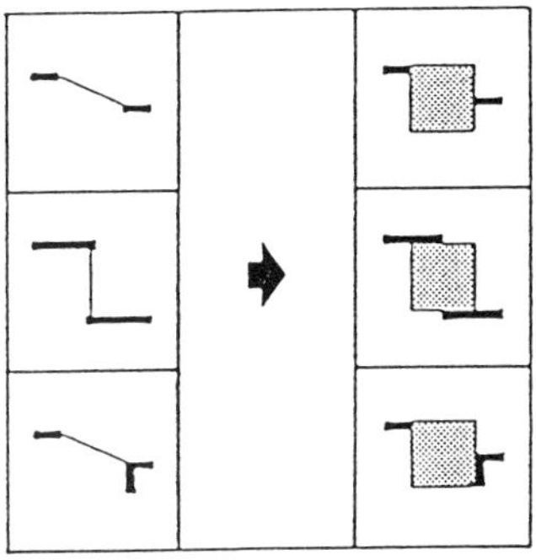

2.4. MATRIX OF GREENHOUSE FORMS AND MICROLOCATIONS

The matrix combines selected forms of greenhouses with their function in a certain ambience and with their orientation. Because of the situation of the longitudinal axis of the dwelling Djakovičeva 9, 11, 13 in the direction N - S the matrix is adapted to the prescribed orientation and a corresponding level of sheltering toward north is assured. A transformation of elementary greenhouse forms which are connected with the first column on the far left side of the matrix is shown in the first line of the matrix. A progressive withdrawal of a greenhouse from an exposed location over halfsheltered location to a position, completely integrated into the building, is represented in first three lines of the matrix. The most suitable forms for defined positions in the wall can so be selected. Greenhouses can be also exposed toward north because of the prescribed orientation. Therefore possible sheltering of different forms is studied in the second part of the matrix. Partial or complete sheltering toward north by a wall, rectangular to the horizontal axis of the building, partial or complete shifting of the wall, lying on the horizontal axis of the building, and a combination of both variants is considered. In the last two lines of the matrix positions of greenhouses in the corner and suitability of an integration of different forms into the building is shown.

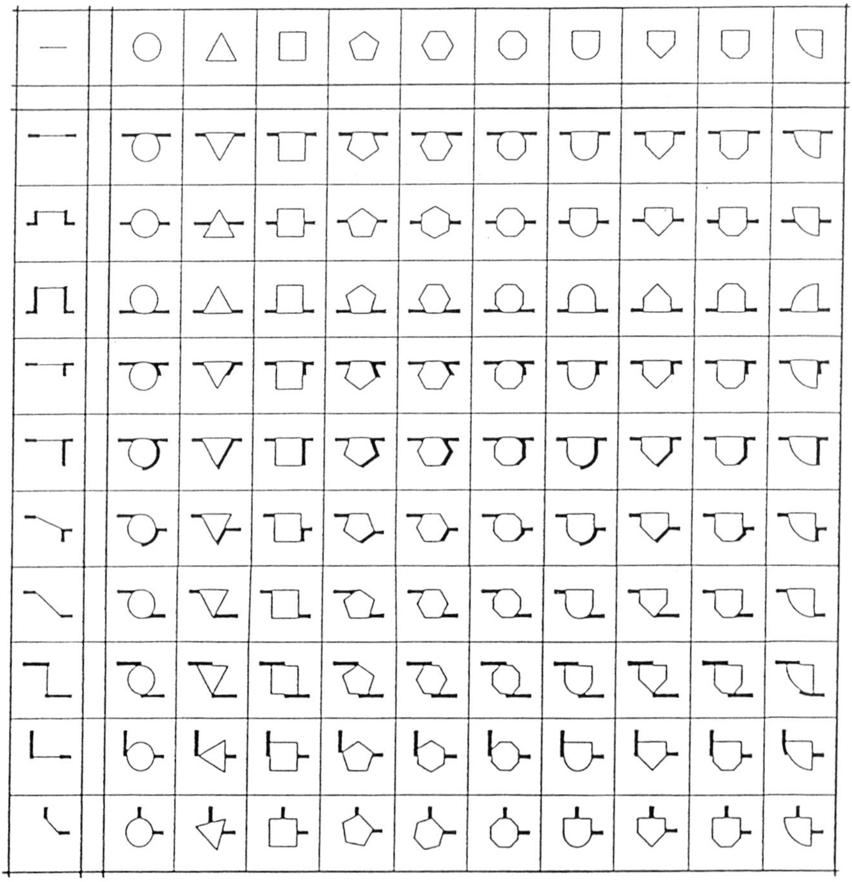

The desired position of greenhouses of different forms and suitable level of sheltering toward north, determinated regarding the function of a greenhouse in the building

3. Presentation of dwelling Djakovičeva 9, 11, 13

Eastern and western facade

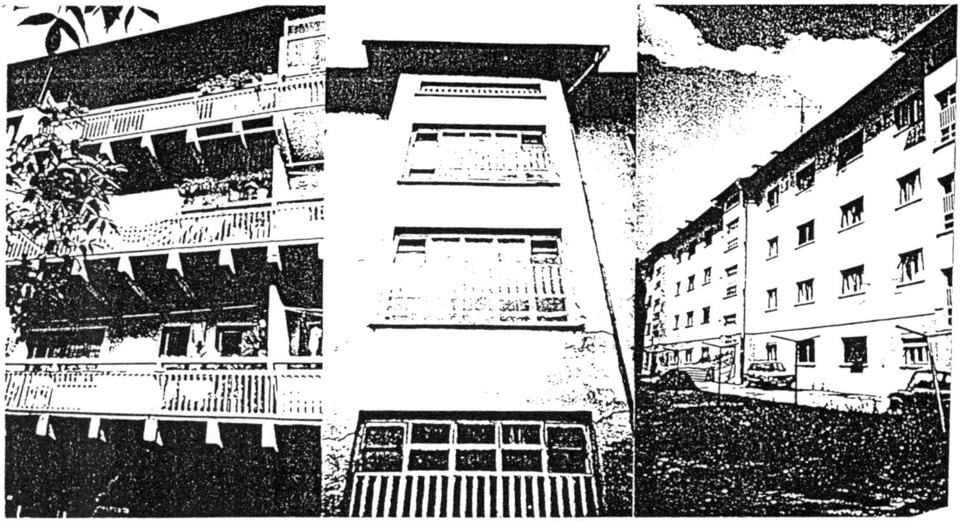

Volume of a building unit - existent state

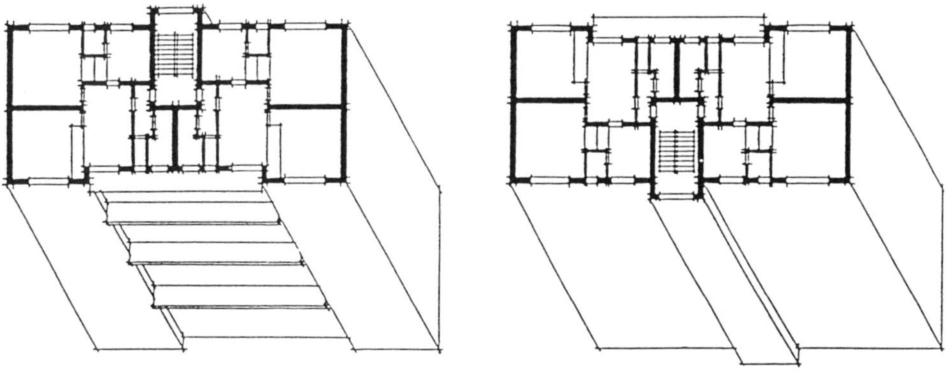

Ground plan of the building - existent state

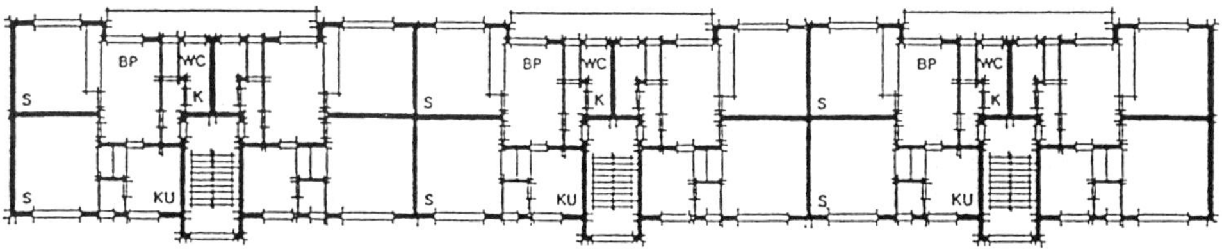

4. Proposals for renovation

Functional analysis of different positions of a greenhouse

Variant 1

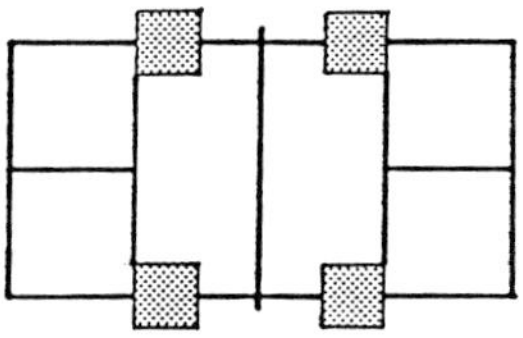

Variant 2

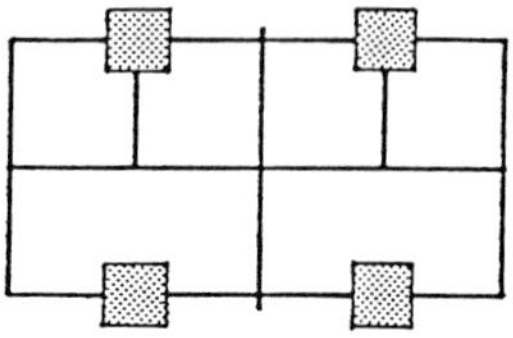

Variant 3

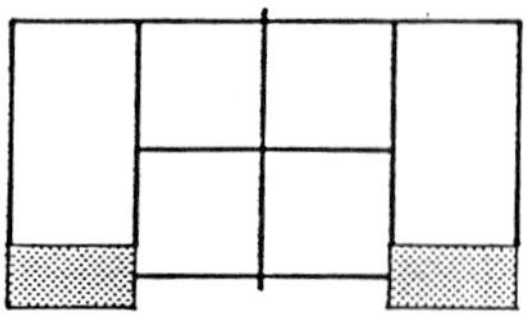

Analysis of a unit ground-plan and proposed solutions with incorporated greenhouses

Existent state

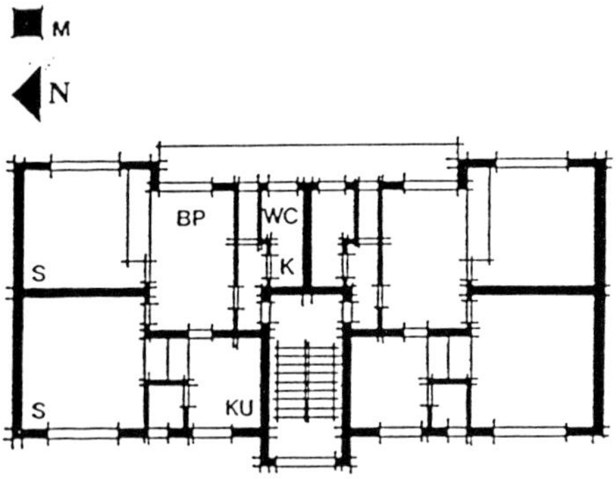

Variant 1

Ground - plan of a building unit with designed transparency of the residential environment and E - W orientation of greenhouses.

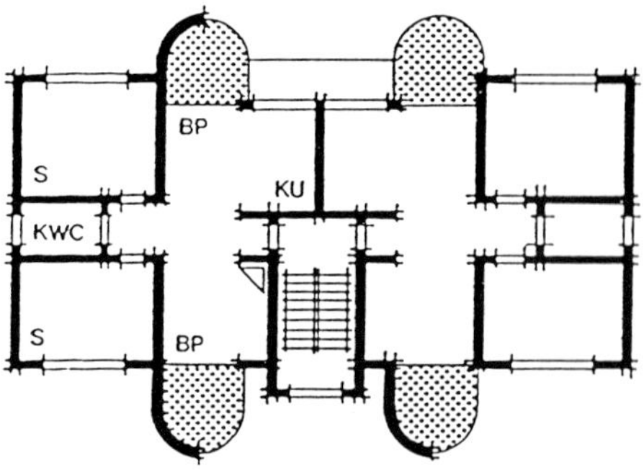

Variant 2

Rearranged ground - plan with rooms along the longitudinal axis and a living space with west - orientated greenhouse.

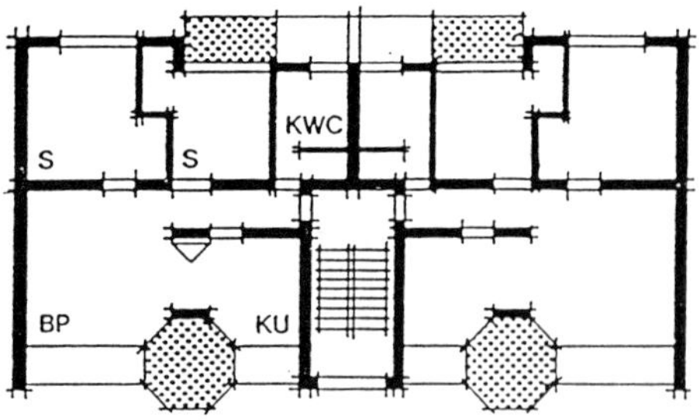

Variant 3

Ground - plan of a building unit with a larger west - orientated greenhouse along the longitudinal axis.

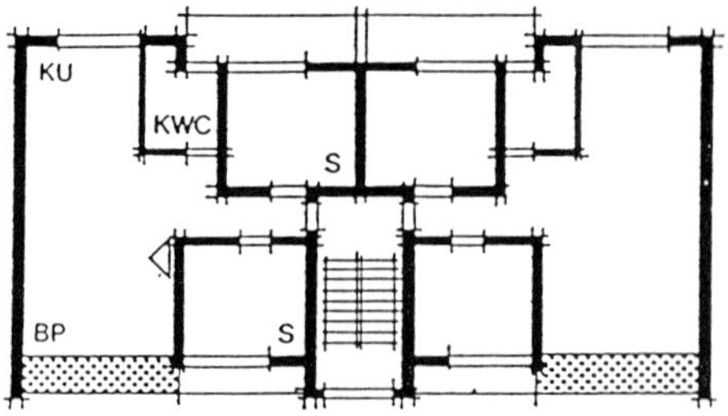

Part 4

DESIGN AIDS

STRATEGIC DESIGN TOOLS FOR NON-DOMESTIC BUILDINGS

NICK BAKER
The Martin Centre for Architectural and Urban Studies
and Cambridge Architectural Research Ltd.
University of Cambridge Department of Architecture
6 Chaucer Road, Cambridge CB2 2EB, England

ABSTRACT. The energy performance of non-domestic buildings varies widely, and is the result of a range of interacting factors. Some of these factors relate to decisions made early in the development of the design. This paper examines the need for strategic guidance at this stage, and describes a simple quantitative energy design tool. The paper also discusses the degree to which early design decisions interact with the performance of servicing systems and occupants.

1. Introduction

Energy consumption in non-domestic buildings, of similar use type, varies over a range of about 10 fold without any obvious correlation between user satisfaction and productivity, and this energy use. We would be surprised to find two equivalent family saloon cars with a fuel consumption ranging from 16 km/l to 1.6 km/l!

How much of this range of energy performance can be accounted for by the design, and can the key design parameters be identified? We can propose that the performance of a building is due to three kinds of factor -

1) that which is inherent in the building design
2) that which relates to services systems design and efficiency
3) that which is due to the effect of occupant behavior

We refer to these as *building, systems*, and *occupant* factors.

2. Independent Factors

First we will consider the building, systems and occupants as independent factors. The final performance of the building would be a product of these three factors.

For example the decision to adopt a deep plan will result in a certain building performance. This creates a demand for useful heating, useful cooling and useful lighting in order to satisfy the conditions for an idealized occupant.

The systems will use energy to provide the environmental conditioning. The delivered energy required will depend upon the system efficiency. This would vary with the choice of system design, e.g. centralized air handling with heat recovery compared with perimeter heating with tempered air and comfort cooling. These efficiencies are relatively predictable, but may carry an uncertainty, e.g. quality control, commissioning etc, the main causes of a system to operate less efficiently than anticipated.

S. Alvarez et al. (eds.), Architecture and Urban Space, 431–439.

Finally, the occupants (and management) will operate the systems in a non-optimal fashion - e.g. having set temperatures too high, heating unoccupied rooms, and leaving lights on. These factors are the least well known, and modellers of building performance have only just begun to recognise their importance.

We could imagine a building with good environmental performance but being equipped with poor services systems and with poor management. This may use more energy than a building of inherently poor design but with highly efficient services and management. Similarly we could imagine other combinations, and only when all three factors are optimised will we get the best performance.

What is the relative magnitude of the building, systems and occupant factors? In studies [1] on energy consumption in relation to urban built form, using an integrated energy model we concluded that if systems and occupant factors are normalised, the range of energy use is unlikely to be greater than about 2.5 fold. This is comparing a fully air-conditioned artificially-lit tinted glass "horror story", with a naturally ventilated and lit building with an atrium. If the system parameters such as lighting loads, ventilation rates, etc are allowed to range over plausible values the energy consumption range increases to 5 fold. This implies that occupant factors typically account for a 2 fold variation, in a total variation of 10 fold. (fig 1).

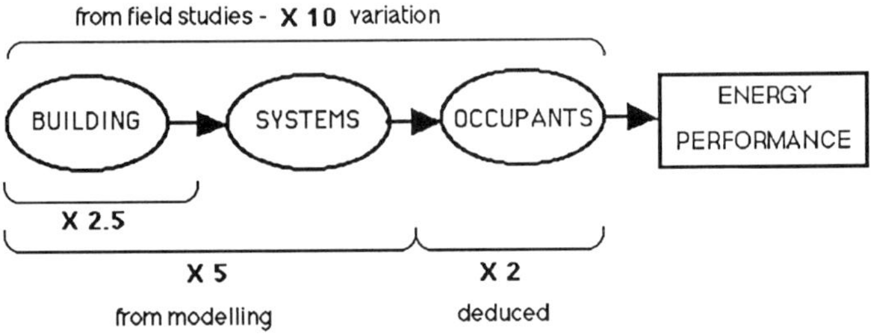

Figure 1. The relative contribution of building, system and occupant factors on energy use.

Thus the building factors influence the building performance little more than half as much as the joint influence of system and occupant factors. This could be seen as rather disappointing for those endevouring to develop strategic design tools, responding to basic building parameters. The designer may question the need to optimise glazing ratio if the energy performance is going to be swamped by the luminaire efficiency or environmental management policy. He may be tempted to resort to stylistic, or short term economic considerations to generate the building concept, believing that the final energy performance will rest in the hands of the engineer and the management.

However, one possible argument for special consideration for the building factors is that these factors are the least easy to change. The building factors could only be changed by major remodelling of the building. Whereas system efficiency could be improved by retrofitting new systems or modifying or simply improved tuning of existing systems, and occupants could be persuaded to operate the building more efficiently.

3. Interacting Factors

However, it can be argued that these factors may not be entirely independent. For example where occupants have access to openable windows they are likely to be much more tolerant to ranges in thermal conditions than in sealed air-conditioned spaces. This has been observed as long ago as 1971 by Humphreys [2]. It is also readily observed that as daylight levels fall at the end of the day, people will happily work at much lower lighting levels than would be acceptable in an artificially lit space. These are examples of where building factors interact positively with occupant factors. It appears then that people are much more ready to accept what mother nature has to offer than what the management thinks is good for them!

There can also be interaction between the inherent characteristics of the building and the performance of the systems. For example a building with low ceilings, all lightweight finishes, and high internal gains, will place much greater demands on system design and control than a heavyweight building with larger room height to depth ratio. The latter may only need systems to intervene under extreme conditions, giving much less opportunity for the system to indulge in gross inefficiency such as similtaneous heating and cooling.

Thus it may be that strategic decisions for design have much wider implications for the actual performance of buildings over their lifetime. This has largely been missed due to the limitations of building performance models.

3.1 BUILDING ENERGY MODELS

The first priority in the development of building models has been in the interaction between the building and climate. This is now achieved to a high level of precision with both simulation and correlation methods. Subsequently, rather fewer models have taken on the modelling of the systems, and to some extent, the interaction between system and building.

It is the area of occupant effects that is almost totally absent. This includes a whole set of topics such as - operation of heating, cooling and lighting controls, window opening, sunshade operation, perception of comfort in relation to activity, dress, spatial location and room layout, and psychological factors.

It is only in the topic of light-switching that we see some progress having been made. In evaluating the energy saving due to daylight, it has been essential to propose some kind of "behavioural" algorithm to describe manual switching performance [3].

3.2 THE RANDOM ELEMENT

Most of our models are causal or mechanistic, and in the case of building and system models, based upon the principles of physics. Mechanisms and processes which are not built into the model will cause differences between real performance and modelled performance, which because unaccounted for will be interpreted as "random error". Each of the sub-systems, building, systems and occupants, will be modelled with this limitation. That is, a given model will only have a limited accuracy due to unforeseen processes which are not included in the model.

But whilst we may not be able to say much about the cause of the error, we may be able to build up a knowledge of the likely random deviation from the predicted performance, not from the conventional error analysis, but from a knowledge of the real subsystem - building, services or occupant - and how they perform and interact in reality. For example, we might be able to say that the performance of a certain kind of office building with a certain air-conditioning system is particularly vulnerable to certain uses - that is it is highly probable that it will perform non-optimally, or at least not in the way that the model prediction assumes. This random element will almost certainly be strongly influenced by interaction between the sub-systems.

We are now in a position to propose our "ideal" integrated model. It models the building, the services systems and the occupants with causal models which interact with one and other. Overlaid upon this is an empirical "random" model which predicts the magnitude and direction of probable departures from the causally predicted performance.

4. Delivered and Primary Energy

The way energy use has been presented, both from model predictions and field studies, has been misleading and has led to incorrect design strategies. Figure 2 shows the monitored energy use of fourteen offices in the UK grouped under naturally ventilated and daylit, and air-conditioned. Note that when presented as delivered energy, heating appears as the largest single demand in non air-conditioned offices. This is misleading since when presented in terms of primary energy*, we see that even in the daylit group, heating is less than lighting by 45%.

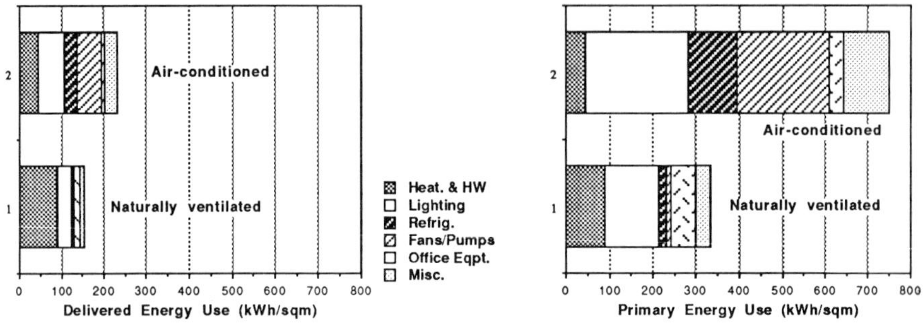

Figure 2. Annual Delivered and Primary Energy use in 14 UK offices. Source BRECSU (UK)

But the greatest distortion occurs when we look at the data for air-conditioned, artificially lit buildings. In this case, even the delivered energy for heating is no longer the largest single component. But when presented as primary energy, heating is twelve times less than the primary energy for lighting, ventilation, pumps and fans.

Historically, this distortion has resulted in a focus of attention towards heating. Technical studies published in the late sixties and early seventies promoted the adoption of small windows and deep plans as a means of minimising surface area to volume ratio, and hence heat losses.

We also see a decade of interest in applying passive solar (thermal) techniques to large non-domestic buildings. We now realise that the returns are very small [4] and the passive design should have concentrated upon natural ventilation and daylighting as the main measure to save auxiliary energy.

**Although primary energy presents difficulties of definition in rigorous scientific terms it is far more meaningful than delivered energy. For example it relates well to CO_2 production, where power is generated from fossil fuel, and recent global environmental concerns have focussed attention to this. Primary energy also relates well to money cost. On the other hand summing delivered energy units of the calorific value of fuel to delivered electricity, is meaningless. It is rather like trying to define the distance fro m London to Paris by adding the miles in England to the kilometres in France!*

5. Strategic Design Decisions

We now address the question - are there design tools and design guidelines which can assist in making strategic decisions which will increase the probability that a building will perform well as a low-energy user? The difficulty with this kind of tool is that early in the design stage there are few parameters available. But if the tool responds to too few parameters, the results may be trivial. If on the other hand the tool requires a large number of parameters both from the building and the services design, the very act of establishing these values will tend to have the effect of freezing the design, even if it is a long way from an optimal solution.

5.1 THE LT METHOD

An example of a design tool which attempts to satisfy such a brief, is the LT Method (Lighting and Thermal). This tool responds to parameters available early in design development - plan depth and section, orientation, facade design, the incorporation of an atrium, and the presence of adjacent buildings. It predicts the primary energy consumption for heating, cooling and lighting. It was originally developed for the EC Working in the City Architectural Ideas Competition [5] and it has been described in more detail elsewhere [6].

In order to introduce the method here, we describe a scenario where it is used by a designer at the beginning of a project.

The Brief

The building is to be a company headquarters of about 4000 m^2. The location is in the UK, a restricted urban site about 50m x 60m overshadowed to the east and north. There is a presumption that it will be air-conditioned and deep plan. However the architect wishes to explore other possibilities.

The first proposal

First he proposes an L shaped plan of four storeys and proposes a 70% double-glazed facade to the south, 30% to the east and west and 15% to the north, under the impression that the passive solar strategy used in housing, could be extended to this building.

The predicted energy performance

Step One *is to designate the passive and non-passive zones of the building as indicated in fig 3. Passive zones are defined as those areas which potentially can benefit from daylight and natural ventilation, and passive solar gains where appropriately orientated. They have a plan depth of twice the floor to ceiling height from the window wall - 6m in this case. The area of the top floor could be designated passive if rooflights were to be incorporated. These areas are entered on the worksheet, fig 4.*

Step Two *The appropriate set of LT Energy Curves is identified by reference to the map of climatic zones fig 5. Annual energy consumption is influenced by the combination of parameters such as temperature, solar radiation, sky luminance, day length, and its variation between winter and summer.*

Step Three *From the set of curves for Mid-European Coastal fig 6. the specific energy consumption for lighting, heating and cooling is read off for the appropriate orientations and glazing ratios and entered on the worksheet. The non-passive specific consumption is read of at zero glazing ratio. Multiply the zone areas by the appropriate consumption values and the total energy can be evaluated. The result is 1148 MWh/y or 265 kWh/m^2y*

Optimisation *Looking at the curves we can see that the glazing areas are not the best. The north shows a rather flat minimum at about 25%, east/west and south are only slightly higher at about 30%. The 70% south glazing was clearly an error; the energy use breakdown indicates that cooling energy is likely to be greater than heating or lighting. But note that if single glazed, heating would remain the largest component. Also the non-passive zone on the top storey could be eliminated by roof lighting. So the proposal is changed to 25% glazing on the north, 30% on*

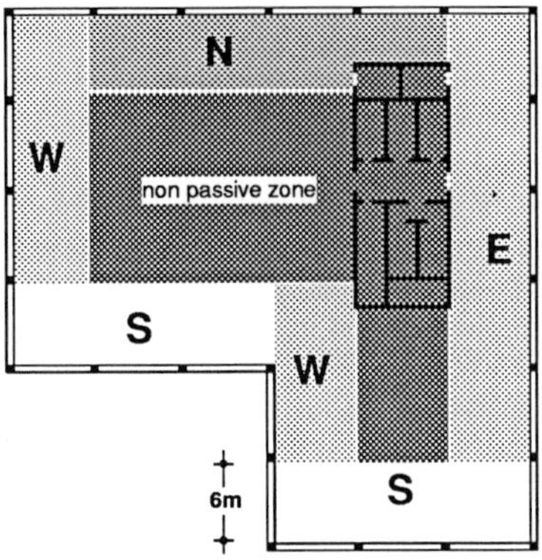

Figure 3. Plan showing passive zones.

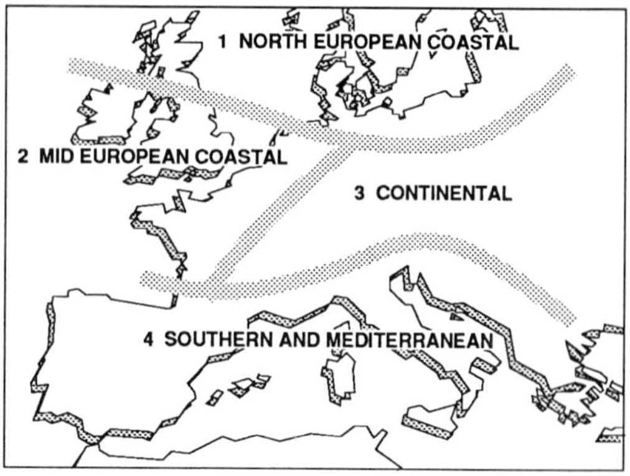

Figure 5. LT Climatic Zones.

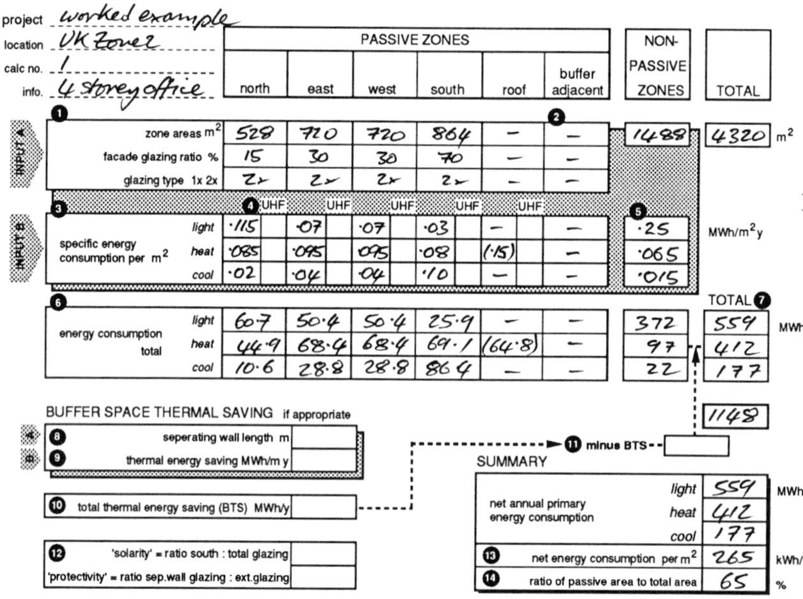

project: worked example
location: UK Zone 2
calc no.: 1
info.: 4 storey office

		north	east	west	south	roof	buffer adjacent	NON-PASSIVE ZONES	TOTAL	
INPUT A (1, 2)	zone areas m²	528	720	720	864	–	–	1488	4320	m²
	facade glazing ratio %	15	30	30	70	–	–			
	glazing type 1x 2x	2x	2x	2x	2x	–	–			
INPUT B (3, 4 UHF, 5) specific energy consumption per m²	light	·115	·07	·07	·03	–	–	·25		MWh/m²y
	heat	·085	·095	·095	·08	(·15)	–	·065		
	cool	·02	·04	·04	·10	–	–	·015		
(6) energy consumption total	light	60·7	50·4	50·4	25·9	–	–	372	559	MWh/y
	heat	44·9	68·4	68·4	69·1	(64·8)	–	97	412	
	cool	10·6	28·8	28·8	86·4	–	–	22	177	
									1148	

TOTAL (7)

BUFFER SPACE THERMAL SAVING if appropriate

A (8)	seperating wall length m	
B (9)	thermal energy saving MWh/m y	
(10)	total thermal energy saving (BTS) MWh/y	
(11)	minus BTS	

(12) 'solarity' = ratio south : total glazing	
'protectivity' = ratio sep.wall glazing : ext.glazing	

SUMMARY

net annual primary energy consumption	light	559	MWh/y
	heat	412	
	cool	177	
(13) net energy consumption per m²		265	kWh/m²y
(14) ratio of passive area to total area		65	%

Figure 4.

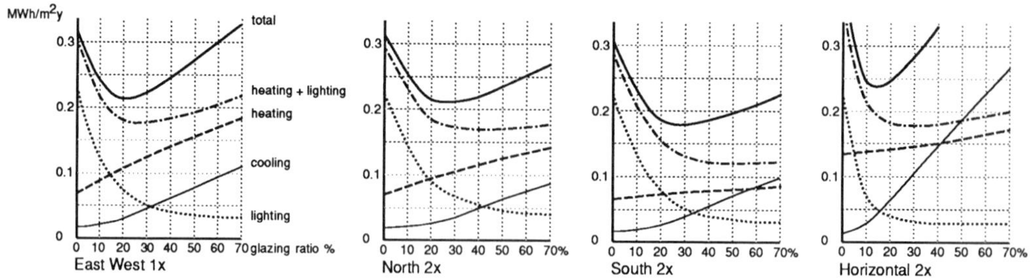

Figure 6. Sample of LT Energy Curves for Zone2.

the east, west and south, and 15% for the central roof lit zone. A re-calculation shows an improvement down to 230 kWh/m^2y or 199 kWh/m^2y if cooling is omitted (see below).

Avoiding air-conditioning *If the cooling energy is below a certain level, as advised in the supporting documentation, and if certain conditions relating to thermal mass, finishes, shading and openings, are met, air conditioning can be avoided and its energy consumption eliminated. Non-passive zones are always considered to need cooling. This takes account of fan power as well as refrigeration.*

Overshading *The north and east zones are heavily overshaded, that is the average horizon angle is greater than 45°. The table shown in fig 7. gives correction factors for lighting, heating and cooling. As we would expect the obstruction has a serious effect on lighting energy, and slightly reduces cooling energy. Rather more surprising is the reduction in heating, until we realise that this is due to the heating effect of the extra lighting. This demonstrates the complexity of the energy interactions.*

This now shows a 7% increase in energy use and suggests that things might be better if the glazing area is increased on the facades which are obstructed. Here we need an inspired guess - the designer tries 45%. A recalculation now reduces the total back to 203 kWh/m^2y.

Atria *The effect of an atrium on the lighting in the atrium-adjacent zone is to reduce the passive zone depth as indicated in the table in fig 8. The reduction in passive zone results in an increase in the non-passive zone and hence an increase in lighting energy. This does not mean that an atrium building will have an overall greater energy consumption than a non-atrium equivalent, because there may be considerable savings in heating energy. The LT Method calculates this effect by applying a Buffer Space Thermal Saving which depends upon the configuration of the atrium, and on the ventilation exchange between the atrium and the heated building. An atrium option was not evaluated in this example.*

orientation	UHA deg.	glazing ratio %	correction factors		
			lighting	heating	cooling
north	15 - 45	15.0	1.3	0.9	1.2
		30.0	1.3	0.9	0.7
		45.0	1.3	1.0	0.6
		60.0	1.4	1.1	0.6
	> 45	15.0	1.3	0.9	1.0
		30.0	1.6	0.8	0.9
		45.0	1.9	0.9	0.5
		60.0	2.1	0.9	0.5

Figure 7. Sample of LT Urban Horizon Factors.

characteristic	score zero	score 1
shading	fixed	movable
glazing	tinted or refloat	clear
internal finishes	dark coloured	light coloured
sky angle*	greater than 45°	less than 45°
total score	**buffer adjacent passive zone depth m**	
0 or 1	0.0	
2 or 3	3.0	
4	4.5	

Figure 8. Passive zone depth for spaces adjacent to an atrium.

The non-passive equivalent. *The preconception of the client would probably have lead to a deep plan air-conditioned building, with total artificial lighting. In spite of the latter, in common with many modern buildings, it would probably have at least 60% glazing. To calculate the performance of this base case, the lighting energy is calculated for zero glazing ratio since no daylight is utilised. Heating and cooling however are calculated using the actual glazing ratios to account for the large heat losses in winter and cooling loads due to solar gain in summer. The result is 427 kWh/m²y.*

6. Developments of the LT Method

We have earlier established a brief for an ideal strategic design tool - one which not only responds to the initial building design parameters, but would also respond to the interactions between systems and occupant factors as well, and quantify the uncertainty that the particular combination of building, system and occupant will perform as predicted.

At the moment the LT Method goes only a little way in meeting this brief. With the manual solution it is possible to handle only a very few parameters. However, the model which generates the curves, has about 30 input variables, most of which have to be given assumed values. Thus including the parameters which are input, we have architectural parameters such as room depth, glazing ratio, orientation etc, system parameters such as U-values, boiler efficiency and lighting efficacy, and finally occupant parameters such as occupied hours, and lighting and thermal comfort targets.

The LT Model already contains a very rudimentary behavioural element. There is an assumption that as soon as there is no net heating load, shading is deployed so that it eliminates the direct part of the solar radiation. This is in order to keep cooling loads to a realistic value. The model in this version does not, but could, assume that ventilation rate is increased under the same conditions, corresponding to the opening of windows. These are very simple behavioural algorithms but we can speculate about more complex ones.

For example in a study on simulating the spatial distribution of comfort conditions in a room [7] we have found that the number of hours of overheating can be reduced from 514 to 104 per cooling season by simply allowing the occupant to move closer or further from the window. The LT Model could take account of occupant mobility when evaluating overheating risk.

If the LT Model itself (as distinct from the manual Method) were presented as a design tool, how would the user handle this large number of parameters? The designer would have to provide values for a large number of variables and this would demand that a number of design decisions had already been made. This tends to defeat the object of a strategic model.

A solution to this problem which we are developing is the presentation of results from parametric runs which include the values of the design proposal. For example a three-dimensional graph of energy consumption versus glazing ratio and lighting datum value would be displayed with the design proposal represented as a point on the surface if a lighting value had been input, or a range of points corresponding to default values of lighting datum if no value had been provided.

There would still remain the problem of selecting which variables to display. This could be handled by an expert system which would steer the designer through a number of parametric displays which were appropriate to the particular building use and context. For example on an urban site, it would be useful to see the relationship between lighting energy use, urban horizon angle, and lighting datum level. This would help the designer to zone the use of spaces in the building to relate to lighting requirements.

The three dimensional graph has a useful analogue. If the vertical axis always represents a reduction in energy, then we are always trying to move our design to the highest point on the surface. But as in mountain climbing, high places are often dangerously steep - small mistakes mean disaster. Perhaps it is better then to look for a high plateau as a compromise. This is where the random element comes in. Rather than representing our design proposal as a point it should be a fuzzy area. If this area takes us over a precipitous edge, we are in trouble.

7. Conclusion

This leads us to a final concept. It seems that for some time we (that is building modellers) have been dreaming of the ultimate model which allows us to optimise every aspect of the design. Due to the complexity of the interactions this is difficult, but the random element, particularly due to occupant influence, makes this impossible. Thus we should stop aiming for optimisation and instead aim for *safe territory*. Our ideal solution is not a point in solution space but rather a volume.

This is not a new concept - that is the concept of a robust solution. What is perhaps new, is the challenge to predict and quantify the robustness from analysis. It would be predicted by modelling the interaction between the building, the systems and the occupants, that is the building as a total system.

The admission of the occupant onto the modelling scene will achieve another important benefit. It will give an opportunity for the health and well-being of the people in response to the building design, to be considered strategically. Fortunately, it is already emerging that the buildings which attain good energy performance by passive means, tend to be those with the least problems of sick building syndrome and other kinds of occupant stress.

This will promt valuable support to passive strategies from clients. Improvements in employee productivity and reduction in absenteeism may well be far more valuable than the reduction in energy cost.

8. References

[1] Steemers K, Baker N, Hawkes D. Parametric Energy Modelling for Urban Situations. Proc. Conf. 2nd European Conf. on Architecture, Paris 1989. Kluwer 1990

[2] Humphreys M, Nicol J. An investigation into the thermal comfort of office workers. Building Research Station Current Paper CP14/71

[3] Hunt D, Predicting lighting use - A behavioural approach. Lighting Res. Technol. 13 (1981)

[4] Dijk H, Arkesteijn C. Windows and space heating requirements. Netherlands Nat. Report from STEP 5 to IEA, Annex XII. 1990.

[5] O'Toole S, Lewis J. Working in the City. CEC EUR 12919 EN. Eblana Editions 1990.

[6] Baker N, Fontoynont M, Simplified Design Tools tools for Daylighting and Thermal Design. Proc. Conf. 2nd European Conf. on Architecture, Paris 1989, Kluwer 1990.

[7] Newsham G, Baker N. The role of the occupant in the assessment of building energy performance. Proc. World Renewable Energy Congress, Pergamon, 1990.

LESOSAI-X : A SOFTWARE PACKAGE WITH EXPERTISE

A. FAIST, F. LERESCHE
Ecole Polytechnique Fédérale de Lausanne
Bâtiment LESO
CH-1015 Lausanne
Switzerland

ABSTRACT. A completely new version of the software package LESOSAI has been developed in order to fulfill the following objectives : more interactiveness, better screen management, improvement of the graphics, heat gains and losses expertise, automatic parameter sensitivity calculation, evaluation of overheating periods.

The last three items are the more interesting novelties: the original expert system built in the programme evaluates both the heat loss and solar heat gains levels of the project. It detects undesired heat losses (bad U values for example) and points out also low overall insulation levels. The potential solar gains are compared to the potential heat losses in order to evaluate the probable thermal behavior of the building. The parameter sensitivity calculation computes automatically the effect of a change of $\pm$ 20 % of any building parameter. The overheating danger is evaluated according to the Method 5000 algorithms. Daily hours of probable overheating are given month by month.

All these informations are translated into direct advices given to the user that help improving the project in the right direction.

1. Introduction

The first versions of the software LESOSAI were dedicated to the computation of solar houses heat gains and losses [1], [2]. LESOSAI is a fast and reliable analysis tool that gives the winter heat requirements of buildings but doesn't give any information about the relative performance of the building or suggest possible improvements. The new software package LESOSAI-X overcomes these weaknesses in the following way :

- An expert system evaluates the insulation level of the building as well as its solar gain level.
- A parametric calculation routine allows fast evaluation of the impact of building parameters change.
- Overheating risks are computed monthly.

This is a first step towards more interactive and more helpful programms adding design help to the common analysis tool [3], [4].

S. Alvarez et al. (eds.), Architecture and Urban Space, 441–446.

2. The expert system

Building a good solar house implies firstly to lower the heat losses and secondly to raise the solar gains. It is also necessary to provide sufficient heat storage as well as an adequate heating control system.

The new SIA Recommendation 380/1 [5] adopted as a national basis for heat load calculation states allowed and recommended (limit and target) heating energy requirement per unit heated floor area. This heating energy requirement is climate and altitude dependant. For example at an altitude of 500 m above sea level the following limit and target values are allowed [Tab. 1].

TABLE 1. Allowed heating energy requirement per unit heated floor area [MJ/m^2] over the heating season of various buildings. Swiss climate, 500 m above sea level (SIA Recommendation 380/1).

Building category	Limit value	Target value	Type of building
I	330	280	One and two family houses, week-ends.
II	300	250	Apartment houses, hotels, homes for elder people.
III	270	220	Administratif buildings, schools, libraries, etc.
IV	240	-	Industrial buildings, warehouses, sport facilities.
V	330		Special buildings : research laboratories, lecture halls, restaurants, etc.
	[MJ/m^2]	[MJ/m^2]	

To evaluate the overall insulation level, the expert system first calculates the parameter "a" defined as :

$$a = 0{,}086 \cdot DJ \cdot H_1 / HFA \qquad [MJ/m^2] \qquad (1)$$

where

$0{,}086 \cdot DJ$ are the SI converted degree · days	[Ms · K]
H_1 the heat loss coefficient to the outdoor air	[W/K]
HFA the heated floor area	[m^2]

Heat losses to the ground are not taken into account in this procedure. If "a" is not smaller than the limit value (heating energy per unit heated floor area) the program is testing the solar gain level (see following paragraph), the adequate choice of windows and the adequate choice of Uvalues. The following information and recommendations are then appearing on the screen :

If a > Limit value and m > 13 (see below)	"The insulation level of your building is too low" "Can you improve the quality of your windows" (only appearing if a bad choice of window type is detected) "Is it possible to improve the envelope U-value ?" (only appearing if inadequate U-values have been chosen) "Is it possible to improve the sunspace glazing quality ?" (only appearing if inadequate glazing quality has been chosen)

The solar gain level is more difficult to estimate. To achieve this an equivalent south oriented collecting surface is first computed through an adequate weighing of all the collecting surfaces in all orientations and tilt (using the local global solar radiation on these surfaces).

$$A_S = \frac{1}{G_S} \sum_j A_j \cdot G_j \qquad [m^2] \qquad (2)$$

A_S equivalent south oriented solar collecting surface $[m^2]$
A_j solar collecting surface in orientation and tilt "j" $[m^2]$
G_j global solar radiation in orientation and tilt "j" over the heating season $[MJ/m^2]$
G_S global solar radiation on south vertical surface $[MJ/m^2]$

The variable "m" is the test parameter used to evaluate the solar gain level [6].

$$m = H_1 / A_S \qquad [MJ/m^2] \qquad (3)$$

If $m \leq 8$	solar gains are very high, leading to high overheating dangers
$8 < m < 13$	high solar gains are expected together with probable overheating periods
$m \geq 13$	moderate solar gains, low overheating dangers.

The expert system carries out these tests very rapidly and displays the following information and recommendations on the screen :

If $m \leq 8$	"Very high solar gains can be expected. Special attention must be paid to overheating dangers that can be overcome through adequate shading or increase of the air change rate"
$8 < m < 13$	"Consistent solar gains can be expected. Do not forget to pay attention to overheating dangers that can be overcome through adequate shading or increase of the air change rate"
$m \geq 13$	"Compared to the heat losses, the solar gains of your building are low." "Wouldn't it be possible to raise the insulation level ?" "Could the solar gains be increased ?" "Wouldn't it be possible to lower the air change rate ?" (only appearing if the air change rate exceeds one per hour.

3. The parametric calculation routine

One of the main advantage of the computer is to compute very rapidly any modified version of the initial building [7]. To take advantage of this, the parametric calculation routine implemented in LESOSAI-X computes automatically the effect of a ± 20% variation of any chosen data including the building data, meteo data, control data, etc. For example if the pointer is located on the air change rate, the following information is going to appear on the screen : the change in specific heating energy (over the heating season) due to a ± 20% variation of the air change rate (Tab. 3) or (Tab. 4) the according change in heat demand over the heating season.

TABLE 3. Parametric calculation : Change in specific heating energy (over the heating season) due to a ± 20% air change rate variation.

Air change rate [1/h]	Specific heating energy [MJ/m^2]	Variation [%]
0,4	435	- 6,4
0,5	465	0
0,6	494	+ 6,4

TABLE 4. Parametric calculation : Change in the heat demand (over the heating season) due to a ± 20% air change rate variation. Solar gains are not affected by this change.

Air change rate	Heat demand		Solar gains	
[1/h]	[MJ]	[%]	[MJ]	[%]
0,40	122 300	- 7,8	26 100	0
0,45	127 700	- 3,8	26 100	0
0,50	132 700	0	26 100	0
0,55	137 600	+ 3,7	26 100	0
0,60	142 800	+ 7,6	26 100	0

If the pointer is located on the window type (labeled A, B, C, D, etc. from single pane to more elaborate window systems) two results are also available. Table 5 illustrates the heating season specific heat loss change corresponding to the choice of different window types (B, C or D) and Tab. 6 shows the change in heating energy and solar gains corresponding to five different window types centered on type C.

TABLE 5. Window type parametric study. Specific heating season heat losses change according to window type.

Window type	Specific heat losses [MJ/m^2]	Relative variation [%]
B	231	+ 12,6
C	205	0
D	183	- 10,7

B : s. gl. + heat mirror; C : doub. gl.; D : as C plus pyrolitic select. coating.

TABLE 6. Window type parametric study. Change in heating energy and solar gains according to window type.

Window type	Heating energy		Solar gains	
	[MJ]	[%]	[MJ]	[%]
A	86 150	+ 17,8	123 700	+ 4,7
B	82 350	+ 12,6	112 300	- 5,0
C	73 150	0	118 200	0
D	65 290	- 10,7	113 800	- 3,7
E	65 990	- 9,8	109 900	- 7,0

A : single glazing, E : as C plus cathodic selective coating.

4. Overheating

High solar gains combined with low heat losses and reduced thermal mass lead to consistent overheating dangers. In Switzerland most buildings (one and two family houses included) are of the medium to heavy construction type. Light weight curtain walls are used on some administrative buildings.

To attract attention on possible overheating problems a routine similar to Méthode 5000 [8] has been implemented. The probable daily hours of indoor temperatures overshoot is given month per month. The indoor maximum acceptable temperature can be set from 24 to 26° C.

5. Using the programm

Special care has been paid to the user friendliness of the programm. Window type screen management together with keyboard or mouse command makes the programm very easy to handle. However the information and recommendations given by the expert system seem to be somewhat rough and not always appropriate to the particular project.

The sensitivity routine is very effective when going through any optimization procedure and speeds up the whole process. It is a very valuable tool.

To get a complete and useful information on overheating dangers it is necessary to run the programm over the whole heating season with various air change rates. It would be more convenient to be able to run the programm with monthly changing air change rates, according the known relationship between windows opening rates and outdoor temperature [9].

6. Conclusion

LESOSAI-X proves to be a very user's friendly solar building design tool. The elementary expertise it features together with the parametric calculation possibility are of valuable help when optimizing a solar project. The overheating routine gives valuable information on overheating dangers.

The next improvements of the programm will be to refine and broaden the expert system as well as the overheating routine : the monthly air change rates could be chosen as a compromise between low heat losses and reduced overheating hours.

LESOSAI-X makes an important step towards real design tools giving not only the energy budget of a given project but helping the user to optimize the project and improve its sensitivity to Energy problems.

7. References

[1] Faist A., Eggimann J.-P. (1985) Programme LESOSAI : Optimisation de constructions solaires passives, 5. Symposium Solaire, Lausanne, 1-497.

[2] Eggimann J.-P., Faist A. (1987) LESOSAI User's Manual, LESO-PB / EPFL, CH-1015 Lausanne.

[3] Faist A. (1987) European methods : Results and experiences, ISES Solar World Congress, Hamburg, 8 W4.03.

[4] Faist A. (1988) Design aids : present and future, Energy and buildings for temperature climates, PLEA Conference, Porto, 1-609.

[5] SIA, Recommandation SIA 380/1 L'énergie dans le bâtiment, Société des Ingénieurs et des Architectes, CH-8039 Zurich.

[6] Gay J.-B., Eggimann J.-P., Faist A. (1985) Comparaison de six systèmes passifs dans des conditions climatiques identiques, 5. Symposium Solaire, 1-107.

[7] Faist A., Niclass A. (1989) The sensitivity of solar passive buildings to construction parameters : a parametric study, ISES Congress, Kobe, 1-353.

[8] Claux P. et al. (1982 et 1985) Méthode 5000, PYC Edition, Paris.

[9] Dubrul C. (1988) Inhabitant behaviour with respect to ventilation - A summary report of IEA ECB Annex VIII - AIVC Technical Note 23, Bracknell, UK.

HYPERCOEN

ir. Anke van Hal and ir. Kees Duijvestein
Delft University of Technology;
Faculty of Architecture
Kabinet 11.04
Berlageweg 1
2628 CR Delft

HyperCoen: A **Hyper**card-program that gives information to architects about living **co**mfort and **en**ergy saving.

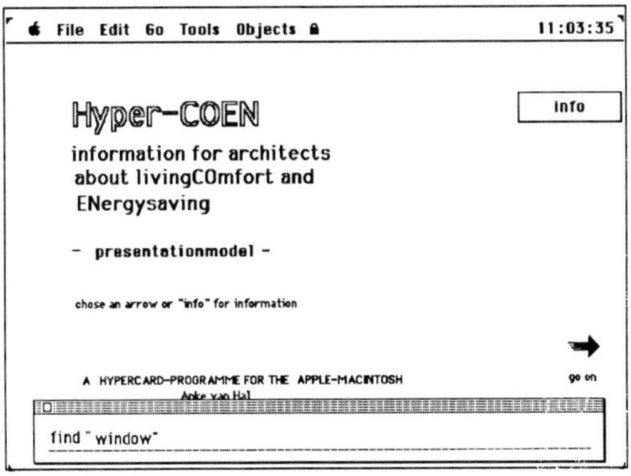

Architecture and environmental care are no longer strangers to each other. Simply said building with environmental care comes to building with the ambition to limit the damage for the environment. Energysaving forms one of the most important gaugers of this form of architecture. However, the architects knowledge on energysaving is mostly rather small. HyperCoen strives to limit this lack of knowledge in the field of house-building. Experience has shown that the success of measures to save energy is, for a great deal, dependent on the behaviour of the occupants. That's the reason why HyperCoen also gives information about what living-comfort contains. The measures to save energy that also improve the living comfort are the most successful ones. To make the big amount of information as accessible as possible, a computersystem has been chosen as information-source. The fact is that research has shown that most architects not often consult a book.

S. Alvarez et al. (eds.), Architecture and Urban Space, 447–450.

HyperCoen wants to make architects aware of the consequences of their design-decisions on the energy-use of their building. Maybe the program can, in this way, make a contribution to the realisation of a low-energy and comfortable living-climate.

- HyperCoen is made with the program HyperCard, version 1.2. of Apple Macintosh.
- It takes 464 K memory.
- In april 1989 HyperCoen formed the final project of Anke van Hal for the Delft University of Technology, faculty of Architecture. In this second version the software has been improved by Jelle Andela and Jaap Schokkenbroek.
- On this moment HyperCoen is in Dutch; we consider a translation to English.

Some questions

What does HyperCoen?

- HyperCoen brings the design consequences to the architects notice during the design proces.
- HyperCoen can be used as a checklist when a design is ready.
- HyperCoen gives answers to architects' questions.
- HyperCoen calculates.

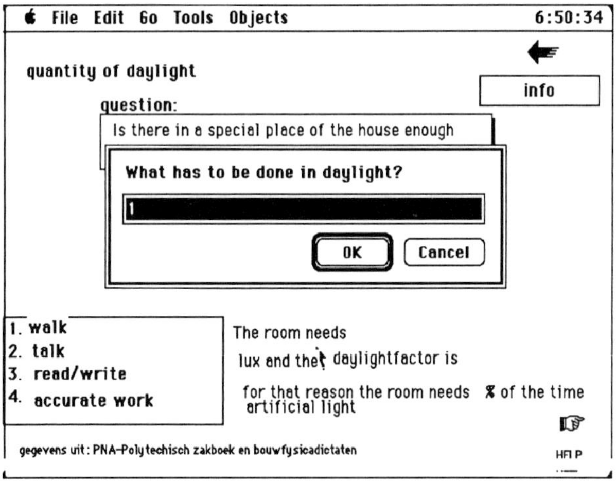

Why a computerprogram?

Giving architects information about energysaving and living-comfort formed the starting point of the development of HyperCoen.
However, research from the University of Eindhoven has shown that this

isn't as simply as it sounds. The research shows that it is very important for architects to look up things easy and that there has to be a lot of drawings. The researchers recommended the use of a computer.

Why the program HyperCard?
Experts Systems, an application of artificial intelligence, lend theirselves very well to this kind of information-transfer. However, to prevent the need of an external program, we choose the program HyperCard of Apple Macintosh. In this program all data can be stored and related to eachother.

The structure of the program
HyperCoen is made with the Macintosh-program HyperCard. This program has to be regarded as a big card-tray in which the cards are entered into relation with each other.
HyperCoen forms one stack of 386 cards. Also cards from other stacks can be related. By this possibility HyperCoen can extend in the future.

Buttons form the links between the cards. By clicking them you start a program that does something. It brings you to an other card or it makes a calculation. The programs are written in the HyperCard-language "HyperTalk".

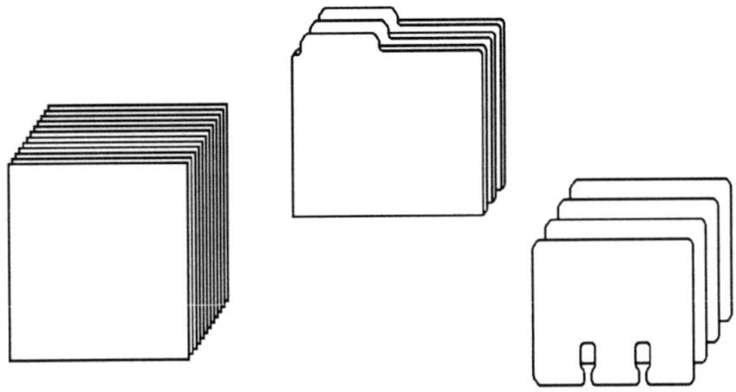

The filling up of the structure

The red line
To make the program as accessible as possible for architects a red line has been chosen that approaches the reality the most. Of course, it isn't possible to create a structure that always fits. There will always be a forced classification. Despite this, we expect the architect to use this equipment succesfully.

The program is split up in several chapters:
1. housing-surroundings
2. linking/opening up
3. groundplan
4. installations
5. materials

Each chapter has the following grouping:

- energysaving
- passive solar-energy
- living-comfort

This grouping has been chosen because of the accessibility of this program. Architects aren't always interested in energysaving. Many designers of houses however, are interested in living-comfort. By this grouping the architect can make a choice in subject. And we hope of course that the program makes the architect also curious about energysaving when he's looking for information about living-comfort.

-passive solar-energy-
Passive solar energy has a lot to do with energysaving. Because of the dependence of the orientation however (mostly South) this subject is separated from the others. Using solar-energy isn't always possible.

-living-comfort-
There's a lot of discussion about the meaning of the word living-comfort. It has reference to temperature etcetera but also to words as "fine" and "easy".
A psychologist from the University of Nijmegen, W. Heijst, did research on this subject. The subject "living-comfort" of HyperCoen is based on his description.

Anke van Hal
is civil servant of the municipality of Delft, the division of environment and energy. She is specialised in energysaving and building with environmental care.

Kees Duijvestein
works in the SOMgroup (Urban Design & Environment) Faculty of Architecture, Delft University of Technology, and in BOOM, (Office for Environmental Research & Design) Delft, Holland

SETIS : AN INTELLIGENT SYSTEM FOR THERMAL DESIGN OF BUILDINGS

C. ROBIN (1)(2) J. BRAU (1) P. DEPECKER (1)

(1) Centre de Thermique de l'I.N.S.A. de Lyon
URA CNRS 1372
20 avenue Albert Einstein BP 168
69621 VILLEURBANNE Cédex - FRANCE
Tél. : (33) 72.43.83.19 FAX. : 72.44.08.00

(2) O.C.G.R. Ingénierie
129, chemin Moulin Carron,
69132 ECULLY Cédex - FRANCE
Tél : (33) 72.20.62.06 FAX. : 78.47.53.13

ABSTRACT. SETIS is aiming at developing a computer support for thermal design of buildings. It deals both with the envelop of the building and with its HVAC system. This project is managed by OCGR, a french civil engineering enterprise, and a laboratory of the Institut National des Sciences Appliquées de LYON. Such a collaboration is very important because SETIS is defined as a tool that integrates both algorithmic programs like a building simulation unit, and knowledge-based parts. They both work on the same database where a thermal description of the building to design is stored. The knowledge bases make data input easier : SETIS is thus available from the initial stage of design. They select HVAC principles too. Then, each air-conditioning system can be evaluated on ten thermal, technical or economical criteria. The following stages of this project are double : install SETIS in OCGR and INSA, and build a *supervisor*, that is a knowledge base which suggests modifications or actions to the designer.

1. A Short Analysis of Thermal Design

A short survey of softwares which the engineer in charge of thermal science can use, has two main drawbacks.

Firstly, the softwares already available in the market often split the thermal design in the study of the outside structure (walls, roof, insulation, glazing...) and the study of the HVAC systems [Deringer 89].

Secondly, most of these softwares are devoted to a calculating and quantitative study (energy analysis, load calculation, pipe analysis, HVAC plants sizing...). Besides the calculating study is very restrictive, and limits the scope of the design process.

Thus the purpose of SETIS is to eliminate these two points. The aim of this project is to develop a complete computer system in Thermal Design, dealing both with the building envelope and HVAC system. SETIS hopes to model the design process in integrating standard techniques of computer science and recent tools of Artificial Intelligence (Expert Systems). This modelling takes into account the main features of the design process.

Because it is *one-of-the-kind* object, each building requires a specific design. To examine the particulars of design, answering the question "What does an expert do when he designs ?", leads us to notice that common computer technique, using procedures and algorithms, is not suitable to solve an ill-structured problem [Simon 74]. Furthermore, the incompleteness of the design activity [Ayrle 91] emphasizes this point.

S. Alvarez et al. (eds.), Architecture and Urban Space, 451–456.

2. Aims of SETIS

As a designer must think up the whole building, it is necessary to guide him and to evaluate as early as possible the effects of choices about architecture and structure, on the thermal quality of the future building.

The complexity and imaginative side of design compels us to focus our attention on the thermal point of view and to define the well-suited tool not as a designer system, but as a *system for building thermal design support*.

The assistance brought to the designer is consistent only in the case where the computer system can run in the initial stages of the design process. Consequently, the system must be able to work with few data. This is the main feature of design : to generate information with few inputs. On the other side, the main outcome of our computer system is a set of HVAC systems.

3. Layout of SETIS

To achieve the aims we have just set, we are going to try to solve each problems related to our model.

Computer machine haven't got any capacity for imagination ; nevertheless, thermal building design activities, like selecting building material [Bourdeau 87] or choosing HVAC systems to ensure a given level for indoor comfort [Doheny 86], need knowledge we can not express with algorithms. Our computer system must be able to deal with formalized symbolic knowledge which describe both qualitative (comfort...) and quantitative dimensions (thickness of window system...) of the cognitive space. Added to the stage where the system must be consistent, we recognize the favoured area of expert systems, one of high-performance developments of Artificial Intelligence science.

To implement an expert system is one part of the solution : we also have to deal with the numerical side of design. Thus, our system has calcuting tools built on standard algorithms.

When an expert system is reasoning or a program is recknoning, we work on only one object, on the description of the object. The database (see 4.1) becomes the main part of the system. So the layout of SETIS looks like a blackboard architecture.

Furthermore, the efficiency of a computer system is linked to a good user front-end. This part of a system is very important in matter of cognitive ergonomics and C.A.D. [Begg 84].

The five main parts of our system called SETIS are shown on Figure 1 which describes its structure.

The integration of calculating tools and of an expert system which collaborate and exchange data describing only one object, and which work behind only one user interface, makes SETIS a DSS (Decision Support System), aimed to be integrated in the line of thought of the designer.

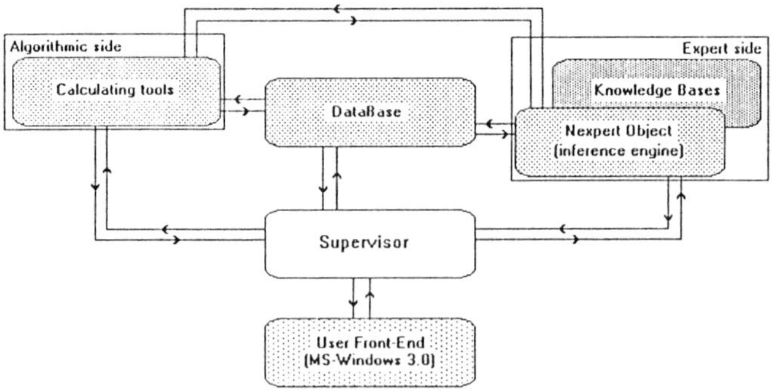

Figure 1 : SETIS Structure.

4. Main Components

This section deals with the four parts now operational. The fifth one, the supervisor, will be discussed in the chapter devoted to the future improvements of SETIS.

4.1. DATABASE

The word database means the set of files where the description of the building is stored. Open to the different kinds of tools, this set is made of dynamic files and static files.

4.1.1. *Dynamic Files.* These files hold the description of a building, that is to say, all the information needed to the design of the walls and the HVAC system of the building, all thermal design variables. That information is dynamic because its number and its value change during the design process. They are modified either by the designer or by the expert system or the calculating tools. Open both to the user and to the system, dynamic files have got a fixed ASCII format.

The description of the building relates to the conceptual and calculating aspects [Björk 90] of the thermal point of view. It also reduces the importance of the topological side of standard description of the building : SETIS looks at a building as a set of "blocks", of which the position in the entire volume of the building may be unknown. Consequently, a block is a concept, an artefact, intended to solve as easily as possible the problems of the designer : it takes the physical form of a room, or a group of rooms or the whole building. To do so, a block is defined by its faces in contact with the outside climate or other blocks. All the adjoining blocks are not necessary : for lack of datum, the adjoining blocks are neutral. So, the designer is not imposed to define all the blocks so as to deal with his problem.

As SETIS only selects its layout, the only information stored about HVAC system is :
- functions it will ensure (heating, ventilating...),
- list of the potential principles which ensure these functions,
- multi-criteria evaluation for the HVAC system choosen by the designer in the previous list.

4.1.2. *Static Files.* This kind of file holds information we can consider time independent. We call DATA, normative knowledge laid down in the regulations, LIBRARIES, traditional description of a block regarding its purpose of use, and DIAGRAMS, set of graphical representation of HVAC principles.

4.2. KNOWLEDGE BASES

Bases Before setting out the knowledge bases the user can trigger in SETIS, the characteristic features of the expert componant of the system are described.

4.2.1. *Expert system technique.* Unable to imagine or to create or to reproduce human mechanisms accurately, expert systems allow us only to model some reasoning of an expert, that is to say, to get the same result as the expert.

Building a prototype of the expert component of SETIS, enabled us to choose an expert system. Nexpert Object of Neuron Data has been selected. The reasons for our choice are the following :
- inference engine based on predicate logic.
- two formalisms for knowledge representation, rules and objects, that make the modelling of the cognitive space very efficient.
- integration with standard relational data bases and spreadsheets.
- SQL-like queries to all kind of database files
- external subroutines written in C-language can be called (you can call execute library that implements list manipulation or multi-values..., or your own functions).
- knowledge bases are supported on many computers, in our case on a MS-DOS micro-computer.

Knowledge acquisition required to build the bases, progresses in an iterative way using interview techniques. From the account given by the experts of OCGR, a french civil engineering office, concepts are found by abstraction, then these concepts make up a model that experts check and test. At the same time, theorical knowledge has been collected by literature review. After the structuration step where knowledge are modelled with the object and rule formalisms provided by Nexpert Object, knowledge bases are written.

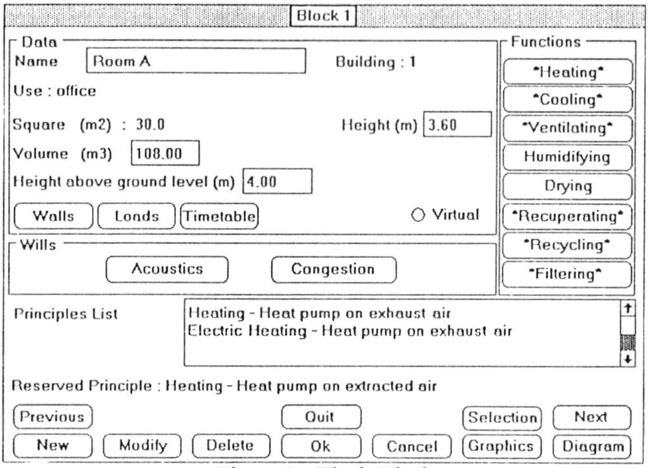

Figure 2 : Block Display.

4.2.2. *Description.* Right now, SETIS integrates three knowledge bases. The first two support the generation of objects fitting the constraints relating to the first stages of the design process. At this step, the designer knows only few data, but wants to trigger calculating tools to enable him to think over his problem, to test solutions he can imagine. Therefore, each knowledge base helps him to create a building and a block. To do that, they ask the user a limited number of data already known in the early stage of design, and generate in an expert way (using both LIBRARIES and heuritic knowledge) the lacking information. For example, to create a building requires only three data (country, department and altitude of the construction), a block only four data (purpose of use, basic geometrical sizes) (see Figure 2). This saves the user from a boring data input, and ensures the right number of data to use the calculating tools of SETIS.

The third knowledge base selects the principles of HVAC system ; a principle is defined as a layout of functional units. At this step, HVAC system design is supposed to be divided into three weakly connected parts : the choice of principle, the choice of the plants ensuring the functions of the layout, and the choice of the controls of the plants [Camejo 89]. The first part, the only operational one at present, is linked to the functional models of a set of 68 principles (see Figure 3). The change from principle to plants will be made in substituting plants for functional unit (it is not necessarily a bijective operation : a plant can ensure more than one function). This step is linked to a structural model, while the choice of controls is related to a behavioral point of view. Using multi-modelling knowledge bases with various aggregation levels is a key to cope with the problem of technical system design.

4.3. CALCULATING TOOLS

SETIS puts a "thermal toolbox" made up of ten tools with different complexity levels, at the disposal of the user. Among the simple ones are unit calculating U-value and capacities of the walls, evaluating the air flow due to infiltration, or generating the meteorological files. We are just going to describe the multicriterion rating unit. Among the complicated ones is the tool for the simulation of thermal and hygrometric behavior of a block.

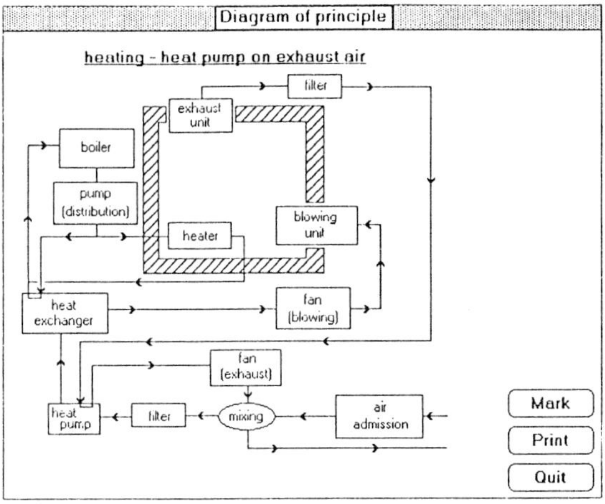

Figure 3 : Diagram Principle.

4.3.1. *Thermal and hygrometric simulation.* This simulation model is based on the enthalpy balance of a block, divided into its dry and moisture part. This last one is quite simple, so we focus our attention on the thermal model only. This model is an adaptation from the Codyba code implemented by the Laboratoire Equipement de l'Habitat [Brau 87]. In SETIS, this code is not aiming at exact recknons, but at valuing as fast as possible the effects on indoor comfort of choices about architecture, structure of the building or functions of the HVAC system. Our model is built by analogy with electric phenomena and used a finite-difference method to solve a constant set of 12 linear equations. Furthermore, it splits the radiative and convective parts of heat transfer. Thus, this simple model is a right solution in the length and the accuracy of the calculating process, and suits the assumptions made for our design problem.

4.3.2. *Multi-criteria rating system.* This unit allows the user to compare the HVAC principles contained within the list of solutions choosen by the third knowledge base described before (see 4.2.2).
This unit tries to deal with the concept of the quality of a HVAC system : it gives a mark from 0 to 10 for criterion characteristic of the principles. These qualitative marks are given for ten criteria, making a right compromise for detailed information required by project managers, and the first stage of design : thermal comfort, hygrometric comfort, air quality, acoustics, congestion in the block and in the boiler room, technical nature (easiness to install), investment cost, maintenance cost and energy cost. These criteria are not only related to thermal science, but to other viewpoints like the economic one. These marks are built by comparing each principles with the others : in fact, a mark is relative to the position given for the principle by the thermal engineers of OCGR. With this tool, the designer can calculate an average value weighed in accordance with the main goals he wants to achieves. To give a mark is interesting because of its simplicity to compare the different solutions proposed by SETIS. Furthermore, some of these marks can be detailed using a particular tool : for example, the user can evaluate comfort with Predicted Mean Vote, to detail marks given for thermal and hygrometric comfort.

4.4. FRONT-END

Like previous tools, the front-end is implemented with C-language and uses graphical means of MS-Windows 3.0. The user interface wants to be ergonomical and user-friendly as shown in the following section.

5. Future Improvements

To enlarge its efficiency and improve the support it brings to thermal designers, we are working on the supervisor of SETIS. This unit is expected to provide the system with an intelligent behavior.

The first requirement is to memorize all events occured in the design process : the journal already achieves this goal. It automatically records the work session.

The second but not the least requirement is to help the user :

- to make his choice (modifications proposal), to answer the question "Which design variable value must be modified to get the choosen level for the performance variable ?"
- to keep the building description consistent (actions proposal), to answer the question "What reckoning or reasoning must be done after each modification ?".

To bring the user this kind of support, a knowledge base using thermal meta-knowledge will be build in the next months [Robin 91].

In our long-term plans, SETIS will be connected with standard C.A.D. softwares, in order to achieve its integration in the range of the usual tools of thermal designers.

6. Conclusions

This paper describes the thermal design support system called SETIS available today. It is not only a prototype, neither is it entirely operational, because its knowledge bases are not completely validated yet. Nevertheless, SETIS is being progressively used by the thermal engineers of OCGR. This first version will be developed so as to improve its actual limits relating to the countries and the purposes of use of block for example, and to comply with the needs of its users. The implementation of the stages of equipment and controls selection will be delayed ; our work is now focused on the development of the supervisor. The SETIS project reveals that the integration of calculting tools and knowledge - based parts is very important for a design support system. Consequently, the development of such a system requires theorical and heuristic knowledge. Within this context, the collaboration between a laboratory and a civil engineering office is determining.

References

[Ayrle 91] **Ayrle**, H. Intelligent tools for intelligent buildings. *Proceedings of MICAD 91*. Paris : Hermès, 1991. p 481-492.

[Begg 84] **Begg**, V. *Systèmes experts en CAO*. Paris : Hermès, 1984.

[Björk 90] **Björk**, B.C. Issues in the development of a building product model standard. *Building Research and Practice*, 1990, N°1.

[Bourdeau 87] **Bourdeau**, M. Systèmes experts : des logiciels aident à les bâtir. *CSTB Magazine*, 1987 (avril), p 37-40.

[Brau 87] **Brau**, J.; Roux, J.J.; Depecker, P.; Croizer, J.; Gaignou, A.; Finas, R. Micro-informatique et comportement thermique des bâtiments en régime dynamique. *Génie Climatique*, 1987 (octobre - novembre), N° 11, p 15-23.

[Camejo 89] **Camejo**, P.J. An expert system for the design of heating ventilating and air-conditioning systems. *ASHRAE Transactions*, 1989, Vol 95, part 1, p 379-386.

[Deringer 89] **Deringer**, J.J.; Hall J.D. Computer software invades the HVAC market. *ASHRAE Journal*, 1989 (july), Vol 31, N°7, p 32 - 44.

[Doheny 86] **Doheny**, J.G. *Preliminary development of an expert system for the conceptual design of energy system*. Forthcoming masters thesis : University College Galway (Ireland), 1986.

[Robin 91] **Robin**, C. *SETIS an intelligent system for thermal design of buildings*. draft paper. Energy and Buildings, 1991.

[Simon 74] **Simon**, H.A. *La science des systèmes, science de l'artificiel*. Paris : Ed. EPI, 1974.

AUTOMATIC DETERMINATION OF SOLAR ACCESS AND SHADING OF BUILDINGS IN COMPLEX SITUATIONS. GRAPHIC AND ANALYTIC APPROACHES.

L. BONANNO, M. PIETRAFESA and G. RIZZO

Institute of Civil Engineering and Energetics
Engineering Faculty
University of Reggio Calabria
89100 Reggio Calabria, Italy.

A. NUCARA

Civil engineer
89100 Reggio Calabria, Italy.

ABSTRACT. We present through the paper a computer code able to define the graphic spatial and temporal distribution of the shadows on a given surface. First aim of the analysis is to produce information concerning the solar access of a site, even in complex geometric situations.
The program also generates a matrix containing a numeric representation of the shadows conditions of the building or of the examined space.
The program can be easily linked with a computer code performing the energy analysis of the building, while the graphical results are stored in proper files that enables the utilization of CAD packadges.
An example application of the program to a complex urban situation is also included in the paper.

1. Introduction

The current wide diffusion in architecture of the bioclimatic concepts, makes the utilization of the solar energy in climatization of buildings a commonly applied strategy. Moreover, the increasing density of the urban areas rises the problem of the solar access and of the overshadowing of the living spaces. Finally, as the building design becomes more sophisticated, the limits between open and confined spaces appear vanishing in the urban context.
Solar radiation plays an important role in the thermal balance of such a kind of buildings and spaces, strongly affecting the thermal comfort (or discomfort) sensations of people living inside them. Thus, it is of paramount relevance the knowledge of the area of the external surfaces of the envelope affected by shadows, due to overhangs and fin openings or brought by the obstructions present in the optical skyline.
This obviously requires more powerful tools when analyzing urban spaces. Infact, other than a graphic representation, designers need numerical results to be utilyzed in the thermal-energy balance of the examined environment.
Through the next sections the analytic description of the most frequent shadowing situations will be investigated and a grahic application will be finally reported for a typical urban context.

S. Alvarez et al. (eds.), Architecture and Urban Space, 457–462.

2. Shadows from horizontal spots

Figure 1 presents a graphic determination of the shadows produced on the building's façade by an horizontal spot, as usually occurs by effect of overhangs.
In this specific case a flat device is shading a single rectangular glazed opening. The shaded part of the façade is represented by the position of the points A", B", C", D", E" and F". Angles α_o (projected direction of the sun beam on the horizontal plane) and α_v (projected direction of the sun beam on the vertical plane) and the characteristics dimensions of the spot, allow to analitically determine the coordinates of the above cited points. The relevant geometric dimensions of the overhang are s_m (height), p_m (depth) and L_m (width).
With reference to the figure, it's then possible to write:

$$X1 = \frac{p_m}{\tan\alpha_o} \qquad\qquad Y1 = X1 \cdot \tan\alpha_v$$

Coordinates of the chacteristic points will be:

$$x_{B''} = x_{A''} + X1 \qquad y_{B''} = y_{A''} - Y1 \qquad x_{C''} = x_{B''} + L_m \qquad y_{C''} = y_{B''}$$

$$x_{D''} = x_{C''} \qquad y_{D''} = y_{C''} + s_m \qquad x_{E''} = x_{A''} + L_m \qquad y_{E''} = y_{A''} + s_m$$

$$x_{F''} = x_{A''} + L_m \qquad y_{F''} = y_{A''}$$

By connecting consecutively these points, the shadow of the overhang will be determined.
The computer code executes the above illustraded operations on the time-step selected by the designer, typically on a hourly base. It stores both the graphic representation and the numeric results. Those are represented by a matrix constituted of a gride having as elementary dimensions the linear extensions of the subsurfaces in which the façade has been devided.

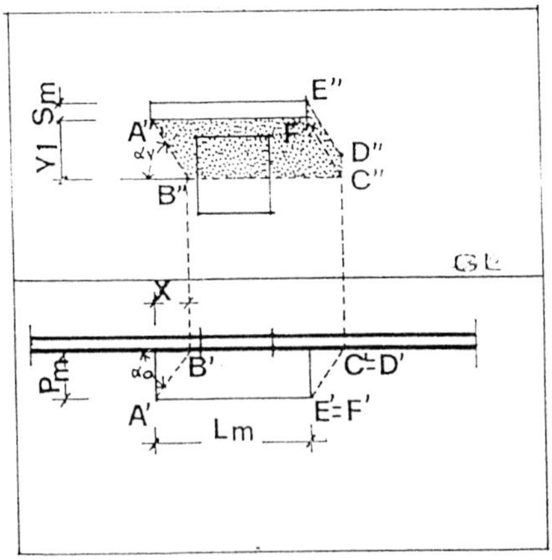

Figure 1

3. Shadows from surrounding buildings

Figure 2 reports the graphic determination of the shadows brought on the examined building's façade.
This case also includes the presence of vertical fins, here assumed as adjoining buidings with a common edge with the building under study.
The shaded part of the façade is represented by the position of the points A_o, A, B, C, C_o. Angles α_o and α_v are defined as in the previous section.
The geometric characteristics of the system are here to be considered **de** (distance from the buildings) and **he2** (vertical coordinate of the point A").
Referring to figure 2, it's possible to analytically determine the coordinates of the point A:

$$X_A = \frac{de}{\tan\alpha_o} + x_{A'} \qquad\qquad Y_A = he2 - \frac{de}{\tan\alpha_o} \cdot \tan\alpha_v$$

A similar procedure can be usefully adopted for the singling out of the remaining characteristic points.
By connecting consecutively the point A_o, A, B, C and C_o the shadow produced by the building 2 on the façade of the building 1 will be determined.
As reported in section 2, the building code is then able to store both the graphic representation and the numeric results also in this case.

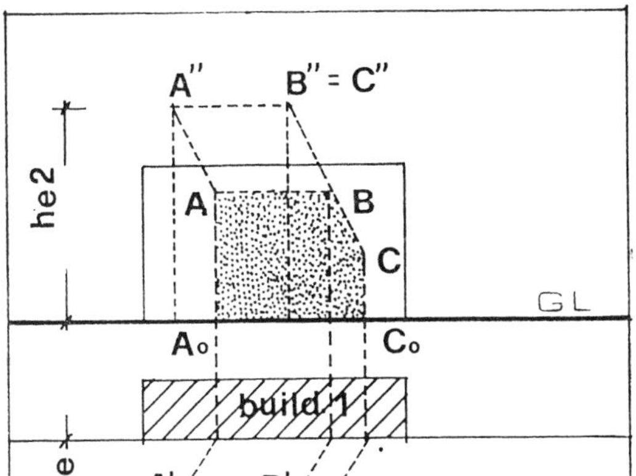

Figure 2

4. Application to an urban context

In order to demonstrate the potentiality of the computer code (written in BASIC language), we have developed an exercise repying some typical interactions among buildings, in a ficticious urban context. Absolute dimensions of buildings and relative distances are here not important and will be omitted for the sake of simplicity.

The selected dislocation of the buildings is that illustrated in figure 3, where the examined volume is the only one shaded with a different grid. Orientation of the façades is also indicated. The urban situation has been supposed to operate, with respect to the sun path, in the geoastronomic conditions of Messina, belonging to the southern part of Italy, in a mediterranean mild climate. For each of the main expositions the percentage of the shaded façade has been calculated, as a monthly mean.

Figure 4 reports a set of graphic results, in terms of shadow percentage over the total examined surface. Percentages are calculated on hourly base for the monthly mean day of some relevant months.

Along with the figures, obviously, the computer program produces a group of tables containing the same kind of data. These tables can be easily used by the common codes that analyze the thermal behaviour of buildings: as it is well known, the hourly percentage of shaded surfaces allows a better evaluation of the energy balance on the external part of the buildind envelope.

The graphic results, on the other hand, can be usefully employed to draw indications about the solar path of the site, in the aim of recognizing the limit situations for the solar access.

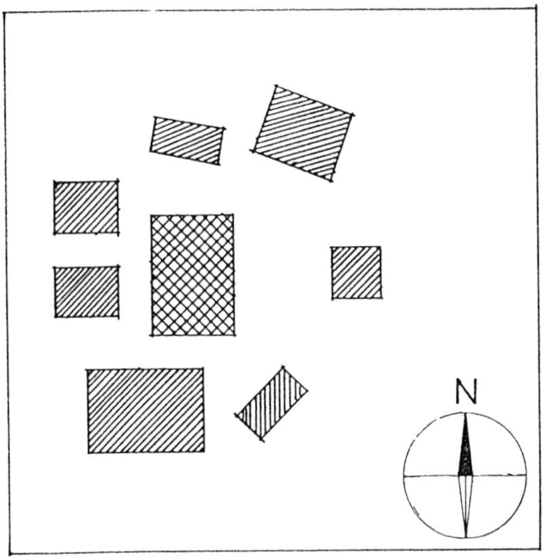

Figure 3

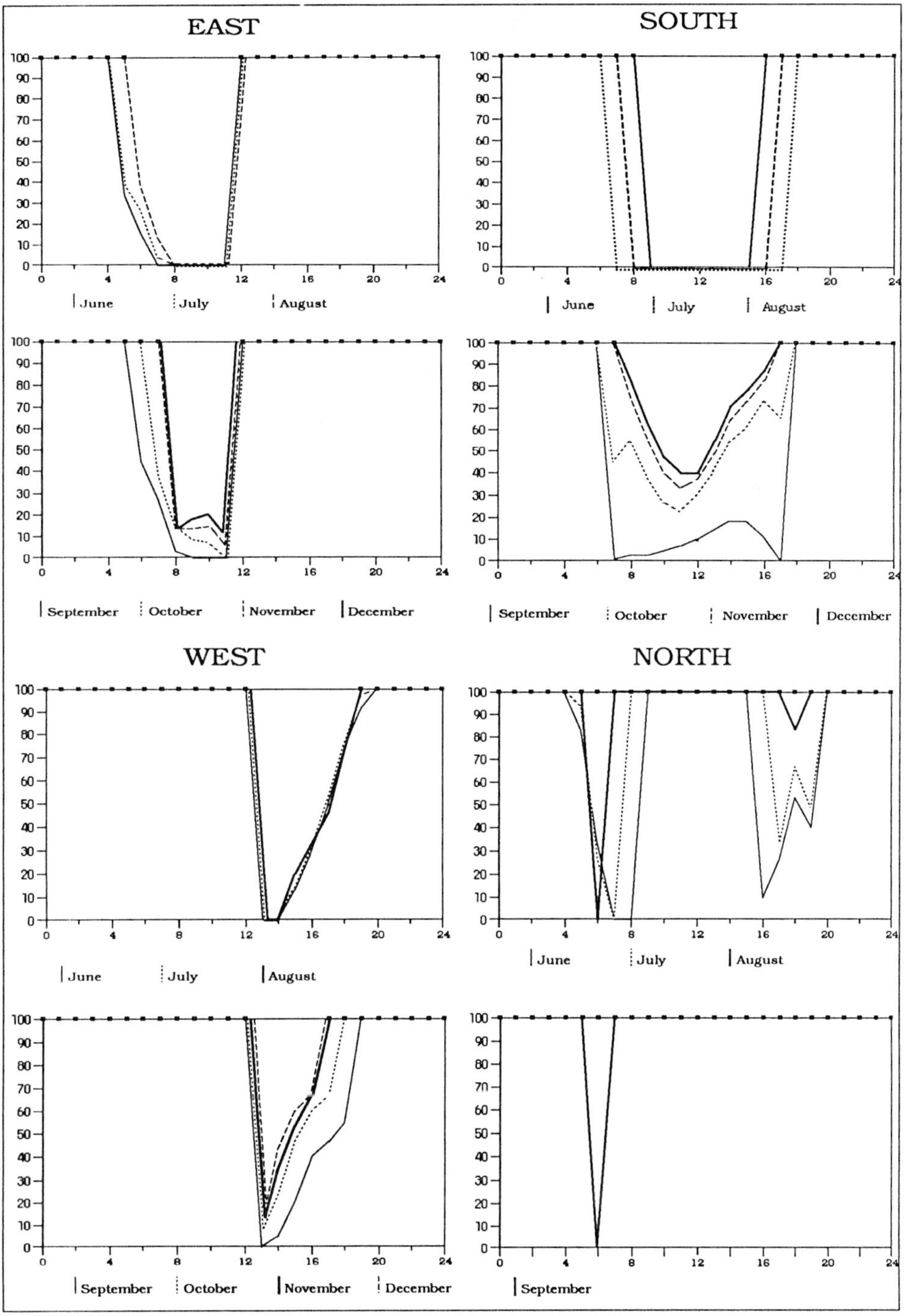
EAST
SOUTH
WEST
NORTH
June
July
August
September
October
November
December

Hr.	June	July	August	Sept.	Oct.	Nov.	Dec.	Yerly Mean
07	100.0	100.0	0.0	0.0	45.0	100.0	100.0	57.5
08	100.0	0.0	0.0	1.7	54.4	74.3	83.3	37.0
09	0.0	0.0	0.0	1.7	37.2	53.7	63.5	20.7
10	0.0	0.0	0.0	3.9	26.3	40.7	47.6	15.8
11	0.0	0.0	0.0	7.2	23.7	34.6	40.0	14.3
12	0.0	0.0	0.0	10.0	30.0	36.7	40.0	16.1
13	0.0	0.0	0.0	13.1	40.0	49.4	55.0	21.7
14	0.0	0.0	0.0	16.7	53.7	64.8	70.0	28.4
15	0.0	0.0	0.0	16.7	61.1	72.2	77.8	31.5
16	100.0	0.0	0.0	11.1	72.2	83.3	88.9	43.5
17	100.0	100.0	0.0	0.0	66.1	100.0	100.0	61.0

TABLE 1. Mean monthly and mean yearly percentage values of the hourly shaded part of the south-facing surface of the example building through selected months.

5. Conclusions

The main features of a computer code, written in BASIC language, able to determine the graphic and analytic representation of the shadows produced on a building's façade, have been presented. The program is essentially concerned with the analyses of solar access and solar path of complex urban situations. It can investigate the shading effects produced on a given surface (showing any orientation and inclination) by surrounding buildings and/or by overhangs and fins belonging to the building structure.

The graphic representation, available on the selected time-step (generally, hourly data computed for the mean monthly day), can be usefully employed within an usual CAD program, in order to study the effects of geometric modifications imposed to the envelope of the examined building, in terms of solar availability. Typical "sun-right" analyses could be performed with this simple tool. The analytic results are constituded by the (mean) percentege of shaded area of the examined surface of the envelope. These matrices can be easily coupled with the computer programs that compute the thermal analysis of the buildings, in order to reach a better energy balance, particularly referring to the contribute of the solar radiation onto the external surfaces of the buildings.

A DESIGN TOOL FOR DIRECT SOLAR GAIN EVALUATION

M. A. Herrero and M. Rodríguez
Depto. de Fisica Aplicada ETSI Telecomunicación
Ciudad Universitaria. 28040 Madrid. Spain

ABSTRACT. To design a passive building a careful assessment of the direct solar gain, so that, the amount of solar energy collected by transparent surfaces is required. With this aim a computer package has been developed by means of which is able to perform an hourly computation of the solar beam radiation coming into the building through oriented windows. The program also obtains the solar energy absorbed on the interior walls, floor and ceiling. The procedure followed is based on solar geometric relations and takes into account the transmission coefficient of the glass and the absortance of the walls.

1. INTRODUCTION

A careful assessment of the amount of solar energy gained during the year begins with the computation of the energy transmitted through transparent surfaces and the portion absorbed by floor, ceiling and walls.

The control of the solar radiation coming into the building implies the correct sizing of windows and overhangs. To this aim a number of different graphical procedure exist (1) (2) which provide the duration of sunshine calculation taking into account the solar obstruction by surrounding buildings. In this paper we provide computer tool to calculate the amount of solar energy transmitted through the windows and absorbed by each interior surface of a rectangular room after being once reflected on the other surfaces.

There are numerous of computer programs for the analysis of thermal behaviour and energy consumption calculation, most of them estimate the solar energy distribution over the walls by an approaching method, computing the absorbed energy multiplying the coming solar energy by a constant factor chosen by the user. The software here presented is able to obtain by hourly outputs of the amount of solar energy transmitted through the glass, reflected on the interior walls and absorbed on them.

S. Alvarez et al. (eds.), Architecture and Urban Space, 463–467.

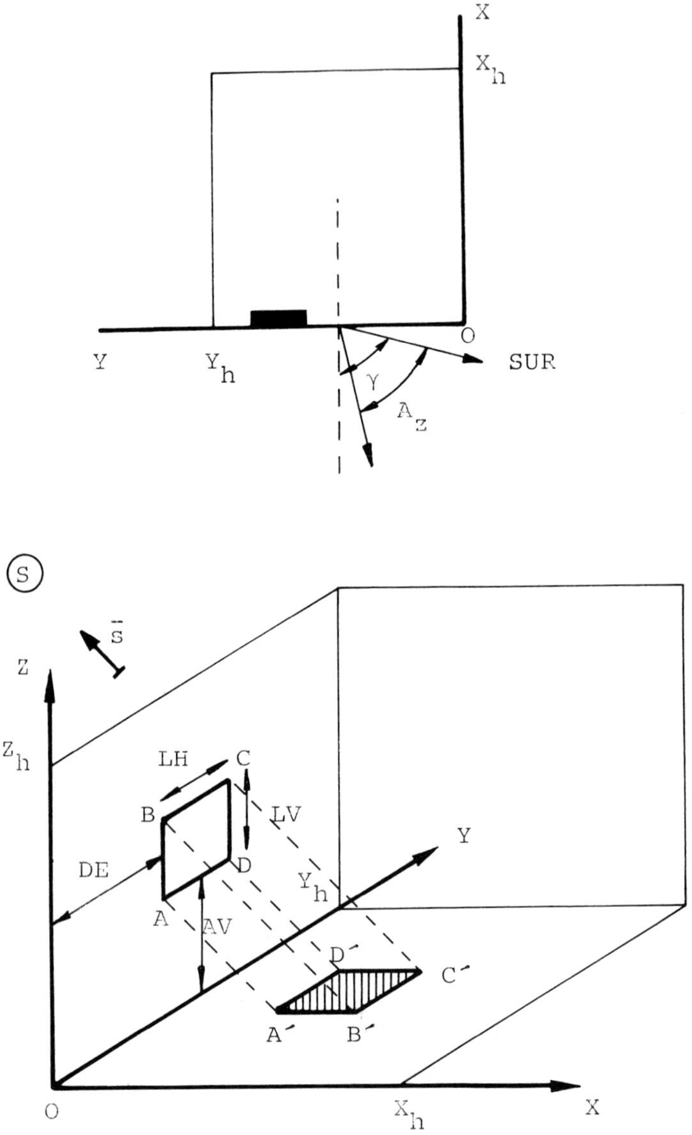

Figure 1. The impacted area

2. THEORETICAL BASES

2.1. Impacted Area

Once it has been defined the size, position of the window on the wall and its orientation, the impacted area on the floor and walls by the sunshine is obtained by geometrical relations developed defining the projected window area on the surfaces as the Figure 1 shows. The direction of projection is given by the sun's rays represented by the unit vector $\bar{s}$. This vector can be expressed in terms of basic angles (3) as follows:

$$\bar{s} = (\cos\delta \sin\Phi \cos\Omega - \sin\delta \cos\Phi,$$

$$\sin\Omega \cos\delta,$$

$$\sin\Phi \sin\delta + \cos\Phi \cos\delta \cos\Omega)$$

where Ω denotes the hour angle, Φ is the site latitude and δ the solar declination. Taking into account the different surfaces in which the projected area can be obtained, eighteen cases have been considered.

2.2. Solar Energy Transmitted

From the hourly beam solar energy on horizontal surfaces and following the method described on (4), the energy by unit area arriving to the vertical plane of the window has been computed and multiplied by the transmission coefficient to obtain the energy coming into the room.

$$Ei = Eh\ Rb\ Ft$$

Where, Ei is the energy per unit area transmitted through the glass, Eh is the energy per area on horizontal plane, Rb is the ratio of beam radiation on the tilted surface to beam radiation on the horizontal surface, Ft is the transmission coefficient of the glass.

2.3. Solar Energy Absorbed

The calculation of the amount of solar energy absorbed on each wall implies the addition of the energy received directly from the window and that reflected from the rest of the surfaces due to short wave radiation. In order to perform this computation it must be used the Angle Factor (5) Fij. This factor must be applied between horizontal surfaces and between perpendicular surfaces, like two adjacent walls. The mathematical relation of this factor includes geometrical parameters based on dimensions and relative positions of the impacted area and the planes.

The energy reflected on the surface j toward the surface i can be expressed:

$$Er)j\text{-}i = Ei)j\ (1 - \alpha j)\ Fji$$

where, αj is the absortance coefficient of the surface j.

The energy absorbed every hour on the surface i is the amount of energy coming from window and the total energy added from the rest of the surfaces by reflection. The energy absorbed is expressed in the following equation,

$$Ea)i = (\ Ei + \sum_{j} Er)j\text{-}i\ )i\ \alpha i$$

The above procedure is applied to all windows adding the energy every hour. Daily and monthly values have been computed.

SURFACE	DAILY ENERGY (KJ)	
	incident	absorbed
FLOOR	20547.75 (45%)	8219.10 (40%)
WALL 0	7106.53 (15%)	5187.77 (73%)
WALL 1	3603.52 (7%)	2630.57 (73%)
WALL 2	4796.23 (10%)	3501.25 (73%)
WALL 3	4975.65 (10%)	3632.23 (73%)
CEILING	4302.13 (9%)	3957.96 (92%)

FLOOR	HOURLY ENERGY (KJ)	
TIME INTE.	incident	absorbed
4-5	0.00 (0%)	0.00
5-6	0.00 (0%)	0.00
6-7	0.00 (0%)	0.00
7-8	180.33 (0%)	72.13
8-9	1096.44 (5%)	438.57
9-10	3257.17 (15%)	1302.87
10-11	3121.41 (15%)	1248.56
11-12	2481.56 (12%)	992.62
12-13	4066.17 (19%)	1626.47
13-14	3012.31 (14%)	1204.92
14-15	2365.47 (11%)	946.19
15-16	688.01 (3%)	275.20
16-17	278.01 (1%)	111.56
17-18	0.00 (0%)	0.00
18-19	0.00 (0%)	0.00
19-20	0.00 (0%)	0.00

Figure 2. Example of the computer program output.

3. THE PROGRAM

The input data are the latitude, the time period of the energy computation, size, position and orientations of the windows and type of glass. The program allows to store and introduce other different types of glasses. The language used has been QUICKBASIC and requires 218 Kbytes to be executed.

An example of the output is shown if Figure 2 and provides the daily energy absorbed by each surface and hourly values by the floor. It has been supposed two windows in different walls with the orientations, $\gamma = -45.46°$, and $\gamma = 44.54°$. The dimensions and situation of the window on the former wall (see, Figure 1) are: LH = 1.76 m., LV = 1.45 m., DE = 2.97 m., AV = 1.00 m. The type of glass is "Parsol gris" 4 mm. thick. On the latter wall: LH = 2.10 m., LV = 0.78 m., DE = 1.00 m., AV = 1.45 m. The type of glass is "Pink" 5 mm. thick. The site considered is Madrid (40° N) and the date, the 11th of February. The dimension of the room are 3.6 m. high, 5.0 m. long and 5.0 m. wide. The absortance of the walls is 0.73, of the ceiling is 0.92 and the floor 0.43.

4. CONCLUSIONS

The software here presented consist of a computer program for the design of passive solar buildings. The hourly amount of energy transmitted through a window and absorbed by the interior surfaces of the building is computed. The program allows a more realistic and accurate method up to now of calculating than most of the existing programs in use.

5. REFERENCES

(1) Ne'eman, E. and Light, W. Availability of Sunshine. Building Research Establishment. Current Paper CP75/75. August 1975.

(2) Herrero, M.A. Construction and Applications of a Sunshine Diagram. International Journal of Ambient Energy. v. 9, n. 4, 171-176. 1988.

(3) Duffie, J. A. and Beckman, W.A. Solar Engineering of Thermal Processes. Wiley Interscience, New York, 1980.

(4) Alvarez S. Análisis Dinámico del comportamiento térmico. Tesis. Sevilla, 1986.

(5) Procedure for determining Heating and Cooling Loads for computering Energy calculations. ASHRAE. 1976

THE RESPECT OF CONSTRAINTS IN URBAN SITES IN CAD

P. FERNANDEZ, A. CHATELET, M. GERBER
Laboratoire d'Architecture Bioclimatique,
Ecole d'architecture de Toulouse
83, rue A. Maillol, 31100 Toulouse
France

ABSTRACT. Our previous work on the integration of thermal design tools into an overall CAD environment has led us to realise that these tools, ranging from measurement to consulting, should be enhanced by a module to help the designer right from the initial phases of the project. We have participated in the development of a "Constraint Supervisor" model for the sitting of the project in its site, particulary in an urban context. These constraints may be regulations, or may result from the schedule of conditions or simply from the prime contractor's intentions. The fonction of this tool is to enable the designer to explore the different options open to him within a given context and also to subsequently check that none of the constraints fixed at the outset is disobeyed as the project progresses

Introduction

Our previous work on the specifications of a tool to aid design in hot countries (1) has been integrated into a prototype of a CAD system enabling all the participants in a project to be brought together (2).

Our guiding principle was to make it possible to pass from measurements to recommendations. Experience showed us that the additional help this gave the architect was essentially at fairly advanced stages of the project. Indeed, whether we consider the bringing into play of application programs or the use of expert systems, they only intervene when the definition of the project is already fairly elaborate.

But in the overall design process, the architect is confronted with a wider context right from the initial phases, which makes it necessary to take account of constraints which are expressed even more strongly in an urban environment when energy saving is a preoccupation.

The list of constraints to be verified in a construction project is indeed very long and designers often mention how difficult it is to make sure that their project is correct at each stage of its progress.

To the numerous constraints imposed by regulations must be added the program set by the prime contractor, the intentions of the designers (self-imposed constraints), and the context.

The problem lies not only in the complexity of the verifications, many of which can be done "de visu", but also in the risk of forgetting a constraint and realizing too late that it has not been verified. This process becomes all the more complex when we consider that

S. Alvarez et al. (eds.), Architecture and Urban Space, 469–473.

any change made to the project may entail the non-respect of a constraint that had been verified previously. It thus becomes practically indispensable to call upon the computer.

It was this observation that led to the study of a "constraint supervisor". This project was supported by the AFME, the "Direction de la Construction" and the FRME of Midi-Pyrenees. Those associated with the project were ID.BAT (official representative), the CSTB (Cognitive Engineering Division), the professional association Logiciel 88, and the LAB of the Toulouse School of Architecture. Their work had two main objectives:

- to propose a way of structuring the constraints resulting from the confrontation of regulations with the point of view of the profession,
- to define the functions of the constraint supervisor, using the prototype of the Krépis CAD system, developed in the framework of Improbat by the CAOMIP and the LAB and Li2A laboratories of the Toulouse School of Architecture (3), as the experimental platform.

The Constraint Supervisor prototype:

A complete presentation of the prototype has been given in a previous report (4). We present here the elements directly concerning our study from constraints concerning the implantation in an urban setting. The method used for developing the prototype consisted of:

- choosing a few representative constraints
- analyzing these constraints and hence deducing a structure
- imagining operating modes for a supervisor and trying them out with the prototype.

FIRST PHASE:

An initial inventory needed for trying out the prototype led us to select a few constriants considered to be representative of each type and able to be exploited directly.

Certain of these constraints are regulations:

- "The height of the building situated at the edge of a road must be less than or equal to the width of the road".
- "To maintain conformity with the POS[1], the SHON[2] should not exceed the COS[3] x area of the plot.
- "The building must be situated at the edge of the property or at least 4 m from it".

Others are consequences of the schedule of conditions or the designer's intentions:

- "In order to reconcile solar energy contributions with comfort in summer, it is preferable that the building should face south"
- "The building must not have a masking effect on the neighbouring buildings to the north".
- "Situating constructions in the part of the land having a slope greater than 15% will lead to increased building costs".

---

1 : POS : Plan d'Occupation des Sols (zoning plan)
2 : SHON : Surface Hors Oeuvre Nette (total floor area including walls)
3 : COS : Coefficient d'Occupation du Sol (building area coefficient)

SECOND PHASE

These constraints were analyzed and stated in a functional diagram according to an expression logic allowing the system to operate.

Expressing a constraint thus involves two types of structuring:

- one making it possible to move around in the constraint base using key words that the software can recognize,
- the other expressing the constraint in a form that can be evaluated by the software.

Taking these modes of structuring into consideration, a constraint is input in successive steps identifying:

- the constraint itself ("To maintain conformity with the POS, the SHON must not exceed COS x area of the plot")
- the field of application ("Nature of project/Destination of project)
- the target concerned ("Building") with any filtering conditions ("Name of plot connected to Building in East Situated on")
- the conditions that can or cannot be directly evaluated by the software ("SHON of building <= COS for plot connected to Building in East Situated on x Area of plot linked to Building in East Situated on")
- the type(s) of action requested when the constraint has been evaluated, depending on whether it is verified or not (object labelling, characteristics of the object, change to the object).

THIRD PHASE:

Once the constraints have been input, the Supervisor evaluates the project in two modes:

- the assessment mode, evaluating the project at a given instant with respect to all the designated constraints, and producing a report.

- the reflex mode, consisting of evaluating the constraints concerned by a change, and those constraints only. Making a change is automatically followed by the evaluation, which is a reflex reaction to the change (fig.1).

Conclusions

We now have a new module in Krépis: the "Constraint Supervisor". It allows the designer to explore a greater number of possibilities, right from the initial phases of project, knowing that he always has a "guard rail" to stop him going too far if the constraints can no longer be respected.

We have shown here the application of constraints in an urban environment by taking the energy approach into consideration but all areas of the project are concerned and a certain number (safety, acoustics, structures, etc.) have also been tested, which allows us to envisage making a complete system.

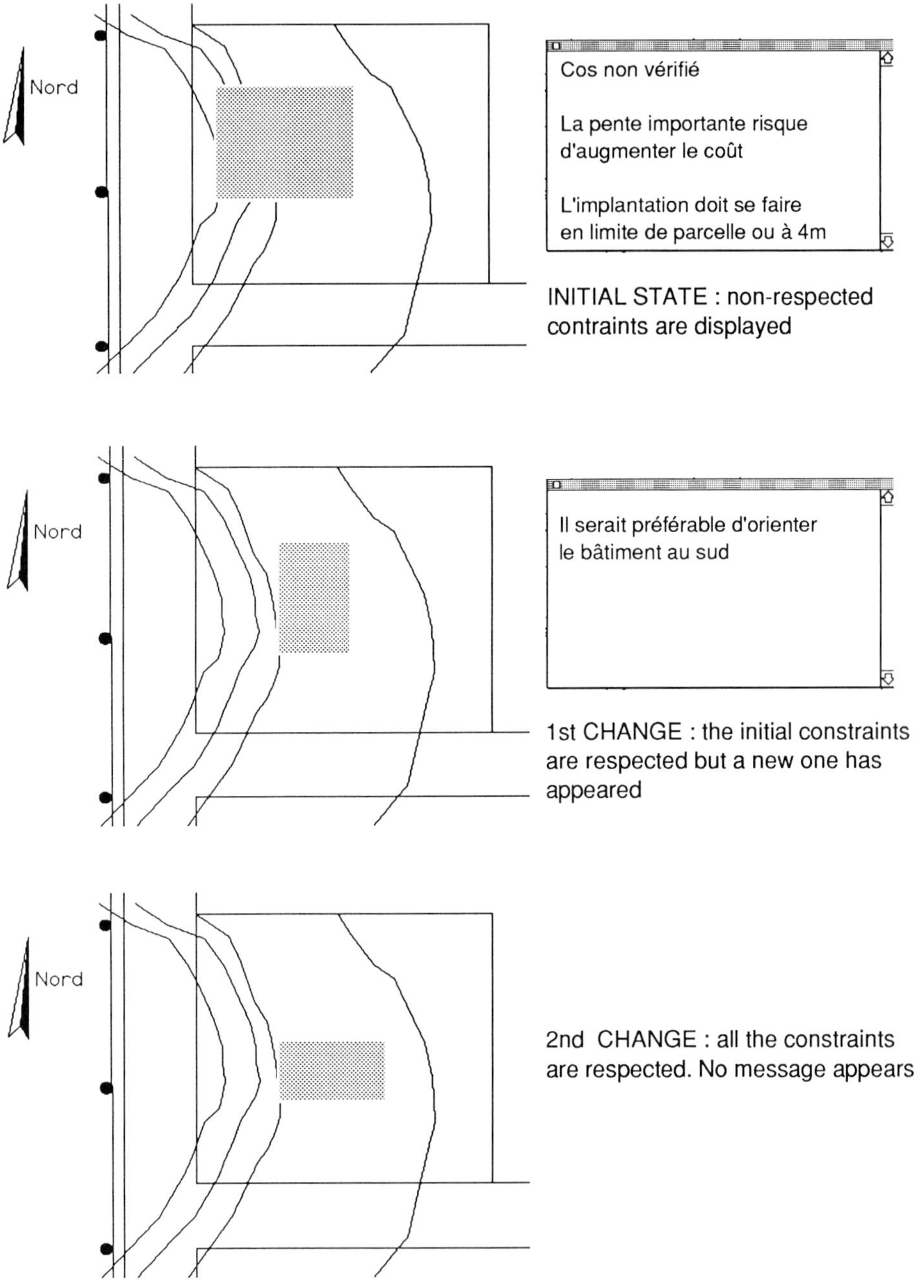

Figure 1. Change in building implantation and configuration in reflex mode.

REFERENCES

(1) - FERNANDEZ P., CHATELET A. : "Spécifications d'un outil d'aide à la conception architecturale en pays chaud", CIB 89, XIe congrès international, Paris, 19-23 Juin 1989, thème I, vol. I, pp. 219-226.

(2) - FERNANDEZ Pierre, CHATELET Alain : "Spécifications du poste thermicien" in "Spécifications du système Krépis". Rapport intermédiaire d'une recherche financée par le M.E.L.A.T.T. dans le cadre du programme IN.PRO.BAT. - p. 55-76, CAOMIP, Toulouse, Juin 1987

(3) - CAOMIP : "KREPIS, Rapport final de recherche", MELATT - AFME, Toulouse, 1988

(4) - ID.BAT : "Définition d'un Superviseur de Contraintes. Intégration au système Krèpis et expérimentation", MELTM - AFME, Toulouse, Mai 90.

FIRST DEVELOPMENTS OF ADLINE* : A LINK BETWEEN A CAD PROGRAM AND DAYLIGHTING DESIGN TOOLS

R. COMPAGNON, C. GREEN, J.-L. SCARTEZZINI
Laboratoire d'Energie Solaire et de Physique du Bâtiment (LESO-PB)
Ecole Polytechnique Fédérale de Lausanne
CH-1015 Lausanne
Switzerland

ABSTRACT. A link between a CAD program and two daylighting design tools is presented. This link dramatically reduces the effort required of a user to prepare the input data for these tools. An application example is also included.

1. Introduction

The growing interest in daylighting among architects stimulates the need for new computerized design tools. At the present time, some programs are available but only few of them are closely coupled to CAD tools commonly used by architects. Each program requires specific data in a particular format. A lot of time must be spent at the input stage by anyone wishing to work with these tools. This fact tends to discourage the use and the spread of these new design tools within practice [1].

The development of links between existing daylighting design tools and CAD softwares is therefore a key part of an integrated daylighting infographic environment such as ADLINE. This paper gives an overview of ADLINE and focuses on the link programs between a CAD software and two daylighting design tools.

2. Description of ADLINE

2.1 OVERVIEW

Basically ADLINE is built around two programs developed at LBL (USA) : SUPERLITE [2] and RADIANCE [3]. The former computes the daylight illuminance in buildings by using a radiosity algorithm. Its limitations (due to the geometry supported and to physical assumptions) make it best suited to the early stage of design. The second program is far more sophisticated and is able to render realistic views of a model. It works with a ray tracing method which is much more computationally expensive. Figure 1 shows the different programs currently linked together within ADLINE.

---

* ADLINE : an **A**dvanced **D**ay**L**ighting **I**ntegrated **N**ew **E**nvironment.

S. Alvarez et al. (eds.), Architecture and Urban Space, 475–480.

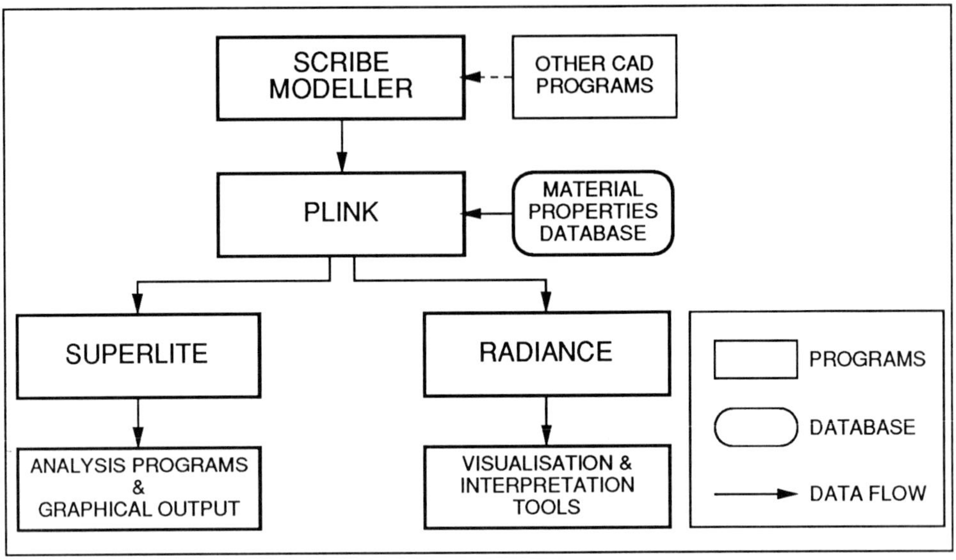

Figure 1. Schematic of ADLINE. Except the RADIANCE package which runs only on UNIX systems, all other softwares runs under the MS DOS system.

Anyone wishing to use one of these programs faces the problem of preparing the input data describing both the geometry of the model and the photometric properties of the surfaces. These input data are stored in text files usually prepared using a common text editor. This means that the user has to give numerical coordinates to describe the geometry and also numerical photometric and colorimetric parameters for each material. Even for very simple models with only a few (say 10 to 20) surfaces, this process is very time consuming and seems very unnatural for designers who are used to working with graphical interfaces. Moreover, errors corrupting the input data are not easily detected before sending the data to their target program.

To replace this tedious process of describing a model, a CAD program should be used to introduce the geometry of the model in a much more efficient and reliable way. Besides other intrinsic advantages the SCRIBE MODELLER CAD software [4] has been chosen at this stage because of its easy way to associate properties data (e.g. colour, reflection and transmission coefficient) to the building elements. Several links have already been developed between SCRIBE MODELLER and other simulation programs in the field of thermal simulation [4] and daylighting [5] confirming the appropriateness of this choice. The program has also the ability to load models prepared on other CAD softwares through the standard DXF file format.

2.2 THE CONVERSION PROGRAM AND ITS PHOTOMETRIC DATABASE

The PLINK program was developed within the framework of this project. It translates the model generated by the modeller in the required input format for the chosen daylighting design tool. An essential task of the PLINK program is to associate the photometric properties to each surface within the model. To relieve the user from giving numeric data for each different surface, a complete set of photometric parameters for common architectural materials are stored in a database. The user can refer to these materials by their unique names. The database can contain an unlimited number of materials descriptions, but the user has to choose a subset of 100 of them that can be used simultaneously for a model. At the present time, the database contains the photometric parameters of nearly 200 materials.

All these photometric parameters are taken from measurements performed over a set of samples of each material. The set includes raw opaque materials (eg. beton, brick, wainscot, linoleum, plaster), transparent and translucent materials (eg. simple and double glazed windows), common light sources and a large amount of tint commonly used for wallpapers or other buildings elements (eg. window frames).

Colours are referred by their names and a Swiss colorimetric standard notation (VSLF). Other standard colorimetric systems will be added in the future (CIELab, Yxy, NCS).

PLINK also does a lot of error checking which helps the user to prepare a model according to the requirements of the target program. Some limitations of the target programs are circumvented directly by PLINK. For example, SUPERLITE only supports a geometry made of trapezoids, but PLINK accepts any quadrilateral planes and generates trapezoids from them. This enhances the flexibility and user friendliness of the design tools incorporated into ADLINE.

3. An application example with RADIANCE

Figure 2 shows a perspective wire frame view of a solar house as it appears in SCRIBE MODELLER. The complete model of this house comprises nearly 2700 lines defining planes and solids. The user gives each line a code number that refers to the material of the surface the line substends. Within PLINK, the user defines the mapping between these code numbers and the name of materials picked from the database.

Because both SUPERLITE and RADIANCE require each surface to have an orientation (ie. a directed surface normal), the user has to follow some simple rules when drawing his design with SCRIBE MODELLER. For example, drawing a planar polygon counter clockwise defines a normal pointing outward the plane.

After the translation performed by PLINK, the model suited for a RADIANCE run comprises nearly 1500 polygons. The text file describing these polygons with their optical properties is 251 kbytes large ! Making the same description file by hand using a simple text editor would require a tremendous amount of work.

Figure 3 shows an outside view of the house rendered by RADIANCE. A clear sunny sky was simulated. More interesting for daylighting study are views from the interior of the house. Figure 4 shows a view inside the sunspace and figure 5 gives a view form the living room at the first floor. Some furnishings are visible on these pictures.

Besides providing detailed and realistic views of a building model, numerical luminance or illuminance values can be extracted from the scene. New developments are underway to compute visual comfort indicators from pictures generated by RADIANCE.

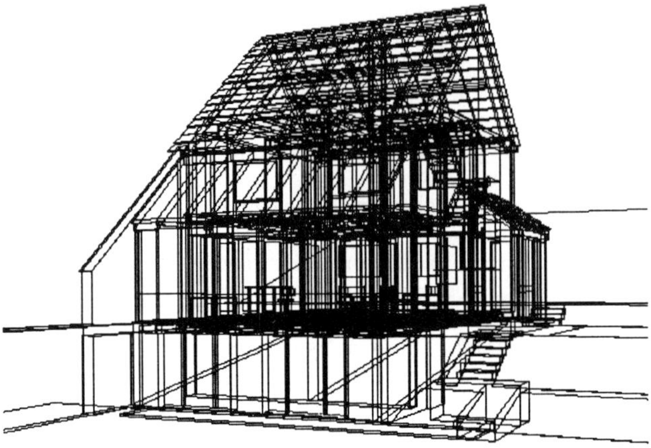

Figure 2. Wire-frame view of a solar house modelled using SCRIBE MODELLER.

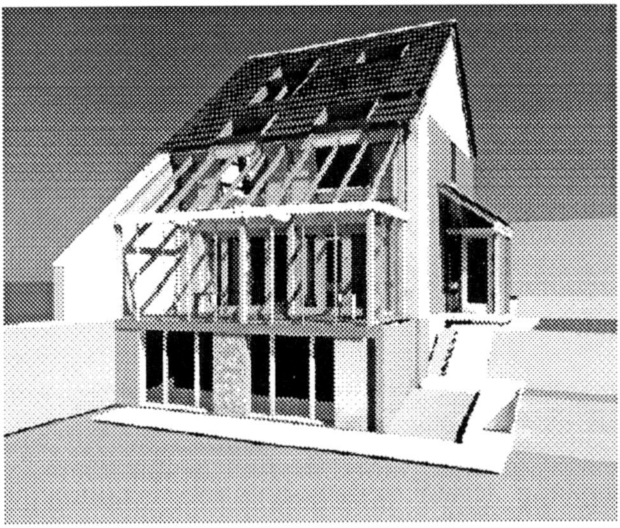

Figure 3. The solar house rendered by RADIANCE. The original picture appears in colour.

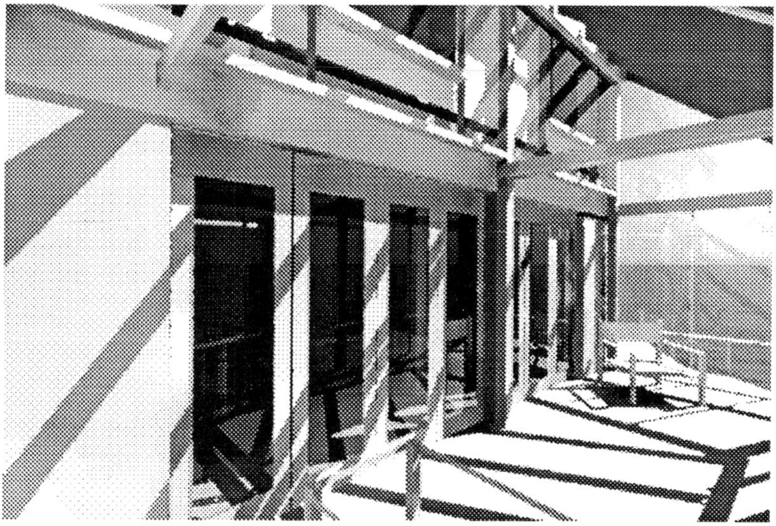

Figure 4. View inside the sunspace. Note the specular reflection in the windows.

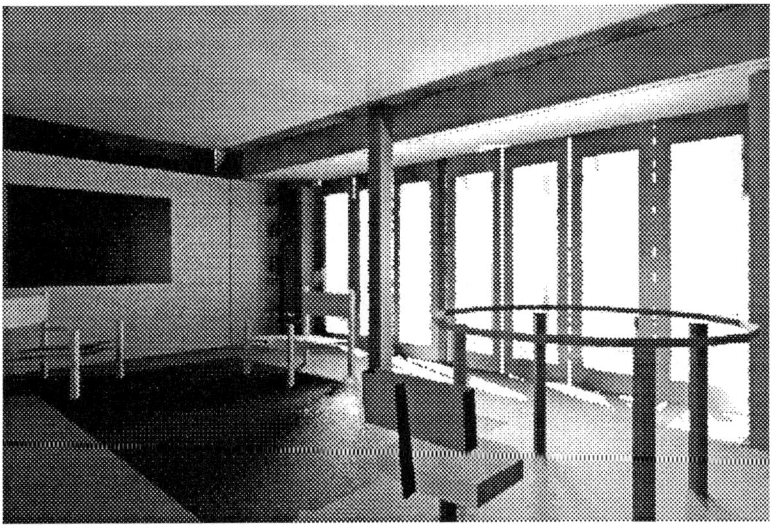

Figure 5. View from the living room. Furniture are also included in the scene.

3. Conclusions

Two daylighting design tools (SUPERLITE and RADIANCE) have been chosen to be part of an integrated infographic design tool called ADLINE. At this early development stage, efforts have concentrated on the following objectives :

- development of a link software (PLINK) between a CAD tool and the two daylighting design tools;
- elaboration of a database containing photometric and colorimetric data for common architectural materials;
- first testing of the usability of the whole system.

By enhancing and simplifying drastically the process of data input, PLINK is an essential part of the integrated tool. It is expected to push further the use of these modern design tools within the design practice.

4. Acknowledgements

This work is done in the framework of a multidisciplinary research project called "LUMEN". This part concerned with the development of computerized design tools in daylighting is funded by the Swiss national energy research fund (NEFF).

5. References

[1] Green C. (1988), CAD and thermal design methods, Proceedings of the sixth PLEA conference, Porto, Portugal, July 1988, pp. 713-723.

[2] Windows and Daylighting Group (1985), Superlite 1.0 manual, LBL, University of California, Berkeley.

[3] Ward G.J. (1990), Visualisation, LD + A, Vol. 20 N° 6, June 1990.

[4] Green C., Haves P. (1986), Generation of data for passive solar building simulation from a three dimensional architectural modelling system Scribe / Seri-Res = Serilink, Proceedings of the fifth PLEA conference, Pècs, Hungary, September 1986, pp. E1 - E10.

[5] Green C., Modlux : final report on program development and user manual version 1.4, Ecotech design ltd, Sheffield, UK.

Part 5

PASSIVE AND LOW ENERGY TECHNOLOGIES

Chapter 1

Progress in Model Development and Simulation

COUPLING VENTILATION NETWORK AND HVAC SYSTEMS TO MULTIZONE INFILTRATION MODELS

A.C. MEGRI, F. ALLARD, G. KRAUSS
CETHIL - URA CNRS 1372,Bât 307 - INSA
69621 Villeurbanne Cedex (FRANCE)

ABSTRACT. Air flow distribution in buildings is due to combined effects of three main driving forces, the effect of wind on the building, the buoyancy effect related to temperature or more generally air density gradients, the effect of mechanical ventilation systems. The ventilation network and HVAC system in general are significant ways by which air, pollutants and thermal energy are transfered from one part of a building to another or to outside. Our contribution here is at first to describe the coupling procedure we use between the pressure network representing the multizone building and the ventilation network. Then we put forward a general proposal dealing with most of ventilation or HVAC systems, and finally we discuss the results obtained on a case study.

1. INTRODUCTION

As a consequence of the better control of building insulation leading to a higher quality of building envelopes, ventilation becomes a more and more significant way by which energy but also pollutants are transfered from a building to outside or between zones of a building.

During the last decade, an important modelling effort has been made ([1]-[7]). Nevertheless, most of the existing codes consider infiltration to be the first source of air transport. In fact, in most of existing buildings, air flow rates between rooms result from the combined effects of wind, stack effect and mechanical systems. The objective of our work is to propose a way to predict the combined effects of these three phenomena on air flow and pollutant patterns in a multizone building.

2. MODELING PRINCIPLES

The basis of our study is the COMIS multizone airflow model developped within the frame of an international joint project[7]. Presently, the COMIS structure includes more than 150 modules representing most of the flow path in a building.

Schematically, a multizone building can be represented by a series of zones at thermodynamical equilibrium and represented by their state variables(temperature and reference pressure). These zones are connected by different kind of links(large openings, cracks...) through which air flows are represented by non linear equations giving the flow as a function of pressure drop through the link.

S. Alvarez et al. (eds.), Architecture and Urban Space, 485–490.

By writing the mass conservation equation of air in each zone, one can represent the multizone air flow model of a building by a network of pressure nodes connected together by non linear flow equation.

HVAC systems(Heating Ventilating and Air Conditionning systems) are made of elementary systems such as ducts, duct fittings, fans, air filters, coils, flow controllers etc. In order to study the behavior of a working ventilation network coupled to a multizone building, the first step consists in modelling each component, then a complete network can be built, integrating the whole ventilation network and the building network.

3. COMPONENT MODELLING

3.1. Flow controllers

Flow controllers represent most of the available dampers or regulators used in a ventilation network. The basic premise of flow controllers is that they have a flap or valve which may throttle the flow by gradually closing the opening when the pressure increases.

A flow controller is called ideal if after a first pressure threashold it delivers a constant flow rate when the pressure increases.

Furthermore, a flow controller is called symmetrical if it has exactly the same behavior for both positive and negative pressure drops.

Figures 1 and 2 present the 4 combinations we can find between this two definitions.

F1: Symmetrical ideal flow controllers.

F2: Non symmetrical ideal flow controllers.

F3: Symmetrical non ideal flow controllers.

F4: Non symmetrical non ideal flow controllers.

For each category we can represent the behavior of any flow controller by power law functions (range 1+ and 1-) and polynomial laws(range 2+, 2- and 3+, 3-).

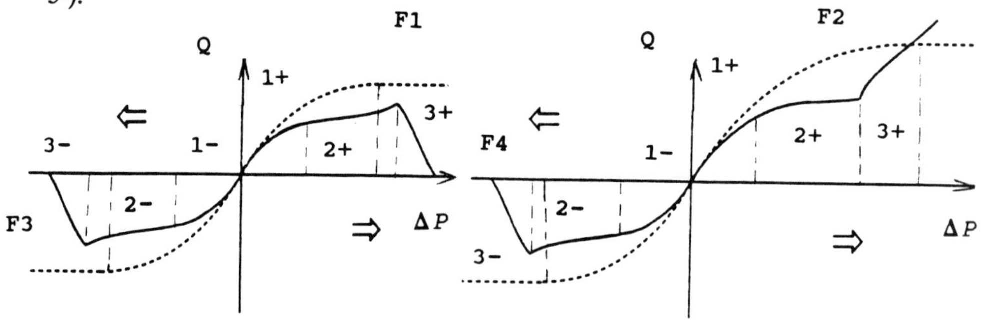

Figure 1: Characteristic curves of flow controllers.

3.2. Fans

Integrating fans in a general network requires at first to define a flow function given the flow rate through the fan as a function of the pressure difference. Polynomial functions are usually used.

Furthermore, in order to avoid any numerical unstabilities if the pressure in the network goes out of the the working range of the fan, we assume that out of this range we can consider a linear behavior of the fan.

Figure 2 gives an example of fan model.

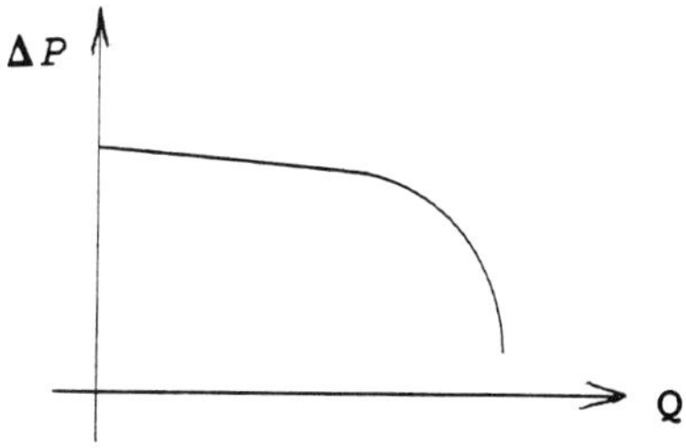

Characteristic curve of a fan.

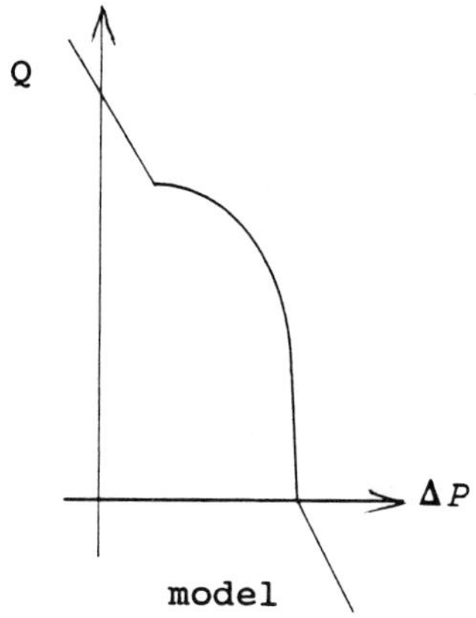

model

Figure 2: Fan model.

3.3. Ducts and Duct Fittings

We consider two types of ducts:

- Straight ducts characterized by their dimensions and a friction factor.
- Duct fittings characterized by their type, their geometry and a local loss coefficient.

In order to integrate ducts and duct fittings in the general network, their behavior is represented by power law functions (orifice laws).

$$Q=C(\Delta P)^{N} \tag{1}$$

C and N are identified by simulating at first the behavior of the duct for various flow regimes, using its characteristic friction factor(for ducts) or local loss coefficient(for duct fittings) to calculate the pressure loss.

Then, the results can be drawn as a flow function and C and N are obtained by numerical regression.

4. CASE STUDY

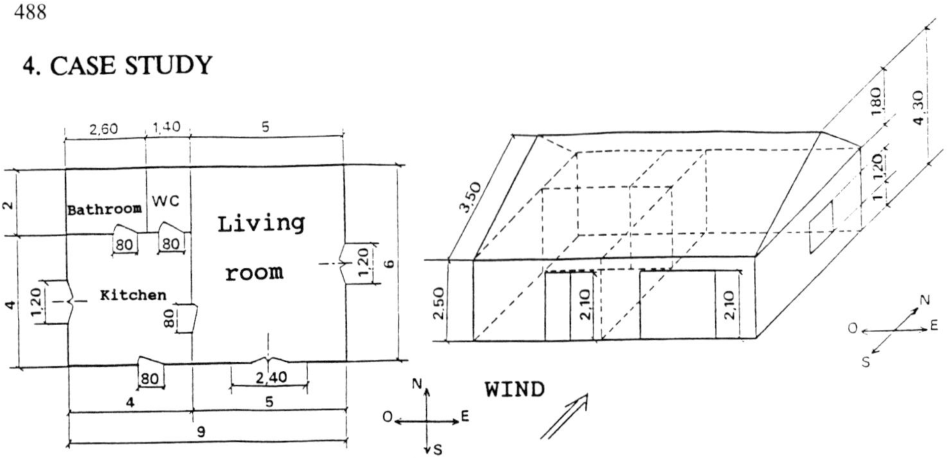

Figure 3: The test building

The test building is a 5 zone single family house drawn on Figure 3.

This building is characterized by 31 flow paths(doors, windows, cracks of internal and external walls...), 4 identical autocontrolled air inlets (F1) located on living room walls and 3 identical exhaust systems located in toilets, bathroom and kitchen. Figure 4 and 5 give the characteristic curves of these components [8],[9]. The Cp-values used for the calculation are given in Table 1.

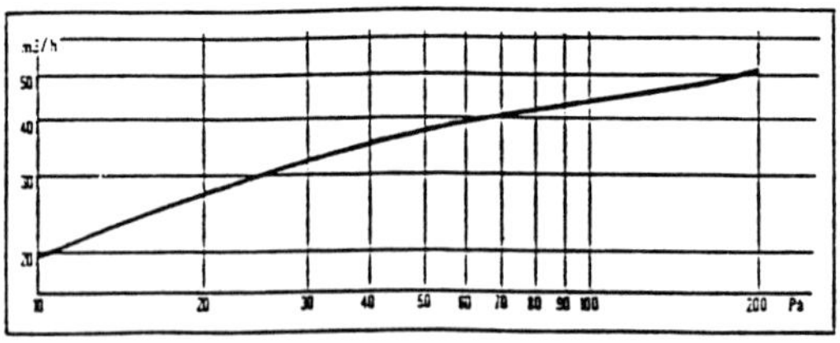

Figure 4: Characteristic curve of the autocontrolled inlet used

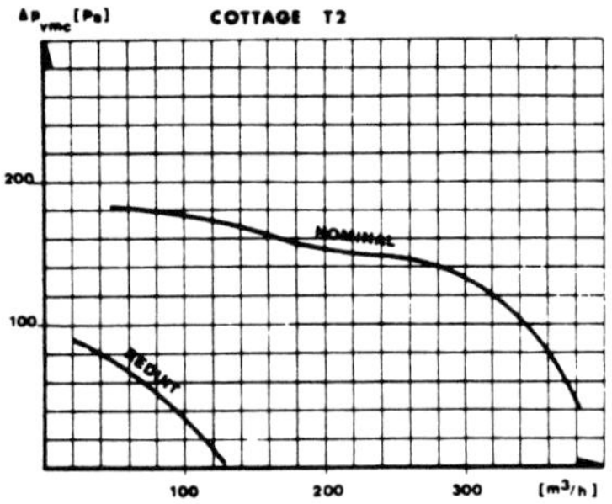

Figure 5: Characteristic curve of the fan used

South façade	West and East façades	North façade	South roof	North roof	End of exhaust duct
+0.25	-0.70	-0.50	-0.55	-0.60	-0.75

Table 1: Cp-Values for each wall

4.1. Combined influences of wind velocity and stack effect

In order to study the influence of each parameter on the infiltration rates in the building, we consider at first the influences of wind velocity and stack effect without any ventilation system(neither flow controller nor fan). The indoor temperature is 20 °C and outdoor temperature is either -20°C or 0°C. Figure 6 shows the evolution of air change rate with wind velocity for the different external temperatures.

4.2. Coupling with autocontrolled air inlets

In order to ensure a better control of infiltration, we change the natural ventilation grids by autocontrolled inlets of same area (50 cm^2).

Figure 7 represents the evolution of the ventilation rate of the building. The efficiency of flow controllers is shown clearly by a decrease in flow rate of 50 % at 7 m/s.

4.3. Influences of wind effect, stack effect and mechanical exhaust systems

If we consider only autocontrolled inlets, the ventilation and by consequence air quality, is poor for low wind velocities. The ventilation is improved by adding mechanical exhaust systems in the kitchen, bathroom and toilets. At first, fans are not operating, they behave as natural outlets. In a second step, the fans are operating; the air change rate increases, evidently, within utility rooms. This increase influences the ventilation within the living room due to a decrease of the crossing flows (Figure 8).

5. CONCLUSION

For low energy buildings, ventilation becomes an important energy load. On the other hand, good air quality conditions require an efficient removal of used air. Furthermore, air flow patterns within a building result from the combined effects of wind, stack effect and mechanical ventilation systems.

The prediction of air change rates in each room of a multizone building is a necessary task, COMIS' code offers an efficient structure to develop such prediction or design tools for complex configurations.

The case study presented in this paper is only given as an illustrative example, it demonstrates the feasability of our project developed within the frame of an international effort coordinated by the International Energy Agency, Annex 23.

REFERENCES:

1. H. FEUSTEL et alii, " Air Flow Models for Multizone Structures,an Updated Review", à paraître Energy and Building 1991.
2. B. MALO, P. VALTON, " Etudes des modèles de ventilation et simulation du enouvellement d'air dans les locaux". Rapport final, convention A.F.M.E/ Co.S.T.I.C, Juin 1987.
3. M.K. HERRLIN, "MOVECOMP a multizone infiltration and ventilation simulation program. " Air Infiltration Review, Vol.9, No.3, 1988.

4. A. ROLDAN, "Etudes thermiques et aérauliques des enveloppes de bâtiment: influence des couplages intérieurs et du multizonage". Doctoral thesis, National Institute of Applied Sciences, Lyon, France, p.310, 1985.
5. G.N. WALTON, "A computer algorithm for predicting infiltration and interroom airflows. "ASHRAE Transactions, Vol.90, Part 1, 1984.
6. G.N. WALTON. " Air flow Network Models for Element-Based Building Airflow Modeling" ASHRAE Transactions, Vol.95, Part 2, 1989.
7. F. ALLARD, V. DORER, H. FEUSTEL, M. HERRLIN, L. MINGSHENG, H.PHAFF, H. YOSHINO, "Technical Note AIVC 29. Fundamentals of the multizone.Air Flow Model-COMIS" May 1990.
8. F. ALLARD, M. HERRLIN, "Wind-Induced Ventilation", ASHRAE Transactions,Vol.95, Part 2, 1989.
9. A.C. MEGRI, G. KRAUSS, F. ALLARD, " Modélisation des réseaux de distribution d'air couplés à des bâtiments multizones" Seminaire GEVRA 1991. INSA de Lyon, Mars 1991.

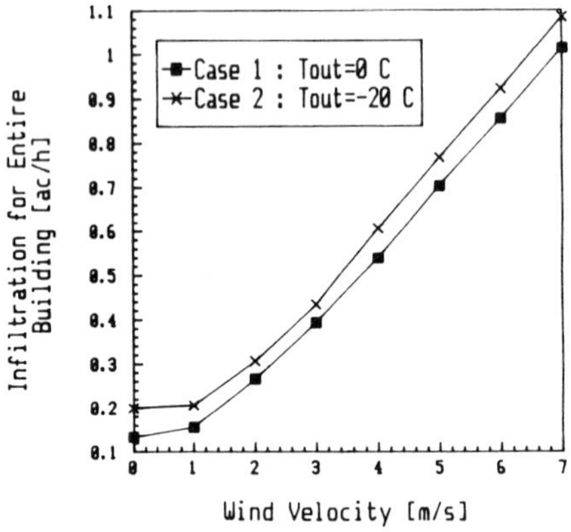

Figure 6: Influence of stack effect.

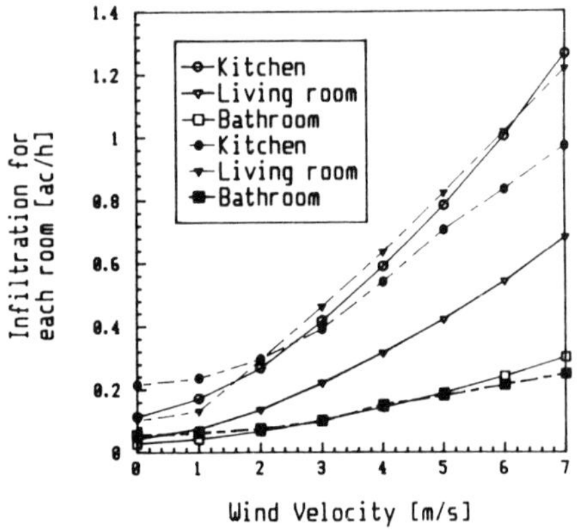

Figure 7: Influence of autocontrolled air inlets

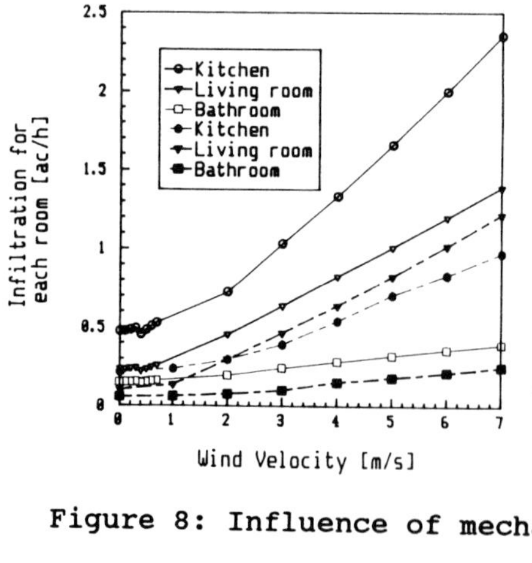

---- Case without any mechanical ventilation

—— Case with Mechanical ventilation

-- — — Case without any autocontrolled air inlet

—— Case with autocontrolled air inlet

Figure 8: Influence of mechanical exhaust system

EVALUATION OF PASSIVE SOLAR COMPONENTS BY IDENTIFICATION TECHNIQUES: TESTS FOR MODEL VALIDATION PURPOSES.

E. Palomo; F.M. Téllez; J. Marco; M.R. Heras
Instituto de Energías Renovables - CIEMAT
Avda. Complutense, 22
28040 Madrid
Spain

ABSTRACT. One of the major goals of the CEC research project PASSYS (1986-1991) is the development of common test procedures for the thermal and solar characterization of building components tested in the PASSYS outdoor testcells. In the present stage of the project, identification techniques have been choosen as the most promising. A special activity of the CIEMAT-IER in PASSYS concerns to the development and evaluation of several tests to be applied for selecting adequate testcell+component models. This paper contents a brief description of such a tests and the information about the model performance that could be achieved using them. As an example, the LUMPA 1.3. model developed by TNO has been fitted to the testcells datarecords supplied by CEA-Cadarache and its residual analysed using the methodology for model validation that we are proposing here.

1. INTRODUCTION.

The system identification procedure involves three basic entities:

a) The data: The input-output data are recorded during a specifically designed identification experiment where the user may determine which signals to measure, when to measure them and how to command, if possible, the inputs signals. The object with experiment design is thus to make these choices so that the data become maximally informative.

b) The set of models: A set of candidate models is obtained by specifying within which collection of models we are going to look for a suitable one. This is the most important and, at the same time, the most difficult choice of system identification procedures.

c) The rule by which candidate models can be assessed using the data: Determining the 'best' model in the set, guided by the data, is the identification method. The assessement of model quality is typically based on how the models perform when attemp to reproduce the measured data.

After having settled on the preceding three choices, we have, at least

S. Alvarez et al. (eds.), Architecture and Urban Space, 491–497.

implicitly, arrived at a particular model: the one in the set that best describes the data according to the chosen criterion. It then remains to test whether it is valid for its purpose. Such tests are known as **model validation.** They involves different procedures to asses how the model relates observed data, to prior knowledge, and to its intended use. Deficient model behaviour make us to reject the model, while good performance will develop a certain confidence in the model. **The aim of this paper is to propose and to discus a set of tests that could be applied for model validation purposes.**

2. NON-PARAMETRIC TESTS ON THE RESIDUALS FOR MODEL VALIDATION PURPOSES.

It is well known that if the considered model were of the correct functional form, the true parameters values were substituted and the assumption about noise were also correct, the residuals from the model would be white noise and its estimated autocorrelation would be distributed mutually independently about zero with variance 1/m (m= degrees of freedom). Serious model inadecuacy can usually be detected by examining: a) The autocorrelation function or the power spectrum of the residuals from the fitted model. b) Certain cross correlation functions, in the time and frequency domains, involving inputs and residuals.

Suppose that the functional form of the system model is correct, but the noise model is incorrect. Therefore, the residuals would not be cross correlated with the inputs. However, the residual process would be autocorrelated. Its autocorrelation function and its power spectrum will show marked correlation patterns. In the other hand, if the functional form of the system model is incorrect, not only the residuals would be autocorrelated, but also they would be cross correlated with inputs. Thus, the set of tests that we are proposing here for model validation purposes are oriented to discover any kind of autocorrelation in the residuals or cross correlation between inputs and residuals. In other words, we are considering the q+1 input-output stochastic processes related linearly by the general dynamic model

$$X_{q+1}(t) - \mu_{q+1} = \int_{\infty}^{\infty} h_{q+1,1}(u) \{X_1(t-u) - \mu_1\} du + \ldots\ldots \qquad (1)$$

$$+ \int_{\infty}^{\infty} h_{q+1,q}(u) \{X_q(t-u) - \mu_q\} du + Z(t)$$

where $X_1(t)$, ..., $X_q(t)$ are the input processes to a linear system, $X_{q+1}(t)$ the corresponding output (in our case the residual from the model to be validated), and Z(t) an independent and white noise (in the following, 'noise of the process'). μ_i (i= 1,2,...,q,q+1) is the mean value of the i process and $h_{q+1,i}$ (i+ 1,2,..,q) is the impulse response function from the input i to the output q+1.
Three steps can be distinguished in the validation procedure:

STEP 1. Tests for white noise on the residuals from the model: Two kind of tests are needed here. One, in the time domain, to detect "local correlations" in the residuals. Other, in the frequency domain, to detect departures of whiteness caused by periodicity effects. The first one is based in the analysis of the autocorrelation function of the residuals, and the second one in the analysis of its autospectrum. After that, if the hypothesis of the residuals whiteness is acceptable we can conclude that both the system model and the noise model are correct. If not, a second step of analysis is needed.

STEP 2. Test for non-zero squared multiple coherency: The following questions must be answered in this step of the analysis: Are the residuals from the model, $X_{q+1}(t)$, and the noise process, $Z(t)$, in equation (1) identical? Or, in the contrary, can inputs explain part of the residuals?. Two tools are available to anwser these: The named residual spectrum of the multivariate process and the squared multiple coherency. If the hypothesis that both residuals and noise processes are identical is acceptable, we can conclude that the system model is of the correct functional form but the noise model is incorrect. If not, a third step of analysis is needed.

STEP 3. Tests for non-zero cross correlation between inputs and residuals: In this step we are trying to measure the existing correlation between each input and the residuals from the model. As in step 1, tests in the time and frequency domains are needed. The first one is based in the cross-correlation function analysis. The second one is based in the analysis of several spectral functions. When inputs are uncorrelated, the spectral estimators of the phase spectrum and of the squared coherency could be used to detect deviations from the hypothesis of non-zero correlations. In the other hand, when inputs are cross-correlated, the specral estimator of the partial coherency spectrum is needed. After that, we will be able to anwser the following question: What inputs are the main responsables of the serious inadecuacy detectected in the system model functional form ?. And consequently, what physical processes are not well represented in the model ?.

3. APPLICATION: 'LUMPA 1.3.' VALIDATION USING A TESTCELLS DATA SET.

3.1. The data.

One of the testcells data sets supplied by CEA (Commissariat a l'Energie Atomique, Cadarache, F) has been used in this paper. Its main characteristics concerning design experiments topics are:

Physical system =	Testcell with callibration wall.
Output variable =	Air temperature of the testroom, T_{tr} (C).
Input variables =	External air temperature, T_e (C). Solar irradiance on south vertical surface, G_{sv} (W/m^2). Air temperature of the service room, T_{sr} (C). Heating power, P_{he} (W).
Sampling interval =	60 min

Record length = 720

Test estrategy = 15 days with a 150 W Pseudo Random Binary Sequence (PRBS) for P_{he} and free floating conditions for T_{sr}.
15 days with 50 W PRBS for P_{he} and 40 C constant T_{sr}.

3.2. The fitted model.

The set of candidate models within which we are going to look for a suitable one is the collection of deterministic models proposed by TNO (Institute of Applied Physics, Delft, NL) (LUMPA 1.3 model). An escheme of the electric network representing it is shown in figure 1. The parameter identification has been performed using the software package MRQTv4.0 developped by TNO. The residuals from the fitted model are represented in fig. 2.

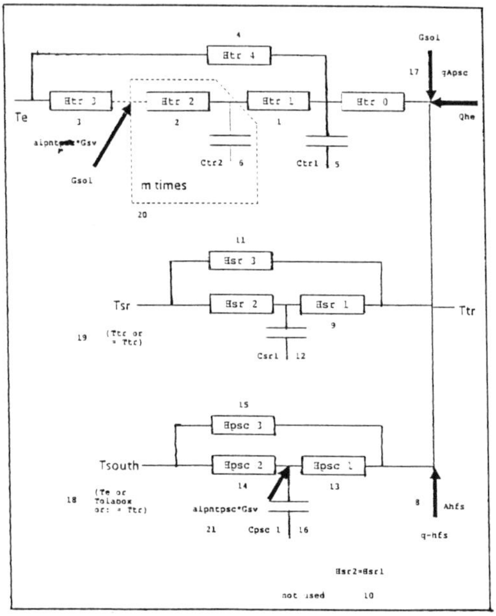

Fig.1. LUMPA 1.3. model representation.

3.3. Model validation results.

Following the methodology for model validation that we are proposing, the results achieved when applied to the LUMPA 1.3. fitted model are:

STEP 1: Tests for white noise on the residuals.

Time domain: The sample autocorrelations estimated has been analyzed. As it can be seen in Figure 3 (squares), they start to be small from lags values greather than 30. Then, it is clear that the residuals from the model are not white noise.

Frequency domain: The normalized smoothed autospectrum of the residuals (see Fig. 4-squares) shows a clear peack at low frequencies and several smaller ones at high frequencies. In an other hand, study of the Fig.5 shows inmediately that there are marked departures from linearity in the normalized cumulative peridiogram. These departures are very pronounced at low frequencies and are what it might be expected if seasonal effects are inadequately accounted for.

The residuals show both "local correlation" and periodicities. They are not a white noise process. Thus, it is expected that the assumption about noise in LUMPA 1.3. and/or the functinal form of the system model were incorrect. To investigate system model inadecuacies a second step of analysis is needed.

STEP 2. Test for non-zero multiple coherency.

As it can be seen in figure 4, where the normalized smoothed autospectra estimated for the residuals from the model and for the noise of the multivariate process has been represented, the residual and noise processes are not identical. The same conclusion leads from the inspection of the squared multiple coherency estimated (see fig.6-squares), in which it could be appreciate that a great portion of the residual spectrum (>.7 for many frequency points) can be predicted from the inputs.

STEP 3. Test for non-zero correlation between each input and residuals.

In order to know which of the four considered inputs is the main responsable of the detected inadecuacy in the system model functinal form, a third step of analysis is needed. Here the hypothesis of non-zero correlation between each input and the residuals is tested.

Time domain: When estimating the cross-correlations between each input (T_a, G_{sv}, T_{sr} and P_{he}) and the residuals the most significant correlations appear in the sample ccf estimated for the heating power and the residuals and for the service room air temperature and the residuals, showing both curves pronounced patterns. However, as no filtering operation has been carried out on the time series to convert them into white processes before to estimate the ccf, this analysis is only orientative.

Frequency domain: At this step, two kind of analysis can be performed to measure the correlation existing between each input and the residuals from the model. If inputs are independent processes, a bivariate analysis on each couple input-output is enougth. If not, the whole multivariate process must be considered in order to measure the above mentioned correlations after allowance is made for the effect of the other input processes. Thus, first of all, a test for non-independent processes on inputs is needed. Non strong correlations are detected in the sample ccf estimated for the bivariate processes (T_a, P_{he}), (G_{sv}, P_{he}) and (T_{sr}, P_{he}). However, significant correlation exist between T_a and G_{sv} and between T_a and T_{sr}. Thus, the frequency domain test for non-zero correlation between each input and the residuals from the model must be performed on the partial coherency spectra.

Squared partial coherencies achieved have been represented in figure 6 jointly with the squared multiple coherency. The analysis of this figure allow us to conclude that: 1) As pointed in STEP 2, the squared multiple coherency indicate, in this case, that an important proportion of the residuals spectrum can be predicted from inputs. 2) The squared partial coherency for P_{he} indicates that at high frequencies (f>0.15) the heating

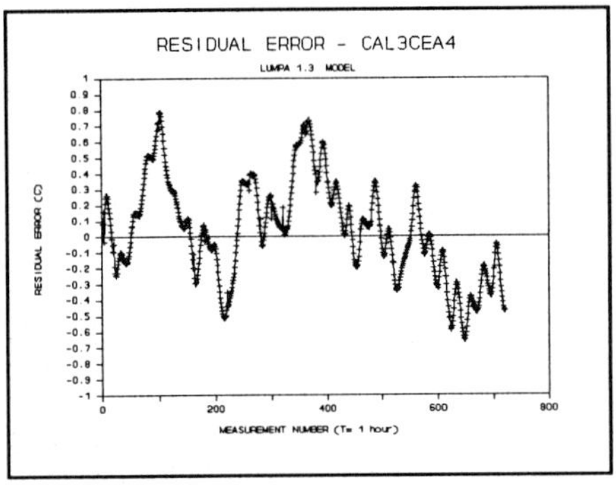

Fig.2. Residual error from the model.

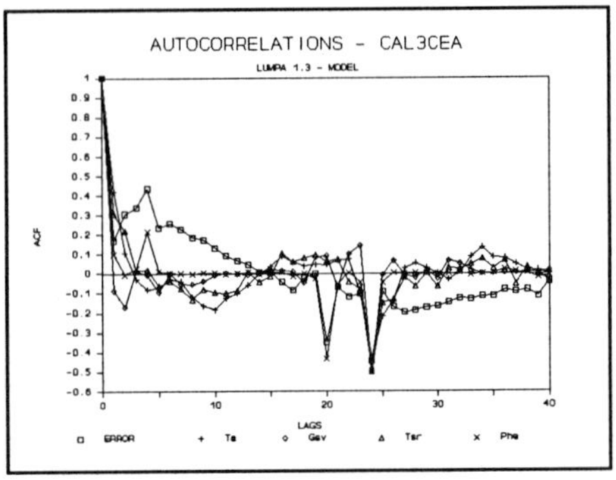

Fig.3. Autocorrelation functions.

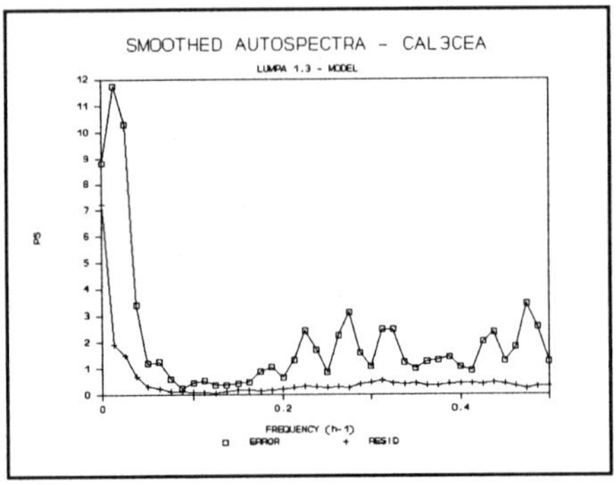

Fig.4. Smoothed power spectra.

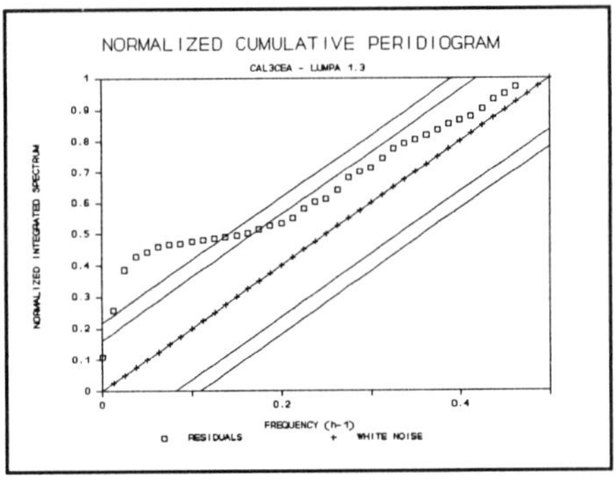

Fig.5. Cummulative pridiogram.

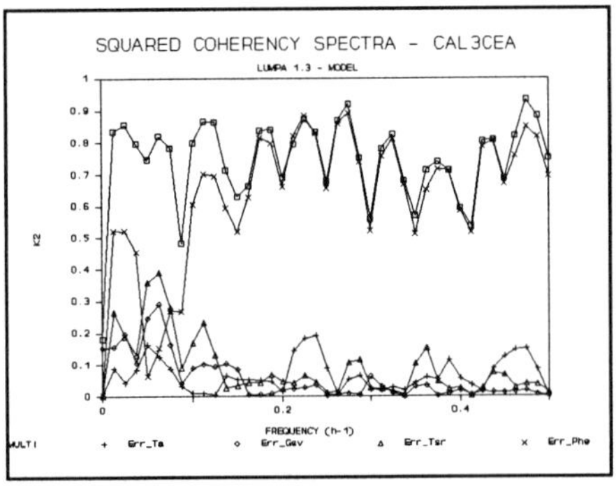

Fig.6. Multiple and partial squared coherency spectra.

power can explain a great proportion of the residuals from the model. At these frequencies, the squared multiple coherency and the squared partial coherency for P_{he} are more or less equal. 3) At low frequencies (f<0.15), the highest values of the squared partial coherency correspond to P_{he} and T_{sr} inputs. Althogh significant, smaller squared partial coherency values have been achieved for solar irradiance and outdoor air temperature. As in the case of high frequencies, a great proportion of the residuals from the model could be predicted from inputs, especially from P_{he} and T_{sr}.

Main conclusions from the prior analysis are:

- The assumption about noise made by LUMPA 1.3. (noise-free model) is not addecuate. A modification of this hypothesis could be interesting in other to improve the model and, possibly, to avoid bias in the fitted parameters.

- Several inadecuacies in the functional form of the model have been detected, mainly related to the heating power and the service room air temperature. The physical processes related to both variables are not sufficiently well represented. The detected inadecuacies of the model related to the other inputs (solar radiation and outdoor air temperature) are no so relevant.

- Using the information contents in the frequency response functions of the analysed multivariate stochastic process, could be a way to proposed interesting modifications of the functional form of LUMPA 1.3. to improve it. (No analysis about this topic has been included in this paper).

4. REFERENCES.

Box, G.E.P. and Jenkins, J.M. 1976: Time series analysis: Forecasting and control. Holden-Day, San Francisco, California.

van Dijk, H.A.L and van der Linden, G.P. 1991: Instructions on MRQT version 4.0, LUMPA version 1.3. and MRQPOST version 1. (PASSYS Ref. xxx-91-PASSYS-TME-WD-zzz)

van Dijk, H.A.L. 1990: User guide on the parameter identification method MRQT version 3.3. (PASSYS Ref. xxx-90-PASSYS-TME-WD-zzz)

Jenkins, G.M. and Watts, D.G. 1968: Spectral analysis and its applications. Holden-Day Inc., San Francisco, California.

Ljung, L. 1987: System identification. Theory for the user. Prentice-Hall, englewwod Cliffs.

Palomo, E. and Téllez, F.M. 1991-a: PAMTIS. Package to Analise Multivariate Time Series. (PASSYS Ref. xxx-91-PASSYS-MVD-WD-zzz)

Palomo, E. and Téllez, F.M. 1991-b: Mathematical and numerical methods used in PAMTIS. (PASSYS Ref. xxx-91-PASSYS-MVD-WD-zzz)

Palomo, E. and Téllez, F.M. 1991-c: Non-parametric tests on the residuals for model validation purposes. (PASSYS Ref. xxx-91-PASSYS-MVD-WD-zzz)

Priesley, M.B. 1981:Spectral analysis and time series. Academic Press.

Vandaele, W. 1983: Applied time series and Box-Jenkins models. Academic Press, Inc.

WATER DROPS AS A NATURAL COOLING RESOURCE. PHYSICAL PRINCIPLES

E.A. Rodríguez**, S. Alvarez*, R. Martín*

** *CIEMAT-IER*
División Termosolar
Av. Complutense 22
E-28040 Madrid. Spain

* *Universidad de Sevilla*
Departamento de Ingeniería Energética y Mecánica de Fluidos
Av. Reina Mercedes s/n
E-41012 Sevilla. Spain

ABSTRACT. This paper deals with the use of small water drops for cooling air, either directly or indirectly through a previous water cooling. A model of the different phenomena involved between a drop and the surrounding air is described. Finally, two examples of applications use the model to select the most suitable drop size for each.

1. Introduction

Most of the natural air cooling techniques developed within the frame of the work on climatic control of outdoor spaces at EXPO'92 [1] are based on the transfer phenomena that take place when water drops and air are put into contact. The cooling can be achieved by two different ways:

- In the first way, water drops directly cool the air of the space being conditioned. Total evaporation of the water droplets is required, so that their air cooling capacity is maximized while preventing people from getting wet.
- A second possibility is an indirect cooling of the air by using cool water as an intermediate medium. In that case, water drops cool the water in a pond and the aim is to obtain the maximum reduction of the temperature of the drops with the minimum water loss. The resulting cool water is used in the precooling coil of air handling units.

The physical equations governing the transfer phenomena on water drops are presented and solved for some particular cases. In complementary papers ([2], [3]) more comprehensive models couple the above equations with those pertaining to the surrounding air, providing the basis for the design, sizing and control of different cooling techniques.

2. Qualitative analysis of the physical phenomena

A single water drop in a non-saturated environment experiences coupled heat, mass and momentum transfer processes:

- The temperature gradient at the drop-air interface provokes a net heat transfer from the air to the drop surface (if the air is hotter than the drop).
- The concentration gradient of water vapor near the water surface gives way to a water diffusion from the drop surface to the non-saturated air. This water diffusion needs a

S. Alvarez et al. (eds.), Architecture and Urban Space, 499–504.

previous evaporation of water from the drop and so, due to the latent energy of this phase-change, this mass transfer is strongly coupled to the heat transfer.

- There exits a momentum transfer between the drop and the air, due to the relative movement of the drop within the surrounding air. This momentum transfer contributes also to increase the effect of both heat and mass transfer.

The first two phenomena show different effects on the thermal state of the drop. The heat transfer tends to increase the drop temperature whereas the latent energy absorbed by the evaporated water causes the drop to cool down. As a result of these two opposite tendencies, an equilibrium drop temperature is reached (the wet bulb temperature of the air). Once the drop has reached the wet bulb temperature, the latent energy needed to evaporate more water is supplied only by the surrounding air and so the latter is cooled.

From the former considerations, two different periods in the evaporation of a single water drop in air can be distinguished [4]:

1.- The drop temperature is changing from its initial temperature to its equilibrium temperature (the air wet bulb temperature).

2.- The drop is at this equilibrium temperature and the radius decreases with the evaporation.

The relative length of each period within the airborne time of the drop depends mainly on the initial drop size and so this will be the most important design variable of the two above-mentioned applications.

3. Model Description

Assuming a single spherical drop at uniform temperature, exposed at uniform boundary conditions and without any radiant exchange, the transport phenomena mentioned above are coupled in the following balance equations:

a) Momentum balance: The equilibrium of the different forces over the drop (friction, buoyancy, inertial, etc) gives rise to the equations corresponding to the horizontal and vertical downwards velocity:

$$\begin{aligned} \frac{dv_x}{dt} &= \frac{-3}{8}\frac{\rho}{\rho_w} f v\, v_x \frac{1}{R} \\ \frac{dv_y}{dt} &= g\,(1-\frac{\rho}{\rho_w}) - \frac{3}{8}\frac{\rho}{\rho_w} f v\, v_y \frac{1}{R} \end{aligned}$$

b) Mass balance: The decreasing rate of the drop size equals the mass of evaporated water.

$$\rho_w \frac{dR}{dt} = -\rho K_m \frac{Y - Y_\infty}{1-Y}$$

c) Energy balance: The variation of the internal energy of the drop equals the net heat flux over it.

$$R\rho_w C_{pw}\frac{dT}{dt} = 3h(T_\infty - T) - 3\rho K_m h_{fg}\frac{Y - Y_\infty}{1 - Y}$$

Converting the above equations in a dimensionless form, the associated transfer coefficients, Nusselt and Sherwood numbers and the friction factor, can be obtained from the correlations [5]:

$$\mathrm{Nu} = \frac{h2R}{\lambda} = 2 + 0.6\cdot \mathrm{Pr}^{1/3}\cdot \mathrm{Re}^{1/2}$$

$$\mathrm{Sh} = \frac{K_m 2R}{D_A} = 2 + 0.6\cdot \mathrm{Sc}^{1/3}\cdot \mathrm{Re}^{1/2}$$

$$f = 24\,(1 + 0.15\cdot \mathrm{Re}^{0.687})/\mathrm{Re}$$

where Pr, Re and Sc are the Prandtl, Reynolds and Schmidt numbers, respectively. Now the problem is closed and it is possible to integrate the dimensionless equations (see below). This integration gives us the evolution of the dependent dimensionless variables of the drop (V_x, V_y, ε and Θ) as a function of the dimensionless time (τ)

Dimensionless variables:

$$V_x = \frac{v_x}{v_{term}} \qquad V_y = \frac{v_y}{v_{term}} \qquad \varepsilon = \frac{R}{Ri} \qquad \Theta = \frac{T - T_{wb}}{Ti - T_{wb}}$$

$$\tau = \frac{t}{t_{evap}} \qquad \Upsilon = \frac{Y - Y_\infty}{1 - Y}$$

where:

$$v_{term} = \frac{2\rho_w Ri^2 g(1 - \frac{\rho}{\rho_w})}{9\mu(1 + 0.15\,\mathrm{Re}_{term}^{0.678})} \qquad t_{evap} = \frac{Ri^2\rho_w}{2\rho D_A \Upsilon_{wb}}$$

Dimensionless equations:

$$\begin{aligned}
\frac{dV_x}{d\tau} &= -\frac{\mathrm{Sc}}{4\Upsilon_{wb}}\left(\frac{3}{8}\mathrm{Re}f\frac{1}{\varepsilon^2}V_x\right)\\
\frac{dV_y}{d\tau} &= -\frac{\mathrm{Sc}}{4\Upsilon_{wb}}\left[9\left(1 + 0.15\mathrm{Re}_{term}^{0.687}\right) - \frac{3}{8}\mathrm{Re}f\frac{1}{\varepsilon^2}V_y\right]\\
\frac{d\varepsilon}{d\tau} &= \frac{\mathrm{Sh}}{4\varepsilon}\frac{\Upsilon}{\Upsilon_{wb}}\\
\frac{d\Theta}{d\tau} &= \frac{3}{4\varepsilon^2}\left[\frac{\mathrm{Nu\ Le}}{\Upsilon_{wb}}\frac{C_p}{C_{pw}}(\Theta_\infty - \Theta) - \mathrm{Sh}\frac{h_{fg}}{C_{pw}}(T_i - T_{wb})\frac{\Upsilon}{\Upsilon_{wb}}\right]
\end{aligned} \tag{1}$$

4. Some results (Applications)

The model previously stated must be coupled with the surrounding air equations in order to describe the actual interactions between the two fluids.

However, if we assume independent surrounding air conditions, it is possible to derive useful conclusions providing the basic requirements of the droplests in order to achieve the two cooling applications envisaged.

For direct cooling purposes, total evaporation of the droplets is required. Figure 1 shows the evaporation time as a function of the initial radius of the droplets for different conditions of the air, expresed by the parameter Υ_{wb}.

From this figure and assuming, for instance, a required lifetime of the droplets of 5 seconds and surrounding air conditions whose Υ_{wb} is the corresponding to 37.5^{0}C, 30 % relative humidity, it can be seen that the drop diameter for such case must be in the order of 75 μm. As the air with water droplets will increase progressively its humidity content, the diameter previously obtained is a maximum value, so that the use of bigger drops will not guarantee their total evaporation.

For indirect cooling, the aim is to reduce the temperature of the water drops. Figure 2 plots the cooling time of the drops against the initial diameter with the cooling percentage as parameter $(100*(1-\Theta))$ This figure helps us to choose the combination of drop diameter for an airborne time in order to get the desired cooling.

Assuming that, the airborne time is fixed at, for instance 2 seconds (fountain of 1,5 m high), the initial drop temperature is 30^{0}C and the ambient air conditions are 37.5^{0}C, 30% relative humidity, the temperature of drops whose diameter is 1 milimetre would be 24.8^{0}C when they return to the pond. For bigger diameters the efficiency of the cooling decreases dramatically. For instance, the temperature of drops whose initial diameter is 6 milimetres would be 28.7^{0}C when they return to the pond.

The above results lead us to select sprayers producing droplets from 0.5 to 2 milimetres, in cases where cooling of water is required [6].

Acknowledgements

The authors would like to acknowledge the EXPO'92 State Corporation for the financial support of the research and development work on which this paper is based.

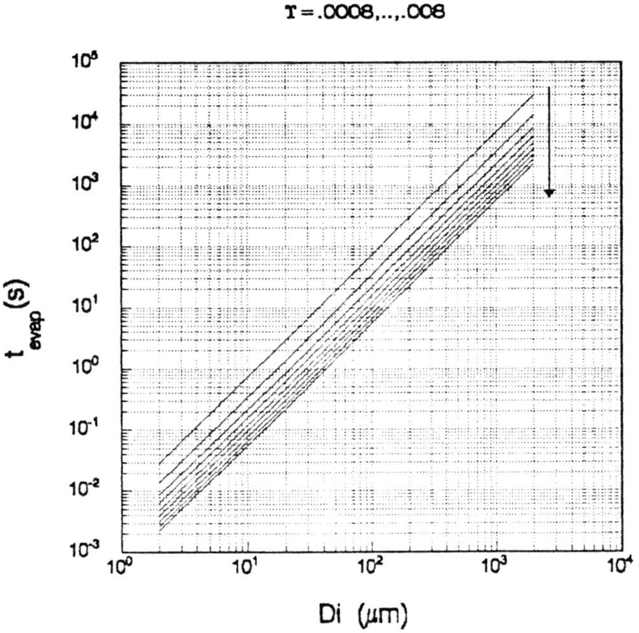

Figure 1: Evaporation time of drops

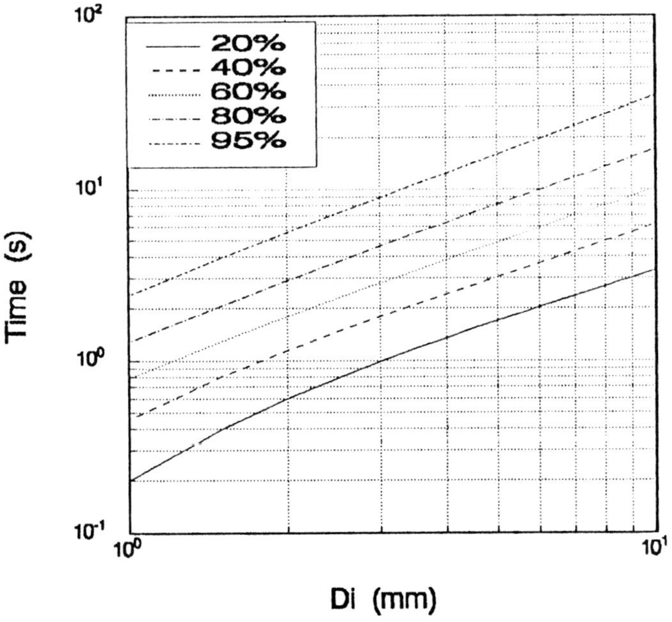

Figure 2: Cooling time of drops

List of symbols

C_p	Specific heat of air
C_{pw}	Specific heat of water
D_A	Binary diffusion coefficient of water in air
f	Fricction factor
g	Gravity acceleration
h	Convective heat transfer coefficient
h_{fg}	Water latent heat of vaporization
K_m	Convective mass transfer coefficient
R	Drop radius
R_i	Initial drops radius
T	Drop temperature
T_∞	Ambient temperature
T_i	Initial drops temperature
T_{wb}	Drop equilibrium temperature
v	Relative velocity drop/air
Y	Saturation mass fraction of water at drop temperature
Y_∞	Ambient mass fraction of water
λ	Air conductivity
μ	Air viscosity
ρ	Air density
ρ_w	Water density

References

1. Velázquez R., Alvarez S., Guerra J., Rodríguez E.A. and Cejudo J.M. (1991) 'General Aspects of the Work on Climatic Control of Outdoor Spaces at EXPO'92' in Architecture and Urban Spaces, PLEA'91, Sevilla
2. Rodríguez E.A., Alvarez S. and Martín R. (1991) 'Direct Air Cooling from Water Drop Evaporation' in Architecture and Urban Spaces, PLEA'91, Sevilla
3. Alvarez S., Rodríguez E.A. and Molina J.L. (1991) 'The Avenue of Europe at EXPO'92: Application of Cool Towers' in Architecture and Urban Spaces, PLEA'91, Sevilla
4. Lefebvre A.H., Atomization and Sprays. Hemisphere Publishing Corporation, New York, 1.989
5. Hewitt G.F., Delhaye J.M. and Zuber N., Multiphase Science and Technology, Vol. 1. Hemisphere Publishing Corporation, New York, 1.982.
6. Velázquez R., Guerra J., Alvarez S. and Cejudo J.M. (1991) 'Case Study of Outdoor Climatic Confort: The Palenque at EXPO'92' in Architecture and Urban Spaces, PLEA'91, Sevilla

LABORATORY MODELLING OF NATURAL VENTILATION VIA CHIMNEYS

G.F.LANE-SERFF, P.F.LINDEN, D.J.PARKER and D.A.SMEED

CERC Ltd	*DAMTP*
3D King's Parade	*Silver Street*
Cambridge CB2 1SJ	*Cambridge CB3 9EW*
U.K.	*U.K.*

ABSTRACT. In this paper we describe a study of the design for a new Engineering School at Leicester Polytechnic, a large building containing a range of spaces including auditoria, drawing studios, laboratories and classrooms. Natural ventilation is the use of naturally occurring pressure differences caused by the wind or by temperature differences to drive a ventilation flow. It is the latter type of natural ventilation (using the "stack effect") that will be considered in this paper. A particular feature of the proposed design is that it uses chimney vents to enhance the stack effect and one of the goals of our study was to establish the effectiveness of this type of system. The building design was studied by a laboratory simulation using a scale model. The experiments showed that the basic design of the ventilation system was sound and indicated areas that needed modification. The results demonstrated that it is generally better to have isolated systems with separate vents for each space rather than having different spaces sharing the same vents. This is because the flows between connected spaces are sensitive to small differences in the occupancy and vent areas of each space. Isolated spaces are also better for the containment of fires and smoke.

1. Introduction: Natural Ventilation

In this paper we describe the results of an investigation into naturally driven ventilation in the proposed building for the School of Engineering at Leicester Polytechnic. The expected heat gains in parts of the building are large, and it was necessary to investigate whether natural ventilation would provide adequate flow rates to maintain comfortable conditions. The study is confined to an investigation of buoyancy driven natural ventilation, ie where the flow is driven by temperature differences (the "stack effect"). In particular, the effect of the wind is not taken into account. The basic principle is to use the temperature differences to induce a flow which will bring cool ambient air into the space.

The ventilation flows found in naturally ventilated spaces can be split into two broad categories: mixing and displacement flows. If the incoming cool air is introduced near the top of a space containing sources of heat then it will descend into the space, mixing up the air and giving a fairly uniform temperature profile. If, however, the cool air is introduced near the floor, and warm air is extracted near the ceiling, a displacement flow will be established with a high degree of stratification. Above the heat sources (such as occupants, equipment, etc) warm air will rise as turbulent plumes, and spread out within the building on

S. Alvarez et al. (eds.), Architecture and Urban Space, 505–510.

reaching the ceiling to form a warm upper layer. This layer will drive flow out of the upper vents and draw ambient air in through the lower vents. (See figure 1.) A steady state will develop with the lower part of the room containing air at close to ambient temperatures, and with warm air confined to the layer in the upper part of the room (and to the rising plumes). There will be an interface between the cool and warm air, and it is clearly desirable to have this interface above the occupied zone. For more details of this type of flow, including a mathematical analysis, see Linden *et al*, 1990.

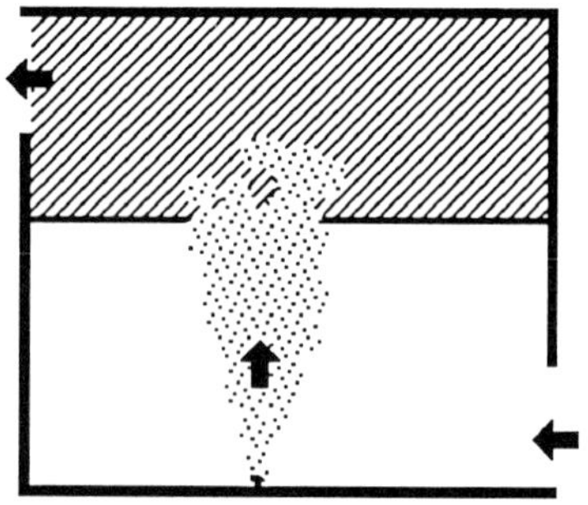

Figure 1. Natural ventilation: displacement type. Warm air rises from the heat source as a turbulent plume, entraining cooler air as it rises. The warm air from the plume forms a layer in the upper part of the space which drives a flow out of the upper opening and in through the lower opening.

2. Case Study: Leicester Polytechnic School of Engineering

The new Engineering School at Leicester Polytechnic is a large building containing a variety of spaces including auditoria, drawing studios, laboratories and classrooms (see figure 2). The building was designed with energy conservation in mind and so makes extensive use of natural light and natural ventilation. Some of the spaces are on the perimeter of the building and fairly shallow so that it is expected that natural ventilation through openable windows will be adequate for these spaces. Many spaces have large gains and little outside wall area and it was not clear that all of these could be adequately ventilated by natural means. The ventilation outlets from these spaces were in the form of tall chimneys to increase the effective stack height. In addition there were low level vents to allow outside air to flow into the spaces near the floor, and with the warm air leaving via the chimney vents this gives a displacement system. The use of tall chimneys is a radical new approach to natural ventilation design, and our work was aimed at assessing the performance of this system.

With the exception of the laboratories most of the heat gains within each space came from occupants. The spaces investigated were the auditoria (gains of 15kW each), laboratories (20kW), drawing studios (3kW), classrooms (3kW) and the central concourse (variable gains of up to 40kW). The effects of a fire in the concourse were also studied (5MW).

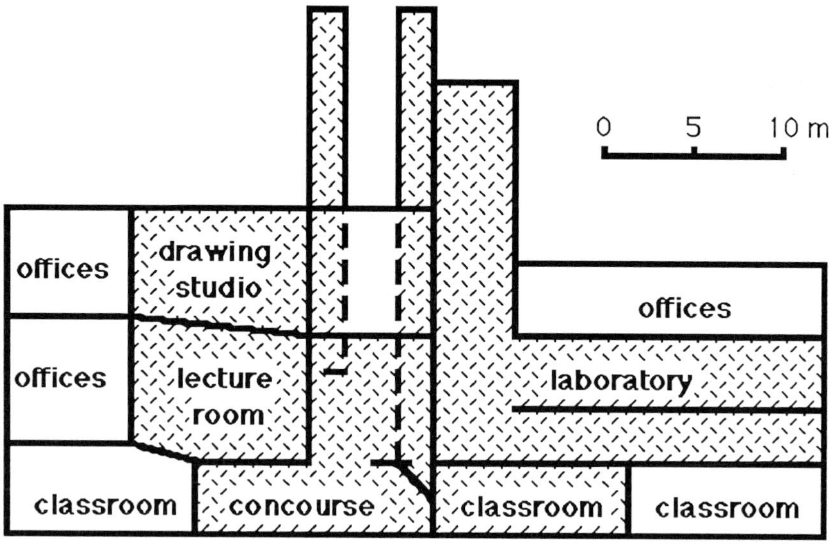

Figure 2. Simplified section of the Engineering School. The shaded rooms were those modelled.

3. Laboratory Modelling

3.1 EXPERIMENTAL TECHNIQUES

A 1:75 scale model of part of the building was used for these tests. The model was made from clear perspex to allow visualisation of the flows. Some features of the building were simplified but the larger spaces and the vents connecting them to one another and to the outside were included. The model was suspended upside down in a large tank of water. Heat sources were represented by sources of salt solution. The falling of dense salt solution in the model represents the rising of warm air in the real building.

The salt water ("hot air") was dyed and a side elevation filmed using an inverted video camera. Samples of fluid were taken during the experiments and the density of these samples determined. These densities can be related to the equivalent temperature for the full scale building.

3.2 SCALING LAWS

For the ventilation flows studied the effects of heat diffusion and viscosity are small compared with advection, as is evident from the high Péclet and Reynolds numbers (of 1000 and above). To simulate these flows accurately in the laboratory it is necessary to ensure that the Péclet and Reynolds numbers in the simulation are also large. This can be achieved in small scale models using water as the working fluid with salt providing density differences, since the viscosity of water is less than that of air and because larger density differences may be achieved without compressibility effects becoming important.

In general lengths, times, velocities and density differences are all different in the model compared with those in the real building. In the experiments described below the flows occurred five to ten times quicker in the model than they would in the real building. Length scales are obviously governed by the scale of the model; other scales are determined by this in conjunction with ratio of buoyancy fluxes between the model and full scale (the buoyancy flux is proportional to the heat flux). The buoyancy forces depend on the difference in densities between different parts of the fluid. A parcel of fluid of density $\rho+\Delta\rho$ surrounded by fluid of density ρ will experience an acceleration of approximately $g\Delta\rho/\rho$, where g is the acceleration due to gravity. This acceleration is known as the "reduced gravity", and will be denoted by g'. For air it is useful to note that $\Delta\rho/\rho$ is approximately equal to $\Delta T/(T+273)$, where T is the temperature in °C. It is not necessary to use the same density differences in the model as would occur in the real building, rather they are chosen to give high enough Reynolds numbers. The scales for times (t), velocities (U) and buoyancy fluxes (B) can be constructed from the length scales (L) and g' as shown in table 1. The model scales are denoted by the subscript M, the full-scale ones by F.

		model	full-scale
scale:	times	$(L_M/g'_M)^{1/2}$	$(L_F/g'_F)^{1/2}$
	velocities	$(L_M g'_M)^{1/2}$	$(L_F g'_F)^{1/2}$
	buoyancy fluxes	$L_M^{5/2} g'^{3/2}_M$	$L_F^{5/2} g'^{3/2}_F$

Table 1. Scales for other variables in terms of the length-scale and the reduced gravity ($g' \equiv g\Delta\rho/\rho$).

Thus, for example, the ratio of velocities in an experiment to those at full-scale will be $(L_M g'_M)^{1/2} : (L_F g'_F)^{1/2}$. The buoyancy flux due to a heat source of 1kW in air at normal temperatures is approximately 0.028 m^4s^{-3}. It should be noted that while the temperatures and flow rates depend on the strength of the heat sources, the basic flow patterns depend only on the geometry and positions of the sources and vents.

As advection provides the dominant part of the heat fluxes in this building, the laboratory model gives reliable estimates of the fluid flow. The details of the fluid flow can then be used in a thermal model to give heat flows between massive parts of the building and also to give the temperature perceived by occupants.

3.3 EXPERIMENTS

Most of the flows observed were displacement type flows with cool ambient air entering through vents near the floor of a space and warm air leaving through vents near the ceiling. There were, however, cases with some mixing caused by incoming fluid entering at high levels or due to high inlet velocities disturbing the stratification in the space. The flow in various spaces was investigated, with both isolated spaces and sets of interconnected spaces being considered. The model incorporated a range of ducts and vents for each space, so that the best strategy for each space could be determined, and detailed recommendations about the size and position of vents could be given. In addition to tests under normal loads, some fire tests were carried out, simulating a fire of 5MW in the concourse.

4. Results

4.1 VENTILATION EXPERIMENTS

A variety of vent areas and positions was tried for each space, and some examples of the results for isolated spaces are given in table 2. Adequate flow rates could be achieved in all spaces, indeed the classrooms proved to be overventilated with uncomfortably large air speeds. However, flow rates do not give the complete picture. From studying the video and from the samples taken during experiments it was clear that there would be a problem with the mezzanine floor in the laboratory. Due to the stratification that built up the temperature on the mezzanine could reach 8°C above ambient even with all the available vents open. In a space ventilated by natural displacement ventilation there is always a problem in keeping the upper part of the space cool, and it is generally necessary to have a "dead space" above the topmost occupied zone that can be allowed to get hot.

Room	Occupancy (people)	Gains (kW)	Vent area (m^2)	Stack height (m)	Flow rate (l/s)
Drawing studio	30	3	4.5	6.5	1300
Lecture room	150	15	3	19	2200
Laboratory	50	20	6	16	5400
Classroom	30	3	1	22	4000
Concourse	100	10	3	22	2100

Table 2. Examples of results from the modelling study for various rooms. The stack height is the distance from the floor of the space to the upper vents. This height is generally bigger than the height of the room because of the use of chimney vents.

As well as experiments with isolated systems, experiments were conducted where different spaces shared the same vents. Problems were found especially when the concourse and lecture room shared the same chimney vents. Hot air flowed from one space to the other depending on only slight changes in the heat gains and vent areas of the two spaces. Given this, and the implications for smoke movement in the case of a fire, it was recommended that the concourse and lecture room have separate vents.

4.2 FIRE TESTS

The results of the fire tests depended on the location of the fire. If the plume from the fire rises close to a wall then entrainment is impeded. This results in a hotter plume but of smaller volume. In all cases there was a rapid build up of a smoke layer at the top of the space which would reach the occupied zone in about 3 minutes. We recommended that there would need to be additional emergency vents (of area at least $15m^2$) to clear the smoke by natural means.

5. Conclusions

The experiments showed that the basic design of the natural ventilation system was sound, and indicated areas that needed modification (design modifications have since been made, particularly to the laboratory and for coping with fires in the concourse). The results demonstrated that it is generally better to have isolated systems with separate vents for each space rather than having different spaces sharing the same vents. This is because the flows between such spaces vary significantly for small differences in the occupancy and vent areas of each space, making the flow effectively unpredictable. Isolated spaces are also better for the containment of fires and smoke.

Laboratory modelling is a very useful tool in the design of naturally ventilated buildings, it is possible to model complicated geometries that are beyond the scope of present computing power. The amount of quantitative information that can be recovered from a laboratory experiment is increasing with the advent of powerful image processing facilities. This type of modelling is also useful in giving a qualitative understanding of the flows, and is especially useful for conveying information about the basic fluid dynamics to non-specialists.

6. References

Lane-Serff, G.F. (1989) "Laboratory models of transient, buoyancy-driven flow", in J.H. Vincent (ed), Ventilation '88, Pergamon, Oxford, pp391-396.

Lane-Serff, G.F. (1989) "Heat flow and air movement in buildings", PhD thesis, Cambridge University.

Lane-Serff, G.F., P.F. Linden, M. Galtry, N.V. Baker (1990) "Laboratory and mathematical modelling: the design of a naturally ventilated building", in F. Lunau and G.L. Reynolds (eds), Indoor air quality and ventilation, Selper, London, pp45-54.

Lane-Serff, G.F., P.F. Linden and D.A. Smeed "Laboratory and mathematical models of natural ventilation", in proc. Roomvent '90, Oslo, 13-15 June, 1990.

Linden, P.F., G.F. Lane-Serff and D.A. Smeed (1990) "Emptying filling boxes: the fluid mechanics of natural ventilation", J. Fluid Mech. 212:309-335.

THERMAL BEHAVIOR MODELLING OF BUILDINGS :

THE NEED OF ADAPTED MODELS

J.J. ROUX, C. INARD, G. KRAUSS
CETHIL, Laboratoire Equipement de l'Habitat
URA CNRS Bât. 307, 20 avenue Albert Einstein
69621 VILLEURBANNE CEDEX

ABSTRACT :

Sub-structuration, reduced order models, "well suited" models of components (walls, ground...) are often used to produced an accurate enough model of the whole system (building) because of all the advantages due to the modularity of this way of modelling. We show in this paper that it's possible to justify using modal analysis some of the pratices used in thermal behaviour modelling of buildings. But, some justifications are still necessary to ensure the robusness of the numerical procedures due to the modelling by composition, particularly in large scale adaptable modelling.

1. INTRODUCTION

The need of adapted models to predict the behavior of a component or to evaluate a proposal has not to be demonstrated anymore. It's particularly true in the field of thermal modelling of buildings.

Various models of numerous components exists [1], but it's not so easy to re-use them to produce the model of the global system.

Encountered difficulties are of differents natures : numerical one, coupling of components models is not always mathematicaly satisfying, the choice of the "best model" for the user is not so evident.

Louis Laret discuss these problems in several articles [2], [3] and propose a methodology to produce linkable, "well-suited" models.

In the following lines we underline some of these difficulties and particularly these due to numerical or mathematical problems and try to justify some usually used technics of resolutions.

S. Alvarez et al. (eds.), Architecture and Urban Space, 511–518.

2. SUB-STRUCTURATION AND COUPLING METHOD

2.1. Modal synthesis and error evaluation

A lot of authors have produced reduced order models of components [4] to [7] using state space approach (set of a few ordinary differential equations) or reduced transfer functions (leading to a set of simple algebraic equations betwen temperature and fluxes).

We used to connect these models of components (walls, ground, thermal bridges...) in order to predict the behavior of the whole building. Is this set of reduced or agragated models really equivalent to the hypothetical original model of the whole building ?. Answers, with mathematical justifications, have been proposed by some authors [8] à [10] ; All these works rely on modal analysis of the thermal system.

CHEN in is thesis [9] show that it is possible to express eigenvalues and eigenvectors of a wall with several slabs of materials knowing eigenvalues and eigenvectors of each slabs of the wall. Futhermore it is possible to obtain a good model of the wall combining reduced order models of each wall (modal synthesis). Moreover an evaluation of the error due to model synthesis is given. It seems that, on the example of a cell (7 walls + 1 air volume) used by Neirac in his thesis [8], NEVEU numerically obtains the same results [10] ; that it to say, it is possible to built a good model of the thermal behavior of the cell using reduced order models of each wall ; but no evaluation of the error is proposed in that case. Moreover, Neirac shows relation ships between eigenvalues of the cell and eigenvalues of each wall of this cell and evaluate their contribution in the response of the whole system.

In his model Neirac uses a finite difference method to discretize the one dimensionnal conduction heat transfer equation. He considers a global "convective + Radiative" heat transfer coefficient between the surfaces of the walls and the "air temperature" supposed to be homogeneous in all the volume of the cell (MODEL1). For n spatials nodes, a set of n equations that can be condensed into a matricial form is obtained :

$$[C]\,\{\mathring{T}\} = [A]\,\{T\} + [B]\,\{u\} \tag{1}$$

The hypothesis for radiative transfer is not very satisfaying ; so in order to evaluate the influence (on the eigenvalues of the system of such an hypothesis we built a second model (MODEL 2) using a "linearized" radiosity model for radiative heat transfer, we obtain :

$$[C]\,\{\mathring{T}\} = [A1]\,\{T\} + [A2]\,\{TS\} + [D]\,\{u\} \tag{2}$$

$\{TS\}$: Internal Surface temperatures vector

Surface balances $[\phi conv] + [\phi rad] + [\phi cond] = 0$ give :

$$[A2]^t \{T\} + [K2] \{TS\} - [RAD] \{TS\} = 0 \qquad (3)$$

$$\text{With } \{\phi rad\} = [IFES]^{-1} [IF1] \{\mathring{M}\}$$

$$= [IFES]^{-1} [IF1] [HRI] \{TS\} \qquad (4)$$

[HRI] includes non linear terms $\sigma\sigma$ $(Ts_i^2 + Ts^2)(Ts_i + Ts)$; if we suppose this terms to be egals and constants, for each surface i, we obtain a linearized form of (4) :

$$\{\phi rad\} = [IFES]^{-1} [IF1] [HRI] \{Ts\}$$
$$= [RAD] \{Ts\} \qquad (5)$$

then, (3) in equation (2) gives :

$$[C] \{\mathring{T}\} = [A'] \{T\} + [D] \{u\} \qquad (6)$$

Where :

$$[A'] = [A1] + [A2] [[RAD] - [K2]]^{-1} [A2]^t$$

If we compare eigenvalues (λi) (or time constants $\tau i = 1/|\lambda i|$) of $[C]^{-1}$ [A] for MODEL 1 and of $[C]^{-1}$ [A'] for MODEL 2 we obtain some differences (Figure 1).

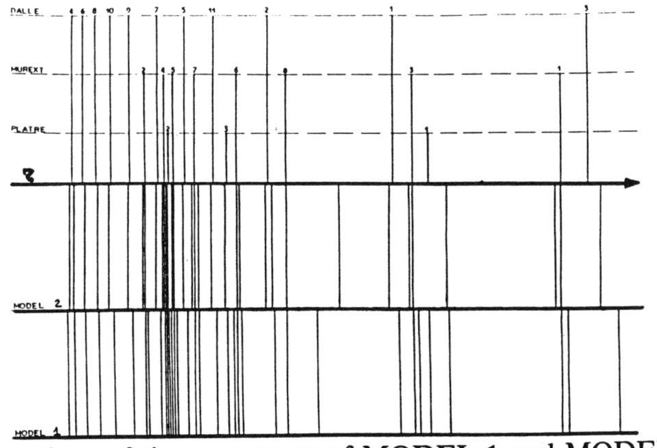

Figure 1 : Comparison of time constants of MODEL 1 and MODEL 2

These differences are normal because MODEL 1 and MODEL 2 are rather differents, but what is interesting is that the observed perturbations in time constants distribution for MODEL 1 and for MODEL 2 are not so important and let us think that modal synthesis and analytical evaluation of errors using MODEL 2 could be possible.

2.2 Numerical Sensitivity Evaluation

In this section, we evaluate the numerical sensitivity to the choice of the reduced order models used for conduction heat transfer. To do so, we compare three models of a cell obtain with non linear radiative heat transfer model + non linear convective heat transfer model and for conduction heat transfer :

DIFF.F : Finite differences [15]

AMABAT : Reduced order models using modal analysis [11]

TONY : Reduced transfer functions using Z - transform method [16]

Results of these comparison are given, for differents cases, on figure 2 (definition of all the cases in [11]).

These results are maybe (depends of the goal) rather good (no more 10 % difference between temperature evolution of internal temperature evaluations).

3. CONCLUSION AND PROSPECTS

It is possible, using modelling by composition (connecting or coupling components models [12] [13]), to obtain well suited adaptable models for buildings and HVAC ; but some efforts are still necessary to evaluate errors due to connection of reduced order models (modal synthesis, physical agregation or reduced transfer function) and to distinguish them those due to numerical technics and then obtain accurate enough global models as described in figure 3, as we are intending to in our project and development of design or simulation tools.

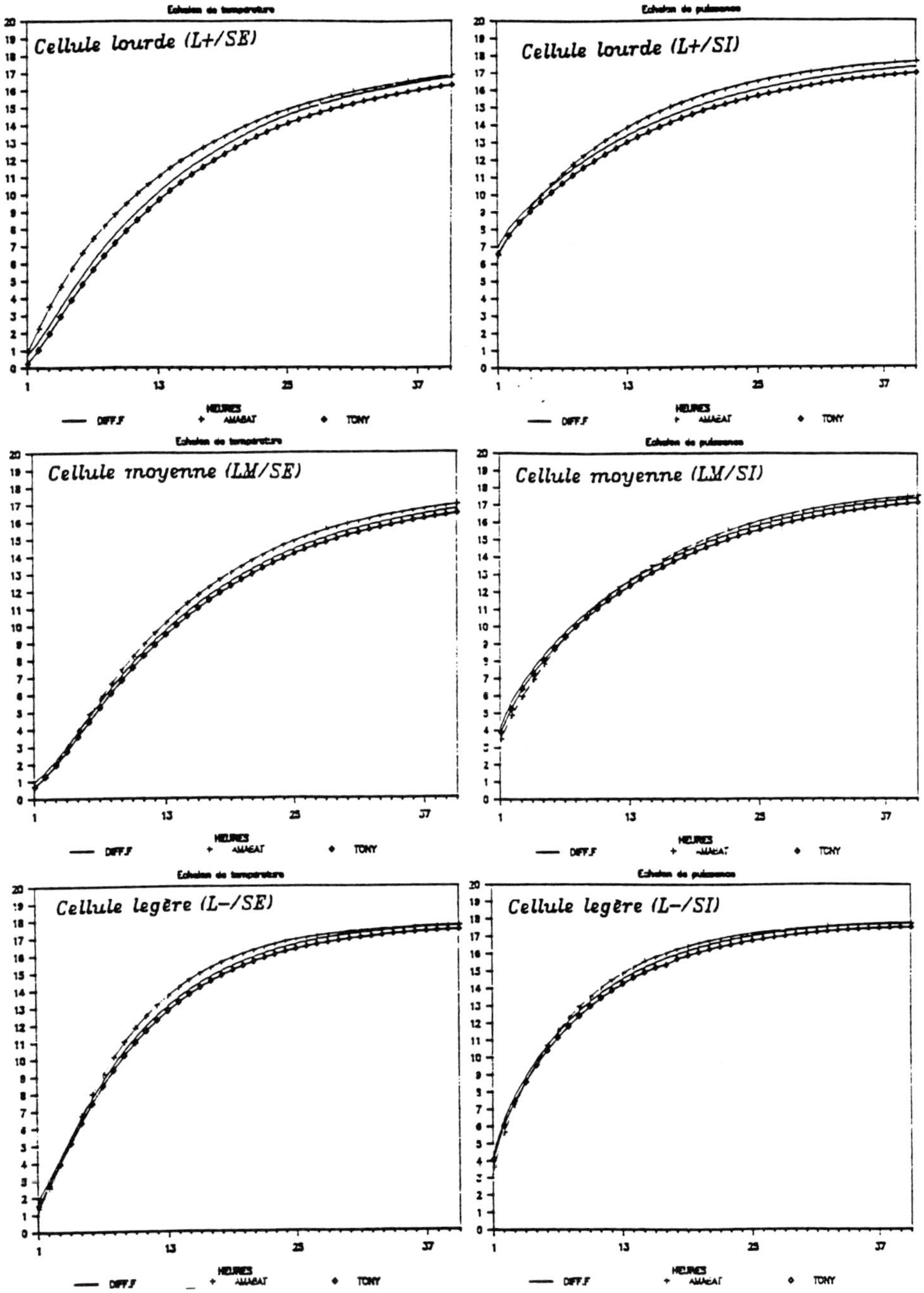

FIGURE 2 : Comparison of coupled reduced order models (AMABAT,TONY) to reference model (DIF.F).

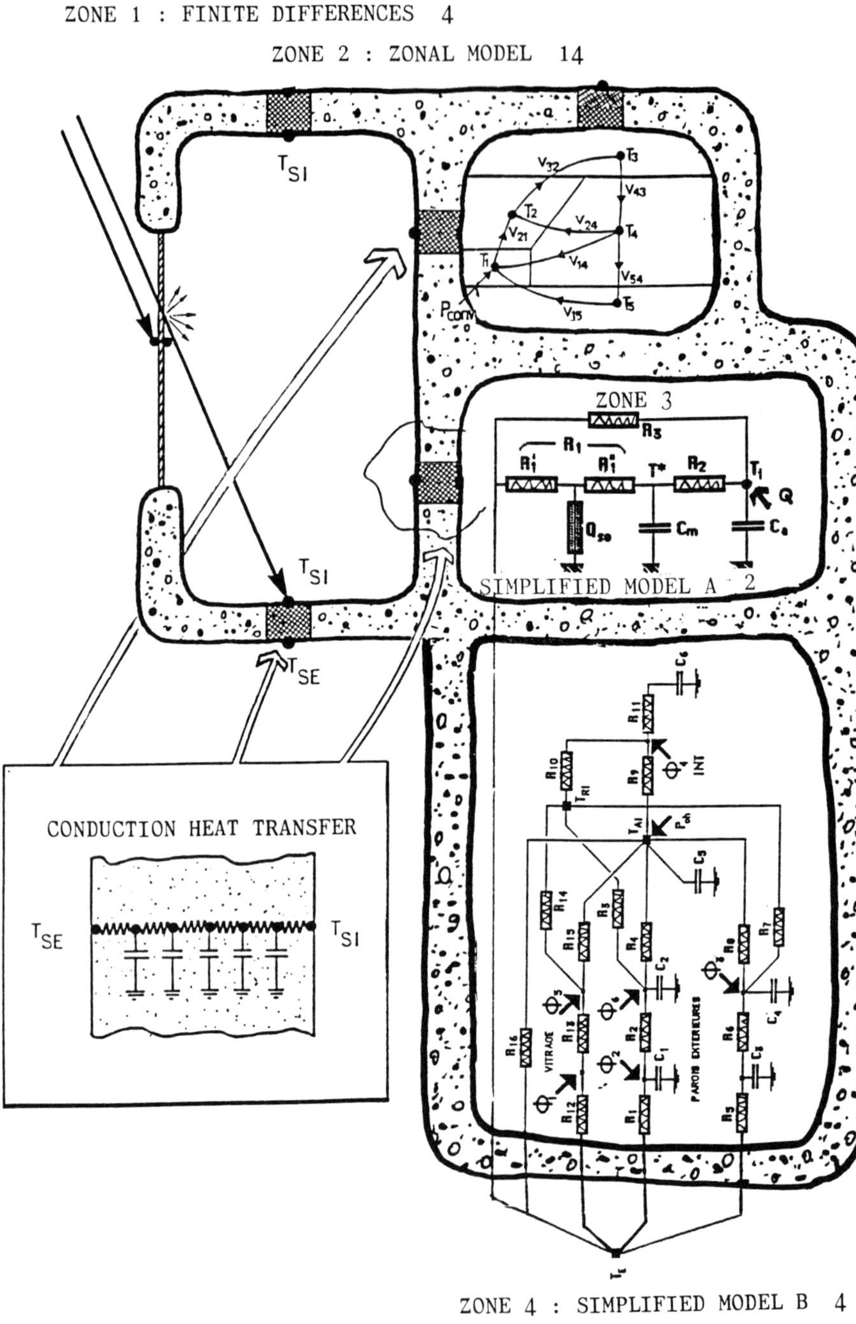

FIGURE 3 : Coupling of different models

BIBLIOGRAPHY

[1] PEREZ Sanchez Maria
"Typologie et Uniformisation Syntaxique des Modèles de transfert de chaleur dans le contexte de la thermique du bâtiment".
Thèse de doctorat INSA de Lyon Janvier 1989.

[2] LARET Louis
"Le Concept de Modèle Adapté"
Séminaire Spécialisé" Outils d'Aide à la Conception et à la Gestion"
Sophia Antipolis 26-27 octobre 1989.

[3] LARET Louis
"Building and HVAC Simulation : The need of well-suited models"
Building Simulation 89 VANCOUVER June 23 - 24 1989

[4] J.J. ROUX
"Proposition de modèles simplifiés pour l'étude du comportement thermique des bâtiments"
Thèse de docteur Ingénieur INSA LYON - Avril 1984

[5] NGENDAKUMANA Philippe
"Modélisation Simplifiée du Comportement Thermique d'un Bâtiment et Vérification Expérimentale"
Doctorat en Sciences Appliquées Université de Liège 1987

[6] SALGON Jean-Jacques
"Analyse Modale par Discrétisation Spatiale. Application à la Modélisation des Ponts Thermiques dans les bâtiments"
Thèse de doctorat Paris VI - 30 Juin 1987

[7] SEEM J.E, KLEIN S.A., BECKMAN W.A., MITCHELL J.W.
"Transfer function for efficients calculation of multidimensional transient heat transfer"
Journal of heat transfer - February 1989 Vol III, P 5-12

[8] NEIRAC F.P.
"Approche theorique et expérimentale des modèles réduits du comportement thermique des bâtiments"
Thèse de doctorat Ecole des Mines de Paris - Mars 1989

[9] CHEN M.
"Nouvelle solution modale pour les parois des bâtiments"
Thèse de doctorat Ecole des Mines de Paris - Octobre 1989

[10] NEVEU A.
"Reduction de modèles"
Séminaire "Outils d'aide à la conception et à la gestion"
Sophia Antipolis - 26-27 Octobre 1989

[11] MOKHTARI A.
"Etude du comportement thermique des bâtiments en contact avec le sol par l'analyse modale"
Thèse de doctorat INSA de Lyon - Septembre 1988

[12] NEVEU A.
"Etude d'un code de calcul d'évolution thermique d'une enveloppe de bâtiment"
Thèse de 3ème cycle - Paris VI Juin 1984

[13] RODRIGUEZ E.A.
"Sistematizacion de acoplamientos termicos y termoaeraulicos en la simulation de edificios"
Tesis ESII de Sevilla - Octobre 1990

[14] INARD C., ROUX J.J., GERY M.
"Anisothermal steady state modelling of heated dwelling-cell with zonal models application to radiation and heating floor"
Third International conference "System Simulation in buildings"
University of Liège - December 3-5 1990

[15] BRAU J.
"Modélisation thermique des bâtiments - Validation des modèles de calcul par une méthode expérimentale en ambiance climatique simulée.
Thèse de docteur d'état es Sciences INSA de Lyon - Avril 1980

[16] ROLDAN A.
"Etude thermique et Aéraulique des enveloppes de bâtiment. Influence des couplages intérieurs et du multizonage
Thèse de docteur Ingénieur INSA de Lyon - Décembre 1985.

A FREQUENCY-DOMAIN IDENTIFICATION METHOD APPLIED TO SIMPLIFIED MODELLING OF BUILDING ELEMENTS

J.M. Cejudo, J. Guerra, J.L. Molina

Universidad de Sevilla
Departamento de Ingeniería Energética
y Mecánica de Fluidos
Av. Reina Mercedes s/n
E-41012 Sevilla. Spain

ABSTRACT. This paper presents a method to obtain simple models of thermal systems by means of a frequency-domain identification technique. This is applied to onedimensional and bidimensional transient conduction which are common in building simulation.

1. Introduction

The importance of using simple models has always been recognized. The less powerful are computer resources the more important are simple models. Nowadays this kind of models are still interesting in order to reduce simulation time. Simple models are also suitable to obtain significant parameters in order to catalog thermal systems.

Generally, models of thermal systems are complex. Even though physical laws which govern heat transfer in buildings are well known, the models obtained combining these laws have important drawbacks: they are time consuming, thermal properties and other transfer coefficients are not always known, etc. The other approach to model building is "the determination on the basis of input and output, of a system within a specified class of systems, to which the system under test is equivalent". This is the commonly accepted definition of identification (Zadeh [1]). It is clear that the system can be physically existing or not. In the last case simulated outputs must be used instead of real outputs.

This paper presents a method to obtain simple models based on the second approach to model building, that is, identification. Application to onedimensional and bidimensional transient heat transfer conduction are presented although the method is appropiate to any kind of linear time-invariant system.

2. Description of the method

The fundamental hypothesis of the method is that the dynamic system under study is linear and time-invariant. Validity of this hypothesis has been discussed elsewhere in the case of heat transfer in buildings (see Molina [2]).

For these kind of systems there are many useful results (see for instance Ljung [3]). Response of a linear time-invariant system to a sinusoidal excitation is also sinusoidal (magnified and shifted), provided the system is stable and the initial transient response vanishes. The quotient between output and input is called frequency transfer function. For these systems if an excitation is composed by different frequencies, the response can be

S. Alvarez et al. (eds.), Architecture and Urban Space, 519–524.

obtained by superposition of responses to each frequency. The proposed method consists of:

a) Selection of a model structure. In the cases presented in this paper, regressive models of order 2 have been selected. From now on we will refer to these kind of models particularly suited for the method. For a discussion about other models refer to Cejudo [4].

b) Imposition of the equality between two transfer functions for some frequencies, one of then obtained from experimental (or detailed simulation) data and the other with the simple model. The former is accomplished for the appropriate number of frequencies in orden to close the system, having the same number of parameters and equations. By this way a linear system is obtained. Solving this system the parameters of the regressive model are calculated. In the following sections two application cases are presented.

3. Onedimensional transient conduction

A simple model for one dimensional transient heat transfer conduction will be obtained for a non-homogeneus wall. Similar applications exist, for instance Virgone [5] using experimental data and Seem and others [6] with simulated data. In order to allow comparison between simple models obtained via identification or model reduction of transfer function (Seem and others [6]) the same case defined in [6] will be considered. The wall under study is named 25 in ASHRAE [7] and inputs are either a unit outside or inside temperature step. Figure 1 shows the wall considered.

a) The class of model:

The following model has been choosen:

$$q(t) = a_1 \cdot q(t-\Delta t) + a_2 \cdot q(t-2\Delta t) + b_0 \cdot T(t) + b_1 \cdot T(t-\Delta t) + b_2 \cdot T(t-2\Delta t)$$

where $q(t)$ is heat flux in time t (W/m^2), $T(t)$ is air temperature (exterior or interior) in time t (C) and Δt is time step . The model order is in accordance with many previous works (for instance Seem and others [6], Neveu and others [8]) and has been verified with other input signals.

b) Calculating the parameters of the model:

In the case treated in this paper the step response is looked for. Then we select frequencies in an appropiate way in order to reproduce, with the simple model, the heat flux obtained with the detailed model.

b.1 Steady state regimen:

In the steady state regimen the following condition must be fulfiled:

$$(b_0 + b_1 + b_2) = Q_{rp} \cdot (1 - a_1 - a_2)$$

where $Q_{rp}(W/m^2)$ is the steady state heat flux for a unit step temperature.

b.2 Choosing the frequencies:

Figure 2 shows the Bode diagram of amplitudes for the heat flux inside when the wall is excited outside. Providing the steady state has been imposed, two frequencies in the descending part of the Bode plot are selected (1 cycles/day and 4 cycles/day). In the

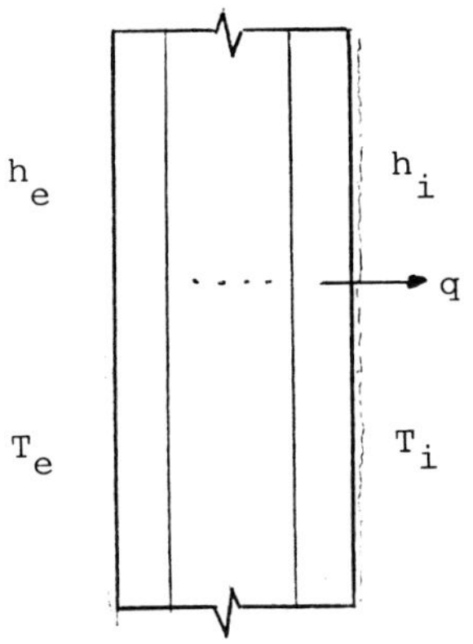

Figure 1. Onedimensional problem

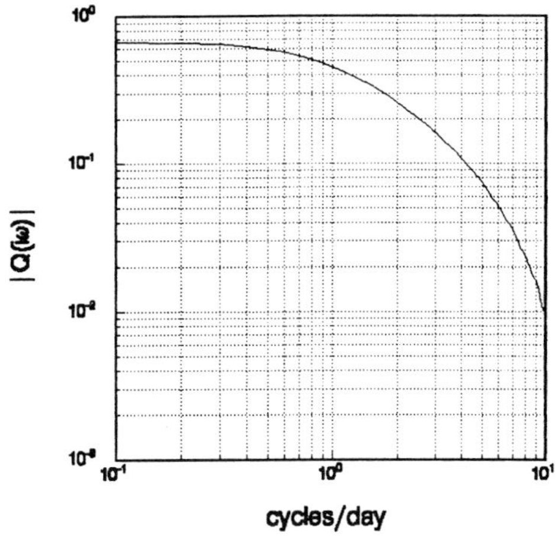

Figure 2. Bode plot of amplitudes

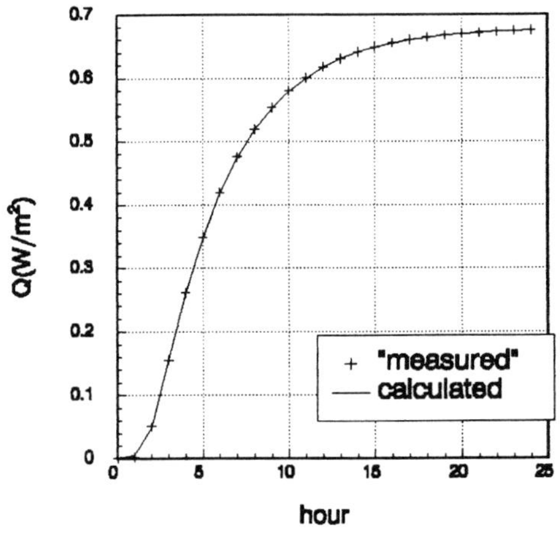

Figure 3. Response to unit step in outdoor temperature

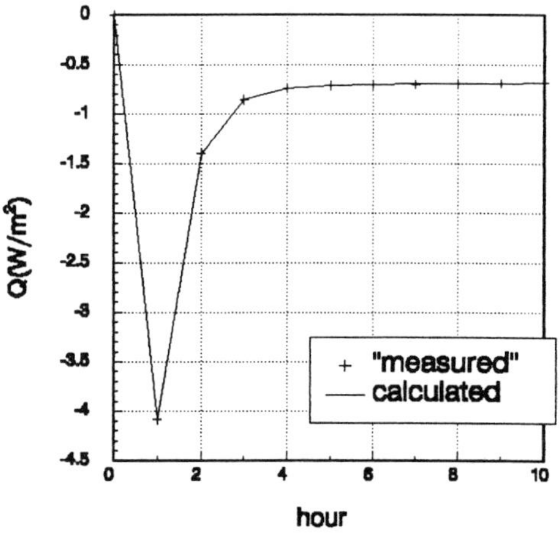

Figure 4. Response to unit step in indoor temperature

	a_1	a_1	b_0	b_1	b_2
Case 1	1.0456	-0.2034	$4.478 \cdot 10^{-3}$	$3.893 \cdot 10^{-2}$	$6.380 \cdot 10^{-2}$
Case 2	0.9813	-0.1547	-4.0830	6.6929	-2.7277
Case 3	0.9839	-0.1553	-4.0877	6.7063	-2.7360

Table 1: Results for onedimensional problem

Excitation in face	a_1	a_2	b_0	b_1	b_2
2	1.2977	-0.3658	3.6778	-5.6425	2.0384
3	1.3746	-0.4309	$-7.596 \cdot 10^{-2}$	$2.250 \cdot 10^{-2}$	$2.615 \cdot 10^{-2}$
4	1.7873	-0.8046	$-2.584 \cdot 10^{-2}$	$6.026 \cdot 10^{-2}$	$-4.480 \cdot 10^{-2}$

Table 2: Results for bidimensional case

case of heat flux produced by an inside unit step the Bode plot is different. Following similar criteria 0.3 cycles/day and 2 cycles/day have been selected. Imposing equality between simulated and "measured" heat flux complex amplitudes, parameters of the regressive model are calculated.

b.3 Results

Following the procedure described above the calculated parameters are shown in table 1 where case 1 refers to heat flux produced by outside step, case 2 refers to heat flux produced by inside step and case 3 refers to parameters obtained using a dominant root method [6].

Figure 3 shows heat flux simulated with detailed and simple models in the case of outside excitation. An excellent agreement between both is obtained. Figure 4 is equivalent to Figure 3 but in this case heat flux is produced by an inside unit step temperature. The agreement is equally excellent.

Different excitations produce different values of a_1 and a_2 parameters (which determine time constants). This is physically coherent because of different behaviour against inside and outside excitations. In this case the whole model have 10 parameters instead of traditional transfer function which have 8 parameters.

4. Bidimensional transient conduction

A typical example of thermal bridge will be modeled with the same methodology discussed above. The problem consists of a plaque, one face isolated (see figure 5). Face 2 is excitated by an air temperature which simulates nocturnal ventilation. Figure 6 shows inside temperature. Faces 3 and 4 are exposed to solar-air temperature corresponding to south and east orientations respectively. Frequencies selected to carry out identification are 1 cycles/day and 2 cyles/ day. In this case a 15 parameter model is obtained, that is, 3 simple models 5 parameters each. Table 2 shows results obtained for the parameters which describe heat flux in face 2 when faces 2, 3 or 4 are excited.

Figure 7 shows heat flux obtained with detailed model [9] versus simple model. A start-

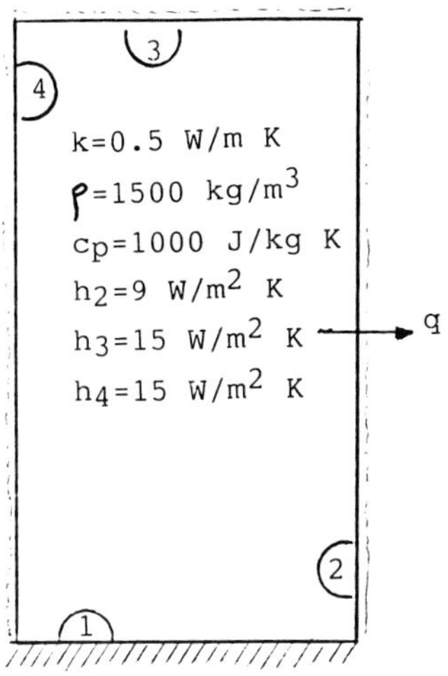

Figure 5. Bidimensional problem

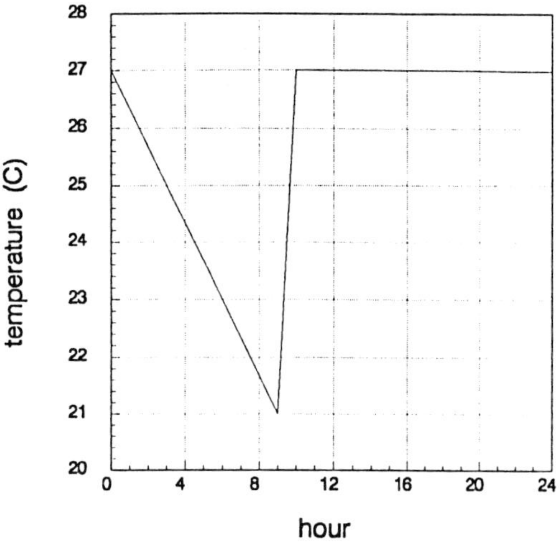

Figure 6. Inside air temperature

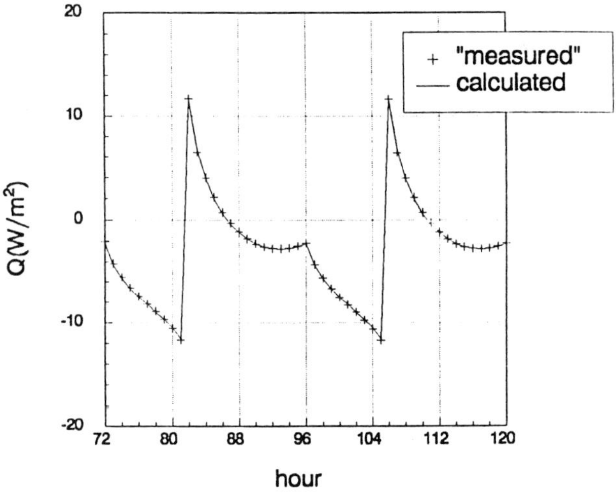

Figure 7. Heat flux in face 2

up period of 3 days has been considered. An excellent reproduction of the "exact" heat flux is obtained with the simple model.

5. Conclusions

A methodology allowing us to obtain simple models for heat transfer phenomena has been presented. This technique can be considered as a frequency-domain identification method. Two succesful applications are shown: onedimensional and bidimensional conduction in building components (non-homogeneus wall and thermal bridge). These simple models permit dynamic simulation of thermal systems with high accuracy and important savings in computing time. Furthermore, interesting parameters like response factors, time constants, etc, can easily be obtained.

References

1. L.A. Zadeh (1962) 'From circuit theory to system theory' Proc. IRE 50, pp 856-865
2. J.L. Molina (1987) 'Simulación energética de edificios mediante un modelo de respuesta en frecuencia'. Tesis Doctoral Universidad de Sevilla.
3. L. Ljung (1987) "System identification. Theory for the user" Prentice Hall, Inc.
4. J.M. Cejudo.(1991) 'Identificación. Aplicación a la obtención de modelos simplificados de muros'. Dpto. Ing. Energética y Mecánica de Fluidos. Universidad de Sevilla.
5. J.Virgone (1982) 'Etude théorique et expérimentale du comportement thermique des parois de bâtiment par les facteurs de réponse'. Tesis INSA Lyon.
6. J.E. Seem, S.A. Klein, W.A. Beckman and J.W. Mitchell. 'Model reduction of transfer functions using a dominant root method'. J. Heat Transfer. Vol. 112, pp. 547-554. (1990)
7. ASHRAE (1977) Handbook of Fundamentals.
8. A. Neveu, P. Bacot and R. Regas (1986) 'Modeles d'évolution thermique des bâtiments. Conditions pratiques d'identification'. Rev. Gén. Thermique. n° 296-297, pp. 413-420.
9. E.A. Rodríguez (1990) 'Sistematización de acoplamientos térmicos y termoaeraúlicos en la simulación de edificios' Tesis Doctoral Universidad de Sevilla.

A SIMPLE MODEL FOR THE ESTIMATION OF POTENTIAL COOLING BY NIGHTTIME VENTILATION

J. VAN DER MAAS, C.-A. ROULET
Ecole Polytechnique Fédérale de Lausanne
Laboratoire d'Energie Solaire et de Physique du Bâtiment
EPFL, LESO-PB
CH-1015 Lausanne
Switzerland

ABSTRACT. We present a simple model to calculate the energy loss during nighttime ventilation. The time dependence of the exhaust air and wall surface temperatures is predicted by a simplified dynamic model that couples air flow, heat transfer, and wall temperature. For given ventilation rate the model predicts that the total heat extracted from the building during the night can be maximized by increasing the heat exchanging surface area and the thermal effusivity, $\sqrt{(\lambda\rho c)}$, of the wall materials. The influence of ventilation rate on the heat removed by nighttime ventilation is discussed. The model is tested in field measurements where the air flow is known from the stack effect.

1. Introduction

Convective air cooling is useful in heavy buildings located in climates where large diurnal temperature variations occur. Large thermal capacity structures cause the indoor temperature to be close to the average outdoor temperature. Nighttime ventilation removes heat from the building mass, resulting in a reduced average indoor air temperature.

Key issues of energy calculations of ventilative cooling are [1,4] the evaluation of the exhaust air temperature (dependence on heat transfer and flow rate) and the estimate of the mass flow rate (dependence on internal flow resistances), but to exploit nighttime ventilation, architects still have to dimension the openings, relying more on their feeling than on calculation. In this paper we present a model which can help in developing detailed guidelines for ventilative cooling.

In the following we first describe the algorithm for calculating the temperature of the exhausted inside air during cooling, then we present a comparison of the model with field measurements and finally, we discuss the parameters influencing the cooling potential of nighttime ventilation [3,4].

2. Model for ventilative cooling

From the mass flow rate $\dot{m}$, and the temperature of the exhausted inside air T_{in}, the heat loss rate (taken positive from inside to outside) is given by

$$Q = C_p \dot{m} (T_{in} - T_{out}) \quad (1)$$

S. Alvarez et al. (eds.), Architecture and Urban Space, 525–530.

where T_{out} is the temperature of the outside air entering the building.The model couples the ventilative heat loss rate with the heat transfer between the air and the walls and a decreasing wall surface temperature. The mass transfer and thermal models form a four-node network (Figure 1).

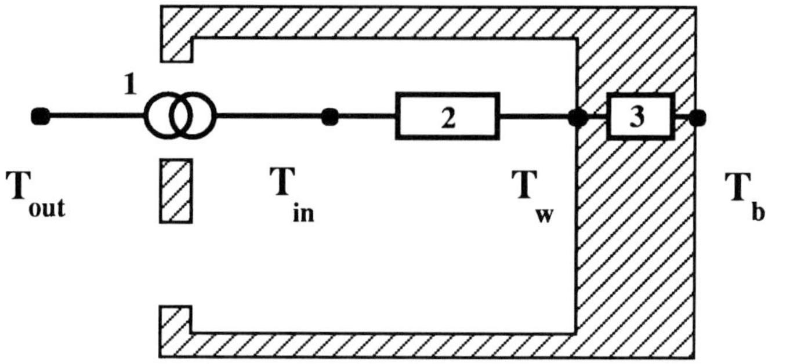

Figure 1. The model for ventilative cooling can be represented by a four-node network. (1) current source for ventilation heat loss (2) boundary layer resistance, (3) dynamic wall resistance. T_b is a reference wall temperature.

2.1 MASS FLOW RATE

The mass flow due to the stack effect, can be calculated for various opening configurations. Our test case is a volume with two openings (area A_t at the top and A_b at the bottom). The influence of wind (in general an enhancement of the air flow) was not considered. For openings A_t and A_b much smaller in height than the distance H between the middle of the two openings, the mass flow rate is given by [2]

$$\dot{m}= \rho_{in}\, A_t\, C_1 \sqrt{2g\, (\frac{T_{in}-T_{out}}{T})}\ \sqrt{\frac{H}{1+(T_{in}A_t/T_{out}A_b)^2}} \qquad (2)$$

where a single discharge coefficient C_1 ($\approx$0.6) is assumed. The ventilation heat flow is represented in Figure 1 by a heat source (1) , its value being given by Equations 1 and 2.

2.2 HEAT TRANSFER

The temperature difference between the wall surface and the air is described by a convective heat transfer coefficient h_c, and equals $T_w - T_{in}=q/h_c$, where q is the heat flux density. A fixed value of $h_c = (6\pm1)W/m^2K$ appeared to be consistent with the measurements [4]. The value of q is obtained by dividing the total heat loss Q by the heat exchanging wall surface area

$$T_w-T_{in}=q/h_c=Q/C_2S_i\, h_c \qquad (3)$$

The coefficient C_2 is the fraction of the total internal wall surface area S_i, which is active in the heat transfer process. The heat transfer is represented in Figure 1 by a thermal resistance (2). The heat capacity of the air adds a small time constant [4] which is here neglected.

2.3 WALL SURFACE TEMPERATURE

We are interested in the time dependence of the surface temperature of the wall, T_w. The walls are considered to be in thermal equilibrium with the building until ventilation starts at t=0, $T_w(t=0)=T_b$. For a constant heat loss density q, the wall surface temperature of a semi-infinite solid is given by

$$T_w(t) = T_w(0) - \frac{2q}{b}\sqrt{\frac{t}{\pi}} \tag{4}$$

where $b=\sqrt{\rho c \lambda}$ (the square root of the product of thermal conductivity, density, and specific heat) is the thermal effusivity of the wall material, see [4]. In the present model, this expression is used even when q does vary with time. This appears to be an acceptable approximation, as long as the actual value of q(t) is close to its time averaged value.

2.4 SOLVING

After substituting Q and $T_w(t)$ in Equation 3, one solves the resulting non-linear expression for T_{in} in a few iterations, and the cooling rate Q is evaluated with (1).

3. Experimental Test

The model has been tested for the case of the cooling of office rooms (volume 50-100 m^3) by a single open window. The agreement was quite good [3].

In this paper we consider another configuration, i.e. the high mass staircase of a three level office building (volume 4.5 by 8 by 15 = 540m^3) with ventilation openings at the roof and ground level ($H \approx 10$m, C2=1). The walls are characterized by an average thermal effusivity of b=1000 [J/(m^2 K.$s^{0.5}$)] with a 20% uncertainty, as discussed in [4]. The total wall surface area of the staircase is S_i=700 m^2 with a 10% uncertainty. Measured are the temperatures of the in- and outflowing air, and of the wall as a function of time.

Figure 2 shows an example of nighttime ventilation lasting from 10 p.m. to 8 a.m. The openings are A_b=1 and A_t=1.8m^2. Are plotted both the calculated and measured temperature of the outflowing air (T_{in}), the measured outdoor temperature and a few values of the wall temperature as a function of ventilation time. The heat loss rate Q is given in the upper part of the plot. The total cooling energy calculated over the 10 hour period is 90kWh.

The calculated exhaust air temperature, T_{in}, is about one degree too low which is nearly 10% of T_{in}-T_{out} and Q is therefore underestimated by up to 15%, showing the kind of precision we can expect from such a simplified model.

We note that both the inside and outside air temperature lowered at about the same rate, and the ventilation rate, being almost constant over the night, was 10 air changes per hour.

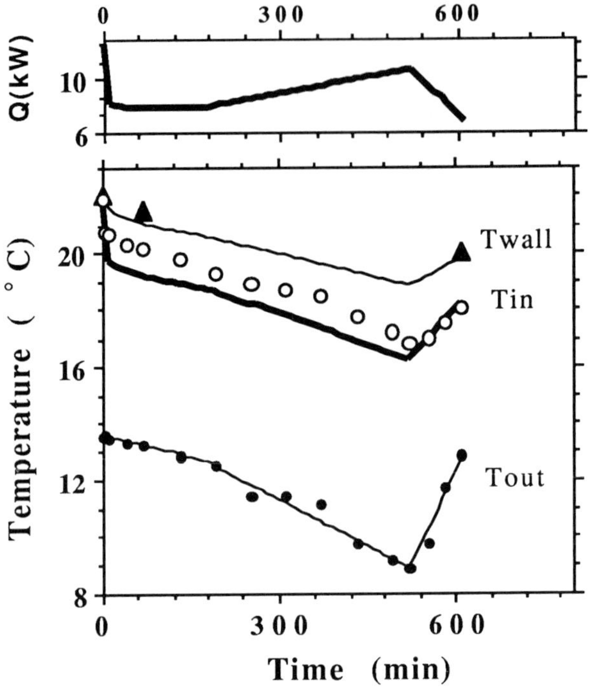

Figure 2. Night ventilation of a $540m^3$ staircase from 10 p.m. to 8 a.m. The measured and calculated exhausted inside air temperature, T_{in}, the outside temperature, T_{out}, the wall temperature, T_{wall}, and the calculated heat loss rate, Q, as a function of ventilation time. $A_b=1m^2$, $A_t = 1.8\ m^2$, H=9.6m. The calculated energy loss over 10 h is 90kWh.

Figure 3. Simulation of energy loss by night time ventilation. Calculated is the total energy loss integrated over 10 hrs, for two values of S_i and b and as a function of initial inside-outside temperature difference, $T_w(t=0)$ -T_{out}; T_{out} = constant. b= 1000 (thick curve), 250 (thin curve) $J/(m^2Ks^{0.5})$. $A_{in}=2\ m^2$ [20 ft^2]. A_{out} = 2 m^2 [20 ft^2]. H=10m [33 ft].

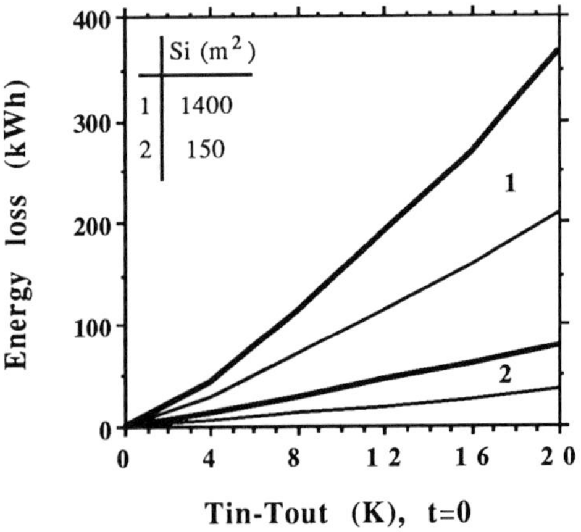

4. Discussion

The simplified model has two main parameters which determine the cooling potential (i) the active heat exchanging wall surface area C_2S_i and (ii) the thermal effusivity b of the wall material, characterizing the rate at which the inside surface temperature decreases.

The cooling rate increases with ventilation rate but is not proportional to it as Equation (1) would suggest. This is because the inside air temperature decreases further below the wall surface temperature.

The influence of S_i on the relation between cooling and mass flow rate is simple when we take $\dot{m}$ and T_w constant. After elimination of T_{in} , Equations (1) and (3) give then the following expression for the heat loss

$$Q = \frac{C_p\, m\, (T_w - T_{out})}{1 + \frac{C_p\, m}{S_i\, h_c}} = \frac{S_i h_c\, CF\, (T_w - T_{out})}{1 + CF} \qquad (5)$$

The maximum value for the heat loss $Q = S_i h_c (T_w - T_{out})$ is approached when the cooling factor $CF = C_p\dot{m} / S_i h_c$ is large. For CF=1, Q is at 50% of its maximum value, and to increase Q further one has to increase the mass flow $\dot{m}$ in a disproportionate way. In conclusion, for a given value of the wall surface area S_i, we estimate an airflow rate of $\approx (S_i/200)\ m^3/s$ to correspond to CF=1. For the case of Figure 2, the air flow rate was nearly constant at 1.6 m^3/s and $CF \approx 0.5$.

To show the relative importance of wall area and material choice, we calculated the total energy loss over a 10 hours period, for a few values of S_i and b and varying initial inside-outside temperature difference, $T_w(t=0) - T_{out}$. The outside temperature was set constant. The results are plotted in Figure 3. Thick curves correspond to a heavy building structure and the thinner lines to a light-weight structure with a wall material four times lower thermal effusivity.

5. Conclusion

This paper concerns a model to estimate the ventilative cooling potential of buildings. Tested on a high mass staircase cooled by stack ventilation, the time dependence of the air and wall temperatures is correctly explained without the need for adjusting coefficients. The main parameters of the model are the heat exchanging wall surface area and the thermal effusivity b of the wall material.

6. Acknowledgement

This research is sponsored by the Swiss Federal Energy Office (OFEN).

7. References

[1] Antinucci, M., B. Fleury, J. Lopez d'Asiain, E. Maldonado, M. Santamouris, A. Tombazis, S. Yannas (1989). 'Horizontal study on passive cooling for buildings', 1st draft. Brussels: Commission of the European Communities, Building 2000.

[2] ASHRAE. (1985). 'ASHRAE handbook -1985 Fundamentals' Atlanta: American Society of Heating, Refrigerating, and Air-Conditioning Engineers, Inc.

[3] Van der Maas, J. and Roulet, C.-A. (1990), 'Ventilation and energy loss rates after opening a window', Air Infiltration Review 11(4), 12-15

[4] Van der Maas, J. and Roulet, C.-A. (1991), 'Nighttime Ventilation by Stack Effect', ASHRAE Winter meeting New York, Symposium NY-91-05-3, ASHRAE Transactions 97, Part 1.

NOMENCLATURE

A_b = bottom opening area [m^2]
A_t = top opening area [m^2]
b = thermal effusivity, $(\lambda \rho c)^{0.5}$ [$J/(m^2\ K.s^{0.5})$]
c = wall specific heat [J/kg.K]
C_p = specific heat of air at constant pressure [J/kg.K]
C_1 = flow discharge coefficient
C_2 = wall surface fraction for heat transfer
g = gravitational acceleration [m/s^2]
h_c = convective heat transfer coefficient [$W/m^2 \cdot K$]
H = vertical distance between centres of openings [m]
$\dot{m}$ = mass flow rate [kg/s]
q = density of heat flow rate [W/m^2]
Q = ventilation heat flow rate [W]
S_i = internal wall surface area [m^2]
t = time [s]
T = mean absolute air temperature[K]
T_{in} = inside air temperature [°C]
T_w = wall surface temperature [°C]
T_{out} = outside air temperature [°C]
ρ = density [kg/m^3]

Chapter 2

Components Performance

EXPERIMENTAL CHARACTERIZATION FOR DAYLIGHTING INNOVATIVE COMPONENTS

F. ALEO, M. ANTINUCCI, P. CARBONARO
Conphoebus S.c.r.l.
95030 Piano d'Arci - Catania
Italy

ABSTRACT. Daylighting control allows an electrical energy saving because limits the utilization of the artificial light and reduces the cooling loads in Summer. A fixed solar shading component located in front of the glazed surface produces a strong decrease of the summer cooling loads whereas an automatic computer controlled shading system allows both the reduction of the overheating and good visual conditions. In this work some suitable components for daylighting applications are described. The difficult problem of the thermal and daylighting performance experimental assessment is then afforded. Illuminance, radiative and thermal measurements have been performed in test cells equipped with different solar shading components.

1. Introduction

1.1 PURPOSE OF THE WORK

Daylighting control allows an electrical energy saving because limits the use of the artificial light and reduces the summer cooling loads and consequently the management of daylighting systems constitues an important strategy in order to obtain energy efficient buildings.
As component and system performance information are very important for the designer, the specification of reliable test methodologies is an essential topic.
The main aim of this work is to show how different types of solar shading components designed for the Mediterranean climate may be characterized. As this climate is characterized by long hot summers, the selected components should contemporaneously shade the solar radiation and secure good visibility conditions during the working hours.
Optical, radiative and thermal measurements have been performed on the components under testing to assess thermal and daylighting performances.

1.2 THE COMPONENTS UNDER TESTING

The first component tested is an automatically controlled external venetian blind installed outside a glazed façade on the South side of a test cell.

S. Alvarez et al. (eds.), Architecture and Urban Space, 533–539.

The control system designed and carried out by Conphoebus consists of an illuminance sensor and a control routine running on an IBM Personal Computer.
The other systems under testing are three "between glass shading system" single window elements and two 3.2 x 3.2 m curtain walls.
The structure of the single window elements is described below:

element 1: 5 mm thick external tempered glass - black metallic screen (23 lamellae per inch) - 6 mm thick low emittance glass - 9 mm air gap - 6 mm thick internal glass ;

element 2: identical to the component 1 but with white screen;

element 3: identical to the component 1 but without the air gap and the internal glass.

Each curtain wall is composed of nine single elements. Type 1 elements constitute the first curtain wall while type 2 elements the second one.
The curtain wall frame utilizes thermal break Aluminium profile.

2. Test methodologies and experimental apparatus

2.1 QUALIFICATION OF THE AUTOMATIC VENETIAN BLIND

The tests were performed using a test cell as controlled ambient having on the South face the automatic venetian blind. A sensor to measure and control the internal diffuse illuminance was placed on a horizontal plane located almost at the center of the test room 90 cm above floor level.
The tests have been carried out in two different operating conditions of the control system. In the first condition the venetian blind was completely raised and without any automatic control, whereas in the second one a control to keep the internal illuminance between 340 and 410 lux was programmed. Internal illuminance and free-floating air temperature comparative measurements have been performed during these two tests.

2.2 SURFACE TEMPERATURE MEASUREMENTS

Internal glass surface temperature measurements on the elements 1, 2 and 3 have been performed.
The measurements were performed using Platinum Resistance Thermometers (Pt 100) glued to the glass surface by means of conductive silicone.

2.3 LUMINOUS AND SOLAR TRANSMITTANCE

The transmittance measurements have been carried out on the photopic and solar spectrum.
To perform the solar transmittance measurements two radiometers have been contemporaneously used. The two instruments were placed on vertical planes, the first inside the test cell, immediately behind the sample, the second one outside, completely exposed to the solar radiation.
During the test, the internal radiometer was placed inside a black box, in order to avoid the radiometer sensor be hit by the reflected solar radiation.
The solar irradiance was measured using two CM11 Kipp & Zonen pyranometers.

The luminous transmittance was measured by a single spectroradiometer equipped with fiber optic light guide. The measurements were performed in a very clear day.

2.4 OPTICAL AND THERMAL PERFORMANCE ASSESSMENT OF TWO CURTAIN WALLS

The experimental characterization of the curtain walls has been carried out applying the "Integrated Comparative-Subtractive" method based on the use of two test cells, previously described in [1].
The wall steady-state energy balance leads to the following linear relationship:

$$\frac{Q}{H} = (gA) - (UA) \bullet \left[\frac{(\bar{T}_i - \bar{T}_e) \bullet \Delta\tau}{H}\right] \bullet \ (1)$$

In this expression Q is the net thermal energy which passes through the wall, H is the solar energy density on a vertical South surface, $\bar{T}_i$ and $\bar{T}_e$ are respectively inside test cell and outside mean temperatures and finally $\Delta\tau$ is the integration time.
The identification of the performance curve (see equation 1) can be obtained by a linear regression of the experimental data. The y-axis intercept and the slope of the steady state performance straight-line represent the solar aperture and the overall heat loss coefficient of the component (respectively gA and UA).

2.5 ILLUMINANCE MEASUREMENTS INSIDE THE TEST CELLS EQUIPPED WITH THE CURTAIN WALLS

Illuminance comparative measurements have been performed inside the two test cells equipped with the two different curtain walls. In the center of the test room a standard daylighting parameter called "window factor" was also calculated. The window factor is defined as the ratio of the internal illuminance to the external.
The internal photometer was placed on a horizontal plane located almost at the center of the test cell 90 cm above floor level.
The external photometer was placed on the same plane as the curtain wall so that it "sees" the same portion of the sky as the façade.

3. Results

3.1 AUTOMATIC VENETIAN BLIND TEST RESULTS

Figure 1 shows the trend of the illuminance inside the cell room in the test days.
The diagram shows a satisfactory operation of the computer controlled shading system that is able to control the illuminance within the selected limits (340 ÷ 410 lux).
Moreover the automatic shading system reduces strongly the overheating inside the test cell. Infact the unshaded test cell reached a peak temperature

of 46°C whereas under controlled illuminance conditions a maximum temperature of 36°C was measured.

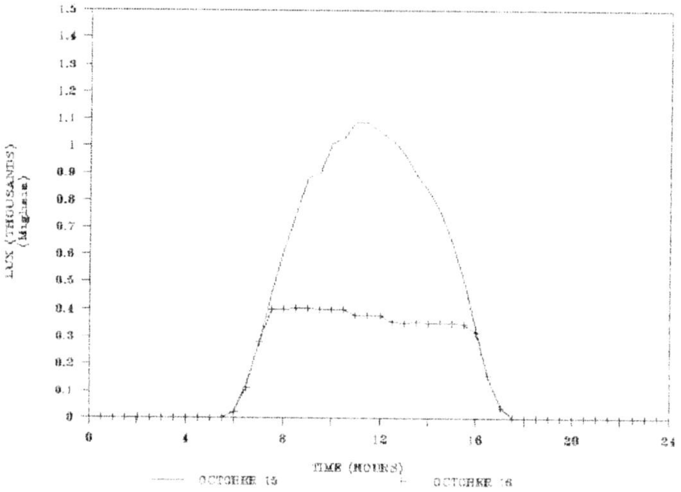

Figure 1. Illuminance control at the test cell center.

3.2 SURFACE TEMPERATURE MEASUREMENT RESULTS

The results of the surface temperature measurements performed on the window elements show that the addition of an internal air gap limits strongly the heat trasmission caused by the solar heated screen. Infact from 8 a.m to 4 p.m the average surface temperature of the elemen 3 (with single internal glazing) is about 8°C higher than the average surface temperature of the element 1 (with double internal glazing).
The surface temperature comparative measurements carried out on elements 1 and 2 show that the different colour of the metallic screens produces the trasmission of sensibly different thermal fluxes.
In this case infact from 8 a.m to 4 p.m element 1 (Black coloured screen) has an average surface temperature 3.5°C higher than that measured on element 2 (white coloured screen).

3.3 PHOTOPIC AND SOLAR TRANSMITTANCE MEASUREMENT RESULTS

The transmittance measurement results are shown in Table 1.
Element 2 is characterized by photopic and solar transmittances considerably higher than those measured on element 1.
This different behaviour is caused by the large difference existing between the white and black metallic screen solar reflectances.
Measurements were performed at angles of incidence from 47 to 57 degrees.

TABLE 1. Transmittance measurement results

ELEMENT	TEST DAY	AVERAGE INCIDENCE: ANGLE(DEGREES)	τsol 0.3-2.5 μm %	τphot. 0.38-0.78 μm %
1	26-02-91	45	-	1.8
	27-02-91	45	2.4	-
2	03-03-91	47	9.0	-
	27-03-91	56	8.9	-
	29-03-91	57	-	13.8

3.4 CURTAIN WALLS TEST RESULTS

The results of the "Integrated Comparative Subtractive" tests performed on the two curtain walls are synthetically showed in Table 2.

TABLE 2. Thermal performance parameters of the curtain walls

CURTAIN WALL COMPOSED OF TYPE 1 ELEMENTS (BLACK COLOURED SCREEN)	CURTAIN WALL COMPOSED OF TYPE 2 ELEMENTS (WHITE COLOURED SCREEN)	GEOMETRICAL PARAMETERS
$(gA)_1$ = 1.89 sqm g_1 = 0.24 $(UA)_1$ = 24.0 W/°C U_1 = 2.34 W/sqm*°C	$(gA)_2$ = 1.62 sqm g_2 = 0.20 $(UA)_2$ = 24.1 W/°C U_2 = 2.35 W/sqm*°C	A = 10.24 sqm A_g = 7.95 sqm A = Total surface of the curtain wall A_g = Glazed surface of the curtain wall

The linear best fits (see figure 2) have a very high correlation coefficient.
The solar factor "g" of the glazed components only, may be obtained, neglecting the solar aperture of the frame, by the following expression:

$$g = \frac{(gA)}{A_g}$$

where A_g is the glass area (here : 8 m^2).
The two elements are characterized by quite similar solar factors because the black one transmits less solar radiation that the white one but re-emits a lot of absorbed radiation.
The solar factors are both approximately 20%. Similar values are shown by "low ε-reflecting" glasses, that are typical competitors of shading devices for curtain walls.
The thermal transmittance "U" is referred to the whole curtain wall (frame and glazed components): the two walls have the same U-value in the range 2.3 ÷ 2.4 W/m^2°C, slightly lower than the U-values of standard double glazing.

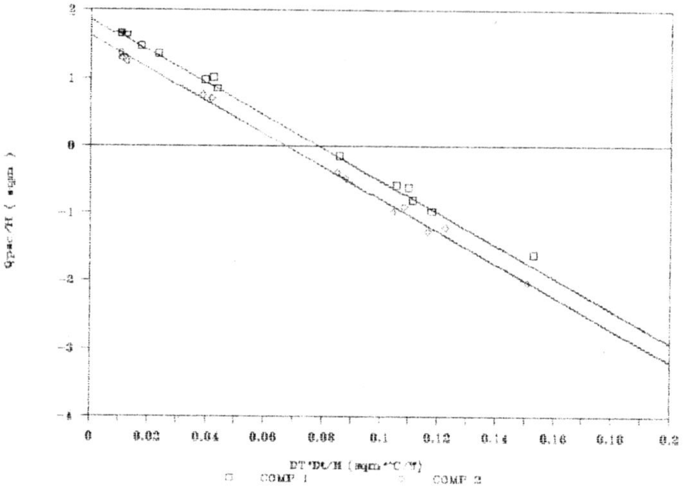

Figure 2. Thermal performance curves of the curtain walls.

3.5 ILLUMINANCE MEASUREMENTS RESULTS

The illuminance comparative measurements performed inside the two test cells show that the illuminance level in the cell equipped with the component 2 (white coloured screen) is much higher than that measured inside the other test cell (see figure 3).

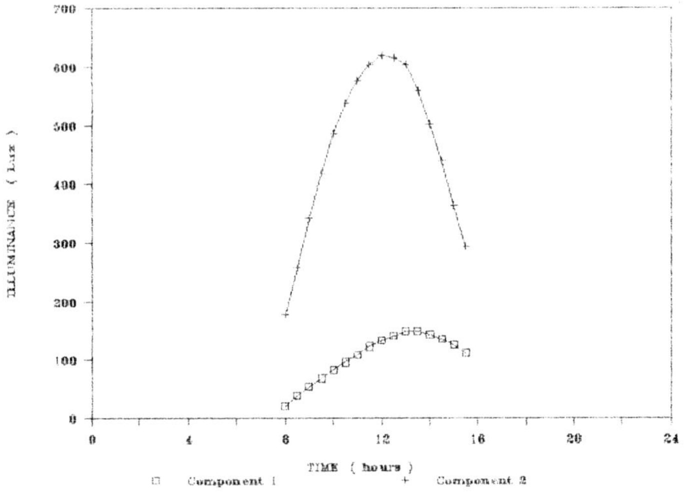

Figure 3. Illuminance inside test cells.

Moreover the white curtain wall gives in the test room center an average window factor of 2%, against 0.2% for the black one.
In most industrial nations the illuminance level to guarantee good visibility conditions for the normal office work has been defined between 250 and 300 lux. The visibility conditions were therefore insufficient in the cell with "black" curtain wall, at least in our test conditions. The version of the same screen with 17 lamellae per inch is preferable.

4. Conclusions

The test results give interesting information on the thermal and daylighting performance of the selected components.
This type of information is useful for the development of new components, before introducing them into the market, to verify performance behaviour in conditions very similar to the actual ones.
Moreover the obtained performance parameters may be used by the designer to better predict building performance.

5. References

[1] Aleo F., Antinucci M., Carbonaro C. (1989)
Performance assessment of sun-shading components for Mediterranean buildings. Second European Conference on Architecture (SECA) Paris 4-8 Dec.1989.

SOME CONSIDERATIONS ON THE ENVIRONMENTAL PROPERTIES OF DOMES.

JOSÉ MARÍA CABEZA LAINEZ
Architect
C/ Castillo de Aroche, 1
Edificio Octógono Aptmto. 122
41013 Seville. SPAIN
Tph. (5) 4233953

ABSTRACT. With this issue we intend to promote a different attitude towards architecture and especially bioclimatic architecture. It is time to react against the increasing trends for standardization of "right angles" in building construction. If bioclimatism has come of age for architectural practice, it should resume the search for richer and more complex new forms so constant, on the other hand, in nature and in the proposals of the past.

1. INTRODUCTION.

1.1. Conceptual properties.

Much has been written about the geometrical, aesthetic and even mystic features of the dome; in words of Giedion /1/ the dome is a typical phenomenon from the second space conception in architecture that of the radiance of interior space. The Roman Pantheon can be a relevant example of this conception.

Inside that building for instance daylighting reveals, according to Burckhardt /2/, visions of the unattainable. Then, the domes of Bernini, Borromini, Guarini and Vittone added important precisions to the original notion and as stated be Wittkower /3/ those building became the measure of themselves for they had no equal.

Nowadays, the use of this typology has seriusly declined mainly with the integration of "urban" concepts in building design and the ebb of spiritual conters.

1.2. Environmental properties.

Historically, the dome belonged to the basic repertoire of most vernacular cultures. Its outstanding structural capacities constituted the basic for the membrane and shell theory. Its thermal properties also, have become rather well known in the last few years, the low surface to volume ratio whith prevents thermal dissipation, the maximum exposure to night cooling in arid regions and the enhanced air-flow through top-vents are typical examples.

Should all this meet with the former conceptual properties we would enter the third idea of space, the radiance between interior an exterior; the building itself, from its spaciousness, reaches

S. Alvarez et al. (eds.), Architecture and Urban Space, 541–547.

out for the environmental.

However, in the aim to retrieve the new dome typology, a major difficulty arises whem analizing the daylighting potential as the available tools are unsuitable for curvilinear geometries. That extreme we will try to solve from now on.

2. CONFIGURATION FACTORS.

2.1. Rectangular surface source.

Let us consider the luminance due to a Lambertian surface:

$$L = \frac{dI_\theta}{dS*\cos\theta}$$

and the illuminance formula:

$$dE = \frac{dI_\theta*\cos\theta}{1^2}$$

Substituting into this expression dI_θ we obtain:

$$dE = \frac{L*dS*\cos^2\theta}{1^2}$$

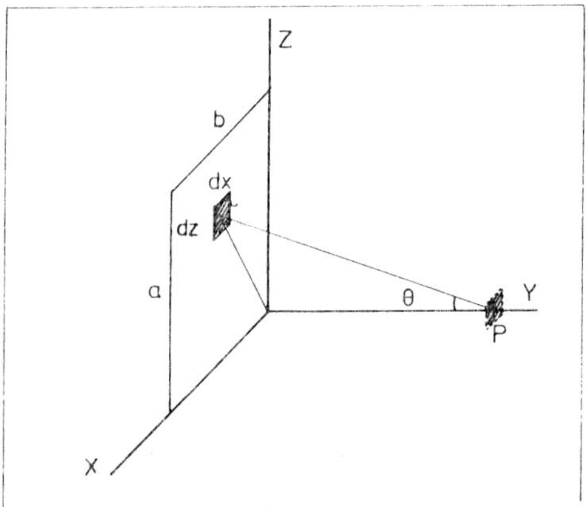

Fig1.rectangular surface.cartesian coordinates

Using Cartesian coordinates, as illustrated in figure 1, this can be written:

$$dE = L * \frac{y^2 \, dx \, dz}{(x^2 + y^2 + z^2)^2}$$

Which over the domain of the rectangular surface becomes,

$$E = L \int_0^a \int_0^b \frac{y^2 \, dx \, dz}{(x^2 + y^2 + z^2)^2}$$

Integration that yields

$$\frac{b}{y^2} \quad \frac{1}{\sqrt{y^2 + b^2}} \quad \text{arctg} \quad \frac{a}{\sqrt{y^2 + b^2}} \quad +$$

$$+ \quad \frac{a}{y^2 \sqrt{a^2 + y^2}} \quad \text{arctg} \quad \frac{b}{\sqrt{a^2 + y^2}}$$

And we arrive at the final result

$$E = \frac{L}{2} \left[\frac{b}{\sqrt{b^2 + y^2}} \quad \text{arctg} \quad \frac{a}{\sqrt{b^2 + y^2}} \quad + \quad \frac{a}{\sqrt{a^2 + y^2}} \quad \text{arctg} \quad \frac{b}{\sqrt{a^2 + y^2}} \right]$$

Thus the sought-for integral is the Higbie configuration factor for horizontal rectangular surfaces. Accordingly we can use the same procedure to find the configuration factor for sloped rectangular surfaces or nonrectangular surfaces. The configuration factor must be multiplied by the luminance from the surface to establish the illuminance at a definite point.

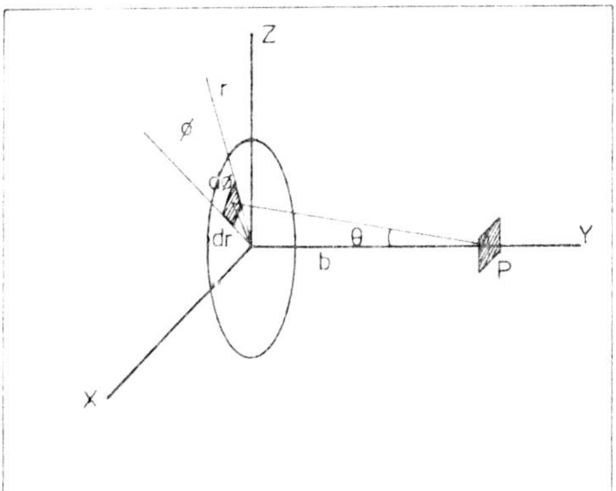

Fig 2. circular surface. polar coordinates

2.2. Circular surface source.

The station poind lies directly under the contre of the circle. Using polar coordinates, as illustrated in figure 2, we have

$$dE = \frac{L*dS*\cos^2 \theta}{l^2}$$

That yields

$$dE = \frac{L*r*dr*d\phi*b^2}{(r^2 + b^2)^2}$$

Integrating over the circular domain

$$E = L \int_o^{2\pi} \int_o^{a} \frac{b^2*r*dr*d\phi}{(r^2 + b^2)^2}$$

And then we obtain

$$E = L \int_o^{2\pi} \left[\frac{-b^2}{2(r^2 + b^2)} \right]_o^a d\phi$$

Substituting

$$E = L \int_o^{2\pi} \frac{a^2 \, d\phi}{2(a^2 + b^2)}$$

And performing the outer integration we get

$$E = \frac{L\pi a^2}{a^2 + b^2}$$

Which is much simpler than the Higbie factor. However, for points other than the formerly described, the integral becomes complicated and is not suitable for the designer's use. Neverthless, the precedent formula shows that illuminance at the station point can be conceived as a ratio between the projection onto the same plane of the whole hemisphere taken as a unit. This is, in synthesis, the projected solid angle principle and many daylighting protractors are based on it though for rectangular apertures.

For circular openings or the like, the problem reduces to intersection of a symmetric elliptical cone and a sphere with common centre and plane of symmetry. This is a usual question of descriptive geometry which can be easily solved over the very desing drawings. Unfortunately, mathematical proofs show that the projection of such intersection is often a fourth degree curve whose area is difficult to determine via analytical methods but, for practical purposes, we can assimilate it to an ellipse. In this way, the illuminance due to direct component of daylighting canbe found graphically with a high degree of accuracy. All this is illustra ted in figure 3.

3. INTERIOR REFLECTED ILLUMINANCE FROM CURVED GEOMETRIES.

3.1. The integrating sphere formula.

With the internally reflected component we could invoke other approaches as, for instance, the integrating sphere formula /4/. There, the average interior reflected illuminance depends on an infinite number of reflections inside the sphere and then

$$IRE = E \frac{W}{A} (R+R^2+R^3+\ldots+R^n)$$

Where W is the total window area, A is the internal surface of the sphere and R its mean reflectance. The sum of this series gives

$$\lim_{n \to \infty} \frac{R^{n+1}-1}{R-1} -1 = \frac{R}{1-R}$$

Whence

$$IRE = E \frac{W * R}{A(1-R)}$$

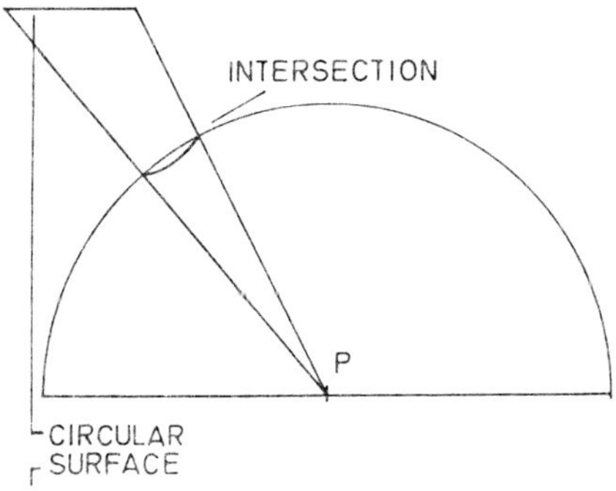

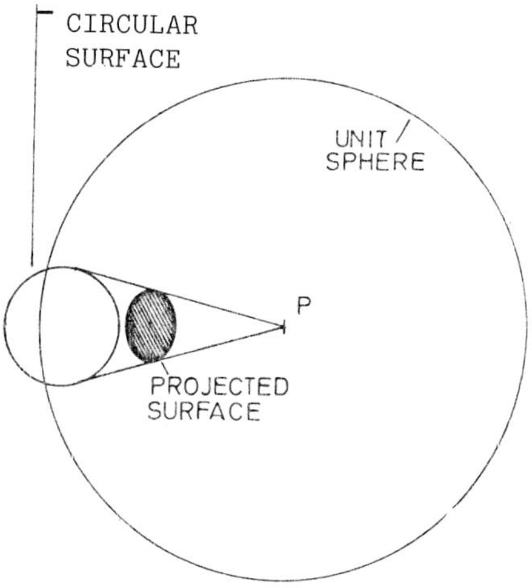

Fig.3 circular surface. determination of direct component of daylighting

4. ILLUMINANCE FROM DIRECT SUNLIGHT IN CURVED ROOMS.

4.1. The secondary source approach.

With direct sunlight we can work as if we had two circular surface sources. In fact, the cylinder defined by beam daylighting through a circular aperture depicts onto the interior surfaces a curvilinear figure often assimilable to an ellipse or circle. Of course, the luminance of this second source depends upon the intensities of solar radiation and its relative position changes with the sun as well. Also the sky and internal components have tobe modified to take into account the circumsolar brightness and the internal reflection of direct sunlight.

5. CONCLUSIONS.

Further generalization of the method with the help of Computer Aided Design will considerably enlarge the possibility of analyzing complex systems in the design stage. Also, curvilinear finite elements will be available. Another important conclusion is the fact, that under symmetric skies, apertures posessing axial symmetry yield a vectorial illumination field with axial symmetry as well. This can be a step forward in the aim to create a new environment in the daylighting field.

6. REFERENCES.

/1/ Giedion, S. 1971. Architecture and the Phemomena of Transition. The three Space Conceptions in Architecture. Harvard University Press. Cambridge, Massachusets.
/2/ Burckhart, J. 1933. Gesammelte Werke, vol 10: Der Cicerone. Berlin.
/3/ Wittkower, R. 1972. Vittone. Atti del Covegno internazionale dell'Academia delle Scienze de Torino.
/4/ Phillips, G. and Littlefair, P. 1988. Avarage Daylight Factor Under Rooflights. Proc CIBS National Lighting Conf., Cambridge.

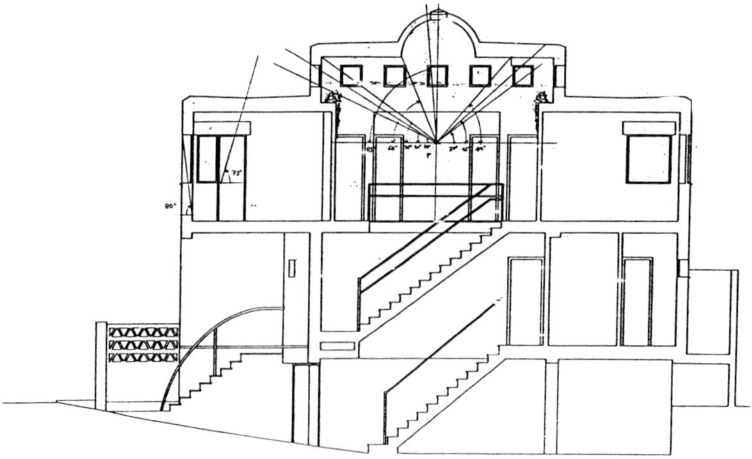

Figure 4. Bioclimatic House with dome for daylighting and cooling. Project by the author now under construction. Finishing July 91.

Figure 5. Bioclimatic library with central dome for cooling and daylighting. Project by the author.

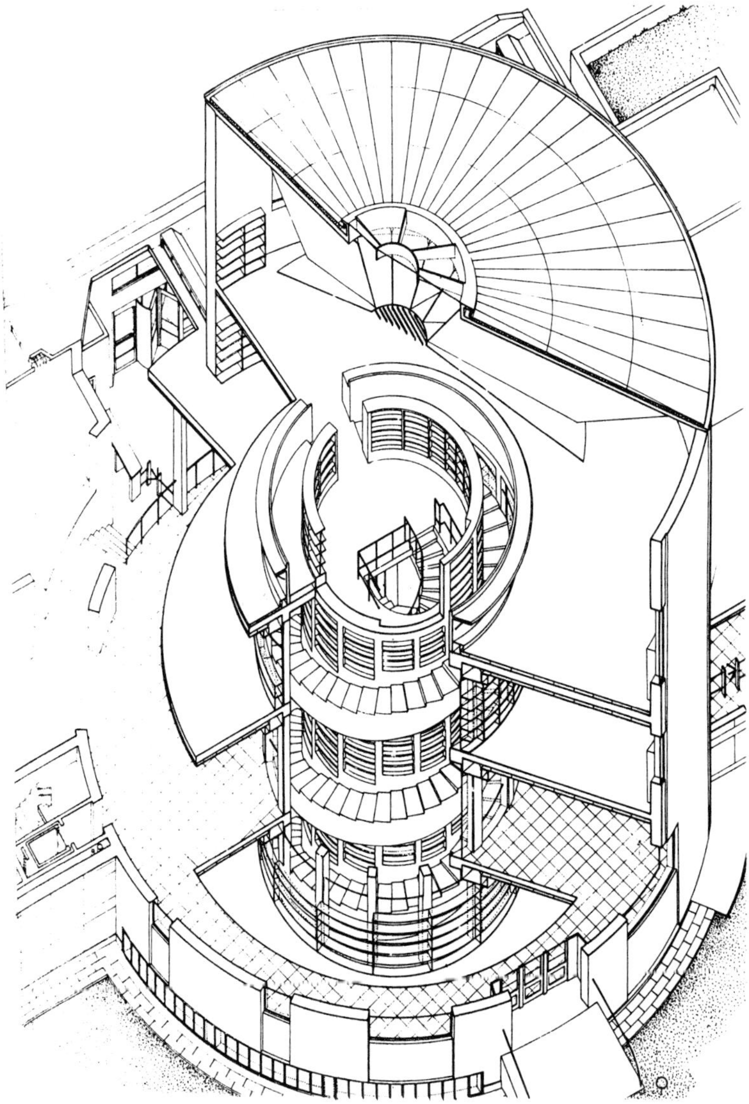

PROCEDURE TO EVALUATE THE EFFICIENCY OF SHADING DEVICES

G. GOMEZ AZPEITIA
Universidad de Colima
Aldama 512 - 5
Colima, Col. CP 28000
México.

ABSTRACT. A procedure to calculate the relative efficiency of shading devices in sunny windows is presented, based in the comparison betwen sun time intercepted by devices and sun time and shade time required, according to a comfort diagnose. The most important aplication of this procedure is in composite climates, where several requeriments of solar control, often opposing, must be pondered.

1. INTRODUCTION.

The architectural designer often must find solutiion for two environmental conditions which are opposite. Both conditions need an entire satisfaction, to achieve suitable climatic performance of buildings.

This situation is most critical in locations whith composite climates where different requeriments of solar control are presented in the same window: solar gain in underheated periods and shading devices in overheated periods.

If we expect to resolve both requeriments whith an unique design shape, we must find the way to permit solar gains across the window while this is necessary and shade on the window only in the season which demand that protection.

This evaluating procedure is developed on this scope. It tries to inform the optimum size of shading devices, according to a comfort diagnose, saving time of trials and calculations, and achieving most efectiveness in device performance.

This method is based in the assuption that the specific gravity of underheated and overheated requeriments are founded in the relation of the hours that need sun or shade and the real sunny hours over the window. The local conditions are stablished at beginings as hourly shade or beam requeriments. Thus the heat transfer calculations is innecessary, but the performing simulation is suitable.

S. Alvarez et al. (eds.), Architecture and Urban Space, 549–554.

Consequently this procedure is most easy to execute that the Olgyay's one (1957), cause only the sun or shade time (in hours) is considered, and not the quantity of solar beam (in BTU's).

2. VOCABULARY.

There are three kinds of shading devices:

a. Overhanging: horizontal devices standing out at the upper part of window, which intercept the vertical component of solar beam (altiude angle). The protection angle of overhanging is measured from window plane to an imaginary line which unite the most standing edge of overhanging whith lowest point of window (see figure 1).

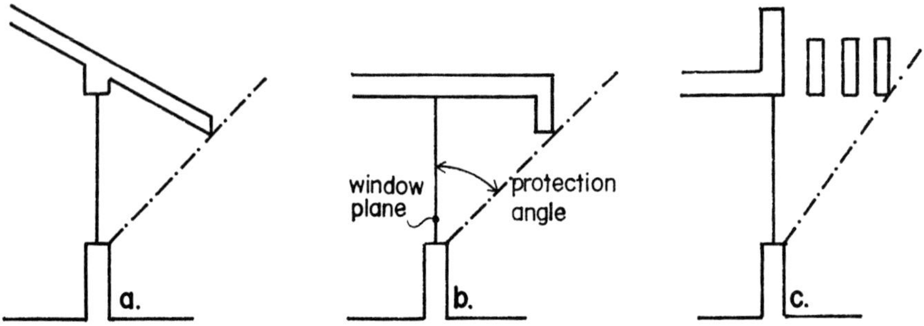

Figure 1. SIDE VIEW OF WINDOWS SHOWING SEVERAL KINDS OF HORIZONTAL OVERHANGINGS AND THE WAY OF MEASURE THE PROTECTION ANGLE.
a. Eaves. b. Solid overhanging. c. Louvered overhanging.

b. Sunbreaker: vertical devices placed at either sides of window, which intercept the horizontal component of solar beam (azimuth angle). The protection angle of sunbreaker is measured from window plane to an imaginary line that unite the most standing edge of sun breaker whith opposite extreme of window, either left or right sides (see figure 2).

c. Lattice: a mix of vertical and horizontal devices, which intercept both horizontal and vertical components of solar beam. Obviously the protection angles of lattice are measured like two anteriors devices, one by one.

3. MATERIALS AND METHOD.

This procedure will be done only once for each location, because it gives precise guidelines to recommend optimunm size of shading devices for any window orientation, according to claims of climatic situation.

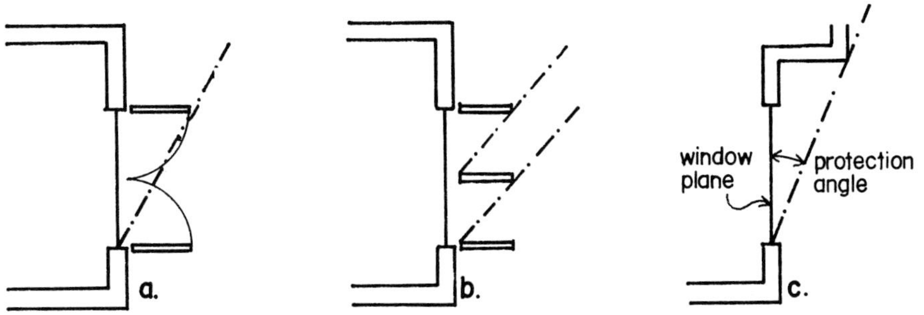

Figure 2. UPPER VIEW OF WINDOWS SHOWING SEVERAL KINDS OF VERTICAL SUNBREAKERS AND THE WAY OF MEASURE THE PROTECTION ANGLE.
a. Fins. b. Vertical louvers. c. Edge oblique to the wall

To effect this procedure, we need the following information:

a. Hourly output temperatures of one day each month, from sunrise to sunset.

b. Comfort diagnose indicating hourly requeriments of shade or beam. This diagnose is achieved by Neutrality Temperature, Effective Temperature or Uncomfort Index models.

c. Two sun path diagrams, one for each half year, preferably equidistant or stereographic (ortographic deform sun path projections making difficult in account of sun or shade hours).

d. Shading masks to place upon sun path diagrams, simuling several kinds of devices and several protection angles. They show the solar obstruction product of overhangings and sunbreakers over the window.

Information of points c. and d. can be substituted by computer pack named "Grafisol" developed at University of Colima.

For this sample, comfort diagnose is achieved by Auliciems (1981) Neutrality Temperature model:

$$Tn = 17.6 + (0.31 * \bar{T}o) \qquad [1]$$

where:
Tn = Neutrality Temperature.
$\bar{T}o$ = Monthly average output temperatures in °C.

This model accept a comfort range of ± 2.5 °C. Output temperatures upper Tn+2.5 indicate overheated period; output temperatures lower Tn-2.5 indicate underheated period; output temperatures into the range of Tn±2.5 indicate comfort period.

The first step of process is to find hourly requeriments of heating or cooling each month, by the preceding comfort diagnose. If we have heating requeriments, we must permit solar beam across the window, that is, direct heating. In exchange, if we have cooling demand, we must shade the window as indirect cooling.

In other hand, if output temperature is into comfort range, then we must to shade the window too.

The second step of process consist in locate this hourly requeriments over the sun path diagrams. This step must be effected twice, one for each half year diagram. Each frame defined by sun path of two consecutive months, and by two consecutive hour patterns, is considered like one "hour-month". Each one of these will contains the respective shade or beam requeriments (see figure 3).

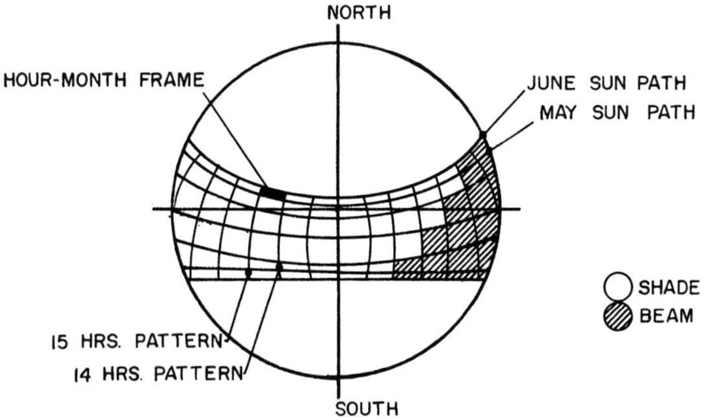

Figure 3. SUN PATH DIAGRAM INDICATING HOUR-MONTH FRAMES AND SHADE OR BEAM REQUERIMENTS.

The third step and following must be done several times, one for each window orientation. In this case we execut the south orientation. Therefore north sector of sun path diagrams is hide, cause the window is exposed only to the sun beam arise from south sector. Now we must count the total hours-month that need solar beam (T1), and the total hours-month that need shade window (T2) staying unhide in south sector. For others window orientations, the hide sector of sun paths is the opposite of those (see see figure 4).

The fourth step consistes in putting shading mask over the sun path diagrams, which will hide another share of sun paths. We must count how many hours-month whith shade requeriments stay unhide (K), and how many hours-month whith beam requeriments remain unhide (L). Therefore, now we can evaluate the device efficiency both underheated and overheated periods:

EC = Overheated period efficiency:

$$EC = 1 - ((K/T2) * 100) \quad [2]$$

EF = Underheated period efficiency:

$$EF = (L/T1) * 100 \quad [3]$$

Nevertheless, the total year efficiency, that is, the ponderate performance of device, is:

ED = Total year efficiency:

$$ED = ((T1/T1+T2) * EF) + ((T2/T1+T2) * EC) \quad [4]$$

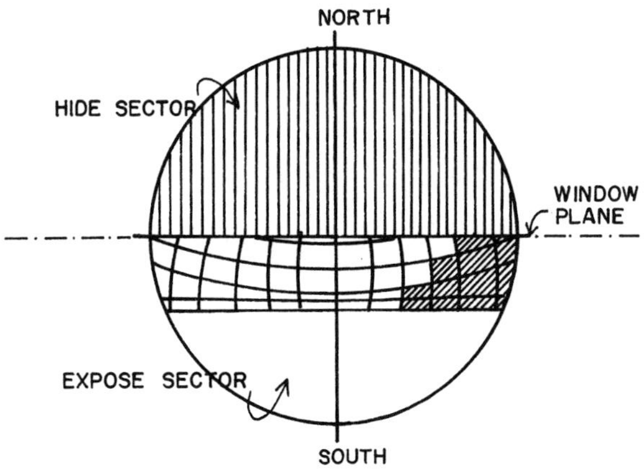

Figure 4. HIDE AND EXPOSE SECTOR OF SUN PATH DIAGRAM ACCORDING TO WINDOW LAY.

This step must be done several times, one for each protection angle, both overhangings and sunbreakers. Is laudable have masks each 10° of protection angle.

If we plot those values over a diagram which relate protection angle values whith efficiency values, we will have three patterns:

a. Overheated period efficiency line (EC)

b. Underheated period efficiency line (EF)

c. Total year efficience line (ED)

The best protection angle for this window orientation, is founded in the point which coincide oll three patterns. Also the best protection angle may be the greatest value (see figure 5).

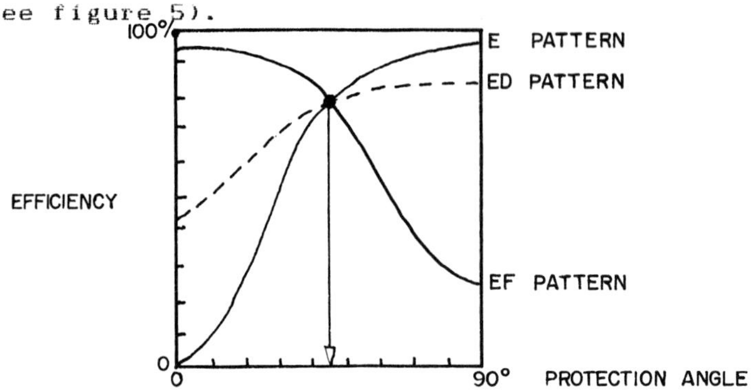

Figure 5. DIAGRAM TO RELATE PROTECTION ANGLE WHITH EFFICIENCY VALUES AND EFFICIENCY PATTERNS.

4. APPLICATIONS.

When these four steps have been executed for each window orientation and for each angle protection, we will have very important information for several aplications, by example:

a. To help in decision process about measurement of both solar gain and shading devices, in architectural, engineering and urban planning tasks.

b. To help to teach the design, cause this information let see that any lay variation of window plane, origine changes in solar gain conditions, and therefore in requeriments of solar control devices.

c. To help to define building and planning regulations of every sites, about solar control devices, to achieve suitables energy savings in buildings, and best comfort conditions for building habitants.

REFERENCES:

Auliciems, A. (1981). Global differences in indoor thermal requeriments. ANZAAS conf., Brisbane.

Olgyay & Olgyay. (1957) Solar Control and Shading Devices. Princetton University Press. Princetton N.J.

Gómez Azpeitia, G. (1990) Método para el Diseño Bioclimático. Universidad de Colima. México.

ACKNOWLEDGEMENTS:

I am greateful to Leandro Sandoval (M. in Sc.) and to Marce de Anda (translator) for revising this paper.

PLASTIC MULTILAYER RIBBED SHEETS FOR BIOCLIMATIC BUILDINGS

G.Raffellini[+] **and S. Piva**[*]

(+) Dipartimento P.M.P.E. - Universita' di Firenze
Facolta' di Architettura - 50100 Firenze
(*) Dip. Ing. Industriale - Universita' di Parma
Facolta' di Ingegneria - 43100 Parma

ABSTRACT

Transparent plastic sheets with double layer of various forms and types (with perpendicular or Y-shaped ribs) are widely employed in passive solar buildings, especially in schools, offices and commercial structures. The solar energy input through these transparent plastic sheets was examined and compared with single or multilayer flat homogeneous slabs, by means of an especially designed experimental apparatus . The results were numerically elaborated in order to collect information on the energetic implications induced by the ribs. The purpose of this paper is to point out some practical relationships in order to correctly use these new materials as direct gain passive elements or for greenhouses.

1. ENERGETIC BEHAVIOUR OF RIBBED PLASTIC SHEETS

New twin-walled plastic sheets are characterized by the presence of longitudinal ribs, connecting two (or more) thin transparent plates. Many different kinds of sections are commercially available (Fig.1). While these sheets possess high mechanical strength, low weight, high thermal insulation and good transparence, the visual connection between indoor and outoor cannot be as perfect as the one offered by glass plates or compact plastic transparent sheets. Both the mechanical and thermal properties of these new transparent surfaces are well known, but there is a lack of information about their sun transmittance. As a matter of fact, the presence of the ribs introduces an anisotropical response into the transmission of an incident sunray, due to the more complicated reflection and refraction occurring when a sunray meets a rib [1].

To assess the particular behaviour of this kind of surfaces, some experimental investigations were executed by the Authors at Bologna University. Many different surfaces were tested by means of an experimental apparatus especially designed to reproduce the effects of the relative position of the ribs with reference to the sunray direction on the solar transmittance. A detailed description of this apparatus and some typical results were presented in [1].

S. Alvarez et al. (eds.), Architecture and Urban Space, 555–560.

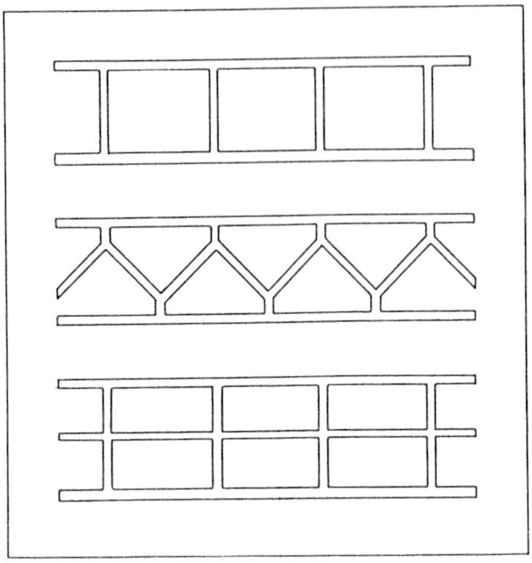

Figure 1 - *Typical shape of multilayer ribbed plastic sheets.*

As the shapes available on the market are very different, many of the experiments were carried out on the simplest one, the rectangular shape, so as to assess the experimental results and to infer from them a general statement. In this set of trials, the ratio between the overall thickness of the sheet and the ribs interaxis as well as the number of ribs per unit of length were the most significant parameters. A physical interpretation of the solar transmittance shown by the materials can be suggested on the basis of these two parameters. The first parameter seems to be related to the incidence angle at which every sunray hitting the surface is to pass through a rib; so the sheet behaves as a three layer plate. Starting from this limit angle, the response becomes similar for every orientation of the ribs . The number of ribs per unit of length is significant up to the limit angle just described. For high numbers, the behaviour is similar to the one shown by a homogeneous twin-walled plate, while for low numbers it is irregular, with strong ripples and oscillations.

The energetic implications of the tridimensional behaviour due to the presence of the ribs, were assessed after calculating the daily and hourly beam, diffuse and total solar radiation transmitted through a ribbed sheet, with different orientation of the ribs. The results related to the beam solar component were reported in [2], while those for the diffuse one in [1]. The data obtained are very interesting and point out firstly that a significant relationship exists between the ribs orientation and the extent of the solar input and secondly that the effect is maximized for south facing surfaces during summer. A horizontal orientation produces the reduction of the total solar energy transmitted during summer and it is able to enhance the energy input at sunrise and sunset during winter. Furthermore the findings suggest that a cleverer design of the shape will allow a bioclimatic usage of the ribbed plastic sheets, enhancing the solar input during winter and avoiding overheating during summer.

2. BIOCLIMATIC UTILISATION OF RIBBED PLASTIC SHEETS

Transparent surfaces are extremely important architectural design. They satisfy the human need for views and contact with the outdoor environment. As referred in the previuos paragraph, twin-walled ribbed plastic sheets are characterized by good thermal and mechanical properties, but they allow only a partial visual connection between indoor and outdoor. For this reason they are used only in the portion of building envelope designed to admit daylight, without any vision of the external landscape, like shed or top windows. On the other hand, this characteristic allows a thorough utilisation of the sheets in bioclimatic buildings: in fact here very large transparent surfaces are needed to obtain passive heating of the inner spaces. This surface is often larger than necessary to assure the natural lightning and a balanced relationship between the user and the external environment. So in these buildings many of the transparent surfaces can be only partially transparent.

However the added south-facing windows needed in passive solar buildings may increase heat losses and cause overheating and glare. Therefore in passive solar design great care has to be taken to balance the useful solar energy transmitted through transparent surfaces against their heat losses. In Tab.I typical U-factor values are reported for 3 mm single and double layer glass plates and for 6 and 16 mm commercial PCO ribbed sheets. It is evident that with ribbed plastic sheets heat losses through windows are comparable with and often lower than the ones with glass plates of the same cost and thickness.

Tab.II shows the average monthly total transmittance (beam, diffuse and reflected) for the same sheets listed in Tab.I The values refer to a south-facing vertical installation at a latitude of 44°12 LN. For the PCO sheets the values corresponding to vertical and horizontal ribs are reported. It is evident that a single layer glass plate always presents the higher transmittance. If compared with a double-glazing plate, the PCO plates always present higher transmittance in winter and lower one in summer. The transmittance of a 16 mm PCO plate is always higher than a 6 mm PCO plate. This is due to the larger ribs interaxis of the first plate, that introduces a lower reduction of the solar transmittance. From the same results it is evident that a horizontal installation of the ribs assures higher transmittance during winter and lower one during summer.

From the analysis of Tab.I and II it is evident that PCO ribbed sheets can be very useful in bioclimatic applications because given the same U-factor, and then the same heat losses, the solar gain is always higher during winter and lower during summer. Furthermore the hourly behaviour, presented in [2], shows that with these sheets it is possible to reduce the risk of overheating.

After dividing the passive solar solutions into three large groups:

- direct heating;
- solar walls with greenhouse effect (open, heavy, Trombe);
- sunspaces;

it is possible to evaluate the best features assured by the new materials examined in the application listed below.

Material		U-factor ($W/m^2 °C$)
Glass	Single layer (3 mm)	5.5
	Double layer (3 + 3 mm)	3.5
PCO	6 mm	3.8
	16 mm	2.8

Tab.I - *Typical values of the U-factor.*

Month	Glass		6 mm PCO		16 mm PCO	
	S.L.	D.L.	Vert.	Hor.	Vert.	Hor.
Jan	0.829	0.691	0.749	0.755	0.765	0.779
Feb	0.809	0.665	0.712	0.719	0.731	0.732
Mar	0.779	0.628	0.668	0.666	0.692	0.684
Apr	0.761	0.598	0.604	0.585	0.625	0.601
May	0.731	0.567	0.524	0.492	0.547	0.507
Jun	0.725	0.557	0.472	0.437	0.491	0.449
Jul	0.719	0.548	0.498	0.464	0.521	0.479
Aug	0.734	0.571	0.571	0.547	0.592	0.561
Sep	0.753	0.602	0.646	0.638	0.669	0.653
Oct	0.804	0.654	0.703	0.706	0.724	0.721
Nov	0.828	0.689	0.743	0.749	0.758	0.769
Dec	0.833	0.699	0.752	0.761	0.767	0.788

Tab.II - *Total transmittance of different transparent plates.*

To produce direct heating, these ribbed plastic sheets are very useful in certain solutions devised for commercial or industrial buildings, where the transparent surfaces are often placed on the cover (sheds) or on the top of the walls (continuous windows). In these cases large surfaces of twin-walled PCO sheets reach an interesting optimisation due to the low U-factor values and high transmittance shown. Furthermore the particular behaviour of the solar transmittance can help reduce overheating during summer, thanks to the correct choice of the ribs orientation. This solution was tested in the design of industrial buildings, where an economical balance shows appreciable pay-back time if compared with glass [3]. In multi-family buildings, where the south exposed surface is usually limited by the floor height, the transparent surface often reaches the floor. In this case the PCO ribbed sheets can be used in the lowest side of the window (from the floor up to the window-sill), reaching very high levels of thermal insulation. This solution was tested in a multi-family building realized in Italy [4].

The ribbed plastic plates can be used in solar walls with greenhouse effect, thanks to the higher thermal insulation they can assure if compared with glass plates of the same daily solar transmittance. Some precautions have to be adopted before choosing them, particularly for heavy closed solar walls, in order to avoid damaging the plastic sheets as a consequence of the high temperatures risen.

Finally, these new materials can find interesting applications in solar spaces. In fact their lightness and pliancy allows interesting solutions. The requirement for letting natural light enter from the top and for using the same transparent surfaces as direct gain systems, is perfectly met by these materials. This solution was tested firstly in an elementary school realized in Italy [5]; in this case large transparent surfaces were created in the sloped roof of the corridors, that worked as solar spaces. Another application was tested in the design of a large bioclimatic office building [6].

3. CONCLUSIONS

The solar transmittance of ribbed plastic sheets has been analyzed on the basis of the experimental results obtained with an especially built apparatus. The tridimensional effect due to the presence of the ribs introduces sensible modificationsin the transmittance, if compared with twin-walled compact sheets, . This particular behaviour can be usefully exploited when the ribbed sheets are used in passive solar elements, as the results presented in the paper clearly show. Some examples of application are reviewed.

ACKNOWLEDGEMENTS

This paper has been sponsored by the National Research Council (CNR-PFE2). The Authors wish to express their gratitude to the Director of the Istituto di Fisica Tecnica of Bologna University, where the experimental apparatus was built.

REFERENCES

[1] Di Luzio N., Piva S. and Raffellini G., The transmission of solar radiation through transparent alveolar plastic sheets for bioclimatic buildings, ISES International Conference on Evolution of Bioclimatic Architecture of External Perimetral Components in Bioclimatic Architecture, Milan 5-6 april 1990, 47-50.

[2] Raffellini G. and Piva S., The utilisation of different transparent plastic sheets in buildings: experimental results and correlations about the transmission coefficient, Second World Congress on Heating, Ventilating, Refrigerating and Air Conditioning - CLIMA 2000, Sarajevo (YU) August 27 - September 1,

1989, 38-43.

[3] Cuppini G., Bezzi L., Mingozzi A. and Piva S., Illuminazione naturale e apporti gratuiti di energia (Daylighting and free gains), Quasco, 2, 42-46 (1987).

[4] Raffellini G. and Piva S., Un intervento sperimentale per il risparmio energetico: valutazione e primi risultati (Energy saving experimental intervent: evaluation and first results), Edilizia Popolare, 193, 56-61 (1986).

[5] Cuppini G., Raffellini G. and Piva S., Energy saving and solar heating Regione Emilia Romagna demonstration program: the elementary school in Massa Finalese (Modena, Italy), Proc. of the CLIMA 2000, Copenhagen (DK), 25-30 august 1985, 277-282.

[6] Cuppini G. and Mingozzi A., Control techniques of direct solar radiation. A case study: the new ENEL premises of Bologna, ISES International Conference on Evolution of Bioclimatic Architecture of External Perimetral Components in Bioclimatic Architecture, Milan 5-6 april 1990, 213-216.

"APPROACH TO TECHNOLOGIES OF SOLAR CONCENTRATORS BY REFRACTION TO ARCHITECTURE AND URBAN PLANNING"

JOSÉ IGNACIO PÉREZ CALERO
Doctor Architect and D. Physics.
University of Seville
Escuela Técnica Superior de Arquitectura
Avda. Reina Mercedes s/n
41012-Seville
Spain

***

ABSTRACT.- This Paper wishes to show the important contribution of certain kind of refraction solar concentrators (R.S.C.) to the development of ecologic solutions in the use of solar energy for architecture and urban planning, establishing unpolluting thecnologies in agreement with the bioclimatic design.

The desired designs will be basically simple, cheap and light systems, that in many cases will be built in large series, generally in plastic materials, and using normally fluids (especially water) as refracting element, working in low temperatures. The methodology followed is multiple, from theoretical analytical study with computarized simulation for design of desired prototypes, to statistical methodology on the basis of data bank and correlations.

The satisfactory results in this work, show the way of many original configurations in inmediate use of solar energy for architecture and urban planning.

S. Alvarez et al. (eds.), Architecture and Urban Space, 561–568.

1. INTRODUCTION

Almost all research on solar energy concentration is done on the basis of the physical phenomenon of the reflection, where a strong technological development has been obtained. However, the refraction as a source of the phenomenon of concentration has not been studied much, that is why it is worth tackling it as an alternative resolution of this proposed problem, since the validity of this hypothesis has been amply checked.

The research done on this Paper aims at the search of a methodology for the desing of solar concentrators, exclusively based on the refractive phenomenon and essentially directed to the fields of architecture and urban planning. The desired designs will be basically simple, cheap and light systems, that in many cases will be built in large series, and using normally fluids, especially water, as refracting element. Above all, we will tackle spherical designs (although we will also deal with the aspherical), thin or thick, whether working on the paraxial zone or not, and with as much opening as needed. we will consider any captivated plane we may be interested in, wheteher focal or not, and where the energetic image that will be mobile, because of the solar focus trasmitter, will be found.

2. METHODS

The methodology followed in this research is multiple, on the basis of the part of work being studied. That applied to the analysis of general models is mixed. On the one hand, we will do a theorical analytical study on the optic-mathematical field of the tracking of tridimensional beam of light incident on the system, for any relative position among them, up to their intersection with the desired captivated planes. On the other hand, the process iis tracked by means of computarized simulation on the basis of the actual diverse parameters, that let us know the more general behaviour of any system of our interest.

Also referring to climatic and radiation parameters, the methodology is statistical on the basis of data banks, once the most complete and recent ones, have been selected.

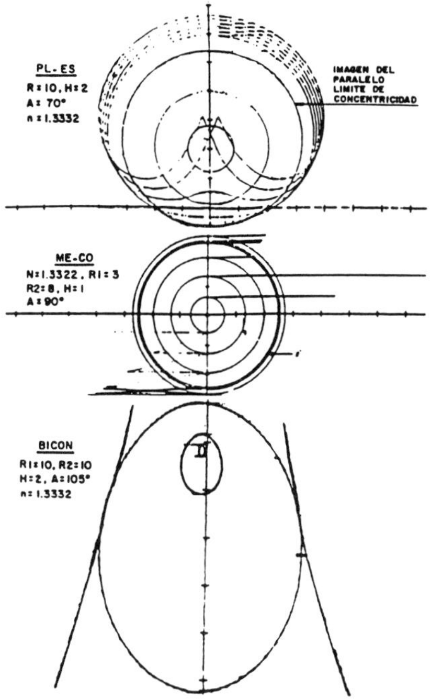

ANALISIS COMPARATIVO DE DIFERENTES ESTRUCTURAS INTERNAS EN TRES SISTEMAS

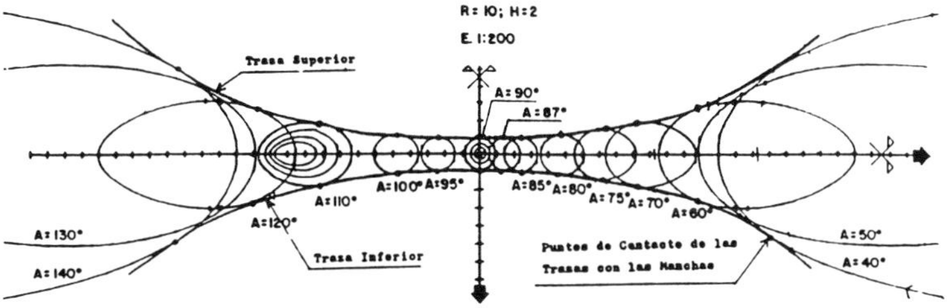

ANALISIS DE TRAZAS EN EL SISTEMA ESFERA-PLANO SOBRE PLANO FOCAL IMAGEN PARA DIAS EQUINOCCIALES

3. RESULTS, DISCUSSION AND CONCLUSIONS

The research work carried out through this Paper leads us to clear results and conclusions, which can be summarized as follows:

a) Given the complexity and diversity of the investigated problem and the difficulty for its general formulation by any other methods, the validity and fitness of the methodological hypothesis used here in the analysis of general models has been verified and reaffirmed. As a result, the models and treatments dealt with show a total feasibility as fundamental basis for the design of real prototypes of the searched concentrators.

b) With the idea of obtaining the maximun performance in a program of concrete applications, in the designing of a determined concentrator and for the evaluation of its efficiency or for the analysis of its fitness in comparison to others, we should take into account not only their simple definite parameters and other determinations contained in these conclusions, but also the comparative analysis of the corresponding parameters.

c) In all the cases, this research work enables the determination of the different positions of the paraxial focal planes in the spherical Systems and the later complete analysis of their behaviours (increasing, decreasing, extreme positions, etc.) on the basis of the different simple parameters of the desing considered, and as an important reference point for the election and definition of the captative planes chosen.

d) In the totality of assumptions we can detec, by means of numerical definition and graphic visualization, the images formed on the basis of the different parameters of the dealing problem (for any relative position of the solar beams and the concentrator, as well as on the desired captativated plane) with the delimitation of the general structures and their internal sub-structures, whose later analysis will show us different behaviours, depending on whether they are axial or not. The first ones are always circular and concenrtric, as opposed to the second ones which are not and are subjet to distortion, with the characteristic of not being conical, bisymmetrical, with lengthening or flattening, and with centripetal or centrifugal poles, depending on the concrete design analyzed.

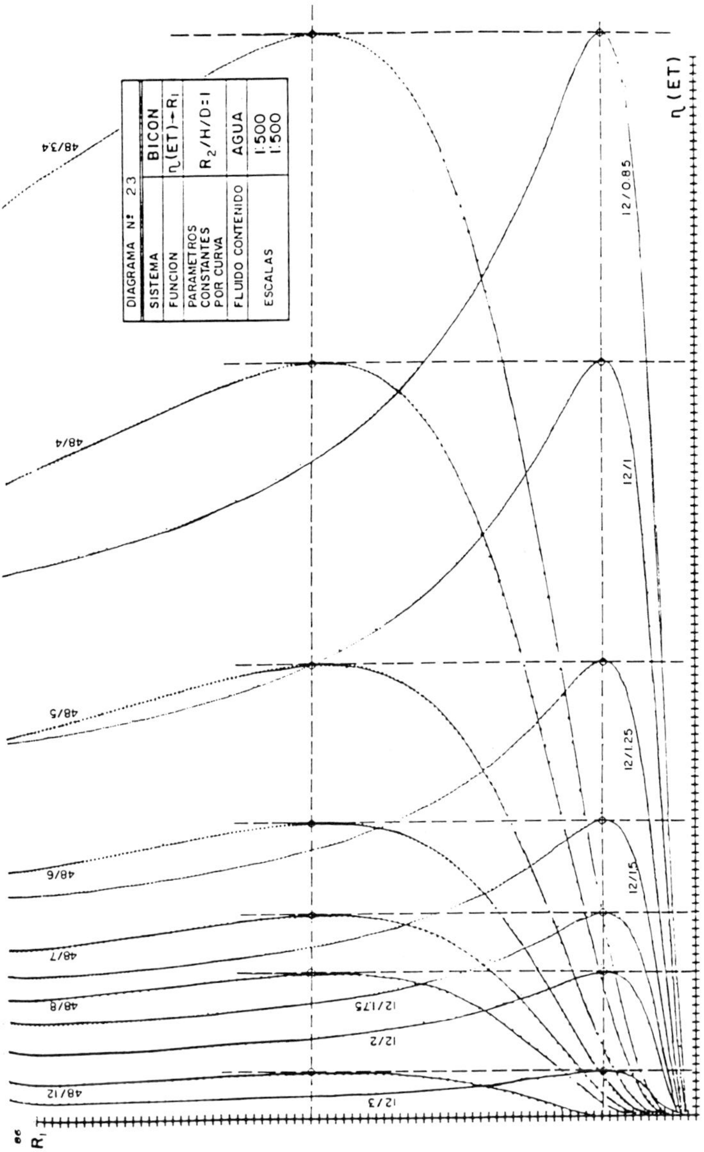
DIAGRAMA Nº 23
SISTEMA
BICON
FUNCION
η(ET)-R1
PARAMETROS CONSTANTES POR CURVA
R2/H/D=1
FLUIDO CONTENIDO
AGUA
ESCALAS
1:500
1:500
48/3.4
48/4
48/5
48/6
48/7
48/8
48/12
12/0.85
12/1
12/1.25
12/1.5
12/1.75
12/2
12/3
η(ET)
R1

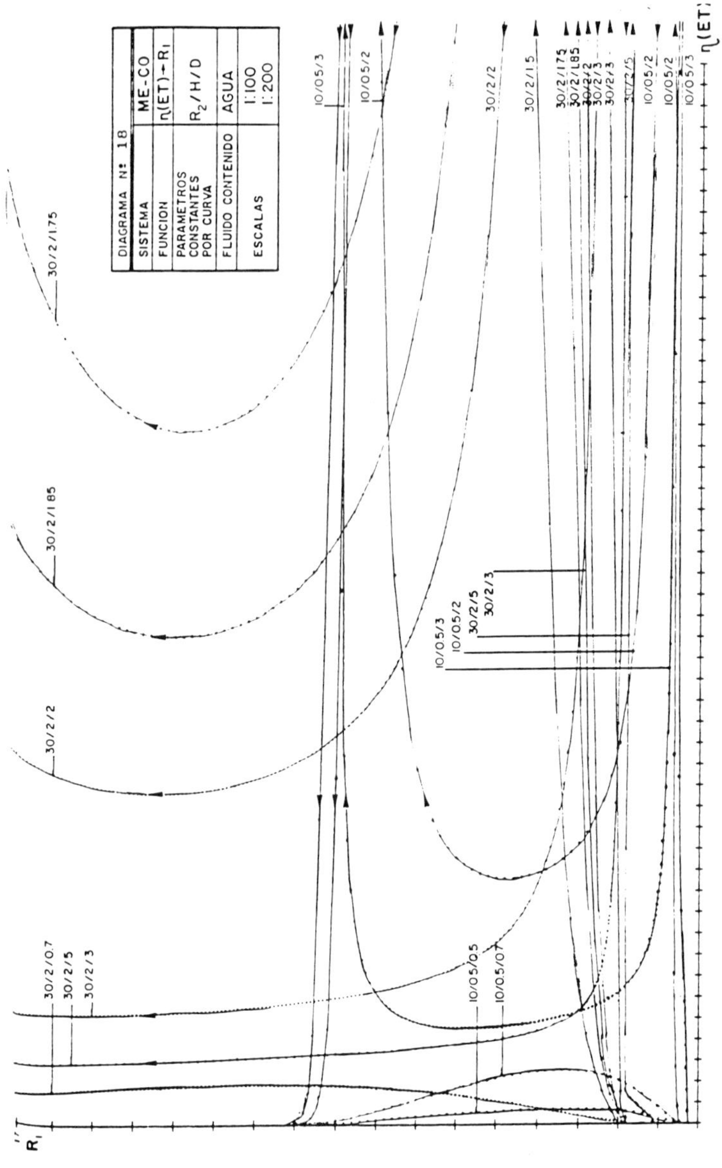
DIAGRAMA Nº 18
SISTEMA
ME-CO
FUNCION
η(ET) → R1
PARAMETROS CONSTANTES POR CURVA
R2/H/D
FLUIDO CONTENIDO
AGUA
ESCALAS
1:100
1:200
30/2/1.75
30/2/1.85
30/2/2
30/2/0.7
30/2/5
30/2/3
10/05/0.5
10/05/07
10/05/3
10/05/2
30/2/5
30/2/3
30/2/2
30/2/1.5
30/2/175
30/2/1.85
30/2/2
30/2/3
30/2/3
30/2/5
10/05/2
10/05/2
10/05/3
R1
η(ET)

e) The study of family of images, layouts and speed movements, allow us to find the angle limits of concentration and the operative time of a system, which are essential for the construction of real prototypes.

f) An important achievement of this research is the delimitation and analysis of factors of concentration, see figures, which quantify in the different Systems the obtained gain in process of refraction. The evolution of their behaviours is achieved with the detection of extreme values that the best designs imply, wheter focal or not, on the basis of the different simple parameters present in each case.

g) Within the framework of this research, the spherical systems prove to be of a higher interest than the aspherical ones.

h) In the study of the materials for the accomplishment of models, we can state that the present technological development in this field offers positive answers for the real construction of prototypes.

In the study of the applications derived from this research, we arrive to the clear conclusion of the existence of ample possibilities, above all, in the field of architecture and urban planning.

Among the architectorical aplications are:

a) Houses with roofs or walls made of transparent refractive concentrative materials in order to collect the solar energy upon prefixed zones to obtain convective heating in air or in tank-stored fluids for posterior using in closed conditioning circuits, as the process of passive bioclimatic architecture.

b) Independent winter quarters-concentrators, or in conexion with family houses.

c) Refractive concentrator-collector units inserting into plane glasses with several original instalment designs.

d) Fixed or moving solar refractive concentrator-skylights, including auxiliaring mirror systems, grouped in service galleries.

e) Ribbons or tubes with inserted "refractive concentrator-cells" in surface with bottom circulating channel for fluids, generally water, even using black body properties and vacuum thecnologies, to install in flat roofs, gardens, etc. -

f) Optimal solar refractive concentrators inciding over circulating fluids in net, to get the instant warming, for use in compact units with solar following sensors.

g) Integral conditioning for large buildings or public spaces of high spans, to get regulable and reversible solar refractive concentrator systems with prefixed confort zones.

Among the urban planning applications are, fundamentally, the "solar parks", centralized conditioning systems in several levels in the urban planning (block of house, quarter or district, little town, etc.), with large equipment of production and storage, made with solar refractive concentrator-collectors, aided with conventional equipment, located in "free urban zones" or in the roofs of "car parkings", and with underground stored tanks, to serve cold and hot water.

REFERENCES

Butti K. and Perlin J. (1981) "A Golden Thread. 2500 yerars of Solar Architecture and Technology". Marian Boyars Publishers Ltda. London.

Hawkes, D. (1981) "The Architecture of Energy". Director of the Martin Centre for Architectural and Urban Studies, Cambridge. Longman Group Limited. London.

Lebers R.M. (1981) "Passive Solar Heating Design" Applied Science Publishers Ltda. London.

Pérez Calero J.I. (1988) "Approach to the studiy of static solar concentrators by refraction: Application to the Cuenca Baja del Guadalquivir". Doctoral Dissertation. Politecnic University. Madrid.

Pérez Calero J.I. and Garcia Moris A. (1985) "Integral System of Bioclimatic Architecture for Social Housing" Proceedings XXIII COMPLES Congress. Seville.

Rubio Requena, P.M. (1979) "Urban Installations. Technology and Territorial Infrastructure" Ambiental Control Publishers. Madrid.

BIOCLIMATIC HEALTH CENTER WITH SOLAR ROOF

JOSÉ MARTÍNEZ GALÁN
ENERCOM, S.A.
c/ Santiago de Compostela, 64
28034 - MADRID
SPAIN

ABSTRACT. The building is a Health Center placed in Jumilla (Murcia), SPAIN; the following architectural and energetic improves have been incorporated in the building: 1) Big skylight with a clerestory design, orientated to the South and with an eaves for protection in summer. 2) Porch in the South. 3) South facade with extensive crystal surfaces protected in summer with a porch. 4) East, North and West facades with reduced crystal cavities. 5) Solar roof on conventional roofing of the building. 6) Radiating flooring system for heating and cooling. 7) Ventilation from the basement (adjustment of temperatures in the months of transition summer/winter/summer). The instalation was put into gear at the end of March, 1991, and it is being supervised for the analysis of the working operation during at least two years.

All the modifications of architectural bioclimatic design, made in the initial proyect, has allowed to obtain a good behaviour of the building as for to the use of skylight to reduce artificial ilumination in the hours of normal use of the Health Center. Also we have obtained notables improves in the climatic conditions of the building by natural ways (thermic inertia of the building, passive saves through walls and crystal surfaces, reduction of solar win in summer, etc.), so in spite of the temporary use of auxiliar energy, the confortable climatization of the building with a considerable reduction of energetic consumption opposite of conventional ways is possible.

1. INTRODUCTION

When the Health Center was designed, it was decided that the building must be energeticly efficient, using renewal and clear energies, lowering the price of the rest of conventional energies, in this way.

To aim to make better results, the energetic load needed to maintain the building in a zone of confort between 20 and 26 ºC along the year, considering all possibles sourcers and sumps of heat, was studied. Among architectural resorts used for guarantee a good bioclimatic design we point up an study by wich we opted for replacing the previous big skylight of half canyon bove for another with a clerestory design.

S. Alvarez et al. (eds.), Architecture and Urban Space, 569–574.

The building is placed in Jumilla (Murcia), SPAIN, where according to data of last ten years, the minimum average temperatures have been 2.2 ºC in January and the minimum absolute temperature have been 10 ºC low zero in February, and maximum average temperature has been 33 ºC in July with picks of 41 ºC also in July. It is a tipically Mediterranean clime with averages moderate temperature values in winter (from 401 to 800 degrees-day/year) and summer, and it is also a high value of solar radiation zone (between 1570 and 1700 Wh/m2 year).

2. TECHNIQUE AND METHODS

2.1. General description of the installation

It is an installation of climatization with water-water heat pumps supported witn solar energy. We intent to take advantage of the Thermic energy get in winter by solar roof in an acumulator and transfer this energy to another acumulator, from wich heat water will be distribute into radiating flooring. In summer the circuit of heat pumps is reversed and the water from the tank of the radiating flooring net is cooled, give in its heat to the tank connected to the solar collector; during the night, the water from this tank is spilled on the solar roof though two mechanisms: nocturnal radiation and evaporating power, and in this way the water is refrigerated given in its energy outside.

The auxiliar system has electric resistance that heat the acumulator of the circuit of the radiating flooring, in case of water temperature of the acumulator is lower than specified heat (cloudly days, increase of heat demand in winter, etc.) and will use only nocturnal acumulation, in valley hours, lower the price of working of this auxiliar equipment.

2.2 Descrption of the solar roof

A solar roof is made up of several photothermic cells. The quantity and shape of these photothermic cells will depend on the service for wich they are required.

A photothermic cell has several reflecting surfaces, made of a good sunlight reflecting material and one or two absorbent surfaces (where solar energy is transformed into thermic energy).

The absorbent surfaces are similar to flat panels of glass and, due to the reflecting surfaces, never reach temperatures above 100 ºC. This relatively low temperature allows for the use of less versatile material, such as certain types of plastic.

The reflecting surfaces are arranged in such a way that, during the time of year when most energy is needed, they send the maximum concentration of sun rays to the absorben surface; whereas, during the time of year when energy is not needed, they prevent sun rays from falling upon the absorbent surfaces.

The reflecting surfaces also serve as a containment area for the hot air produced by the absorbent surface, in order to minimize the loss of energy throug natural convection.

The absorbent surface is placed so that the solid angle, which is exposed to the cold focus (the sky), is the minimum possible, and therefore considerably reduces the loss of energy by radiation.

The back of the absorbent surface also has a thermic isolator in order to reduce heat loss throug conduction. However, any loss taking place goes to the next cell or to the building on wich the solar roof is installed.

The thermic energy is collected in the absorbent surface by a heat-transporting fluid and is sent to all points of consumption by means of a pump, or is stored in a acumulator.

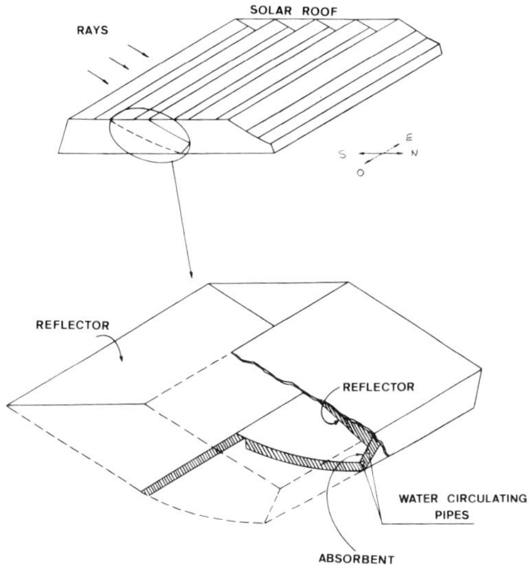

FIGURE 1 – **Phototermic Cell.**

As can be seen in Fig. 2, in winter the rays of the sun fall directly on the absorbent surface, thus achieving greatest efficiency during the season of highest energy demand.

In spring ans autum the rays of the sun fall first onto the reflectors and most are reflected onto the absorbent surface. As in each reflection some energy is lost, the efficiency in spring and autunm is less than in winter.

In summer, the season in wich thermic energy is hardly needed, the rays of the sun fall onto the reflectors, but, due to the position of the sun, no rays fall on the absorbent surface. The energy absorption in this season is low.

Figure 3 shows the temperature curve reached by the phototermic cell absorbent compared with the absorbent of a solar panel with glass, when the heat transporting fluid is not circulating.

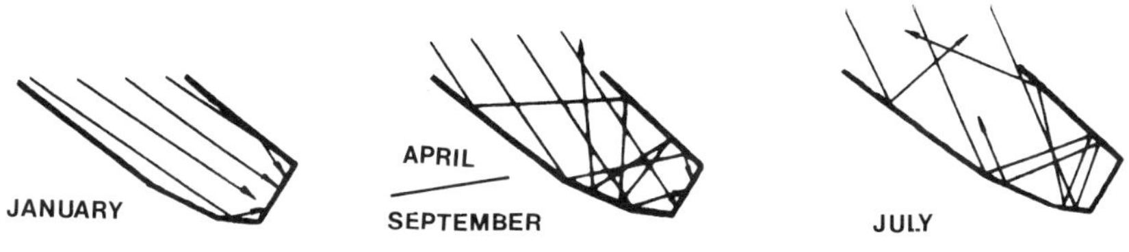

FIGURE 2 .- **Course of the solar rays inside a cell (Latitude 40º N).**

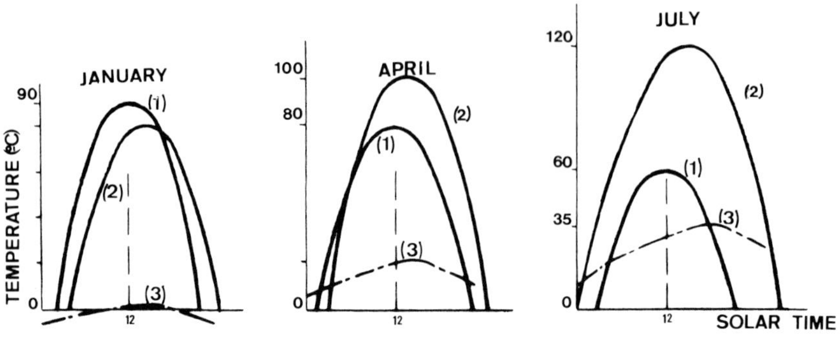

FIGURE 3.- Graphics showing the maximum temperature reached by the absorbent surface (when the water is not circulating).

From Fig. 3 the following can be deduced:

1) The maximum temperature reached in any season is below 100 ºC.

2) The highest temperature is obtained in January and the lowest in July.

These characteristics give the following advantages:

a) As the temperature in the absorbent surface does not exceed 100 ºC, the installation of safety valves is unnecesary.

b) The installation functions at atmospheric pressure. (The glass panels must operate at 2 Kg/cm2 of pressure in order to avoid the formation of vapour on reaching 120 ºC in summer).

c) If in summer a fault occurs in the control system and the circulation pump does not stop, the maximum temperature reached in the acumulator would be 60 ºC wich would not damage the installation. In a conventional system of glass panels, however the temperature of the acumulator may rise above 100 ºC, with the subsequent deterioration of the installation and an increase in the risk of injuries (burns, etc).

d) As the temperature of the installation does not rise above 60ºC in summer, the human safety factor is high (no burns,etc.), and repairs and maintenance work can be carried out without special precautions.

It is clear from Figure 4 that a system of hanessing solar energy using photothermic cells adjusts almost perfectly, in terms of energy obtained (curve 1), to the demand for (central) heating (curve 5);

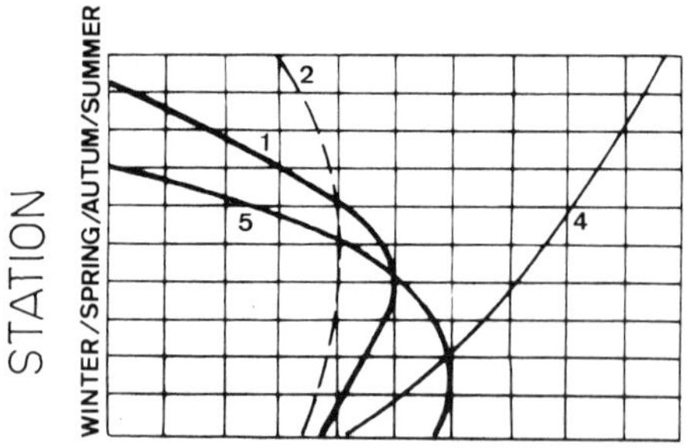

FIGURE 4.- Curves of absorbed solar energy.

KEY:
1) Photothermic cell with no selective black paint.
2) Flat panel with 45º leaning glass and with no selective black paint.
3) Environmental temperature.
4) Solar energy over 1 m2 of horizontal plane.
5) Energetic demands for heating.

whereas a conventional glass panel system harnesses too much energy in summer (curve 2), reaching excessive temperatures of up to 120 ºC. The photothermic cell system, however, obtains more solar energy than the conventional system in winter.
Materials of the solar roof:
Absorbers.- Sheet of copper painted in black on the previous side.
Reflecter.- Anodizated aluminium of 0.5 mm. of thick.
Structures sheet.- Sheet of galvanized steel of 0.6 mm. pleated.
Metallics structures of steel.

2.3. Architectural bioclimatic elemnts

The modifications from original architectural proyect of the building had the following improvements:

- Initial cylindrical bove was replaced, on central courtyard, by a big skylight that allow direct solar earnings in winter, with reduce of loss though crystal surfaces, and in summer impede directs solar earn but allow to extract hot air in upper zones of the building, preventing dazzlings by the use of diffusers of direct light. Energetic conditions get a biggest thermic passivity of the system, that allow extend confort conditions longer time. All them makes an energetic save of around 80000 Kw/h each year.
- The solar roof has been raised to the front fringe, in a way that in winter the rooms orientated to the South have sunlit. In this way the in let of direct solar radiation in winter days is improved, obtaining bigger solar earn in cold days, as in summer the dimensions of the porch, continues impede sumlit on South facade.
- Cavities had been specially reduced in North, East and West facades, for reducing loss and earns of heat through crystal surfaces, raising the inertia of the building. Besides this, a good level of ilumination in the building is maintained, while economic cost of the building is reduced. The material used in the crystal surfaces is grey polished glass of 6 mm. in the facade with a solar factor of 0.7, that allow protection in summer, and colourles mild glass of 4 mm. in the big skylight, that allows to take a very best advantage of it.

3. RESULTS

An electric power of 100 Kw has been installed in the building; 45 Kw have been destined for climatization. In Health Centers with similar dimensions and conventional climatization would be installed electric power of 200 Kw and more than half of the power would be destined for climatization. It is deduced that electric price and consumptions in our Health Center are lower thain in the others Health Centers, with a cosequent energetic and economic saving.
Figure 5 shows some temperature variations, variables of our system, along the same day in April. Curve 3 shows us a narrow rank in which environment temperature has oscilated in spite of sudden variations of exterior temperatures, (curves 1 and 2); this shows a effectiveness in the adjust of the climatization system in adverse climatic exterior

conditions. Curve 3 shows the high passivity of our system, because acumulator temperature is almost constant along the whole day and around 25 ºC. Curve 4 shows the possibility of fast adjust of room temperatures with cool air in case of the temperature would raise too much and would be necessary a control of temperatures faster than radiating flooring could give us, because there is a big difference between air temperature in basement and temperature of acumulator tank.

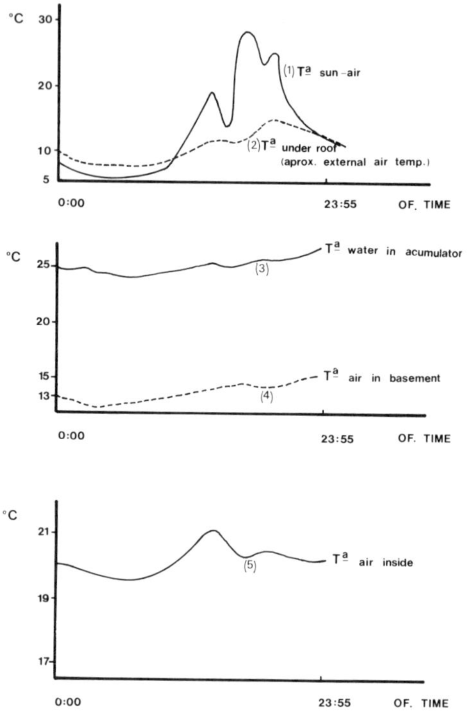

FIGURE 5 – Curves showing the variation of several interesting temperatures in a typical day of April.

4. DISCUSSION

Photothermic cells allow to take advantage of solar energy, using lower temperatures in heat transffer fliud, this raise the efficiency of the system and besides offers another advantages already described, as: the using of lower atmospheric pressures, is not necessary the use of sequrity valves, equpments support lower mechanic requests and so they are more available and cheaper, etc. Besides this technical advantages, the instalation is really interesting from a human security point of view because allows to reduce the level of risk of another installation in this way:

1) Electric risk level.- Low, as the only dangerous points are supply points for the back-up electric resistances and the circulation pumps, which are in protected places.

2) Thermic risk level.- Nil, because de maximun temperature that the water in the circuit can reach on a hot summer day and in the event of a control failure is below 60 ºC.

3) Mechanical risk level.- Nil, because the system operates entirely at atmospheric pressure.

4) Toxic risk level.- Nil, the only heating-refrigeration fluid is ordinary water.

5)Risk of fire.- Nil, no inflamable material are used.

5. ACKNOWLEDGEMENTS

The proyect has been carry out with the collaboration of Tomás Menor, Architect, INSALUD and Commission of the European Communities Directorate General for Energy (DG XVII).

SAV SYSTEM : TOWARDS THE IDEAL COMPONENT

X. TRAVER, M. MIRO, F. BONVEHI[+] , A. TRIAS[*]

Ecotecnia sccl	+ Asolba	*BCN Cambra Logica
Demostenes 6	Pintor Pahissa 27 4º	Sn. Joan 8
08028 Barcelona	08028 Barcelona	08629 Torrelles de Llobregat
España	España	Barcelona España

SUMMARY : **Parietal Ventilation, or the integration of the ventilation of a building in its skin facilitates the optimisation of the functioning of the building upon the criteria of comfort and energy rationalisation. The application of Parietal Ventilation to opaque or transparent** walls presents great flexibility for its **adaptation to diverse climatic conditions and also to various forms of construction.**
We present here the characteristics and basic properties of the SAV System which we have developed, and its analysis in relation to the climatology of southern Europe.
The optimisation of the functioning of a building on the basis of criteria of comfort, hygiene and energy rationalisation inevitably rests on a greater integration of the components, systems and installations in the global conception of the building.
The technology of Parietal Ventilation constitutes a clear example of this principle.
The maintenance of the air in appropriate conditions, especially in public places, must be a fundamental requirement. The fulfilment of these conditions requires the use of controlled mechanical ventilation.
The integration of the ventilation with windows under solar radiation permits at the same time an adaptative **control of the thermal flows through the windows and a low temperature use of solar radiation. In energy terms the resulting balance is positive**

1. Parietal Ventilation

The ventilation of a building through its skin provides a means of ensuring air renewal and increasing the control of the thermal flows through the external walls with a satisfactory energy balance.

The technology of parietal ventilation can be applied to either transparent or opaque walls.

A parietal ventilation project must combine the conditions imposed for its use with the climatic conditions of the location and the exposure of the facades. The known systems are : parietodynamic, permeodynamic and thermodynamic.

Basically two modes of ventilation are available : with renewal air or with exhaust air.

In the second case there is a distinction between direct extraction to the exterior or integration into air/air recovery circuit of the building's own structure.

S. Alvarez et al. (eds.), Architecture and Urban Space, 575–580.

Working with exhaust air we achieve the effect of dynamic insulation with which the temperature differences between the internal wall and the ambient air are minimised. If renewal air is used, we achieve a pre-heating of the new air in winter. It is obvious that needs vary from one season to another, in spring and autumn, even during the same day. Parietodynamic systems are adaptable to these changeable conditions.

It is also possible to connect to the system opaque solar collection components integrated into the facade.

Parietal Ventilation permits the treatment of special buildings like the pyramidal roof of the Aquatic Park in Le Touquet, Normandy.This roof consists of 1000 m² of double glazing with the enclosure ventilated by means of external air pre-heated in an air/air recovery unit.

This technology facilitates the elimination of surface condensation, avoiding the fall of drops on the park over the users, and the cold-wall effect. Optimal comfort conditions are maintained in winter with energy saving included. The downward circulation of air limits thermal stratification with the result of good conditions of higrothermic homogeneity in the useful space.

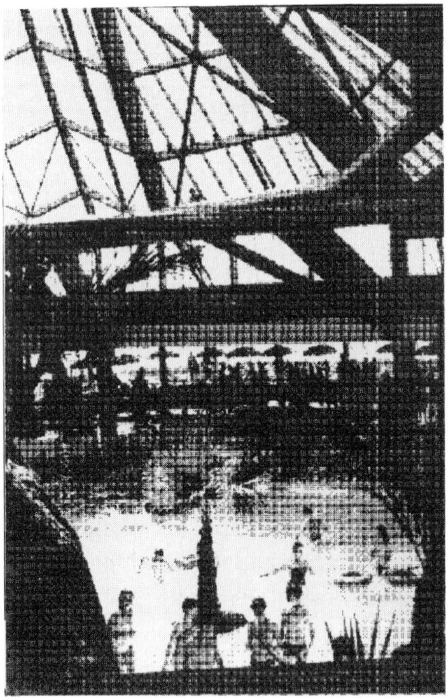

Figure1. Aquatic Park in Le Touquet, Normandy.

2. Air Flow Window : SAV System.

The windows present, in comparison with opaque walls, energy flows of grater amplitude and with faster variations.The Air Flow Windows permit an effective control of the various flows.

The SAV system consists basically of two glass panes with a venetian blind in between and air circulation inside this chamber.

Naturally, the characteristics of the components can be varied in accordance with the needs of the application : in the blind, the colour and the slat width can be varied; the outer glazing can be double, with chamber or laminar; if required, the system can incorporate low-emissivity, reflecting or absorbent glass

As mentioned previously, the ventilation air can be exhaust or renewal air. As for the volume, in winter the prime consideration will be the needs involved in air renewal : in summer it can be increased without creating a sensation of discomfort because of air movement.

Air Flow Windows began to be developed and installed in the Scandinavian countries in the 1950's with the aim of improving comfort conditions in winter and economising on heating energy.

In these cases, ventilation is done by direct extraction or by renewal. In this way a dynamic insulation is established in the glazed wall, approximating the surface temperature of the outer pane to the ambient air temperature of the room.

In the warm climates of southern Europe the demands of thermal comfort require solutions for winter heating and also for summer air conditioning, which may coincide in intermediate seasons.

The functions involved in the design of air-flow windows for warm climates with marked variations are :

* Natural lighting
* Solar Protection
* Thermal Insulation
* Solar Collection
* Acoustic Insulation

The SAV System has been thoroughly researched, theoretically and experimentally, by our team. Its basic properties have been characterised and quantified, and pilot and demonstration installations have been set up in houses, schools and special buildings.

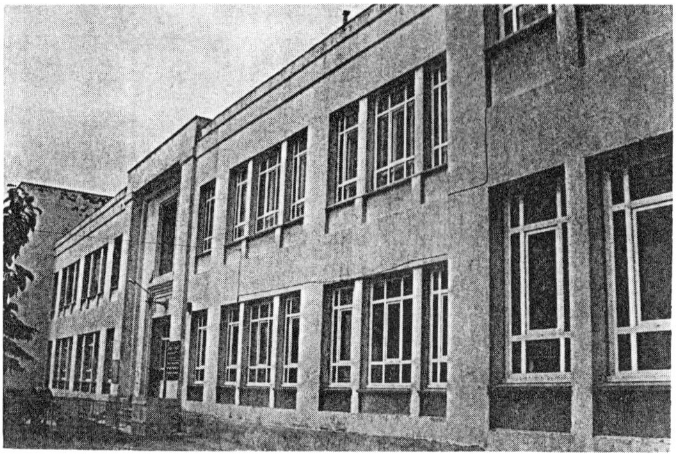

Figure 2. SAV System in the Primary School of Agramunt (Lleida)

3. Properties of Air Flow Windows

The basic system of the Air Flow Window which has been described, when appropriately developed and adapted to each set of circumstances, permits the optimisation of the functioning of the windows.

The air flow window is easily adaptable to the occupants needs, guaranteeing an optimum level of comfort in very varied circumstances and producing a considerable saving of energy.

An appropriate volume of ventilation and the correct positioning of the blind slats allow for an efficient control of natural lighting, solar irradiation an thermal transmission . Also, because of its double glazed construction, this component offers an effective control of acoustic transmission.

The venetian blind allows for the adjustment of the level of internal natural lighting, and it acts as an absorber of solar radiation.

The circulation of air under solar radiation permits the recovery of the solar energy absorbed by the blind, by virtue of which :

* the solar protection effect is reinforced;
* usable warm air is produced
* the radiant wall effect is attenuated

Independently of solar irradiation, the air circulation gives the Air Flow Windows the characteristics of a dynamic insulation, the coefficient of surface thermal transmission will vary according to the volume of ventilated air and the internal/external thermal interval.

One of the consequences of this property is that, by ventilating with exhaust air, it will be possible to maintain the surface temperature of the window close to that of the ambient air.

3.1 NATURAL LIGHTING

"With the blind slats inclined us to intercept sunrays, the entry of direct solar radiation is prevented, and the level of internal lighting at the horizontal plane 70 cm. from the floor is 600 lux.

The daylight factor, the relation between internal and external illumination at one point, has a very homogenous distribution.

3.2 SOLAR PROTECTION

This property is characterised by the Solar Factor, the relation of the transmitted energy flow to the incident one.

With the blind folded up and without circulation of air, the solar factor is 78%, corresponding to two 4 mm. panes of float glass.

With a classic white venetian blind with the slats inclined at 45° the solar factor is 20% : ventilating with 25 m^3/hm^2 of extracted air the solar factor is reduced to 14%.

The external protection devices with double-glazed windows give factors of the order of 10%.

3.3 COLLECTION OF SOLAR ENERGY.

The Air Flow Window under solar irradiation functions as an air collector.

With a ventilation flow of 20, m^3/hm^2 the solar collection performance is 20% with a black blind and 15% with a white one.

On a sunny day the increase of air temperature on passing through a 2 meters high window is 15-25 °C.

3.4. DYNAMIC INSULATION.

The surface transmission coefficient K of the windows without ventilation is 2 W/°Cm^2.

Ventilating with a flow of 20 m^3/hm^2 of extracted air the value of the coefficient k is reduced to 1 w/°Cm^2.

At the same time, with an internal/external thermal gap of 18 °C, the temperature difference between the internal ambient air and the glass surface is reduced on average to 1.5 °C.

Table 1. Technical specifications of the SAV System

Thermal transmission coefficient	1.9	$W/m^2 °C$
Performance of solar collection		
Ventilation	20	%
Direct	15	%
Total	35	%
Solar Protection		
Static Conditions	20	%
Summer ventilation	14	%
Acoustic damping index	46	dBA
Luminous dispersion index	20	%

INTELLIGENT BUILDING : LOW CONSUMPTION AND INCREASED COMFORT

X. TRAVER, M. PUJOL, A. TRIAS*, X. MILLET*

CCFC sccl
Via Laietana 54
08002 Barcelona (España)
Tf. 268.20.87 Fax. 268.01.39

* BCN Cambra Logica scp
C/ San Joan 8 - Torrelles de Llobregat
Barcelona (España)
Tf. 689.01.36 Fax. 689.05.45

SUMMARY The object of this project is to integrate energy-saving technology (with spacial emphasis on the use of solar energy) with integral control and management technology, with the aim of designing and constructing in the present day homes with levels of comfort which will be in demand in the near future.

In the design of the dwellings the concept of integral comfort is applied : thermal, luminous, acoustic, visual and hygienic. to this end the following technologies are used :

* SAV System : Solar Acoustic Ventilated Window
* Microporous Brick
* ICM System . Integrated Control and Management System

The form chosen is that of a linear block with a ground floor plus four, and it will be built in Torrent dels LLops in Martorell (Province of Barcelona). The promotion is under the charge of Incasol (The Catalan Land Institute, a public company of the Generalitat, the autonomous government of Catalonia) and is included in its operating plan for the period 1991-1992.

In addition the CCFC (Construction Cooperative Consortium) has proposed its presentation to the Thermie programme of DGXVII of the CE with the participation of Italy and Spain.

The SAV System consists of ventilation and acclimatization systems controlled by the integral management system.

The microporous brick has been developed in Italy, and due to its alveolar and microporous design it possesses very attractive thermic characteristics while preserving the features of mass and transpiration typical of traditional bricks.

The ICM System has been developed by Bems (Mondragon Group) and is applied to the use of facilities of ambiental comfort, safety and communications which are incorporated in this project.

1. The SAV System

The SAV system is composed schematically of a venetian blind between two panes of glass between which air circulates. This system, appropriately developed and adapted to each set of circumstances, permits optimised operation by modulating the functions of :

* Natural Lighting
* Solar Irradiation
* Thermal Transmission
* Acoustic Transmission
* Ventilation

S. Alvarez et al. (eds.), Architecture and Urban Space, 581–586.

The SAV System can be adapted to users needs and guarantees interior comfort in the different climatic conditions, at the same time providing a considerable energy saving.

The venetian blind permits the modulation of natural lighting and at the same time acts as an absorber of solar radiation. The circulation of air under solar radiation allows for the recovery of solar energy absorbed by the blind, by virtue of which :

* The solar protection effect is reinforced;
* Usable warm air is produced;
* The radiant wall effect is attenuated.

The ventilation can be effected with either exhaust or renewal air. Independently of solar irradiation, the air circulation gives the Air Flow Windows the characteristics of a dynamic insulation. The coefficient of surface thermal transmission will vary according to the volume of ventilated air and the internal/external thermal gap. One of the consequences of this property is that, by ventilating with exhaust air, it will be possible to maintain the surface temperature of the window close to that of the ambient air.

1.1 FUNCTIONING OF THE SYSTEM

The SAV System operates in conjunction with the controlled renewal of air in accordance with two basic schemes, in function of the climatic conditions : Winter and Summer.

The openings at the bottom of the outer part permit aspiration in winter or expulsion in summer. The air circulation inside the chamber is upwards or downwards In winter the new air is aspired through the window and expelled to the outside by the north facade by means of electroventilators situated in the false ceiling. The renovation rate of the air is one volume per hour.

RI : Incident Solar Radiation
RR : Reflected Solar Radiation
RT : Transmitted Solar Radiation
V Thermal Energy Evacuated By Ventilation
RE : External Infra-red Radiation
Ri : Internal Infra-red Radiation
CI : Interior Convection
CE : Exterior Convection
Te : Exterior Temperature
Ti : Interior Temperature.

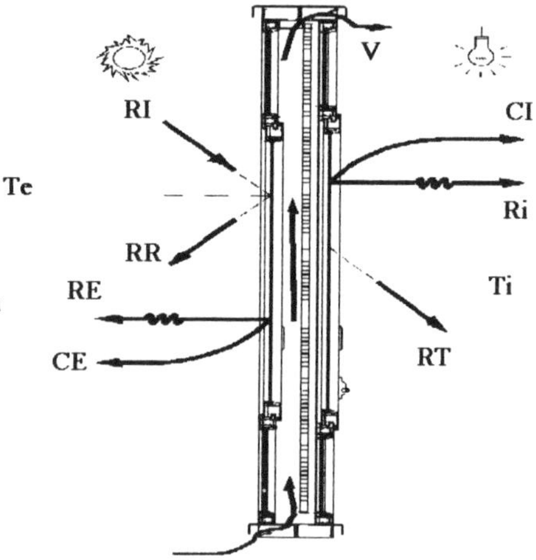

Figure 1. Diagram of functioning of the SAV System

In summer the air circuit is inverted, the electroventilators aspire the new air through the north facade and expel it through the windows. The renewal rate is three volumes per hour.

In static conditions without forced ventilation, the functioning has the same characteristics. The free convection of the air in the chamber is activated exclusively by the action of the incident solar radiation on the windows.

1.2 CHARACTERISTICS OF THE SYSTEM

The follow table shows the characteristics of the SAV System. The measurements are taken from the SAV System installed at the Primary School of Agramunt (Lleida).

Table 1. Technical Specifications of the SAV System

Thermal Transmission Coefficient	1.9	$W/m^2°C$
Efficiency of the Solar Collection		
Ventilation	20	%
Direct	15	%
Total	35	%
Solar Protection (Solar Factor)		
Static Conditions	20	%
Summer Ventilation	14	%
Acoustic Damping Index	46	dBA
Luminous Dispersion Index	20	%

2. Microporous Brick

Since the 1960 Italy has been producing bricks of microporous clay with vertical and horizontal ribs which also serve to withstand the stresses caused by settling processes.

In this project, this technology is used with the following variations :

* Use of the brick both as conventional closure and as ventilated wall. This allows an improvement in thermal insulation performance, a reduction of the summer load by means of air circulation, and also a faster evacuation of possible humidity in the external wall and their protection from atmospheric agents.

* Improvement of the performance as thermal insulation by means of the modification of the geometry of the alveola.

* Improvement of the thermal insulation performance through the selection of the basic mixture in the manufacture of the brick.

The project visualises the use of the brick both as an external closure and as a ventilated wall. In the latter case two lugs are added to the brick to support the transparent closure.

The brick is made of microporous clay, that is to say, clay lightened with air pores. The dimensions of the brick are length 40 cm., height 39 cm. and width 40 cm. The thermal transmission coefficient is 0.58 kC/hm^2

Table 2. Thermal flows of the normal and modified microporous brick

Brick	w	w/m^2
Normal	0.553	14.9
Modified	0.464	12.3

External surface area : 0.037 m^2

3.Internal Management System IMS

This is a technological system which is applicable to the operation of installations and which achieves a reduction in operating costs while improving the user's conditions of comfort.

Continuously tests, evaluates and reports on the building's status.

This system supplies the building with a set of services which satisfy user needs and optimise comfort, being based primarily on :

* Decentralised system of integral management;
* Internal communications system by local network
* CCC : Control, communications and computer

This system controls the following services :

* Heating and ventilation
* Control of electrical energy and cosinus _
* Optimisation of starting/stopping of installations
* Energy savings
* Lighting
* Security : monitoring of accesses and notification to guards. Control of closure of automatic doors with access by card.
* Protection against fires.
* Lifts
* Maintenance.

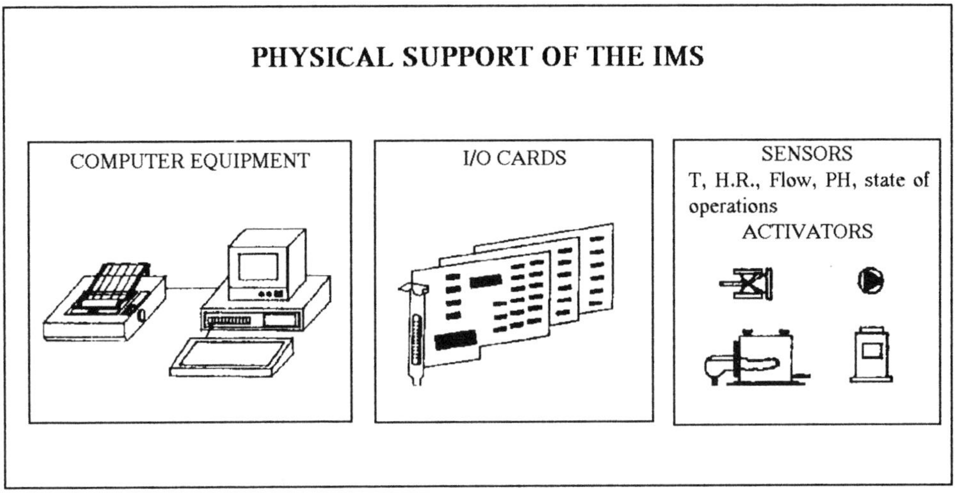

Figure 2. Physical Support of the IMS

4. Architectural Aspects

From the architectural point of view, the project uses energy saving design criteria, with advanced technologies for an increase in comfort and an integral control of installations : these technologies ensure the optimum functioning of the services supplied by the building.

The drafting of the project has coherently used popular criteria of distribution and orientation of buildings, and technological tools to simulate the building's behaviour such as energy balances and computer-aided synthesis images.

Given the planning regulations of the district, the orientation of the site and the others variables concerning planning and regulations, it has been decided to construct a building consisting of ground floor and four storeys, with all the principal rooms of the dwellings (bedrooms, living-rooms) oriented towards the south. Only the kitchens, bathrooms, washrooms and accesses are north-oriented. The building's exposure to the sun, solar gains and inertia are under control.

The adaptation of the building to the various climatic changes affecting the exterior has been made possible by intelligent planning work on the skin of the building, and not only on its interior as is customary in conventional architecture.

Special attention has been paid to the aspects of energy saving, security, hygiene and integral comfort in order that all the building's services operate optimally and integrally and can be adapted easily and economically to future systems.

5. Conclusions

The integration of these technologies developed by cooperative firms contributes to a functional and futuristic design of a warm and comfortable home. The most important treatment is that of the skin of the building achieving a harmonic balance between the internal and external atmospheres.

At no time is an excess of technology visible, this remaining always invisible and at the service of the user.

This is the purpose of working with a second skin which insulates thermically from the exterior, modulates the internal lighting by acting on the windows, breaks through controlated ventilation and also transpires, thereby ensuring integral comfort in its interior.

Chapter 3

Cooling

An Illustration of semi-passive bioclimatic Architecture: The Belgian Pavillon Project for the Seville World Fair 1992

A.C. GILLET , J-M de LAMINNE
Atelier "D" S.C.
23, Belle Voie
B 1300 Wavre
Belgium

S. LEFEBVRE
Emeritus Professor
Faculté Polytechnique de Mons
99 Chaussée de Beaumont
B 7000 MONS

ABSTRACT

In the national competition for the conception of a belgian pavillon for the Seville World Fair 1992, our bureau has presented an original project.
The main features of our conception were:
- a building wrapped in an energetic glazed curtain façade, open to natural lighting, designed as to give a passive shading to the premises combined with a semi-passive cooling system.
- a building extremely economical in investment as well as running costs.
All the architectural design has been driven by the concern of avoiding solar gains through the apertures needed for the natural lighting of the floors. The whole building shade itself protects from direct solar radiation an open forum running at basement level.
On this bioclimatic passive conception, we have conceived the opportunity to act on the ventilation flows with a belgian patented evaporative system known as AMAZONE, a direct contact air/water exchanger in which water is vaporized while flowing on a vertical fiber bundle crossflow to air. The latent heat of vaporization is taken from the air which is cooled byabout 5 kelvin with a relative humidity gain of about 20%.
Compared with a classical thermodynamic air conditioning system, our proposal was able to reach an appropriate comfort level with much less investment as well as running costs.

S. Alvarez et al. (eds.), Architecture and Urban Space, 589–594.

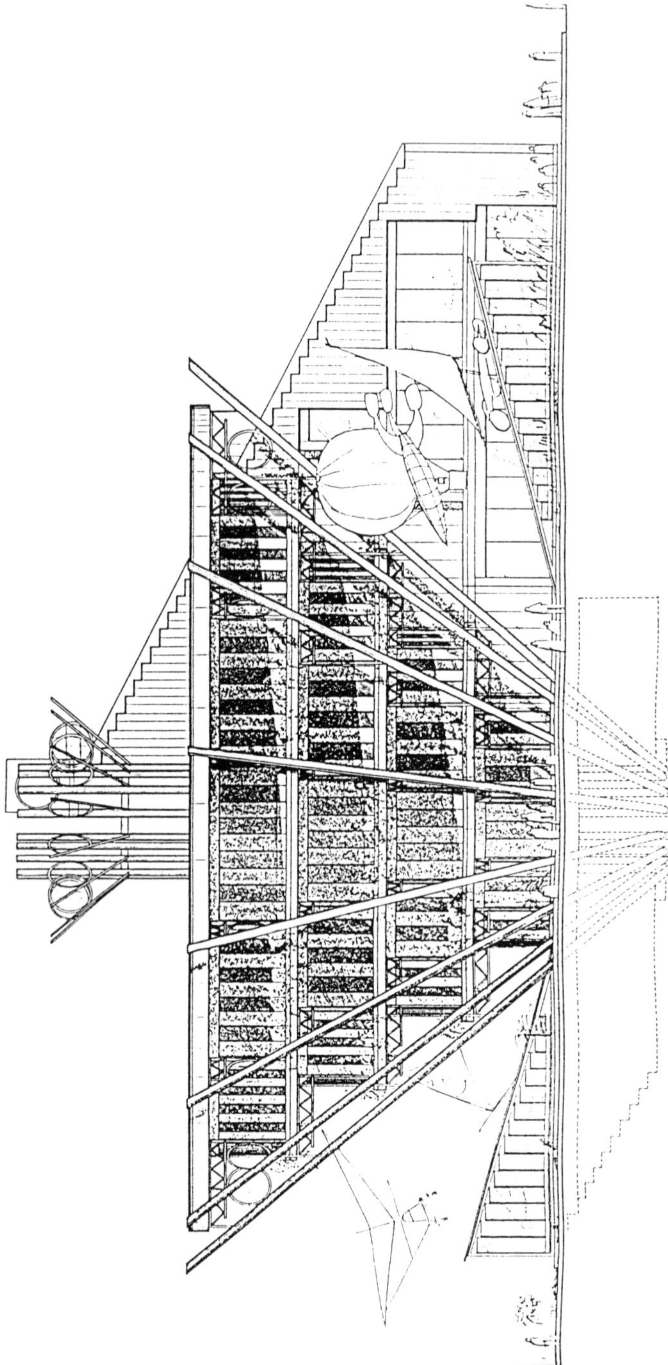

fig.1 General Sketch of the Belgian Pavillon

I Introduction

The building itself (fig.1) is caracterized by an uptodate architecture integrating the more recent experience in bioclimatics :

The project consists of a gigantic metallic top-like tree in a quarter grove of which is set a triangular based pyramid. The later acts as a stabilizing root to the top volume. Oriented south and covered with an opaque roof, it gives the necessary shade to the south part of the building.
In the top-like volume, each floor overhangs the lower glazed façade avoiding any direct solar gain. These façades are provided with openings for forced ventilation of the adjacent floors. Their repartition has been designed as to avoid any distributing air duct system while assuring a good equalized air flow everywhere.

II The Semi-Passive Cooling

Taking advantage of this well designed ventilation system, we have concieved the opportunity to act on the air flow with a belgian patented évaporative appliance known as the
"AMAZONE".
The Amazone is a direct contact gas/liquid exchanger (fig.2). The liquid phase (water in our case) flows downwards on a vertical synthetic fibers bundle streched between floor ans ceiling in front of the openings of the glazed façade. The gas, air in this case, is sucked through these openings by the general ventilation system and flows partly crossflow, partly counter currentwise through the bundle (fig.3). It vaporizes water and gains humidity on an adiabatic or subadiabatic way. The latent heat of vaporization is taken from the air which is cooled.
The installation has been computed as to get a 5 kelvin cooling effect with a relative humidity gain of about 20%.
The Amazone exchangers are modular and removable.

III Complimentary Active Cooling

The water needed for the wetting of the fibers bundles is stored in an underground 600 m3 reservoir, pumped through a pipîng network to every part of the building and flows back to the reservoir. Night running of the system, when wet thermometer temperature falls down to 17°C, provides the necessary cooling of this water volume. As a security back up to the normal cooling load, this is also helped by a water to water

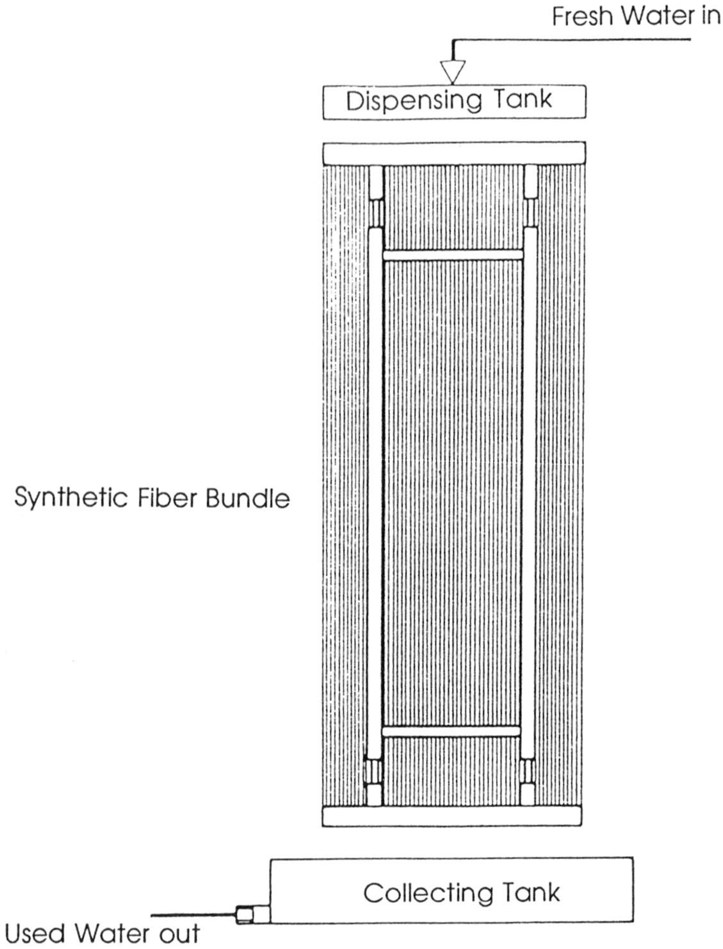

fig 2. Front View of the Amazone

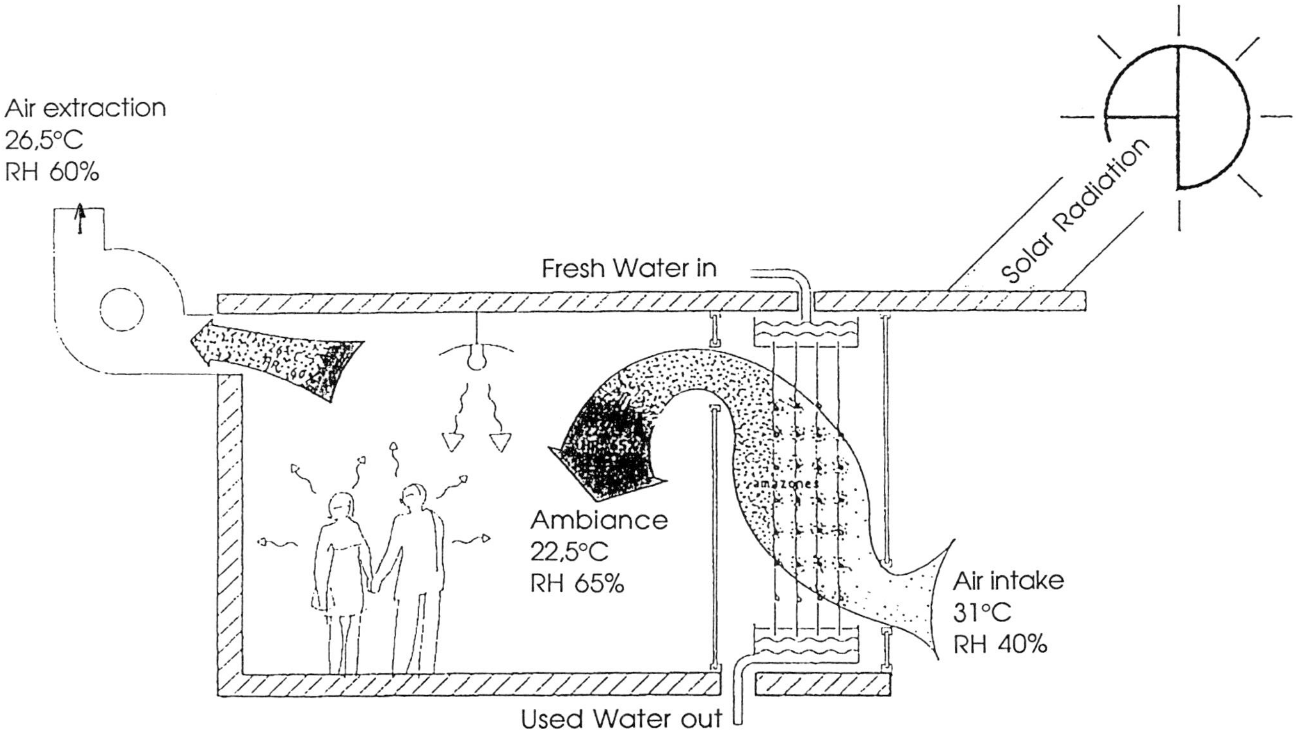

Fig 3. Principle of the Semi-passive Cooling

domestic hot water needed by the kitchen and the sanitary appliances of the building.

IV Discussion

The high volume flow of air (20 vol/h) will provide a steady breeze in the premises and an increased comfort feeling to the inoccupants by evaporative cooling of their own perspiration.
Compared to classical thermodynamic air conditioning system, our proposal is able to reach an appropriate comfort level with much less investment and running costs.

V Conclusion

The project assembled a lot of up-to-date and highly desirable features: **Passive and Low Energy Architecture** combined in an elegant , economical and comfortable building. In spite of these qualities, it was not the winner of the comptétition and will not be built.

INTELLIGENT CONTROL OF NIGHT-TIME VENTILATION

A. C. PITTS and R. S. ABRO
Building Science Unit
School of Architectural Studies
University of Sheffield
Sheffield S10 2TN, U.K.

ABSTRACT. In hot climates various passive cooling strategies may be employed to reduce the need for, or reliance on, air conditioning systems for buildings. One of the strategies which may provide substantial benefits is the use of night-time ventilation. This paper describes research currently being carried out at the University of Sheffield to enhance and improve control over this option. A photovoltaic array is used to generate power which may be utilised to operate fans; these fans are linked to a solar chimney. The power generated can be used immediately or stored for future use, for example during the night-time period. Decisions as to the optimal time periods of use of the energy available must be made in a logical fashion. An "intelligent" control strategy is therefore being developed.

1. Introduction

Interest in passive solar design features has increased in recent years, partly as a result of the improved design methods/techniques being developed, and partly as a result of the incorporation of some form of control system or strategy. In order to increase the limits within which passive systems operate, the level of control, or degree of "hybridization", is also likely to increase. In this paper a Solar Chimney enhancement is descibed which requires more sophisticated control. The dwelling is still able to benefit from the normal ventilation increase from the chimney during the day, whilst also being able to use additional ventilation at a time period of most benefit for comfort.

S. Alvarez et al. (eds.), Architecture and Urban Space, 595–600.

2. Building Bioclimatic Design

It is possible to analyse and assess passive cooling strategies using the method originally suggested by Givoni (1976) and later developed upon by Szokolay (1985). These analyses produce a Building Bioclimatic Chart (drawn onto a psychrometric chart outline). The principal features shown on such a chart are a "comfort zone" and "control potential zones": the second category here represents the range of external conditions which could exist whilst allowing passive features of the building design to control the internal conditions to within the comfort zone. Figure 1 illustrates the use of such a chart for the city of Lahore in Pakistan. The chart shows that by using "mass" and "ventilation" effects/features, that the control potential is almost sufficient to provide comfort even in the relatively hot month of June.

3. Enhanced Solar Chimney

Conventional solar chimneys consist of a static absorbing surface in a near vertical plane. Though improved versions of solar chimney design have been investigated, by for example Ward and Derradji (1987), the static nature limits its use. A particular drawback is the lack of option for night-time ventilation, yet it is such ventilation which can substantially increase cooling potential.

In the enhanced version being developed at the University of Sheffield, a photovoltaic array is used to collect solar energy. This energy is employed as the power source for fans located within the solar chimney. The fans may be operated immediately by the system, or the energy may be stored for future (night-time) use. More details concerning the system may be found in Pitts and Abro (1990). The choice as to whether the energy is used immediately or stored, is a complex one. The inefficiencies of conversion and storage must be set against the extra benefits that night-time ventilation may bring. It has therefore been decided to use a form of "intelligent" controller in an attempt to optimize the system on a day-to-day basis.

4. Control System Features

4.1 General Data

The decision making process is contolled by computer program incorporating three calculation modules which are used to evaluate: the thermal conditions; the solar energy

MAX. = EXTREMES OF EXTERNAL ENVIRONMENTAL CONDITIONS
MIN.

(vertical hatching) = COMFORT ZONE

M = MASS EFFECT CONTROL POTENTIAL ZONE

V = VENTILATION EFFECT CONTROL POTENTIAL ZONE

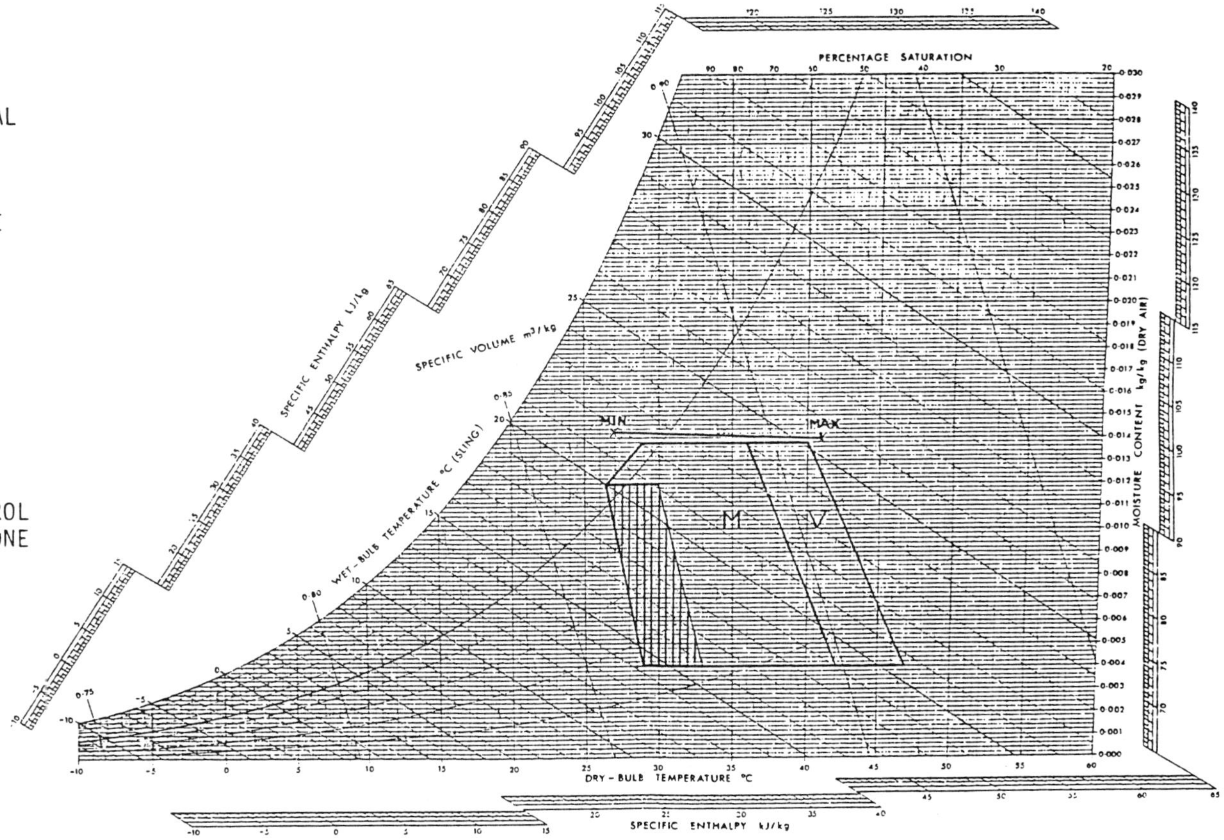

FIGURE 1: BUILDING BIOCLIMATIC CHART: LAHORE, PAKISTAN:JUNE

availability; and the degree of comfort/discomfort. The control system is currently being implemented using a standard PC, though it is anticipated that a single chip could be used for all calculations/decision making when fully developed. A limited database of information is required to enable the calculation modules to function; this includes :

(a) Building thermal parameters
(b) Occupancy pattern
(c) Climate data (general monthly values plus running averages of the previous 3 days)

Other data required and measured are values of internal and external air temperature and solar radiation level.Details of the three calculation modules are given below.

4.2 Solar Energy Availability (SEA) Module

This module uses as its basis the average of previous days' solar radiation levels to predict the subsequent day's levels. The prediction might be modified as the day progresses, however the main use of the enhanced solar chimney is seen as being in climatic zones with well defined seasons and patterns, (thus reducing the need for such "updating" which introduces several problems). The expected solar energy incident on the array is used to calculate equivalent air changes for the building either for immediate use to power the fans or via storage for later use. The ventilation rate provided by the basic solar chimney is also determined.

4.3 Thermal Environment (TE) Module

This module is based on a relatively simple thermal model in which hourly predictions of internal conditions may be made. Factors included are the building thermal parameters, climate data (external temperature and solar radiation levels), casual heat gains associated with the occupants, and ventilation rates.

4.4 Comfort Assessment (CA) Module

This module utilises a modified form of the Fanger Comfort Equation to determine the degree of discomfort. The inputs required are found as follows: air temperature is provided by the TE module; humidity is estimated from the database; radiant temperature is estimated arising from the TE module; air velocity may be calculated from ventilation data; human factors (expected clothing and activity levels) are included in, and thus derived from, the occupancy database. The module provides an hourly calculation of "Predicted Mean Vote" which can be integrated over the relevent time periods to give an index of discomfort. It is this index which must be minimized in the optimization of the control strategy.

5. Control System Operating Strategy

Initially the sequence of control system operation was devised to be as follows:

(i) Use SEA module to calculate potential ventilation rates
(ii) Use TE module three times to calculate internal conditions using ventilation rates: Solar chimney alone; Chimney plus day ventilation; Chimney plus night ventilation
(iii) Use CA module to calculate discomfort under each ventilation option

On the first day of operation in any sequence, it has been found that it is most beneficial to use solar radiation to charge the battery store, unless it is already charged. This then allows scope for night-time ventilation during the first night and thus influence on the "start-point" for the next day. In other words, the initial situation for system operation should be at the beginning of the night with a fully charged store. The decision then becomes one of whether to discharge the battery overnight, (which can improve comfort during the night and give a better starting condition for the following day); or whether to save the stored energy because the following day will not permit recharging.

6. System Testing

There are many complexities involved in the field testing of both enhanced solar chimney and control strategy. At present therefore, the testing has been confined to a reduced scale laboratory investigation. Initially, climatic conditions represented in areas of Pakistan have been chosen for analysis. This arises from that country's continued interest and support for photovoltaic technologies and the potential opportunity to exploit enhanced passive solar design.

Detailed results are not yet available to indicate any general pattern of system use/optimization.

7. Summary

An enhanced solar chimney design has been described which incorporates fans powered from solar photovoltaic derived energy. The option exists to utilise the fans during the day-time or to store the energy for use at night-time. An intelligent control system has been devised to determine the optimum operational period. Though this preliminary design for the controller is large and complex, a smaller scale device could be produced when all factors have been evaluated.

8. References

Givoni, B. (1976) Man, Climate and Architecture, Applied Science Publishers, 2nd Edition

Pitts, A.C. and Abro, R.S. (1990) 'Improved Building Comfort in Hot Climates', Proc. World Renewable Energy Congress, Reading, U.K. Sept.1990

Szokolay, S.V. (1985) 'Thermal Comfort and Passive Design' Advances in Solar Energy, American Solar Energy Society

Ward, I.C. and Derradji, M. (1987) 'Air Movement in Buildings Cooled by Passive Means', Proc. European Conference on Architecture, Munich

THERMAL PROCESSES IN BUILDING ENVELOPES TO IMPROVE SUMMER COMFORT.

by

A.N.AYOOB and R.A.ATTALAGE.

(Centre d'Energétique, Ecoles des Mines de Paris,
rue Claude Daunesse, Sophia Antipolis,
F - 06565 Valbonne.)

Abstract.

This study intends to present the behaviour of the building envelope treated by ambient air ventilation or by evaporatively cooled air ventilation. A mathematical model for a test was developed and simulations were carried out for different stratergies of applications.

1 - Intoduction.

The specific climatic conditions in hot regions, play an important role for the passive tools in building design. The combined action of wide temperature fluctuations between night and day - timeleads to a rapid transient thermal behaviour of the walls.

The need for a more economical method of cooling in summer season than that available at present is becoming necessary with the increase of the need for comfort and the development in the world, especially in hot or desert regions.

The thermal storage in building material for passive solar construction has generated considerable interest in recent years. The use of walls and floors for thermal storage for solar heating has been successfully demonstrated and has gained some acceptance by the public why not cooling also. The object is to make use of the thermal mass, which can be well adapted in hot climates, not only to change the time lag but also to allow for the night time storage cooling.

In general the thermal load of the wall structure in buildings present a considerable percentage of the total thermal load. So, the treatment in the wall construction proves to be an efficient way for reducing cooling load in buildings.

The use of night time ventilation or evaporative cooling may be utilised to reduce the thermal load and to improve the comfort in buildings.

This study intends to present the behaviour of a building envelope in view of reducing the cooling load and improving the comfort inside the building.

S. Alvarez et al. (eds.), Architecture and Urban Space, 601–605.

2 - Modelling and simulation of a test cell.

In order to illustrate the applicability of the treatment of the building walls as a means of improving the thermal comfort in buildings using nocturnal operation of ventlation and evaporative cooling, we have tried to simulate a ficticiuos test cell of which one or several walls are formed as a cooling storage. As a preliminary study the test cell is considered to be comprised of six identical walls (3m x 3m) of which the cross section and the isometric view are shown in Fig.1. In addition, the nothern wall contains a window pane of dimension 3m x 1.5m. The blocks are considered to be filled with rock particles of a certain porosity when the wall acts as a storage.

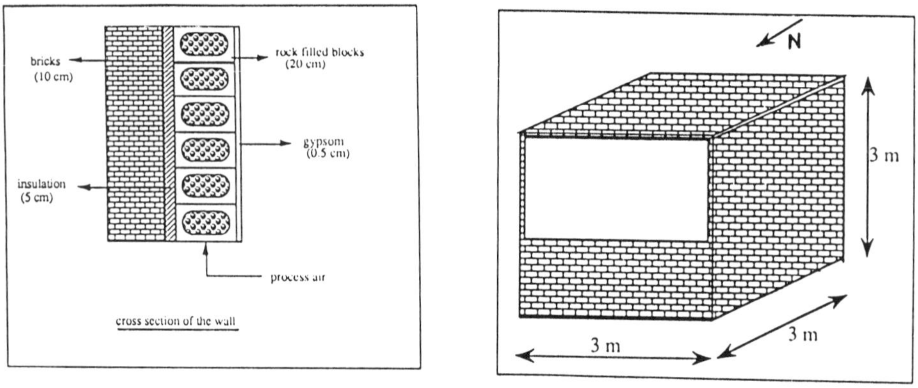

Fig.(1) cross section of the wall being analysed.

We have adapted the technique of nodal discretisation of the space variable by finite differences, in order to solve the set of equations containing partial dervetives of time and space describing the behaviour of building.

For each node, conservation of energy is applied taking into consideration the boundry conditions of imposed temperature or flux type. All nodes of heat transfer condition convection,radiation and transport of heat by mass transfer (only for cooled walls) are included in the analysis. We have assumed

(i) the properties to be isotropic and homogeneous independent of temperature.

(ii) emmission and reflection of long wave radiation to be diffused and isotropic.

In addition the mass of air inside the test cell is represented by single thermal capacity at the centre of the cell.

These yield a system of non-linear algebraic first order differential equations in node temperatures.

for node i, the equation with classical notations is

$$m_i C_{pi} \frac{dT_i}{dt} = \sum_{j=1}^{N} \phi_{ij}(A) + P_i$$

where $\phi_{ij}(A) = \frac{\lambda_i}{\Delta x_{ij}} A_{ij} (T_j - T_i)$ for conduction

$\phi_{ij}(A) = h_{ij} A_{ij} (T_j - T_i)$ for convection

$\phi_{ij}(A) = \varepsilon_i \sigma A_{ij} F_{ij} (T_i^4 - T_j^4)$ for radiation

$\phi_{ij}(A) = \rho_a C_a v_i A_{ij}(T_i - T_j)$ for mass transfer

P_i - external heat gain

The heat transfer coefficient between the air and the masonary block is taken as (2.4 $W/m^2.k$) (3), where is the volumetric heat transfer coefficient between the humid cold air and the rockbed is taken from this formula (4).

$$\frac{h_v.d_e^2}{k} = 1.56 \left(\frac{\rho.v.d_e}{\mu}\right)^{0.7}$$

where d_e is the equivalent diametre of the rock (m)
k is the conductivity of the rock (W/m.K)
v air velocity (m/s)

As the means of solving the system of equations, we have made use of the simulater NEPTUNIX , which is based on the technique of Newton-Raphson as the mode of resolution.

The results concerning the daily maximum and minimum temperatures and relative humidity inside the test cell are presented in Tab.1. for different processes of treating the system for different climatic conditions, treated air ventilation and building inertia.

3 - Discussion and Conclusion.

This study serves as a preliminary presentation of several treatments of the building envelope by air ventilation or by simple evaporative cooling system to improve the comfort.

The research on improving and optimizing a simple air conditioning system which can operate efficiently are carried out and on the other hand the research for new materials such as PCM (phase changing material) which can replace the traditional rocks in the storage are continuing.

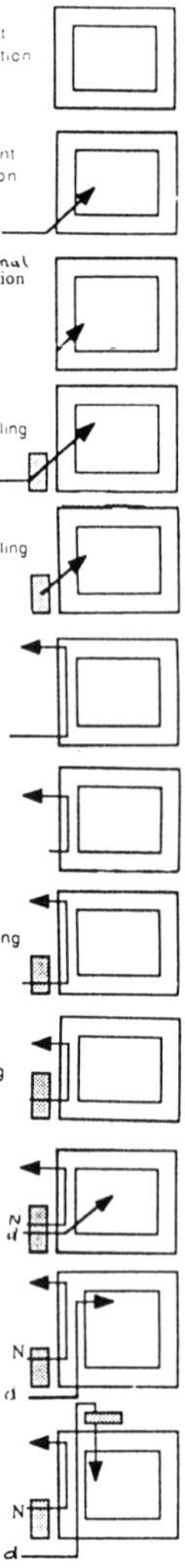

Configuration		HOT - ARID CLIMATE						MODERATE CLIMATE					
		low inertia			high inertia			low inertia			high inertia		
	vol/h	5	10	25	5	10	25	5	10	25	5	10	25
without ventilation	Max. Temperature°C	46.6				39.0			34.6			38.0	
	Min. Humidity %	14.0				20.6			47			39	
	Min. Temperature°C	32.7				37			25.4			26	
	Max. Humidity %	29.0				24.0			71			70	
permenant ventilation	Max. Temperature°C	45.7		45.4		42.7	39.0		31.7	31.0		30.7	30.6
	Min. Humidity %	15.0		14.7		12.7	21.0		52	52		51	51
	Min. Temperature°C	32.0		31.0		39.7	34.0		24.7	24.0		25.7	25.0
	Max. Humidity %	29.7		32.0		22.0	22.4		72	74		71	71
nocturnal ventilation	Max. Temperature°C	45.4		44.7		41.0	40.7		31.7	31.4		31.2	31.0
	Min. Humidity %	15.2		15.0		21.7	21.7		51.6	51.6		51	51
	Min. Temperature°C	29.4		29.7		30.4	30.4		26.2	25.7		26.4	26.0
	Max. Humidity %	27		27.2		26.4	26.4		69	77		69	69
permenant evaporative cooling	Max. Temperature°C				37.7	36.0	32.2				29.0	28.4	
	Min. Humidity %				42.7	50.0	60.0				77.0	81.0	
	Min. Temperature°C				28.4	25.7	25.0				22.4	22.0	
	Max. Humidity %				56.0	60.0	90.0				95.4	96.0	
nocturnal evaporative cooling	Max. Temperature°C		42.7	41.6		39.7	38.4		31.4	31.0		29.4	29.0
	Min. Humidity %		36.0	41.0		19.6	20.0		52	52		61	61.,6
	Min. Temperature°C		27.7	23.0		28.7	24.0		22.7	21.4		23.0	27.0
	Max. Humidity		66.0	91.0		70.0	91.0		77	79		74	62
permenant wall ventilation	Max. Temperature°C	45.2			44.2			30.7			30.4		
	Min. Humidity %	14.7			15.7			57.0			57.7		
	Min. Temperature°C	34.4			40.0			24.0			24.4		
	Max. Humidity %	22.4			21.0			77.0			76.4		
nocturnal wall ventilation	Max. Temperature°C	43			42.4			32			31.7		
	Min. Humidity %	12.7			12.6			54			56.0		
	Min. Temperature°C	34.4			38.6			24.2			25.0		
	Max. Humidity %	22.1			22			77			76.6		
permenant wall evaporative cooling	Max. Temperature°C				37.0	38.0					28.7	28.4	
	Min. Humidity %				41.7	41.4					24.6	24.4	
	Min. Temperature°C				27.0	27.4					24.4	24.4	
	Max. Humidity %				36	36.6					77	77	
nocturnal wall evaporative cooling	Max. Temperature°C				44						31.0		
	Min. Humidity %				15.0						57.0		
	Min. Temperature°C				27.6						26.0		
	Max. Humidity %				36.0						74.0		
nocturnal wall evaporative cooling + daily direct evaporative cooling	Max. Temperature°C				35.6	33.7	31.4				28.1	27.7	
	Min. Humidity %				47.0	53.0	61.0				76.4	77	
	Min. Temperature°C				25.0	23.7	21.0				22	[illegible]	
	Max. Humidity %				57.0	67.0	91.0				94	95	
nocturnal wall evaporative cooling +daily air ventilation through the wall	Max. Temperature°C					40.6	41.2					28.0	28.2
	Min. Humidity %					20.0	21					76.0	77.0
	Min. Temperature°C					28.0	27.6					23.4	23.0
	Max. Humidity %					57.0	91.0					87	87
nocturnal wall evaporative cooling +daily air ventilation through the wall+ evaprative cooling	Max. Temperature°C				31.2	30.0	29.0				27.0	27.9	
	Min. Humidity %				64.0	66.0	70.0				77.0	77.7	
	Min. Temperature°C				24.7	24.6	24.6				23.0	23.0	
	Max. Humidity				34.0	35.0	36.0				86	86	

Tab.1.

References.

1- A.Ayoob, P.Achard and R.A.Attalage."Characterization of parallel evaporative cooling pads". Ecole Des Mines de Paris-France, sent to be published 1989.

2- M.Bordier "Identification de systeme lineaires par la methode de Marquardt, Rapport, Ecole Des Mines de Paris-France,Octobre 1984.

3-Mitchell,R ,and D.Elmer, Aunique use of concrete block for cooling applications. THe Renewable Challenge; Proc 1982 Annual Meeting(Huston,TX) ASES ,

4-P.Achard,A.Ayoob,L.Elegant,Correlation of the volumetric heat transfer coefficient for air-rock bed thermal storage system - considering humid air, The Int. Jnl. of solar energy June 1990.

THERMAL BEHAVIOUR OF ROOF COVERS IN HUMID TROPICAL MICRO-CLIMATES

Cláudia Barroso Krause (*) and Oscar Daniel Corbella(°)[1]
FAU - UFRJ - Cidade Universitária
21.945 - Rio de Janeiro
Brazil - (FAX 55-21-290.66.26)

ABSTRACT. This paper deals with the evaluation of the hydrothermal behavior of buildings when the characteristics of the roof cover are changed. It works with 15 different types of roofs, selected among the most usually found on the Brazilian market for social lodging buildings. They are studied in summer conditions for 3 different sites in the neighborhood of Rio de Janeiro City, representing each one a different summer situation in this climatic region. A reference building, representative of a typical social lodging in the region, was created and modified by simulation in order to be the most effective low consuming house for the local climate. For the last version, more than one hundred computational simulations were made for the three climates, in order to determine the internal conditions of hydrothermal comfort. As a result a hierarchical order of roofs is presented according to their response when submitted to the most extreme external summer climate.

Introduction

Traditionally the problem faced by the environmental conscious design was that of the maintenance of a comfortable environment in the interior of a building, lowering the heating load, and as a consequence, lowering the level of pollution related to corresponding energy generetion.This approach is a consequence of the starting of the discussions and development of these ideas in developed, energy importers and polluted countries.

In tropical countries the problem would be put as that of how to lower the cooling load, and also how to diminish the pollution produced by the energy consumption and the devices used to cool(1). However, the problem seems to be more complex because, thinking about social lodging, it is far more expensive to produce cooling than heating. On the other hand, usually tropical countries are "developing" countries where the lack of low cost houses is enormous and, when built, they are built without any criteria concerning comfort.

---

[1](*) Present address: Ecole de Mines de Paris - Centre d'Energétique - Paris - France - Fax: 33.1.43.54.18.98
(°) Researcher of the CNPq - Brazilian National Research Council

S. Alvarez et al. (eds.), Architecture and Urban Space, 607–612.

So, we ought to think how to produce an internal comfort without mechanical cooling, performing design works considering only the space arrangement, building materials and climate characteristics in order to produce a low cost habitable house.

This work is a contribution to this general idea, studying one of the variables considered in the whole problem: the roof. This building element is the most important from the thermal heating load point of view. Therefore, if we can select a low cost roof, using local market materials, that would produce the minimum heat load, we shall be in a better position to solve the problem.

Basic Elements for the Simulation

A reference building was introduced in order to allow comparisons of the response obtained when the 15 different types of roofs are used. Its basic design follows the one used in Reference 2, similar to those of the majority of the self-built houses found in the South-East zone of the country. It has 12sqm. on the interior, 3m. high and an overhang of 0.75m. in all perimeter, as it is shown on Figure. 1

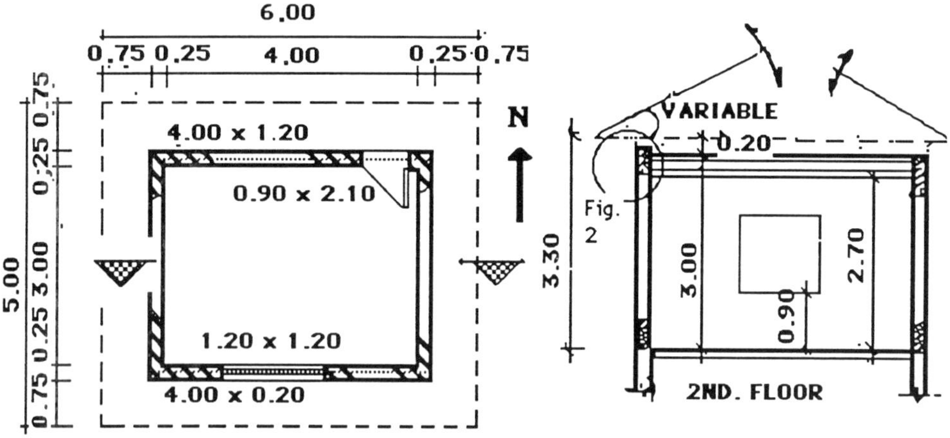

Figure 1 - The reference building

The walls are made of hollow clay bricks, gypsum board finishing on both sides with white washed (using fixing solution) outside.Its top was modified in order to accept several choices for the different roofs and to permit its ventilation The floor is made of wood tables over concrete cement plate and the ceiling of preformed wood..

There is just 1 window with 2 internal wood wings with sheets of glass and 2 external venetian-like sheets, all white painted.

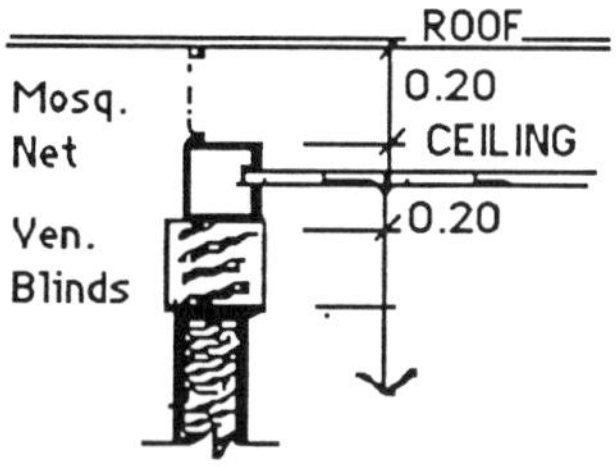

Figure 2

The door, opposite to the window, consists on preformed wood, white painted on both sides.To increase ventilation, there are also 2 openings on the top of the walls with fixed venetian-like blinds. 2 other openings, with the same size and on the same previous walls are placed with mosquito-net, above the ceiling in order to allow attic ventilation like is shown in Figure 2.
As the building had to be located in 3 different micro-climates, the best orientation for each one was determined by simulation. Related climates are discussed lather and here it is only quoted the result of these optimizations.
By analysis of the sun path and local winds, for the climate of "Baixada de Jacarepaguá" (location 1) and for that of the "Baixada da Penha" (location 3), the best orientation was E-W for the major axis, and for the climate of "Teresópolis" (location 2), it was 45° NW.
In order to reproduce a tipical family daily routine in this type of house, the occupational and load profile of the reference building was fixed in one person between 6 and 8 p.m. and two persons between 8 p.m. and 7a.m..The energy load was distributed on 50W of illumination load from 4 to 7a.m., 6 to 7 p.m. and 11 to 12 p.m., 100W from 7 to 11 p.m. For other energy loads 50W was assumed from 5 to 7 a.m. and 6 to 10 p.m..

ROOF TYPE SELECTION

The selection of roofs was made after consulting the local material building suppliers and based on IPT Report (3). 15 different types of roofs were then selected among the most usually found on the Brazilian market for low cost housing, and their thermal behavior studied on the reference building.
Two main groups, light and heavy, were established. The heavy one has consisted of concrete plates, with and without insulation sheet and a pre-fabricated plate (using pre-fabricated concrete beams and hollow clay bricks). The 12 light covers selected have included 3 metallic covers, (aluminium sheet alone, plus preformed wood and in a sandwich with polyurethane insulation), 4 asbestos-cement sheets, with 2 different thickness, with or without white-washed for 2 slopes, and finally 4 different types of clay tiles. Physical properties of these materials are quoted below.

TABLE 1 - PHYSICAL PROPERTIES OF CHOSEN ROOF MATERIAL

Note: R = thermical resistance,in $m^2/W°C$

Roof material N°	Slope °	e mm	(ext)	(int) (ext)	W/m°C	kg/m3	c J/kg
1.Aluminium	15	1	0.2	0.04	138,16	2680	880
2.Aluminium/wood	15	21	0.2	0.04 0.95	138.16/ 0.14	2680/ 800	880/ 1214
3.Al./Polyur./Al	15	30	0.2	0.04	138.16/ 0.14/ 138.16	2680/ 50/ 2680	880/ 1590 /880
4.Asbestos-cement	15	6	0.7	0.95	0.407	1600	1010
5.Asb.w/white-wash	15	10	0.7	0.95	0.407	1600	1010
6.Asb.w/white-wash	15	6	0.1	0.95	0.407	1600	1010
7.Asb.w/white-wash	25	6	0.1	0.95	0.407	1600	1010
8.French Tile, cream	25	15	0.4	0.5	0.85	1700	921

9.French Tile, red	25	15	0.6	0.9	0.85	1700	921
10.n°8 + air + n°8	25	50	0.4	0.5	0.85/ R=0.49/ 0.85	1700/ 1 / 1700	921/ /921
11.n°9 + air + n°9	25	50	0.6	0.9	0.85/ R=0.49/ 0.85	1700/ 1 / 1700	921/ /921
12.Concrete Tile	25	20	0.3	0.5	0.319	1438	780
13.Concrete plate + imperm.	0	55	0.9	0.95	1.6/ 0.17	2200/ 1200	1005 /1675
14.n°13+ insulation.	0	105	0.6	0.95 0.9	1.6/ 0.58/ 0.17/ 0.035	2200/ 1000/ 1200/ 30	1005 1047 1675 1210
15.pre-fabr.plate +imperm.(thick)	0	85	0.9	0.9 0.95	R=0.11/ 0.17	1850/ 1200	1005 1675

The influence of the roof geometry was analyzed by simulation. Horizontal roof, 1,2 or 4 slopes(15° and 25° slope) were considered. Only 2 types of roof material, light and heavy ones, were studied in order to diminish the spread of results. Best results were obtained for horizontal heavy covers and for 4 slopes (25° slope) covers.

LOCAL CLIMATE CHARACTERISTICS

The reference building, with their different roof possibilities, was studied for 3 different sites in the neighborhood of Rio de Janeiro City, covering every summer situation in this climatic region. Climatic data is available from 1986 to 1990, showing that February is the month with the most extreme summer conditions. Corresponding data for each zone is shown in Table 2.

The zone 1, "Baixada de Jacarépaguá", located between 22°54'S and 23°02'S latitude and between '43°13'W and 43°18'W longitude, is a region limited by the Atlantic beaches in the South and by mountains in the other sides. It is a flat zone with low urbanization density, that has many lakes and channels.

The zone 2, "Teresópolis", is located at mean latitude of 22°26'S and 48°58'W of mean longitude. It is an alluvial valley at 870 m altitude in its urban zone, with high altitude differences within the city and its surroundings, having also large tropical forests. Its climate is characterized as tropical-altitude one(5), having, in a yearly basis, a temperature of 17.7°C, precipitation of 1671 mm and relative humidity of 84%.

The zone 3, "Baixada da Penha", located between 22°45'S and 22°52'S of latitude and 43°12'W and 43°18'W longitude, is part of the plain area surrounding the Guanabara Bay. Many industries and low cost rent houses are installed and there are no forest or representative group of trees. Mean annual data are 24.9°C (temperature) and 74%(relative humidity).

TABLE 2 - CLIMATIC DATA FOR FEBRUARY

		ZONE 1	ZONE 2	ZONE 3
temperature:	minimum(°C)/hour	23.4° - 5:30	18.2° - 5:30	25.1°- 5:00
	maximal(°C)/hour	31.4° - 13:00	28.6° - 15:00	34.7° - 16:00

relative humidity: minimum/hour	63% - 13:00	56% - 15:00	53% - 16:00
maximal/hour	91% - 5:30	87% - 5:30	92% - 5:00
wind: day (m/s - orientation)	3.5 - S	3.0 - E	5.1 - 150°SE
night(m/s - orientation)	3.1 - 30°NE	0.1 - 135°SE	2.6 - 300°NO
surrounding albedo	0.25	0.20	0.30
Linke (turbidity factor)	2.0	5.5	2.0

Simulations

The simulations of the hydrothermal behavior of the reference building, well located in each climate and with the chosen roof, with or without attic, ventilated or not, was performed with the software Casamo-Clim, a computational design aid developed and adapted to work in tropical humid climates by the Ecole des Mines de Paris(4). More than one hundred simulations have been made, for extreme summer conditions, obtaining graphical and numerical results. The graphical ones include plots of solar radiation in each external surface showing the shadow effect, external, internal and resultant temperatures, relative humidity evolution, and bioclimatic charts showing the position in the comfort zone. Numerical results have allowed to define a comfort index from which it was possible to rank the studied roofs .

Results and Conclusions

In order to analyze the results obtained from the simulations, showing the relative efficiency of each roof to produce a situation as close as possible to Brazilian summer human comfort, an index (expressed as a percentage) was defined that shows the swing over 27°C, the highest limit temperature usually accepted in summer by Brazilians (5), and T_m, the average temperature found by the simulation during the period of occupation, that is,

$$S = (27 - T_m) \times 100/27$$

Note that $S=0$ implies a satisfactory result, being improved as S increases positively (within certain limits), and meaning a bad result if S is negative, worse as S increases in the negative sense. Therefore the values of S are indicators of the design quality and they also indicate in what way the design has to be changed. Table 3 shows the values of S for every roof and for each climate studied.

TABLE 3 ROOF COVERS EFFICIENCY - VALUES OF *S* FOR EACH CLIMATE

N° roof	Zone 1	Zone 2	Zone 3	RANK	N° roof	Zone 1	Zone 2	Zone 3	RANK
1	1.52	8.33	-9.78	12	9	2.03	9.66	-8.96	6/7/7
2	2.15	9.77	-8.82	3/5/4	10	2.30	10.22	-8.56	2
3	1.63	8.37	-9.63	11	11	2.56	10.85	-8.07	1
4	2.00	9.63	-9.00	9/8/8	12	2.00	9.48	-9.04	8/9/9

5	2.03	9.74	-8.93	7/6/6	13	1.37	7.85	-10.00	14
6	2.11	9.81	-8.82	4/3/5	14	1.44	7.96	-9.89	13
7	2.11	9.81	-8.78	5/4/3	15	1.33	7.67	-10.04	15
8	1.85	9.22	-9.22	10					

First conclusion is that materials with less thermal inertia work better than heavy ones, as it is expected from the theory. In general, clay tiles behaved better than covers of asbestos-cement, that are better than metallic ones, even with insulation. One exception was the aluminium/wood cover(n°2), with a good behavior also on the hottest zone. Unfortunately, it will have to be improved in its finishing, because it deteriorates faster than the other materials.

It is also interesting to note that red clay tiles(n°11) behave better than cream ones(n°12), due to their good emissivity that increases heat losses during the nightthat are greater than the gains produced by the absorption of solar energy during the day.

As a final conclusion we would like to remark that Table 3 is a first trial to establish a hierarchical order of roofs, from the point of view of their hydrothermal behavior, that allows architects or designers to choose better among the less expensive possibilities in the Brazilian market. Better results should be established when better computational programs are available for hot or tropical and humid climates and, fundamentally, when systematic experimental work is done in order to validate them.

Acknowledgements

One of us, CBK, wishes to thank CNPq for the fellowship received during her MSc. studies, period during which part of this work was done(6), and the Centre d'Enérgetique(5) for the use of the software CASAMO-CLIM.

References

(1)Corbella O. and Cavalcanti E., Bioclimatic Architecture in Hot Countries and Atmospheric Pollution. Proceedings. of the 1st WRE Conference, Reading, UK, Pergamon Press, 1990

(2)Akutsu, M. and Sato N., Comparative Analysis in Thermal Behaviour of Some Prototypes of Conventional Buildings (in Portuguese), Proceedings of the Latin American Symposium on Building Rationalization, IPT, São Paulo, Vol III, 1981

(3) IPT - Instituto de Pesquisas Tecnológicas de São Paulo(Brazil).Assessment on Thermal Behaviour of Unifamiliar One Floor Houses(Report in Portuguese)., IPT n° 6277, 1981

(4)CASAMO-CLIM, Centre d'Energétique d'Ecole de Mines de Paris, 1990.

(5)IBGE - Instituto Brasileiro de Geografia e Estatística , Annual Report, (in Portuguese) 1989

(6)Krause,C.B. ,Roofs, Higro-thermal Comfort, Buildings: Thinking Over and Propositions for Tropical Humid Climates During Summer Situations. Master's thesis, (in Portuguese),FAU- Rio de Janeiro Federal University,1990

THE EFFECT OF CONVECTION ON A ROOF POND COOLED BY RADIATION IN A HOT-ARID CLIMATE.

E. ERELL, Y. ETZION
The Desert Architecture Unit
J. Blaustein Institute for Desert Research
Ben-Gurion University of the Negev
Sde-Boqer Campus, Israel 84990

ABSTRACT. Roof ponds have been universally acclaimed as passive climate control systems for moderate to hot-arid climates. However, for mostly architectural reasons, their application so far has been limited to a few test or demonstration houses. Experiments were conducted to overcome some of the most common failings of roof ponds, such as the need for wind screens and movable insulation to eliminate unwanted convective heat exchange. Under certain climatic conditions, convection can assist in the nocturnal cooling of a roof pond. The effects of convection on the performance of the system described are discussed: while the average nightly cooling rate on a test room in the Negev desert of Israel was 70 - 80 watts*m^{-2} of roof area, up to one fifth of this was due to convection, depending on the meteorological conditions and mode of operation.

Introduction

Conventional roof ponds may be classified into two major categories:

- In *"Wet" roof ponds*, the water is exposed to the surrounding air, and cooling is achieved mainly through evaporation. The temperature of the water and the daily swing depend on the depth of the water, but a shaded roof pond may be expected to be about 1-2 ^{0}C above the ambient wet bulb temperature. This type of roof pond is simple and cheap, but has a number of limitations: daytime evaporative cooling is more than offset by radiative and convective heat gains; if the surface of the pond is shaded to reduce daytime gains, night-time radiative cooling is reduced significantly; and water consumption may be high.

- In *"Dry" roof ponds*, water is contained within sealed, infra-red transparent bags, and is not directly exposed to the surrounding air. Cooling occurs during the night by convection and by longwave radiation, not by evaporation. This type of roof pond is sealed and insulated during the daytime, and exposed during the night, requiring complex systems of movable insulation. While Hay (1978, 1980) claimed great success for his "Skytherm" system, the rigid geometry required and the difficulty of installing a reasonably airtight movable insulation system reduced the attractiveness of such systems considerably.

This research was funded by the Israeli National Council for Research and Development (NCRD) and by the German Federal Ministry for Science and Technology (BMFT)

S. Alvarez et al. (eds.), Architecture and Urban Space, 613–618.

Windscreens have been proposed to reduce unwanted convective heat exchange between a surface cooled by radiation, often to well below ambient air temperature, and the relatively warmer air. However, dust accumulation reduces the transmissivity of so-called transparent windscreens considerably. And, unless the air is extremely dry, the radiator surface cools down to the dew point, whereupon condensation forms on it. At this point, the film of water collected on the windscreen effectively becomes the radiating surface itself. Therefore in *practice*, windscreens, while being expensive and difficult to install and maintain, do not improve the performance of roof ponds significantly.

The approach adopted by the authors is based on the "dry" roof pond concept. However, instead of moving the insulation as required to control the heat exchange between the water and the environment, water was circulated from a roof pond to a fixed heat exchanger which was installed on the outside of the insulated envelope of the building.

Experimental Methods

Fig. 1: Section through test cell.

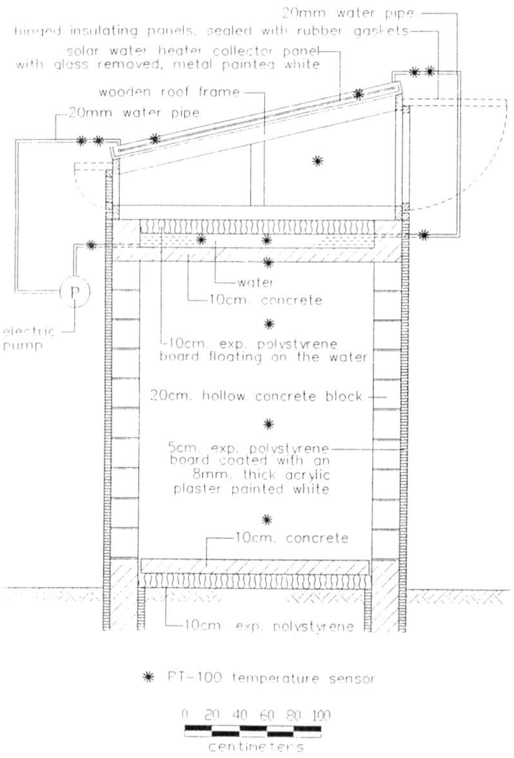

A room-size chamber (Fig. 1) was designed to test methods of cooling on a larger structure, as a step towards establishing a cooling technology that could be used in a full-size building. The system integrated a cooling mechanism with a storage mechanism, and a means of transferring energy between them.

Water was circulated at night through a radiator with a hollow section, by means of a small pump controlled by a timing device. The water reservoir, in the form of a shallow roof pond, served as a heat sink which absorbed heat from the building (through a thin concrete roof). During daytime, the collector functioned as a roof shading device. The mass of the structure and its insulation helped maintain the low interior temperature throughout the daily cycle.

Analytical Methods.

DETERMINING THE COOLING POWER OF THE RADIATOR.

Total cooling power of the *radiator* was calculated using the temperature difference measured between water entering and leaving the radiator:

$$Q_t = (T_{in} - T_{out}) \cdot c \cdot \frac{dV}{dt}$$

where Q_t is the momentary cooling power [watts], $T_{in} - T_{out}$ is the temperature difference [^oC], c is the specific heat of water [watts*hrs*m^{-3}*$^oC^{-1}$] and $\frac{dV}{dT}$ is the rate of flow of water [m^3*hr^{-1}]

HEAT EXCHANGE PROCESSES AT THE RADIATOR SURFACE.

Heat exchange between the radiator and the ambient involves two mechanisms - convection and radiation. The two mechanisms may be either complementary or opposing, depending on the temperature of the radiator and on the ambient meteorological conditions.

Convective heat exchange at the surface of the radiator (Qc) is given by:

$$Q_c = h_c \cdot (T_{rad} - T_{air})$$

where Q_c is the convective heat exchange [watt*m^{-2}], h_c is the convective coefficient [watt*m^{-2}*$^oC^{-1}$], and T_{rad}, T_{air} are the radiator and air temperatures [^oC]

The convective coefficient of the boundary layer under various conditions was evaluated using the expressions developed by Clark and Berdahl (1980), and is a function of air velocity and temperature.

Net radiative cooling was calculated following Givoni (1982):

$$R_{net} = \sigma \cdot \epsilon_{rad}(T^4_{rad} - \epsilon_{sky} \cdot T^4_{air})$$

where R_{net} is the net radiative cooling [Watt/m^2], σ is the Stephan-Boltzman constant [5.67*10^{-8} Watt/m^2/$^oK^4$], ϵ_{rad} and ϵ_{sky} are the emissivities of the radiator and of the sky, and T_{rad} and T_{air} are the absolute temperature of the radiator and the ambient air [^oK].

Sky emissivity was calculated using the expression developed by Clark and Berdahl (1980), based on dew point temperature data.

DETERMINING CONVECTIVE COOLING AS A PROPORTION OF THE TOTAL COOLING.

Fig. 2: Calculated vs. measured cooling rate - covered radiator, Aug. 24, 1990.

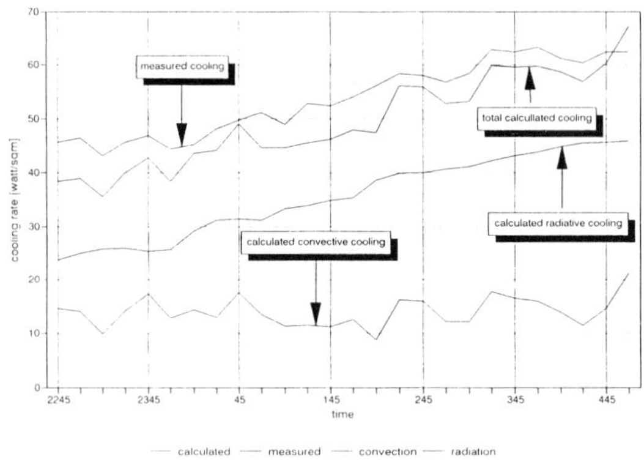

Since the surface temperature of the radiators was generally higher than that of the ambient air, they were cooled by both radiation and convection. However, radiation to the sky, which is affected by meteorological conditions and cloud cover, is difficult to measure accurately. To determine the proportion of cooling due to each mode of heat transfer, the sky was eliminated as a heat sink by placing a plywood sheet parallel to the radiator, about 20 cms above it. The radiator was therefore cooled by convection and by radiative heat exchange with the plywood sheet. The radiative heat exchange with the plywood sheet was assessed fairly simply, following Givoni (1976):

$$q_r = 5.6987E \cdot \left[\left(\frac{T_2}{100} \right)^4 - \left(\frac{T_1}{100} \right)^4 \right]$$

where q_r is the net radiative heat exchange per unit area, T_1, T_2 are the absolute temperatures of the radiating surfaces [^oK] and E is the effective emissivity of the two surfaces [watts*m^{-2}], given by

$$E = 1/(1/\epsilon_1 + 1/\epsilon_2 - 1)$$

where ϵ_1 and ϵ_2 are the emissivities of the two parallel radiators.

Fig. 3: Net cooling rate on a typical night - Aug. 8, 1990.

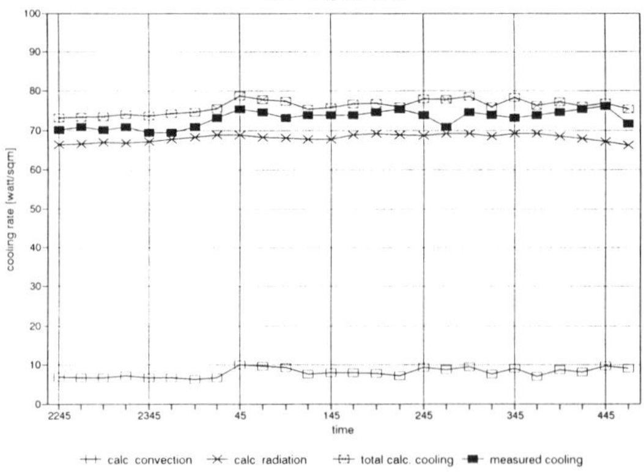

The total cooling power calculated for the covered radiator using the above procedure shows good agreement with the cooling power calculated using the temperature difference between the entry and exit from the radiator (Figure 2). It was therefore concluded that the expression used to calculate the convective heat exchange at the surface of the radiator was valid under the test conditions.

Results

CALCULATED COOLING RATES.

The system described here consistently produced cooling rates of approximately 80 watts*m^{-2}. Each 2.78 m^2 radiator therefore provided approximately 1.5 kWh of cooling per night. *The cooling rate was nearly constant throughout the night, so that the average nightly cooling rate reported here is comparable to peak cooling rates reported elsewhere* (See fig. 3, showing the net cooling rate for one radiator on a typical night). The following table summarizes the results for a three-week measurement period in August 1990. Data are *net* values for the *radiator*, averaged over 7 hours of operation each night (22:00 hrs - 05:00 hrs):

	cooling rate [watt*m^{-2}]	convection as % of total
Highest average nightly value	104	17.3
Lowest average nightly value	59	-2.2
Mean value for 3-week period	81	5.7

FACTORS INFLUENCING THE SIZE OF THE CONVECTIVE COMPONENT OF THE COOLING.

When the *water flow rate* was reduced, the temperature difference between the radiator entry and exit increased. At extremely low rates of circulation, the radiator temperature dropped below the ambient air temperature, thus reversing the direction of the convective heat exchange. It was therefore concluded that the water flow rate had to be sufficiently high to prevent this occurring.

The effect of the *Wind*, measured at the height of the radiator surface, was to increase the cooling rate, since the temperature of the radiator was generally higher than that of the ambient air for most of the night.

Other meteorological factors monitored, including ambient dry bulb temperature and relative humidity, had little effect either on the average nightly cooling rate of the radiator or on the proportion of cooling due to convection. However, since the meteorological conditions in Sde-Boqer are fairly constant throughout the summer, this finding should be viewed as valid only under similar, fairly constant, conditions.

The difference between roof pond temperature (which is in effect the inlet temperature of the radiator) and the ambient DBT, reflects the relation between the prevailing meteorological conditions and the thermal inertia of the building. While this factor was found to have a significant influence on the overall cooling rate of the system, it did not appear to alter the balance between convective and radiative cooling.

Discussion of Results

Perhaps the greatest advantage of the proposed system is the close coupling between the radiator and the thermal storage mass of the building. Due to this coupling, the temperature of the radiating surface remained fairly high - generally higher than that of the ambient air - throughout most of the night. This has several advantages:

a. When the radiator temperature is *higher* than the ambient air temperature, convective heat exchange increases the rate at which energy is removed from the system, rather than reduce it. This obviates the need for wind screens, which previous research (Givoni 1982) has shown are essential for achieving low radiator temperatures.

b. High radiator temperatures increase radiative cooling as well.

c. Since the radiator temperature is not only fairly high but is also nearly constant - the temperature of the roof pond is only reduced by about 3-4 ^{O}C on a typical night - the cooling rate is nearly uniform, at a level similar to *peak* cooling rates reported by other researchers.

Conclusion

A system consisting of a roof pond in which the water is circulated through a radiating roof, is able to maintain the radiator at a relatively warm temperature, thus avoiding convective heat gains. It avoids the need for movable insulation, and does not require wind screens.

References

Clark, E., Berdahl, P. (1980), "Radiative cooling: Resource and Applications" In: Miller, H. (Ed.) *Proceedings of the Passive Cooling Workshop*, Amherst, Mass. Oct. 20, 1980. Center for Energy Efficient Design, Berkeley, Ca. pp. 177-212.

Givoni, B. (1976) "*Man, Climate and Architecture*", 2nd. Edition, Applied Science Publishers Ltd., London.

Givoni, B. (1982), "Cooling Buildings by Longwave Radiation - Review and Evaluation", *Research Report to the Israeli Ministry of Energy and The US Dept. of Energy*. J. Blaustein Institute for Desert Research, Ben-Gurion University of the Negev, Sde-Boqer Campus, Israel. 62 pp.

Hay, H. (1978) "Skytherm Design Evaluation; Skytherm Production Research", *Proc. 3rd. Annual Solar Heating and Cooling R&D Branch Contractors' Meeting*.

Hay, H. (1980) "Skytherm Natural Thermal Control - Design, Production and Commercialization", *Proc. U. S. Department of Energy Passive and Hybrid Solar Energy Program Update Meeting*, Washington, D. C., Sept. 21-24, 1980, pp. 3-80 - 3-83.

DIRECT AIR COOLING FROM WATER DROP EVAPORATION

E.A. Rodríguez**, S. Alvarez*, R. Martín*

** *CIEMAT-IER*
División Termosolar
Av. Complutense 22
E-28040 Madrid. Spain

* *Universidad de Sevilla*
Departamento de Ingeniería Energética
y Mecánica de Fluidos
Av. Reina Mercedes s/n
E-41012 Sevilla. Spain

ABSTRACT. The present paper deals with the interaction between the drop evaporation model and the conservation equations of the air. Depending on the nature of the air flow (natural convection or forced convection), two different applications are considered in terms of their theoretical treatment as well as in their practical issues.

1. Introduction

One of the most important applications of producing water drops is to cool the surrounding air through the latent energy absorbed in the evaporative process. As pointed out in [1], the expected lifetime of a water drop and the objective of obtaining a total evaporation, defined the maximum drop size usable for this application. Thus, for a lifetime of between 2 and 5 seconds, the maximum drop diameter ranges from 60 to 100 μm for very dry air (10% relative humidity) and from 15 to 25 μm for very humid air (90% relative humidity). For average ambient conditions we can assure that a drop with less than 50 μm diameter will evaporate totally in almost all applications.

2. Drop Generation. Atomizers

The process of drop generation, more commonly known as atomization, consists of a liquid jet or sheet which is disintegrated by the kinetic energy of the liquid itself (pressure atomizers), or by exposure to a high-velocity air stream (twin fluid atomizers), or as a result of a mechanical energy applied externally through a rotating or vibrating device (rotary atomizers). Essentially, all that is needed is a high relative velocity between the liquid to be atomized and the surrounding air.

For the thermal conditioning of Open Spaces in EXPO'92 and, after experimental evaluation of the different types, pressure nozzles have been chosen as the most suitable atomizing devices. The specific type adopted is the impaction type nozzle (Fig. 1), in which the water exits from the orifice as a cylindrical jet and then spreads into an expanding hollow cone after striking the impact pin. This cone breaks into small particles or drops at a given distance from the orifice.

The random and heterogeneous nature of the atomization process that takes place in the nozzles, produces sprays which are a spectrum of drop sizes distributed around some arbitrarily defined mean value. We use a based on volume of mass distribution and the corresponding mean value is the mass median diameter (MMD) or the volume median

S. Alvarez et al. (eds.), Architecture and Urban Space, 619–624.

diameter (VMD) defined as the drop diameter, such that 50% of the total atomized liquid is in drops of smaller diameter. Figure 1 also shows a typical cumulative distribution based on volume or mass with the volume median diameter arrowed.

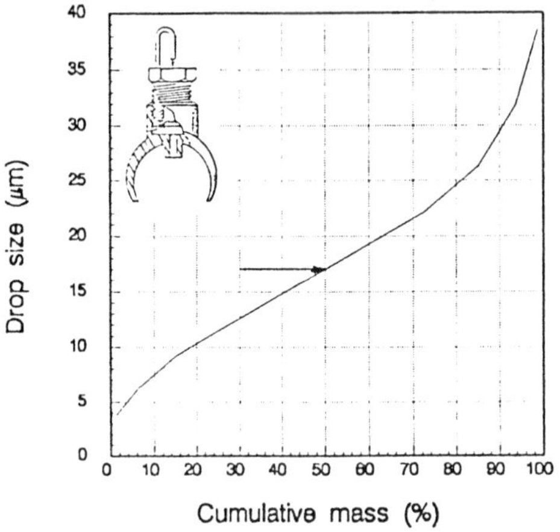

Figure 1: Impact Pin Nozzle Used and its Cumulative Drop Size Distribution

3. Modelling the Interaction Between Water Droplets and Surrounding Air

The calculation of the cooling effect on a certain air stream due to the drop evaporation process, requires the coupling of the drop evaporation model [1] and the energy balance on the air. Two additional assumptions are made:

- The drops are at their equilibrium temperature: This hypothesis is equivalent to neglecting the cooling time of the drops and the corresponding sensible energy exchanged (the cooling period is in the order of $\frac{1}{1000}$ of the total evaporation time).
- The relative drops/air velocity is zero. This is almost true for small drops because the deceleration time is much lower than the evaporation time and the terminal velocity is very low.

Assuming that the drop size distribution can be characterize by taking N_S segments, each with representative radius R_K, the energy equation applied to a control value of air is written as:

$$\rho v C_p \frac{dT_\infty}{ds} = \underbrace{(\sum_{K=1}^{N_S} N_K 4\pi R_K^2 \frac{dR_K}{dt}) \frac{h_{fg}\rho_w}{v}}_{E_{evap}} \tag{1}$$

where E_{evap} is the latent energy of the evaporated water from the drops.

According to reference [1] and using the two hypotheses above:

$$\frac{dR_K}{dt} = -\frac{\rho D_A}{\rho_w R_K} \frac{Y_{wb} - Y_\infty}{1 - Y_{wb}} \tag{2}$$

From (1) and (2) the final formulation of the energy balance equations of the air is:

$$\frac{dT_\infty}{ds} = -(\sum_{k=1}^{N_S} N_K 4\pi R_K)\frac{D_A}{v^2}\frac{h_{fg}}{C_p}\frac{Y_{wb}-Y_\infty}{1-Y_{wb}} \tag{3}$$

which gives the temperature evolution of a stream line.

In an analogous way, the water vapor balance in the same control value results in:

$$\frac{dY_\infty}{ds} = (\sum_{K=1}^{N_S} N_K 4\pi R_K)\frac{D_A}{v^2}\frac{Y_{wb}-Y_\infty}{1-Y_{wb}} \tag{4}$$

Concerning momentum balance, it strongly depends on whether the air flow is provoked by an external force (forced convection) or caused by density changes (natural convection).

4. Forced Convection

Two applications of air cooling using water drops lay in this category:

- Wet barriers: These are rows of atomizers (or water curtains) placed upwind the treated zone and the prevailing wind direction.
- Fan-driven cool tower: A fan blows air downwards through a tower in which nozzles atomize water. Cool air is released at the bottom to treat a certain zone.

As the air velocity is known or can be fixed, the integration of equation 3 and 4 gives us the spatial variation of both air temperature and mass fraction of water. A comparison between the measured and calculated results in the cool tower of the rotunda [2] (see Figure 2) revealed the consistency of the proposed formulation and hypothesis, for very different water flow rates and outside air conditions. It was also demostrated during the validation process that a number of segments (N_S) equal to 9 represents the best compromise between the accuracy required and the calculation effort.

By way of example, Figure 3 shows a sensitivity analysis performed for a cool tower in order to evaluate the influence of the height of the tower (distance between the base and the atomizers) and the VMD of the atomizers on the outlet air temperature. The inlet air conditions are 37.5^{0}C, 30% relative humidity. The two parameters examinated together with the water flow rate constitute the relevant design variables of the fan-driven cool tower.

5. Natural Convection

The density changes provoked by changes in the air temperature act as the driving force of the air movement.

The formulation of the problem is now more complex due to the necessity of include the momentum equation for the air:

$$\rho\frac{D\vec{v}}{dt} = -\nabla p + \mu\nabla^2\vec{v} + \rho\vec{g}$$

The solution of the resulting system of equations depends on the specific characteristics of the application. A particular example of solution can be found in [3].

Apart from the increasing complexity concerning the modelling viewpoint, the design and sizing involves more parameters and the control strategies are more critical.

The two main configurations used in the outdoor spaces at EXPO'92 are atomizers in trees or pergolas and cool towers.

5.1. ATOMIZERS IN TREES OR PERGOLAS

Atomizers placed in the foliage of trees or on the lower side of vegetation pergolas create an artificial fog. Droplets evaporate in contact with the surrounding hot air which becomes cooler, and, as a consequence heavier. A continuous descending flow of cool air is obtained.

Trees and vegetation pergolas are elements that will be pleantiful in the landscape of EXPO'92 and so, the will be profusely used to integrate atomizers.

The most important factor that guides the sizing of these configurations is the total evaporation of the atomized water. The problems of having an excess of water in the air are:

- The non-evaporated drops may wet people
- Those drops which do not wet people, strike onto the floor sooner or later and their air cooling potential in lost.
- This wetness deposited on the floor absorbs heat from it to evaporate, so that the humidity level is affected in the same manner as if this evaporation had happen in the air. As a consequence, the humidity may increase without the corresponding sensible cooling given by an adiabatic process, and the humidity control may stop the systems at a higher air temperature than expected.

The chosen nozzles produce a drop size distribution whose VMD is around $17 \mu m$. Modellig reveals that this size is small enough for evaporating before striking the floor whenever the droplets are released higher than 4 metres. In fact, in the rotunda experiments [2] it was observed that a continuous working of the nozzles (placed in the foliage of tress, 2.5 m high) formed a humid plume that avoided the total evaporation of the water. The reason for this plume formation was that the higher cooling point (3 m) was not high enough to induce the necessary air flow for evaporating the plume. The solution for this trouble was found through an intermittent working of the nozzles following short ON/OFF periods. The ON time span must be shorter than the airborne time of a drop so as to break the plume.

Besides the avoidance of the plume formation, this ON/OFF strategy permits a very fine partial load regulation of the atomized water, while keeping the uniformity of the water distribution. This partialization is made through a progressive increase of the OFF time span.

5.2. COOL TOWER WITHOUT FAN

This configuration will be used as an air cooling system in the Avenue of Europe, one of the main conditioned outdoor space of the EXPO'92 site. Twelve conical towers representing the CEE countries will blow cool air downwards to the entrances to this Avenue.

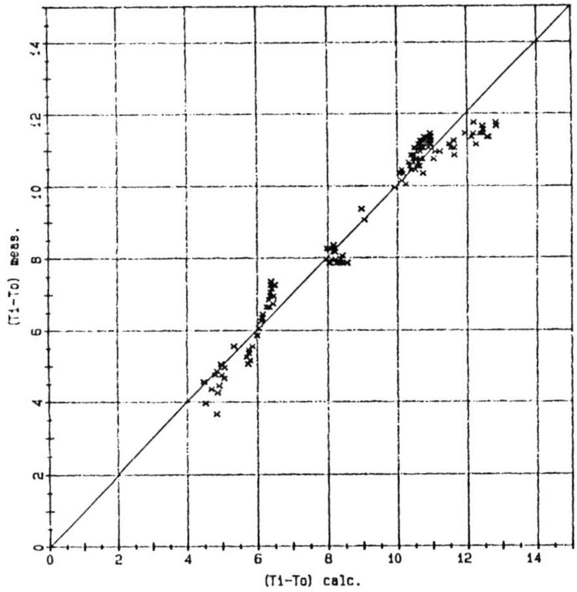

Figure 2: Measured (y axis) and Calculated (x axis) Temperature Drop in the Tower at the Rotunda

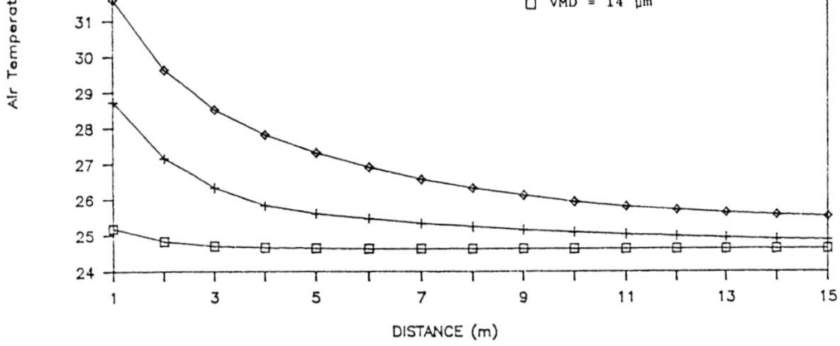

Figure 3: Temperature Evolution of the Air within the Cool Tower of the Rotunda as a function of the Volume Median Diameter of the Drops

The modelling of these kind of cool towers together with all the design and operation issues are extensively described in [3].

Acknowledgements

The authors would like to acknowledge the EXPO'92 State Corporation for the financial support of the research and development work on which this paper is based.

List of Symbols

C_p	specific heat of air
D_A	binary diffusion coefficient of water in air
$\vec{g}$	gravity acceleration
h_{fg}	water latent heat of vaporization
N_S	number of drop sizes that models drop size distribution
N_K	number of drops of size K released per unit time
p	static pressure of air
R_K	radius of each drop of size K
s	position variable
T_∞	air temperature
v	air velocity
Y_{wb}	saturation mass fraction of water in ambient air
Y_∞	mass fraction of water in ambient air
μ	dynamic viscosity of air
ρ	air density
ρ_w	water density

References

1. Rodríguez E.A., Alvarez S. and Martín R. (1991) 'Water Drops as a Natural Cooling Resource. Physical Principles' in Architecture and Urban Spaces, PLEA'91, Sevilla
2. Alvarez S., Cejudo J.M., Rodríguez E.A. and Guerra J. (1991) 'Full Scale Experiments in EXPO'92: The Bioclimatic Rotunda' in Architecture and Urban Spaces, PLEA'91, Sevilla
3. Alvarez S., Rodríguez E.A. and Molina J.L. (1991) 'The Avenue of Europe at EXPO'92: Application of Cool Towers' in Architecture and Urban Spaces, PLEA'91, Sevilla

THERMAL PERFORMANCE OF WATER PONDS: MODELLING AND COOLING APPLICATIONS

J. Guerra, J.L. Molina, J.M. Cejudo

Universidad de Sevilla
Departamento de Ingenieria Energética
y Mecánica de Fluidos
Av. Reina Mercedes s/n
E-41012 Sevilla, Spain

ABSTRACT. This paper deals with the termal behaviour of ponds applied to the outdoor spaces thermal conditioning. The thermal performance of the pond is examinated by numerical solution of the dynamic governing equations. The model developed is validated in an experimental installation at EXPO'92 and the basic design parameters are optimized through a sensitivity analysis.

1. Introduction

The pond considered is composed of a mass of water with a depth of the order of 1m which may include a system of sprayers on its surface. The use of ponds in the conditioning of outdoor spaces is carried out on the basis of two fundamental aspects: low reflectivity, and water temperatures below the exterior ambient temperature.

The reflectivity of a pond is low, of the order of 3% during periods of maximum solar radiation. The solar radiation absorbed is however appreciable and may reach values of up to 80% depending on the depth of the pond and the reflectivity of the bottom. This high level of absorption of short-wave radiation is compensated by the evaporation that takes place on the surface and temperatures inferior to the exterior ambient temperature are obtained. The inertia of the mass of water also gives rise to a time lag of 1 to 2 hours in the maximum temperature in relation to the maximum air temperature.

The inclusion of fountains, cascades and sprayers in the pond increases losses through evaporation and an additional temperature reduction is achieved. The operation of sprayers gives rise, in addition, to a direct evaporative cooling of the air which comes into contact with them. This fact may be taken advantage of for conditioning the areas close to the pond, provided the layout of the sprayers and the wind direction promotes displacement of the air towards the area to be treated [1].

The use of ponds for conditioning outdoor spaces thus satisfies the following basic conditioning criteria [2]:

a) Reduction of the reflected solar radiation.

b) Reduction of the surface temperature and thus a decrease in the long-wave radiant exchange.

c) Reduction of the air temperature due to the operation of spray system and the evaporative cooling process which takes place.

S. Alvarez et al. (eds.), Architecture and Urban Space, 625–630.

d) Reduction of the air temperature using the cool water in the precooling coil of air handling units.

This paper studies the thermal behaviour of a pond. The model developed was validated in the experimental setup at EXPO'92 and, finally, a sensitivity analysis was carried out of the fundamental design parameters (depth, flow rate and operating times of the spray system, type of covering, etc).

2. Model Description

The following hypoteses are considered for the thermal analysis of the pond:

- For practical sized ponds, the lateral dimensions are large compared with their depth and hence it is possible to consider one-dimensional heat flow in the vertical direction;
- The water is a selective absorber and scatterer. The phase function (i.e. the angular distribution of scattered radiation) is independent of wave length and water clarity, and is assumed to be peaked in the forward direction. Then, attenuation of radiation in water can be considered to proceed as if there is no scattering since all the non-absorbed energy propagates in its original direction;
- The air-water interface roughness affected the reflectance very little [3], so Fresnel's equations for reflectance are applicable on both sides of the air-water interface;
- The water-air interface reflects specularly and the pond bottom surface reflects diffusely;
- The refractive index of water varies very little with wave length in the range of interest, using an average value of 1.333 for all wave lengths. The absorption coefficient, however, depends strongly on wave length and is treated as a constant over each wave length band considered. The absorption coefficient is assumed not to vary with location and the temperature in the range encountered in ponds;
- The boundary conditions are uniform;
- The diffuse radiation incident on the water surface is represented as direct radiation incident at 60°[3].

The heat balance in the pond expresses the variation of the internal energy of the pond as:

$$M\,C_p\,\frac{dT}{dt} = Q_s + Q_c + Q_r + Q_b + Q_e + Q_{sp}$$

where Q_s, Q_c, Q_r, Q_b, Q_e and Q_{sp} correspond to heat fluxes due to absorption of solar radiation, convection, long wave radiation, conduction from bottom, evaporation and cooling by the spray system.

A complete description of all the terms of the above balance is in reference [1]. Once all these terms have been expressed as a function of the water temperature, the meteorological variables and the geometrical and functional parameters of the pond, we obtain an equation like this:

$$\frac{dT}{dt} = -C_1 T(t) + C_2$$

where C_1 and C_2 are time-dependent variables. However, this time-dependency can be neglected in each integration step. Thus C_1 and C_2 remain at a constant value equals to that of the precedent time step, and the water temperature follows the evolution given by:

$$T(t) = \frac{C_1}{C_2} + (T(0) - \frac{C_1}{C_2}) \exp(-C_1 \Delta t)$$

where $T(0)$ is the water temperature at the begining of the time step (Δt).

3. Validation

The pond used for validation of the former model is a 37 m long, 15 m wide and 0.4 m deep. It is placed within the EXPO'92 site, incorporates spray and jet systems and its three-level configuration originates a water circulation along its longitudinal dimension.

Experiments took place during summer of the years 1988 and 1989 to provide data to characterize the thermal performance of the pond and to validate the theoretical model. The water pond temperature has been measured by a PT-100 probe located in the centre of the pond at an aproximate depth of 0.2 m. The air temperature and relative humidity, horizontal global radiation, wind velocity and direction are measured in a meteorological station located 5 m high in a non-perturbated area.

Figure 1 shows the evolution of the air and water pond temperatures during 4 days, starting on 25th August 1988. Only during the two first days the spray system is operating. The water pond temperature reduction due to the spray system is clearly observed. The daily range of the water temperature is lower than that of the air temperature, due mainly to the pond inertia. Practically in all the sunny hours, the pond is cooler than the ambient air registering up to 16 °C temperature difference.

Figures 2 and 3 show comparisons between the measured and calculated water temperatures of the described pond. The agreement is very good both in the case without sprays (Figure 2) and in the case with the sprays working (Figure 3).

4. Sensitivity Analysis

The influence of the design parameters on the termal behaviour of the pond are analyzed. We have choosen for this analysis the following initial data:

- Pond depth: 0.4 m
- Spray system flow rate: 1.3 pond volume circulations per hour, working from 8 a.m. to 8 p.m.
- Bottom reflectivity of the pond: 0.2
- Outdoor conditions: Seville, 1 % percentile refered to the summer months. The solar position corresponds to the 15th of July.

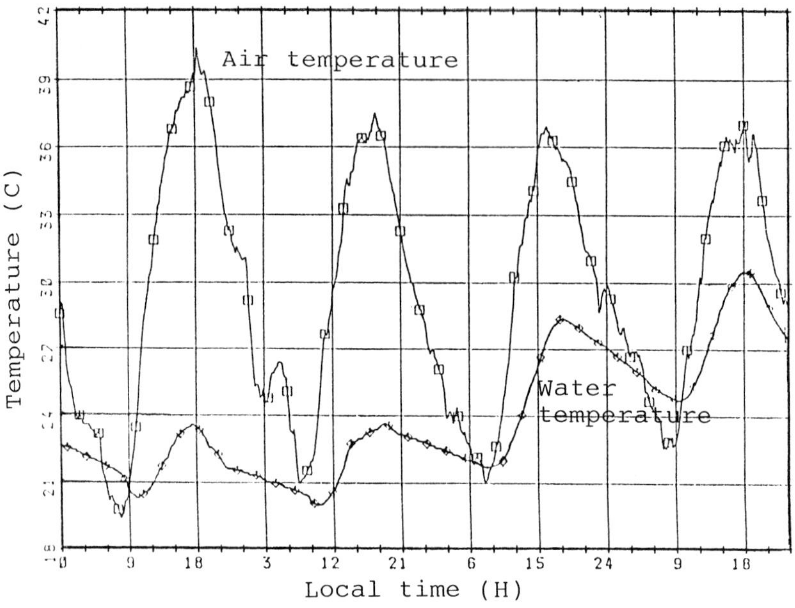

Figure 1: Water pond and air temperatures

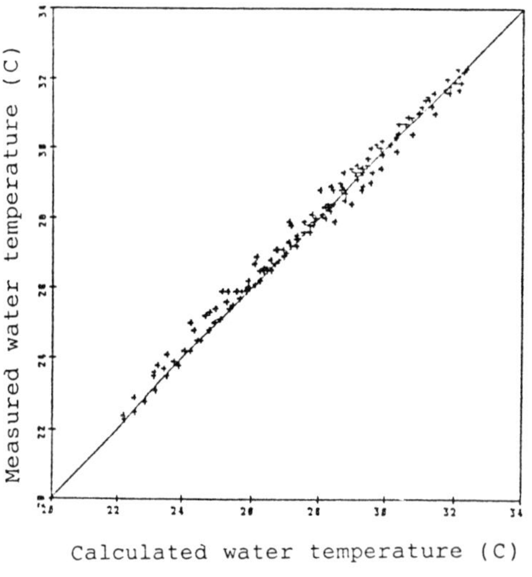

Figure 2: Measured versus calculated water temperature without sprays.

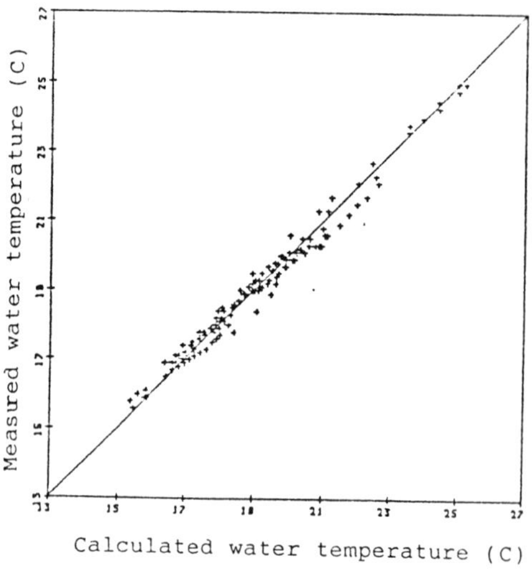

Figure 3: Measured versus calculated water temperatur with sprays.

In each sensitivity analysis, only one parameter is varied within the prescribed range, remaining the other parameters at the initial value. The design parameters analyzed are:

- Spray system water flow rate (in pond volume circulations per hour);
- Nozzle type;
- Depth of the pond;
- Bottom reflectivity;
- Air velocity above the pond;
- Covering type;
- Working schedule of the spray system.

Two examples of these analysis are shown in figures 4 and 5. In both the plotted variables are the water temperature of the pond and the dry and wet bulb temperatures of the air.

Figure 4 shows the influence of spray system water flow rate. From this figure it is observed a considerable decrease in the water temperature of the pond when the spray system is working. This effect is improved as the spray flow rate grows. However, the marginal improvement of the cooling effect decreases with this flow rate, reaching an asymptotic behaviour from 1.5 pond volume circulations per hour.

The value of the transmissivity of a covering placed over a pond affects directly the amount of solar radiation absorbed by it. Figure 5 shows the way in wich a shading element over the pond reduces sensibly its water temperature specially in the absence of sprays working. For example a pond covered with a 10 % transmissivity covering reduces its water temperature in 5 °C without sprays and only 1.5 °C with these working. The temperature levels reached by a covered pond (10 % transmissivity) without sprays are similar to those reached by a spray-cooled pond.

Acknowledgements

The authors would like to acknowledge the EXPO'92 State Corporation for the financial support of the research and development work on which this paper is based.

References

1. Departamento de Ingeniería Energética y Mecánica de Fluidos (1989) 'Acondicionamiento Climático de los Espacios Abiertos en EXPO'92. Modelización' Vol. I and II. Final Report. Sevilla
2. Velázquez R., Alvarez S., Guerra J., Rodríguez E.A. and Cejudo J.M. (1991) 'General Aspects of the Work on Climatic Control of Outdoor Spaces at EXPO'92' in Architecture and Urban Spaces, PLEA'91, Sevilla
3. Payne R.E. (1972) 'Albedo of the surface' J. Atmos. Sci. 29, 959
4. Cengel Y.A. Özişik M.M. (1984) 'Solar Radiation Absorption in Solar Ponds', Solar Energy, Vol 33, pp 581-591

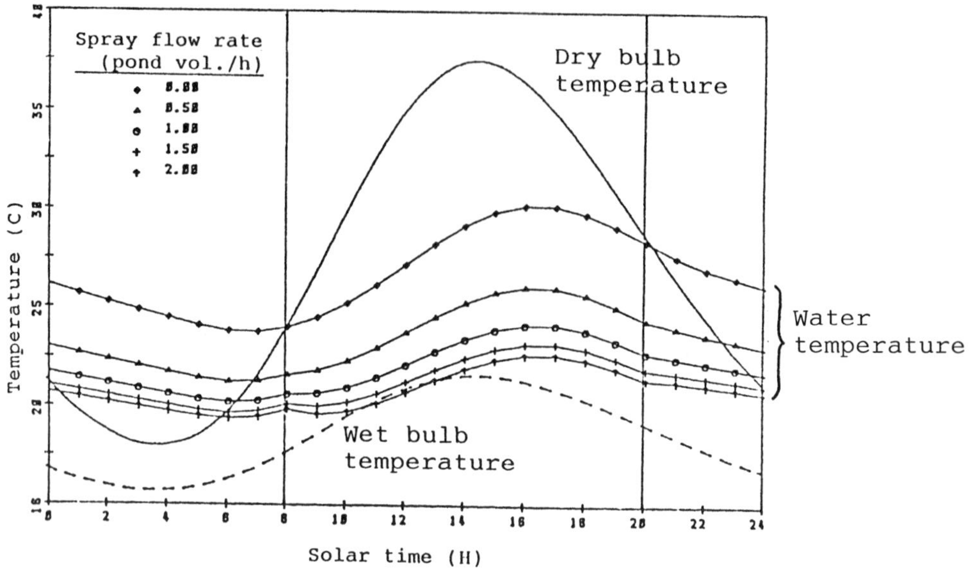

Figure 4: Sensitivity analysis of the sprayers flow rate

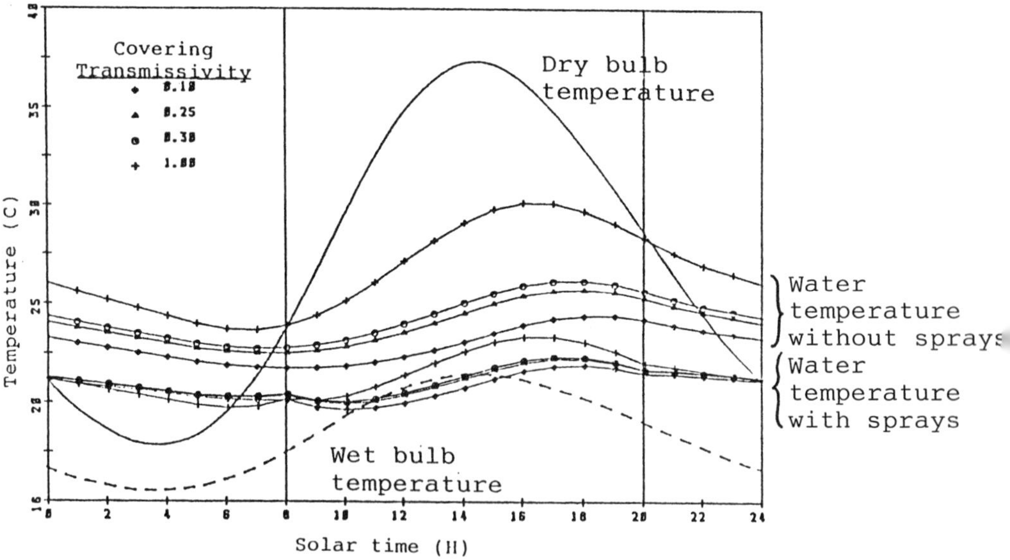

Figure 5: Sensitivity analysis of the covering type

EVAPORATIVE COOLING EFFECTS IN HOT AND HUMID URBAN SPACES

KEN-ICHI KIMURA
Professor
Department of Architecture
Waseda University
Okubo-3, Shinjuku, Tokyo 169, Japan

ABSTRACT. Two different natures of evaporative cooling effects on indoor and outdoor spaces in hot and humid regions are described in view of the devices in traditional architecture with some examples of measured results. One of the evaporative cooling means can be found in a traditional way by spraying water in the garden or fountains to make the air cooled through vegitations. This will give the air of one or two degrees lower in temperature than the ambient air, which was measured with a house in Japan. This will contribute to the improvement of thermal comfort to a considerable extent owing to a certain air movement along the skin surfaces of human body, though the relative humidity of the air becomes higher. The other can be seen also in traditional houses with thatched roofs or tiled roofs. The rain water soaked by those roofs will be evaporated by strong solar radiation whereby the inside surface temperature becomes much lower than the case with the roof which does not soak rain water. This was examined by laboratory experiments and a field study with an earth sheltered building.

1. Introduction

PLEA has been playing an important role in promoting bioclimatic architecture under the serious environmental conditionds where many people are much indulged in abundant lives of so called industrialized age. One of the most dangerous situations in view of the energy supply and demand is a high peak demand of electricity for cooling for a short duration of hot spells in the daytime. The increase rate of this peak demand in the whole households of Japan is equivalent of four nuclear power plants to be constructed every year. To relieve such grave situations architects, urban planners, scientists and engineers must strive for creating new ideas of bioclimatic designs, though policy decision makers should impose very high rate charges for electricity consumption during the summer peak hours.

Bioclimatic solutions are being expected to be effective against the problem concerned. There are several ways of passive cooling to be applied to different situations. From the viewpoint of thermal comfort for people inside or outside with a certain clothing and activity level, it is desirable to have lower dry bulb temperature, lower wet bulb temperature, lower mean radiant temperature and greater air movement.

In the urban spaces where people walk, stand, sit down and work, it is generally difficult to have dry bulb temperature and relative humidity changed by direct means of passive cooling. Radiative cooling does work effectively under hot and dry conditions to have the surface temperatures of ground or building lowered, but less effectively under hot and humid conditions as long wavelength radiation is absorbed by moisture in the atmosphere. Shading is of primary importance, but sometimes avoided in modern buildings where cooling load naturally turns out quite high.

Evaporative cooling is one of the most effective ways of passive cooling for architecture and urban spaces from ancient times in hot regions. It is more effective in hot and dry regions than hot and humid regions in terms of total amount of cooling, but may be

S. Alvarez et al. (eds.), Architecture and Urban Space, 631–636.

equally effective in both regions in terms of the enhanced level in thermal comfort against severe summer conditions.

There are two different natures in the evaporative cooling effects in hot and humid urban spaces: use of evaporatively cooled air for urban or inside spaces and sensible cooling of inside surfaces by evaporative cooling on outside surfaces.

2. Evaporatively Cooled Air

2.1 OUTLINE OF THE PROCESS

When the air is cooled by evaporation as we experience the breeze, the adiabatic change occurs in every bulk of air which absorbs moisture from sprayed water or evaporated water vapor from wet surfaces in the surroundings. Then the temperature of the air can be lowered down to the wet bulb temperature when perfectly saturated.

This process is called ventilative cooling, because the moist but cool air enters into the space to deprive the sweat from the skin surface of a human body. Thus the heat loss associated with mass transfer from the skin surface to the air makes the body feel cool. Of course this effect is more pronounced when the air velocity is higher and when the skin surfaces is wet. If the air is stationary, the evaporation takes place only slightly in the form of skin diffusion and the sweat tends to be accumulated on the skin surface. As the air comes in and flows along the skin surface, the evaporation will take place to a large extent and the person feel it a comfortable breeze. It can be maintained that the intermittent air movement makes the person feel cooler than the constant air movement. Fig. 1 shows one example of air movement in velocity recorded at the center of the room in one of the Japanese vernacular houses on a hot and humid summer day.

Owing to the moisture absorption the temperature of the breeze through the garden with a lot of moist foliage gets lower than in the field by 1 - 2 °C as shown in Fig.2. This was measured by Kimura et al (1987) at one of the rural houses facing the garden in Yamanashi Prefecture in Japan.

2.2 THEORETICAL CONSIDERATION

It is rather difficult to estimate the air temperature after evaporatively cooled through vegitation and evaporation from water surface. From the energy balance the decease in sensible heat should be equal to the increase in latent heat with ragard to the air flow Q[m^3/h] between before and after the moisture absorption. Thus,

$$C_p \gamma(\theta_1-\theta_2)Q = \gamma r(x_2-x_1)Q \qquad (1)$$

where

θ_1, θ_2 : air temperature before and after the moisture absorption respectively [°C]

x_1, x_2 : humidity ratio before and after the moisture absorption respectively[g/kg]

C_p : specific heat of the air [J/kg K]

γ : specific weight of the air[kg/m^3]

r : latent heat of water [J/g]

Then the air temerature after the moisture absorption becomes

$$\theta_2 = \theta_1 - \frac{r}{C_p}(x_2-x_1) \qquad (2)$$

The higher the humidity ratio x_2, the lower the air temperature θ_2 after moisture absorption, which is to be called secondary air temperature hereafter. The value of x_2, however, cannot be higher than saturated humidity ratio for the temperature θ_2. The lower the humidity ration of primary air x_1, the greater the difference in humidity ratio x_2-x_1, and the lower the temperature of secondary air θ_2, which people would feel. It can be explained clearly that evaporative cooling works more effectively in dry regions than in wet regions.

Having the relationship between temperature and humidity ratio linearlized for the temperature range around 30°C, saturated humidity ratio against air temperature may be approximated by

$$xs = 22.6 + 1.55\,(\theta w - 27) \qquad (3)$$

where θw is wet bulb temperature of the primary air.

If the state of secondary air is assumed to reach a half way between the initial and saturated conditions, i.e.,

$$x_2 = \frac{x_1 + xs}{2} \quad \text{and} \quad \theta_2 = \frac{\theta_1 + \theta w}{2} \qquad (4)$$

then the temperature of the secondary air may be estimated by substituting equations(3) and (4) into equation (2), namely,

$$\theta_2 = \theta_1 - 2.08\left(\frac{x_1 + 22.6 + 1.55(\theta w - 27)}{2} - x_1\right)$$

$$= \theta_1 + 1.04x_1 - 23.5 - 1.6\,(\theta w - 27) \qquad (5)$$

This is an engineering approximation and may be used by architects or urban planners for their practical estimation.

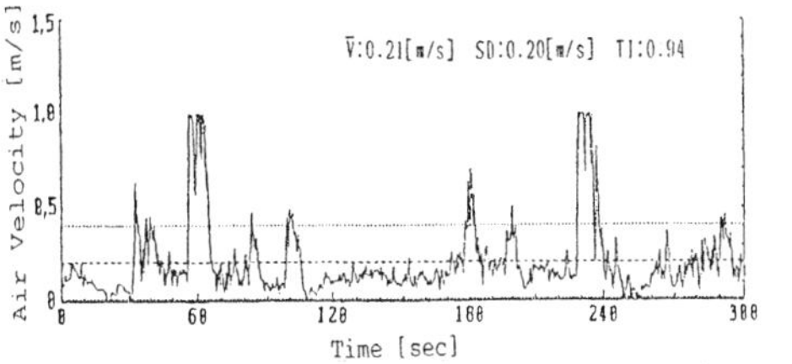

Fig.1 Air velocity in the second floor of a vernacular house in Kawagoe

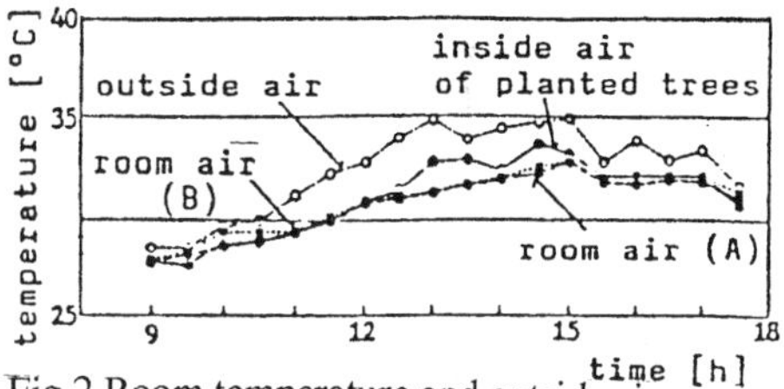

Fig.2 Room temperature and outside air temperature in a rural house

3. Surface Temperature Suppression by Evaporative Cooling

3.1 HEAT BALANCE AT THE SURFACE

Evaporation from a wet surface is accompanied by heat loss from the surface to the ambient air to be resulted in the decrease in temperature of the surface. This is explained by the following equation of heat balance at the surface:

$$qr_1 - qr_2 + qd = qc + qe \qquad (6)$$

where

qr_1 : incoming radiation absorbed by the surface [W/m2]

qr_2: outgoing radiation emitted from the surface [W/m2]

qd : conduction to the surface from behind [W/m2]

qc : convection from the surface to the ambient air [W/m2]

qe : heat loos by evaporation from the surface to the ambient air [W/m2]

The term(qr_1- qr_2)is large when strong solar radiation is flaling on the surface and becomes negative during the night with greater value for drier atmosphere. The radiation exchange (qr_1-qr_2) is negligible when the surface is in the shadow and when the sky is overcast. Thus the convection loss qc can be positive or negative depending on the temperatures of ambient air and the suface, as expressed by

$$qc = \alpha c(\theta s - \theta a) \qquad (7)$$

where

αc : convective heat transfer coefficient along the surface [W/m2K]

θs : surface temperature [°C]
θa : ambient air temperature [°C]

Under the summer conditions the ambient air temperature is higher than the surface of the streets or roofs in the daytime and lower in the nighttime. Namely, the convection loss is positive in the nighttime and negative in the daytime. The absolute value of the convection loss is minimal for the shaded surfaces, where evaporation loss dominates accordingly.

Heat loss by evaporation is associated with mass transfer between the wet surface to the ambient air to be expressed by

$$w = kx(xs - xa) \qquad (8)$$

where

w : rate of mass tranfer [g/m^2h]
kx : coefficient of mass transfer based on humidity ratio [$g/m^2h(g/kg)$]
xs : saturated humidityratio [g/kg]
xa : humidity ratio of ambient air [g/kg]

Then the heat transfer by evaporation is given by

$$qe = rw \qquad (9)$$

The surface temperature should be determined by the conditions so that the heat balance as expressed by equation (6) can be maintained. For simulation unsteady state heat transfer through the solid with which the surface temperature is concerned must be involved to give the rate of surface heat flow qd.

3.2 IMPERFECTLY WET SURFACE

If the surface is not perfectly wet, it may be assumed to be partially wet. Namely, the surface is assumed to be evenly distributed with partially wet surface and partially dry surface. The ratio of wet surface area to the total surface area may be comparable to the relative humidity very close to the surface, which can be measured by hygrometer or through humidity transducer.

It is quite difficult to identify the correct figure of the relative humidity for partially wet surface like soil surface or grass covered surface. It is possible, however, to infer the value by assumption to a certain degree.

In fact evaporation loss especially during the night is so great that the surface temperature can be suppressed remarkably. This effect must be regarded significant in the hot and hmid regions where radiation loss cannot be expected substantially. This resembles with the case that a person sweat very much under the hot and humid conditions when the air temperature is as high as skin surface temperature with no air movement.

Figure 3 shows the results of comparison of the soil surface temperature of an earth sheltered building in Japan made by Kimura et al (1988). This verifies that the evaporative cooling effect can be approxiamted by mass transfer formula where the humidity ratio of the soil surface was inferred by the measured relative humidity close to the surface multiplied by the saturated humidity ratio for the surface temperature. Conduction is relatively easy in estimation, as the soil temperature under the surface can be easily measured. The temperature of the ceiling surface can be

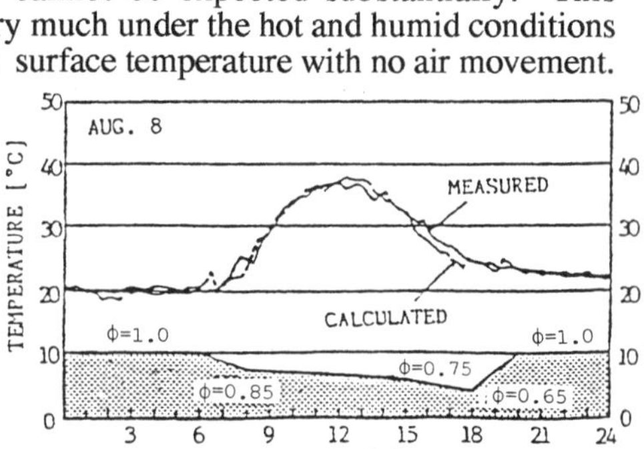

Fig.3 Comparison of roof surface temperature between calculated and measured results in the room of an earthe sheltered building

lower than outside and stable throughout the day.

This principle may be applied to the buildings and urban spaces where earth covered surface can be used. It is generally desirable to have about 30 cm of soil thickness for obtaining good evaporation effect, but the soil of 20 cm thick even can offer quite stable, low temperature underneath.

Thatched roof has much the same effect as soil. The rain water soaked by thatch will be evaporated by strong solar radiation afterwards. The evaporation loss continues after sunset and the upward heat flow through roofs will make the inside surface temperature lower. This is the reason why the houses with thatched roof is generally cool as being popularized by people at large.

The porous roof tiles have similar effects of evaporative cooling, though the amount of soaking rain water may be much smaller than in the case of thatched roofs. Figure 4 shows the results of measurement made by Kimura (1984) with experiments on roof tiles to be compared with steel plate roof. The obvious difference was found between them. The effect of the small air spaces between the roof tiles and seathing board underneath was found to be ventilated as it was warmed up by strong solar radiation. This is the device of avoiding excess heat of solar radiation and may be somewhat effective in storing solar energy in the winter daytime within the roof tile.

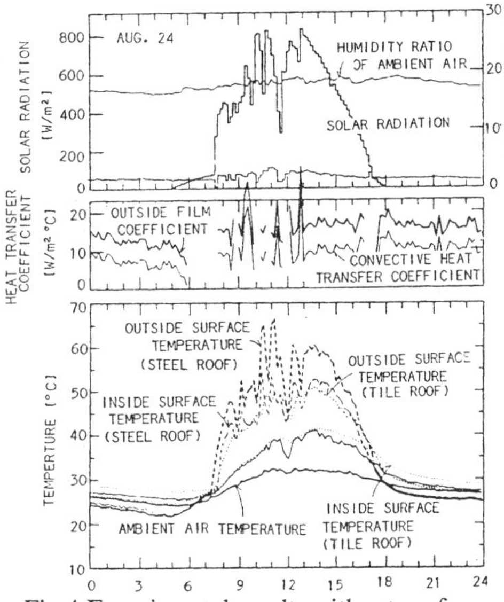

Fig.4 Experimental results without roof spray in comparison between tiled roof and steel plate roof

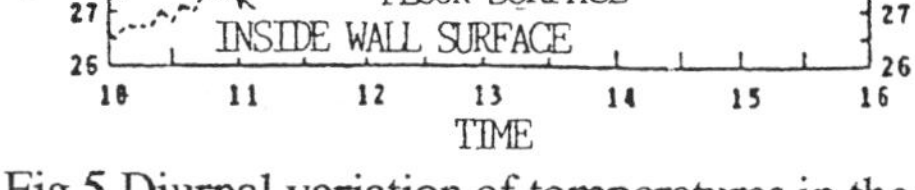

Fig.5 Diurnal variation of temperatures in the 2nd story of a massive house in Kawagoe on Aug.5, 1986

Kimura (1988) presented the results of a field measurement made at the massive house in Kawagoe located at 40 km north-west of Tokyo as shown in Fig.5. It was found that the room air temperature during the daytime on a hot day in mid summer was 1 - 2 degrees lower than the air temperature outside and more interestingly the mean radiant temperature in the room of the second floor turned out 2°C lower than the room air. This is very much owing to lower temperature of the ceiling surface which must be cooled by evaporation.

3.3 INTERIOR SURFACE TEMPERATURE

Further implication of evaporative cooling effect of the surface may be extended to the interior or shaded exterior surface which is made of relatively porous material so that moisture may be absorbed and desorbed at the surface concerned.

The porous material such as clay wall embraces a lot of minute pores where moisture is absorbed in its hygroscopic state. This moisture may be desorbed when drier air comes along the surface. This process is another phase of mass transfer and the surface temperature can be lowered by desorption of water vapor. This is not the evaporative cooling and may be called desorption cooling, but the effect is not so much great as evaporative cooling as it does not involve latent heat loss.

In some cases the water vapor absorbed by a porous material may change to the form of water droplets. If the water state within the pore continues, mold and mildew may develop in natural course. The smaller the pore, the more the mold and mildew, because the probability of water formations is much greater than wood or straw matt floor of large pores.

On the contrary the interior surfaces with vinyl cloth or oil paint do not absorb and desorb moisture and no cooling effects can be expected.

4. Concluding Remarks

Two types of evaporative cooling for hot and humid urban spaces are described: the air cooled by evaporation and the surface cooled by evaporation. Physical background for both cases are explained with some measured results.

In the modern urban spaces under hot and humid climate, green areas are being minimized with the result that the ambient and temperature tends to rise. It is recommended, therefore, to provide foliages and water surfaces to increase the heat loss by evaporation for avoiding excess accumulation of heat within the massive ground surfaces.

Roofs of buildings occupying a large area of urban spaces also had better be covered with evaporative surfaces by soil or equivalent. Heat loss by evaporation from the roof would in turn makes the interior cooler as the ceiling surface temperature becomes lower.

Various arrangements of evaporative cooling process in the urban spaces for hot and humid regions will help make the peak demand of electricity reduced considerably owing to the passive approach.

References

1) Kimura, K. & Tanabe, S. (1987), 'Traditional system of complex use of natural energy in vernacular houses', Reports of Special Project Research on Energy under Grants Aid of Scientific Research of the Ministry of Education, Science and Culture, Japan, Vol. 20, 213 - 218

2) Kimura, K., Nagasawa, K., Sugiura, S. & Oba, K.(1988) 'Field investigation of thermal environment and passive cooling efects in an earth sheltered building', Proc. of Solar '88, 13th National Passive Solar Conference and the Annual Technologies Conference, American Solar Energy Society, 359 - 363

3) Kimura, K. (1984) ' Studies on the architectural planning with complex use of natural energy', Reports of Special Project Research on Energy under Grant in Aid of Scientific research of the Ministry of Education, Science and Culture, Japan Vol. 8, 1 - 16

4) Kimura, K. (1988) 'Passive cooling effect of a massive vernacular house examined by field measurements under sumer conditions in Japan, Proc. of the 6th International PLEA Conference, Porto, Pergamon, 765 - 770

PERFORMANCE EVALUATION OF BUILDINGS ASSOCIATED WITH EARTH TO AIR HEAT EXCHANGERS. THE COOLING POTENTIAL

M. SANTAMOURIS [1], G. AGAS[2], T. MATSAGOS [2] T. ARGYRIOU [3] and K. THEOFILAKTOS 3

1. University of Athens, Physics Department, Athens, Greece.
2. TEI Pirea, Department of Mechanical Engineering, Aegaleo, Greece.
3. PROTECHNA Ltd, Themistokleous 87, 106 83 Athens, Greece.

ABSTRACT. This paper deals with the evaluation of the summer performance of buildings associated with earth to air heat exchangers. A thermal model coupling the operation of the pipes with the building is created and simulations have been performed for the Reference Greek building. A sensitivity analysis concerning the more important parameters of the exchangers has been carried out and is presented.

1. INTRODUCTION

Earth to air heat exchangers have gained an increasing acceptance during the last years. Earth to air heat exchangers, better known as buried pipes, can be used to provide preheating of the space during winter and cooling during summer. Although there are several set of data concerning the performance of the exchangers there is very few information concerning the impact of the system to the overall building performance.

Successive applications of earth to air heat exchangers are described in (1-2), without however to discuss the sensitivity of the various parameters characterizing the exchanger.

The present paper aims to provide information on the influence of the buried pipes technique to a building as well to investigate the sensitivity of the more important parameters to the overall performance of the earth to air heat exchangers.

For this purpose a computer program linking the performance of the earth to air heat exchangers with a building has been created. The program is

S. Alvarez et al. (eds.), Architecture and Urban Space, 637–642.

used to estimate the influence of a reference Greek residence defined in (3) when earth to air heat exchangers are used.

The building is an ordinary one storey house situated in the Athens region, designed by A. Tombazis. The house consists of two bedrooms, a living room together with a kitchen and a bathroom. The building is well insulated. All windows are single glazed and well shaded during the summer period. More details on the reference building are given in (3).

For the simulation procedure the Test Reference Summer weather data for Athens are used. Test Reference Summer weather data are created for 18 Greek locations, (3), in order to provide hourly values of all the necessary climatic data used for cooling purposes, i.e. Ground Temperature at various depths, Sky temperature, wet bulb temperature, etc.

2. THE CALCULATION PROCEDURE

The thermal behaviour of the building was simulated using as a basic code the program CASAMO-CLIM, developed by the Ecole Nationale des Mines de Paris, France, (4). This program is developed especially for cooling purposes and is validated against real buildings.

In order to calculate the impact of the ground cooling system to the building a subroutine providing the thermal performance of the buried pipes has been developed and integrated into the whole program. For this purpose the algorithms proposed in (5) have been used. Comparative analysis of 13 different algorithms proposed to calculate the efficiency of such systems, repported in (6), has shown that the used algorithms is the more complete while it is characterized by high accuracy.

The simulated system consists of an underground horizontal PVC pipe, whose inlet sucks air from the environment using an electric fan. The air is cooled by being circulated underground and then injected into the building.

The basic configuration of the system consists in a 50 m long pipe, having an internal diameter of 0.2 m, placed at a depth of 4 m. The thickness of the pipe wall is of 0.005 m. The air velocity inside the pipe is 5 m/sec.

In order to investigate the impact of the depth to burry the exchanger, a series of simulations have been performed, considering that the depth of the exchanger varies from 1.5 m to 6.5 m.

The obtained variation of the indoor temperature of the building when no earth to air heat exchangers are used as well as when the exchanger is placed in various depths is reported for July and August in Figures 1 and 2. The daily variation of the ambient ait temperature has also been reported. As shown, a 2-5 C reduction of the peak indoor temperature can be obtained when an earth to air heat exchanger buried at 1.5 to 6.5 m is used correspondingly.

The effect of changing the lenght, the internal diameter and the air velocity inside the exchanger, on the indoor temperature of the reference building has also been investigated. During these simulations the depth of the exchanger has been fixed at 4 m. The lenght has been increased from 50 to 70 m, the diameter from 0.2 m to 0.22 m and the air velocity has been decreased from 5 m/ sec to 3 m/sec. Te impact of these variations to the indoor temperature of the building is illustrated for July and August in Figures 3 and 4.

As it can be observed, when the lenght of the exchanger increases from 50 to 70 m, the corresponding indoor temperature drop is about 0.5 C. This is because that the temperature decrease of the air during the last 20 m in the exchanger is not very important compared to the temperature decrease at the first 50 m.

The increase of the diameter of the exhanger from 0.2 m to 0.22 m leads to a more important decrease of the indoor temperature of the building, which is of about 1.5 C. This is because increasing the diameter and maintening constant the air velocity in the exchanger, increases the cooled air flow and consequently the cooling energy injected to the building. On the contrary, when the air velocity is reduced, the cooling energy offered to the building, is also reduced, resulting in higher indoor temperatures.

3. CONCLUSIONS

The impact of the earth to air heat exchangers to the thermal performance of a reference building has been studied.

The obtained results have shown that when earth to air heat exchangers are used, the pipe has to be buried at a depth ranging between 3.5 m and 5 m. Regarding the sensitivity of other parameters of the system, it has been concluded that the diameter air air velocity variations affect significantly the indoor temperature of the building. Icrease of the lenght of the exchanger over a certain value has not an important impact on the buildings performance.

4. REFERENCES

1. M. Antinucci, B. Fleury, J. Lopez d Asiain, E. Maldonado, M. Santamouris, A. Tombazis and S. Yannas : Horizontal study on Passive Cooling. M. Santamouris (ed), Building 2000 Research Project, E.E.C., D.G. 12, 1990.
2. A. Tombazis, A. Argiriou and M. Santamouris : Performance evaluation of Passive and Hybrid Cooling components for a hotel complex. Int. J. Solar Energy, 9, 1-12, 1990.
3. PROTECHNA Ltd and MELETITIKI Ltd : Potential of Passive and Hybrid Cooling of buildings in Greece. CRES- EEC Valoren Program 1991.
4. Dialogic : Manual Description of CASAMO-CLIM, Paris, France, 1988.
5. G. Schiller : Earth Tubes for Passive Cooling. The Development of a transient numerical model for predicting the performance of earth to air heat exchangers. M. Sc. Repport, MIT, 1982.
6. A. Tzaferis and D. Liparakis : Comparative study of 13 different algorithms predicting the performance of earth to air heat exchangers against experimental data. Dissertation Thesis. TEI Piea, Greece, 1990.

Figure 1.
Indoor Air Temperature Variation

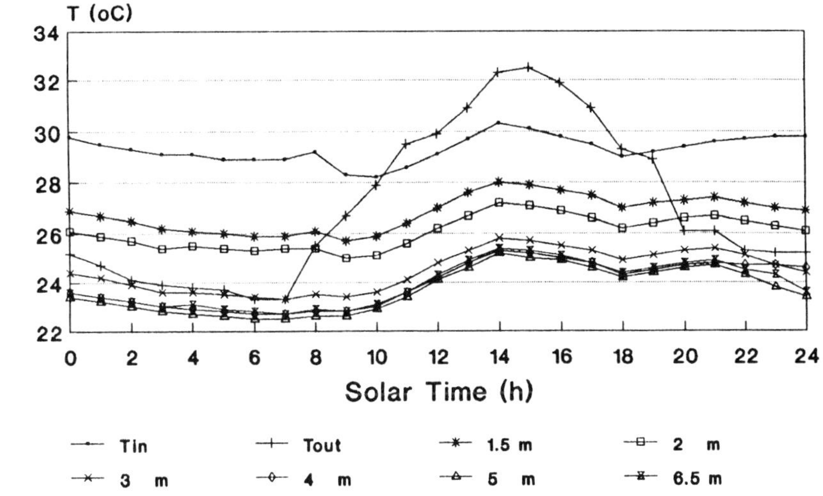

July

Figure 2.
Indoor Air Temperature Variation

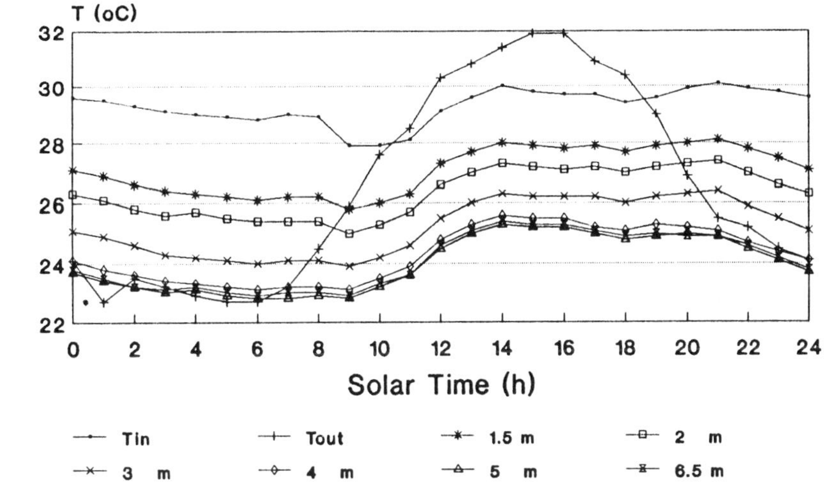

August

Indoor Air Temperature Variation

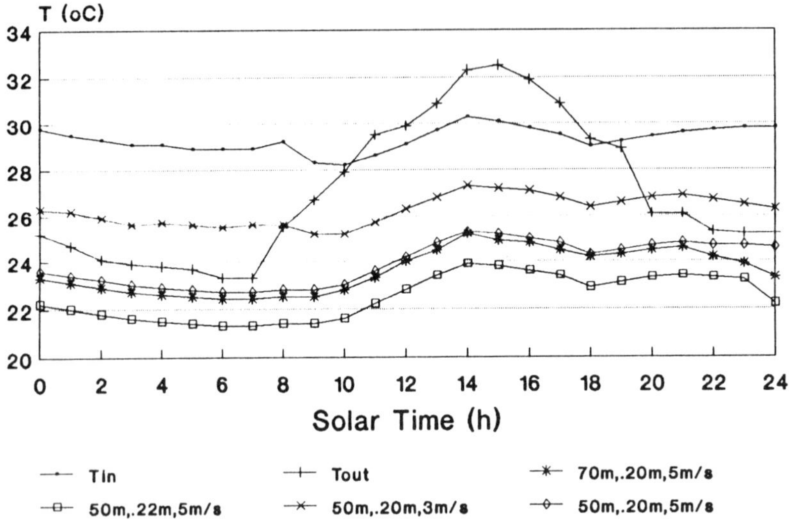

July

Figure 4.
Indoor Air Temperature Variation

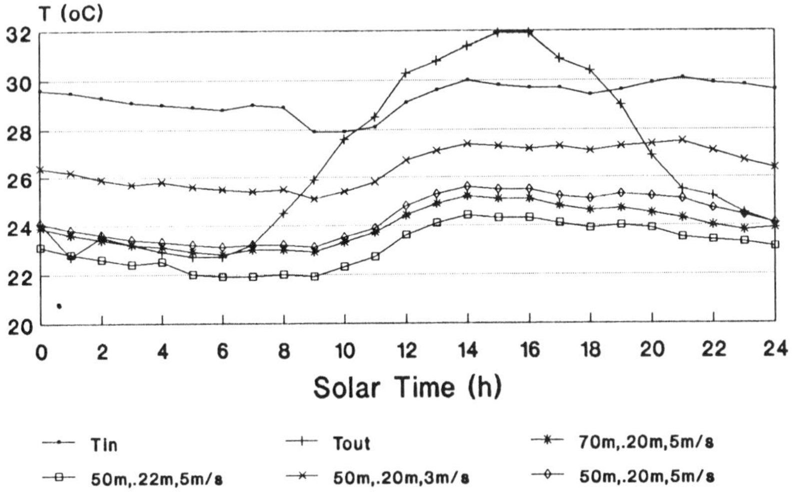

August

PRELIMINARY INVESTIGATION OF THE EFFECT OF A PASSIVE DIRECT EVAPORATIVE COOLING SYSTEM ON A COURTYARD AT UCLA

N. A. ALHEMIDDI
Graduate School of Architecture and Urban Planning
University of California, Los Angeles
405 Hilgard Avenues,
Los Angeles, California 90024,
U.S.A.

ABSTRACT. This paper reports the preliminary stage of a research project to study the performance of a passive direct evaporative cooling system in the model of a courtyard space at a UCLA Laboratory. Two experiments were conducted to evaluate the proposed design, namely: (1) quantify the system's average air velocity, ventilation air flow, cooling and humidifying efficiency, (2) observe the effect of the system on the thermal variations of the courtyard. Results indicate that the new design is capable of providing thermal comfort in outdoor spaces. Preliminary data were recorded during the period of December 1990 - February 1991. Statistical evaluation of the recorded data was carried out for the proposed design. Finally, the results are presented graphically.

1. INTRODUCTION

Passive cooling and natural ventilation systems have been studied by various scholars, but their work was mostly theoretical and limited to indoor spaces (Bahadori,1985; Wu and Yellott, 1987; Cunningham and Thompson, 1986). However, passive cooling and natural ventilation systems can also be used effectively for thermal comfort in <u>outdoor</u> spaces (yards and courts, or in "free standing" areas such as public rest areas, walkways, etc.) through the lowering of air temperature and increasing air velocity. Givoni (1989) has designed an evaporative cooling system for an outdoor space at the '90 EXPO in Spain. This system consists of an open vertical shaft at the top of which water is sprayed. The falling fine drops draw a large volume of air, resulting in an inertial air flow. Evaporation of the fine drops cools the air close to the ambient WBT. In the design used in the present study, evaporative cooling is increased by adding to the system a wind catcher, allowing more air to circulate downward. In other words, the new tested system is an enhanced version of Givoni's model. Therefore, this study presents two experments to evaluate and analyze the preformance of the new design: (1) quantifying the system

S. Alvarez et al. (eds.), Architecture and Urban Space, 643–648.

influence of the system on the thermal conditions of the courtyard. Statistical analysis of the recorded data shows that the system is efficient in providing cool air.

2. SYSTEM DESIGN DESCRIPTION

The proposed design of a passive direct evaporative system is illustrated in Fig. 1. It consists of three main parts: a wind catcher, an evaporative column, and a water pond. The design of this system is based on the concept that a wind catcher can force a large volume of air downward in an open vertical column at the top of which water is sprayed. The falling thin drops come in contact with the hot air and cool it through latent heat transfer to a temperature close to the Wet Bulb Temperature (WBT) of the ambient air. Water is collected in a pond located at the bottom of the evaporative column. It is then recirculated by means of a pump. In addition to the evaporative cooling, there is a sensible cooling of the air when the water temperature is lower than the ambient WBT. The difference in temperature between the ambient air and the air in the shaft increases the density of the air and thus it creates inertial flow and a higher flow rate. The wind catcher allows the wind to enter the column from any direction. This part has cross vertical barriers (+) to deviate the airflow downward without allowing it to leave through the opposite openings. The second part, the evaporative column allows full utilization of the airflow evaporative cooling potential by spraying very tiny water drops. At the top, four water spraying humidifiers (shower heads) spray the water vertically downward. A plastic sheeting covers this part. The third part is a water pond which consists of a bucket and a submersible electrical pump. The pump is connected to the head showers by a plastic pipe to recirculate the water. A galvanized metal sheeting surrounds the water pond and rises vertically to channel the flow of cooled air to the entire space of the courtyard.

3. THE COURTYARD CHARACTERISTICS

Figure 1 shows a plan and section of the proposed system and the courtyard dimensions. The courtyard was set up in the UCLA Laboratory in 1989. It is constructed of wood framing with plywood and insulation material (4" thick and R=11). The exterior and interior are painted in white to minimize the effect of solar radiation. The north half of the courtyard is open to the sky with a movable door at the south side. The system is located at the center of the courtyard.

4. INSTRUMENTS AND MEASUREMENT METHOD

Two instruments for temperature and air velocity measurements were used in this study. The first was Bacharach Sling Psychrometer used to read Dry Bulb Temperature (DWT) and Wet Bulb Temperature (WBT) outside the courtyard 4' above ground level, at the point where the air exits from the system, and inside the courtyard. The water temperature was recorded continuously. The second instrument is a digital air velocity recorder with an accuracy of $\pm$ 0.5 m/sec. Figure 1 shows the reading points of the DBT and WBT and air velocity (V).

5. EXPERIMENT #1: QUANTIFYING THE AVERAGE AIR VELOCITY, VENTILATION FLOW, COOLING LOAD, AND HUMIDIFYING EFFICIENCY OF THE SYSTEM

On December 3, 1990, the weather was clear, the ambient air temperature was 80°F and the wind speed was moderate. The system was operated and two experimental methods were used to quantify the average air velocity and total ventilation flow of the system.

The First Method: The air exit opening of the system was divided into 8 equal areas (A, B, C, D, E, F, G and H). Air velocity readings were recorded at the center of each area with the probe tip placed in the plane of the opening. This allows the air to flow perpendicularly through the duct probe tip. It was found that the air velocity values were non-uniform and the average air velocity was 1.18625 m/sec. Figure 5.a shows the recorded air velocity and total air flow ventilation.

The Second Method: Each one of the 8 equal areas was divided into 10 subopenings. Air velocity readings were recorded at the center of each subopening area with the probe tip placed in the plane of the opening. The collected data presented in Figure 2.a shows the location of the reading points. The average air velocity was 1.182 m/sec. The findings of both methods show that air velocity was almost the same. Threfore, it is possible to minimize the number of air velocity measurements, but average air velocity at the four openings should be multiplied by a ratio of the average air velocity and the total average air velocity 1.186 m/sec.

Having found the air velocity, we can calculate ventilation flow and cooling using the following equations:

$$Q_i = A \cdot V_i \cdot V_c \qquad (1)$$

where

Q_i = air flow rate at second i in F^3/sec or m^3/sec
A = area of the exit air in F^2 or M^2
V_i = exiting air velocity the at second i in F/sec or M/sec
V_c = velocity correction ratio of average air velocity of a given opening = 1.182 MPH

In order to calculate the cooling of the system, the difference in temperature between the inlet of the wind catcher and open air exit (outlet) should bc known. Therefore the following equation can be used to find the amount of heat extracted from the courtyard air at a given time (Givoni, 1991):

$$C_i = Q_i\ (2(T_{in} - T_{out})) \times 0.33 \times 3600 \qquad (2)$$

where

C_i = cooling load (extracted heat) (heat loss)
T_{in} = inlet air temperature (F°)
T_{out} = outlet air temperature (F°)

0.33 = volumetric heat capacity of air (ws/C·m^3)
(Note: Calculations were converted from Britsh unites to US units in this study).

The hidifying efficiency of the system is a measure of the completeness of the humidifying process using a spray chamber. It (E_h) can be calculated using the following equation (Clifford, 1984):

$$E_h = \frac{T_{in\text{-}D} - T_{out\text{-}D}}{T_{in\text{-}D} - T_{out\text{-}W}} \qquad (3)$$

where

E_h = humidifying efficiency
$T_{in\text{-}D}$ = DBT of air entering the system (F°)
$T_{out\text{-}D}$ = DBT of air leaving the system (°F)
$T_{out\text{-}W}$ = WBT of air leaving the system (°F)

Air velocity, ventilation air flow, cooling load and humidifying efficiency will be used for calculation in the evaluation of the system based on thermal variations in the courtyard.

6. EXPERIMENT #2: THE EFFECT OF THE SYSTEM ON THE THREMAL VARIATIONS OF THE COURTYARD

DBT & WBT readings were taken outside and inside the courtyard, as well as at the open air exit at the base of the tower. The water temperature was also taken. This was done on February 4, 1991; the sky was clear and wind speed low. Fig. 3 represents 4 time periods of one and a half hours each. The systems performance is shown in phases 2 and 4. In phases 1 and 3, the system was not operating. The graph shows that DBT inside the courtyard in phases 2 and 4 are substantially lower than in phases 1 and 3. Furthermore, the DBT inside the courtyard in phase 4 are lower than those in phase 2. This is due to the fact that they were recorded at different periods of the day. In phase 2, which is the phase where cooling is most needed, the graph shows an average DBT reduction of 4°F inside the courtyard. The exiting air DBT is close to the outdoor air WBT, with a range 0.0 to 0.75°F. This demonstrates the cooling efficiency of the system. On the other hand, phase 2 shows that the water temperature rises by an average of 4°F, due to heat exchange between the air's sensible heat and the latent heat of the water droplets. Fig. 4 represents average air velocity measurements through 4 exiting air openings (C, D, E and F) and ambient air. It was found that air velocity averaged 3.872 Ft/sec. in phase 2 and 3.937 Ft/sec. in phase 4. The air velocity value changes from one opening to another because air velocity depends on two factors: the speed of downdraft air captured by the wind catcher and the water pressure sprayed in the column (inertial air flow). Fig. 5 shows the graph of ventilation air flow generated by the system with an average 32.500 Btu/Ft^3 per hr in phase 2 and 4. This air flow increases the ventilation potential in the courtyard which is characterized by lower pressure. Thus, the air flow leads to a comfortable living condition. Fig. 5 shows the cooling effect produced by the system in the two phases 2 and 4. The heat extracted from the courtyard (cooling) depends on the difference between DBT of the air entering the system and DBT of the exiting air. The higher the difference the more effective the cooling of the system. The average cooling load recorded was 35000 Btu/hr in phase 2 and 15000 Btu/hr in phase 4. Fig. 7 shows the humidifying efficiency of the system. Humidity fluctuates between 63% and 94% in phase 2 and between 50% and 88% in the phase 4.

7. CONCLUSION

The proposed system can supply air to the courtyard space at higher flow rates. Furthermore, the air delivered to the courtyard space is cooled by evaporation. Two experimental studies conducted to evaluate the proposed system were carried out successfully and may be used for designing passive direct evaporative cooling systems. Additional applications are being investigated such as quantify the system's water and energy consumption, the effect of wind catcher on the system performance, and the effect of the water quality (brackish and sea water) on the performance of the proposed system.

8. ACKNOWLEDGEMENTS

The author is grateful to professor B. Givoni for his advice and helpful suggestions. The advice of Proffesors M. Milne, D. Schoen, is gratefully acknoledged, as are the suggestions of Professors C. Harris, S. Aroni, K. Iyengar, and graduate students in the course of "298 Research Practicum in Architecture Technology" in Winter quarter 1991. This study was in part funded by a grant from the UCLA.

9. REFERENCES

Bahadori, M. N. (1985) An Improved Design of Wind Towers for Natural Ventilation and Passive Cooling. Solar Energy Vol. 35, No. 2, pp. 119-129.

Clifford, G. E. (1984) Heating, Ventilating, Air Conditioning. Reston Publishing Company, Inc., Reston, USA.

Cunningham, W. A. and Thompson, T. L. (1986) Passive Cooling with Natural Draft Cooling Towers in Combination with Solar Chimneys. submitted to: PLEA 1986, Pecs, Hungary, September 1-5.

Givoni, B. (1989) Cooling of Outdoor Spaces. Proceedings Workshop Interaction between Physics and Architecture in Environment Conscious Design, International Center for Theoretical Physics. Trieste, Italy, p. 1.

Wu. H. & Yellot J. I. (1987) Investigation of a Platetype Indirect Evaporative Cooling Systems for Residences in Hot and Arid Climates. ASHRAE 93:1.

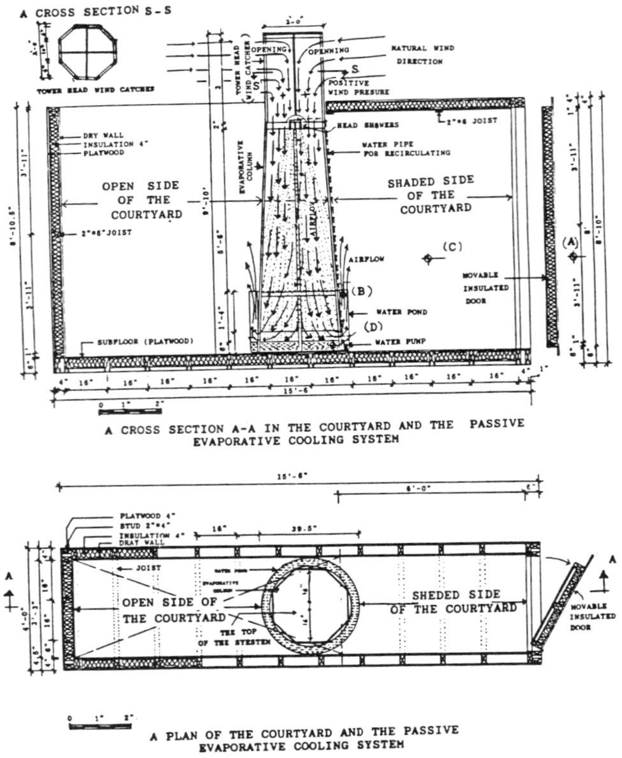

Fig. 1 Detaild plan and section of the passive direct evaporative cooling system in the model of a courtyard space at UCLA Laboratory.

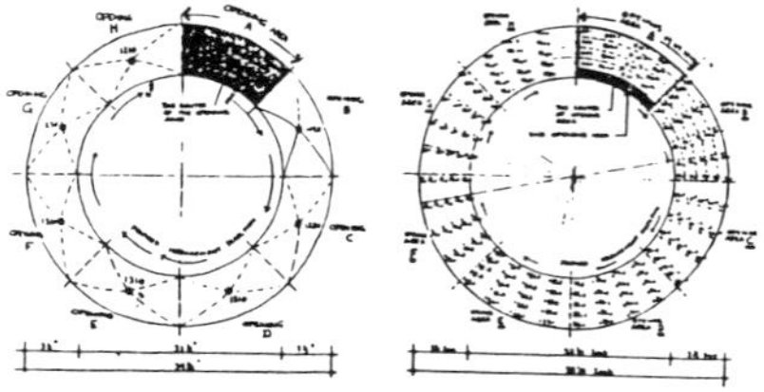

Fig. 2.a The recorded air velocity and total ventilation air flow for experimant method #1.

Fig. 2.b The recorded air velocity and total ventilation air flow for experiment method #2.

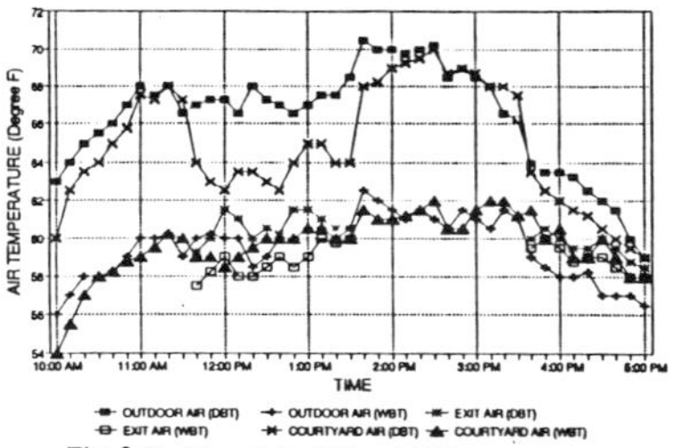

Fig. 3 Profiles of the DB and WB temperature of the outdoor air, tower exit air and courtyard air, and water.

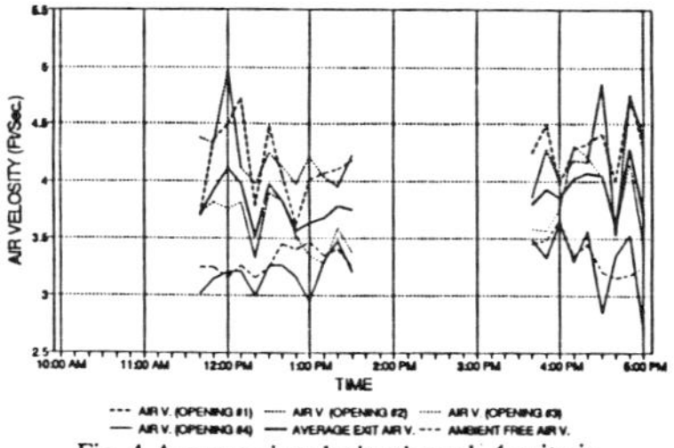

Fig. 4 Average air velocity through 4 exit air openings and ambiant free air velocity.

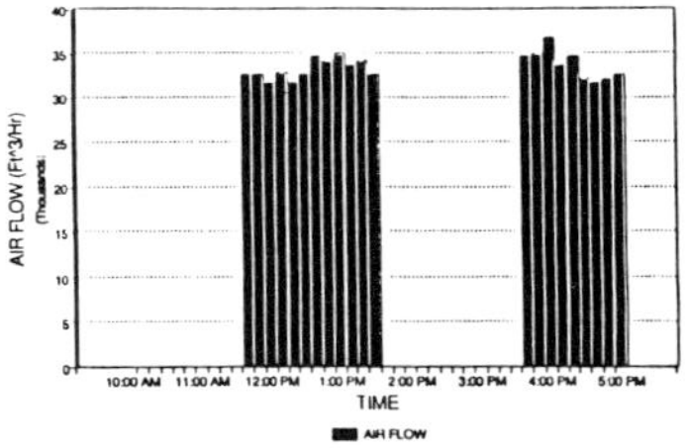

Fig. 5 Profile of the ventilation air flow generated by the system.

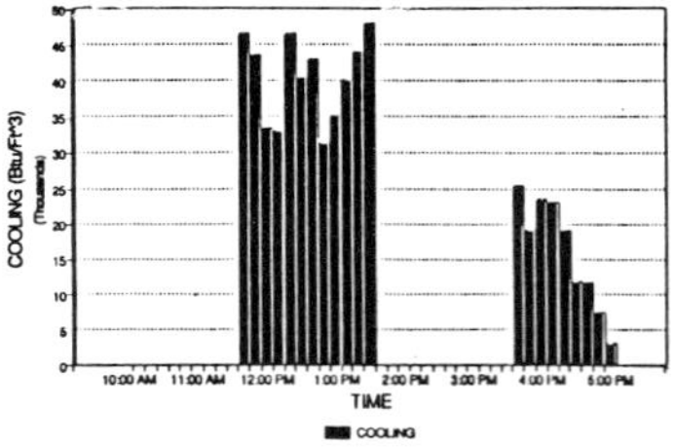

Fig. 6 Graph of the cooling load.

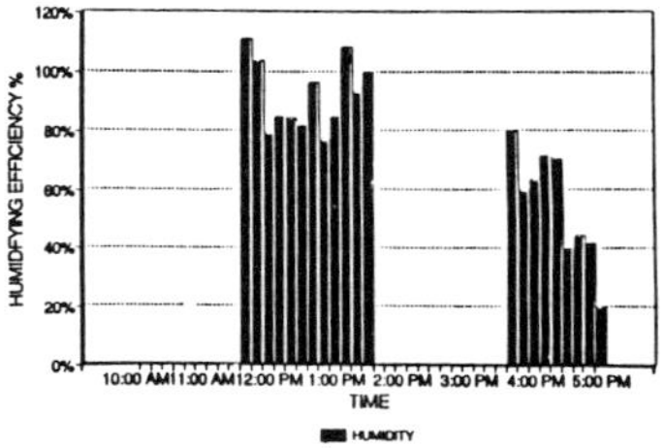

Fig. 7 Graph of humidifying efficiency.

Part 6

BUILDING ENERGY ANALYSIS

Chapter 1

Climatic Data

APPLYING REDUCED WEATHER DATA TO TEMPERATE CLIMATES FOR ENERGY ANALYSES OF BUILDINGS

M. COLUMBA, S. BARBARO
and G. CANNISTRARO

Department of Energetics and
Application of Physics
University of Palermo
Viale delle Scienze
90128 Palermo, Italy

G. RIZZO

Institute of Civil Engineering
and Energetics
Engineering Faculty
University of Reggio Calabria
89100 Reggio Calabria, Italy

ABSTRACT. Using a methodology recently proposed for cold climates, we develop a reduced weather data set for the spanish location of Almeria.
Reduced data are constituted by a simple periodic temperature distribution, while solar radiation is computed as a function of the mean monthly clearness index.
A "regional" mediterranean correlation is also proposed between the diurnal variation of the air temperature and the mean clearness index.
The reliability of the produced data set has been verified by means of a computer code, comparing the energy requirements of a building module, chacterized by different mass and insulation and located in the climatic situation of Genoa.

1. Introduction

The energy requirements for climatisation of buildings are usually predicted by means of mathematical models, numerically solved with the aid of computer codes. Notoriously, weather data are among the most important parameters in this context.
Two problems do arise at this regard. First, reliable set of meteorological data are available only for a very limited number of sites, essentially belonging to the industrialized countries. Second, even though they were available, the use of extended series of weather data often turns out into unbearably time spending computations, particularly when using personal computers.
On the other hand, architects and designers push for the knowledge of climatic data even at regional and urban level, in the aim of producing more and more sophisticated and detailed building analyses.
As an attempt of avoiding these constraints, we have studied the reliability of a method, recently proposed in the literature, able to build-up reduced set of weather data starting from simple typical climatic information, on a monthly base. The method, originally proposed for cold northern locations, has been extended to mild mediterranean sites.

S. Alvarez et al. (eds.), Architecture and Urban Space, 653–658.

2. Structure of the reduced data set

The reduced data set here applyed (Bahadory et al., 1986) essentially consists of a periodic temperature distribution, as a function of the mean clearness index, k_T:

$$\frac{T_a - T_m}{A_t} = \sum_{i=1,4} a_i \cdot \cos(i \cdot t' - b_i) \tag{1}$$

where $\mathbf{T_a}$ is the hourly ambient temperature (to be determined); $\mathbf{T_m}$ is the mean monthly temperature; $\mathbf{A_t}$ is the temperature amplitude of the diurnal variation; $\mathbf{a_i}$ and $\mathbf{b_i}$ are suitable constant coefficients; and **t'** is given by:

$$t' = 2\frac{\pi}{24}(t-1) \tag{2}$$

where **t** is the time, in hours (12 at solar noon).
For the constant coefficients we have adopted here those proposed by Erbs (Erbs et al., 1983), since the diurnal variation of $\mathbf{T_a}$ is only a weak function of location and time of year.
Unfortunately, the maximum and minimum value of the daily temperatures are not easily available, so that the computation of the $\mathbf{A_t}$ parameter is rarely well suitable. But the "peak to peak" diurnal variation has been found to present a linear dependence (Erbs et al., 1983) by the clearness index, that is:

$$A_t = A_1 \cdot K_t + A_2 \tag{3}$$

being $\mathbf{A_1}$ and $\mathbf{A_2}$ parameters essentially site-depending.
The method has been initially applyed to northern cold climates, but we have verifyed its reliability for the site of Genoa (Cannistraro et al., 1990), characterized by a mild temperate climate.

3. Correlation for the mediterranean area

In the aim of generating a correlation like that described by equation (3), but valid for the spanish site of Almeira, we have used the Test Reference Years (TRY) produced within the Commission of the European Communities (CEC, 1985) for 29 stations, showing a latitude spreading from 37°55' (Trapani, Italy) to 55°26' (Valentia, Eire). An extensive application of the data set allowed us to find the $\mathbf{A_1}$ and $\mathbf{A_2}$ coefficients for nine southern Europe locations, as shown in table 1. We have also produced a "regional" correlation, on the base of the above cited locations: the $\mathbf{A_1}$and $\mathbf{A_2}$ coefficients are respectively values of **14.823** and **-0.245**.
This correlation should apply to mediterranean sites and should allow the determination of the $\mathbf{A_t}$ parameters for any location belonging to the area. A validation for a case-study is reported through the next section.

4. Validation of the algorithm

The validity of the distribution built-up with the "regional" values of the coefficients has been checked for the italian location of Genoa.

In this aim, we have driven a computer code based on the finite differences method (Butera et al., 1984) with the following sets of meteorological data:

1) **TMY** A typical meteorological year, statistically (Dogniaux et al., 1977) built-up using a six-years (1965-1970) group of data;

2) **R** A reduced weather data set characterized by the temperature distribution of equation (1) and a "local" correlation for the $\mathbf{A_t}$ computation;

3) $\mathbf{R_m}$ A reduced weather data set characterized by the temperature distribution of equation (1) and a "regional" correlation for the $\mathbf{A_t}$ computation.

TABLE 1. Coefficients of the correlation between amplitude diurnal variation and clearness index for nine mediterranean locations.

Station	Country	Latitude	A_1	A_2
Nice	France	43° 39'	- 0.15	6.50
Carpentras	France	44° 05'	38.57	- 6.72
Cagliari	Italy	39° 15'	11.29	1.64
Crotone	Italy	39° 00'	14.93	- 0.34
Foggia	Italy	41° 32'	20.62	- 0.69
Genova	Italy	44° 32'	5.61	2.64
M.te Terminillo	Italy	41° 48'	2.93	3.27
Ciampino	Italy	41° 48'	17.29	0.51
Trapani	Italy	37° 55'	9.93	1.52

Two sample modules (showing a volume of 75 m^3) have been chosen for the comparison: a room with a light insulated frame (overall mean thermal resistance of 0.474 $m^2{\cdot}K/W$) and a room with a typical massive insulated frame (overall mean thermal resistance of 0.675 $m^2{\cdot}K/W$).
Figure 1 reports the energy requirements for the climatisation of the sample modules for each of the above described climatic data set.
It's evident that the energy consumptions of the regional-correlated distribution (R_m) presents a quite good agreement with the line representing the comsumptions predicted by means of the Test Referenc Year (TRY) data set. Only for a reference to a real situation, in figure 1 is also reported the behaviour of the SYA distribution, that is the six-years average (1965-1970) of the recorded data in the station of Genoa.
As a general remark, it's possible to observe that the regional distribution remains both under the local distribution and the Typical Metereological Year through all the simulated period. During the most sunny months the regional (as well the local) distribution seems to underestimate the energy requirements: this evenience has not to be ascribed to the methodology of generation of the reduced data set, since its sistematically occurs when using monthly average days instead of the extended serie of climatic data. Another significant feature related with the use of reduced data is that the difference between massive and light frame structures are in some way smoothed, as it's possible to see in table 2. It concisely reports the percentage errors of the seasonal energy consumptions respectively predicted with the "local" distribution (R) and with the "regional" distribution (R_m). The long-term agreement appears fairly good.

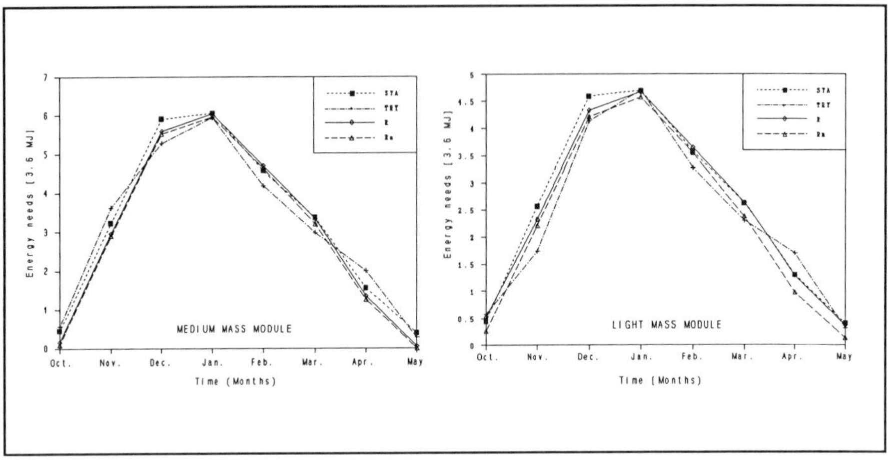

Figure 1

TABLE 2. Percentage errors of the seasonal energy consumptions.

Level of the module insulation	Local distribution	Regional distribution
Medium	-2.76 %	-5.30 %
Light	-2.00 %	-5.67 %

5. Temperature distribution for Almeria

In order to demonstrate the suitability of the above proposed method, we have applyed it to the generation of a reduced data set of temperatures, to be employed for the energy analysis of building, for the spanish location of Almeria.

This city (36 51' N latitude) belongs to the southern iberian region of Andalusia and is chacterized by external conditions during the heating season inside or nearly the comfort area, although, usually, winter early mornings are slightly cold.

Using equation 1 and the "regional" parameters for equation 2, we have calculated the reduced set of the external air temperature on monthly base. Results are reported in Table 3. Last row of the table contains the mean daily (monthly) global solar irradiation values, G, published on the European Solar Atlas (CEC, 1984), from where have been also taken the mean values of the clearness indeces, k_t. Values of the mean monthly air temperature have been found on a work produced within a IEA international exercise.

We would to remember here that the presented values of temperature have to be utilized along the hourly values of the radiation data. These can be usefully computed by means of the usual well known relationships, allowing the generation of the hourly data starting from the daily (or monthly) total value.

6. Conclusions

A periodic distribution of external air temperatures has been here presented as a method for generating reduced sets of climatic data for the energy analysis of buildings. The only required parameters are the monthly air temperatures and the mean monthly clearness indeces.

TABLE 3. Hourly external temperature (°C) and daily (monthly) global irradiation, G (Wh/m^2) for the site of Almeira, Spain.

Hr	Jan.	Feb.	Mar.	Apr.	May	Jun.	Jul.	Aug.	Sep.	Oct.	Nov.	Dec.
1	8.3	9.3	12.7	13.2	17.5	21.0	22.4	22.0	20.6	15.2	13.4	10.1
2	7.9	8.9	12.4	12.8	17.1	20.6	21.9	21.5	20.2	14.8	13.0	9.7
3	7.5	8.5	12.0	12.5	16.7	20.2	21.5	21.1	19.8	14.4	12.6	9.3
4	7.2	8.2	11.7	12.1	16.3	19.8	21.1	20.7	19.4	14.0	12.3	8.9
5	6.9	7.9	11.5	11.9	16.0	19.5	20.7	20.4	19.1	13.8	12.0	8.7
6	6.9	7.9	11.4	11.8	16.0	19.4	20.7	20.3	19.1	13.7	12.0	8.6
7	7.2	8.2	11.8	12.2	16.4	19.9	21.1	20.8	19.5	14.1	12.3	9.0
8	8.2	9.2	12.6	13.1	17.4	20.9	22.2	21.8	20.5	15.1	13.3	10.0
9	9.5	10.5	13.8	14.4	18.8	22.4	23.8	23.3	21.9	16.4	14.6	11.3
10	10.9	11.9	15.2	15.8	20.3	24.0	25.5	25.0	23.4	17.9	16.0	12.8
11	12.2	13.2	16.4	17.1	21.9	25.4	27.0	26.5	24.8	19.3	17.3	14.2
12	13.2	14.2	17.3	18.1	22.9	26.6	28.3	27.7	26.0	20.3	18.3	15.2
13	14.0	15.0	18.0	18.8	23.7	27.4	29.1	28.5	26.8	21.1	19.1	16.0
14	14.5	15.5	18.5	19.3	24.2	28.0	29.8	29.1	27.3	21.7	19.6	16.6
15	14.7	15.7	18.7	19.6	24.5	28.2	30.0	29.4	27.6	21.9	19.8	16.8
16	14.6	15.6	18.6	19.4	24.3	28.1	29.9	29.2	27.4	21.7	19.7	16.6
17	14.0	15.0	18.1	18.9	23.7	27.4	29.2	28.5	26.8	21.2	19.1	16.1
18	13.1	14.1	17.2	18.0	22.7	26.4	28.1	27.5	25.8	20.2	18.2	15.1
19	12.1	13.1	16.2	16.9	21.6	25.2	26.8	26.3	24.7	19.1	17.2	14.0
20	11.1	12.1	15.4	16.0	20.5	24.2	25.7	25.2	23.6	18.1	16.2	13.0
21	10.3	11.3	14.6	15.2	19.7	23.3	24.8	24.3	22.8	17.3	15.4	12.2
22	9.7	10.7	14.1	14.6	19.1	22.6	24.1	23.6	22.2	16.7	14.8	11.6
23	9.2	10.2	13.6	14.1	18.5	22.1	23.5	23.0	21.6	16.2	14.3	11.1
24	8.7	9.7	13.2	13.7	18.0	21.5	22.9	22.5	21.1	15.7	13.8	10.6
G	2724	3492	4273	5468	6702	7215	7436	6577	5264	4006	2894	2541

Using the Test Reference Years produced within a CEC programme, we have proposed a regional correlation, valid for mediterranean climates, between the mean monthly clearness indeces and

the "peak to peak" diurnal variation of the air temperatures. This expands the application of the methodology, originally proposed only for cold climates. We have also applyed the method to the spanish location of Almeida.
The comparisons presented through the paper show that the method can be properly employed particularly for long-term analyses, that is a typical feature of the monthly based climatic data sets. Since the needed climatic parameters are generally easy to be find, the method candidates itself as an useful tool for buildings energy analyses, even until regional and urban levels of application.

7. References

Bahadori, M. and M. Chamberlain (1986) 'A Simplification of Weather Data to Evaluate Daily and Monthly Energy Needs of Residential Buildings', Solar Energy, vol. 36, no. 6, pp. 499-507.

Erbs, D., Klein, S. and W. Beckman (1983). 'Estimation of Degree-Days and Ambient Temperature Bin Data from Monthly-Average Temperatures. ASHRAE Journal, no. 6.

CEC (1984). 'European Solar Atlas. Volume I: Global radiation on horizontal surfaces'. Verlag Tüv Rheinland.

IEA (1985). 'Passive and Hybrid Solar Commercial Buildings'. International Energy Agency: Solar Heating and Cooling - Task XI.

CEC (1985). 'Test Reference Years TRY. Weather data sets for computer simulations of solar energy systems and energy consumptions in buildings'. Directorate-General XII for Science, Researche and Development (Tape Format).

Cannistraro, G., Franzitta, G. and G. Rizzo (1990). 'Comparison of extended and reduced weather data in computing the energy requirements for climatisation of buildings'. Proceedings of the 7. Internationales Sonnenforum, Frankfurt, 9-12 october.

Butera, F., Farruggia, S., Rizzo, G. and G. Silvestrini (1984). 'Il codice di calcolo SMP per la simulazione del comportamento termico di moduli edilizi solari passivi e convenzionali: algoritmi usati e logica funzionale'. Quaderni dello IEREN/CNR, no. 1, Palermo.

Dogniaux, R. and R. Sneyers (1977). 'Methodologie d'analyse statistique des donnes metereologique'. Rapport des Journees Nationales d'Etudes sur le Chauffage Solaire dans le Batiment. AIM, Liege.

A METHOD TO DETERMINE A SET OF STATISTICALLY INDEPENDENT CLIMATIC ELEMENTS. APPLICATION TO 44 URBAN STATIONS ALL OVER PENINSULAR SPAIN

M. PARDO GONZáLEZ-NANDíN; C. GóMEZ CAMACHO (*)

Departamento de Análisis Regional y Geografía Física
C/ Dª María de Padilla, s/n
41004 Sevilla (Spain)

(*) Departamento de Ingeniería Energética y Fluidomecánica
Avda. Reina Mercedes, s/n
41012 Sevilla (Spain)

ABSTRACT. The number and nature of possible independent climatic elements of a geographical region is obtained by Spearman's Rank Correlation techniques. The most meaningful of them are chosen as the independent ones, thus characterizing by themselves the different regional climates. Monthly averages of global horizontal solar exposure, precipitation, and relative humidity become representative variables for climates of 44 urban stations covering peninsular Spain.

1. INTRODUCTION

The aim of this paper is to give a method to check the relationship between pairs of climatic elements that could affect urban climates, allowing to determine their number of independents variables; once this number is known, the most adequate selection of independent climatic elements can be made in order to establish climatic zones, design guidelines, and develop algorithms useful for urban architecture.

The method should be as general as possible, avoiding any unnecessary apriori hypothesis about distribution of data and relation between them. The criteria to fix independent variables depends on the objective: this paper developpes criteria emphasizing cause-effect, even if some of selected variables are not the most available. In this way, mechanisms of urban climate can be better understood and a most efficient control can be achieved.

See Landsberg (1984), to a detailed study of urban climate, and Budyko, (1974) and (1982), which states a climatic classification based on radiation balance and radiative index of dryness, by means of rather complex physical models and experimental measurements, whose conclussions are close to ours, which are statisticaly model-free obtained.

S. Alvarez et al. (eds.), Architecture and Urban Space, 659–664.

2. STATISTICAL PROCEDURE

Most of statistical tests performed to correlate two variables are based on the assumption that data to analyze have a well defined distribution function, usually Gaussian. If data do not follow such function, results are at best approximate, more often erroneous; besides that, product-moment correlation coefficients are very sensitive to extreme values of data.

Distribution free tests, or non-parametric tests, are valid regardless of distribution functions of observations, and become especially useful to deal with climatic data used in this work, whose Gaussian distribution is sometimes doubtful.

Spearman's Rank Correlation Coefficient tests monotonous relationships between two variables using the ranks of the data, instead of their values; if variables are independent and continous, the test is exact: see Moses (1986). Thus, Spearman's test will be used to check statistical independence of couples of climatic elements, month by month, in a wide sample of stations, covering most of urban climates of peninsular Spain.

3. DATA SET FOR 44 URBAN STATIONS IN PENINSULAR SPAIN

Climatic variables selected to carry out this study has been the three geographical coordenates -latitude, longitude and altitude- which are also climatic factors, monthly average of the five climatic standard normals -air temperature, °C, precipitation, mm, relative humidity, %, atmospheric pressure, mbar, and sunshine hours, h-, and monthly daily average of global horizontal solar exposure, kWh/m^2; this last six quantities are climatic elements.

To cover the disparity of climates in penisular Spain, these climatic variables have been recorded at 44 urban stations; see their location in Figure 1.

Climatic standard normals and solar exposures data used are the ones published by Spanish National Institute of Meteorology: see López Balaguer (1989) and Font Tullot (1984), respectively.

4. RESULTS OF SPEARMAN'S RANK CORRELATION

The table gives the significance levels for the null hypothesis of independence between variables, of a two-sided test of Spearman's Rank Correlation Coeficient. Variables are the six climatic elements stated in the previous point.

The table is ordered in 12 groups of 5 rows each, and 5 columns. Each group represents a month, ordered from January, and the associated rows and columns stand consecutively

Figure 1. Location of Urban Stations used in this paper

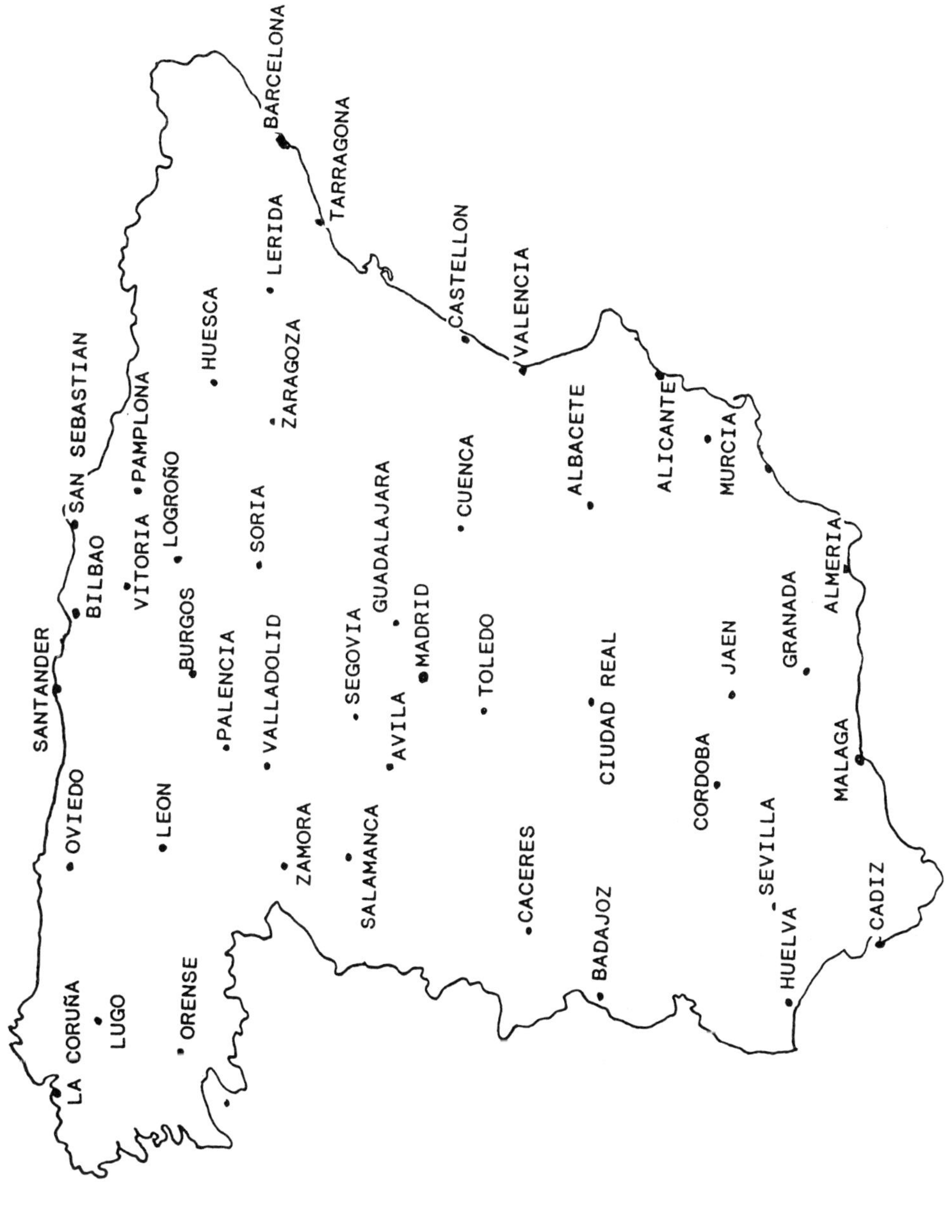

for monthly averages of air temperature, te; precipitation, pc; atmospheric pressure, pa; sunshine duration, hs; and global horizontal solar exposure, ex.

TABLE

Monthly Significance Levels of Spearman Rank Correlations

Monthly	pc	hr	pa	hs	ex
te1	.6287	.0000	.0000	.0001	.0011
pc1		.2267	.3322	.0063	.0329
hr1			.0004	.0012	.0476
pa1				.0132	.1133
hs1					.0000
te2	.5959	.0053	.0000	.0003	.0010
pc2		.0029	.2652	.0013	.0235
hr2			.0250	.0036	.0266
pa2				.0273	.1376
hs2					.0000
te3	.9836	.5159	.0000	.0005	.0001
pc3		.0002	.1975	.0008	.0121
hr3			.6112	.0002	.0105
pa3				.1121	.0854
hs3					.0000
te4	.0863	.0157	.0000	.0002	.0000
pc4		.0022	.7402	.0000	.0003
hr4			.5299	.0000	.0001
pa4				.3780	.0470
hs4					.0000
te5	.0000	.0585	.0000	.0000	.0000
pc5		.0087	.1210	.0000	.0000
hr5			.0179	.0011	.0017
pa5				.5096	.1789
hs5					.0000
te6	.0000	.0062	.0034	.0000	.0000
pc6		.0130	.0459	.0000	.0000
hr6			.0036	.0001	.0001
pa6				.8473	.5773
hs6					.0000
te7	.0000	.0036	.0544	.0002	.0000
pc7		.0061	.1889	.0000	.0000
hr7			.0002	.0000	.0001
pa7				.0401	.5834
hs7					.0000

Monthly Significance Levels of Spearman Rank Correlations

	pc	hr	pa	hs	ex
te8	.0000	.0164	.0138	.0012	.0000
pc8		.0002	.9378	.0000	.0000
hr8			.0001	.0000	.0000
pa8				.0654	.2063
hs8					.0000
te9	.0001	.3623	.0000	.0001	.0000
pc9		.0000	.9174	.0000	.0000
hr9			.0005	.0001	.0003
pa9				.9910	.7359
hs9					.0000
te10	.2134	.0849	.0000	.0005	.0001
pc10		.0619	.0049	.0043	.0147
hr10			.5989	.0007	.0004
pa10				.2570	.2366
hs10					.0000
te11	.5508	.0001	.0000	.0006	.0007
pc11		.2547	.1111	.0318	.0142
hr11			.0011	.0095	.0127
pa11				.1206	.1969
hs11					.0000
te12	.0957	.0000	.0000	.0002	.0004
pc12		.9390	.0388	.1225	.0702
hr12			.0000	.0020	.0033
pa12				.0204	.0267
hs12					.0000

5. SELECTION OF CLIMATIC ELEMENTS AS INDEPENDENT VARIABLES

The null hypothesis of independence between variables will be rejected if significance levels exceed 0.01, at least for a month. If two variables are related, criteria for assigning the independent one are two: first, his nature as a cause of climatic behaviour, rather than as a response; second, the continuity and quantification of his recording, better than discontinuity, qualification, or both.

Besides that, if a climatic element is continously related, for every station, to a climatic factor -that is to say, it is nearly constant for a given station, but changes between them- it will be considered as parameter, rather than a variable, to define local climates.

This last fact is applied to atmospheric pressure, that shows an excellent correlation with altitude, which is obviously constant in each station; thus, effects of atmospheric pressure in local climate will be taken into account through its altitude, been discarded as an independent variable.

Sunshine duration, hs, and solar exposure, ex, have always significance levels lesser than 10^-4. Consequently, null hypothesis of independence is rejected, and a relationship between them is accepted. Exposure is considered independent variable, because is recorded quantitatively, and because solar irradiance causes sunshine.

Monthly average values of air temperature, te, and exposure, ex, have a maximum of their rank correlation significance level of 10^-3, so they are taken as interrelated climatic elements. Exposure is selected as the independent one, because it can be considered cause of air temperature, while the reverse can not be stated.

Monthly values of precipitation, pc, and relative humidity, hr, show maximum significance levels clearly over 0.01, and the same occurs between precipitation and exposure, and relative humidity and exposure.

In consequence, monthly average values of daily solar global horizontal exposure, precipitation, and relative humidity are selected as a set of independent climatic elements, defining the climate of a peninsular spanish urban station, together with its constant climatic factors.

This fact means that, for a given station, climatic elements, others than exposure, precipitation and relative humidity, are function of these ones, as well as climatic factors. Such functions can have, station to station, very different expressions and parameters, but their independent variables are the tree before said.

Physical reasons to explain it is to consider the climate of a station, then with fixed climatic factors, as the response to an input of energy absorbed by ground -exposure-, exchanged with its atmosphere through water -precipitation- evaporation, which is controled by athmospheric composition -relative humidity-.

6. REFERENCES

Budyko, M. Y. (1974) Climate and Life, Academic Press, New York.

Budyko, M. Y. (1982) The Earth's Climate: Past and Future, Academic Press, New York.

López Balaguer, J. L. (coord.) (1989), Calendario Meteorológico, Instituto Nacional de Meteorología, Madrid.

Font Tullot, I. (1984) Atlas de la Radiación Solar en España, Instituto Nacional de Meteorología, Madrid.

Landsberg, H. E. (1983) 'City Climate', in Flohn, H. (ed.), General Climatology, Vol. 3, Elsevier, Amsterdam.

Moses, L. E. (1986) Think and Explain with Statistics, Addison-Wesley, Reading, Massachusetts.

CLIMATIC SEQUENCES IN URBAN ENVIRONMENT

Rafael Serra, Xavier Solsona, Jaume Roset
Escola d'Arquitectura de Barcelona, U.P.C.
Av. Diagonal 649, 08028 Barcelona

ABSTRACT. Any assessment of building performances, and specially of its thermal behaviour, requires climatic data as one important input. However, these data are not usually available for all places at suitable time intervals and reliable statistics over the years.

Moreover, it is well recognized that the urban environment produces special microclimatic effects, such as the 'thermal island' in large cities.

The climatic effect of cities is not only important on average conditions, but it also has to be studied in time dependent changes. In fact, day-of-the-week weather dependence had been already reported in Paris. Therefore, it would be interesting to analyze the existence of common sequences in the daily evolution of weather, and some work has been done in this field.

We will analyze in this paper the sequences using radiation data from different Spanish cities. Effects on the urban environment are searched for at this stage.

1. INTRODUCTION

Large urban concentrations have considerable influences on climate; the "thermal island" effect, which gives rise to appreciable differences between the city centre temperature and that of the outlying area, is particularly well known. Similarly, it is widely accepted that urban concentrations may increase rainfall considerably [1].

However, the influences that they exercise over periodical changes of weather is not so widely known. For example, a study carried out on Paris revealed that, statistically, rainfall has its own weekly behavioural period and that the average Friday precipitation is greater than that of every other day of the week [1].

Similarly, the application of the Fourier analysis, on the basis of readings of solar radiation taken hourly over several years, suggests the existence of behaviour periods of approximately seven days [2].

S. Alvarez et al. (eds.), Architecture and Urban Space, 665–670.

Likewise, for daily radiation data, a system of classification into three different types of days has been established: Cloudy (C), Medium (M) and Clear (D). Applying this classification, it has been demonstrated that the unrelated probability hypothesis for the type of day (i.e. that the probability of any particular day being type C, M or Cl is not affected by the surrounding days) is not borne out, and that furthermore neither is the Marcovian transition probability (i.e. that the probability of a particular day being of a particular type depends solely on the type of the previous day) borne out [3].

In view of the results above commented, which tend to demonstrate the existence of a periodical co-relation between day types of a longer duration, we have studied the relation between said day types and their average frequency in weekly sequences. This study has been conducted for urban centres of different sizes and geographical locations.

2. DESCRIPTION OF THE INVESTIGATION

Our research was carried out using daily readings of global radiation measured over a horizontal surface, provided by IER-CIEMAT [4]. Madrid, Toledo and Ciudad Real have been chosen for study as central zones of the Iberian Peninsula and Barcelona, Valencia and Castellón de la Plana representing the Mediterranean coast.

In determining the readings selected for use, we discarded the readings for weeks in which there was a day without reliable readings, so that although the number of weeks which we could use in the study was reduced, as this investigative method provided us with more reliable results.

Firstly, we made three different classifications of day type: Cloudy, Medium and Clear, corresponding to the greater or lesser proportion of solar radiation recorded with respect to the theoretical average for this day within the astronomical year. The proportion of radiation between the different day types has been adjusted so as to provide data which correspond with the statistics of the I.N.M. for the three types of days which appear in its findings [5].

With the data thus obtained, we have developed a classification of all the different possible types of weeks according to the number of times each particular day type occurs in each week.

Furthermore using the statistics on the occurrence of each different day type, we have calculated the theoretical probability of different weekly compositions on the basis of daily radiation not being affected by that of the previous day or days (the process of unrelated phenomena).

3. RESULTS

The following graphs, shów the results obtained for Madrid (1979-1984), Toledo (1979-1984), Barcelona (1975-1980) and Valencia (1975-1984).
The following information may be obtained from the graphs:
-The number of cloudy days Nc from the X axis
-The number of clear days Nd from the Y axis
-The number of medium days, being $Nm = 7 - (Nc + Nd)$
-The probability of occurrence (weeks or calculated with reference to the total number of weeks considered) from the basis of axis Z.

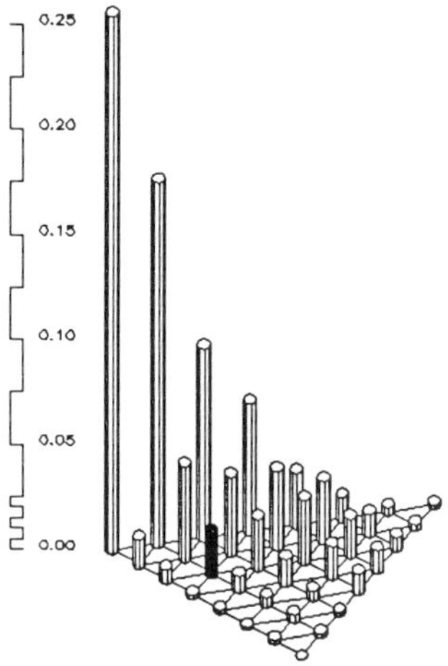

Figure 1. Example: if we consider that the probability of occurrence of a week with one cloudy day, 5 clear days and, hence one mixed day , we arrive at point A and the level of the corresponding column over the scale of probabilities is 0.021. In this graph, we have used 1938 possible weeks, so that the margin of error has been estimated as being 0.003.

In evaluating the margin of error, we assumed the standard deviation to be the square root of the observations. The estimated margin of error is less than the usual differences which exist between the values obtained from the actual occurrence of data and those calculated on the basis of the unrelated probabilities hypothesis for each day.

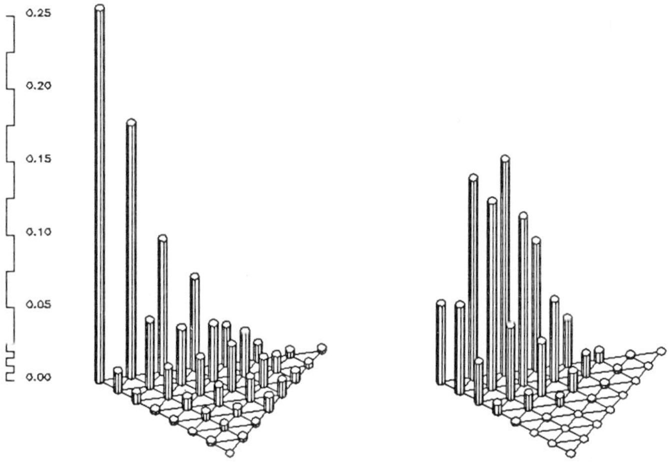

Figure 2. Probabilities of occurrence of weekly sequences obtained from meteorological data for Madrid (1979-1984). Left, actual readings; right, readings calculated on the basis of unrelated probability.
Number of weeks for which readings were considered: 1938.

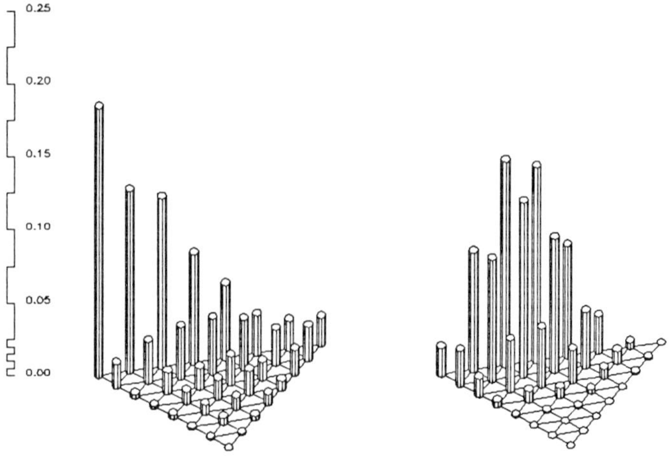

Figure 3. Probabilities of occurrence of weekly sequences obtained drom meteorological data for Toledo (1979-1984). Left, actual readings; right, readings calculated on the basis of unrelated probability.
Number of weeks for which readings were considered: 1909.

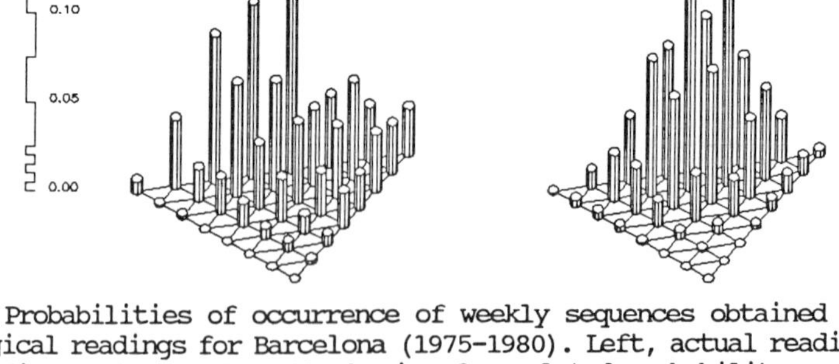

Figure 4. Probabilities of occurrence of weekly sequences obtained from meteorological readings for Barcelona (1975-1980). Left, actual readings; right, readings calculated on the basis of unrelated probability.
Number of weeks for which readings were considered : 943.

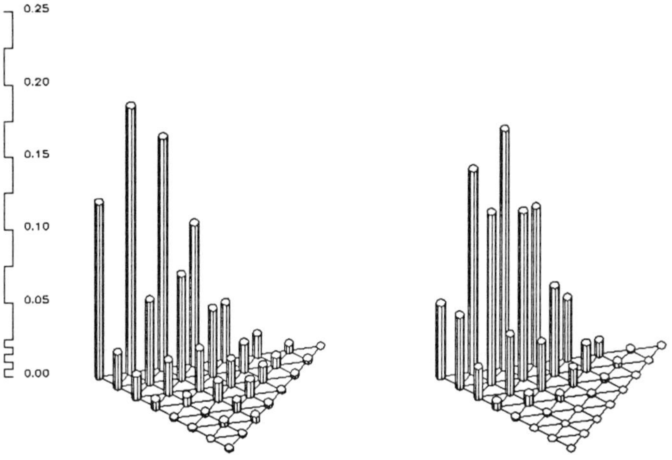

Figure 5. Probabilities of occurrence of weekly sequences obtained from meteorological readings for Valencia (1975-1980). Left, actual readings; right, readings calculated on the basis of unrelated probability. Number of weeks for which readings were considered: 2176

4. DISCUSSION

The results of our investigation on the cities we have studied bears out what what we already stated [3]: that the hypothesis of unrelated probability does not apply.

As regards the different geographical situations, the graphs for Madrid and Toledo (and those of Ciudad Real, which are similiar to those of the latter) are representative of the climate of the central zone of the Iberian peninula, with an abundance of weeks of many clear days, as opposed to other day types.

The data obtained from the cities of the Mediterranean coast (Barcelona, Valencia and Castellón de la Plana) provide us with diagrams, both representative as well as actual values similar to each other but different to those of the cities of the centre of the peninsula, with an abundance of weeks of many medium days, characteristic of the Mediterranean coastal climate.

As regards the effect of the size of cities, our investigations show that the differences between the experimental values and those calculated on the basis of unrelated probability for each day, show a marked dependance on the size of the city in the case of the central zone and somewhat less of a difference, but existant nevertheless, in the case of the coastal zone. This dependance shows that the greater the size of the city, the greater is the difference between the occurrence of weekly sequences observed and the probability caculated on the basis of the unrelated daily behaviour hypothesis.

This leads us to deduce that the size of the city affects not only the average parameters of the climate (temperature, humidity, etc.) but also the dynamic sequences, leading towards a different duration of periods of constant day types.

The results of this investigation and of the multiple aspects which remain to be resolved, lead us to propose that this investigation be continued in order that our undertstanding of meteorology be widened, and that such research should consider in particular the area of periodical behaviour.

REFERENCES

(1) Ph. Duchene-Marullaz: Climatologie locale. Climatologie. REEF-vol. 2,page 52. Ed. CSTB (Paris)

(2) J. Roset, A. Isalgué, R. Serra: (1989) Secuencias climáticas para la evaluación térmica de los edificios. Proc. 22th Bienal Conference R.S.E.F., vol 3, page 294. Palma de Mallorca.

(3) J. Roset, A. Isalgué, R. Serra: (1989) Seqüències Climàtiques en Arquitectura Bioclimàtica. Proc. Jornades científiques de la Mediterrània: Energia, Medi ambient i edificació, page 17. Maó (Menorca).

(4) Meteorological data IER-CIEMAT (Spanish Ministry of Industry).

(5) Guia resumida del Tiempo en España. (1968) Instituto Nacional de Meteorologia, Madrid.

Chapter 2

Concepts and Methods

TERMOGRAPHIC PROJECTIONS FOR EVALUATING ENERGY PERFORMANCES.

Arch. D. NICOLETTI
Dipartimento di Configurazione e
Attuazione dell'Architettura
via Tarsia , 31
80135 Napoli
ITALY

ABSTRACT. One of the most interesting systems of energetic control is the thermographic analysis. This is an effective diagnostic tool providing a lot of information through a not-hurting investigation.
This communication will concern the possibility of building thermic balance control using datas collected through thermographic investigation.

1. Introduction

Studies about building control systems are becoming very important for the determination of the parameters enabling the intervention choices.
One of the most interesting systems of energetic control is the thermographic analysis. This is an effective diagnostic tool providing a lot of information through a not-hurting investigation.
Using thermographic analysis it is possible to control all the thermic irregularity of buildings which provoke particular negative effects on thermic balance.
The new software supplied by "Dipartimento FARE ENEA" and designed by "Matec-Modelli Matematici s.r.l.", allows to verify different energy performances of elements and

S. Alvarez et al. (eds.), Architecture and Urban Space, 673–678.

parameters which contribute to energy demand for each room. The software is based on the so called "free energy utilisation factor method". This software enables the user to determine the energetic needs of heating or cooling of inner spaces of buildings. The user have to determine the over- and the under-value of temperature , with constant values of set-point.

1. Methodology

The contribution of termographic investigation is important to determine energetic requirements both in terms of quality and quantity.
This kind of investigation gives to the software information on three different levels:

a- building characteristics
b- degradation conditons
c- thermo-physical characteristics of used materials.

a - For what concerns the building characteristics, the contribution regards the behaviour of elevations in respect to the "heat quantity transmitted across the walls". This method offers an interesting map of dispersions and disorganic heat transmissions of wall surfaces.

b - Concerning the degradation condition of the building,the termogaphic method allows a precise definition of zones having a particular degradation conditions both in terms of environment (umidity, thermic bridges) and technology (inflations, disconnectedness, fissures,...).

c - In this study, the thermo-phisical characteristics of used material are analyzed for what concerns the definition of evaluating criterias for the losses of the materials in terms of performances.In these parts of the buildings that are characterized by dramatic conditions of degadation the investigation gives more information about the "heat quantity capacity of elevation walls", and about the "heat quantity capacity of solar radiations of the structure".

3 Conclusion

This investigation is still in progress,and it will give specific indications about several parameters which will contribute to define the energetic balance of buildings.
With the following images,we are going to illustrate an application of this semplificated simulation software on an ancient convent.
On this ancient convent we are actually doing the thermographic analysis.

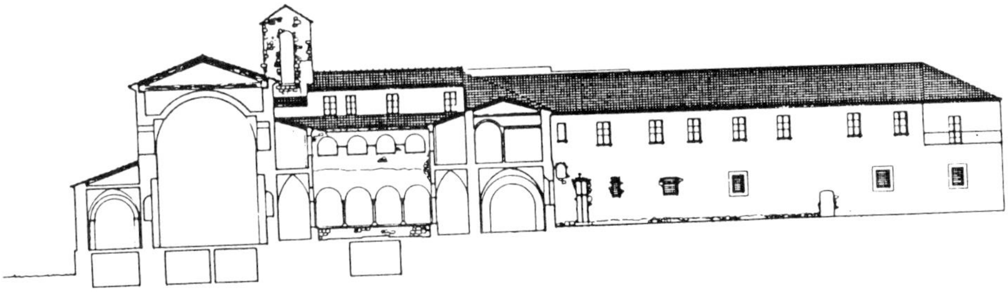

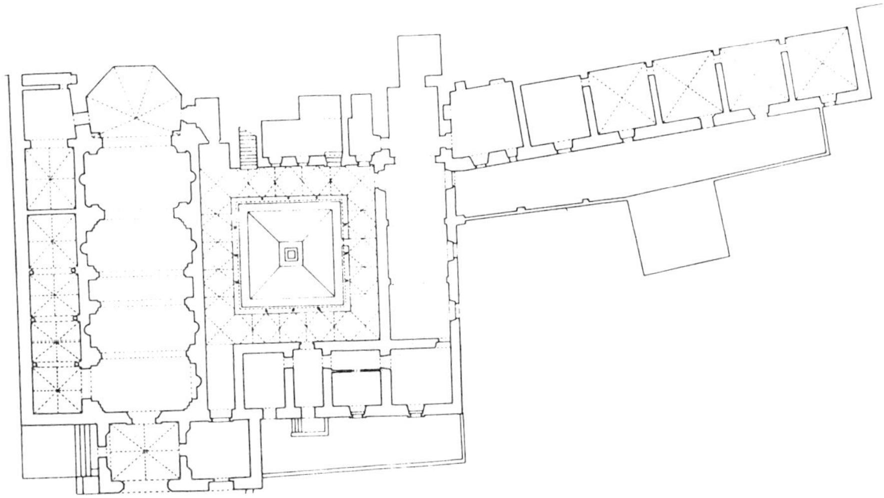

CALCOLO DEL FABBISOGNO ENERGETICO DI UN EDIFICIO

Metodo del Fattore di Utilizzazione degli Apporti Gratuiti di Energia

Nome del file di input : CELLA 1

R I S U L T A T I D E I C A L C O L I

Calcolo Trasmittanza media e Coefficienti di Riduzione

PARETE SUD	s =	14.82	k =	2.459
FINESTRA SUD	s =	1.80	k =	5.708
TELAIO FINESTRA SUD	s =	0.36	k =	2.613
PORTA	s =	2.76	k =	2.613
PARETE OVEST ESTERNA	s =	6.84	k =	2.042
SPIGOLO S-W	s =	3.80	k =	0.050
PONTE TERM SOFFITTO	s =	15.40	k =	0.060
PONTE TERMICO PORTA	s =	7.00	k =	0.230
PONTE TERMICO FINEST	s =	5.60	k =	0.230

Trasmittanza media (km) [W/m2/K] = 0.826
Coefficiente di riduzione Fc = 0.854
Coefficiente di riduzione Fr = 0.756

R I S U L T A T I D E I C A L C O L I

Scambi Energetici attraverso Pareti / Elementi

	Nome Elemento		ang.	sup.	trasm.	tsa	tsa'	Ea	Em+Ef
1	PARETE SUD	S	180	14.82	2.459	12.94	14.51	163389	-1163
2	FINESTRA SUD	S	180	1.80	5.708	8.98	11.83	71907	-121325
3	TELAIO FINESTRA SUD	S	180	0.36	2.613	8.98	11.83	6583	-2
4	PORTA	N	0	2.76	2.613	8.98	11.83	50469	-13
5	PARETE OVEST ESTERNA	W	270	6.84	2.042	8.98	11.83	97748	-1082
6	SPIGOLO S-W		0	3.80	0.050	8.98	11.83	1330	0
7	PONTE TERM SOFFITTO		0	15.40	0.060	8.98	11.83	6466	0
8	PONTE TERMICO PORTA		0	7.00	0.230	8.98	11.83	11267	0
9	PONTE TERMICO FINEST		0	5.60	0.230	8.98	11.83	9014	0
	Totale							418173	-123584

R I S U L T A T I D E I C A L C O L I

Calcolo Massa Specifica della Struttura

PARETE SUD	s =	14.82	dens. sup. =	1531.0	
FINESTRA SUD	s =	1.80	dens. sup. =	7.5	
TELAIO FINESTRA SUD	s =	0.36	dens. sup. =	28.8	
PORTA	s =	2.76	dens. sup. =	28.8	
PARETE OVEST	s =	7.60	dens. sup. =	900.0	non disperdente
PARETE EST	s =	14.44	dens. sup. =	900.0	non disperdente
PARETE NORD	s =	11.68	dens. sup. =	900.0	non disperdente
SOFFITTO	s =	14.82	dens. sup. =	234.0	non disperdente
PAVIMENTO	s =	14.82	dens. sup. =	654.0	non disperdente
PARETE OVEST ESTERNA	s =	6.84	dens. sup. =	1578.0	

Massa Specifica 3734.168
Teta / Pi 2.993
Omega 0.190
Eta 0.940
Consumo .... 342196 Wh per RISCALDAMENTO

Battere un tasto per continuare

```
                R I S U L T A T I   D E I   C A L C O L I

  Ea (QUANTITA' DI CALORE TRASMESSA ATTRAVERSO L'INVOLUCRO
      DOVUTA ALLA DIFFERENZA DI TEMPERATURA INTERNO/ESTERNO)
+ Er (QUANTITA' DI CALORE DOVUTA ALLA RADIAZIONE SOLARE
      SULLE PARETI OPACHE) ................................... 418173 Wh

  Em (QUANTITA' DI CALORE DOVUTA ALL'ACCUMULO NELLE STRUTTURE)
+ Es (QUANTITA' DI CALORE DOVUTA ALL'ACCUMULO DELLA RADIAZIONE
      SOLARE NELLE STRUTTURE) ................................  -2259 Wh

  Ef (QUANTITA' DI CALORE DOVUTA ALLA RADIAZIONE SOLARE
      SULLE SUPERFICI VETRATE) ............................. -121325 Wh

  Ev (QUANTITA' DI CALORE DOVUTA ALLA VENTILAZIONE) ..........  80785 Wh
  Ei (QUANTITA' DI CALORE DOVUTA ALLE SORGENTI INTERNE) ....... -43152 Wh

  TETA     (TOTALE ENERGIA USCENTE) ......................... 498958 Wh
  PI GRECO (TOTALE ENERGIA ENTRANTE) ....................... -166736 Wh

  TETA / PI GRECO ............................................  2.993
  ETA ........................................................  0.940

  CONSUMO .................................................... 342196 Wh
Devo salvare ?
```

CONSUMI STAGIONALI PER RISCALDAMENTO

Mese	En. Uscente	En. Entrante	Rapporto	Eta	Consumo
OTTOBRE	14778	-344867	0.043	0.007	12524
NOVEMBRE	175393	-178931	0.980	0.370	109237
DICEMBRE	361518	-147740	2.447	0.888	230397
GENNAIO	499219	-166843	2.992	0.940	342359
FEBBRAIO	400799	-180603	2.219	0.851	247120
MARZO	327257	-250107	1.308	0.545	191016
APRILE	148546	-293213	0.507	0.136	108809
MAGGIO	26655	-454024	0.059	0.009	22497
Totale					1263959 Wh

A:\>

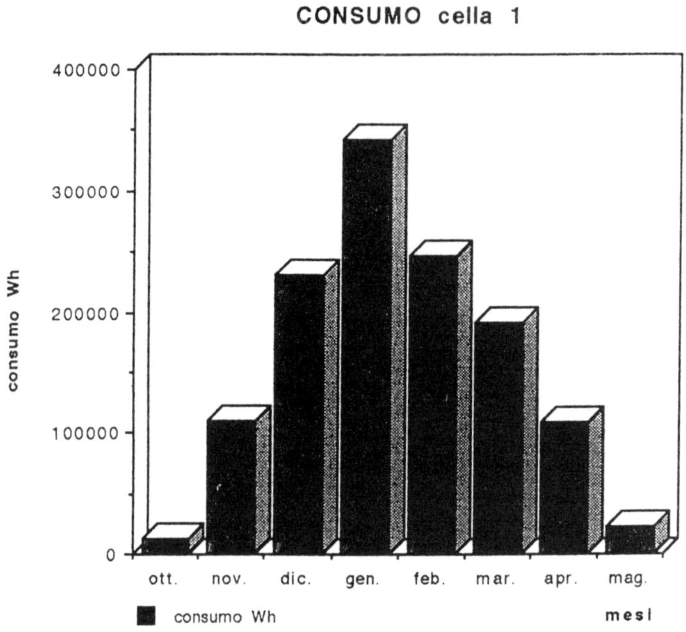
CONSUMO cella 1
400000
300000
200000
100000
0
consumo Wh
ott.
nov.
dic.
gen.
feb.
mar.
apr.
mag.
consumo Wh
mesi

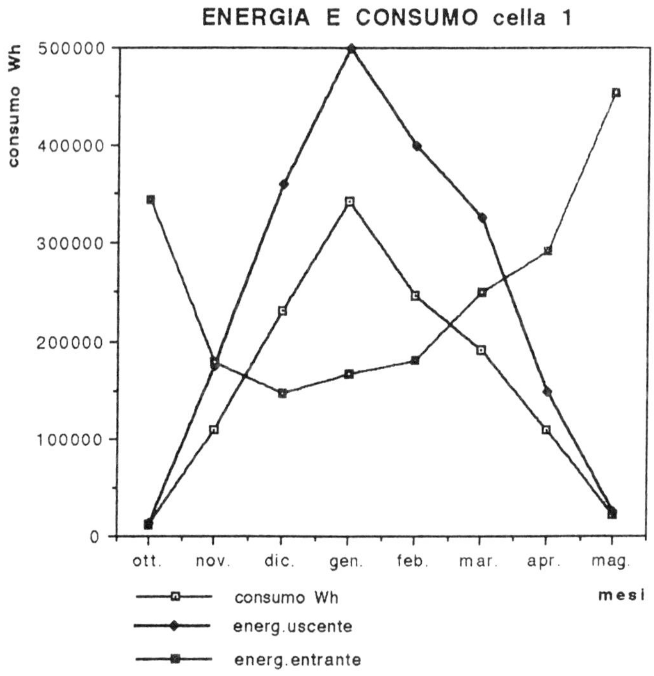
ENERGIA E CONSUMO cella 1
500000
400000
300000
200000
100000
0
consumo Wh
ott.
nov.
dic.
gen.
feb.
mar.
apr.
mag.
consumo Wh
energ.uscente
energ.entrante
mesi

ENERGY CONSUMPTION AND WALL OPENINGS IN COMMERCIAL BUILDINGS

Juan C. Klainsek
Depto. de Construcción y Tecnología Arquitectónicas
E.T.S.Arquitectura-Universidad Politécnica de Madrid
Av. Juan de Herrera, 4 - 28040 Madrid
Spain

ABSTRACT. A proper building design can reduce the amount of operational energy required and windows have an important role in this process, since they allow -if properly designed-, to make use of solar radiation as much and long as possible, and to provide efficient insulation. Due to the fact that non-residential buildings have large glass areas, an office building has been chosen to carry out the analysis.Two different window elements have been considered: a) shape:continuous horizontal window and discontinuous vertical window, both having the same surface; b) type: colorless, bronze colored and reflectant, each of them for insulating glass with different gap thickness, and for single glass. The results of this analysis show the building energy consumptions in summer and winter,to be considered when designing a building.

INTRODUCTION.

An important percentage of the energy produced from processing fossil and fissile fuels is consumed by heating, cooling and illuminating the buildings; this consumption can be considerably reduced if building technology and conditioning techniques complement each other. With regard to this, the choice of both the shape and type of the wall openings is an important decision at the initial design stage of the building. As in non-residential buildings the wall openings influence greatly the lighting, cooling and heating loads, the shape and type of glass areas in an office building have been studied in this paper.

The Weighting Factor Method has been used. This method, which was first introduced by Mitalas and Stephenson, accounts for the important parameters that affect the energy flow of the building. It allows space heat gains at constant space temperature to be determined from a physical description of the building, the ambient weather conditions and internal load profiles. A computer program has been developed

S. Alvarez et al. (eds.), Architecture and Urban Space, 679–684.

to simulate this office building energy needs in a climate region: Madrid.

The purpose of this paper is to study the incidence that certain shapes and types of glass can have on the energy performance of a commercial building.

Base case building.

An office building with 10 floors, located in Madrid: 40m x 26m floor; 37m height, which corresponds to a gross surface per floor of 1036 m^2, useful conditioned of 795 m^2.This building has the most disadvantageous situation, that is, the largest façades face towards the east and west.

The winter season includes the months from November to March, the rest months being included in the summer season.

A constant interior temperature of 20 ºC and RH 50% in winter, 25 ºC and RH 55% in summer is assumed in the calculations; the exterior temperature is variable, so the estimations are carried out hourly.

With regard to the exterior conditions, meteorological variables are generated hourly from reliable meteorological data, and an auxiliary program is used.

The heat transmission coefficients K (U-value) of opaque surfaces are: -horizontal- roof Kr=0.43 W/m^2 K, light color; floor levels above open space Kf=0.66 W/ m^2 K.-Vertical traditional wall composed of seven layers, total thickness 23.5cm, insulating material 4cm, 244.6 kg/m^2 and Kw=0.61 W/m^2 K, and four layer curtain-wall, total thickness 5.3cm, insulating material 5cm thick, 89.6 kg/m^2 and Kw=0.42 W/m^2 K both medium color. Metallic windows with different sorts of glass, which are studied in this paper will be described.

Glass areas will not have any shading devices, except those shadows produced by setback.

Results from the energy consumption calculations are obtained per useful or net surface unit, taking into account that the free height in all floors is 2.80m.

Internal loads are fixed by: occupation 55 pers./floor, equivalent approximately to 14 m^2/pers.; ventilation 2000 m^3/h floor, that is 2.5 m^3/h m^2; and illumination 7.95 kW/floor or rather 10 W/m^2.

Glass areas.

The glassing used in this paper has the following characteristics: colorless, bronze colored-filtering and reflectant, each of them single 6mm with Kg (U-value) of 5.8 W/m^2 K; insulating 6+6mm, gap of 6mm, Kg=4.0 W/m^2 K; and insulating 6+6mm, gap of 12mm, Kg=3.7 W/m^2 K. Each of this glassing is affected by its corresponding corrective coefficient of the solar gain factor (c.c.), which varies from 0.48, for the reflectant insulating, to 0.94 for the single colorless.

ANALYSIS OF RESULTS.

For the base case building with windows of 1.30m height, all round the building perimeter and using the glassing above, the energy consumptions (e.c.) per conditioned surface in summer and winter are calculated. Later, the glass shapes are transformed from horizontal continuous (h.c.) to vertical discontinuous (v.d.), with the same façade surface than the previous ones and the e.c. are calculated. To carry out the change, the glassing height is 2.10m from floor to lintel

In all the cases considered the traditional wall and curtain-wall have been used.

In the first group, Table 1-Fig. 1, for summer and with traditional wall, the values obtained are analysed. The e.c. per conditioned useful surface unit (kWh/year m^2), are represented on the diagram ordinate axis, as a result of the three different types of glassing (colorless, bronze colored filtering and reflectant), and of their corresponding glassing shape transformations.

In Table 1 (T1), it is shown that the e.c. differences between h.c. windows and v.d. windows, with the same surface (Δec_1), for the three types of glassing, change from 0.39% to 0.75%. The h.c. ones have always the least e.c.

For h.c. ones and with regard to the single glass, for each type of glassing (Δec_2), the least variation occurs between the reflectant single glass (s.g.) and the reflectant insulating glass (i.g.) with a gap of 6mm (-7.34%); and the greatest variation between the colorless s.g. and the colorless i.g., gap=12mm, (-9.25%).

Still for h.c. ones and with regard to the colorless glass, for bronze colored glass and for each glass constitution -s.g. and i.g., gap=6/12mm-, (Δec_3), the s.g. shows the least variation (-8.54%), and the i.g. gap=6mm the highest (-12.41%).

The colorless and reflectant glass (Δec_4) are analysed in the same way; the least variation corresponds to the i.g. gap=12mm (-14.34%), and the greatest to the s.g. (-15.81%).

Approach to values shown in Tables T2, T3, T4 are made in a similar way.

CONCLUSIONS.

It would be necessary to discuss the different types of fuel, conditioning systems -together with their operation and maintenance-, and ways of using the buildings, so that a technical economic evaluation could be finally carried out. This has not been made in this paper, so the following conclusions are given separately for summer and winter.

With regard to the glass shapes, the vertical discontinuous window has an e.c. greater than the horizontal continuous window, except in winter with curtain-wall and for all kinds of colorless glass studied, when no difference is noticed. Thus, an important factor to take into account is the glass setback in relation to the wall exterior surface. As a result, the differences of the values obtained are not of great significance.

As for the types of glassing and for horizontal continuous windows, the least e.c. in summer, for both traditional wall and curtain-wall (22.76/23.76 kWh/year m^2) is given by the reflectant i.g., gap=12mm.

In winter, with the same sort of walls (16.22/14.83 kWh/year m^2) the lowest e.c. corresponds to the colorless, also i.g., gap=12mm.

Considering the geographical location and the typology of the building studied, and having obtained more significant e.c. in summer, the reflectant insulating glass should be adopted.

NOMENCLATURE.

c.c.: corrective coefficient of solar factor gain. Glassing without blind or shading
e.c.: energy consumption
h.c.: horizontal continuous windows
i.g.: insulating glass
s.g.: single glass
v.d.: vertical discontinuous windows

REFERENCES.

ASHRAE Handbook 1989 Fundamentals. SI Edition. American Society of Heating, Refrigerating and Air Conditioning Engineers, Atlanta, Georgia.

Bennett, J.R. et al. (1988). "A Study of the Thermal Characteristics and Potential for Energy Conservation at Four London Office Buildings". Building and Environment. Vol. 20, Nº 2, pp. 83-94.

Klainsek, J. (1988). "Energy Saving in Office Building Design and Retrofitting". Energy and Buildings for Temperate Climates. Plea 88. Pergamon Press, pp. 173-178.

McKinley, A.D., and Mitalas, G.P. (1983). "Room Thermal Transfer Functions, Computer Programs to Calculate Z-Transfer Function Coefficients for Rooms". National Research Council of Canada, pp. 1-3.

Selkowitz, S. (1986). "Smart Windows". Glass Magazine. Windows and Daylighting Group, Lawrence Berkeley Laboratory, University of California, pp. 86-91.

Sullivan, D. (1987). "The Influence of Glazing Selection on Commercial Building Energy Performance in Hot and Humid Countries". Lawrence Berkeley Laboratory Report, BS-247.

TABLE 1. Energy Consumptions in Summer-Traditional Wall

Type glass	s. g.			i.g. gap=6mm			i.g. gap=12mm		
	Cont.	Disc.	Δec1	Cont.	Disc.	Δec1	Cont.	Disc.	Δec1
Cless	29.28	29.50	+0.75	26.67	26.84	+0.64	26.57	26.74	+0.64
Δec2				-8.91			-9.25		
Bron.	26.78	26.94	+0.60	23.36	23.47	+0.47	23.28	23.38	+0.43
Δec2				-12.77			-13.07		
Refl.	24.65	24.77	+0.49	22.84	22.94	+0.44	22.76	22.85	+0.39
Δec2				-7.34			-7.67		
Δec3	-8.54			-12.41			-12.38		
Δec4	-15.81			-14.36			-14.34		

TABLE 2. Energy Consumptions in Winter-Traditional Wall

Type glass	s. g.			i.g. gap=6mm			i.g. gap=12mm		
	Cont.	Disc.	Δec1	Cont.	Disc.	Δec1	Cont.	Disc.	Δec1
Cless	22.48	22.54	+0.27	17.22	17.26	+0.23	16.22	16.26	+0.25
Δec2				-23.40			-27.85		
Bron.	23.79	23.85	+0.25	18.85	18.89	+0.21	17.82	17.86	+0.22
Δec2				-20.76			-25.09		
Refl.	24.98	25.03	+0.20	19.10	19.14	+0.21	18.06	18.10	+0.22
Δec2				-23.54			-27.70		
Δec3	+5.83			+9.47			+9.86		
Δec4	+11.12			+10.92			+11.34		

TABLE 3. Energy Consumptions in Summer-Curtain-Wall

Type glass	s. g.			i.g. gap=6mm			i.g. gap=12mm		
	Cont.	Disc.	Δec1	Cont.	Disc.	Δec1	Cont.	Disc.	Δec1
Cless	30.81	30.93	+0.39	28.04	28.13	+0.32	27.95	28.04	+0.32
Δec2				-8.99			-9.28		
Bron.	28.05	28.13	+0.28	24.41	24.46	+0.20	24.33	24.38	+0.20
Δec2				-12.98			-13.26		
Refl.	25.71	25.77	+0.23	23.84	23.89	+0.21	23.76	23.81	+0.21
Δec2				-7.27			-7.58		
Δec3	-8.96			-12.96			-12.95		
Δec4	-16.55			-14.98			-14.99		

TABLE 4. Energy Consumptions in Winter-Curtain-Wall

Type glass	s. g.			i.g. gap=6mm			i.g. gap=12mm		
	Cont.	Disc.	Δec1	Cont.	Disc.	Δec1	Cont.	Disc.	Δec1
Cless	20.93	20.93	+0.00	15.79	15.79	+0.00	14.83	14.83	+0.00
Δec2				-24.55			-29.14		
Bron.	22.19	22.20	+0.04	17.35	17.36	+0.06	16.35	16.36	+0.06
Δec2				-21.81			-26.32		
Refl.	22.36	22.38	+0.09	17.59	17.60	+0.06	16.58	16.59	+0.06
Δec2				-21.33			-25.85		
Δec3	+6.02			+9.88			+10.25		
Δec4	+6.83			+11.40			+11.80		

Δec1: between continuous w. and discontinuous w. [%]
Δec2: among continuous w. always with relation to s.g. [%]
Δec3: colorless-bronze colored glass [%]
Δec4: colorless-reflectant glass [%]

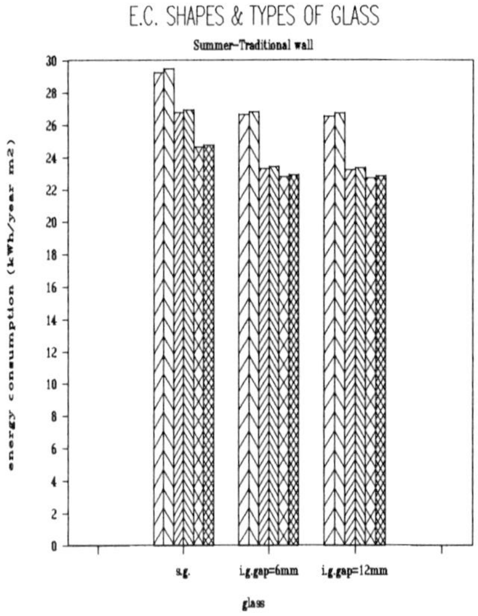

Fig.1

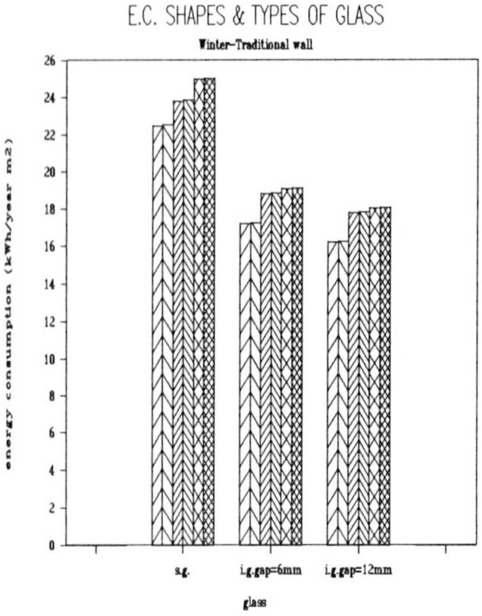

Fig.2

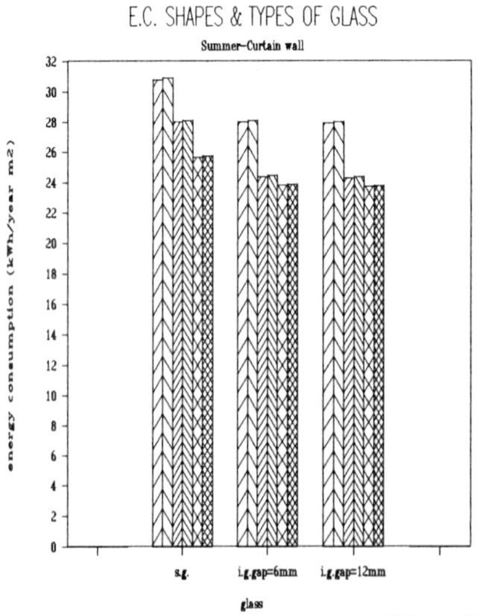

Fig.3

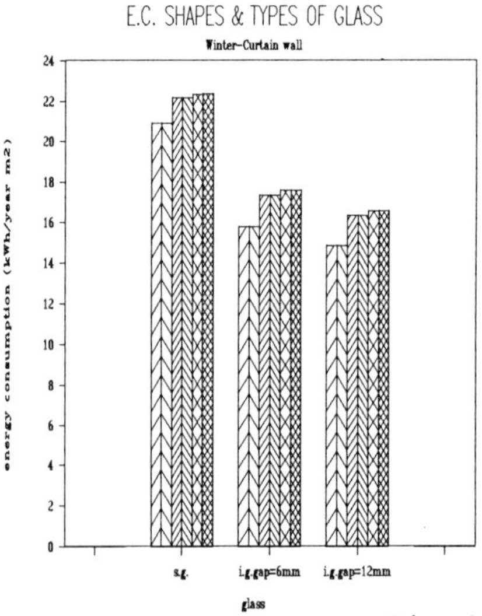

Fig.4

h.c. window; colorless glass
v.d. window; colorless glass
h.c. window; bronze colored glass
v.d. window; bronze colored glass
h.c. window; reflectant glass
v.d. window; reflectant glass

PASSIVE SOLAR ENERGY AS A FUEL 1990–2010

J. V. DOGGART
The ECD Partnership
11-15 Emerald Street
London WC1N 3QL
United Kingdom

ABSTRACT. The purpose of this study was to assess the current and potential future use of passive solar energy in buildings in Europe up to the year 2010. The study was commissioned by DG XII of the Commission of the European Community and a summary report is available. Data on the use of passive solar energy for heating, and passive design for the avoidance of overheating and the provision of daylighting, were collected from published sources and experts in all European countries. The study demonstrates that there is considerable potential for using passive solar design for heating, cooling and lighting in all countries of Europe and that the current equivalent saving of 230 million tonnes of CO_2 could be increased by 103 million tonnes by 2010 if positive action were taken to promote passive design in buildings.

1. Purpose of the Study

The study set out to answer three main questions:-

- How much solar energy is used currently in buildings in the European Community?
- How much solar energy could be used in the future - in the years 2000 and 2010?
- What could be the consequent reduction in pollution, particularly carbon dioxide (CO_2)?

A summary report is available and background material and research details are given in a full report prepared for the Commission of the European Communities (CEC).

2. Using Passive Solar Design

2.1 PASSIVE DESIGN

This study looks at the use of passive solar design for heating, cooling and lighting dwellings and non-domestic buildings, throughout the European Community.

Heating by passive solar energy involves entry of solar radiation into the building either

S. Alvarez et al. (eds.), Architecture and Urban Space, 685–690.

directly through the windows or indirectly through sunspaces, wall panels, etc., and its use or storage inside the building with little or no use of fans or pumps.

Cooling by passive solar design involves shading, natural ventilation and the use of naturally cooled air (from the ground or evaporative cooling) to prevent the building overheating.

Daylighting by passive design means allowing natural light to enter deep into a building and ensuring that artificial lighting is only used when the natural lighting is insufficient.

Good design optimizes the utilization of solar energy and can achieve large savings in the amount of conventional fuel required for heating, cooling and lighting.

2.2 SOLAR DESIGN IN NEW AND EXISTING BUILDINGS, IN NORTH AND SOUTH

All buildings make use of some solar gain for heating and lighting and this use can be increased at any stage in a building's life. The simplest and most effective way is at the design stage for a new building, but some solar features can also be introduced with significant effects during major refurbishments. In addition, it is possible to add certain solar measures onto an existing building at any time.

Passive solar design can be applied in all climates, though naturally in different ways. In the northern latitudes where winters are long and cold most of the available solar gains can be used to reduce consumption of fossil (and nuclear) fuel for heating. In Mediterranean areas, passive design is used to best effect to reduce or prevent overheating and consequent use of air-conditioning; in addition, the small heating loads which exist in these regions can be minimized. Daylighting design is applicable in all climates.

3. Methodology for Estimating Solar Energy Use

3.1 THE "SPREADSHEET" AND INFORMATION SOURCES

The method used was to build up the total usage of solar energy in the European Community for all the relevant sectors in all the various countries. Computer "spreadsheets" were developed - one for each country - and available information, estimates and projections inserted. Much of the information required was simply not available, so estimates were made on the basis of what did exist, comparison with other countries and expert opinion.

3.2 BASE AND "TECHNICAL POTENTIAL" SOLAR CONTRIBUTIONS

The "base case" solar energy usage was defined as the contribution of passive solar design to heating and cooling loads if no action is taken to increase the usage in the future.

The "technical potential" solar contributions to heating, cooling and lighting were defined as the maximum achievable overall in the years 2000 and 2010. They take into account technical limitations (such as the overshading provided by neighbouring buildings, etc.) but otherwise assume that all buildings would be constructed or refurbished according to solar principles.

3.3 DWELLINGS

Using solar energy to heat houses and other dwellings is the best researched and documented area of passive solar design. The current solar usage was built up for each country from the number of dwellings, their gross heating demand and the percentage solar contributions as determined from monitoring projects. New build, refurbishment and demolition rates were then used together with the solar contributions monitored in new and refurbished solar dwellings to estimate the technical potential.

The need for cooling in dwellings was only considered for countries in southern latitudes. Passive or natural cooling design was estimated in this study to currently provide half the demand in dwellings which have no air-conditioning but where cooling is needed.

In dwellings, daylighting was not considered as an area where solar energy was likely to make a contribution to fuel savings.

3.4 OFFICES, FACTORIES, SCHOOLS, ETC.

The non-domestic sector has far more diverse building types and less information is available on them. Broad estimates were made mainly on the basis of current energy usage and predictions of increases in energy consumption available from CEC research. The contributions which solar design might make to heating, cooling and lighting for the years 2000 and 2010 were built up.

3.5 REDUCTION IN POLLUTION FROM SOLAR DESIGN

The reduction in CO_2, SO_2 and NO_X pollution and nuclear waste for each country was calculated from the breakdown of the different types of fuel in the country and the total equivalent savings resulting from the solar design.

4. Results of Study – Solar Energy Use and Pollution Reduction

4.1 USE OF SOLAR ENERGY COMPARED WITH OTHER FUELS IN 1990

Passive solar design at present supplies the European Community with the equivalent of 96 million tonnes of oil equivalent (mtoe) of primary energy per annum. This is 9% of the total fuel used in the European Community (Figure 1). It is larger than the amount of coal directly burnt for heating, which amounts to 6% of the total. In the building sector alone (i.e., excluding industrial process heat and transportation energy use) solar design supplies 13% of the total (Figure 2). Thus it is clear that solar energy is already a very important fuel in Europe.

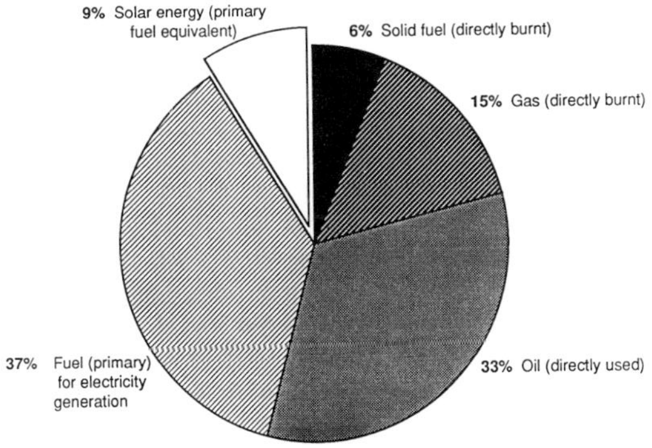

Figure 1. Total use of fuel in the European Community in 1990 (including transport and industrial process use)

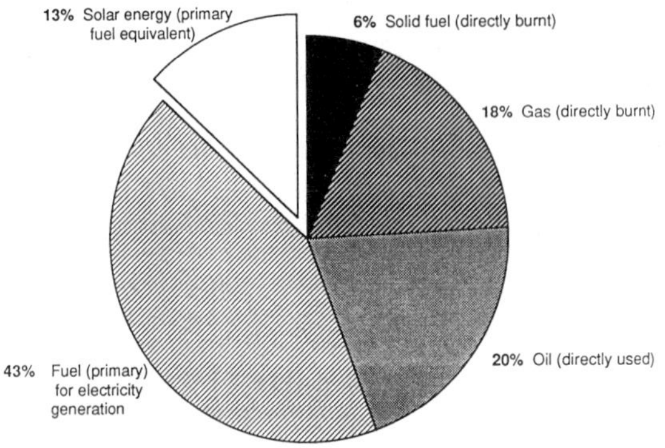

Figure 2. *Use of fuel in dwellings and non-domestic buildings in the European Community in 1990 (excluding transport and industrial process use)*

4.2 SOLAR ENERGY USE IN THE FUTURE

If no specific action is taken to promote the use of passive solar design in the future, a small rise of 8% above the 1990 levels is predicted by the year 2000. It is expected that this will fall back to 6% above the 1990 figure by the year 2010. The reduction is due to the lower heating demand resulting from higher insulation standards.

If, however, specific action is taken, there is the potential for a great increase in the use of solar energy. By the year 2000, the overall amount could be increased by 27% of the 1990 usage - an increase of 26 mtoe. By 2010, the increase could be 54% or 52 mtoe per annum.

These potentials are the maximum technically possible and include allowances for buildings which cannot be oriented optimally, etc. In practice, the potential will be reduced by such factors as low take-up rates and poor design and operation.

4.3 WHERE MOST SOLAR ENERGY IS USED

Figure 3 shows the increase in the contribution from solar energy in each of the five categories (domestic heating, non-domestic heating, domestic cooling, non-domestic cooling and non-domestic lighting) for the European Community as a whole from 1990 to the years 2000 and 2010.

Of the five categories, heating currently receives the greatest contribution from solar energy in absolute terms (it was nearly 76 mtoe per annum in 1990). However, the largest increase is forecast for the use of daylighting in non-domestic buildings; the increase amounts to nearly 18 mtoe per annum by 2010. The predicted growth for solar heating in non-domestic buildings is also large with an increase of over 17 mtoe per annum in 2010 compared with 1990. This reflects the relatively high rates of new construction and refurbishment in this sector compared to housing.

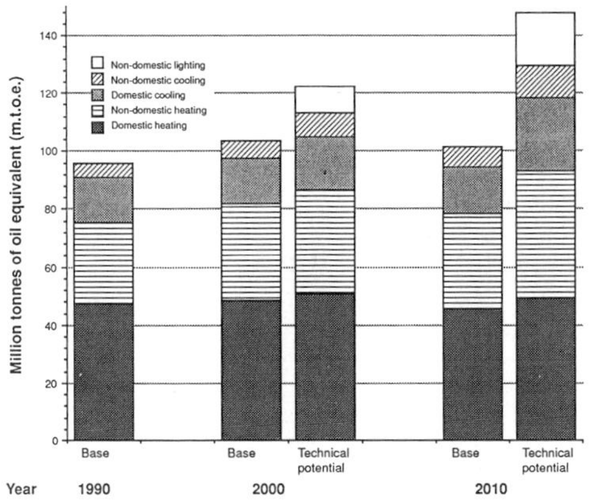

Figure 3. Contribution of solar energy to heating, cooling and lighting buildings in the European Community in 1990, 2000 and 2010

Note: Base Case = no action taken to increase solar usage.
Technical potential = all potential solar usage exploited.

The overall contribution of passive solar design to reduction of cooling loads is smaller than its contribution to the heating requirements of buildings. This is due to the small need for cooling within Europe as a whole. Within individual countries, however, the picture is very different. For instance, the northern countries of Denmark, Ireland and the Netherlands have negligible cooling demand whereas Greece, Spain, Italy and Portugal achieve 50% of their total technical potential solar contribution through passive cooling design. Taking the Community as a whole, an increase of 9 mtoe per annum in the contribution from natural cooling in dwellings is possible by 2010 (compared with 1990) - a significant growth rate.

Whilst the largest use of solar energy, in absolute terms, undoubtedly comes from the countries with the largest populations, the greatest growth rates for passive solar design are likely to come from countries in southern latitudes where there is a potential for increasing the contributions made by solar design to heating, lighting and cooling.

4.4 REDUCING ATMOSPHERIC POLLUTION BY USE OF SOLAR DESIGN

The reduction in use of fossil fuels brought about by the current usage of passive solar energy prevents 229 million tonnes of CO_2 being emitted to the atmosphere per annum. This represents nearly 17% of the CO_2 which would be produced if the buildings in the Community experienced no solar contribution. Emission of 1.3 million tonnes of SO_2 and 0.56 million tonnes of NO_x (oxides of nitrogen) is also avoided by use of solar design. The potential cuts in pollution which could arise from use of passive solar design throughout Europe in the future are shown in Figure 4.

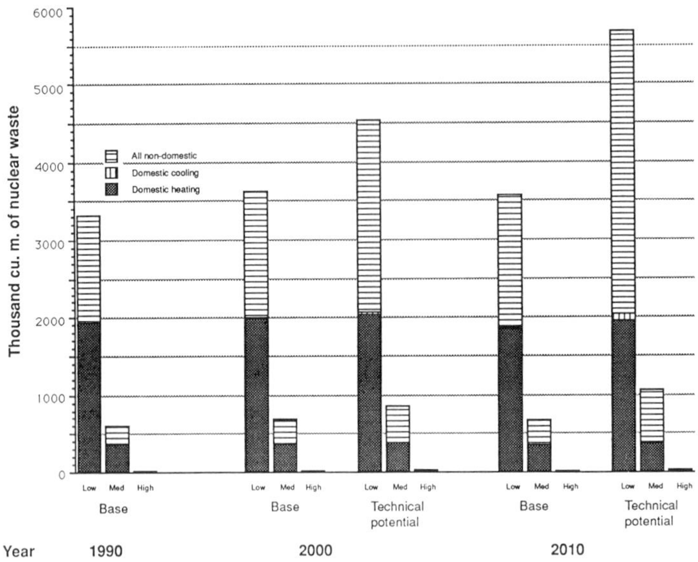

Figure 4. *Savings in atmospheric pollution arising in the European Community through use of passive solar design*

Notes: 1. Base case = no action taken to increase solar usage.
Technical potential = all potential solar usage exploited.
2. SO_2 and NO_x reductions are largely due to implementation of legislation to reduce power station emissions of these gases.

If the technical potential solar contributions were achieved, the amount of CO_2 saved per annum by the year 2000 would rise to 272 million tonnes, an increase of 19% compared with 1990. By 2010, the savings would be 332 million tonnes a year, an increase of 45% over 1990.

In the future, legislation will be implemented to reduce emission of SO2 and NO_x (principally "acid rain") from power stations. If the targets established by this legislation are met, there will be less scope for reduction of emissions of these gases from use of passive solar design.

Three countries - Germany, Italy and the UK - currently contribute two thirds of the total reduction in CO_2 emissions due to solar design. Because of its large nuclear programme, France contributes only 9% to the savings made in the Community as a whole. The picture will remain broadly unchanged up to 2010, assuming the technical potential is achieved.

The increased savings in CO_2 emissions possible by 2010 vary greatly between countries, from 19% in Belgium to 104% in Greece. The differences roughly reflect the changes in solar contributions, the countries in southern latitudes generally showing the highest percentage savings.

4.5 REDUCING NUCLEAR WASTE BY USE OF SOLAR DESIGN

At present, the production of nuclear waste is 18% below what it would be if passive solar design made no contribution to heating, cooling and lighting the buildings in the European Community. Although seven countries use nuclear power, France is the major producer by far and contributes around 70% of the 1990 savings and 57% of the 2010 potential savings. Overall, solar design could, by 2010, increase savings of nuclear waste by around 70% of the 1990 level.

USERS RESPONSE AND ENERGY CONSUMPTION PROFILE IN BIOCLIMATIC ARCHITECTURE

J.A.CUSIDÓ, J.JORGE, J.PUIGDOMÈNECH
Escola Tècnica Superior d'Arquitectura del Vallès
P.O.Box 508, 08220 Terrassa

J.J.ESCOBAR, J.ESTEVE, A.MITJÀ
Institut Català de l'Energia, Generalitat de Catalunya
Diagonal 514, 2nd floor, 08006 Barcelona

ABSTRACT. In the energetic audits of full scale buildings there is an important energetic contribution with a great degree of uncertainity: it is the internal gains due to the occupants and the management that they make of their own bioclimatic buildings. So far, the users of such kind of buildings in Catalonia were persons involved themselves in these technologies. It is not the case of an experimental field of 20 bioclimatic dwellings placed in Begues (40 Km far from Barcelona), in a closed urban nucleus. Significant data about occupation profiles were collected in similar dwellings (four typologies on the same architectural program), which allows us to compare energy consumption results according to the building use, for a generic sample of non specially motivated users.

1. INTRODUCTION

The main objective of this study deals with the analysis of a set of 20 dwellings placed in Begues (Baix Llobregat), 30 Km far from Barcelona. The height over sea level is 400 m, which implies a decrease of 2.5 ° C under the average monthly temperature of its climatic zone (15 ° C) and a considerable increase of rain regime. Included in the usual monitoring tasks of this set of dwellings occupacy profile and users response have been carried out by means of inquiries, similar to what has been done in other countries [1,2].

In this way our study tries to show the users response on passive solar houses and consequently the energy management of their own dwellings as well as to extract from this response the effective validity of internal gains evaluation used in the substractive analysis of energetic behaviour of bioclimatic buildings:

Solar Savings = Heat Requirements - Auxiliar Energy

Heat Requirements = Useful Load - Internal Gains

Obviously the results here analyzed have not a direct extrapolation to other climatic zones, but architectural designs and cultural scope can supply an idea about possible deviations of standard behaviour usually considered in these kind of analysis. In other words, we can try to determine what the error range by means of the users intervention when estimating solar savings.

2. THE BIOCLIMATIC FIELD HOUSING OF BEGUES

The 20 dwellings are grouped in 5 different blocks and every house is composed of a ground floor and a first floor. The garages are built separetelly in the North zone of the urbanistic site (Figure 1). Exterior walls of North facades and exterior lateral

S. Alvarez et al. (eds.), Architecture and Urban Space, 691–696.

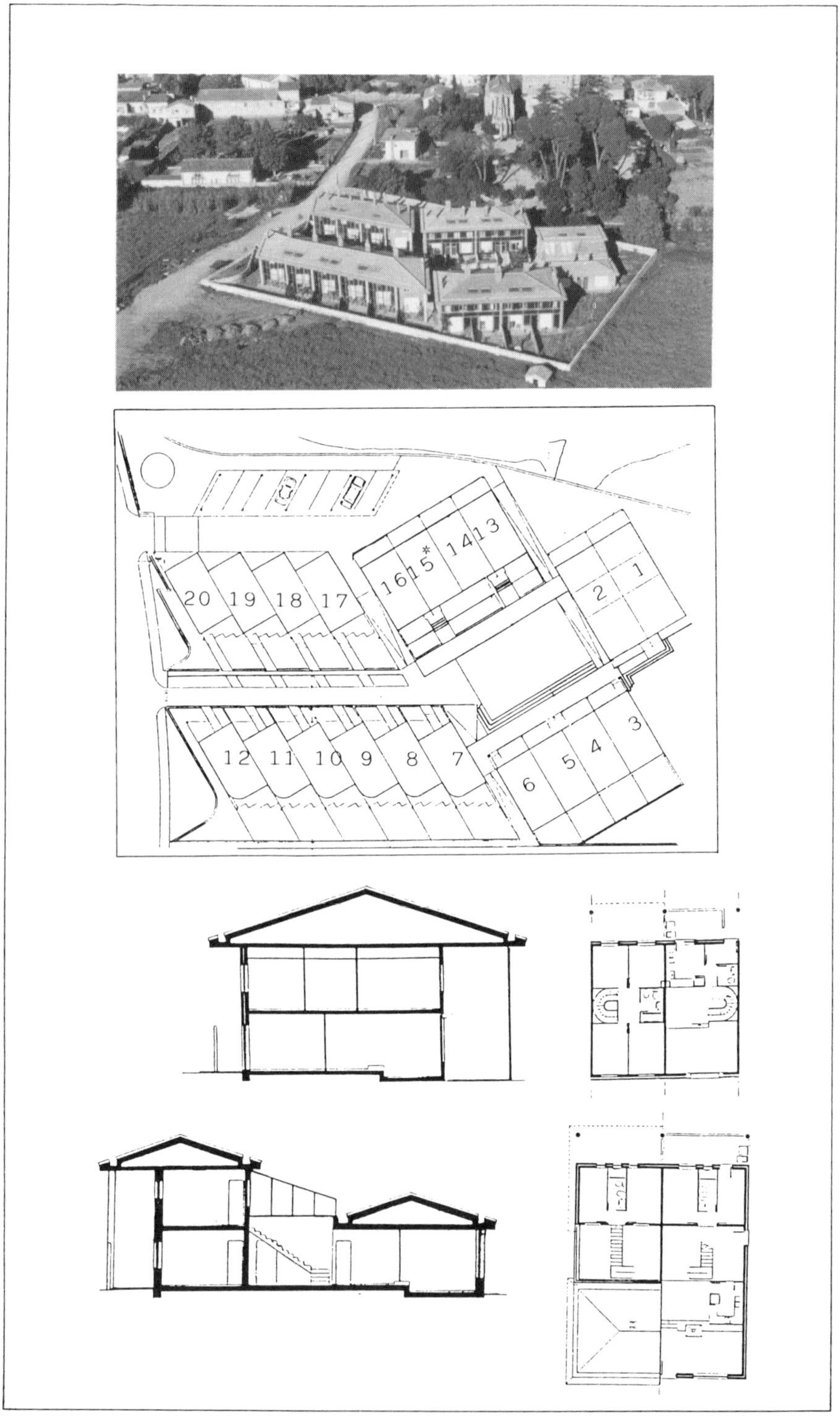

FIGURE 1.- Urbanistic plan overview and sections and plants of two representative dwellings (type S and CC) in the Begues bioclimatic set.

walls are built on 15 cm massive masonry, a 5 cm polstyrene layer and an interior 20 cm layer of massive concrete on the ground floor in order to achieve an important inertial effect, of great importance during winter and so much during summer. The differences in the South facade charactarize the passive systems used: attached sunspace,(S= 8), Trombe wall,(T= 8), covered courtyard,(CC=2) and conventional houses,(C= 2).

Heat transfer from these South Solar spaces to the interior ones are carried out by radiation through the massive separation walls placed between glass surfaces and interior spaces and by convection through the movable openings in the internal frames.

We can remark the existence of evacuation ducts of hot air from the upper part of the captation systems to the North rooms, as their exterior evacuation to the North facade during summer. This allows an efficient thermal regulation in all weathers.

An active solar system has been also included composed of a 2 m^2 solar panel integrated in the house roof and a reservoir of 250 l capacity.

Users of such popular dwellings (with subsidized prices) belong to families with low rents who work, generally, on the service sector.

3. EXPERIMENTAL DATA ACQUISITION PROGRAM AND METHODOLOGY

The whole set of 20 dwellings has been monitored by means of sensors of temperature , humidity and energy flow metters. In addition to this, electric energy consumption and other energy sources (like wood) have been measured. Inquiries to each family about the use and management of their dwellings have been made including the personal (subjective) opinion of each user. These interviews (Table I) per site were conducted in order to know the motivations of the occupants and their role in the effective thermal behaviour of each house and specially daily management of the house and life style of the occupant [2].

Before delivering the dwellings a brief training meeting was carried out in order to explain to the users the bioclimatic concepts included in the house design and the general recomendations to achieve an efficient energy management. It is important to remark that these houses are designed for any kind of occupants but particularly for sedentary ones.

4. DATA ANALYSIS AND DISCUSSION.

The average value among all houses of electric energetic consumption averaged all year through and during summer season are 341.4 and 260.6 Kwh/month respectivelly. Houses 15,18 and 20 have not been considered in the analysis because of anomalous circumstances.

4.1. Tipology correlation (question 2)

There is a greater electric energy consumption in the CC-tipologies (1 and 2) and C-tipologies (7 and 17), which desappears during summer. On the other hand, the buildings with S-tipology show a slight advantage on energy savings in front of any other tipology.

Diferences can reach nearly 20%, specially in the C-typology buildings.

4.2. Correlation with cultural level (question 3)

N	Dwelling Number		1	2	3	4	5	6	7	8	9	10	11	12	13	14	15	16	17	18	19	20
1	Typology (S, T, CC, C)		CC	CC	T	T	T	T	C	S	S	S	S	S	T	T	T	T	C	S	S	S
2	Number of inhabitants		2	2	3	3	4	2	3	2	4	4	2	3	4	4	0	3	2	5	7	1
3	Man/Woman Profession		E/-	W/E	W/D	E/E	B/D	E/E	E/E	E/E	W/D	E/-	W/D	E/D	W/D	W/D	-/-	W/D	W/W	E/E	W/E	W/-
4	Subjective confort level: Winter/Summer		3/3	2/1	3/3	3/3	4/4	3/3	3/3	2/3	4/4	3/3	3/3	3/3	1/2	2/2	-/-	1/1	3/3	3/1	4/1	2/2
5	Occupation profile (mean persons/day)		1.5	1.7	1.8	1.9	2.8	1.7	1.8	1.25	3.1	2.3	1.7	1.4	2.7	2.2	0.1	1.5	1.6	2.1	5	0.45
6	Winter ventilation (opening windows)		1	3	5	2	4	6	4	5	4	4	5	4	4	4	1	1	3	7	7	1
7	Closing North windows and South blinds		Nor	Yes	Sometimes	Yes	Yes	Nor	Yes	Yes	Yes	Yes	Yes	Yes	Somes-times	Yes	Yes	Nor	Yes	Sometimes	Nor	Yes
8	Wood consumption in winter (kg/day)		0	12	6	2	1.7	1.5	23	8.3	5.5	5	1.5	8	22	0	0	3	2	5	1.5	0
9	Electrical mean consumption (kWh/month)		395.7	346.9	323.5	330.8	379.9	352.7	534.1	195.1	294.5	256.0	369.2	242.9	357.7	485.1	203.6	275.2	277.6	72.6	386.6	23.9
10	Electrical summer mean consumption (kWh/month)		339.6	237.1	222.4	208.7	291.6	281.6	376.3	147.8	221.2	224.2	369.9	202	294.1	322.3	194.4	218.7	130.5	10.7	342.5	3.7
11	Variability	November	-11.5	66.2	64.0	78.6	46.9	55.7	74.5	37.0	61.5	43.7	0.9	34.1	32.6	116.2	3.2	30.6	163.5	944.4	20.2	14.1
12	of the	December	56.4	115.2	76.1	83.1	71.3	59.5	105.6	112.5	72.5	33.0	43.5	35.9	28.5	150.2	16.6	91.6	254.7	7146.3	25.8	2613.8
13	electrical summer	January	36.3	101.7	97.3	106.6	85.7	70.7	118.2	113.2	88.2	46.0	16.3	60.2	71.4	89.7	28.6	88.5	170.6	-100	40.5	2301.1
14	mean consumption	February	105.4	103.0	154.2	145.7	117.5	65.9	109.3	127.7	76.6	35.2	8.7	58.5	57.1	97.2	27.3	118.8	88.9	-100	43.6	2794
15	in percentage (%)	March	11.6	104.1	83.5	91.3	62.0	31.3	40.7	48.5	47.3	13.4	-15.0	32.7	17.3	83.0	-10.4	41.9	6.8	-100	16.2	281.7
16	Electrical light and appliances consumption (Wh/day)		1750	750	300	300	2000	800	560	1800	600	800	1500	350	1200	600	20	600	1200	200	1500	500
17	Special remarks		Overheating			No convective loop		Acoustic noises	Dry machine		Acoustic noises				No convective loop	Bad entrance door	Monitoring house			No counter meter	Summer control	No occupied

TABLE I.- Experimental data obtained by lectures in counter meters and inquiries in the Begues bioclimatic set. Notation keys:
Question 1: (CC: covered courtyeard; T: Trombe wall; C: Conventional; S: Sunspace).
Question 3: (E: Office employer; W: Working class; D: Domestic labours, B: Businessman).
Question 4: (0: Very bad; 1: Bad; 2: Normal; 3: Good; 4: Very good).
Question 6: (0: Never; 1: Sometimes 1 h; 2: 15 min for a week; 3: 4 h for a week; 4: 30 min day; 5: 1 h day; 6: 2 h day; 7:3 h day).

Connected with a greater cultural level we can observe a greater demand of comfort level by users (houses number 1, 5 and 7). Probably related with activities of intellectual nature and a higher living standard. This increase in consumption can be estimated up to a value of 10%.

4.3. Correlation with users motivation (question 5)

When we analise houses number 5 and 9, we cannot deduce a clear inferior energetic consumption in this way. We can only remark a little decrease, more obvious during summer when it is more evident the active management of the dwelling by their users. The acceptance of their dwelling is remarkable (3) and is slightly more favourable during summer (related with the appropiate thermal inertia of these dwellings).

4.4. Correlation with wood consumption (question 8)

Obviously electricity consumption in houses number 1 and 4 is higher because their users do not use wood as auxiliar energy. Although this contribution does not seem to be greater than 10% in all the heat inputs of non-solar character.

4.5. Correlation with occupation (question 2 and 5)

Occupation profiles, after beeing evaluated on Man/day ratio, are quite homogeneous. A higher level of occupation that would increase internal gains is balanced with a greater demand of hot sanitary water within other needs, as it can be analised in house number 19. Then we can average when evaluating this energetic contribution.

4.6. Correlation of energetic consumption with lights and appliances (question 16)

This factor do not contribute to establish any kind of tendency in the electric energy consumption, as it can be seen in houses number 4,8,1 and 19.

4.7. Correlation with seasonal variability (questions 11,12,13,14,15 and 16)

We can observe a clear and determined dependence of electric consumption with meteorological variables. On data collected in February 1990, where bad climatic conditions took place, we observe a variation greater than 100% in front of the average of summer consumption.

4.8. Correlation with other variables (questions 6 and 7)

Other variables such as the management of nocturnal closing of shutters and blinds as well as higienical ventilation in winter do not seem to be relevant with average data. The complexity and the diversity of parameters to be taken into account leave large opportunities to chance to have a role in the identification of the optimum, which is a mathematical concept named attractor. One part deterministic, one part chance. Since, the interaction into principles and tools are motivated by the afectation of the technology [3].

For instance, the variability of used patterns of the building -quite unconnected with the instrument- and the variability of the weather are both superimposed on the situation [4].

Considering the influence of different elements which contribute to the auxiliar energy consumption and internal gains as we have discussed, we can establish a priority order taking into account the stadistic weight of these components in monthly or yearly average:

a) Seasonal variation: $\delta(s_v) \approx \pm 50\%$.

b) Typology (architectural design): $\delta(t) \approx \pm 10\%$.
c) Correlation with cultural level: $\delta(c_l) \approx \pm 5\%$.
d) Use of non-electric auxiliar energy: $\delta(a_u) \approx \pm 5\%$.
e) Others: $\delta(o) \approx \pm 2.5\%$.

Although the limitations of the stadistic sample used, with a possible model to correlate the internal gains amount with electric energy consumption, an approach to the variability of each contribution on monthly scale above mentioned can be obtained by the following expression where different weight contribution has been approximately included:

$$\delta(Q_{int}) = 0.35\delta(s_v) + 0.25\delta(t) + 0.20\delta(c_l) + 0.10\delta(a_u) + 0.10\delta(o) \approx 0.22 = 22\%$$

If we consider a house with 50% solar contribution, 30% auxiliary energy used and 20% internal gains, the acumulated errors in the evaluation of the solar contribution is the order of 8%.

5. CONCLUSIONS.

With an appropiate uselfulness behaviour, important energy savings can be reached even in conventional houses. The use of stadistic or stocastic models, as well as the Markov chains method, etc., is not suitable to be used without a more enlarged quantity of data which are not common in bioclimatic architecture research and then it is difficult to deduce an expression of the variability of the evaluation of internal gains of general use.

However the fact that an occupation profile can be obtained and then evaluated in the global energy balance of a building shows a variability from its mean value, which 22% of the mean value usually considered in the internal gains determination. In this way the error which this parameter introduces in the solar energy savings can reach 8% value.

REFERENCES.

[1] Jesch, L.F. (1989), 'Thermal performance of 90 passive solar family houses in Birmingham, UK' in T.C.Steemers and W.Palz (eds.), 2nd European Conference on Architecture, Kluwer Academic Publishers, Dordrecht, pp.600-603.
[2] Fleury, B. and Richalet, V.(1986) 'Solar homes field monitoring: a simplified method', ENTPE, Rue Maurice Audin, 69120 Vaulx en Velin, France
[3] Butera, F.M. (1989) 'Conflicts between physics and architecture in solar systems optimization' in T.C.Steems and W.Palz (eds.), 2nd European Conference on Architecture, Kluwer Academic Publishers, Dordrecht, pp.346-352.
[4] Fisk, D.J. and Morrison, R.C. (1979) 'Energy conservation test with simulated occupancy', Building Research and Practice, Cl/SfB 81 (RS), 149-155.

COMPARISON OF ENERGY BALANCES OF PASSIVE AND CONVENTIONAL PREMISES - DIFFERENTIATIONS IN THE PERFORMANCE OF THE PASSIVE SYSTEMS

Prof. Dr. PAPADOPOULOS M. Dipl. Eng.
Dr. CHRISSOMALIDOU N. Arch. - Lect.
Dr. KOTOULAS K. Dipl. Eng.
ZYGOMALAS M. Dipl. Eng.
Dept. of Civil Engineering Division of Structural Engineering
School of Engineering
Aristotle University of Thessaloniki - 54006 GREECE

ABSTRACT: The study details results and conclusions obtained from 2 apartments which were part of research carried out in 8 apartments of the Solar Village SV3 in Lycovrisi, Attiki. The results refer to a) the comparison of the energy balances between the passive and the conventional apartments, b) the verification of the simulation programme (software package) ESP, c) the evaluation of 6 passive solar systems and d) the effect of the tenants' interference and their possible energy behaviour on the internal rooms' temperature during the winter period.

1. Introduction

Solar Village 3 (SV3) is located at a distance of 18 km from the center of Athens, near Lycovrisi housing, which is an experimental housing estate, to which the principles of bioclimatic architecture were applied, as well as passive and reference conventional components-systems, aiming at the maximum exploitation of solar energy. The whole housing estate consists of 435 residences. 12 apartments and 22 maisonettes out of these, are combined into six distinctive passive solar systems (PSS), to ensure during the heating period, the collection and storage of solar energy, and also its appropriate distribution, over the specified time periods.
On completion of the project, our research team, consisting of 8 scientists of various specializations, was assigned, the evaluation of the energy behaviour of 8 apartments, and in particular, the monitoring and evaluation of the PSS output. This last was achieved by continuous (for two years), short-time, in situ measurements (thermophotographies, U-value, thermal comfort, infiltration measurements etc), by means of the continuous (per five minutes) electronic measuring system DAS, and also by using estimating methods (energy balance of the buildings with conventional method). In this study the results from only two of the residences are indicatively presented: More specifically, from the research carried out in a passive solar apartment (UBB02A2 Fig.1) and a reference conventional one (UDB03A0 Fig.2).

2. Field of the study - Conclusions

The 4 result categories, presented within the confined limits of the study, refer to the same time interval (February) of the heating period, so that comparisons are possible. The temperature data of the period considered, and also the projects data of the buildings are included in Fig.3, Fig.4 and Fig.5 (external temperatures).
The 4 distinctive fields of the study, with their respective results, which are given in the Figures that follow, refer to:

S. Alvarez et al. (eds.), Architecture and Urban Space, 697–702.

A. The comparison of energy balances between passive and conventional apartments (Fig.1,2 and 3). The conventional method was used for the estimations, with data from the measurements and the actual recordings of energy consumption. From the comparison between the two inhabited apartments, it was found that the passive one consumes 48% less energy than the conventional one per m2 of useful surface and real degreeday (Kwh/m2 DD.d). Furthermore, in the first case, the auxiliary heating takes 22.26% of the gross thermal load, which, in the second case is only 53%.
B. Verification of the simulation software-programme ESP. During the period when the apartment UBB02A2 was non uninhabited, temperature recordings were carried out at the four thermal zones of the apartment (Fig.1), by the data acquisition system (DAS) (Fig.4). For the same time period, multiple sequential simulations were attempted using the ESP programme, in order to approximate the actual values of the variable parameters (eg ventilation of the spaces), and adjust the programme in accordance with the real data. The final results (Fig.5) confirm the reliability of the method, to a satisfactory degree.
C. Evaluation, after comparison of output, between passive systems of same size. Alternative solutions of P.S.S. have been studied for apartment UBB02A2 (Fig.6) and by using the ESP program, an equal number of simulations, were carried out, concerning the thermal zone 4 (bedroom). The results show a clear superiority of the Trombe-wall with double glass-panes (curve 1), followed by the Trombe-wall with single glass-pane (curve 2), the mass-wall (curve 3), the combination of mass and Trombe-wall (curve 4), the direct gain (curve 5), and finally, the attached sun-space (curve 6).
D. Control of the tenants' interference, and their possible energy behaviour, concerning the internal temperature of the spaces. The temperatures of apartment UBB02A2, in the following situations, are given in Fig.8: (1) the Trombe-wall has night-time insulation (blinds), (2) without interference by the tenants, (3) the kitchen door (zone 3) is open during the whole 24 hours, (4) as in No 3 but with the external south door of the living-room (zone 2) also open, (5) all the external doors of the apartment are open, (6) all the doors are closed with the southern orientated awnings rolled down (cancellation of the PSS). The results reconfirm that no energy measure can have the expected output, unless the building's tenants attribute positively.

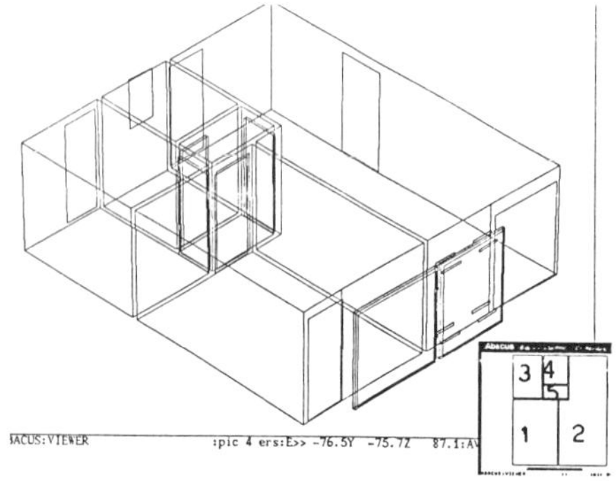

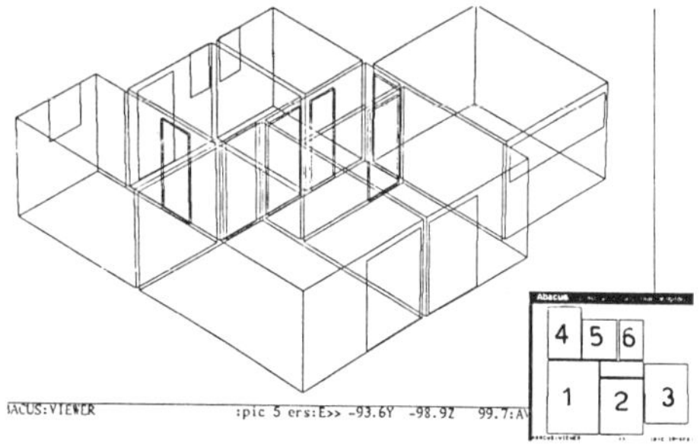

Fig.1. Passive solar apartment (UBB02A2) **Fig.2.** Conventional apartment (UDB03A0)
(Isometrics - Indicative plans with the thermal zones)

3. References

Prof. Papadopoulos M., Chryssomalidou N., Economides G., Kotoulas K., Liveris P., Papadopoulos A., Tsakiris N. and Zygomalas M. "Measurements and evaluation of the passive systems and buildings in Solar Village III (Athens)" Research project 1988-1991

A. Energy balances

GEOMETRIC DATA			
	UBB02A2	UDB03A0	Units
Heated volume	176.79	210.44	m^3
Heated floor area	58.93	72.56	m^2
Ext. wall area	208.12	232.03	m^2
Window area	5.28	7.72	m^2
Trombe wall	4.05	-	m^2
Mass wall	4.73	-	m^2

THERMAL CHARACTERISTICS			
	UBB02A2	UDB03A0	Units
U-value wall	0.354	0.354	W/m^2K
U-value roof	0.225	0.225	W/m^2K
max. int temp.	18.67	20.16	^oC
min. int. temp.	15.37	18.41	^oC
aver int. temp.	17.28	19.91	^oC
aver. ext. temp.	8.64		^oC

A1. Passive solar apartment UBB02A2

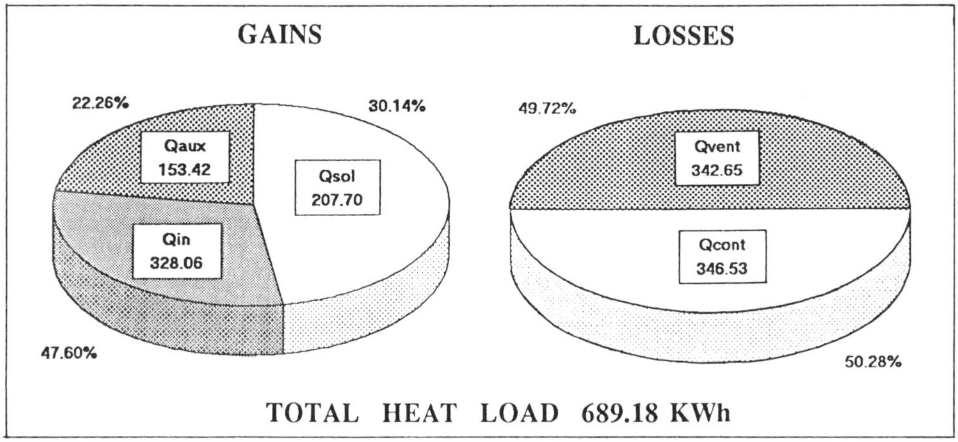

A2. Conventional apartment UDB03A0

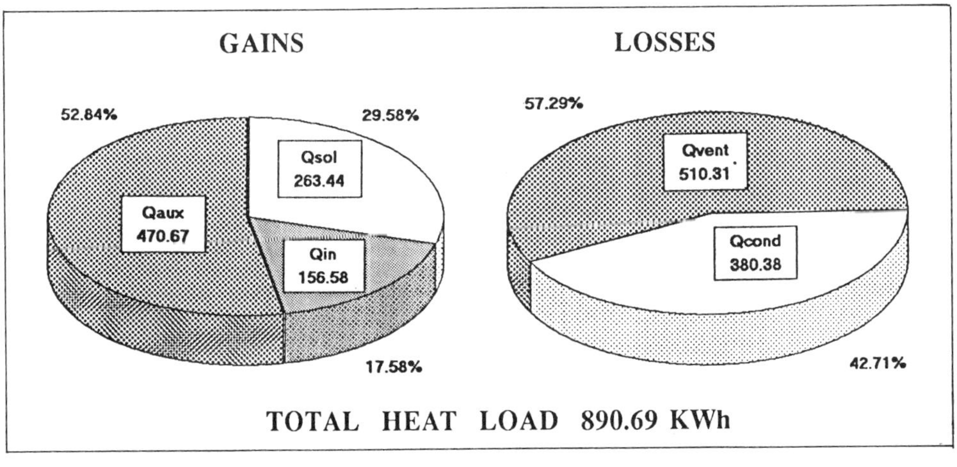

Fig.3. Projects data - energy balances of buildings

B. Verification of the simulation software-programme ESP.

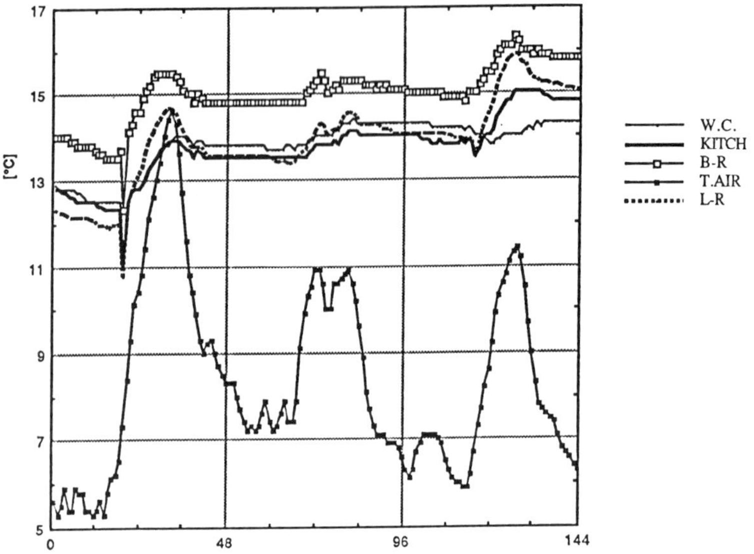

Fig.4. Internal room temperatures (4 zones) from the recordings of the electronic recording system DAS at passive solar apartment UBB02A2 during the time that it was uninhabited (Tair=ext.temp)

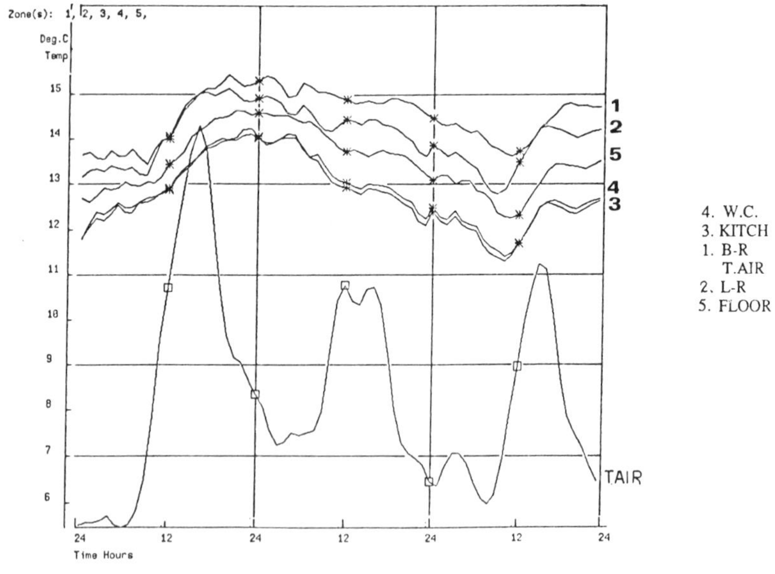

Fig.5. Internal room temperatures (5 zones) as they resulted from the simulation of apartment UBB02A2 using the ESP programme

C. Alternative solutions of passive solar systems

1. Trombe wall - double glazing (T.W.d.g.)

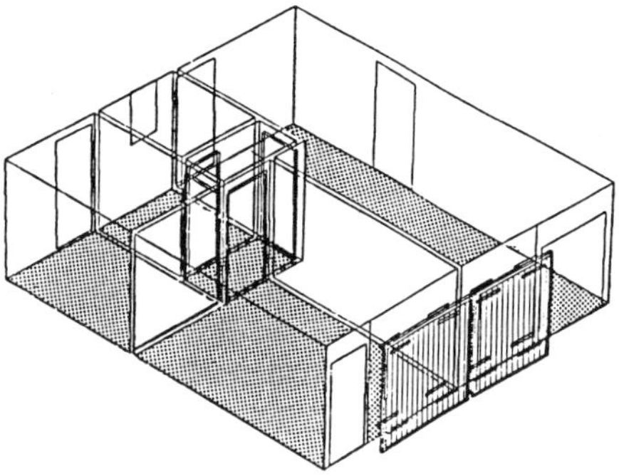

2. Trombe wall - single glazing (T.W.s.g.)

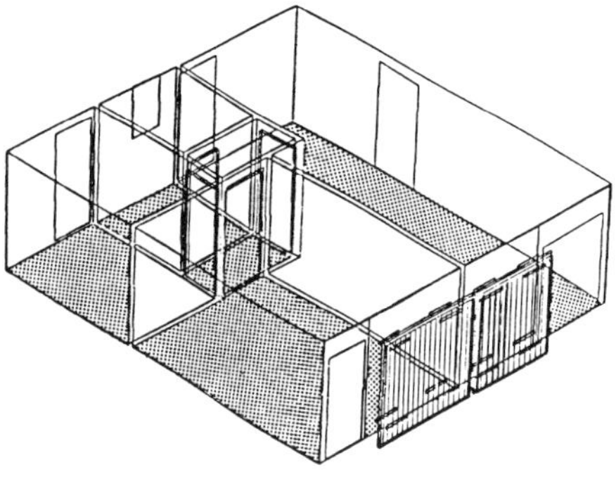

3. Mass wall (M.W.)

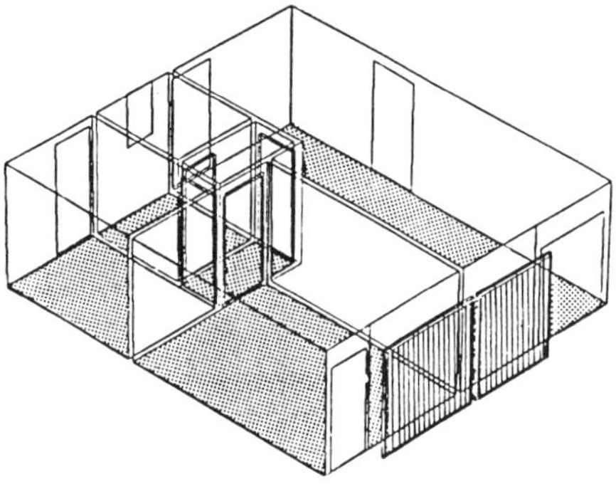

4. Direct gain (D.G.)

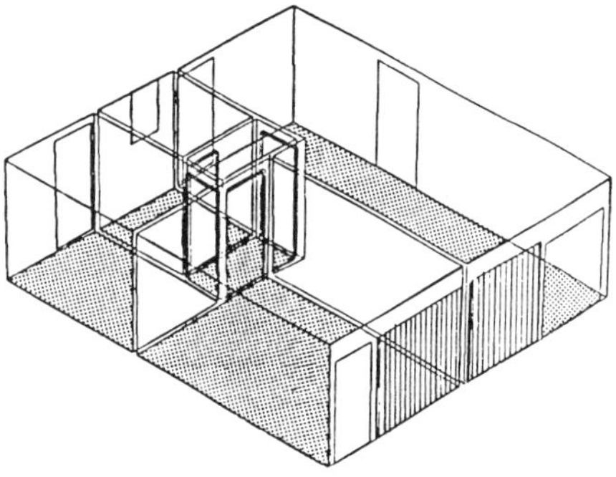

5. Mass and Trombe wall (MW-TW)

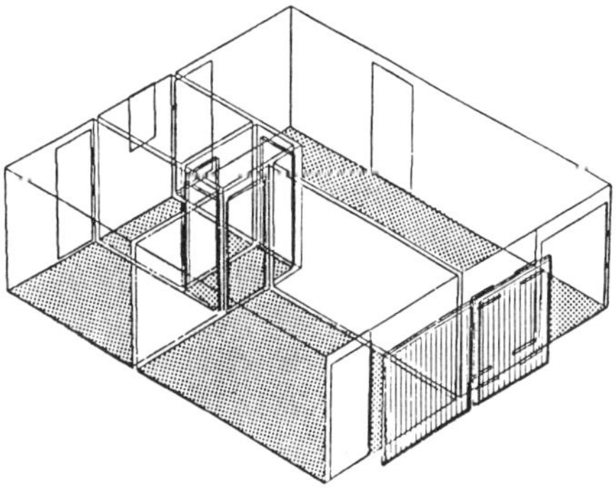

6. Greenhouse (GH)

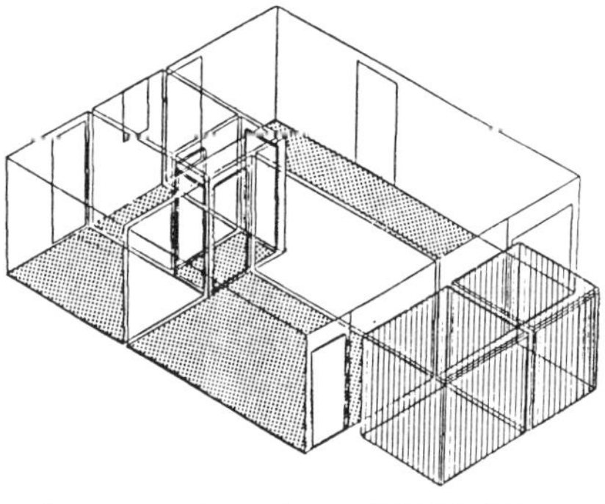

Fig 6. Passive solar systems. Investigation of their performance at apartment UBB02A2

D. Influence of passive solar systems (Fig.7) and the energy behaviour of the tenants (Fig.8) on the internal temperature

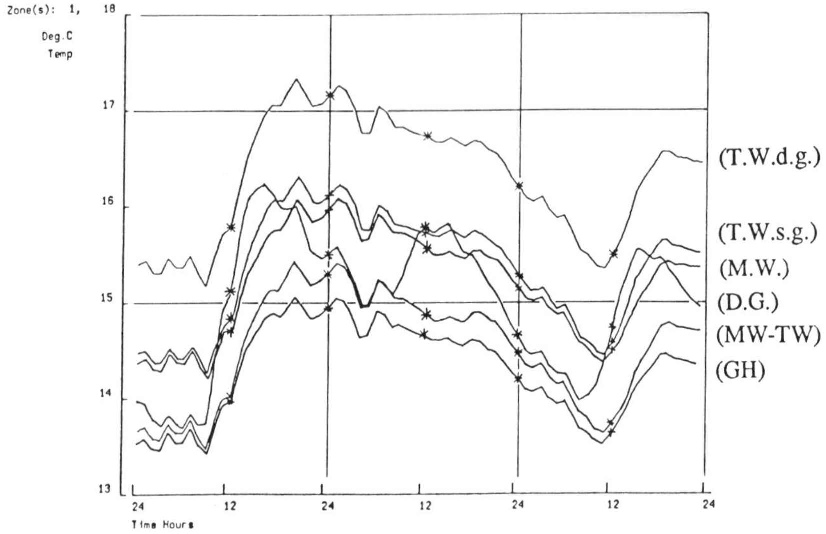

Fig.7. Influence of 6 passive solar systems on the internal temperature of the bedroom (zone 1) at apartment UBB02A2 during the period it was uninhabited

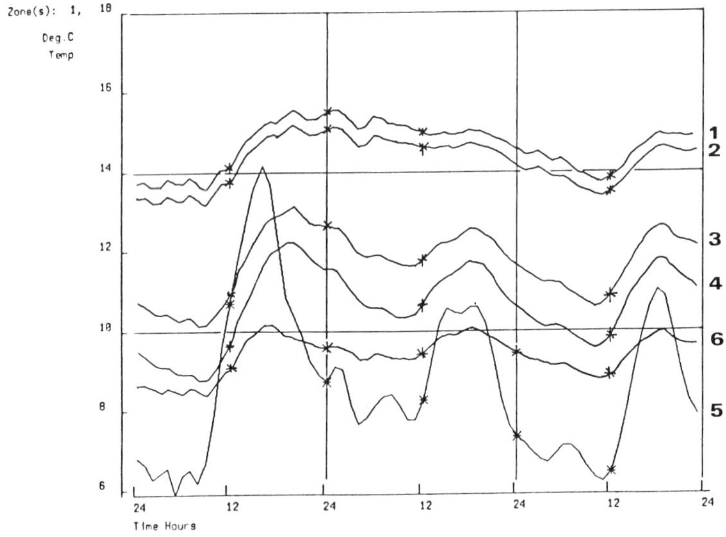

1. TW with night-time insulation (blinds)
2. without interference by the tenants
3. the kitchen door (zone 3) is open
4. as 3. but with south door of L-R also open
5. all the external doors are open
6. the southern orientated awnings rolled down

Fig.8. Control of the tenants' interference, and their possible energy behaviour, concerning the internal temperature of the zone 1 (bedroom-apartment UBB02A2)

THE CLIMATE OF URBAN SPACES AS A DECISIVE ENERGY PARAMETER IN BUILDING DESIGN AND CONSTRUCTION-

ENERGY AND ECONOMIC EVALUATION OF BUILDINGS

Dr. CHRISOMALLIDOU NIOBE -Architect -Lecturer
Dept. of Civil Engineering Division of Structural Engineering
School of Engineering
Aristotle University of Thessaloniki-54006 GREECE

ABSTRACT.This study presents the research on optimum solutions, as far as energy and economic criteria are concerned, for different forms and insulation of the shells of residential buildings in all three climatic zones (C.Z.) of Greece. Measurements in one of the three buildings studied, (one 8- and one 5-storey conventional building and a 2- storey passive solar building) as well as numerous thermal analyses and cost-benefit analyses, constituted the principal methods leading to the verification of the simulation programme and to a series of results and conclusions.
One of the more important conclusions relates to the passive solar building, which verifies the restricted pay-off period (3 years in the cold C.Z.) of the additional investment in the energy saving measures and also evidenced, a very high percentage of energy saving (up to 82%) in comparison with other conventional buildings.

1.Introduction

Numerous energy parameters, based on the principles of bioclimatic architecture design, are considered, which finally influence not only the sensitive thermal balance of the building, but the total life cycle cost as well (cost of energy consumption, construction cost, heating installation cost, maintenance cost). Hence the problem of rational use of energy, a compound problem, is better confronted as an entity, thus evaluating the parameter “energy” as well as the parameter “economy”, if, of course, the aim is to determine the optimum solution in each case, using not only thermal but also economic criteria. Based on the above, and assuming that the climate of each area, is one of the most decisive parameters of energy behaviour of buildings (fig. 2), this study investigates the optimum thermal and economic solution in the three climate zones (C.Z.) of Greece and for three types of buildings' . The buildings that were studied were chosen on the basic criterion of form, expressed by the A/V ratio, as well as the existence or not of an energy design.

2. Field of study - Method used

For the purposes of this study, as mentioned above , three residential buildings were chosen: a two-storey passive solar building (fig. 1), house of the architect M. Chrisomallidis, awarded a prize in a Hellenic architectural competition, with an attached green house of 15 m2, A/V: 0.923

S. Alvarez et al. (eds.), Architecture and Urban Space, 703–708.

(unfavourable form) and dimensions L.7.5*W.7.5*H.7.5 m; a conventional five-storey building (A/V: 0.36 L.20*W.18*H.16.5 m.), and an eight-storey conventional apartment building with A/V: 0.313 (favourable energy form ,L.20*W.14*H.27 m.). These buildings, designed for the cold climate zone of Greece (Thessaloniki) were assumed to be also suitable for construction in the mild climate zone (Athens), and in the warm climate zone (Chania, Crete). The shell of the buildings was considered to have 4,6,8 cm. thermal insulation, (glass-wool) or to have none.The geometric data, the thermal characteristics, and the climate data of the three urban spaces, are presented in table 1.

In the first stage of the study, simulations were carried out for all the climatic cases and all possibilities of thermal insulation of the shell of the three buildings, by using a computer programme in order, to determine their thermal behaviour (fig.2). It should be especially noted that the verification of the programme results was checked by temperature measurements and annual registers of consumption (24 tons oil) in the existing uninsulated 5-storey building. The results from the comparison are considered to be very satisfactory. (deviation in the order of 7%)

The second stage comprised a cost-benefit analysis, for all the alternative cases investigated in the earlier, stage of thermal analysis. The totally uninsulated building was taken as a base of comparison for the calculations. The aim was the evaluation of the solutions also with reference to economic criteria, so that, for each climate zone and building, the optimum solution (table 2) could be proposed. The economic analyses were based on the existing 6 dynamic economic methods, which were applied by means of a special computer programme. In table 2, only the results from the methods " Net Benefits-Benefits Costs- B-C" and "Discount Pay-back D.P.B." are indicatively presented, taking into account all the cash in- and outflows, that come from the energy-saving measures during the life cycle of the building. A sensitivity analysis was also attempted, for the two most unstable parameters ("I" :interest rate, (%),"L": increase of energy cost (%)), by assuming possible alternative values. The shaded areas (table 2) , refer to results, with values of "I" and "L", that are valid today in Greece, based on which the choice of the optimum economic solution could be realized.

TABLE : 1 Geometric data, thermal characteristics of the buildings, and climate conditions of the three climate zones in Greece

	PROJECTS DATA														
	8- STOREY CONVENT. BUILD.					5-STOREY CONVENT. BUILD.					2-STOREY PASSIVE SOLAR BUIL.				
SHELLS THERMAL INSULATION		0cm	4cm	6cm	8cm		0cm	4cm	6cm	8cm		0cm	4cm	6cm	8cm
GEOMETRIC DATA															
			U-VALUE [W/M2K]					U-VALUE [W/M2K]					U-VALUE [W/M2K]		
VOLUME [M3]	7943					5869					353				
FLOOR AREA [M2]	323	3.80	0.54	0.49	0.33	360	3.78	0.54	0.49	0.33	56	2.83	0.59	0.42	0.33
ROOF AREA [M2]	323	2.15	0.52	0.44	0.30	360	2.15	0.52	0.44	0.30	56	2.00	0.54	0.39	0.30
EXTERNAL WALL AREA [M2]	1457	1.78	0.66	0.51	0.39	1044	2.16	0.66	0.51	0.39	188	2.36	0.55	0.40	0.32
WINDOWS TOTAL AREA [M2]	380	5.80	3.02	3.02	3.02	338	5.80	3.02	3.02	3.02	26	5.81	3.02	3.02	3.02
THERMAL CHARACTERISTICS															
TOTAL Um -VALUE [W/M2K]		2.50	0.95	0.85	0.75		2.70	0.97	0.87	0.76		2.30	0.68	0.56	0.49
VENTILATION RATE $n : h^{-1}$		0.60	0.60	0.60	0.60		0.60	0.60	0.60	0.60		0.60	0.60	0.60	0.60
HEATED VOLUME [M3]	7943					5869					353				
HEATED FLOOR AREA [M2]	2647					1956					113				
NET HEAT LOAD [KWh/M2A]	THESSA	84	34	31	28	THESSA	107	39	36	32	THESSA	227	61	50	44
	ATHENS	47	19	18	17	ATHENS	61	22	21	20	ATHENS	137	34	28	26
	CHANIA	27	11	11	11	CHANIA	37	13	13	12	CHANIA	87	20	17	16
SITE CLIMATE															
		THESSA	ATHENS	CHANIA											
LATITUDE		40 37	38 58	36 22											
LONGITUDE		23.00	23 43	24.00											
AVERAGE AMBIENT TEMP.[C]	JANUAR	5.5	9.4	11.9											
	JULI	26.7	27.9	26.9											
DEGREE DAYS [days] [base 18°C]		1725.0	1100.0	782.0											
GLOBAL IRRADIAT. ON HORIZ	[KWh/M2	1403.0	1581.0	1630.0											

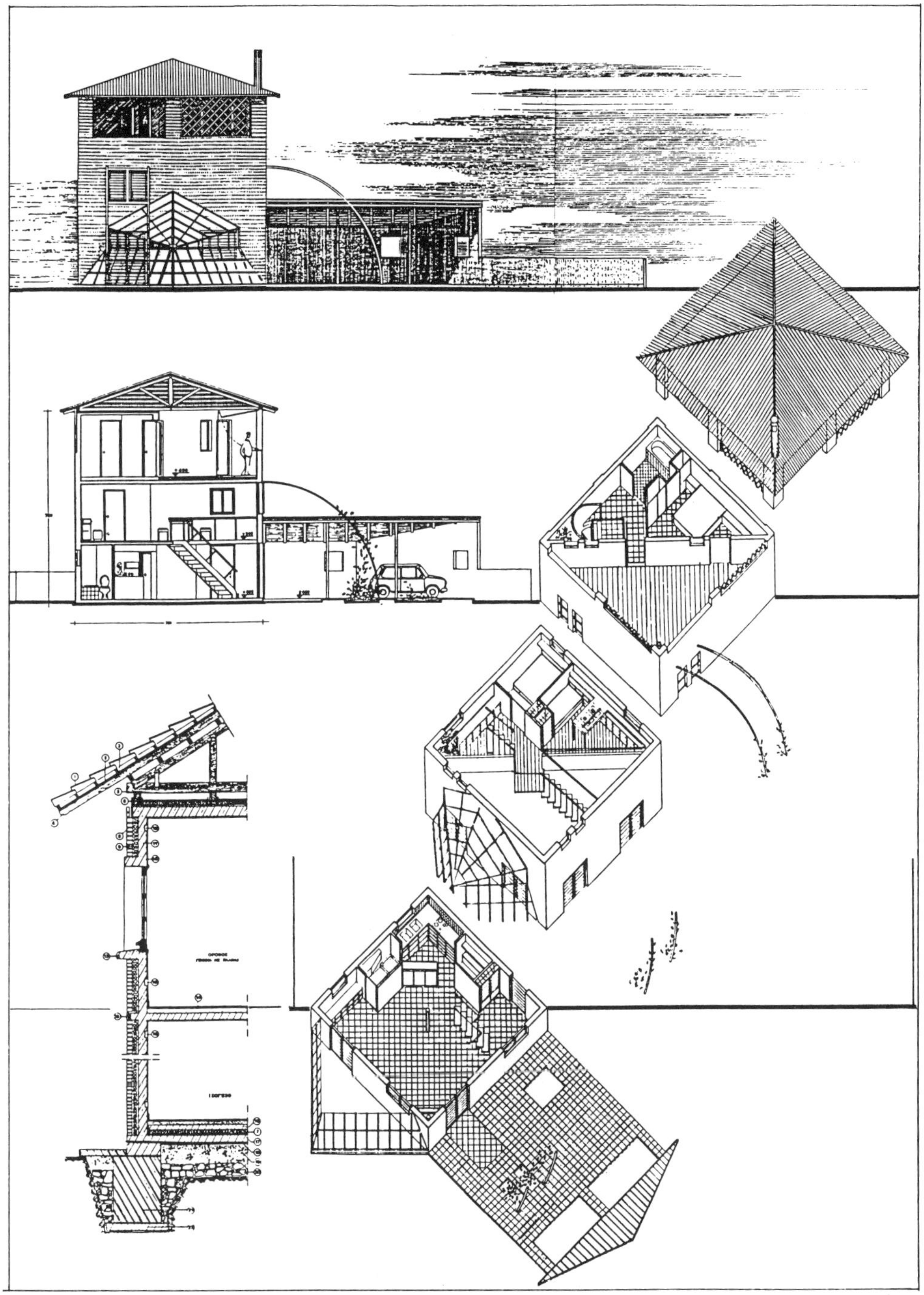

Figure 1. Passive solar 2-storey house, designed for mountain in the cold climate zone (Thessaloniki)

3. Results

3.1 THERMAL ANALYSIS RESULTS.

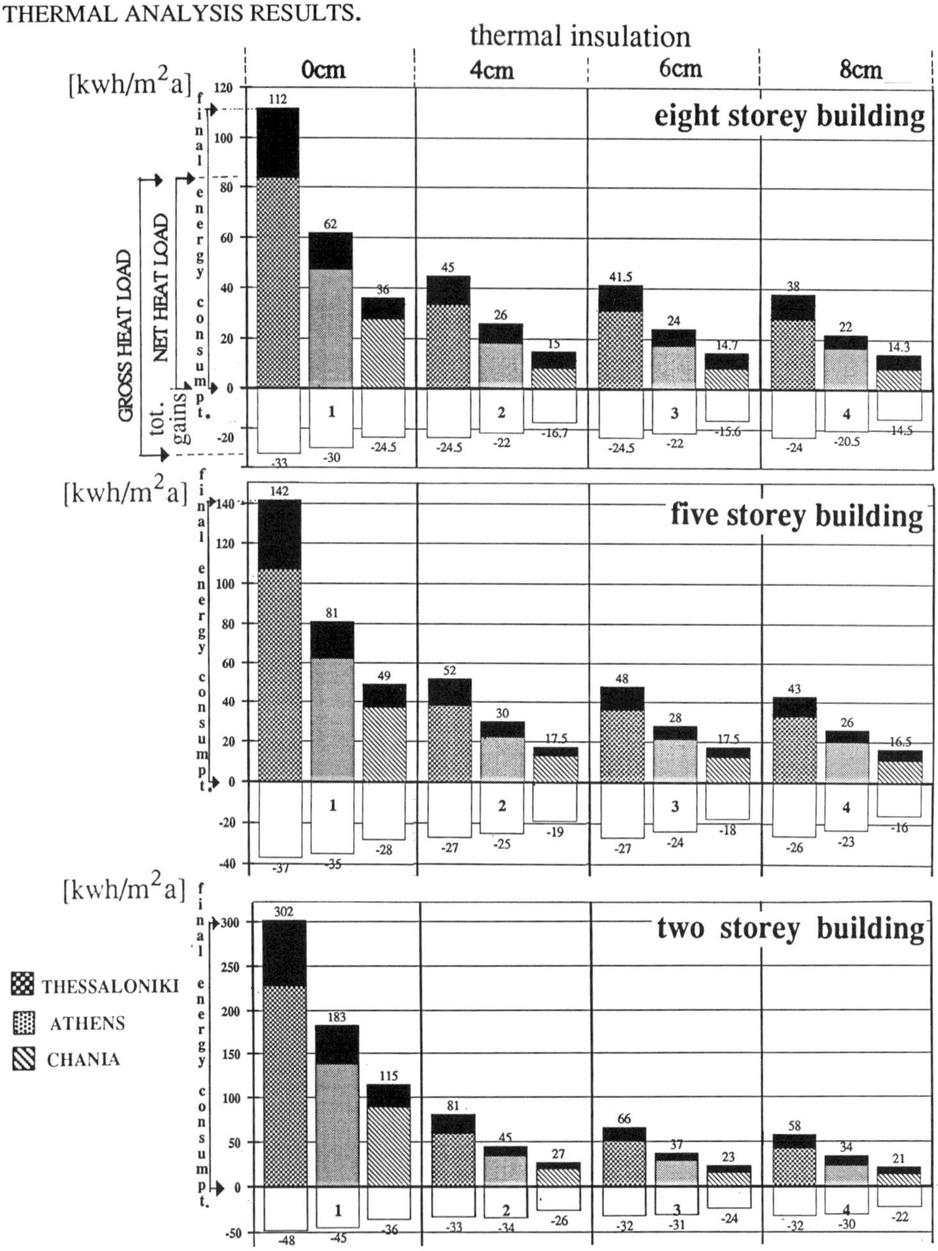

Figure 2. Thermal balance of buildings and final energy consumption, in the three climate zones and with different thickness of shell insulation. (central heating, degree of efficiency of the installation 0.75). Note: The total thermal gains appear higher in the cold C.Z. due to longer heating period (6 months) compared to the mild (5 months) and the warm (3 months) C.Z..

3.2 COST-BENEFIT ANALYSIS RESULTS

8-STOREY BUILDING

		THESSALONIKI			ATHENS			CHANIA		
		4cm	6cm	8cm	4cm	6cm	8cm	4cm	6cm	8cm
I:14%-	L: 8%	11.982.100 [4.0]	12.505.700 [4.4]	12.492.500 [5.0]	3.783.470 [10]	3.776.150 [10.5	3.175.060 [12.4	0	0	0
	L:14%	27.882.100 [3.8]	29.313.600 [4.0]	30.188.700 [4.5]	12.500.500 [7.9]	12.944.900 [8.2]	12.729.000 [9.3]	4.558.030 [15]	4.153.600 [16]	3.272.260 [18]
	L:20%	70.476.900 [3.5]	74.340.900 [3.7]	77.595.500 [4.2]	35.852.700 [6.8]	37.507.200 [7.0]	38.323.400 [7.6]	18.228.600 [11]	18.002.700 [12]	17.383.200 [13]
I:18%-	L: 8%	7.373.860 [4.4]	7.634.250 [4.6]	7.363.620 [5.4]	1.257.030 [12.6]	1.118.790 [13.5]	406.042 [17.5]	0	0	0
	L:14%	16.044.300 [4.0]	16.799.800 [4.2]	17.013.500 [4.8]	6.010.500 [9.0]	6.118.590 [9.5]	5.615.920 [11]	758.743 [20]	304.709 [22]	0
	L:20%	37.243.000 [3.7]	39.209.100 [3.9]	40.607.100 [4.4]	17.632.500 [7.5]	18.372.900 [7.8]	18.353.800 [8.7]	7.562.370 [13]	7.197.180 [14]	6.373.370 [16]
I:22%-	L : 8%	4.698.270 [4.8]	4.805.870 [5.0]	4.385.770 [6.0]	0	0	0	0	0	0
	L:14%	9.836.250 [4.3]	10.237.300 [4.5]	10.104.200 [5.2]	2.607.010 [11]	2.538.730 [11.5	1.885.640 [14]	0	0	0
	L:20%	21.252.500 [3.9]	22.305.400 [4.0]	22.810.100 [4.6]	8.865.840 [8.5]	9.121.890 [8.8]	8.745.410 [10]	2.430.280 [17]	1.998.070 [18]	1.075.970 [22]

5-STOREY BUILDING

		THESSALONIKI			ATHENS			CHANIA		
		4cm	6cm	8cm	4cm	6cm	8cm	4cm	6cm	8cm
I:14%-	L: 8%	13.420.900 [2.9]	13.681.300 [3.3]	13.456.000 [4.0]	5.379.030 [6.8]	5.185.420 [7.6]	4.384.810 [9.7]	1.231.320 [15]	685.574 [18]	0
	L:14%	29.454.900 [2.8]	30.441.100 [3.0]	31.049.500 [3.7]	14.351.700 [5.8]	14.489.000 [6.4]	14.026.000 [7.7]	6.731.450 [11]	6.224.470 [12]	5.245.130 [15]
	L:20%	72.409.000 [2.7]	75.339.300 [2.9]	78.181.200 [3.5]	38.388.900 [5.2]	39.412.700 [5.6]	39.853.900 [6.6]	21.465.900 [9]	21.062.800 [10]	20.406.000 [11]
I:18%-	L: 8%	8.773.740 [3.0]	8.823.890 [3.4]	8.356.880 [4.3]	2.778.490 [7.7]	2.488.970 [8.8]	1.590.530 [12]	0	0	0
	L:14%	17.517.300 [2.9]	17.963.200 [3.2]	17.950.800 [4.0]	7.671.390 [6.4]	7.562.320 [7.0]	6.847.950 [8.8]	2.636.500 [13]	2.100660 [15]	1.031.650 [20]
	L:20%	38.894.800 [2.7]	40.308.300 [3.0]	41.407.500 [3.6]	19.634.300 [5.6]	19.966.400 [6.0]	19.702.100 [7.3]	9.969.600 [10]	9.485.440 [11]	8.577.000 [13]
I:22%-	L : 8%	6.075.590 [3.2]	6.003.610 [3.5]	5.396.300 [4.6]	1.268.600 [9.0]	923.395 [11]	0	0	0	0
	L:14%	11.256.900 [2.9]	11.419.400 [3.3]	11.081.500 [4.2]	4.168.070 [7.0]	3.929.800 [8.0]	3.083.630 [10]	489.014 [17]	0	0
	L:20%	22.769.400 [2.8]	23.453.000 [3.0]	23.713.700 [3.8]	10.610.500 [6.0]	10.609.800 [6.7]	10.006.000 [8.2]	4.438.120 [12]	3.914.980 [13]	2.885.420 [16]

2-STOREY BUILDING

		THESSALONIKI			ATHENS			CHANIA		
		4cm	6cm	8cm	4cm	6cm	8cm	4cm	6cm	8cm
I:14%-	L: 8%	2.220.200 [1.0]	2.202.120 [1.9]	2.097.360 [2.9]	1.226.960 [2.5]	1.127.800 [4.0]	948.594 [6.0]	643.612 [4.9]	491.977 [8.4]	313.006 [12.7]
	L:14%	4.472.400 [1.0]	4.605.460 [1.8]	4.580.470 [2.8]	2.623.060 [2.3]	2.604.430 [3.8]	2.484.460 [5.4]	1.547.350 [4.5]	1.432.810 [6.9]	1.270.020 [9.5]
	L:20%	10.505.900 [1.0]	11.043.800 [1.8]	11.232.600 [2.7]	6.363.100 [2.3]	6.560.220 [3.6]	6.535.590 [4.8]	3.968.380 [4.0]	3.953.240 [6.0]	3.833.800 [7.8]
I:18%-	L: 8%	1.567.450 [1.0]	1.505.560 [1.9]	1.377.680 [3.0]	822.332 [2.5]	699.833 [4.4]	533.957 [6.9]	381.684 [5.4]	219.296 [10.]	35.635 [18.]
	L:14%	2.795.600 [1.0]	2.816.130 [2.0]	2.731.750 [2.8]	1.583.640 [2.4]	1.505.060 [4.0]	1.358.590 [5.9]	874.500 [4.7]	732.343 [7.7]	557.506 [11]
	L:20%	5.798.360 [1.0]	6.020.410 [1.8]	6.042.380 [2.7]	3.445.000 [2.3]	3.473.790 [3.7]	3.374.770 [5.2]	2.079.410 [4.3]	1.986.720 [6.6]	1.833.460 [8.8]
I:22%-	L : 8%	1.188.460 [1.0]	1.101.140 [2.0]	959.830 [3.0]	587.401 [2.6]	451.351 [4.7]	279.485 [7.9]	229.607 [5.9]	60.975 [13]	0
	L:14%	1.916.250 [1.0]	1.877.760 [2.0]	1.762.230 [3.0]	1.038.540 [2.4]	928.515 [4.2]	768.150 [6.5]	521.643 [5.0]	365.001 [9.0]	183.846 [14]
	L:20%	3.533.330 [1.0]	3.603.370 [2.0]	3.545.110 [3.0]	2.040.940 [2.4]	1.988.740 [3.9]	1.853.930 [5.6]	1.170.530 [4.5]	1.040.520 [7.0]	870.986 [10]

Table 2. Net benefit (in drachmas) during the life cycle of the building and pay off time (in years-values in the parentheses) of the extra-cost for the application of energy saving measures.Base of comparison is the uninsulated building. The optimum solution for countries with different values of interest rate "I" and different increase of the energy price "L"can also be determined from this table.(1$=191 Drachmas, ECU=228, 1 Pesetas= 1.8 Drachmas).

4. Conclusions

4.1 CONCLUSIONS FROM THE THERMAL ANALYSES (fig.2 , tab. 3)

Table 3.Indicative summary results.The shaded area refers to the base of comparison.

	8-STOREY BUILDING											
	THESSALONIKI				**ATHENS**				**CHANIA**			
1	**0cm**	**4cm**	**6cm**	**8cm**	**0cm**	**4cm**	**6cm**	**8cm**	**0cm**	**4cm**	**6cm**	**8cm**
2	[103] +0.00	[48] -53%	[45] -56%	[42] -59%	[66] +0.00	[31] -53%	[29] -56%	[27] -59%	[41] +0.00	[20] -54%	[18]-57%	[17] -61%
3	32%	51%	53%	56%	45%	70%	73%	75%	56%	84%	84%	84%
4	0.00	60%	63%	66%	0.00	59%	59%	61%	0.00	59%	59%	61%
	5-STOREY BUILDING											
2	[126] +0.00	[54] -57%	[50] -60%	[46] -63%	[83] +0.00	[35] -57%	[33] -60%	[30] -64%	[55] +0.0	[22] -59%	[21] -62%	[20] -65%
3	29%	50%	53%	56%	42%	71%	73%	76%	51%	85%	85%	86%
4	0.00	63%	66%	69%	0.00	63%	65%	68%	0.00	64%	64%	66%
	2-STOREY BUILDING											
2	[254] +0.00	[79] -69%	[67] -74%	[61] -76%	[167] +0.00	[53] -68%	[45]-73%	[40] -76%	[110] +0.0	[35] -68%	[29] -74%	[26] -76%
3	19%	42%	48%	52%	27%	64%	70%	74%	32%	78%	81%	84%
4	0.00	74%	78%	81%	0.00	75%	80%	82%	0.00	77%	80%	82%
	1.thermal insulation [cm]			2. total losses [kwh/m2a] -[%]				3.total gains % of the tot. losses				
					4. energy conservation [%]							

The thermal analysis results confirm what is already known , that
-the more favourable the outdoor climate conditions are
-the stronger the passive protection of the shell is and
-the more favourable the form of the building is, from the energy aspect,
the greater the percentage cover of the total thermal losses, by the thermal gains,(with conductivity and ventilation). (tab.3 ,8-storey building., row 3, 8cm.therm.insul., in the 3 C.Z.)

Therefore, in mild and warm zones, energy autonomous buildings could be very easily achieved.. In cold zones the total gains can cover 42-56% of the gross thermal load,(total losses and energy consumption for use of warm water), a percentage that can be increased with the use of passive solar systems of southern orientation.

The energy saving, with respect to the degree of insulation of the shell and independent of the climatic zone, is, in the conventional 8-storey and 5-storey building, between 59%-68%, whereas in the passive solar 2-storey building it is between 74%-82%. (tab.3)

4.2 CONCLUSIONS FROM THE COST -BENEFIT ANALYSES .(tab.2)

In many cases, and especially in the multistory buildings, in the warm climatic zone the uninsulated building shows a lower life cycle cost compared to the insulated one, and hence result negative benefit values. (0 in table 2).

Using the "I" and "L" values valid for Greece today, and for all the three buildings, the economical optimum solution for the cold climatic zone is shown to be the strong insulation of 8 cm, for the mild climatic zone 6 cm, whereas for the warm climatic zone 4 cm of shell insulation.

In multistory buildings particularly, the discounted pay-back period increases, successively from the cold to the warm C.Z., hence the optimum solution must be determined with the additional consideration of the results by the "B-C" method.

The passive two-storey building shows a very restricted pay-off period, especially in the cold C.Z. (up to 3 years in the most unfavourable case), an encouraging factor for making additional investment in energy-saving measures in buildings of this scale, and especially for the application of passive solar systems.

SOLAR DESIGN ZONES FOR QUEENSLAND

S.V.SZOKOLAY
Director, Architectural Science Unit
The University of Queensland
Brisbane, 4072, Australia

ABSTRACT. A study is described, which was carried out for Queensland (one of the seven states of Australia) in order to define zones of building climatic design, to select the most suitable constructional elements and to produce a set of building- and solar- design guidelines for each of these zones. The paper describes the computer program used, the methodology followed and presents some of the results and conclusions.

1. INTRODUCTION

In many countries advisory publications are available for the architect, the builder or the home owner on climatically suitable and energy conservative house design. In Queensland we have two problems with this:

a) The state extends form latitude 11° to 28°S, a north-south length of some 1900 km. Its area is some 7 times that of the UK or 3.5 times that of Spain. Its eastern (Pacific ocean) coast tends to be humid, but its western, inland areas (at places more than 1500 km from the coast) are very dry. It includes quite a number of clearly distinguishable climate types. No one set of recommendations will be suitable for the whole state.

b) Many textbooks (explicitly or implicitly) recommend that climatic design should start with an analysis of climatic data, on the basis of which the climate will be classified and a label will be attached, such as 'hot dry', or 'warm humid'. This label would immediately conjur up the image of the stereotype house for this class of climate and then the design would proceed accordingly. Unfortunately the reality rarely corresponds to one of these 'typical' climates. None of the Queensland climates match any of the generally recognised four main climate types. Each location has its own charcteristics, they all have transitional or composite climates.

2. FIRST APPROACH

At the start of the study (Szokolay, 1990) it has therefore been decided to eliminate the intermediate step of climate classification

S. Alvarez et al. (eds.), Architecture and Urban Space, 709–714.

and go directly from data analysis to design recommendations. In the first instance a computerised version of the 'Mahoney Tables' (UN, 1971) was used (the CLIMATE module of the ARCHIPAK program). This is a fairly simple 'expert system', which gives certain design recommendations from an analysis of climatic data.

From multiple runs of this program seven climatic design zones can be distinguished and the extent of applicability of five basic design characteristics can be plotted on a series of maps, as shown in Fig.1 overpage.

3. THE TOOL

In the second part of the study a thermal response simulation program, the HARMON module of ARCHIPAK, has been used (Szokolay, 1983). This is based on the 'admittance procedure' of the British Building Research Establishment and its output is the hourly indoor temperatures.

The ARCHIPAK suite includes a climatic data base (monthly mean data, but including standard deviations of the distribution) for some 70 locations and a building elements data base, which contains the U-value, admittance, time lag and decrement factor of some 500 different building envelope elements. The facility exists to create new entries for multi-layer elements, reading the relevant material properties (conductivity, density and specific heat capacity) from another data file.

4. THE VEHICLE

We took a low-cost typical 2-bedroom, about 100 m^2 suburban house and examined its thermal response on an average winter (July) day and an average summer (January) day, in each of the seven selected locations, with a large number of constructional combinations:

- 2 types of floor (concrete and elevated timber)
- 11 wall constructions
- 6 roof constructions

(because of the practical incompatibility of some roofs and walls, a total of 32 roof/wall combinations were examined, each with both floor types.)

5. THE METHOD

A total of 3584 simulations runs were carried out in the first phase. The criteria used to measure the performance of these variants were the indoor minimum and maximum temperatures in July and January.

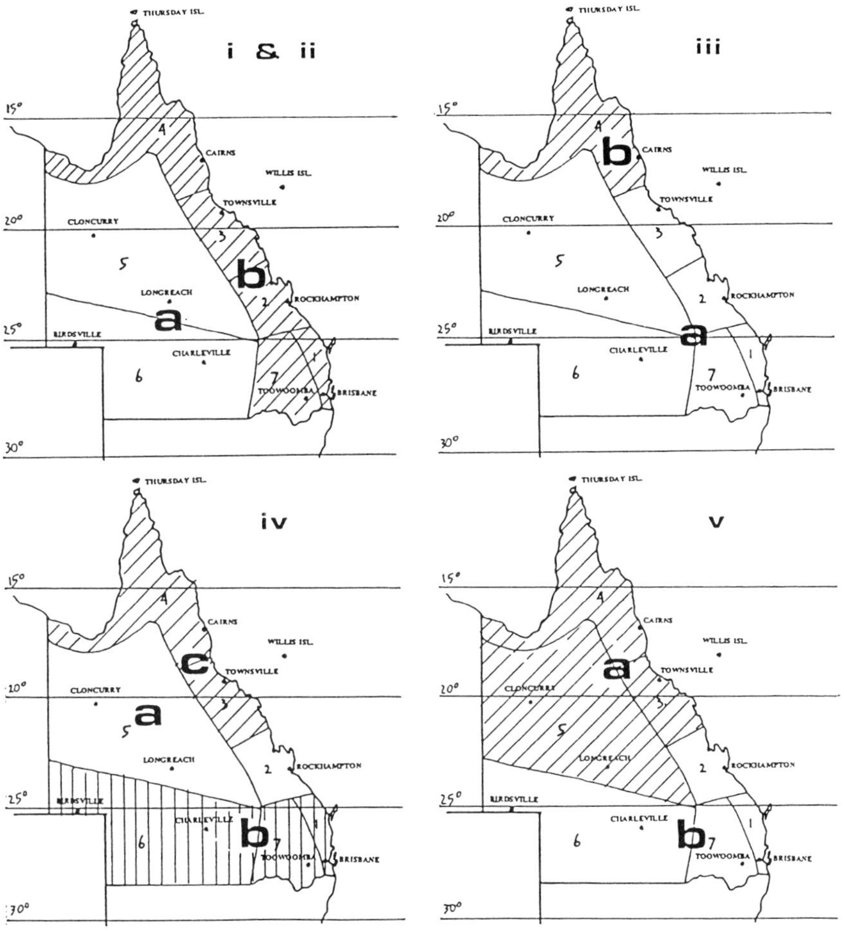

Fig.1. Building climatic zones of Queensland

i. plan shape	a. courtyard
	b. elongated in the east-west direction
ii. roof	a. heavy
	b. lightweight, insulated
iii. walls	a. heavy
	b. lightweight
iv. windows	a. small (<25%)
	b. medium (25-50%)
	c. large (>50%)
v. shading	a. full shading all year
	b. winter sun penetration (passive solar heating)

NOTE: the boundaries of zones are indicative only, these are not sharp lines; the zones tend to merge into each-other and are affected by local variations.

In the second phase a macro program automatically produced 64 runs (for the 32 constructional variants, for January and July) so only 56 of these composite runs were necessary. The measure of performance was the total number of degree-hours of overheating and underheating of the indoors in the above two months.

To produce this second measure, a modified version of the program was used: for every hour when the indoor temperature exceeded the upper comfort limit, the excess temperature was summed and for every hour when the indoor temperature was below the lower comfort limit, this temperature deficit was summed. These sums were then multiplied by the number of days of the month to get the monthly total Kelvin-hours.

6. RESULTS

Floors: In all cases the concrete slab-on-ground floor gives better results than the elevated timber floor according to both sets of criteria, except in zone 4 (Cairns), in January: with most constructions a concrete slab-on-ground floor gives a lower maximum temperature than a timber floor, but in terms of cumulative overheating (Kelvin-hours) the timber floor gives a lower number.

Walls: In terms of cumulative over- or underheating in all cases the best wall construction is the insulated reverse block-veneer: 200 mm concrete blocks, filled with concrete with insulation on the outside. In terms of summer indoor maximum temperature the above and the insulated reverse brick veneer are equal bests, giving the lowest maxima. These are closely followed by filled concrete blocks and insulated cavity brick walls.

Roofs: The insulated lightweight metal roof gives the best year-round performance in all locations, except in zone 7 (Toowoomba), in terms of cumulative over- or underheating. This is surprising for the inland locations, ie. zones 5 & 6 (Cloncurry or Charleville), where conventional wisdom, including our own mini 'expert system', would recommend a concrete slab roof. In the relatively cool climate of zone 7 (Toowoomba) the insulated concrete roof gives a better winter performance.

If the lowest summer indoor maximum temperature is taken as the criterion, then the insulated concrete slab roof is better. The problem in all zones (except zone 7) is that the summer night-time temparatures will also remain high. The concrete slab roof will give satisfactory summer temperatures only with night-time ventilation.

Fenestration: winter solar gain (eg. a direct gain passive system) is necessary in zones 7, 6 & 1 (Toowoomba, Charleville and Brisbane), desirable in zone 2 (Rockhampton), but summer shading is essential in these locations. In zones 3, 4 & 5 (Townsville, Cairns and Cloncurry) full shading of all windows should be ensured all year round.

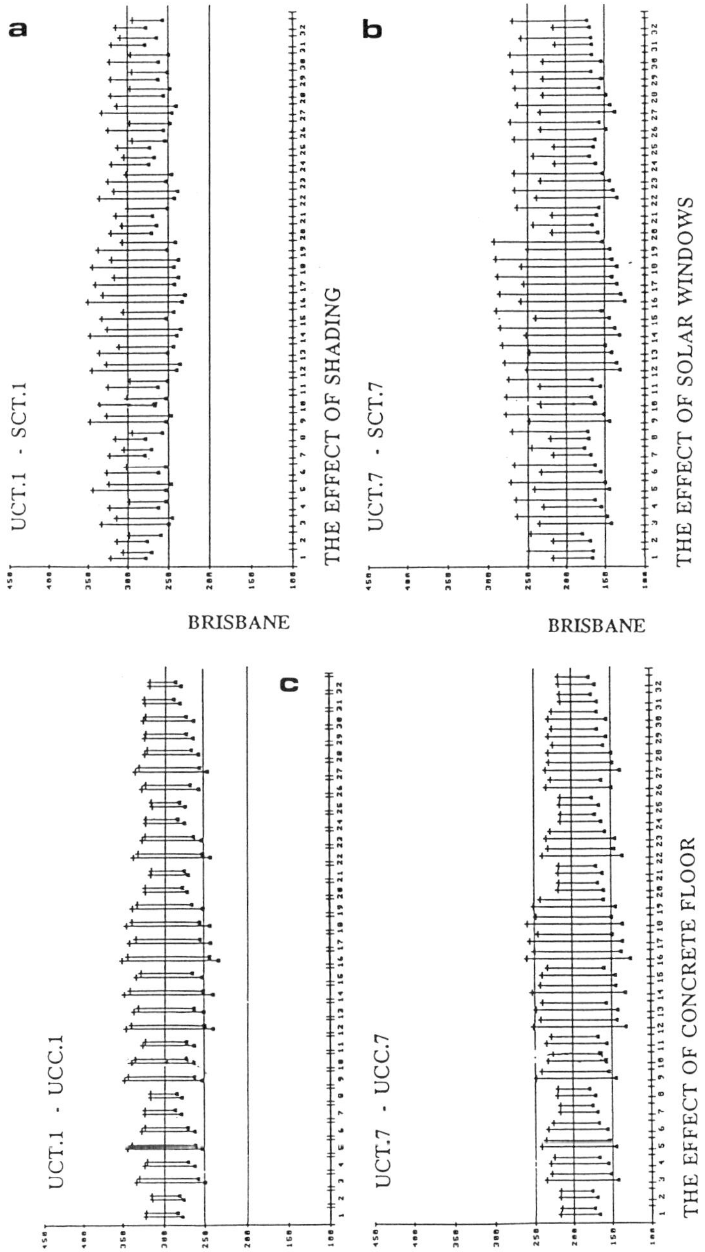

Fig.2. Parametric study: binary comparisons

a. unshaded, constant ventilation, timber floor in January vs. same but shaded
b. normal windows, constant ventilation, timber floor in July vs. same with enlarged solar windows
c. unshaded, constant vent., timber floor, both January and July, vs. same with concrete floor

7. PARAMETRIC STUDY

The results of phase one (indoor minimum and maximum temperatures) were fed into a spreadsheet and plotted in the form of bar-charts, such as that shown in Fig.2. The bars extend from the minimum to the maximum indoor temperature. The 32 pairs of lines represent the 32 constructional (wall and roof) variables, where there is only one difference between the two bars of any pair.

The illustration shows the effect of a concrete slab-on-ground floor, as compared with an elevated timber floor (for Brisbane). It can be seen that the concrete floor in all cases (both in January and July) shortens the lines: elevates the minimum and lowers the maximum temperature. In January shading substantially reduces the maxima and slightly lowers the minima.

The effect of enlarged solar windows in July warrants a closer look. The maxima are increased in all cases, especially for the timber floored variants. In fact some of the lightweight constructions can produce indoor overheating even in this coldest month (reaching 30°C in some cases). However, the minima do not necessarily follow: in some cases the minima are unchanged or even slightly reduced. These are variants 1, 8, 21, 25, 31 and 32, - all of which have concrete roof slabs. Interestingly, the effect disappears if the floor is also concrete.

The study produced a whole series of these "if..., then..." type conclusions.

As the study had a practical relevance, the press became interested when it was released and the public response was so great that we felt compelled to summarise the results in a small popular booklet (Szokolay, 1991). This is available from both the University and the state government's Energy Information Centre at a nominal price.

REFERENCES

United Nations, Dept.of Econ.& Soc. Affairs (1971): Climate and house design, vol.1. Design of low-cost housing and community facilities. UN. New York

Szokolay, S.V. (1983): A thermal design tool for the sketch design stage in Predictive methods for the energy conservative design of buildings, ed. H.J.Cowan, p.17-36, Pergamon, Sydney

Szokolay, S.V. (1990): Climatic design of houses in Queensland. Final report on research project. Arch.Sc.Unit, Uni.of Qld.
also Supplementary report (1991)

Szokolay, S.V. (1991): Climate, comfort and energy (design of houses for Queensland climates). Arch.Sc.Unit, Uni.of Qld.

NEW PORTUGUESE BUILDING THERMAL REGULATIONS: A PASSIVE AND LOW ENERGY APPROACH

VITOR ABRANTES (1)
Depart. of Civil Engineering
University of Porto
Rua dos Bragas, 4099 Porto Codex
Portugal

MENDES DA SILVA (2)
Depart. of Civil Engineering
University of Coimbra
FCTUC, 3049 Coimbra Codex
Portugal

ABSTRACT. The Portuguese building thermal regulations under application, for the first time, in 1991 only establish requirements for the building envelope. The account situation is characterized by the generalized absence of heating or cooling systems in most buildings based on the amenity of the climate and on the accepted discomfort conditions indoors. Therefore, the new regulations are established for nominal conditions inespective of the future use of energy systems.

For Winter requirements, a target value is established. This target value - a nominal label - represents the maximum nominal annual heating needs per m^2 of useful floor area (NI - kWh/m2.yr). For the calculation of this target value reference U-values are defined for each part of the envelope according to the local climatic zone and the maximum allowable window area is 15% of the floor area. For any building the corresponding nominal annual heating needs (NIC) are calculated through a standard procedure in which useful solar gains are taken into account. The building will be in compliance with the regulations when NIC does not exceed NI.

This paper presents the results of the thermal analysis of 130 residential existing buildings in the central region of Portugal and the main conclusions that were obtained:

- No existing residential buildings were in compliance with the new regulations;
- If it is intended to comply with the regulations using only thermal insulation, the increase in construction cost in near 4%;
- However, by taking window areas into account through a more adequate vertical south-projected area, increasing solar gains, then the compliance with the regulations is possible without any increase in construction costs.

These conclusions show that Portuguese thermal regulations can promotte imaginative and diverse architectural solutions in which solar gains can be correctely taken into account.

1. Introduction

For a long period of time typical Portuguese buildings were easily identified with the region where they were built. Using traditional technics and local materials the ancient knowledge often led to high inertia buildings with different glazing solutions depending on the climate. Nowadays most modern buildings are lighter, since concrete frame structure and perforated brick walls are generally used [2] [3].

---

(1) Associate Professor - University of Porto; (2) Lecturer - University of Coimbra

S. Alvarez et al. (eds.), Architecture and Urban Space, 715–720.

Total lack of insulation and thermal bridges lead to a low thermal quality and a low indoor comfort level; the reduced use of heating and cooling systems increase these effects.

Now, for the first time, a thermal regulation is available in Portugal [1]. In order to test its impact several research works have been developed. Since there is no tradition of heating systems, energy savings are only "potential", preventing future energy consumption; the control of construction costs is rather important in a country where more than five hundred thousand new dwellings are necessary; finally, the correct construction technics must be adopted to improve building quality and, particulary, thermal quality.

2. The new Portuguese Building Thermal Regulations (a brief description)

The main goal of the new Portuguese building thermal regulations [1] is to improve the natural climatic response of buildings. The new regulations result of several constraints and compromises establishing goals in terms of energy, comfort level and construction quality, with limited costs and reduced technological changes [3].

The country is divided into 3 Winter regions and 3 Summer regions. The buildings are analysed according to their different inertia.

U-values, thermal bridges corrections and shadding coefficients are recommended for each climatic zone and kind of envelope element ("reference values"). These values, obviously, lead to automatic compliance with the regulations. Designers can use other values providing established limits are respected ("maximum permissible values"). In this case the annual heating and cooling energy needs per m^2 of useful floor must be calculated and compared with the regulations "target-values". In Winter, heating needs can be reduced by solar useful gains.

3. Actual building thermal performance

A set of 130 buildings in the central region of Portugal was subjected to a careful thermal analysis [4] through the new Portuguese building thermal regulations calculation methods. The buildings were selected from all single family buildings in construction stage in six different towns in different climate zones. This procedure allowed both the design analysis and the construction technics identification.

All the buildings were automatically excluded. One or more of the U-values maximum limits, established in the building thermal regulations for each kind of envelope component, were often exceeded.

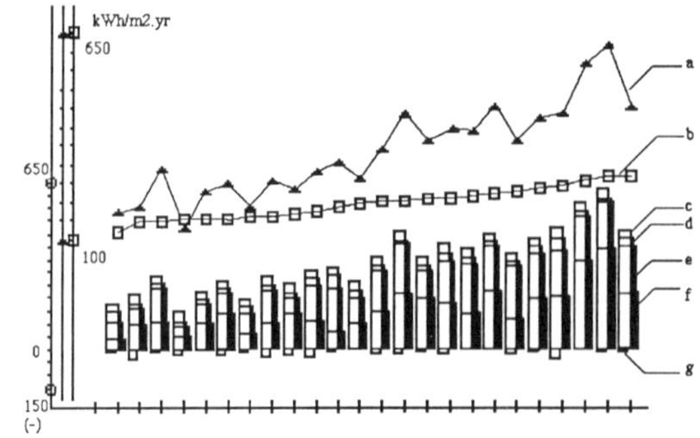

Figure 1.
Actual heating needs -Winter (24 buildings):
(a) nominal needs - NIC
(b) nominal needs regulations limits - NI
- Losses through:
(c) air renovation
(d) glazed surfaces
(e) roofs and floors
(f) walls
- and solar useful gains (g).

Nevertheless, the annual heating and cooling needs were calculated and compared with the corresponding target-value (maximum nominal needs per m^2 of useful floor area established in the code). Only 1.55% of the buildings did not exceed these limits in Winter, while, in Summer, this goal was achieved in a larger number: 62.79%.

The causes of this low thermal performance were studied: comparison of losses (in Winter) and gains (in Summer) through each building envelope component, and technic construction used. Figure 1. illustrates the results of this analysis in Winter in one of the towns; figure 2. illustrastes the results in Summer.

Total lack of insulation, thermal bridges and defficient glazing protection had to be correct!

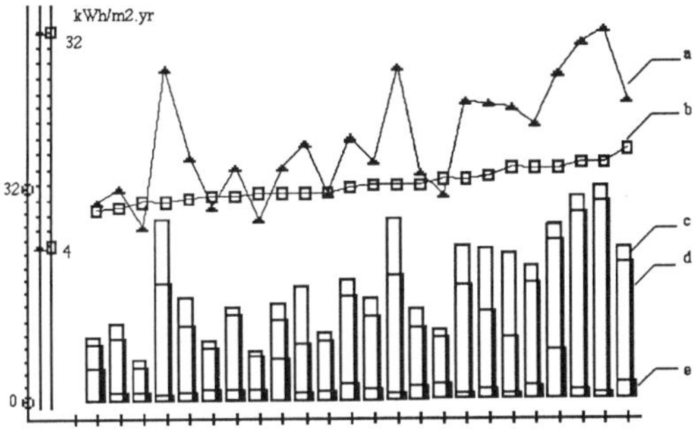

Figure 2.
Actual cooling needs -Summer (24 buildings):
(a) nominal needs - NVC
(b) nominal needs regulations limits - NV
- Gains through:
(c) roofs
(d) walls
(e) glazed surfaces

4. Thermal correction through traditional technical solutions

Every wall, floor and roof has been analysed and modified by simulating a new design, in order to reach the reference level. This procedure had two main constraints: a) each new element should not differ from the initial one in thickness and type, allowing to maintain the architectural and design characteristics and options; b) the insulation should lead to minimum costs, beeing, however, in compliance with the support technical characteristics and the moisture exposure.

Floors and roofs have been completely insulated (with boards 2 to 6 cm thick) in the outside, avoiding the occurence of thermal bridges and increasing the inertia. The inclined roofs have been covered with traditional red ceramic tiles, since Summer climate do not encourage the usage of dark colors.

Single-layered walls have been insulated outside, but double-layered walls have insulated only when necessary, in the air gap, between the concrete structure. The resulting thermal bridges were corrected by using a protective layer of perforated bricks (25 to 70 mm thick) covering the concrete beams.

The impact of this solutions on technological status of Portuguese building industry has been discussed and technical detailed procedures were recommended.

Since no major changes could be made in architectural design, glazed surfaces have not been modified.

All the buildings are now in compliance with the regulations.

Figure 3. illustrates this conclusion and shows the "potential" energy saving in Winter, in each type of envelope component. This performance was achieved with an impact of 4% increasing costs.

A more complete and innovative approach in design phase can, however, be made if a new architecture, dealing with building form, orientation and window placement, is allowed.

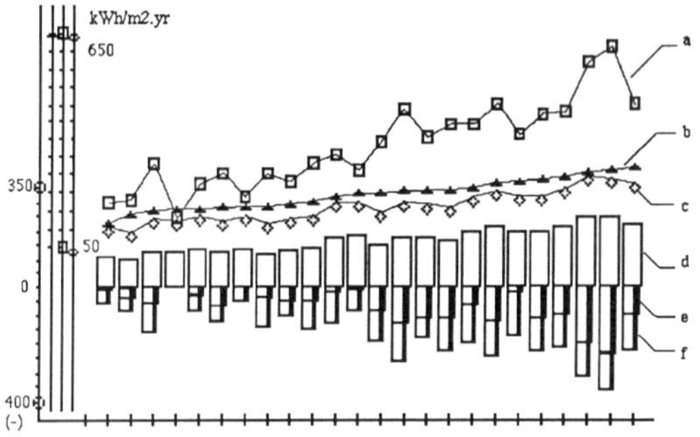

Figure 3.
Winter heating needs before and after insulating correction:
(a) initial nominal needs - NIC
(b) nominal needs regulations limits - NI
(c) or (d) nominal needs after insulation - NIC
Potential energy saving through:
(e) walls
(f) roofs and floors

5. A passive and low energy approach

The new strategy consists in giving a particular attention to building and glazed surfaces orientation, increasing the useful solar gains in Winter , but also avoiding Summer overheating.

The original buildings were sequencially subjected to three transforming operations:

5.1. MINIMUM OPAQUE ENVELOPE CORRECTION

Since many envelope opaque construction elements exceeded maximum U-values allowed, as referred above, it was necessary to reach this limits. They are higher than the reference U-values used in the traditional thermal corrections described in the previous item.

The walls are now non-insulated, double layered (each layer in perforated bricks 10-15 cm thick separated by a 5 cm air gap). This solution takes into account the traditional construction technics, the low costs, the acceptable thermal performance and the rain protection. This solution provides also an easier thermal bridge correction.

The roofs and floors are insulated as well as in the previous approach, because the reference U-values and the maximum limit U-values are too close and the available solutions for each one do not differ significantly. This decision is also based in the knowledge that the roof insulation is the more efficient and lower cost in this kind of building.

After this, studied buildings are no longuer "automatically excluded" by the opaque elements behaviour. Nevertheless, only 45% has an annual "potential" energy comsuption - to reach the comfort reference level established in the code - bellow the target-value.

Reaching this non-excluding thermal skills in opaque envelope increased construction costs 0.7%. It is now possible to begin the solar analysis.

5.2. CHANGING BUILDING ORIENTATION AND PROTECTING WINDOWS

The main goal of this step was to increase the solar useful gains (calculated through the concepts of vertical south-projected area) to reduce the heating needs in Winter.

This was made in the 55% reminder buildings; they have been rotated until the major glazed surface matched South orientation. Additionally, unprotected windows have been corrected with ligth colored roller blinds outside improving Summer shadding effect and reducing U-value at night, in Winter.

Solar useful gains can be taken into account provided there are no shadding obstacles, according to sun elevation in Winter. For this reason, when this obstacles occured, a new dimension was suggested allowing not only the solar gains but also, where possible, a Summer additional shadding. This double goal, often neglected in design, is easy to reach in South orientation in Portuguese latitude.

After this process 80% of the buildings were in compliance with the new building thermal regulations. The other 20% needed a deeper analysis.

5.3. CHANGING PROPORTIONAL WINDOW ORIENTATION

In the previous section was described how the initial buildings were rotated until the major glazed area is South oriented. Nevertheless the glazed areas in other orientations still remain very high. In the buildings in which an upper value of solar gains was desired, all the windows (except South orientation) were reduced to half of their surface, transferring the other half to the South wall, with no consequences except architectural re-design. When this change was not enough, more 25% of the windows surfaces were transferred to South, leaving in the original re-oriented wall 25% of the initial windows.

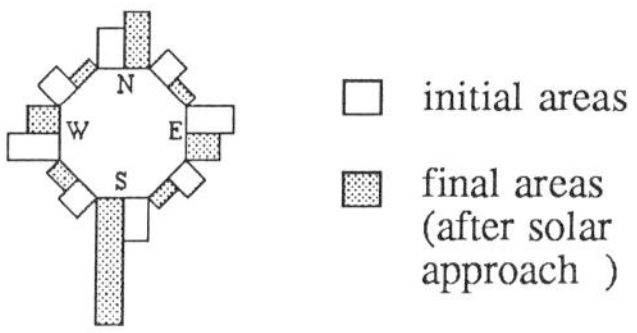

Figure 4.
Comparison of glazed surfaces orientation before and after the solar approach

Figure 4. illustrates the relation between the initial and final glazed surfaces in each oriention. It is important to remember that 45% of the buildings did not need to be rotated and 35% were South rotated without changing the initial window proportions. Nevertheless, an important South glazed area is obtained. The increasing of North glazed area is the result of a traditional architecture with two main opposite façades; in some cases, when rotated to South, an important glazed area was also oriented to North.

In the most severe climate regions this work was completed with an additional correction of thermal bridges.

The final goal is achieved: all the buildings are, now, in compliance with the new Portuguese building thermal regulations, not only in Winter but also in Summer.

5.4. ENERGY SAVING

Comparing the heating needs in the original buildings with the heating needs after the solar approach, calculated by regulations methods, the result is a 40% "potential" energy saving in Winter.

Even if the heating systems have a reduced use, an improved comfort level is expected. Figure 5. illustrates the final heating energy needs for buildings from one of the towns. It shows the losses through opaque and glazed envelope surfaces and infiltration losses, but also, useful solar gains obtained. Comparing these with Figure 1. it is quite clear that in this approach the solar useful gains have a remarkable contribution to improve the thermal quality and energy saving.

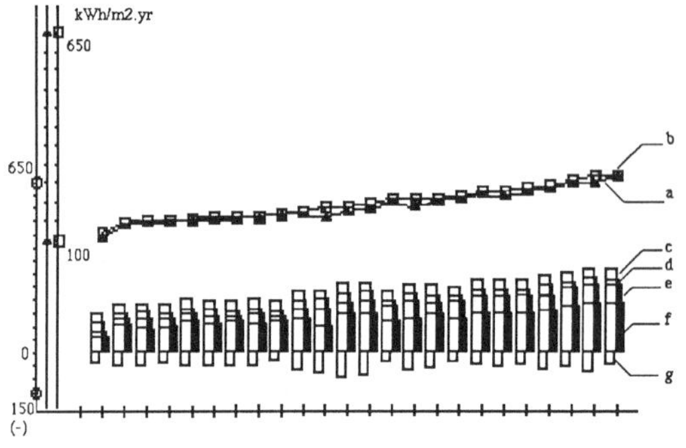

Figure 5.
Actual heating needs after solar approach:
(a) nominal needs - NIC
(b) nominal needs regulations limits - NI
- Losses through:
(c) air renovation
(d) glazed surfaces
(e) roofs and floors
(f) walls
- and solar useful gains (g).

5.5. COSTS

Most of the corrections made through this solar approach do not represent an additional cost provided they will occur in design phase. The final impact in costs is about 0.9%, which means that the solar approach increased costs in a meaningless value: 0.2%

6. Conclusions

It is urgent to improve general building quality in Portugal. One of the more significant recent contribution to this goal has been the new Portuguese building thermal regulations .

The developed analysis shows that a careful insulation of roofs is recommended, as well as a thermal bridge correction, to reach thermal behaviour minimum skills.

To improve thermal quality and to be in compliance with the new regulations, more changes are required to design and construction. Maintaining traditional architecture (i.e. insulating walls, unselecting adequate orientation and windows) it is possible to reach this goal with 4% of increasing cost.

If a passive and low energy approach is made in design phase (i.e. a correct building orientation and a selected window placement) those costs will decrease to a meaningless value of 0.9%. Futhermore, this new design strategy will lead to 40% "potential" energy saving. As the new Portuguese building thermal regulations have been available only since the beginning of 1991, its actual effects on building quality are still unknown.

REFERENCES:

[1] "Regulamento das Características de Comportamento Térmico dos Edifícios" (1990) (Portuguese Building Thermal Regulations), Decreto-Lei 40/90, Diário da República nº31, SÉRIE I, Lisboa, pp. 490-504.

[2] Maldonado, E. and Oliveira Fernandes (1989) "Improving Housing Quality Through Thermal Regulations", Proceedings of IAHS World Congress on Housing - University of Porto, Pergamon Press, Exeter, pp. 211-216.

[3] Maldonado, E. and Oliveira Fernandes (1989) "The Design of Energy-Efficient Envelopes in Mediterranean-Type Climates", Proceedings of "Thermal Performance of Exterior Envelopes of Buildings - IV", Orlando, USA , pp. 135-146

[4] Silva, J.Mendes (1990) "Building Behaviour and The New Portuguese Thermal Regulations" (in Portuguese), Msc Thesis, Depart. of Civil Engineering, University of Porto.

Chapter 3

Case Studies

MEASURED PERFORMANCE OF A PASSIVELY COOLED INDUSTRIAL BUILDING IN MALTA - PRELIMINARY RESULTS

BRIAN FORD
Peake Short & Partners
15 Prescott Place
London SW4 6BS

ABSTRACT. This paper desribes preliminary results from the first few months of monitoring the thermal performance of the passively cooled process building at the Simonds Farsons Cisk Brewery in Malta. Although only limited data is available it can be shown that with daily maximum external air temperatures regularly above 35°C, the process hall air temperature remains below a peak of 27°C for all but extreme days. The process hall floor surface temperature shows a mean of 25.3°C for this period and is very stable, while the 'jacket' air temperature fluctuates more widely between process hall and external air temperature. These results are broadly in line with performance simulations made at design stage with a two zone finite difference model.

1. Introduction

The design of the new process building for Farsons Brewery in Malta has been described previously (1). The building was officially opened on July 12 1990, and the design has been very favourably received (2). The building's performance has been monitored since September 1990. Initial results are very encouraging, and it is anticipated that over the next year analysis will reveal more detail on exactly how night ventilation coupled with internal thermal mass maintains these very stable temperatures.

S. Alvarez et al. (eds.), Architecture and Urban Space, 723–728.

2. The Building

The process building was designed to exploit the high thermal capacity of the fabric coupled with night ventilation during the summer to maintain acceptably cool conditions internally and to minimise the cooling load associated with the vessels inside the building. The process hall is surrounded by a 'jacket' of corridor/circulation space (Fig.1 plan). This has a number of functions: reducing solar gain into the process hall; providing a plenum via which the process hall is ventilated; allowing daylight to be reflected into the process hall via high level glazing; providing a major route for piped services as well as an access corridor for workers in the building.

The process area is ventilated via high level openings in the 'towers' above the jacket (Fig.2). The glazed vent openings into and out of the process hall are automatically controlled, regulated to open at night on a temperature differential and to close in the morning according to ambient light level.

3. Evaluation

The objective of this performance evaluation is to test the design intention. This includes:

1) Achievement of satisfactory summer time temperatures within the process hall by means of
   a) Night time ventilation
   b) The 'buffering' effect of the jacket space
   c) The high thermal mass of the building fabric

2) Achievement of satisfactory daylighting of the interior by means of indirect and reflected light via high level clerestorey glazing.

Appropriate methods include both short term (spot checks) and long term (continuous) measurements. Spot checks of air movements through vent openings and within the jacket have been made using a thermal anemometer, providing an indication of air speed and direction. Continuous recording of temperatures (at ten minute intervals) at 12 different locations within the building allow comparison with external air temperature for basic assessment of performance. The data logger used is a 16 channel 'Squirrel' recorder, Type 1206 by Grant Instruments of Cambridge.

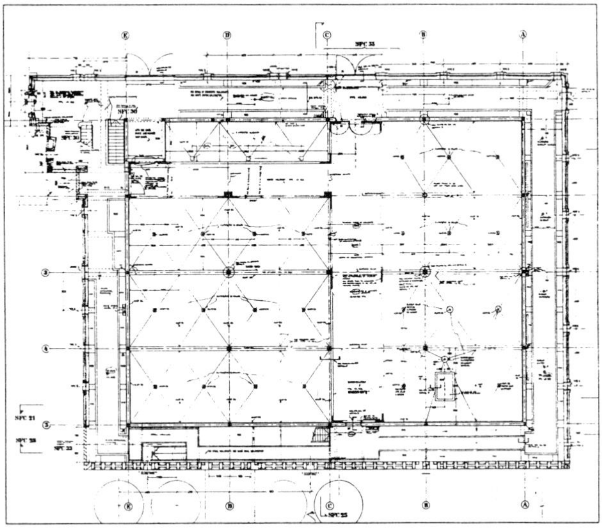

Fig.1. Process Hall Floor Plan

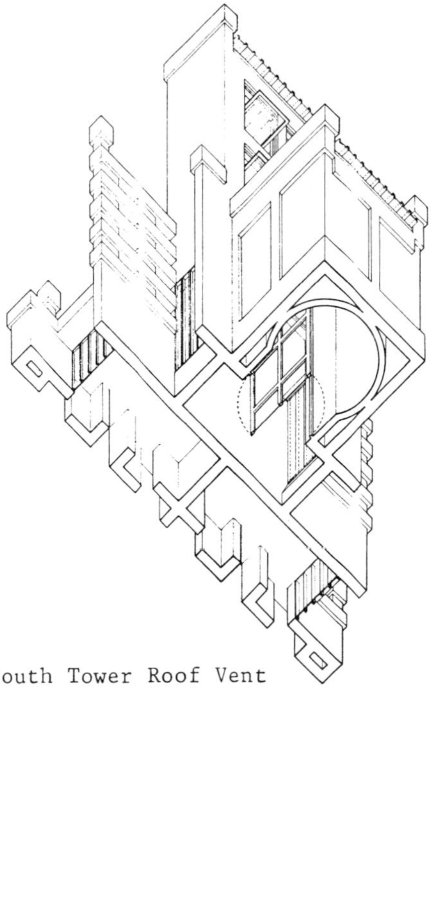

Fig.2. South Tower Roof Vent

4. Preliminary Findings

Data is so far only available for the months of September - November 1990. This is clearly a very limited set of data, but preliminary conclusions can be drawn.

Simulation of performance at the design stage predicted that for all but extreme days in August, the dry resultant temperature within the process hall will stay below 27°C. An 'extreme' August day is characterised by an external air temperature above 40°C. Records indicate that September 1990 was about average (mean monthly external air temperature 23.3°C compared with 24.0°C for the 80 year average). The daily maximum external air temperature was regularly above 35°C, and night time minima were generally above 20°C, with an extreme minimum for the month of 16.8°C.

So, in this context, how does the building perform? For September as a whole the mean air temperature in the process hall was 25.7°C, the minimum 23.1 and the extreme maximum 28.9°C. The stability of the temperatures within the process hall is demonstrated clearly in Fig.3. which shows that for a week (3 - 10 Sept) when daily maximum external air temperatures were regularly above 35°C, the process hall air temperature remained very stable, with a maximum diurnal swing of 2.5°C and a mean of 24.9°C.

The effect of night ventilation can be seen by comparing results for 3 - 10th September with 9 - 15th September (Fig.4). Lower nightime air temperatures coupled with lower daytime solar heat gains result in a considerable drop in the amount of heat stored in the roof slab. It appears that a cool night (minimum below 20°C) has a profound effect on the start temperatures for the following day, and that a succession of cool nights can have a marked effect in depressing surface temperatures within the process hall. This effect is carried through to the following week (15 - 21 September), with process hall temperatures remaining stable around 24.0 C.

For October the monthly mean external air temperature was 23.0°C, with a maximum of 39.8°C and a minimum of 10.2°C. The mean process hall air temperature during this period was 23.1°C, 2.6°C below the previous monthly mean.

It is apparent that cooling associated with night ventilation, the 'buffering' effect of the jacket, and the influence of internal thermally massive floors and walls all contribute to the extremely stable temperatures within the process hall. The building is behaving broadly in line with simulated performance, and it is anticipated that the

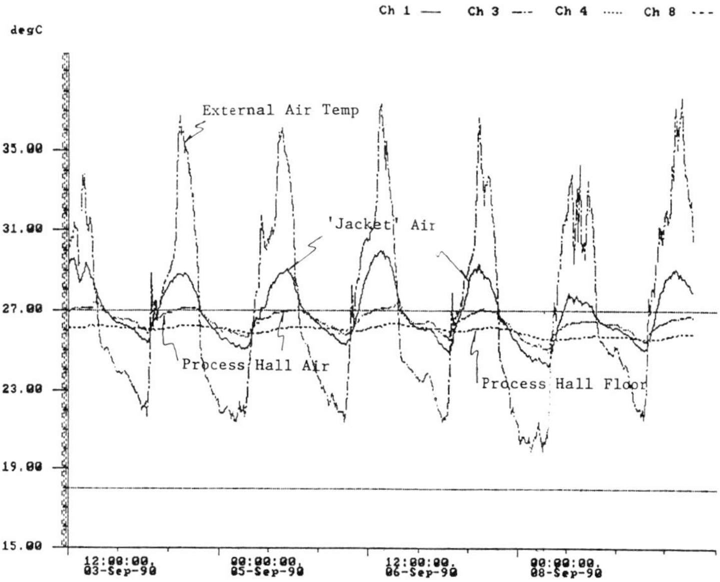

Fig.3. 3 - 9th September 1990 SFC Brewery, Malta.

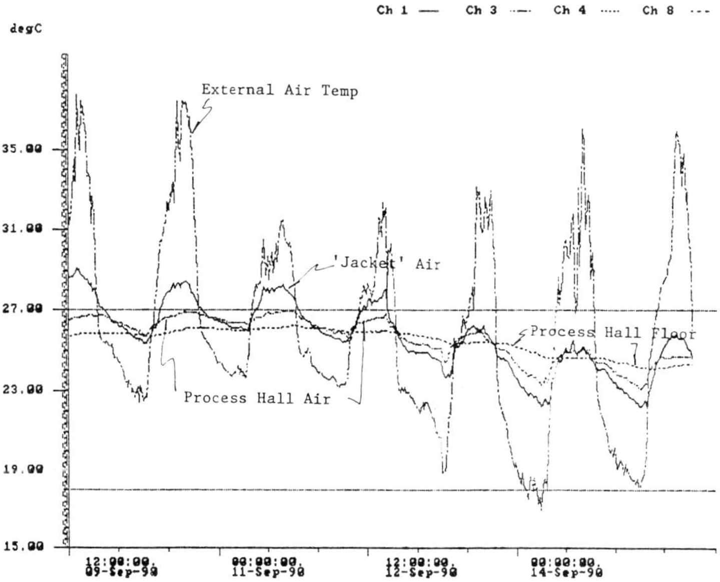

Fig.4 9 - 15th September 1990 SFC Brewery, Malta.

continued programme of measurements will enable disaggregation of the influence of the different elements.

5. Future Measurements

The programme of measurements will continue for at least a further 12 months, and it is intended that additional sensors (to measure relative humidity and solar radiation) will be installed. A series of spot measurements of air movement will be made during the summer of 1991 to obtain information on air change rates. The influence of building geometry on the pattern of air movement will be studied using smoke (to trace movement through vents) and helium filled neutral buoyancy balloons (to trace movement through large spaces). Spot measurements of daylight levels will enable the determination of daylight factors throughout the building.

It is anticipated that the final programme of measurements will enable a detailed investigation of parameters for which little empirical data currently exists, including:

1. Convective heat exchanges to and from the soffit of the roof slab
2. Long term mass effects on the stability of internal temperatures
3. User assessment of the operational success of the building.

This investigation is being supported through the award of a research fellowship at the Martin Centre, Cambridge University, by the Leverhulme Trust.

References

1. Ford B., and Short A. (1988) 'Design of a passively cooled brewery building in Malta'. PLEA Conference Procedings, Porto, Portugal.

2. Rickaby P., '(1991) The art of energy: Peake Short & Partners - Malta'. Architecture Today No 14.

MODULAR SYSTEMS FOR NATURAL VENTILATION-RESEARCH ON NEW TYPOLOGIES FOR A TROPICAL ARCHITECTURE

AGUSTIN ADARVE GOMEZ
Arch. M.Sc.A. U de Montreal
Solaire Ltd-Cra 7 32-33 Of 1703
Bogota, D.E. 1
Colombia S.A.

ABSTRACT. Passive solar cooling for hot humid areas represents an important field for innovation, if we want to solve comfort needs in houses designed to reduce economical, technical and health requirements.

In urban areas or deep valleys which is a common situation in Colombia and many tropical countries (specially those located around equatorial areas), external breeze is not frecuent and air speed patterns are very low in order to produce cooling using a single cross ventilation system (fig.1).

Since 1982 SOLAIRE LTD has developed new designs and constructions using passive cooling techniques. Thus, the purpose of this paper is to show the application of modular cooling systems able to promote indoor air flow, reduction of internal temperatures and to control, both direct and indirect, solar radiation.

The main objectives of the investigation are : a) Comparation between real results and theorical responses within houses designed according to bioclimatic requirements as predicted by main researchers like B. Givoni [1], J. B. Dick [2], Karakalsanis C. [3] and the author [4], [5]. b) Evaluation of the influence borne by the envelope (internal space) and specific cooling elements like openings, walls, ceiling, roof top (solar chimney). c) Description of apropriate architectural typologies (as modular components or the whole building). d) The methodology employed consists in testing three different houses (urban/suburban), taking into account the following parameters. e) Measurement of indoor and outdoor air speed to determine how the cross ventilation and stack effect occurs. f) Measurement of temperature differences between outside/inside, first/second floor and ceiling (chimney or roof top). g) Registration of indoor air flow patterns (laminar flow, turbulent flow, still-air areas).

As a conclusion, a comparative table shows the effectiveness of different spaces and some designs describe improved strategies to assure bioclimatic comfort.

Minimun cooling devices are proposed such as : openings type A-B-C which improve high pressure zones, convective flow or evacuation of hot air. Also, the solar chimney or wind turbine coupled with a sloped ceiling increase the stack effect. Aluminum foil reflective ceiling and

S. Alvarez et al. (eds.), Architecture and Urban Space, 729–735.

overhangs placed over east/west facades control the temperature difference.

1. INTRODUCTION

According to the analysis of the bioclimatic building chart (fig.2), we can locate the comfort zone in hot humid areas (standart temp. 20-35°C and RH above 70%), for interior air speed between 1.20 m/sec and 1.50 m/sec. By placing openings between opposite walls, we can obtain theorically 42% of external breeze when cross ventilation occurs.

By installing a cooling a tower 3. mts over the floor [2] the stack effect could produce a decrease of 5°C (when external air speed is about 1.6 m/sec). Indoor air speed could reach .10 m/sec to .50 m/sec [3].

Finnally, reflective ceiling and clear painted roofs could reduce from 2°C to 5°C. This insulation must be guaranteed if we wish to shift the comfort requirements from the zone V3 thoward the zone V1 (air speed 1.00 m/sec, see fig.2) [4].

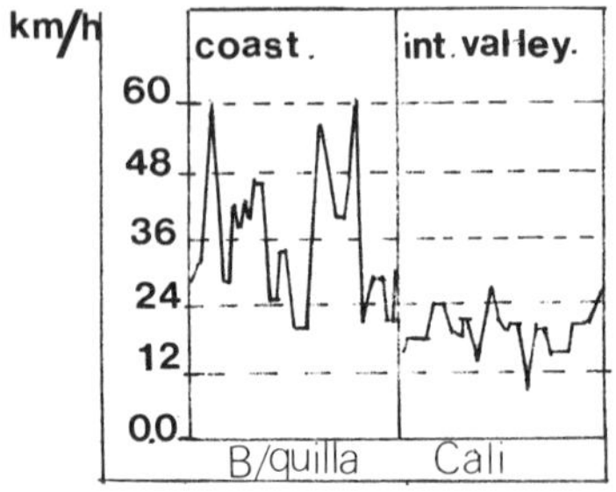

fig.1 . Col. wind pattern

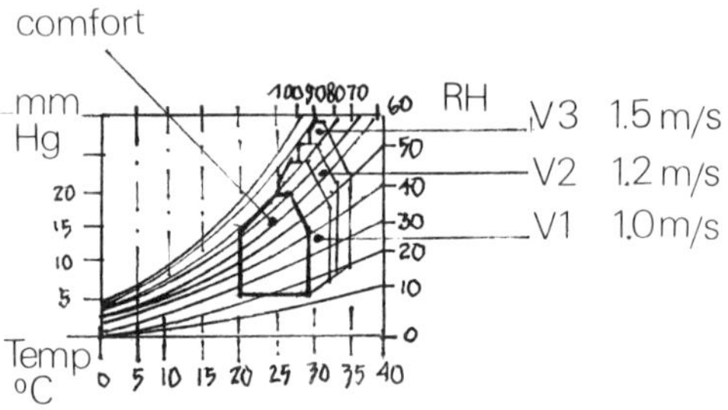

fig.2 . Bioclimatic Chart

2. CASE STUDY No.1

This house is located in the urban area of Honda, 4° Lat N, 200 mts ASL, with climatic data described in fig.6. The construction was designed as a "turbine" to promote maximun cross ventilation effect. A 45° sloped ceiling is coupled with an open top roof to promote a large stack effect. A ventilated double roof assures the insulation. The building is made with standar available materials like asbestos cement for the roof, cement blocks (with air cavity) and wood board for the ceiling.(fig.3).

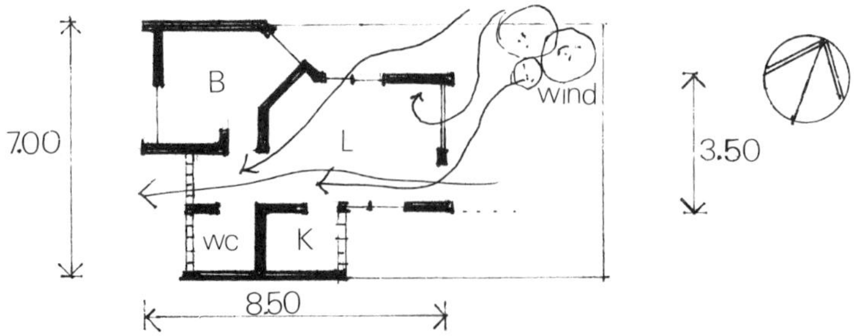

fig.3 . Conjunto Campestre Honda

Anemometer measurements were taken during february (peak period) and the results of the air speed are plotted in figs. 4,5 (first/second level 1.00 mts from opening and 1st/2nd level 1.50 mts from opening). The indoor temperature pattern is shown in fig.6.

The air flow pattern is described in fig.7. It shows the horizontal air flow, (laminar motion) the turbulent flow, the vertical air movement and the interior air speeds, depending on the location within the house.

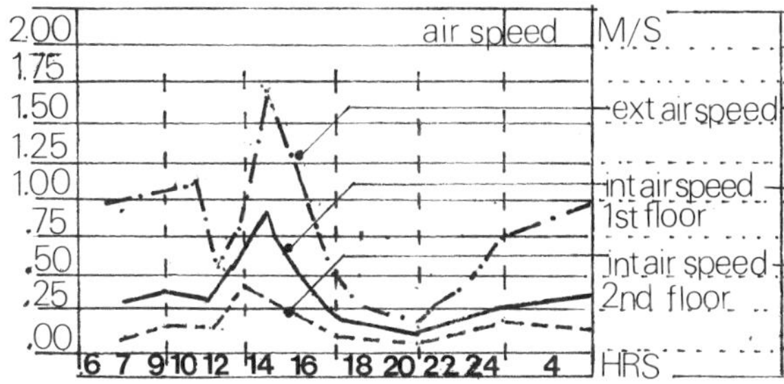

fig.4. Air speed 1.00 mt

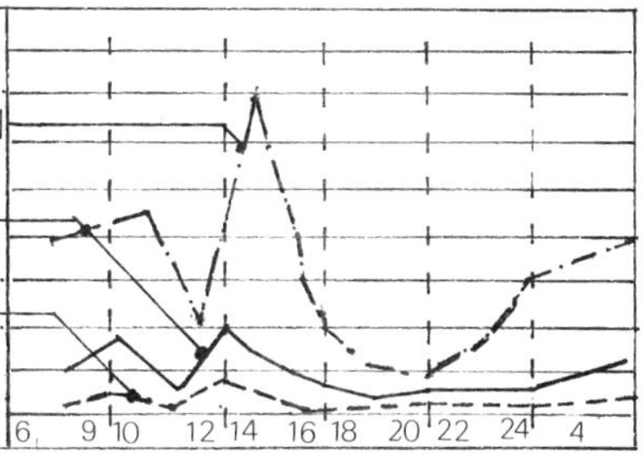

fig.5. Air speed 1.50 mt

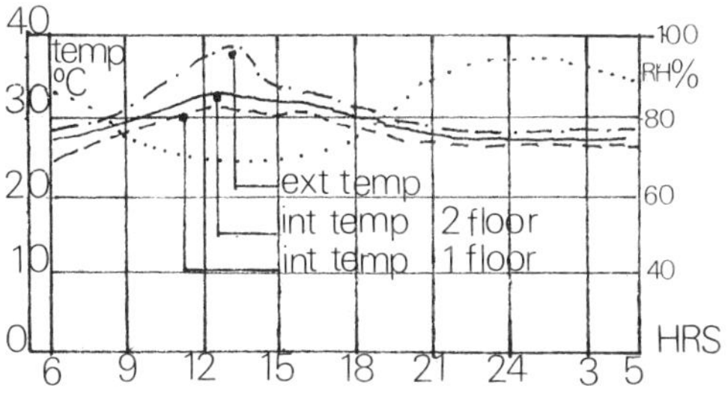

fig.6. Temperature pattern

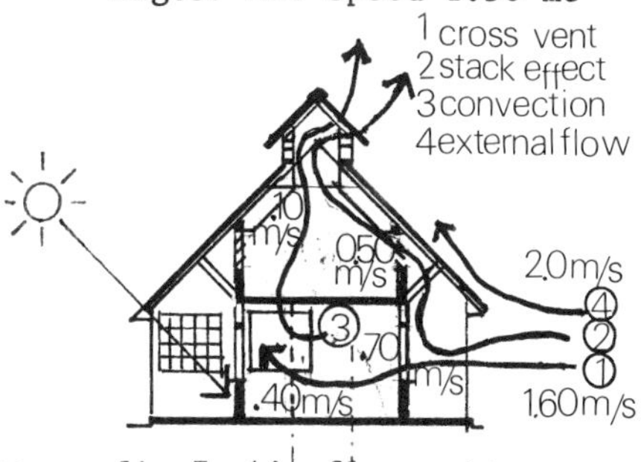

fig.7. Air flow pattern

3. CASE STUDY No.2

This house is located near Bogota (4°N) but 1.000 mts ASL. See climatic data, fig.8. The bioclimatic approach of this house consists basically in improving a combined cooling system by placing a "greenhouse" on the top of the house so that it could receive a greater amount of solar energy in order to produce the largest diference in temperature between bottom and top. In this way the stack effect is improved and the cooling system assured. In fact, a cross ventilation effect is produced by placing a large outlet (jealousy) on the top to exhaust the air comming from the inlet openings. Results of this combined vertical/horizontal air flow pattern is showen in figs. 9, 10, 11. An aluminium reflective double roof with a white painted mortar ceiling was incorporated as a necessity to reduce maximun air temperatures (5°C). A .25 mts cavity wall with white painted plaster (U value 1.43 kcal/hr °C cm2) helps to control heat transmition.

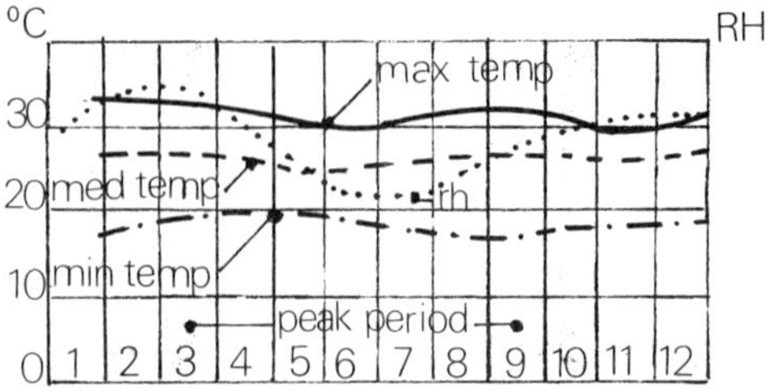

fig.8. Climatic Data

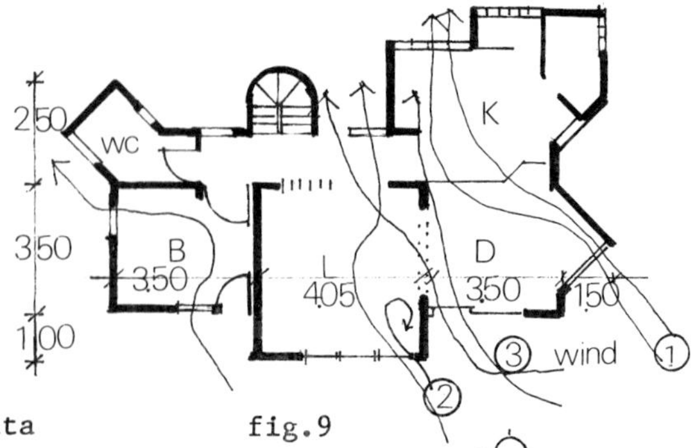

fig.9

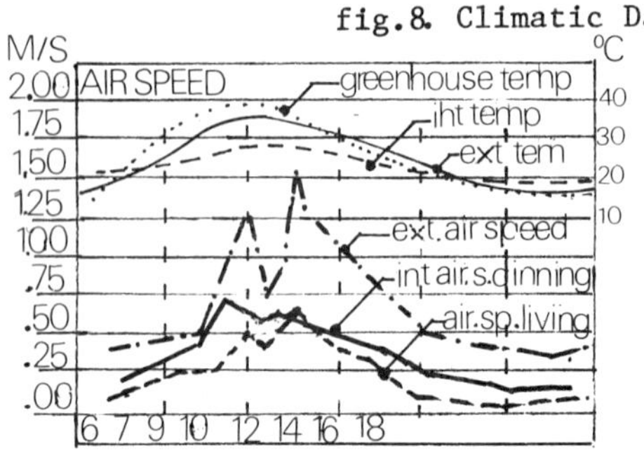

fig.10 . Air speed

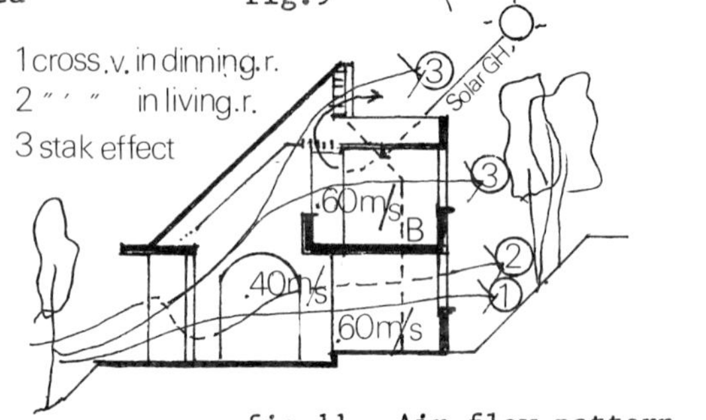

fig.11 . Air flow pattern

4. CASE STUDY No.3

It consists of a special low income house program in a little urban community, Utica (300 mts ASL). Where we find a hot/humid climate with prevailing winds from NE (Rionegro Valley). In 26 selfbuilt houses a new typology take into account a 9.00 x 9.00 mts terrain area (the usual is 6.00 x 6.00 mts), that could make possible to place all living areas facing the prevailing winds, and so create the minimun cross ventilation effect (fig.12). With a very low budget (US$1.500 per house) it was necessary to increase the stack effect through a sloped zinc-foil roof without ceiling. The windows were improved to permit a laminar air flow pattern (with standarized dimenssions) and louvers placed over east/west side made possible (80% of facade) the protection against direct solar radiation in.

It's important to note that the circulation area located on the rear, provides buffer from direct solar radiation. This zone controlles heat flow through the building enveloppe (convective or conductive heat flow). Also, vertical sun shades produce a double cooling effect. First, they controlle direct solar radiation (during the morning or afternoon), and second, they actuates as deflectors that could promote laminar air flow to indoor, that is needed for ventilation.

Results of indoor air speeds and temperatures are plotted in fig.13.

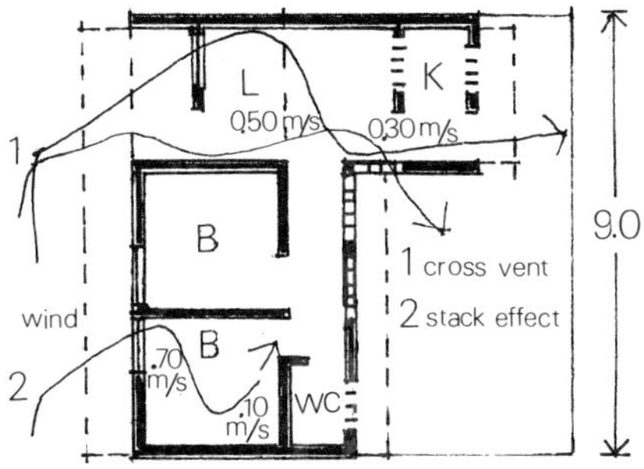

fig.12. Flow pattern

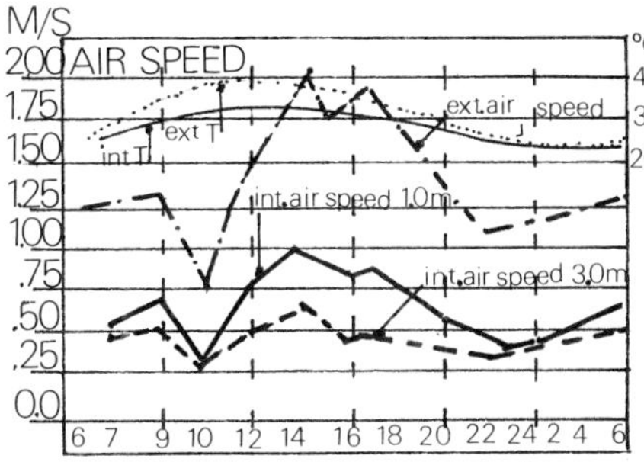

fig.13. Air speeds/temp.

5. RESULTS AND CONCLUSIONS

Looking at the comparative table it is important to note that the following elements make it possible to obtain real effectiveness in a passive cooling system for hot humid areas.

	case study 1	case study 2	case study 3
REQUIRED INDOOR AIR SPEED M/Sec	1.20	1.00	1.20
INT.AIR SPEED. PREDICTED % OF EXT.A.SP.	42	42	42
AVE. MEASURED INT AIR SPEED M/S	.70	.50	.40
INT AIR SPEED. % OF EXT.AIR SPEED.	50	40	27
EFFECTIVENESS %	58	40	34

MAIN DESIGN FACTORS AFFECTING COMFORT
definitivly important : X relative: /

CASE	orientation to pred.wd.	opening/wall distribution	45° pitched double roof	reflective singe roof	roof vent. solar chimney
1	/	X	X		/
2	/	/	X		X
3	/	/		X	X

a) In urban areas landscape elements (trees, hills, neigborhood), decrease eolic potential by 50%. It is prudent to consider 1.00 m/sec to 2.00 m/sec for the DESIGN external wind speed. b) When opossite openings are placed close to an adjacent wall, a laminar air flow take place. The mesured air speed is about 80% of the external air speed (not continuos) Givoni's prediction is 66% to 98% [1]. c) When cross ventilation effect occurs indoors, (2.50 mts from opening) the air speed represents 50% of the external air speed. A standar speed of .70 m/sec is equivalent to 60% of the effectiveness required during peak climatic conditions (1.20 m/sec).d) The stack effect is really efficient (.80 m/sec) when suction exists between the first and second floor. This is possible by the combination of a pressure difference (solar tower) and cross ventilation from bottom to top (diagonal motion). The laminar air flow benefits from the 45° sloped ceiling in order to increase its speed (see case study 2).

On the other hand, vertical anabatic natural airflow could reach maximun air speed of .10 m/sec to .20 m/sec (internal temperature under 30°C). e) Reflective roofs, ceilings and overhangs must be incorporated in all cases. Ceilings .40 mts under the roof, need not be ventilated, except if they actuate as solar collectors to create suction of hot air by natural convection. A decrease of the external temperature by 5°C to 6°C is produced by this means to compensate the impossibility to produce the required air movement.

As a result of the above analysis, the following schemes show some modular systems for natural ventilation in tropical areas. These design criteria have the purpose of establishing an appropriate technology for both developing and industrialized hot countries.

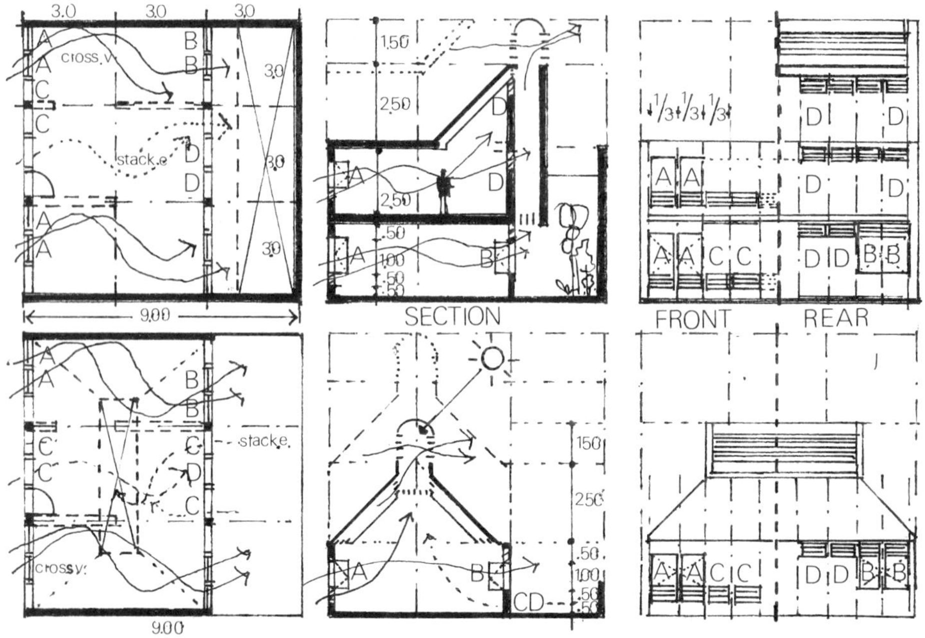

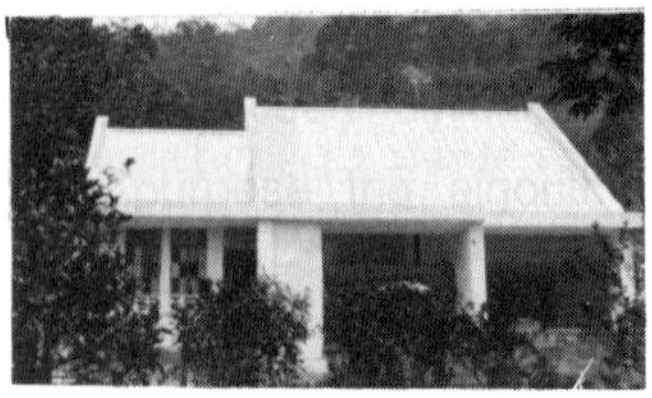

Example 1
Case St. No.2

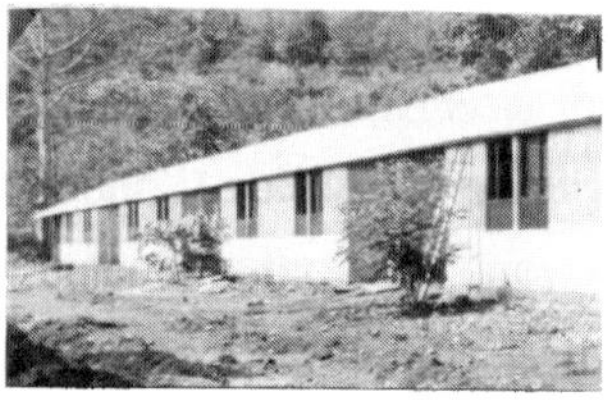

Example 2
Case St 3

6. REFERENCES

[1] Givoni B. (1978)."L'Homme L'Architecture et Le Climat", Ed. du Moniteur, Paris.

[2] J.B. Dick (1950)."The Fundamentals of Natural Ventilation of Houses Journal of Heating and Ventilating Engineers," 123-134.

[3] Karakatsanis, C, M.N. Bahadori, et Al (1985)."Evaluation of pressure coefficients and air flow rates in building employing wind towers," in E. Bilgen and K.G. Hollands (edsi), ISES Congress Abstr. Intersol 85', Montreal.

[4] A. Adarve, E. Mayer, P. Rivet (1987)."Natural Ventilation in equatorial Developing Countries," in W.H. Bloss and F. Pfisterer (eds), 1987 ISES Congress Abstr Pergamon Press, Hamburg, pp.3238.

[5] A. Adarve (1989)."Evaluacion del Circuito de Ventilación Natural en Clima Tropical Ecuatorial,"in H. Rodriguez (ed) VI Congreso Latinoamericano de Energia Solar, Impregraf Hd. Bogota, pp. E59-E66.

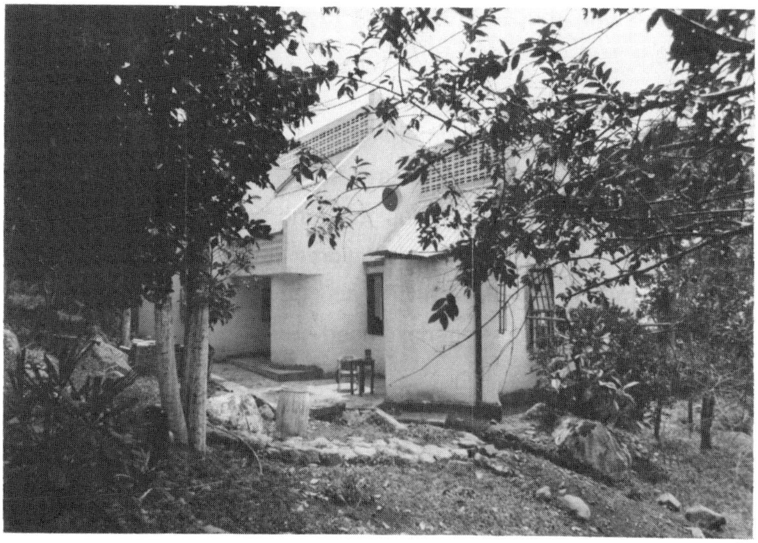

1

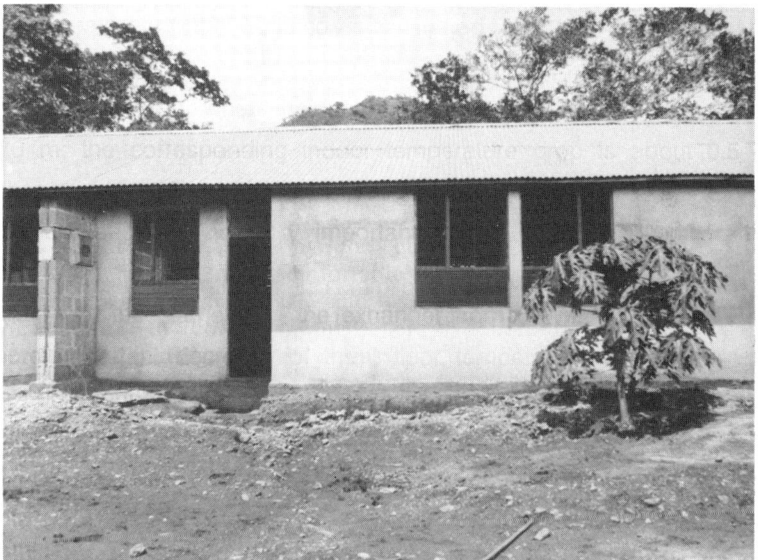

2

ANNEX No.1

1. Casa Santandersito (prop. Emiro Gonzalez)
2. Vivienda Popular Utica (Inscredial)

USE OF BASEMENT IN PASSIVE COOLING OF AN ADMINISTRATIVE BUILDING IN SION (SWITZERLAND)

B. LACHAL (1), R MELDEM (2), W. WEBER (1), C. ANCAY (3), O. GUISAN (1)
(1) CUEPE, University of Geneva, CH-1231 Conches
(2) Meldem Energie SA, Consulting Engineers, CH- 1920 Martigny
Center for Building Science, Lawrence Berkeley Laboratory, Berkeley, CA 94720, USA
(3) Etat du Valais, Département des Travaux Publiques, Service des Bâtiments, CH-1950 Sion

ABSTRACT. This study presents the measurement and the analysis of a passive cooling in an office building. Dividing the cooling plant into three subsystems, we model the whole system in order to determine which parameters are the most influential on the system efficiency.This paper discusses the possibility to use the basement in passive cooling.

1. Introduction

In an urban context, even in countries with a moderate climate like Switzerland, cooling is often necessary in administrative buildings when considering effects like : urban microclimate, solar gains through windows, lighting and internal gains. In a well designed building, cooling needs are not high enough in intensity and time to justify the investissement cost and the electricity consumption necessary to install and to run a standard air conditionning system. Because these buildings generally include several underground floors, we are investigating the possibility of using the fresh air from the underground floors to cool down the offices.

2. Plant description

We have studied a simple case in order to acquire a good know how for such systems. The chosen building is located in the city of Sion, which lies in the Rhône Valley, in the Alps. Built in the 19th century,it was renovated in 1989 and devoted to administrative offices. In summer time the outside temperature can easily reach 30°C and even if the building is properly insulated, the last floor (mainly garrets under a slated roof) can reach more than 28°C. It was then desirable to install a cooling system for this last floor. The building includes wide and deep cellars about 10 meters underground, as seen in figure 1. The air used to cool the last floor was brought from these cellars thanks to a fan. The main characteristics of the plant are :

S. Alvarez et al. (eds.), Architecture and Urban Space, 737–742.

cooling source : cellar

deepth : 10m
volume : 300m3
walls + ground area : 500m²

Air distribution network

fan : 1000m3/h, 400W
ducts : steel, not insulated, 8m before fan and 23m after.

Cooled area

total area : 200 m²

Monitored office

South-West orientation
.wall insulation : 0.10m
.roof insulation : 0.16m
.area : 17.5 m²
.volume : 44.5 m3
.window : double glazed
.skylight window area : 1.1 m²
.shell window area : 0.4 m²
.changing air rate : 4.5 per hour
.internal loads : 78 W (1 pc on 24h/24)
.employees : 1-2

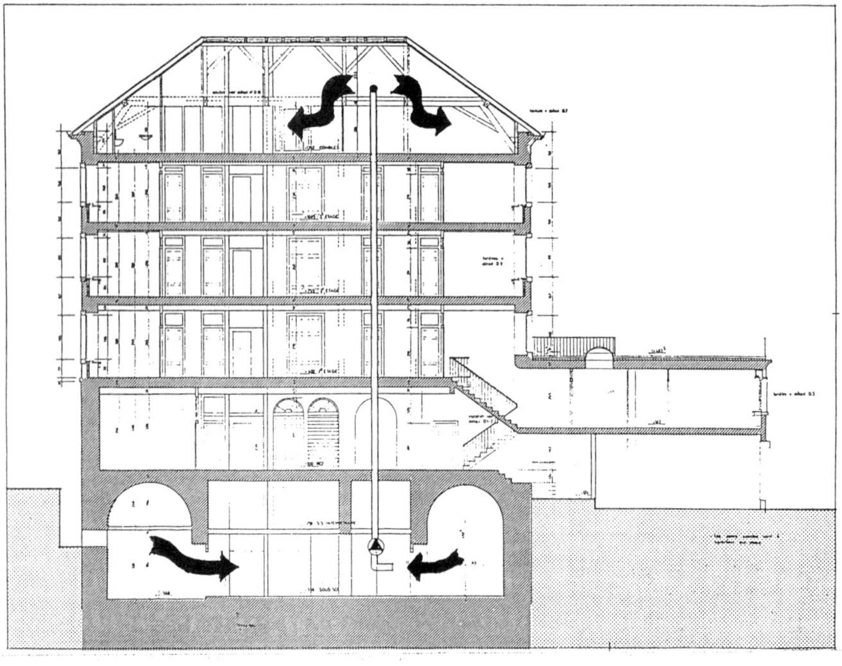

Figure 1. Passive cooling system.

3. Monitoring and results.

We recorded data (mainly temperatures) during the summer 1989, in order to characterize the building without cooling system. The passive cooling plant was completed during the spring 1990. We then started the monitoring of the building to evaluate the performances of the cooling system. The data deal with hourly measurements on 40 channels such as temperatures, humidities, solar radiation, air flow rate, electrical consumption and other parameters. We mainly focus on the largest cellar and the most overheated office (located South-West).

Good comfort was reached during summer 1990 thanks to the passive cooling system. Figure 2 shows the meteor data for two weeks in August and the corresponding South-West office temperatures at three different levels : 0.1 m (close to the floor), 0.9m (working place) and 3m (under the roof). The fan has been switched off during 3 days (from August 11th to 13th) to confirm its effects and an immediate stratification of temperature appears.

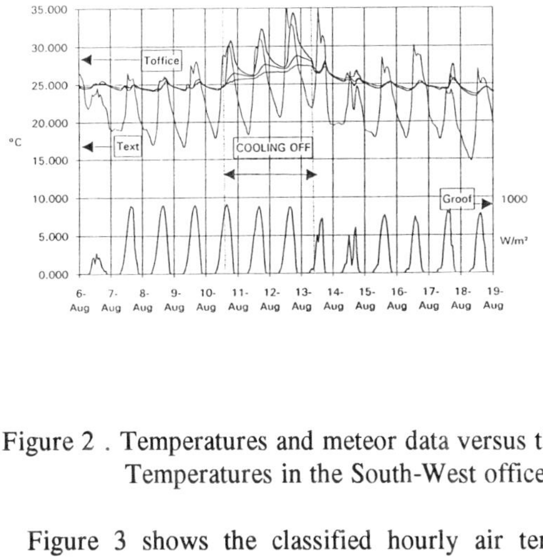

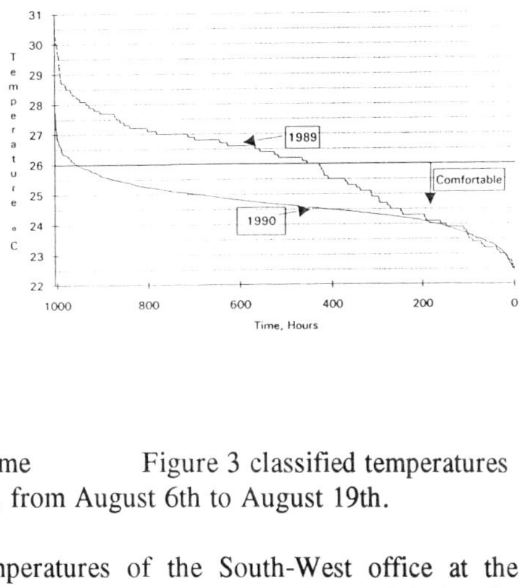

Figure 2 . Temperatures and meteor data versus time Figure 3 classified temperatures
Temperatures in the South-West office, from August 6th to August 19th.

Figure 3 shows the classified hourly air temperatures of the South-West office at the working place during 1989 (without cooling), from July 27th to September 9th, and during the same period in 1990, with cooling but with 70 hours without cooling, as seen in figure 2. In the summer of 1989, which was cooler than summer 1990, the South-West office was inoccupied for most of the month of August.

4. Modelisation of the system

4.1 THE COOLING SOURCE : THE CELLAR

The measurements show that,when applying daily values, the cellar output air temperature can be described as a mixture of two air flows, one at the cellar structure temperature and one at the mean daily outside temperature (see figure 4).

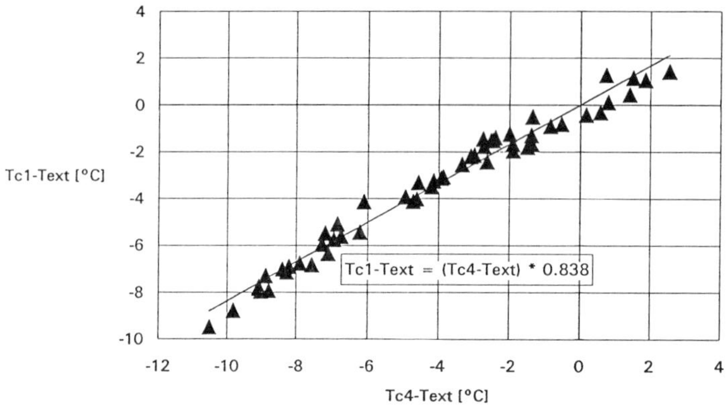

Figure 4. Correlation between the temperatures of the outside air (Tex), of the cellar structure (Tc4, 50 cm under the ground) and of the cellar output air (Tc1). Daily values.

4.2 FAN AND DUCTS

The signifant sources of heat gains were :

. the fan, because the motor is inside the duct,

. the friction losses in the duct,

. the heat transfer through the wall ducts,

. the air infiltration before the fan.

The relative importance depended of many parameters and the modelling of this part was long. An example is given in figure 5.

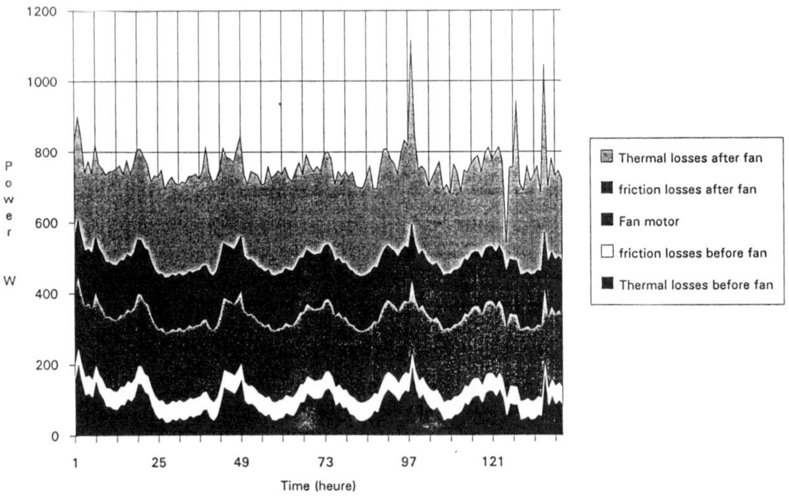

Figure 5. Heat gains in the air distribution network.

4.3 SOUTH-WEST OFFICE

For this office, we used a simplified method with one node thermal diagram. In fact, the most important cause of discrepancies of the model was the bad knowledge of parameters like : incline of internal blind strips and the corresponding solar transmission, heat transfers from the other rooms in relation with the opening of the door, the number of people in the office, air infiltration through the open window,... All these parameters are very difficult to deal with and we preferred to use simple but reliable parameters including all these effects and determined with daily value analysis (see [1]):

. total heat loss factor U =2 to 27 W/K,

. total effective area for solar gains Aeff = 0.3 to 0.45 m² (0.2 m² was due to the skylight window, 0.1 m² was due to the roof transmission and 0.1 m² was due to the shell window).

Dynamic model for the office temperature with this one node diagram gave good results, specially when considering classified temperatures during the summer time. Parameters used in this model are hourly values of outside temperature and of solar radiation on the roof, heat loss factor = 27W/K, effective area for solar gains = 0.4m² and heat capacity of the office but with respect with the air temperature variation = 1500 Wh/K (see [2]). We recall that the main cause of errors was not the simplicity of the model, but was due to the difficulty in estimating the parameters.

4.4 RESULTS OF SIMULATION

Thanks to these models, we were able to describe the whole system and to carrefully study the effects of each parameter. We gave the quantitative results, described in details by [1] and [3] and we only focus on the qualitative consequences in the next section.

6. Discussion : use of basement in passive cooling

The use of basement in passive cooling and in urban context is possible but has several implications :

Site implication :

.In addition to the existence of fresh undergrounds, the air must be clean and safe. Radioactive pollution with radon is to be considered carefully. In our case, we measure a very low content of this gas in the cellar (24 Bq/m3). When a problem of air pollution occurs, a air air heat exchanger will be used, but simplicity and low cost are partially lost.

Architectural implications : such a system needs :

.a very good insulation of the roof (decrease of solar load)

.a very good solar protection

.a good thermal inertia with good thermal coupling between thermal mass and inside air,

.a space to install the duct (from undergounds to the last floor!) in case of retrofitting,.

.a limited area is to be cooled (for instance, the last floor under the overheated roof). For greater areas, it would be more complexe and expensive.

Technical implications

.good fan, with external motor, and low friction losses (well designed ducts)
.no duct in depression in order to avoid warm air infiltration,
.good insulation of the ducts in warm conditions

User implications

.good comfort if good management of solar protection, lighting and electricity use.

7. Conclusions

The passive system we have studied in detail runs very well and the same comfort as a standard climatisation is obtained, but requires half of the investissement and less than 25 % of the operational costs [1]. The use of underground in passive cooling in an urban context is often possible, in new buildings as well as in retrofitted buildings, but needs special attentions. Note that other cooling sources could be used : river, underground, ...

8. References

[1]. R. Meldem, B. Lachal, C. Ançay, W. Weber, O. Guisan, Rapport final : Climatisation passive du Bâtiment Aymon à Sion, to be published in June 1991, Geneva University.

[2]. T. Frank, R. Saggeldorff, "Schweizerenergie Fachbuch", 1989, p117, St Gallen : Kenzler-Bachmann AG, 1991.

[3]. C. Ançay, O. Guisan, B. Lachal, R. Meldem, W. Weber to be published in "ISES 1991", August 1991 in Denver.

A SOLAR PASSIVE HOUSE IN AN URBAN CONTEXT: SIMULATIONS

E. GRATIA: researcher
B. MEERSSEMAN: designer
Architecture et Climat
Bâtiment Vinci
Place du Levant, 1
B-1348 Louvain-la-Neuve
Belgium

ABSTRACT
In the frame of the task 13 of the International Energy Agency,we had to design an advanced solar low-energy residential building. We selected the urban location of Louvain-la -Neuve (Belgium). During the design process of the two pre-projects,simulations have been realized to help the architect to answer a few questions.
The following topies have been studied
-the location of the thermal inertia
-the impact of a greenhouse
-the constitution of the wall separating the living from the greenhouse
-the advantages of a ventilation with air extracted from the greenhouse
-the heating economy if the same house was placed in a well-oriented street
-the determination of the heating power in a well insulated house
-the overheating problems

1. Introduction

When an architect designs in an urban context, he has always to face strict rules imposed by the town planner. Indeed, the latter doesn't think about the restricting impact of the streets layout and of the impositions on the buildings themselves (such as outlines, alignment, roof slope, maximum height..) on a possible passive solar design.
In Louvain-la-Neuve, the urbanistic rules are numerous and very strict. Taking all those rules into account, we intended to establish pre-projects with an objective of passive solar design.
The house is a row house along a street running north-west to south-east. Thus, the street façade is north-east oriented and the garden side is south-west. The greenhouse could make the link between the livingroom and the garden.
Since the house will be build in the frame of the IEA-task 13, it must be an advanced solar low-energy residentiel building.
Simulations have been realized by the programs:
-LPB4 :a multizone static program
-MBDS :a multizone dynamic program
-SOLPA1:a unizone dynamic program adapted for greenhouse calculations.

S. Alvarez et al. (eds.), Architecture and Urban Space, 743–748.

2. Our approach to help the architect

2.1. At the beginning of our research we thought that the location of the **inertia** would have much importance because the house is a solar passive house (with the idea of direct heat storage) and a very well insulated building.

To test this, we first simulated an imaginary house with an outline corresponding to those imposed by the urbanistic rules.The imaginary house was ideal, protected by a non-heated space on the north-east side and closed by windows only at the south-west side.

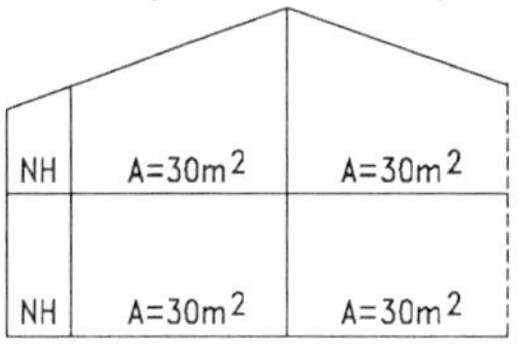

We then reached the following simulation results.

house without inertia	**100%**
house with inertia	**89.4%**
south inertia only	**89.2%**

2.2. The second step took the following constraints into account:
- the strict urbanistic rules,
- the idea of a passive solar design,
- the necessity of a very well insulated house,
- the problematic of the location of the inertia,
- the outlines of the two adjoining not builded houses.

Two pre-projects were established, of which the second was the following one :

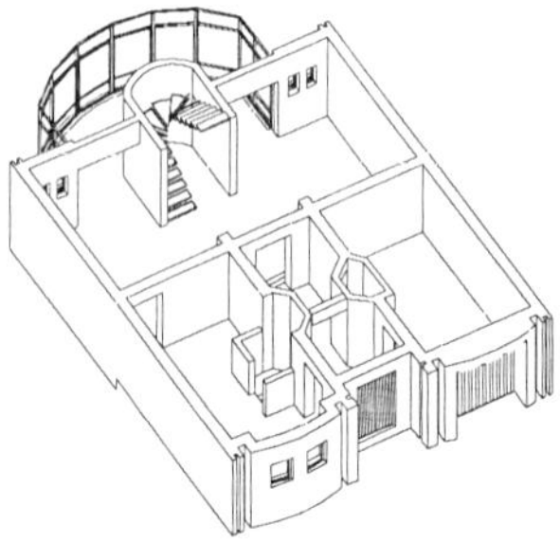

GROUND FLOOR

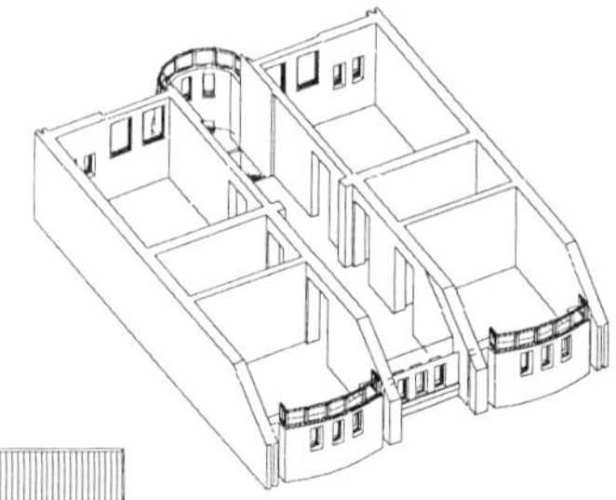

FIRST FLOOR

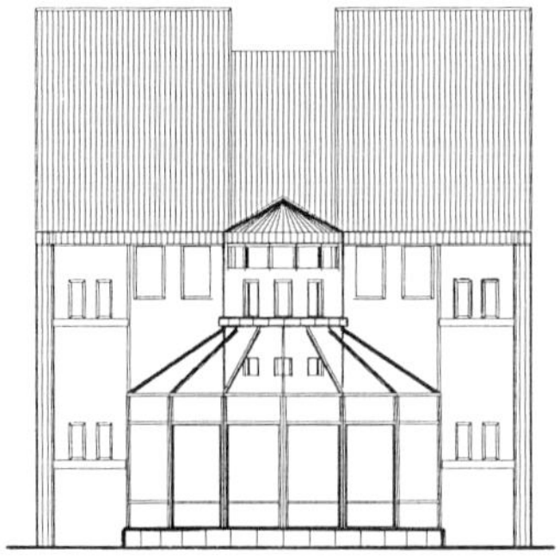

FACADE

Real simulations could then begin.

The chosen assumptions are detailed here under :

- Constitution of the walls : well insulated

  Examples: opaque walls : 12 cm of insulation
  roof : 20 cm of insulation
  windows : k = 1.3 W/m2°C

But : wall between the living and the greenhouse : 20cm of bricks

- Position of the inertia

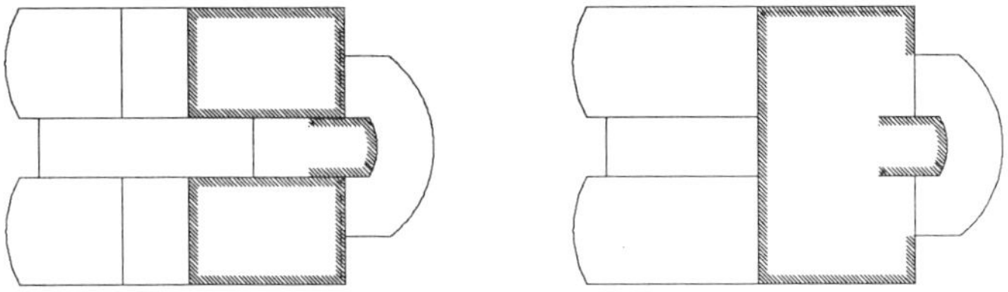

- Infiltration, ventilation

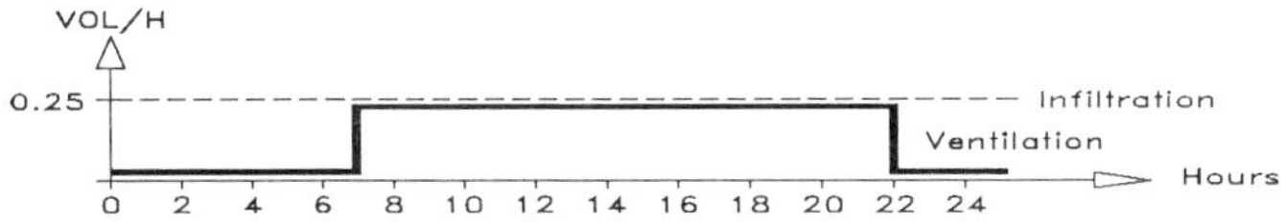

- Free gains for the whole house
(for simulation needs, they are distributed in the rooms)

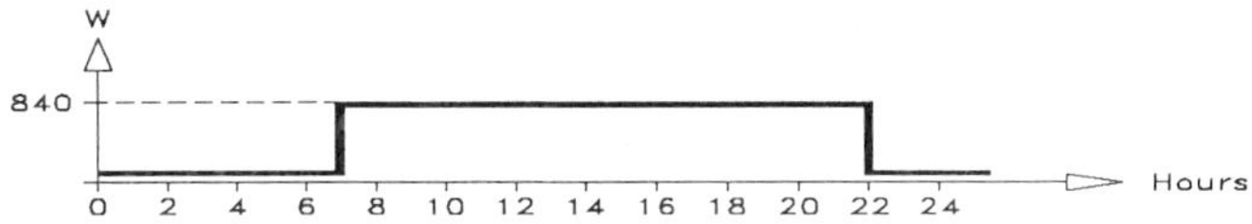

- Regulation of the heating system

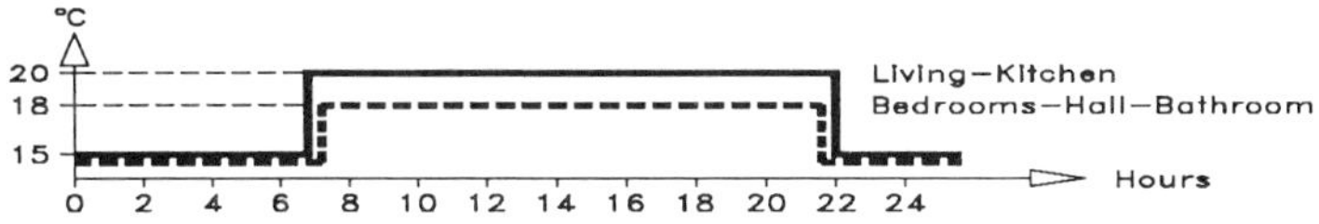

Result of the simulation on that basic case :

3646 kWh or 22.8 kWh/m2 heated

The table below shows the great importance of the losses by infiltration and ventilation.

	infiltration and ventilation rate	consumption
	0.1	28%
	0.25	51%
basic case	0.5	100%
	1.0	198%

2.3. **Orientation** of the house.

If the house was oriented N-S, the consumption would only represent **91.5%** of the basic case.

2.4. **With or without sunspace ?**

Assumptions : the wall living-sunspace which was an uninsulated storage wall (20cm bricks) is replaced by an external insulated wall.

The results of simulation show that the consumption is decreased by about **10%** when the greenhouse is removed.

Why ???

- -Because of the insulation of the wall living-sunspace.
- -Because the window between the living and the sunspace is a very good window and the external windows of the sunspace is only single glazed.
- -Because the solar radiation is filtered by the second layer of glass of the greenhouse.

In this basic case, the advantages of the greenhouse were not used very well. We then decided to run new simulations with new hypothesis :
-the wall living-sunspace is insulated
-the ventilation of the living (O.25 vol/h between 7h and 22h) is provided by the air of the sunspace.

The gains obtained were
12% as compared with the basic case
3% as compared with the case without sunspace.

The energetic gain provided by the sunspace still remains low because of :
-the very good window between the living and the sunspace
-the filtration of the solar radiation.

But the sunspace can't only be seen as an element providing energetic gain : it provides a freely heated space during many months in the year. It is also a privileged space for the senses through the surface aspects, the variations of the light, the transparencies, the presence and the odors of the plants, and the physic sensation of the outdoor climate.

When the sky is clear, the comfort temperature (20°C) is reached in the sunspace during the following periods :

March:	between 12h30 and 18h
April:	between 10h and 21h30
May:	between 07h and 24h
June, July, August:	all day
September:	between 08h and 24h
October:	between 12h30 and 20h30
November:	between 14h and 16h15

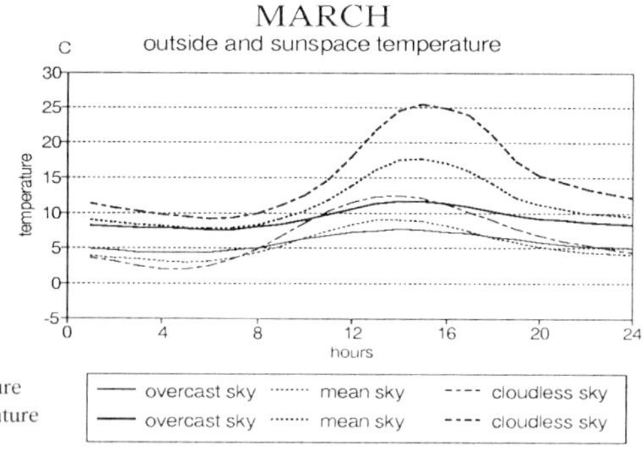

The graph above shows the greenhouse temperature evolution during the month of March for three different sky conditions (overcast, mean and cloudless).

2.5. **Problematic of the heat power determination in a solar low-energy house with inertia.**
Usualy heat power determination in the various zones of a building is calculated using a multizone static program.
So, in MBDS, we used the heat power calculated by the LPB4 program to study the response of the heating system during a cold period.
The chosen period (15/1 - 18/1) includes sunny and clouded cold days. Simulations are made without the sunspace.
The climatic data and the graph wich gives the evolution of the temperature and the heat demand in the livingroom (south zone with inertia) are represented here below. It allows us to see that four hours are needed to reach the desired temperature.
It also shows that the heating system can be stopped during one or two hours in spite of very low outside temperature (-7°C, -8°C) and that thanks to the sun.

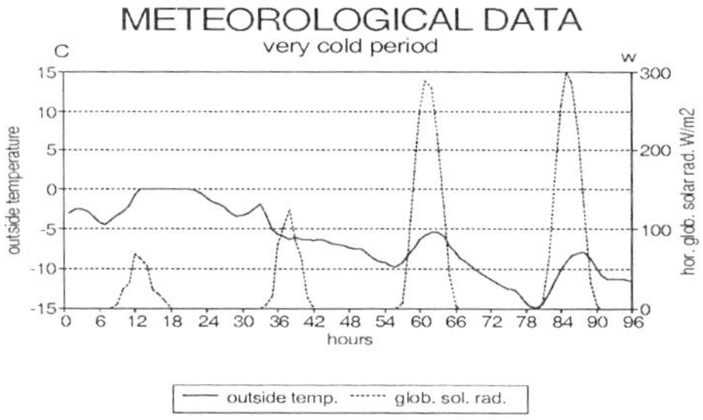

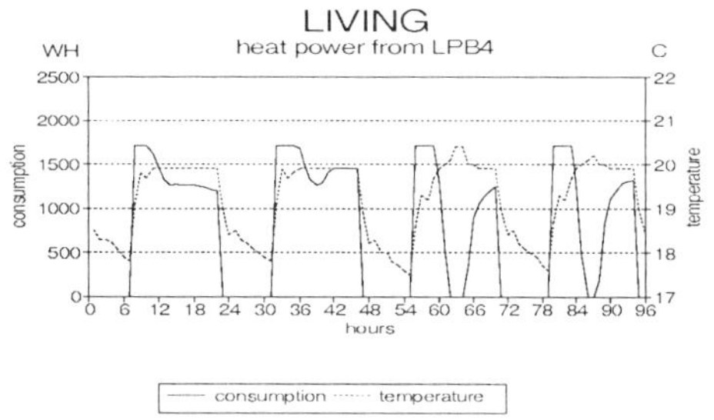

If the heating system is stopped during four days,the living temperature don't decrease under **14°C**.But **36 hours** are needed to reach the chosen temperatures.

When the heating power is increased by about 50%, the wished temperature is reached immediately.

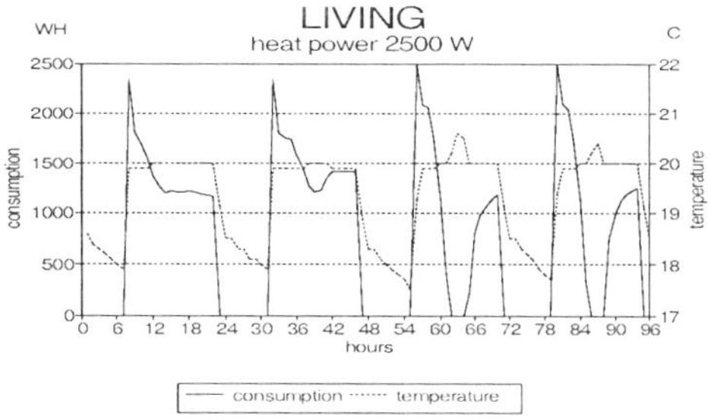

2.6. **Overheatings.**

We have then examined a very hot period to try to solve the overheating problems.The simulations are made without the sunspace.The chosen climatic data are represented here below.

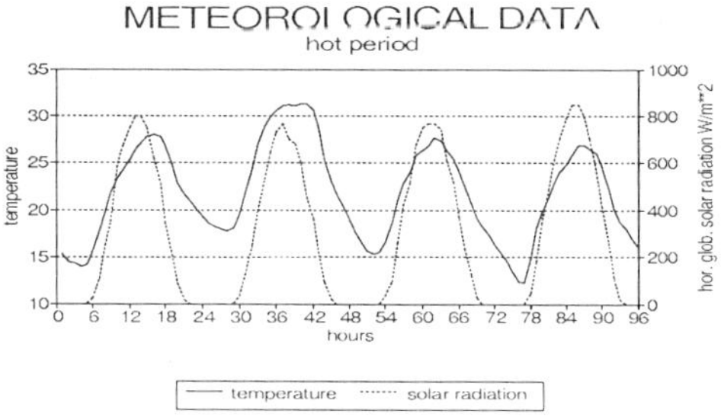

We have analyzed what would hapen if inertia was placed only in south zones or in the whole house and if air movement was created between south and north zones.

Results: When inertia is placed in the whole house, temperature decreases by about **2.5°C** in the south zones and **4.0°C** in the north zones

When air movement is created, temperature still decreases by about **1°C** in the south zones and increases by about **1°C** in the north zones.

We have then examined the influence of an increase in the ventilation rate during the day and during the night (air movement maintained).

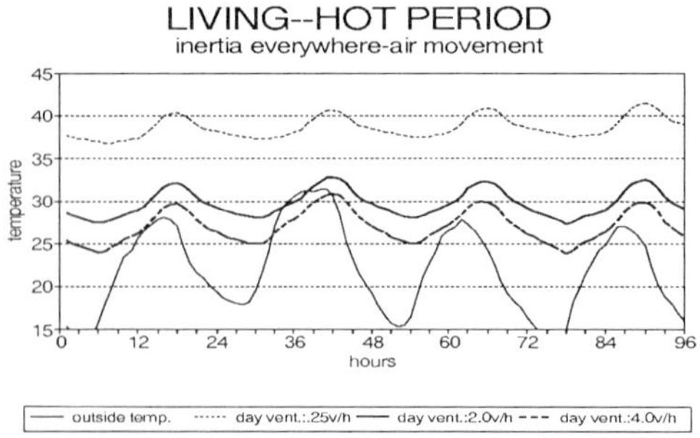

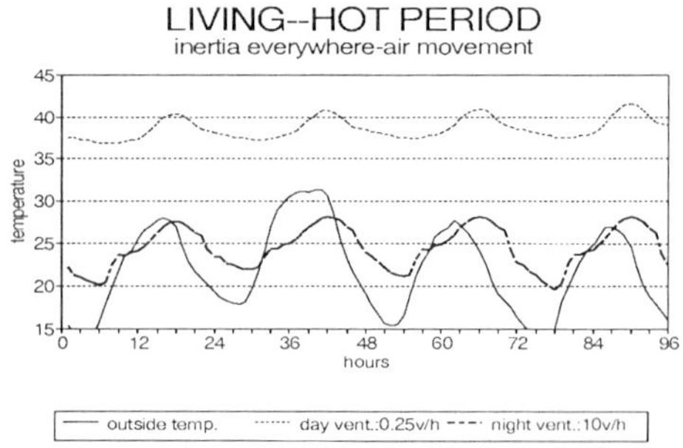

The use of blinds when solar radiation penetrating by the windows is greater than 150 W/m2 is also effective (air movement maintained)

The graphs below compare the basic case with the case using all strategies.

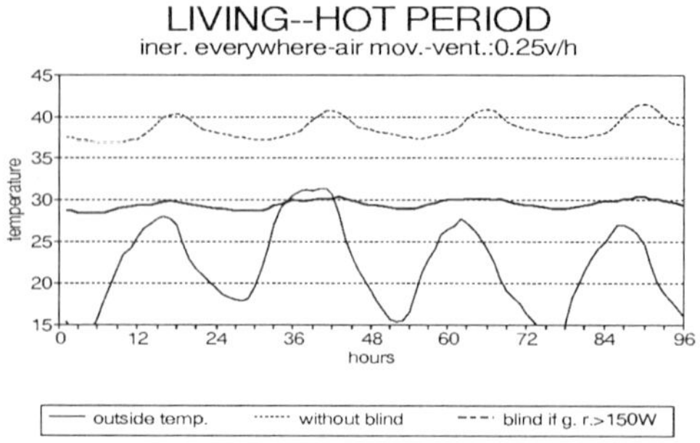

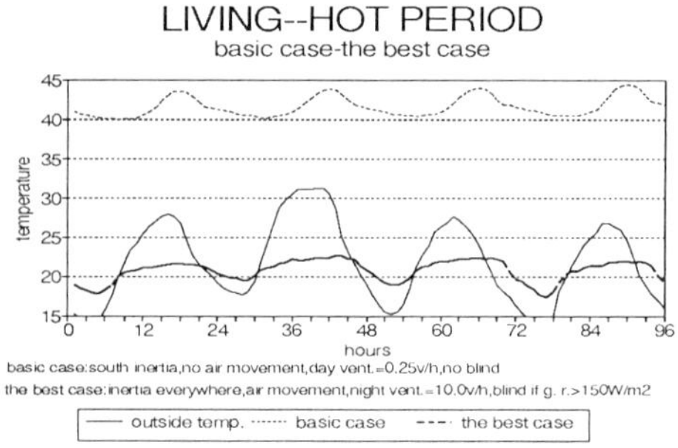

3. Conclusions

The simulations have shown that

-the inertia located in the south zones decreases the energetic consumption

-the inertia located in the north zones decreases the overheatings

-the energetic gain given by the sunspace is low but it provides a freely heated space during many months in the year

-the heat power determined by static program is underestimated

-overheatings are decreased by placing inertia in the whole house,by creating air movement between south and north zones, by increasing the ventilation rate and mainly during the night, by using blinds.

SIMULATING THE THERMAL PERFORMANCE OF NATURALLY VENTILATED SPACES: A CASE STUDY

Herbert Eppel and Kevin J. Lomas
School of the Built Environment
Leicester Polytechnic
PO Box 143
GB - Leicester LE1 9BH

ABSTRACT. In order to be as energy efficient as possible, the new building for the School of Engineering and Manufacture at Leicester Polytechnic was designed to be naturally ventilated, as far as possible. Computer simulation was used to test whether this could be achieved, and design optimisation and performance assessments studies for the auditoria were undertaken. The results indicated that a suitably controlled natural venting strategy could work, and the final design was based on this approach.

1. Introduction

The new building for the School of Engineering and Manufacture at Leicester Polytechnic, which is described in a companion paper (Ford & Short, 1991), was designed to be as energy efficient as possible. To minimise energy consumption for lighting and temperature control, maximum use was to be made of natural light and the building was, as far as possible, to be naturally ventilated.

Research into whether, and how best, the latter could be achieved, was commissioned by the Energy Technology Support Unit of the UK Department of Energy. The Environmental Design Unit at Leicester Polytechnic undertook work using computer simulation and the results are reported in this paper. Complementary work was undertaken by the Department of Applied Mathematics and Theoretical Physics at Cambridge University, using a perspex model suspended in a salt solution; their work is described in another companion paper (Lane-Serff et al., 1991)

2. Research Brief

It was uncertain whether natural ventilation would be adequate to prevent summertime overheating in the two densely occupied auditoria, so mechanical ventilation, and perhaps full air-conditioning, had to be considered. Research at Leicester concentrated on this issue and on design optimisation and performance assessment of the auditoria.

The $870m^3$ auditoria are designed to seat 150 people, and the brief indicated this level of occupation for 8 hours a day, 5 days per week. The heat produced by the occupants (100W each), together with $15W/m^2$ from lighting and 500W from equipment, result in an internal heat load of 18.3kW.

S. Alvarez et al. (eds.), Architecture and Urban Space, 749–754.

The natural ventilation scheme consists of a plenum below the raked wooden staging supporting the seats. This plenum is fed from the outside, and the air enters the auditoria through grills below the seats. The air outlet consists of a 13.3m high vertical chimney, connecting an air extract grill at the front of the auditoria to the outside above roof level (Figure 1). In winter, heating is provided by fin convectors behind inlet grills in the seating risers. To try and prevent excessive temperature fluctuations, the side walls and ceiling were made of heavy-weight concrete.

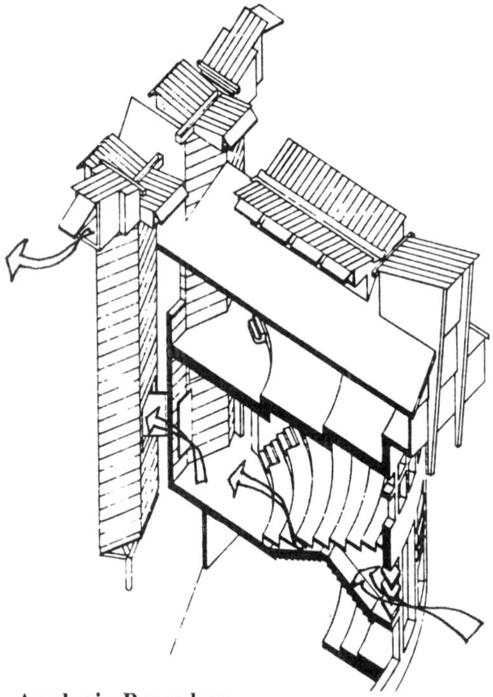

Figure 1:
Natural ventilation scheme for the auditorium

→ N

3. Analysis Procedure

The thermal comfort conditions in the auditoria are dictated by: the convective and radiant heat from people and lights, and the stack (buoyancy) driven airflows which these induce; wind driven infiltration; and heat flow into and out of the thermal mass. These complex time-varying interactions could only be adequately investigated by using a dynamic thermal computer simulation program capable of simultaneously solving both the equations governing heat transfer and the equations describing the air flows; such a program is ESP (Clarke et al., 1990).

Some initial simulations were undertaken to identify the factors, which would have the largest influence on the results. Weather data for a typical hot sunny day was used to study the influence of design changes on the hourly variations in temperatures and air flow rates. In the next stage, design optimisation, the design options were refined and simulation results compared with a reference, or base case. An attempt was made to compare these simulation results with the results obtained from the physical model used at Cambridge. Finally, the occurrence of warm conditions in the auditoria was studied for a typical year.

4. Initial Results and Design Optimisation

The 'summer sizing' weather data used in this stage of the work was devised by analysing the 310 July days in the 10 years of hourly weather data collected at Kew, London, from 1959 to 1968 (Loxsom 1985/86). Only 1% of July days were hotter <u>and</u> sunnier. Analysis of the CIBSE banded weather data (CIBSE, 1986) shows, that no more than 5 July days, and no more than 10 days in the whole year, can be expected to be hotter than this design day.

The computer simulations clearly illustrated the strong relationship between the rate of heat generation and the induced air flow (Figure 2). For the base case, air change rates of over $6ach^{-1}$ occurred during the occupied periods. The progressive warming of the structure during the day leads to peak dry resultant temperatures occurring in late afternoon.

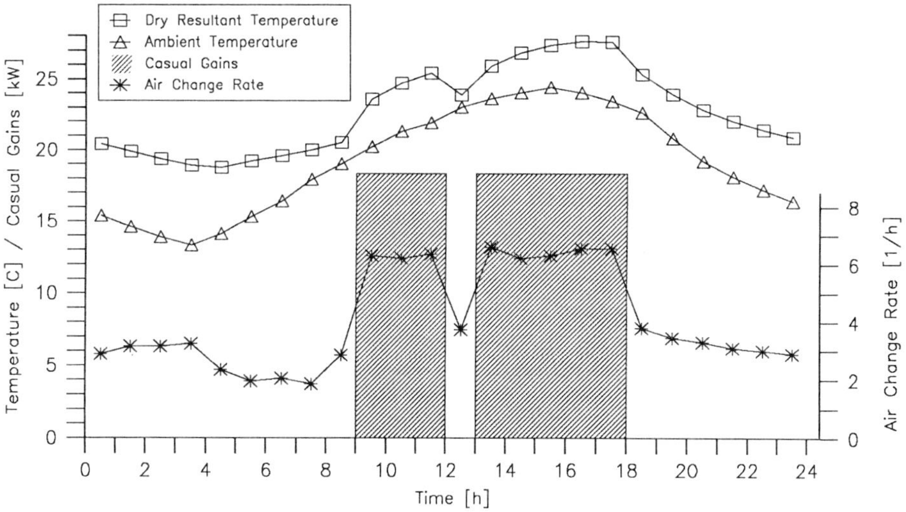

Figure 2: Hourly variation of environmental conditions for the base case auditorium

Nine cases (design and operational variants) were investigated in detail and compared with the base case. Concentrating on the peak dry resultant temperature (Figure 3), the following observations can be made.

(a) Large reductions in the dry resultant temperature would occur during days of reduced occupancy. At 60% of the design value, this reduction was 1.3°C. (At 80% occupancy the reduction was 0.9°C).

(b) Of the design variants studied, the provision of windows was the single most effective temperature reducing measure. A 5 m^2 window opening reduced the peak dry resultant temperature by 0.8°C.

(c) However, if both the plenum inlet and chimney cross-sectional area were enlarged, then a reduction of 1.2°C in peak dry resultant temperature could be realised. Of these two design changes, enlarging the chimney cross-sectional area had the greater effect.

(d) Heating the chimney reduced the peak dry resultant temperature by only 0.5°C, and night time mechanical venting reduced it by only 0.1°C. One must therefore set the complexity and cost of these measures against these marginal environmental improvements.

(e) Mechanically ventilating the structure at night to reduce the temperature of the structure, and increasing the chimney height (results not shown), had little impact on the peak dry resultant temperatures.

(f) Acoustic tiles around the back third of the auditorium raised the peak dry resultant temperature by 0.8°C. However, it should be noted, that this value represents the average for the auditorium as a whole. Locally higher values will occur in spaces surrounded by the tiles, leading to greater discomfort towards the rear of the auditorium. The rear of the auditoria is already higher and so closer to the warmer, stratified, air below the ceiling. Opening windows at the rear of the auditorium, which has an outside wall, would give greater local benefits in this region (but at the expense of the acoustic qualities of the space).

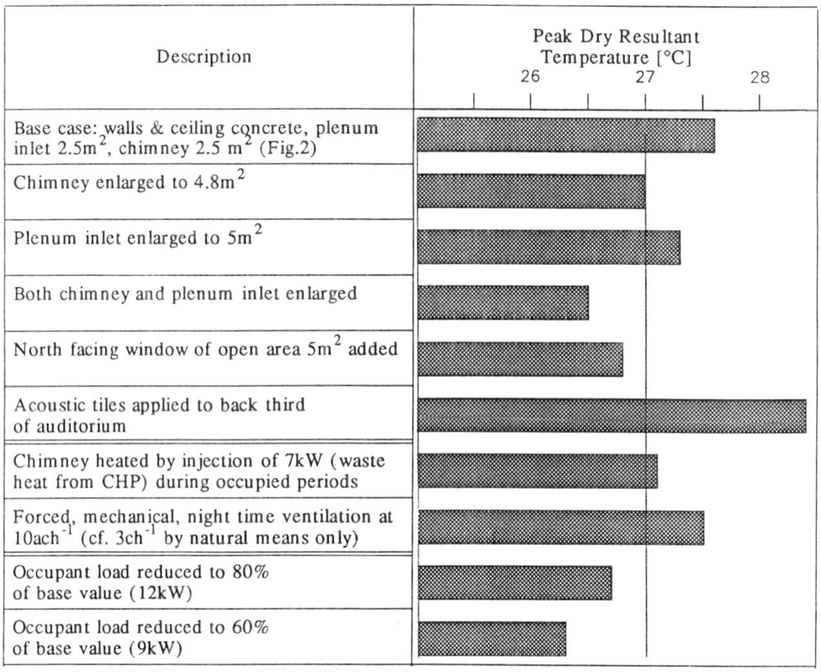

Figure 3: Comparison of peak dry resultant temperatures

A design guide for educational buildings (Design Note 17, 1981) states, that "it is undesirable that the resultant temperature should exceed 27°C during normal working over the school year, but an excess for 10 days during the summer is considered to be a reasonable predictive risk". The peak dry resultant temperatures on the hot sunny design day were close to, or below, 27°C for many of the scenarios studied. It was felt therefore, that a design based on natural ventilation could succeed, and so a more rigorous performance assessment investigation was undertaken.

5. Performance Assessment

The occurrence of warm conditions in a typical year was studied by simulating the 'summer' period from 1st May to 30th September using the weather data for Kew, London, collected in 1967. (This is the single year closest to the average of the 10 years of data collected at Kew between 1959 and 1968.) No corresponding data for Leicester were available, (but it is reasonable to assume it will be cooler if anything, so the results will be conservative).

The design options most likely to be adopted were used in this study. The results were plotted as the cumulative number of occupied hours (lecture periods), for which the dry resultant temperature exceeded 20°C, 21°C, 22°C etc. (Figure 4).

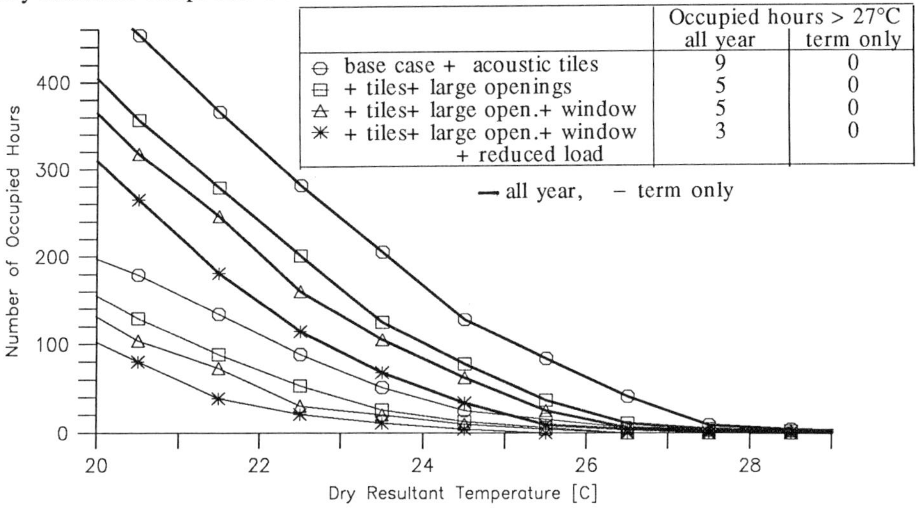

Figure 4: Occurrence of warm conditions on the auditorium

ESP predicted, that for the basic design with acoustic tiles, there would be only 9 hours per year with dry resultant temperatures above 27°C, and none of these would occur during normal academic term time. Increasing the air inlet and outlet areas reduced the number of occupied hours, in which summertime overheating occurs, by 50%. The addition of a window reduced this by a further 20%.

These results gave both the client and the architects valuable insight into the effect on comfort of alternative designs and a sound basis upon which to make design decisions. Partly on the strength of these results, a natural venting strategy was adopted.

6. Comparison of Thermal and Physical Modelling

The explicit finite difference program ESP is inherently a dynamic, time-varying model, which takes into account the thermal mass of the structure, radiant exchange mechanisms and wind effects. However, thermal models such as ESP have difficulties in dealing with temperature stratification and complex geometries. Conversely, whereas the perspex model is well suited to study the temperature distribution within complex spaces, it is inherently a steady-state model, and none of the dynamic processes mentioned above can be treated adequately. Therefore, the two approaches are complementary, since neither method can deal with all the physical processes occurring in a real building.

It is difficult to undertake a precise comparison of the results obtained with the two methods. Nevertheless, by averaging spatially the air temperatures predicted by the salt model and assuming the same surface temperatures, an average dry resultant temperature can be calculated. This was 26.9°C, and the air change rate was 9.2ach^{-1}. Corresponding values predicted by ESP were 27.6°C (Figures 2&3) and 6.6ach^{-1}. This level of agreement is encouraging, but the reasons for the differences are worthy of investigation.

7. Conclusions

Combined thermal and air flow simulations gave valuable insight into the complex dynamic variations of the temperatures inside the auditoria. These studies provided a sound basis for refining the design and for assessing the overheating risk.

In so far as this work made comparisons possible, ESP and the salt model agreed reasonably well. However, little work has been done to assess the applicability and bsolute accuracy of these design tools. Funds for monitoring the environmental performance of the building are being sought.

The results confirmed that a suitably controlled natural venting strategy was viable, even in the densely occupied auditoria. The final design evolved trough close co-operation between the architects and the researchers. This makes the new School of Engineering and Manufacture one of the largest naturally ventilated buildings in Europe.

Acknowledgements

Funding by the Energy Technology Support Unit of the UK Department of Energy under a pilot study for the new energy design advice scheme is acknowledged.

References

CIBSE Guide, 1986, Volume A: Design Data, The Chartered Institution of Building Services Engineers, London

Clarke, J., Hand, J. and Strachan, P., 1990, 'ESP - A building and plant simulation system', Version 6 Series, ESRU Manual U90/1, Energy Simulation Research Unit, University of Strathclyde, Glasgow

Design Note 17, 1981, 'Guidelines for environmental design and fuel conservation in educational buildings', Dep. of Education & Science, Architects & Buildings Branch

Lane-Serff, G.F., Linden, P.F., Parker, D.J. and Smeed, D.A., 1991, 'Laboratory modelling of natural ventilation via chimneys', Proc. PLEA'91 Conference, Sevilla

Loxsom, F., 1985/86, 'Meteorological data for the Passive Solar Programme', Polyt. of Central London, Collection of reports for ETSU model refinement study

Ford, B. and Short, A., 1991, 'Design of a naturally ventilated laboratory building in Leicester, England', Proc. PLEA'91 Conference, Sevilla

A LOW ENERGY DESIGN FOR KING ALFRED SCHOOL, LONDON

J.WALKER*, M.WILSON, Y.DEAN, Low Energy Architectural Research Unit, Polytechnic of North London, Holloway Road, London N7 8DB. *main author

J.VAN HEYNINGEN, B.HAWARD, N.CHOUDHURY, van Heyningen and Haward Architects, 1c Laurier Road, London NW5.

ABSTRACT. This paper describes the low energy design of a new $896m^2$ passive solar re-development at King Alfred School, Hampstead, London. Environmental design input came from the Low Energy Architectural Research Unit (LEARN) to an original design by architects van Heyningen and Haward (vHH). The project consists of groups of steel framed and timber clad modular units allowing completion during one summer holiday period of nine weeks. Energy saving design features include: Daylighting optimization utilising rooflights with daylight insulation and automatic light switching, passive solar ventilation from a buffer space, direct gain with external shading, and improved insulation levels. Estimated savings for all the measures are 50% (£1165 per annum) compared to a conventional design to current UK building regulations, and overall payback time is 7 years. In the school context, daylighting and lighting control provided the largest savings, insulation levels being only moderately improved upon.

1.0 Project Description

The new lower school re-development replaces existing buildings with groups of steel framed and timber clad $150m^2$ modular units to allow rapid completion during one summer holiday period of nine weeks, Figure 1. The frame will sit on a 100mm concrete raft foundation, chosen because of the unstable site conditions, with a floating timber floor, quick to install and unaffected by future cracking of the slab. The pre-painted steel frame, of 100mm hollow square section columns and 175mm x 100mm beams, is bolted together allowing the mill aluminium finish roofing system to be fixed, providing immediate weather cover. Wall construction is 150mm timber stud with red cedar tongue and groove boarding, all dimensions on a 610mm grid to aid rapid internal finishing with pre-painted plasterboard. A covered area at the front of the building has a triple skin polycarbonate sheet roof, with an opaque upper roof for shading. The addition of 'cafe style' concertina opening doors creates a buffer sunspace for pre-heat ventilation in the heating season which can fold completely back in the summer to prevent overheating.

S. Alvarez et al. (eds.), Architecture and Urban Space, 755–760.

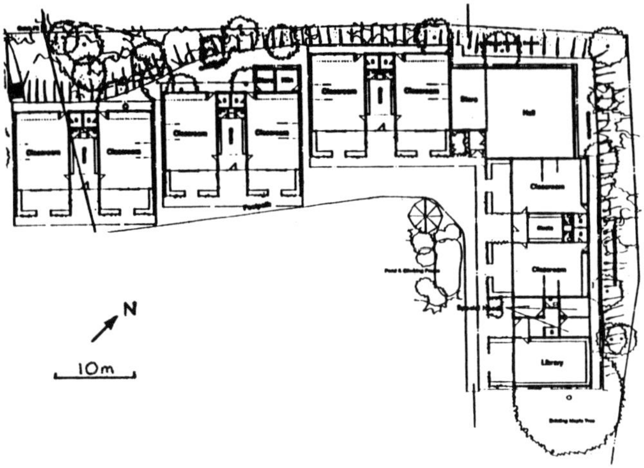

Figure 1. Site plan of King Alfred Lower School Re-development. The buildings are formed in an 'L' shape around an open playing field, the front facades being either SE or SW orientated, the rear overshadowed by trees

Mineral fibre insulation consists of 100mm in the walls, 100mm in the roof and 80mm under the floor, giving U-values of $0.37W/m^2K$, $0.41W/m^2K$ and $0.28W/m^2K$ ($0.49W/m^2K$ to ground temperature).

2. Environmental Design

Environmental studies focused primarily on a typical end classroom, as shown in Figure 2.

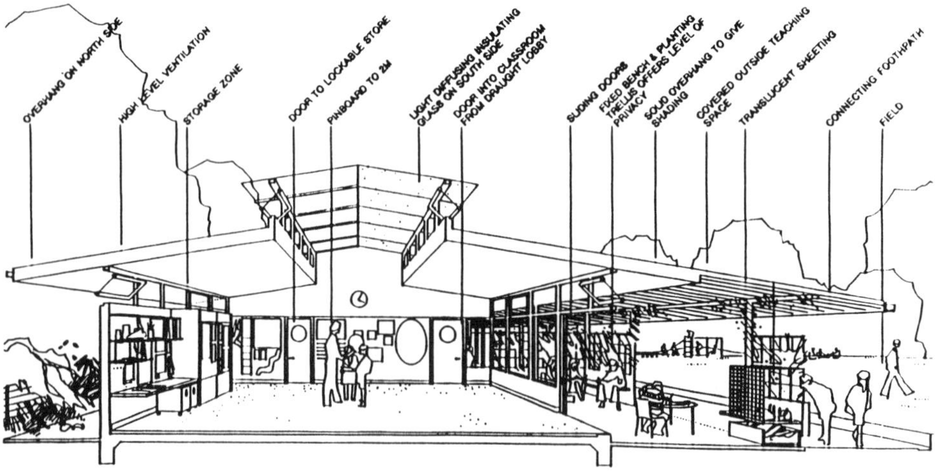

Figure 2. Sectional Perspective through a Typical Classroom. Rooflights have daylight insulation at the front and insulated panels at the rear; vertical lights are opening. Windows are plain double glazing.

2.1 DAYLIGHTING DESIGN

Computer daylight modelling of the original design indicated a poor daylight distribution, with high daylight levels near the front windows and under the rooflight but poor at the rear of the room. Electric lighting use will be based on the minimum daylight factor but will also be increased when daylighting is uneven. Using 1:20 physical scale models in an artificial sky, white venetian blinds were tested under the rooflight but did not increase daylight factors at the rear of the room, instead decreasing daylight factors under the rooflight, as seen in Figure 3. From the rear of the room the rear of the rooflight cannot be seen, has little effect on minimum daylight factor but adds to the unevenness of daylighting. Therefore insulated panels replace glazing at the rear of the skylight reducing thermal losses in areas where daylight is not useful with the resulting daylight distribution in Figure 4. Front window heights above 1.8m were found to provide little additional daylighting due to the shading overhang obstructing sky sight lines. Sill heights below 0.7m were also found to provide little daylighting benefit.

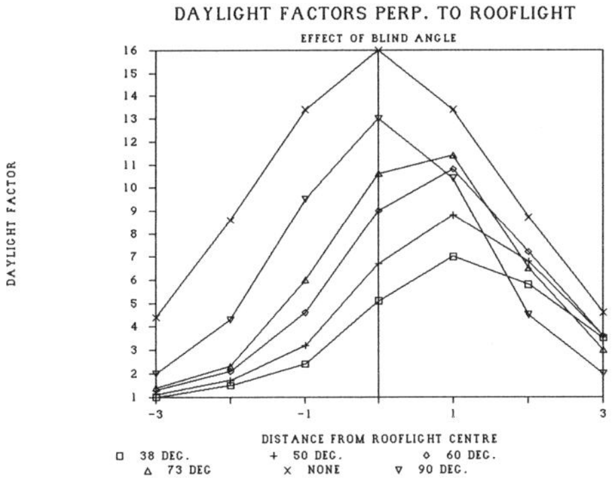

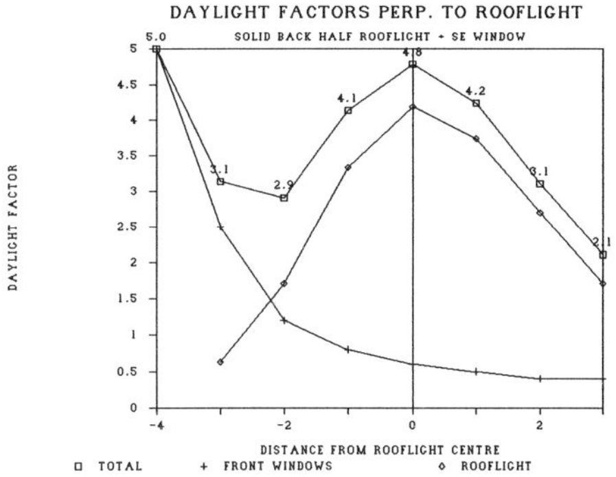

Figure 3. Effect of Blind Angle. Figure 4. Effect of Rooflight.

2.2 SHADING SYSTEMS

Simple venetian blinds could not exclude the sun, due to the non south facing orientation of the buildings, an 'egg crate' type solar shading device with a height to width ratio of 3:1 would be required. However such systems would reduce daylight levels and would give rise to problems and cost to keep clean. Light diffusing 'daylight insulation' double glazing was considered the best option for solar control in the rooflight. Using a 1:20 model of a classroom, the effectiveness of a 1.3m overhang was investigated. Excluding non-term time, the potential problem months for overheating are May, June and July. For these months the whole facade is shaded, for both

SE and SW at noon. The worst solar penetration occurs at 9-10am for SE and 3-4pm for SW, when the sun patch rises up to 1.0m in May. If overheating is a problem increasing sill height will therefore have an effect in solar exclusion.

3.0 Thermal Modelling with SERI-RES

The SERI-RES model was used to look into energy consumption of the building over the 185 day school year; term times 2/9-6/12, 8/1-27/3, 15/5-11/7. Using typical hourly weather data for Kew weather station, an annual simulation has been made based on 8am-4pm heating, 9am-4pm occupation (5 days a week) at 20°C for term times, with 5°C winter frost protection at other times. Economic appraisal aimed for a simple payback less than 11 years. The following aspects were investigated:

i. LIGHTING. Figure 5. shows the variation of electric lighting consumption with daylight.

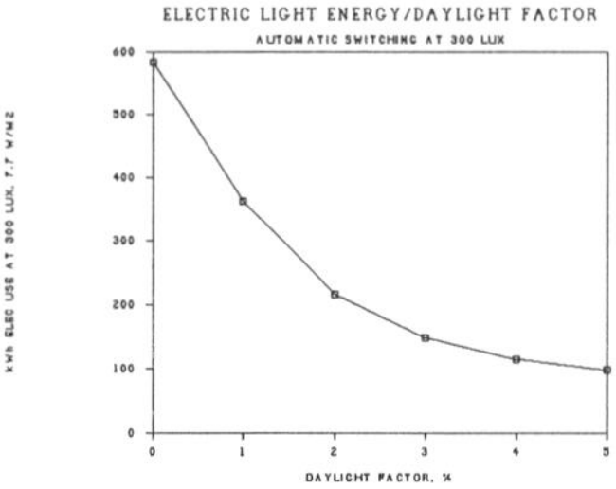

Figure 5. Electric Lighting vs. Min. Daylight Factor. It is clear to see that largest savings are made increasing DF from 0% to 2%. Automatic switching of 300lux lighting (7.7W/m^2) below 300lux daylight.

Automatic electric light switching is by Setsquare Liteminder automatic occupancy and daylight level switching at a unit cost of approximately £40. The units payback in less than one year, excluding installation costs.

ii. INSULATION OPTIMIZATION. Variation of insulation levels in 10mm increments has allowed a life cycle costing of various insulation thicknesses. Savings are based on gas heating at £0.02/kWh and mineral fibre insulation at £0.167/10mm thickness per m^2. This has yielded optimum economic insulation thicknesses for mineral fibre of: Roof 100mm, walls 100mm with the floor remaining at current UK building regulations of 80mm.

iii. WINDOWS. Sensitivity to increased front window area is for a small reduction in energy use due to the solar gain more than off-setting greater heat losses.

iv. DAYLIGHT INSULATION. Daylight insulation has been incorporated in the front half of the rooflights primarily as a means of solar control, in order to diffuse light and minimise glare. Due to the decreased solar transmission the use of 12mm daylight insulation actually increases annual energy consumption by £1.65 p.a. This measure improves comfort and obviates the requirement for further solar control devices, such as blinds, which will require maintenance.

v. SOLAR VENTILATION PRE-HEAT. The required ventilation rate of $30m^3$ per person per hour translates to 3.5 air changes per hour in each classroom. Assuming 3/4 of the ventilation air comes via vents to the buffer zone (2.625 ach, 467 m^3/hour) and 1/4 directly from outside (0.875 ach) an annual saving of $13.6kWh/m^2$ (£$0.27/m^2$) is predicted. This measure is not cost effective unless the amenity value of a covered and glazed area for school activities is assumed to offset most if not all of the cost.

vi. SUMMER OVERHEATING. In order to see the effect on overheating, details of window area and overhang size were investigated, defining overheating as an air temperature of 25°C or more. In the original design the main front SE (or SW) facing windows have outside dimension 1.6m x 6.2m (sill height 0.7m, area $9.9m^2$), with opening lights of 0.53m x 6.2m ($3.3m^2$) above. Venetian blinds were not considered appropriate in the context of a school, being expensive, requiring maintenance and being venerable to abuse by children. From this and the heliodon study the overhang of 1.3m is both effective and appropriately sized, reducing peak temperatures by 1.5-2.5°C. Given the loss of useful solar gain and daylighting penalty of smaller windows, a sill height of 0.7m has the greatest effect on overheating whilst having small effect on daylighting. The overall result of this will be at most 12 days where for one hour or more the temperature is 25°C or more, the maximum temperature being 30.9°C for the worst hour of the year.

4.2 OVERALL BUILDING PERFORMANCE

The energy balance for the heated areas is shown below in Figures 6 and 7.

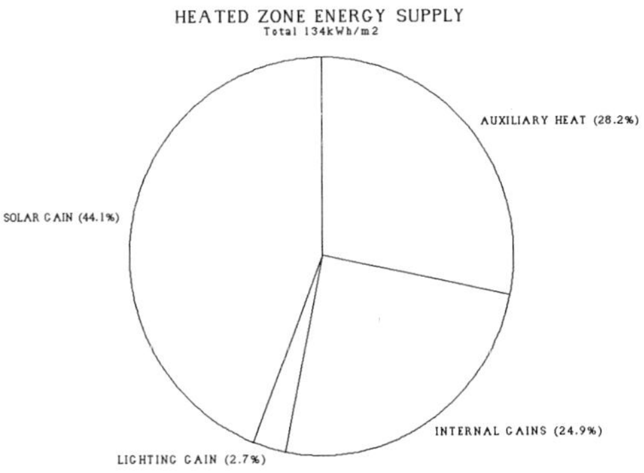

Figure 6. Energy Supply

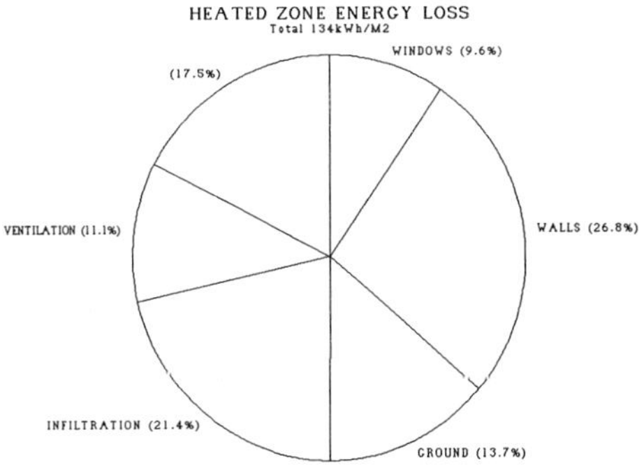

Figure 7. Energy Loss

The main factors giving rise to energy savings are:
i. Optimization of daylight design and use of efficient luminaires has led to lower electric lighting use. Annual electric lighting use is £224 (3.74kWh/m^2, £0.25/m^2) for 300 lux on the working plane at 7.7W/m^2 using high efficacy fluorescent lamps. This is a saving of up to £385 p.a.(£0.43/m^2), the figure compared to continuous lighting while occupied. Compared to present school practice of leaving lights on until approximately 9pm for cleaners, a further saving of up to £352 p.a. (5174kWh) is made, allowing half an hour for cleaning. Total lighting saving is in the order of £737 p.a. Had standard fluorescent lights been specified both running costs would be increased by approximately 42% (£94 p.a.).
ii. Despite the relatively short occupancy period of a school building, economic optimization of insulation thickness has resulted in greater than building regulation standards being used. Wall, roof and floor insulation thicknesses are 100mm, 100mm, and 80mm respectively, providing an annual saving of 2.4kWh/m^2 (£0.05/m^2) compared to UK Building Regulations.
iii. The use of solar ventilation pre-heat (SVP) from an attached glazed space has reduced the ventilation load, leading to heating net savings of £242 p.a. 13.6kWh/m^2 (£0.27/m^2).
iv. Orientation of main glazing to the SE or SW has allowed direct solar gain to contribute significantly. Simulated useful direct solar gain (term time, October-April) is 18kWh/m^2 for the heated zones in addition to indirect gains due to SVP and would be 24kWh/m^2 if the buffer were not in place.

4.0 Conclusions

Total annual heating and lighting energy use simulated for a similar west facing design to building regulations provides a reference of 83.4kWh/m^2, with a net running cost of £2.37/m^2. The chosen design has a predicted annual energy use of 42.4kWh/m^2 and a net running cost of £1.06/m^2, a saving of 55% on compared to building regulations. A more common reference for school energy use is primary energy units (PEU) used in Department of Education and Science design note 17, where ratios of 3.73 and 1.68 are applied to electricity and gas respectively. Allowing 2.6kWh/m^2 for hot water provision and 3W/m^2 for general electrical use, the proposed design has a total energy consumption of 48.0kWh/m^2, that is 95.8kWh/m^2 PEU. This is a 48% saving on the building regulations reference of 185.4kWh/m^2 PEU and a 66% saving on the DES recommended maximum of 280kWh/m^2 PEU for middle schools. On this basis gross running costs are £1.32/m^2 for the chosen design compared to £2.62/m^2 for the building regulations reference, a saving of 50%.
Total estimated net overcost for additional insulation, double glazing, transparent insulation, and controls, less plant saving, is £7,749. Estimated saving are £1165 per annum indicating a project payback of 6.7 years.

PASSIVE COOLING TECHNIQUES IN INDUSTRIAL BUILDINGS: A SIMULATION AND EVALUATION STUDY

J.L. Molina, J. Guerra, R. Velázquez

Universidad de Sevilla
Departamento de Ingenieria Energética
y Mecánica de Fluidos
Av. Reina Mercedes s/n
E-41012 Sevilla, Spain

ABSTRACT. This paper presents an analysis of the thermal conditions in an industrial building, carried out with a view to improving the level of comfort inside the building by means of passive cooling techniques. In its original state, the temperature inside the building at the height of Summer surpassed 40 °C. The S3PAS computer programme enabled us to reveal the source of the problem. At the same time, we were able to simulate the thermal behaviour deriving from each passive cooling measure, prior to final implementation, and thereby select the most suitable one for the usage of the building. Monitoring of the building, once the proposed measure was implemented, reveals a close approximation between the results forecast by S3PAS and those actually obtained. A reduction of the interior temperature to a maximum of 34 °C was achieved.

1. Introduction

In general, industrial buildings are not designed so as to obtain reasonable levels of comfort therein. A particularly important aspect, given its contribution towards discomfort, is the lack of insulation in the building. This factor causes the interior surfaces of the enclosure to reach extremely high temperatures.

Construcciones Aeronáuticas S.A. have, at their Seville factory, industrial buildings inside which the temperature surpasses 40 °C on the hottest days of the Summer. In this paper we study the levels of comfort within one of these buildings and of the possible passive cooling techniques which would lead to a reduction in the resultant inside temperature.

The problem was analyzed with the aid of the S3PAS computer programme [1]. The building was simulated, with the outdoor conditions pertaining on a severe Summer's day, as laid down by Spanish regulations, in its original state and following implementation of a series of passive cooling techniques. The simulations performed revealed the source of the problem and enabled us to select the most suitable technique considering the usage of the building.

2. Description of the Building. Original State

The building under study [2] is an industrial building of approximately 1650 m^2, forming a 31×55 m rectangle. The main axis runs NE-SW. All the enclosing walls are exterior with the exception of the NW one which is adjacent to an air-conditioned office building.

The building is divided into two areas: a Warehouse of approximately 1300 m^2 and a

S. Alvarez et al. (eds.), Architecture and Urban Space, 761–766.

Table 1: Areas by Orientation (m^2)

	SE	SW	NW	NE	Roof	Floor
Warehouse	297^a	175^a+50^b	225^e	225^a	1250^c+125^d	1364^f
Printing Works	75^a	225^e	200^a+25^b	200^a+25^b	300^c+40^d	286^f

Table 2: Constructional Features

Walls	U ($W/m^2\ K$)	Weight (kg/m^2)
^{a} Exterior	1.47	376.
^{b} Metal doors	5.8	
^{c} Roof	0.72	96.
^{d} Skylights	4.30	
^{e} Partition	3.31	122.
^{f} Floor	1.74 W/mK	480.

Printing Works which occupies some 350 m^2. The average height of the building is 7.3 m. The area of each confining surface is shown for each orientation in Table 1. The construction features are shown in Table 2.

The building is occupied by 6 to 8 people when in use, which is from 5 to 14 hours (solar time). During this period the illumination is 0.88 W/m^2 in the Warehouse and 3.52 W/m^2 in the Printing Works, where there is equipment with an installed capacity of 37 kW.

The Warehouse area is naturally ventilated whereas in the Printing Works there is forced ventilation by fans which renew the air 6 times each hour. The doors are open when the premises are occupied.

The S3PAS programme was used to simulate the interior conditions of the building on a design Summer's day in Seville (percentile 1 %), both at a constant interior temperature and allowing this to fluctuate freely. The former conditions enable us to ascertain which constructional aspects influence the thermal load to a greater extent. The latter permit comparison of the simulated and actual interior conditions, given that the building has no air-conditioning equipment. The distribution of the load components as determined by S3PAS is shown in Table 3.

Figure 1 shows the resultant interior temperatures in the Printing Works and in the Warehouse, together with the exterior temperature for reference purposes. A situation of obvious discomfort may be observed (maximum temperatures of 39.7 °C in Warehouse and 41.5 °C in the Printing Works), elimination of which is the purpose of this work.

Table 3: Principal Load Components (kWh per day)

	Printing Works	Warehouse	TOTAL
Conduction (Roof)	218	909	1127
Conduction (other Walls)	217	535	752
Radiation	47	194	241
Occupancy	2	6	8
Illumination	14	15	29
Equipment	370	0	370
Ventilation	81	57	138
TOTAL	949	1716	2665

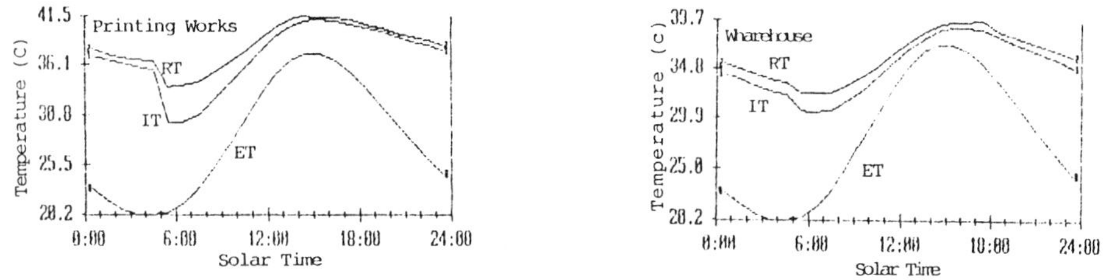

Figure 1: Resultant Interior Temperatures in the Original State. ET: Exterior Temperature, IT: Interior Temperature, RT: Resultant Temperature

3. Selection of Passive Cooling Techniques

Analysis of the thermal loads at constant interior temperature (Table 3) reveals that the principal load component is conduction via the roof. (In the Printing Works, the greatest load is that of the equipment, but this component cannot be reduced). The cause of the high resultant temperatures is the high roof temperature.

In order to reduce heat transfer via conduction in general terms, and particularly via the roof, together with the interior surface temperature thereof, the following techniques may be utilised:

a) Insulation of the exterior walls,
b) Installation of a false ceiling beneath the roof,
c) Insulation of the roof with 4 *cm* of insulation,
d) Installation of an irrigation system on the roof,
e) Installation of a false ceiling with 4 *cm* of insulation and mechanically ventilated.

Each measure was evaluated from the point of view of the thermal load at constant

Table 4: Maximum Values of Total Load and Roof Temperature

Technique Simulated	Max. load (kW)	% Reduction	Max. Temp.
initial situation	223	-	38
a	207	6.8	38
b	186	16.4	33
c	167	24.9	30
d	118	47.2	27
e	110	50.7	26.5

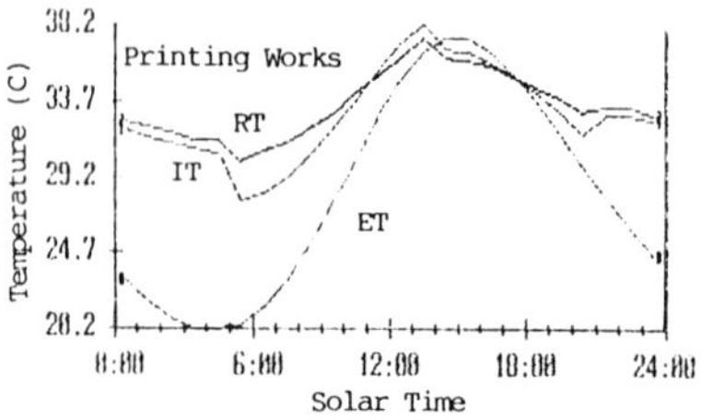

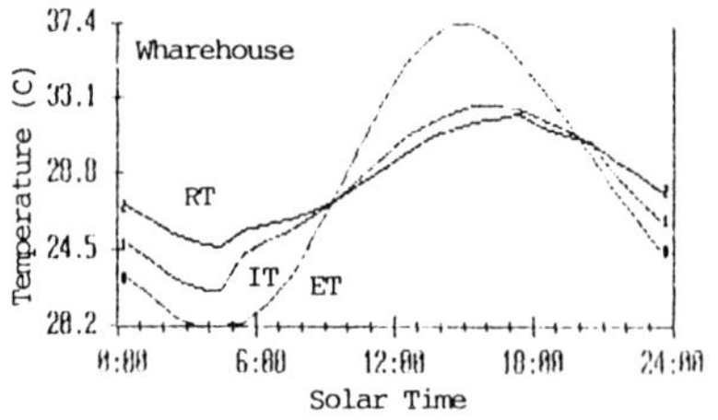

Figure 2: Resultant Interior Temperatures After Simulation of Technique e). ET: Exterior Temperature, IT: Interior Temperature, RT: Resultant Temperature

interior temperature with the S3PAS programme. Table 4 shows the maximum daily values of the total load and the maximum temperature obtained on the interior surface of the roof (or of the false ceiling when present).

The last-mentioned technique, the installation of an insulated and ventilated false ceiling, has the advantage of being simple to install, requires no special maintenance and appears to be sufficient on its own to achieve the desired effects.

4. Free Fluctuation of the Interior Temperature

Figure 2 shows the free fluctuation of the temperature in the two areas following the installation of the proposed technique, technique e), obtained by simulation with S3PAS. With respect to the initial values (Fig. 1), a reduction of 4 °C in the resultant temperature in the Printing Works may be observed, and 6.5 °C in the Warehouse.

Analysis of Figure 2 shows that the outdoor temperature is inferior to the inside temperature at night, so the resultant inside temperature could be reduced still further by ventilation of the building at night.

Figure 3 shows the temperatures obtained by simulation assuming forced ventilation for 6 renewals per hour at night-time. A reduction of the resultant temperature of an additional 1 °C is achieved in the Warehouse and 4 °C in the Printing Works.

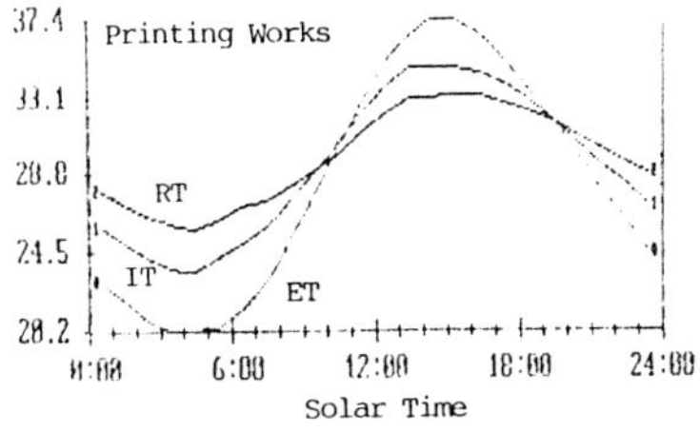

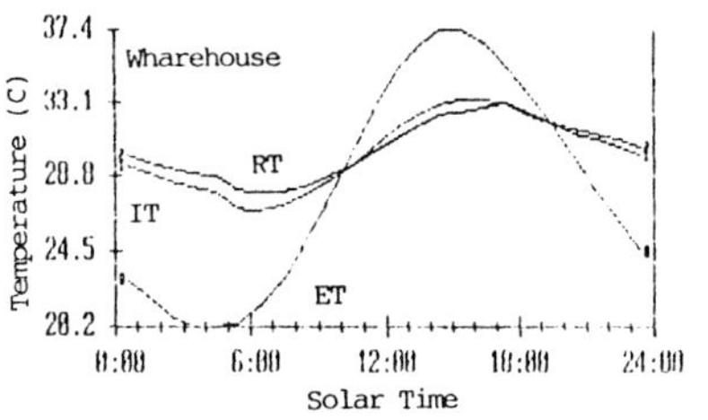

Figure 3: Resultant Interior Temperatures After Simulation of Technique e) plus Night Ventilation. ET: Exterior Temperature, IT: Interior Temperature, RT: Resultant Temperature

This situation is better than the initial technique e, therefore technique e) with night ventilation was recommended for implementation in the building: An insulated false ceiling of 4 *cm* of glass fibre was installed, with spaces for translucent panels beneath the skylights, ventilated mechanically by means of a group of extractors located above the false ceiling on the SE wall. Renewal of the air is carried out via grilles situated at ground level in the SE wall.

The extractors will operate continually in the Printing Works. In the Warehouse, all the extractors will operate when the outside temperature is inferior to the inside temperature; otherwise only two of the extractors will operate to enable ventilation of the false ceiling without causing an excessive increase in the ventilation load of the building.

5. Monitoring of the Building

After implementation of the proposed technique, in the Summer of 1990, the two areas of the building were monitored.

The records obtained are shown in Figures 4 and 5 for the Warehouse and Printing Works respectively. Maximum temperatures of 33 and 34 °C are seen, which are much lower than the original values. Furthermore, the values recorded are fully in accordance with those obtained by simulation with the S3PAS computer programme.

6. Conclusions

This research permits us to draw the following conclusions:

1. In this case, the mere use of insulation and ventilation is sufficient to reduce the temperature inside an industrial non-conditioned building.
2. The results obtained with the S3PAS programme are satisfactorily close to those actually occurring.
3. The S3PAS computer programme is sufficiently versatile to enable its usage as a building design and evaluation tool.

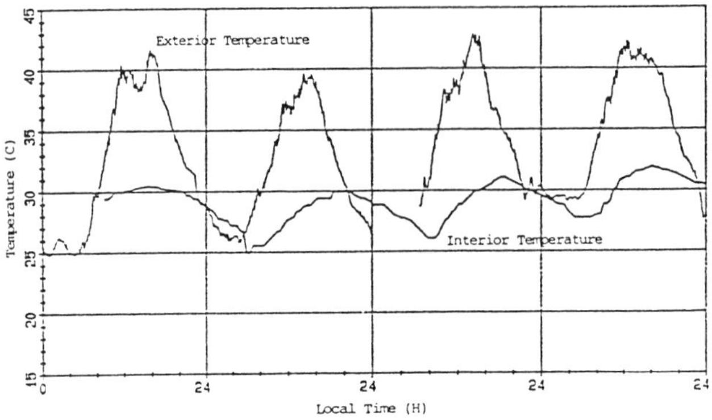

Figure 4: Recorded Exterior and Warehouse Interior Temperatures

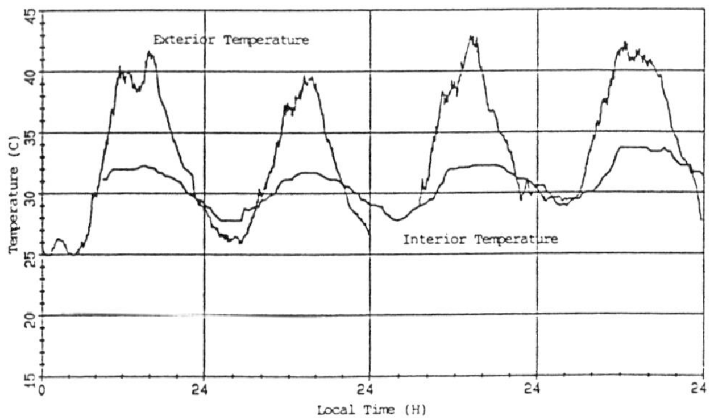

Figure 5: Recorded Exterior and Printing Works Interior Temperatures

References

1 Alvarez, S., Rodríguez E.A., Velázquez R. (1989) 'Thermal Performance of Passive Buildings Modelling and Simulation in the S3PAS Computer Program' in Proceedings of the 2nd European Conference on Architecture. Kluwer Academic Publishers, Paris, pp 416-418

2 Guerra, J.J., Molina, J.L. (1989) 'Estudio de Evacuación de Calor en Almacén Postventa de Construcciones Aeronáuticas S.A. Factoría San Pablo' Final Report. Cátedra de Termotecnia, Sevilla

COMPARISON BETWEEN SIMULATION AND EXPERIMENTATION APPLIED TO REAL BUILDING

M.R.Heras, J.Marco, J.L. Montesinos*
Instituto de Energías Renovables.CIEMAT
Avda. Complutense,22
28040 MADRID

*Escuela Superior de Ingeniería y Arquitectura
AV.IPN Norte, S/N
U.P. Zacatenco
07000 México D.F.

ABSTRACT. In this work, we compare experimental measurements on a house belonging to vernacular architecture with the simulations carried out using a detailed simulation program (S3PAS).

1.- INTRODUCTION

When the design of a house attends to energy demands the two relevant factors (constructive and the weather of its place) must be well linked. So, we can find some answers in the vernacular architecture which the man has given to the different conditions of the place, taking advantage of the climate.

In this work, we have chosen a house belonging to a typology corresponding to spanish central climate zone, where we will analyze and evaluate its thermal response. The parameters and constructive shapes of the popular architecture in that place, correspond to the central system. The selected house is located in San Martín de Valdeiglesias town (Madrid).

2.- TYPOLOGY DESCRIPTION

In the central system, there are a lot of cases of hardwood floor which is a very important part of the building structure. So, in San Martin de Valdeiglesias and all around there are houses with hardwood floor and adobe, granite, which offer their most valuable group.

The coordinates of San Martin (Madrid) are: latitude of 40º 22′ north,

S. Alvarez et al. (eds.), Architecture and Urban Space, 767–771.

length 0º 12' W, altitude 678m.

The house analyzed belongs to popular architecture and was built at the end of the XVIII century.

This house has three floors. The facade is, in the ground floor (outside), built with granite, the first floor was built with clay bricks, without showing the hardwood floor structure, wooden eaves, and it is covered with a arabian roof tile with two needles. (fig. 1).

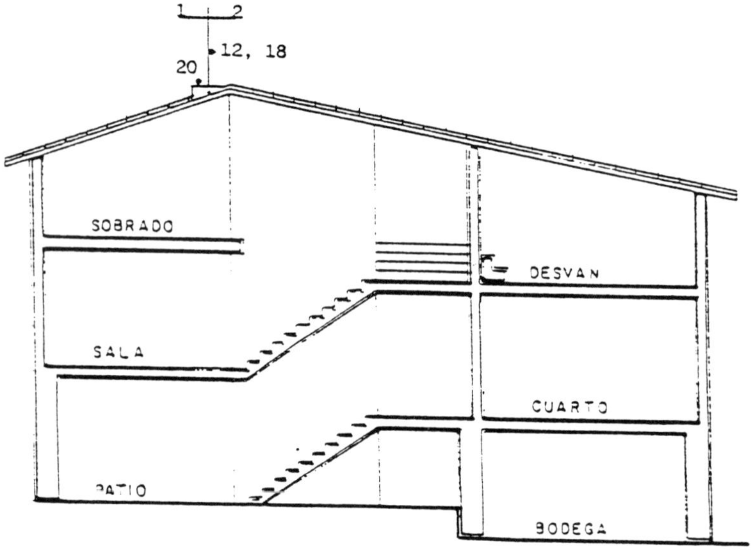

Fig. 1. Cross section (North (left) - South (right)). Sala = living room.

3.- EXPERIMENTATION

In the experimental phase continuous measurements were averaged and recorded over the following magnitudes:

- outside variables: solar global radiation, wind's direction and wind velocity, dry and humid temperatures.

- inside variables: temperature, relative humidity and heat flux throught a envelope wall.

In total, measurementsof 18 variables have been recorded, originating from 18 sensors and averaging every half hour. Each one of the four experiement periods lasted between 12 and 15 days. These four periods were differentiated by controlling power heating and rooms intercinnections as follow:

1st phase. Intermittent heating (10 hours a day) with the isolated living room.
2nd. phase. Continued heating (24 hours a day) with the isolated living room.
3rd. phase. Continued heating (24 hours a day) with the living room connected to the rest of the house.
4th. phase. Free evolution with the living room connected to the rest of the house.

Since there have been more simulations than those which could be experimentally proved, in this work, they are compared as if they had been considered as more characteristics, according to the experiments which have been done.

4.- SIMULATIONS.

To carry out simulations for camparation a detailed program for simulation of thermal and energy conditions in passive systems were used. The selected program was the 'S3PAS' developped in the Escuela Superior de Ingenieros industriales de Sevilla with the finantial support of the CIEMAT-IER.
In this program a simplification of the real house into two zones (ground floor and first floor) was defined with the real dimensions and defining the thermophysical properties of the used materials. The climate conditions were also introducted. And simulations were carried out for sequences of a day chosen as representative of each phase.

5. RESULTS.

In figure (2) both simulated and measured temperatures in the two zones are represented together with the external air temperature. We can observe a good agreement between the temperatures measured and the simulated with S3PAS for the two-zones hause defined. Also we can observe a great damping of the internal temperatures in comparison with the external one. The difference in temperatures between ground floor and first floor is due to the contact with the ground.

In figure (3) a similar comparison, but for a phase with intermitent heating of ten hours per day, can be sown. The power heating (2000 w) was located in the living room of the first floor. We can observe that temperatures simulated for the ground floor fit well the measured ones but there is a dissarrange between the observed and simulated temperature in the first floor where dynamical behaviour is more relevant (due to the dynamical excitation). The real system is heated and cooled slowlier than the simulated one. So, the deficient adaptation of the building represented with S3PAS to the dynamical behaviour of the real building will cause a bad estimation of the dynamical parameters (time constants)

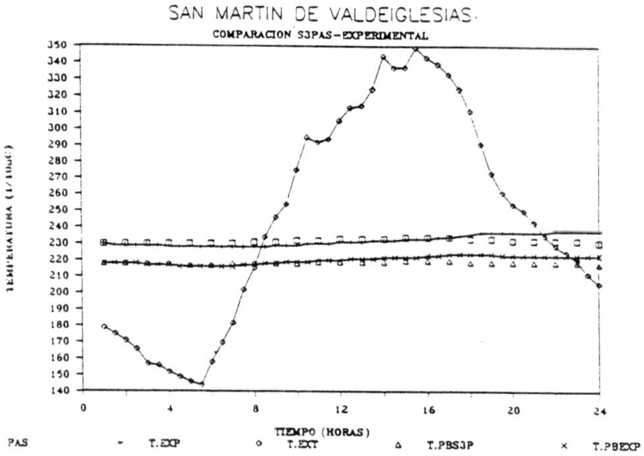

Fig. 2. Comparation free evolution.

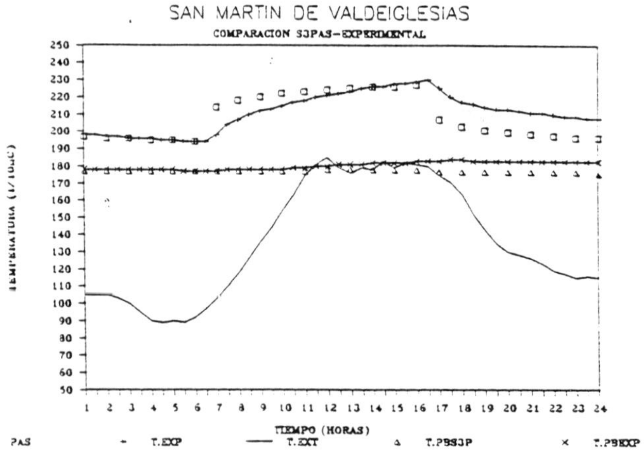

Fig. 3. Comparation intermitent heating.

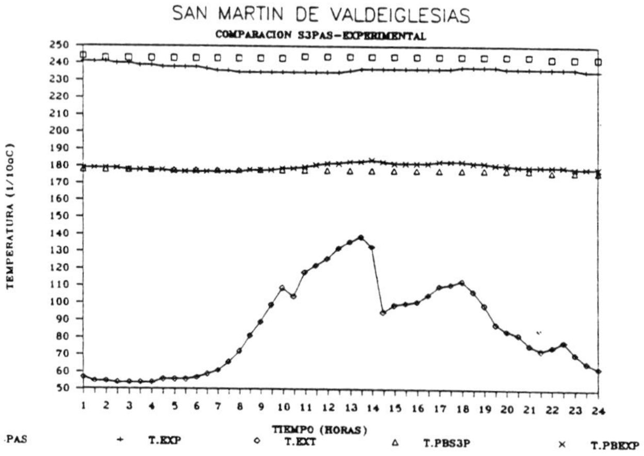

Fig. 4. Comparation continuous heating.

of the building by using this representation. But, on another hand the good fit of the simulated stationari conditions to the measured ones allows to rely in the stationary parameters (global heat loss coefficient) that can be obtained with S3PAS-representation of the building that we are defined.

Finally, in figure (4) we can observe the same comparison but for the 3th experimental phase (2000 W continuous heating in first floor). There is again a good fit between simulation and experimentation. And it shows that for stationary conditions, like these, the simulation with a shuch simplification of the real building is quite satisfactory.

5.- CONCLUSIONS

The usage of detailed models for energetic simulation, is a worth tool for studies of energetic behaviour of 'popular' architecture typologies by identifying a good representation of the building under study, in a first step, which is possible as we have presented in this paper. In a second step this representation could be used to carry out sensibility analysis over the different parameters that vary in a same typology (such as kind and thickness of walls, orientations, hollow surface, etc.), or for evaluating the response of the building to a range of climatic conditions.

6.- BIBLIOGRAPHY

Alvarez,S. et al."S3PAS 1.0. Simulación de Sistemas Solares Pasivos. Manual del Usuario" Informe Técnico, Catedra de Termotecnia. ETSI Industriales de Sevilla. Dic. 1988. (In spanish).

Montesinos, J.L. "Descripción de una metodología de análisis energético aplicada a la arquitectura popular. Influencia energética de la adopción de tipologías españolas en México." TESIS DOCTORAL. ETSAM, Madrid, 1991. (In spanish).

EXPERIMENTAL ANALYSIS OF ENERGETIC BEHAVIOUR IN A SPECIFIC TYPOLOGY OF THE SPANISH VERNACULAR ARCHITECTURE

M.R. Heras, J. Marco, J.L. Montesinos*

Instituto de Energías Renovables. CIEMAT
Avda. Complutense, 22
28040 MADRID (Spain)

* Escuela Superior de Ingeniería y Arquitectura
Avda. IPN Norte, s/n
U.P. Zacatenco
07000 México D.F.

ABSTRACT. In this work, an experimental analysis of a building typology of the vernacular architecture is carried out. It is done from an energetic point of view. This typology belongs to Central Spain.

1.- INTRODUCTION

The popular Spanish architecture presents numerous typologies, somehow related with the great variety of the climate and culture from the country. This architecture has had a great power through the time, from a historic point of view, through the direct solution of the human necessities and through the way of resting the environmental elements.

In this work, a meaningful typology of the popular spanish architecture has been chosen to analyze its energetic behaviour. Thus, we have chosen a climatic zone where we will analize and evaluate the termal response of a typical building of the popular architecture of that place. This zone corresponds to Central Spain, more specificly to the Madrilenian mountain range.

2.- TYPOLOGY DESCRIPTION

This house is one of the Central System. It is located in San Martin de Valdeiglesias, povince of Madrid, with 40º22'N latitud, 0º42'W longitude and 678 m. of altitude.

The climate is Continental. The annual maximum temperature is 23.3ºC and the minimum is 7.8ºC. Winter is cold and July is the month with highest temperature (37ºC). The annual rain precipitation is between 800 and 1000

S. Alvarez et al. (eds.), Architecture and Urban Space, 773–778.

mm.

This house has 3 floors. The first floor is made of granite, with a thickness of about of 50 or 60 cm. The walls of the higher floor are made of wood and adobe. Wood is also used in doors and windows, as well as in the stairs structure. The ceiling is made of arabian roof tile, to 2 watersheds, with the principal beams parallel to the front part of the house and a pendent scale of 30%. This house belongs to those which are called "solid" without outside yard, where all streets are constituted by continued buildings (see next plane).

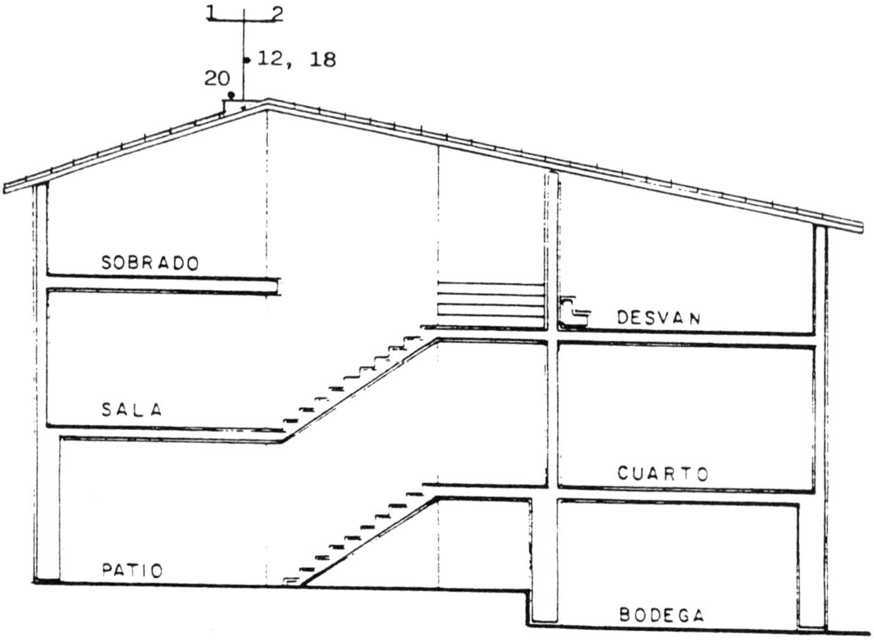

Esqueme 1. Cross section of the house.

3.- EXPERIMENTATION

The indoor air temperature and humidity have been measured, as well as the outside meteorological conditions during more than 4 months.

The performed experiement has been divided in four phases. Each phase ranges over a period of 10 to 12 days, in wich the most relevant variables concerning the thermal behaviour of the building has been recorded with a sampling interval of ½ h.

1st. PHASE. Intermittent heating (10 hours a day) with isolated hall.

The experiment in this first phase was carried out with the doors and windows closed, without allowing direct solar gains through hollows (isolated house). In the hall, there was allocated an auxiliar heating system of 2000 W scheduled on for 10 hours daily, from 8:00 to 18:00 (oficial time). The doors of the hall remain closed during the experiment (isolated hall).

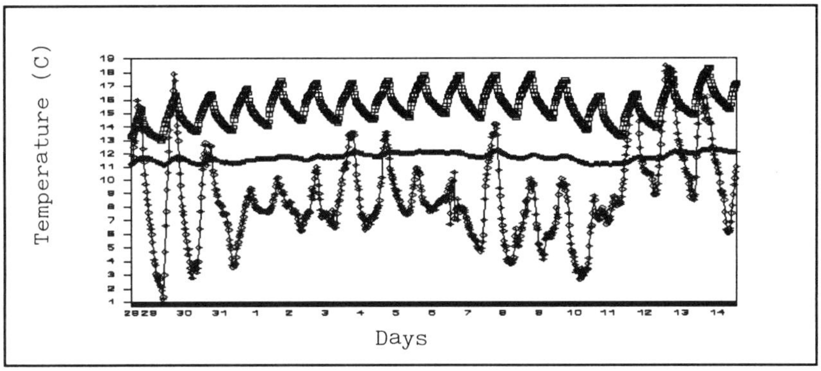

Figure 1. Phase 1: Intermitent heating & isolated hall.

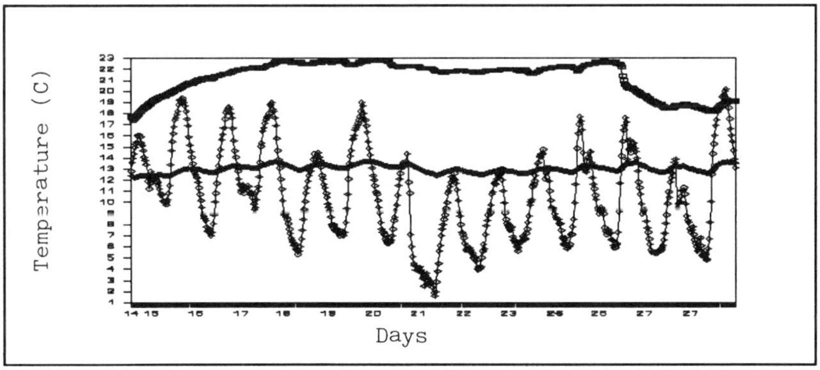

Figure 2. Phase 2: 2000W constant heating power & isolated hall.

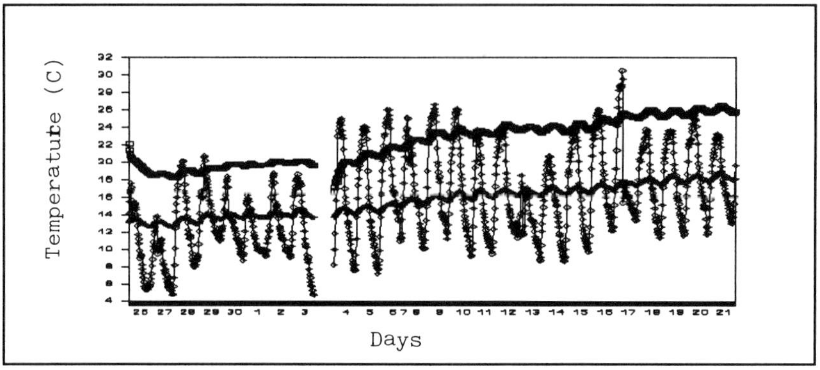

Figure 3. Phase 3: 2000W constant power & connected hall.

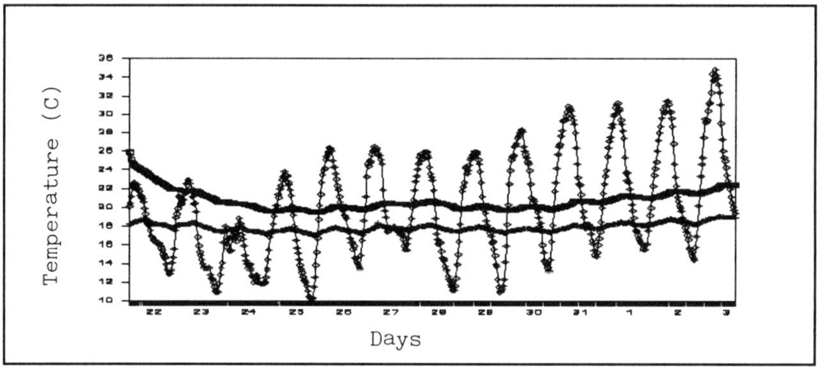

Figure 4. Phase 4: Free-floating conditions.

The experiment ranges over March 28th to Abril 14th 1990 (spring). During this time, the indoor air temperature of the hall (see fig.1) remains between 13ºC (minimum) and 17ºC (maximum), with daily amplitudes of 3ºC, and the outside air temperature oscillates between 12ºC (maximum) and 6ºC (minimum). Due to the high inertia of the building, when we turn off/out the auxiliary heating equipment no relevant variations of the indoor air temperature are observed (< 1ºC)

2nd. PHASE. 2000 W constant heating (24 hours a day) with isolated hall.

In this second phase, the experiment was similar to the first one, except that the auxiliar heating system works during 24 hours in a day (constant heating).

This experiment was carried out from April 14th to 25th 1990.

Looking at figure 2, it can be observed that there is a general decrease of the outside air temperature from April 19th to 21st. However, while the outside temperature oscillates between 12ºC (maximum) and 2ºC (minimum), with a daily amplitude of 10ºC, the indoor air temperature remains between 22ºC (maximum) and 21.5ºC (minimum).

3rd PHASE. Constant heating (24 hours a day) with the living room connected to the rest of the house.

In this phase a 2000 W constant heating power has been used and the living room has been connected with the other zones of the house opening its doors, allowing the air interchange among them. The experiment ranges over April 25th to May 21st 1990.

In the figure 3, we can notice a big degration in the indoor air temperature (May 5st) due to a failure in the electrical system. After that, the indoor air temperature becomes to increase little by little to reach a maximun value of 26.5ºC. During the experiment the outside air temperature shows a tendency to increase from mean values of 10ºC to 16ºC. A similar tendency is observed in the indoor air temperature.

4th PHASE. Free-floating evolution with the living room connected to the rest of the house.

This phase is characterized by free-floating conditions. The heating system was turned off and the hall remained connected the rest of the house (opened doors). It was carried out from May 21st to June 4th.

As you can see in figure 4, when the auxiliar heating system is turned off the inside temperature decrease to 21ºC, begining to increase again, since the outside temperatures begin to increase too, keeping a temperature of 23ºC.

4. - THERMAL GLOBAL CHARACTERISTICS.

The global charactreistics parameter of the house have been obtained fitting macrodynamic models (ARMAX) to the measurements coming from the analyzed system.

In the following table we show the results.

PHASE	LOSS GLOBAL COEFFICIENT	SOLAR APT.	t1(h)	t2(h)	HEATING	CONECTED
1	508	0.77	8.21	0.41	INT.	NO
2	520	1.12	9.80	0.23	INT.	NO
3	1403	1.06	10.0	0.63	CON.	YES
4	1272	1.14	9.15	0.34	CON.	YES

Due to the selected sampling interval and the length of the performed experiments the achieved values for the two first time constants (t1 and t2) are not significative.

5.- CONCLUSIONS

The house has a high power of damping the outside thermal variations, because of the great thermal mass of the house. As a consequence of this thermal inertia the daily outside oscilations do not produce appreciable thermal variations inside the building, while the seasonal variation do.

Focussing our attention in the thermal behaviour of the living room and using an ARMAX model to fit the data, the achieved global thermal characteristics, refering to the solar opening (As) and loss global coefficient (U), are:

As = 0.97 ± 0.15m2
Uisolated = 490 ± 40 W/ºC
Uconnected = 1340 ± 100 W/ºC

6.- BIBLIOGRAPHY

HERAS, Mª del Rosario y Marco, José. Comportamiento Energético de Edificios Solares Pasivos. Plan de Monitorización del Instituto de Energías Renovables. Ed. CIEMAT. Madrid, 1990. (In spanish)

MONTESINOS, José Luis. Descripción de una metodología de análisis energético aplicada a la arquitectura popular. Influencia energética de la adopción de tipologías españolas en México.Tesis doctoral. ETSAM, Madrid, 1991. (In spanish).

A THERMAL IMPROVEMENT OF A 9 STORY BUILDING, BUILT IN 1963, SITUATED IN THE SUBURB OF GENEVA (SWITZERLAND)

W. WEBER, B. LACHAL, A. GSPONER, O. GUISAN
CUEPE, University of Geneva, CH-1231 Conches

ABSTRACT. This study presents an experiment of thermal rehabilitation of a typical prefabricated building of the sixties in Geneva. The rehabilitation of a 126 flat building included a drastic thermal renovation. Insulation of the roof, facades, new windows and blinds permitted to cut by half the heating demand. The building was monitored over four years in order to obtain a detailed case study concerning rehabilitation of buildings from the '60s.

1. Introduction

Between 1960 and 1970 the population of Geneva nearly doubled, creating a need for apartments and leading to the construction of many "satellite towns" around the center of the city.

The rehabilitation of these buildings will represent a large part of architect activity in coming years.

We have the possibility to participate in the thermal rehabilitation of a large building of this type

The CUEPE in collaboration with the "Délégué à l'énergie du canton de Genève", the owner and the architects, decided to take this opportunity and make a detailed case study of this rehabilitation in order to:

- evaluate solar gains before and after renovation
- get some experience in the choice and effectiveness of new windows
- study the behavior of the inhabitants

2. Building description

2.1. GOAL OF THE RENOVATION

"Onex 1", was the first part of the "satellite town" of Onex. It consists of 750 flats, constructed in 1962 with prefabricated concrete elements.

S. Alvarez et al. (eds.), Architecture and Urban Space, 779–784.

Built over very rapidly, these 9 story buildings are badly or not insulated and include large single glass windows (more than 50% of the facades) and many thermal bridges.

After 25 years of use, these buildings are in poor condition and need to be renovated.

One of the buildings (126 flats), oriented north-south, has given us the opportunity to make the case study of thermal rehabilitation [1].

The thermal renovation had three goals:
- improve thermal comfort
- reduce the heat losses
- enhance the use of solar energy with the south facade

2.2. DESCRIPTION OF THE RENOVATION

After two years monitoring of the existing building and thermal simulation (LESOSAI, SERIRES) of different types of improvement, the proposed solution was predicted with a reduction of 40 -50% of the energy demand.

The renovation of the envelope was achieved during spring and summer 1990 (figure 1).

The envelope was fitted with new windows (plastic frames and double glazing), walls and roof insolation with aluminium cladding.

A careful study of the inhabitant behavior led to the choice of shading and ventilation devices in order to optimize the use of possible solar gains:
- outside aluminium blinds
- manual adjustable ventilation valves in the window frames

Figure 1: South facade of the renovated building

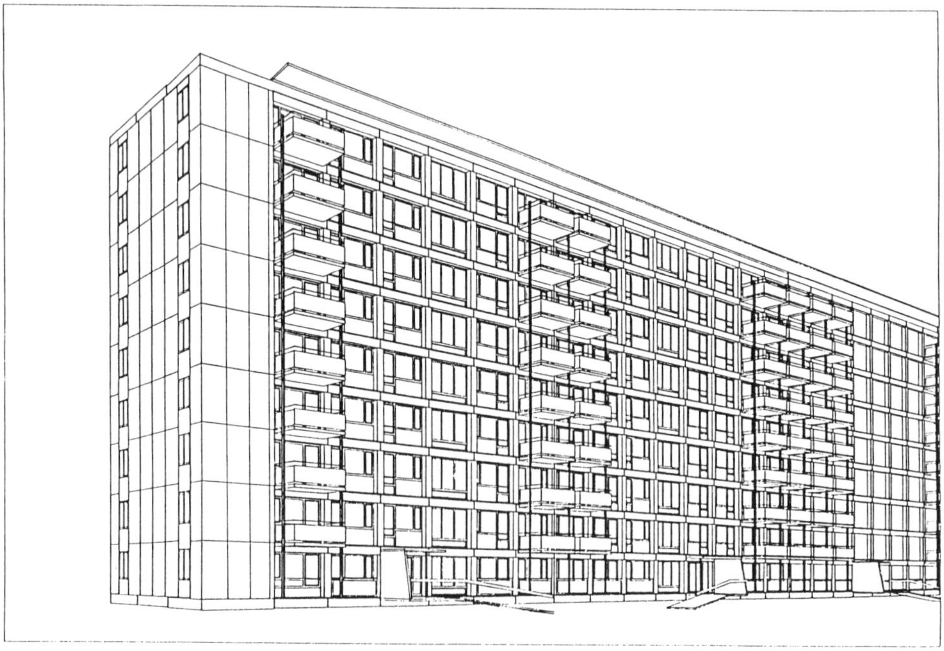

- small windows for ventilation

The heating control was adapted to the new envelope.

The production of hot water was decentralized and 200 m2 of solar panels will be installed at the end of 1992 to preheat this water.

The relation between inside and outside was improved mainly by adding balconies and by requalifying outdoor spaces.

3. Monitoring and results.

3.1. MONITORING

This building is being monitored since 1987.

Only five data are measured:
- heat demand for the two zones of the building (north and south) with two heatmeters;
- external temperature and solar radiation;
- internal temperature in an unoccupied flat in the southern part.

These data are read 20 times per hour, by conventional data acquisition system and stored in hourly values in order to be compatible with software for thermal simulation as SERIRES.

3.2. RESULTS

Results for the heating season 1987-1988 are shown in table 1. The annual consumption is not too high in regards to the bad thermal quality of the envelope. This is due to the temperature of

Table 1: Mesured heating demand between October 1987 to May 1988

Month	Days	Tout	Tint	Qsouth	Qnorth	Qtotal
October	22	12.3	23.6	119727	178751	298478
November	30	6.3	22.3	323994	366183	690177
December	31	4	21.9	410346	426518	836864
January	31	5.2	21.7	366808	395581	762389
February	29	3.4	21.1	324481	409987	734468
March	31	6.4	21.6	279602	351686	631288
April	30	12.1	22.6	133584	205721	339305
May	16	16.6	24.1	41992	56119	98111
YEAR	220.0	7.6	22.2	2000534	2390546	4391080
consumption (MJ/m².an)				373.2	446.0	409.6

Tout : outside temperature, °C

Tint : Internal temperature, °C

Qsouth : heating for south part, MJ

Qnorth : heating for north part, MJ

Qtotal :total heat = Qsouth + Qnorth, MJ

this period that was about 10% higher than average and to the good form factor (volume/envelope surface = 0.36 (1/m).

Results of the last heating season, after the thermal improvement of the facade, show a saving of around 50% (figure 2). For this period no regulation of the heating was working, so night temperature of the water was as high as during the day. It means that the final results should be even better and the building will still have to be monitored during winter 91 to 92.

Figure 2: Daily heat demand for last winter, October 1990 to January 1991

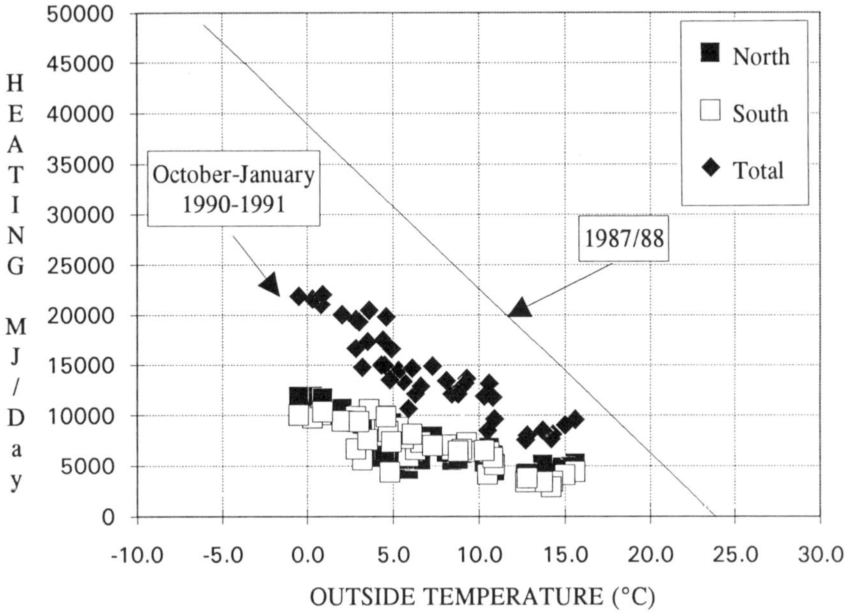

4. Photography

By scanning pictures of the south facade of the building (212 windows), taken every two hours during a week in March, we obtained rich information on how the occupants deal with external blinds, internal curtains and window opening.

These observations combined with the use HM diagrams [2], enabled us to explain the observed value for effective solar gain area (figure 3).

Solar gains are reduced because of different effects:

- the total window area includes frames. We have to consider the net glass area;
- shadings of different kinds;
- use of external blinds (30% are always closed);
- glass transmission, and in the staircase, useful fraction of solar gains (30%);
- use of curtains (75% are usually drawn).

By referring to the temperature measured in an unoccupied flat, we may deduce that occupants react mainly to an internal temperature increase rather than to solar radiation. These reactions are strongly connected to the comfort and highlight the importance of good regulation and thermal inertia in order to obtain a high utility of solar energy in the building

Figure 3: Analyse of the effective solar area

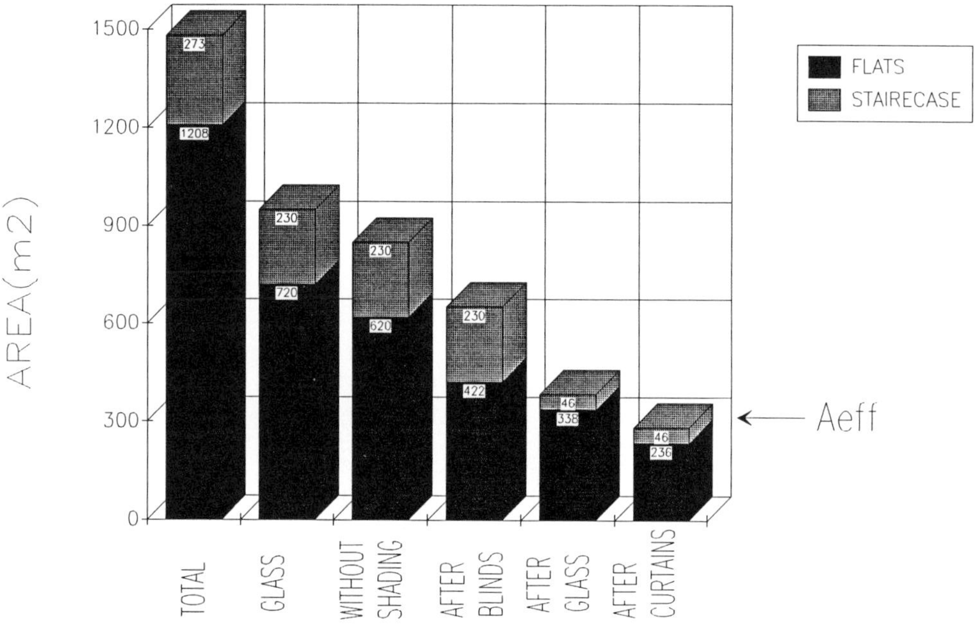

5. Conclusion

This study shows that the three goals of this thermal rehabilitation can be easily achieved with actual technics of building and good collaboration between architects and ingineers:
- thermal comfort was improved;
- the heat losses were reduce by about 50%;
- the part of solar energy was increased in the thermal balance.

However an investment of 4 000 000 SFr for a maximum of 40 000 Sfr savings a year means a pay back of over 100 years with actual fuel prices.

So drastic thermal rehabilitation of buildings of this type, without the help or regulation of the administration, are not profitable and can only be done if owner:
- takes the opportunity of a general renovation that they have to do anyway;
- wants to take into account the thermal comfort and environmental preservation;
- understands that the thermal renovation of the envelope will preserve and give value to his building.

6. References

[1]. D. Baillif, R. Loponte, bureau d'architecte, W. Weber, architecte, mars 1989, Rapport pour le bureau du Délégué à l'énergie: Projet de réhabilitation des immeubles 8 à 14 rue du Comte-Geraud, Rapport interne.

[2]. B. Lachal, W. Weber, O. Guisan, September 1990, Analyse thermique d'un bâtiment locatif des années 60 avant rénovation, 6. Schweizerisches Status Seminar, Energie Forschung im Hochbau, EMPA_KWH.

BIOCLIMATIC DESIGN FOR AN ELEMENTARY SCHOOL IN NORTHERN JAPAN

YUICHIRO KODAMA

Building Research Institute
Tatehara 1, Tsukuba JAPAN 305

ABSTRACT

In northern part of Japan, especially on the Japan Sea side of the main island, they have relatively cold and cloudy days in mid-winter. Solar Insolation in January is almost half of that in Tokyo on the Pacific Ocean side. This is a project of public school that should be built with a very limited budget. The bioclimatic design strategy in this kind of climate are proposed and discussed.

1.Bioclimatic Design Procedure

The principle of bioclimatic design is how to adapt building design to the climate.In the design procedure, it is required for designer to understand the climate at first and then to check the performance of the building which he designed. Authors has proposed following technics: 1) climate analysis with Climate Characteristic Chart and 2) performance check with interactive simulation program for personal computer. This project is developed along with this bioclimatic design procedure. This elementary school has 12 classrooms with multi-purpose space (450 pupils). Its total floor area is 4800m2 except arena. The main structure is of 2 storied reinforced concrete.

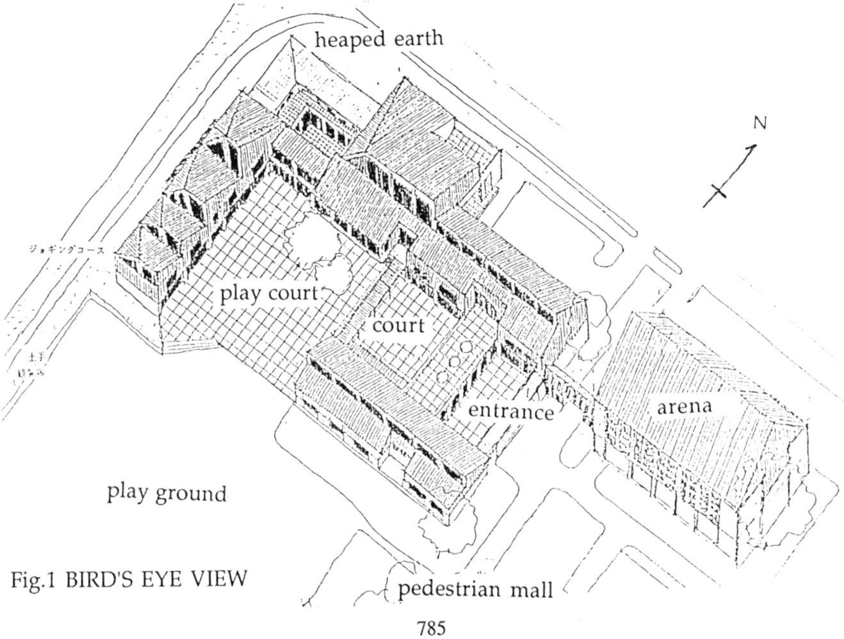

Fig.1 BIRD'S EYE VIEW

S. Alvarez et al. (eds.), Architecture and Urban Space, 785–790.

2.Climate Analysis

For the climate analysis, the regional climate characteristic chart s are referred, which are compiled by authors for 24 major cities in Japan. They are compiled with the objective of simplifying designers work to discover the potentials of natural environment which are endowed to the region.
The charts are compiled for five climate elements; (1) air temperature, (2) relative humidity, (3) wind velocity and direction, (4) solar insolation on horizontal and vertical south surface and (5) could cover. On the chart that has the vertical axis as time and the horizontal axis as the month, one can observe both of the daily changes and the yearly changes. Contour lines indicate the hourly changes over one year period. Furthermore, it is also possible to know the simultaneous generation of some climate elements when these charts are over-layed.
In this case, was referred the chart for Akita city which is 25 km south of the site and has almost same climate condition.

TEMPERATURE

[°C]

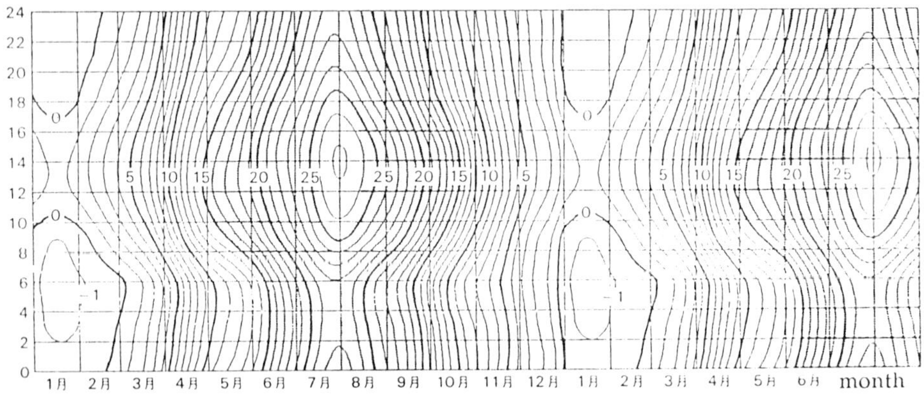

WIND DIRECTION /WIND VELOCITY

[m/sec]

above: north

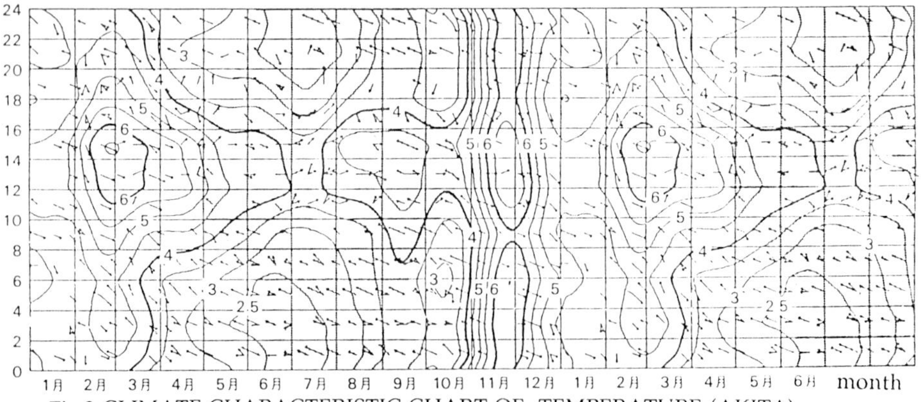

Fig.2 CLIMATE CHARACTERISTIC CHART OF TEMPERATURE (AKITA)

Fig.3 CLIMATE CHARACTERISTIC CHART OF WIND (AKITA)

As the air temperature chart (Fig.2) shows, it is relatively cold winter(average temp. of Jan. is 2.5deg.C) and hot summer (average temp. of Aug. is 28.4deg. C)
The wind chart (Fig.3) shows that strong north-west wind blows whole day long in the mid-winter.This is very characteristic point in this area. On the contrary, mild breeze blows from east-south in the summer but has enough potential to cause cross ventilation in hot day time.

3. Design Strategy

1) wind protection
The site is located at the west end of hills and face to an open rice field which extends to lagoon and there is nothing to prevent cold winter wind.
Buildings are located along with the west and north end of the site and 1.8 depth of earth is heaped up against the buildings. Also the roof shape is decided with the consideration of wind stream. So that the play-ground and courts are protected from cold wind and major openings can faces to sunny side. (Fig.4)

2) Maximum Solar Access
Although solar insolation in mid-winter is not much in this area, south-facing windows have still advantages to inlet solar heat in cool periods (heating season)and to control it easily in hot periods (cooling season). Each class room has large south-facing high-side windows for solar heat gain and daylight , with glare control system to avoid glare problem which is afraid to occur in case of windows except that of north-facing.
Double glazing is used for major widows.

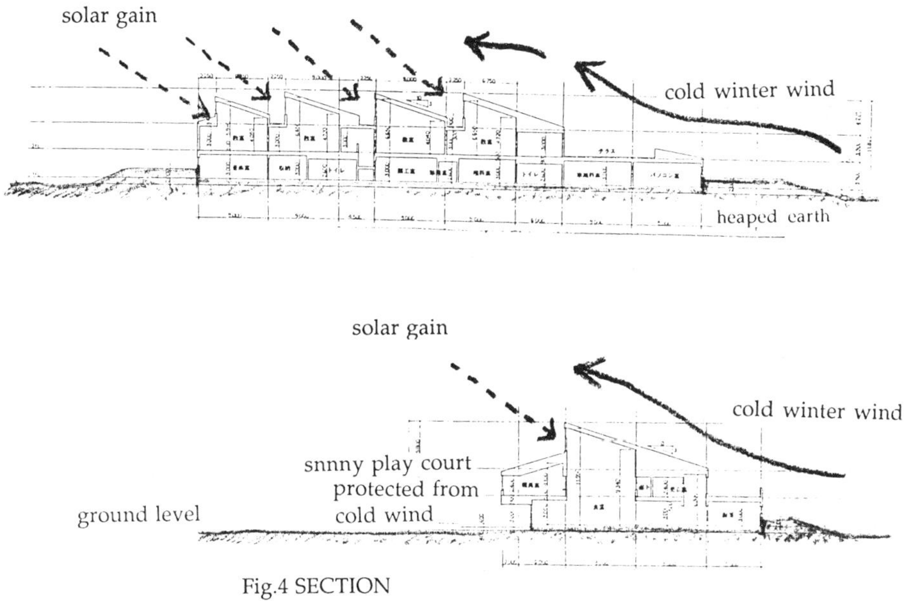

Fig.4 SECTION

3)Exterior Insulation

The main structure is reinforced concrete and exterior surface of concrete wall should not be exposed to the air. Insulation materials are installed at the exterior side and then covered with claddings. This results inevitably an exterior insulation system which add heavy mass of concrete inside. The effect of mass is checked by computer simulation.
(see 4-3))

4)Cross Ventilation

Major cooling systems depend on cross ventilation and natural exhausting. Each classroom bas openings both for inlet and outlet of wind, and high-side windows at the top of the room from which hot air is exhaust. A large air volume of the room promotes ventilating.

4. Performance Check

Fig.5 shows a section of the typical classroom which has south-facing high-side windows in addition to normal windows. Usually north-facing high-side windows are preferred to avoid glare however they can not get solar gain.The effect of south-facing high side window are checked from the view points of heating and daylighting. Also the effect of mass are checked.

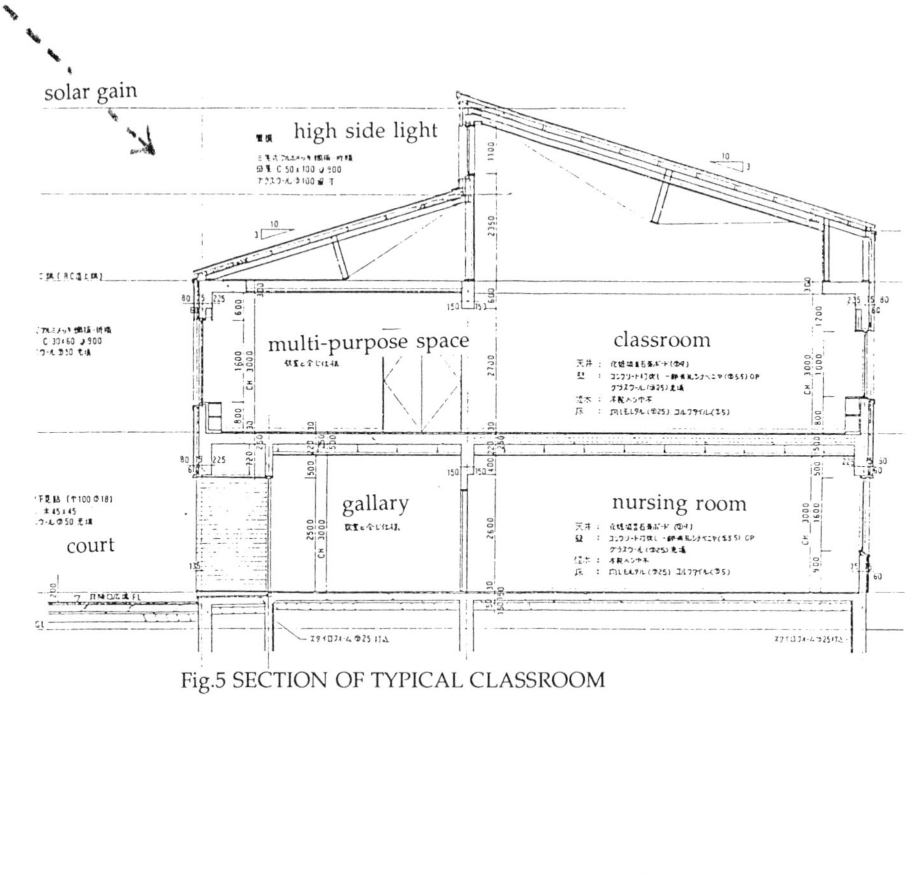

Fig.5 SECTION OF TYPICAL CLASSROOM

1)Thermal effect of south -facing high-side window

When the glazing area increases, the heat loss from the room increases. Fig.6 shows the ratio of heat loss: one is with high-side window and another without it. In both cases, major heat loss causes from glazing and ventilation.
The Aux. heating of three cases are compared:

case A: conventional (without high-side window)
case B: with south-facing high-side window
case C:with north-facing high-side window

It is obvious that case C needs AUX most. (Fig.7) However there is not so big difference between case A and case B. This is because of the climate of this area where they have a little solar insolation.
For instance in Tokyo where they have much insolation, the advantage of case B is obvious.(Fig.8)

A without high-side light

B with high-side light

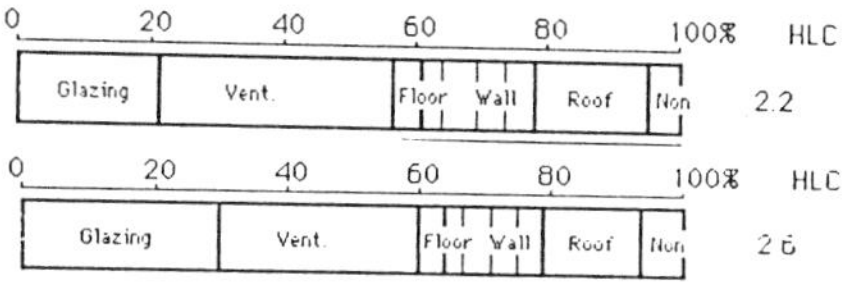

Fig.6 RATIO OF HEAT LOSS

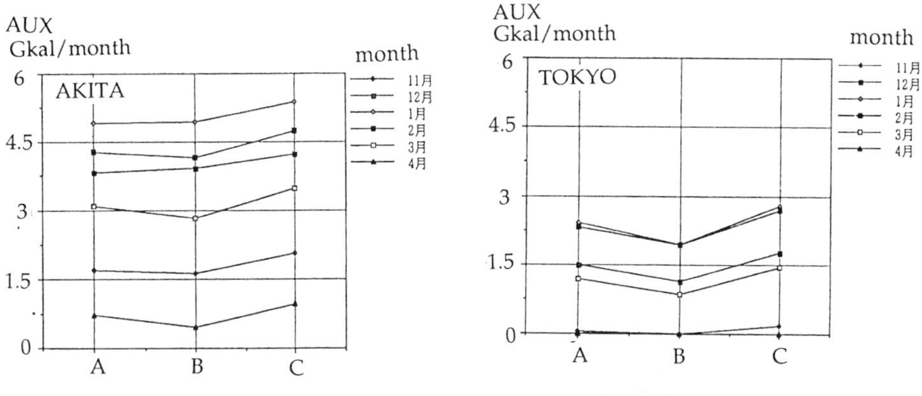

Fig.7 EFFECT OF SOLAR GLAZIC SIZE ON AUX

2)Day lighting effect of south -facing high-side window

A typical classroom with multi-purpose space is 7m in width and 15m in depth. The distribution of illuminance is shown in case A (Fig.8) and in case B (Fig. 9). The day lighting condition in case B is better than that in case A. (These calculations do not consider about direct beams but only diffusive beams.)
In conclusion, the thermal effect of adding south -facing high-side window is not clear in this climate however the effect to improve day lighting condition is obvious. In the sunny climate like in Tokyo, both effect are obvious.

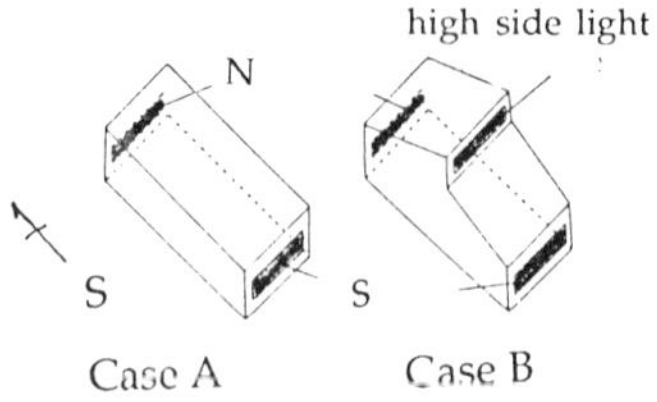

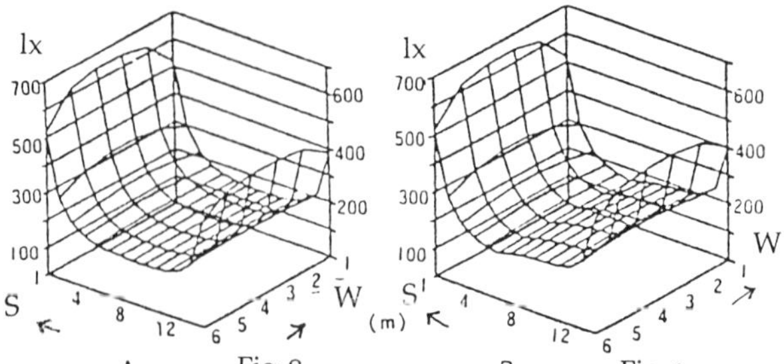

Fig. 8 Fig. 9

DISTRIBUTION OF ILLUMINANCE

3) Thermal effect of mass

The classroom is heated periodically (usually 8:30-15:00). In such a heating mode, it is generally expected that the thermal effect of mass is not clear in the cloudy climate.

Fig.10 shows the comparison of AUX among three building type:

case B0 : heavymass in side of the classroom

case B1 : light mass in side of the classroom

case B2 : no mass in side of the classroom

The shape of the room is same to that of case B above mentioned.

In mid-winter (Dec., Jan. and Feb.), the room with heavy mass needs more AUX than others in this local climate.

On the contrary, in the sunny climate like in Tokyo the effect of mass saving AUX is obvious in any month of heating season.(Fig.11)

It is concluded that the heat capacity is considered too much when all of the concrete elements are exposed to the interior space of the classroom in case of this school.

In order to reduce the influense of the mass, the floor is finished with cork tile and some part of interior wall is finished with wood. So that the mass come to medium level as same as case B1. Those finishes are desirable also from acoustic point of view.

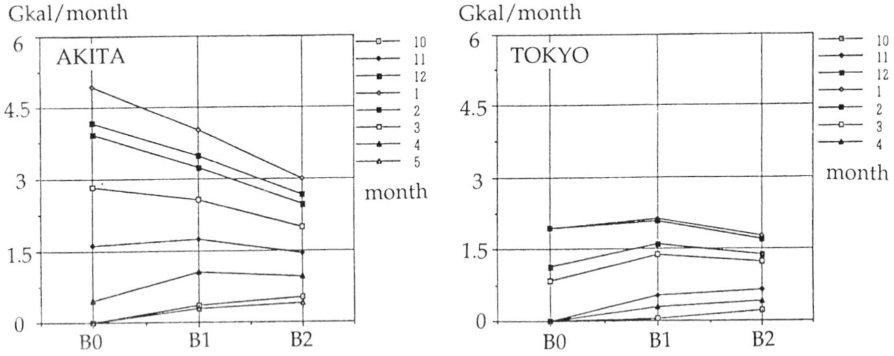

Fig. 10 EFFECT OF MASS ON AUX Fig. 11

Part 7

EDUCATION AND TECHNOLOGY TRANSFER

THE NEGEV DESIGN AND CONSTRUCTION WORKSHOP: AN ALTERNATIVE EDUCATION PROTOTYPE

I. A. MEIR
Desert Architecture Unit
J.Blaustein Institute for Desert Research
Ben-Gurion University of the Negev
Sede Boqer Campus 84990, Israel

ABSTRACT. The paper discusses the problems of architectural information dissemination in the sparsely populated Negev Desert, Israel. It presents the Negev Design and Construction Workshop, a "nomadic" framework of professional seminars, its activities and some of its accomplishments, and argues the necessity and viability of alternative education frameworks in the fields of planning and design.

1. Introduction

The Negev Desert, the southern part of Israel, is an arid region constituting about 65% of the country's area (some 17,000 km^2). Having been sparsely populated so far (less than 7% of Israel's population), it has been given little attention on matters of planning and design. Standard settlement and building types designed for other, climatically different parts of the country, have been introduced to this arid region in the past. This practice still goes on regardless of the Negev's division into four distinctly different climatic zones, ranging from the hot and humid desert of the coastal plain to the cold and dry mountain region (Meir,1989; Meir,Etzion,Faiman,1990) (Fig.1-a). The distribution of the Negev population to a number of conceptually different settlement and social organization types (cities, development towns, cooperative villages and collectives) complicates planning and design matters even more, since the spatial organization of each one of these settlement types is different than the others' (Fig.1-b).

2. The Negev Design and Construction Workshop

2.1 BACKGROUND

The Desert Architecture Unit (DAU) of the Blaustein Institute for Desert Research is a research unit concentrating on the investigation of planning, design and construction appropriate for deserts, and particularly appropriate for the Negev. However, the knowledge developed by DAU in these fields does not always reach the target population, namely the people actively involved in the relevant fields

S. Alvarez et al. (eds.), Architecture and Urban Space, 793–798.

in the Negev. The need for an information dissemination framework that would deal with the specific problems concerning these specific fields of interest was identified in the late 1970s. A first attempt to establish such a framework was undertaken in the years 1981-82 and took the form of a *Workshop*. Its activities covered mainly the *kibbutzim* (collectives) in the close vicinity of the Institute. Eight seminars were held, each one concentrating on a design or construction topic of unique desert-influenced character. However, several changes inside DAU, as well as various external influences, brought the *Workshop* activities to an end.

Between the years 1982 and 1988, DAU was contacted by individuals and groups seeking advice and assistance in appropriate planning and design. As a consequence, the author of this report decided to renew the *Workshop* activities. A number of decisions were to influence the form of the renewed framework.

2.2 AIMS AND TARGETS

The main aim of the *Workshop* is the creation of a framework within which communication and information transfer will be enabled through continuous contact between the academe and decision makers, planners, architects and the people directly involved in construction and supervision in the field. The subjects to be investigated, be they theoretical or practical, have to be closely related to the design and construction activities in the field. Applicability of the theoretical work presented is an imperative.

A second, but not less important, aim of the *Workshop*, is the creation of an informal forum for the people connected to the fields of design and construction to meet and exchange information and ideas. The importance of this aim will be better enhanced if one considers the long distances between the various settlements, the distinctly different socio-economic, demographic and political background of the different settlements in the region creating a gap between them, and the fact that contacts between the planning and settling authorities and the architects, builders and users are usually loose and infrequent.

2.3 MEANS AND METHODS

To achieve these aims, it was decided to give the *Negev Design and Construction Workshop* a unique character. The potential forum addressed by the *Workshop* includes professionals from all the settlements in the Negev region. The topics investigated are selected by an *ad hoc* committee formed by representatives of the different groups (settlements, authorities, the Federation of Architects and DAU). The *Workshop* is a "nomadic" forum which holds seminars, each one in a different settlement. Each seminar concentrates on one or more projects in the settlement, includes *in situ* investigation of the project, lectures by the planners and designers, background lectures on theoretical aspects of the projects investigated (by members of DAU or other academic institutions) and open discussions. The proceedings of each seminar are published and mailed to all the participants in an attempt to form a reference base on the topics investigated.

To enable the unobstructed activities of the *Workshop*, and to ensure the participation of all people interested, financial support was sought from the Municipalities, Local and Regional Councils in the Negev, the Ministry of Housing and the J.Blaustein International Center for Desert Studies. This financial support, which covers the basic operational expenses, enables the participation of people interested while removing the obstacle (sometimes more psychological than real) of participation fees.

3. Activities and Evaluation

So far, the *Workshop* has held eight seminars on various topics, such as appropriate design for the desert; appropriate details for buildings and clusters in hot areas; bio-climatic architecture; evolution of residential areas and the urban fabric; land reserves as a policy in desert settlements; landscaping and its effects on micro-climatic conditions in the desert; soil-related problems of structures; and solar energy applications in arid zones.

The seminars were held in nine different locations, covering most of the settlement types and climatic sub-regions of the Negev (Fig.1-c). They were attended by a total of over 200 professionals. Their professional background differed (architects: 48.0%; landscape architects: 8.2%; planners/engineers: 7.8%; construction related professions, including foremen, supervisors, contractors, builders and project managers: 25.9%; others, such as geologists, geographers technicians and students: 10.1%), as did their affiliation (academic institutions: 21.6%; municipalities/local and regional authorities: 38.0%; settling authorities: 16.9%, government: 11.3%; private: 12.2%) and settlement type basis (city: 40.7%; town: 6.2%, rural: 39.2%, others, including Sede Boqer Campus and regional services centers as defined by the Population Census: 13.9%). These data were collected by means of attendance records, as well as from a questionnaire mailed to all participants who had attended at least one seminar. The number of participants per seminar has grown gradually from 15-20 in the first two seminars, to an average of 30-35.

It is interesting and important to note the relatively high participation rate of people from rural settlements: 39.2 of the Workshop participants came from the rural sector which represents less than 9% of the target population (State of Israel,1984). It is assumed that the direct involvement of people from *kibbutzim* (collectives) and *moshavim* (cooperative villages) in decision making processes causes higher awareness to various aspects of administration, management and budget allocation and, therefore, stirs interest in the topics of appropriate design and construction. This awareness may also be attributed to the fact that, through this direct involvement in decision making processes, these people can influence decisions concerning the formation of their built environment much more than the urban population can.

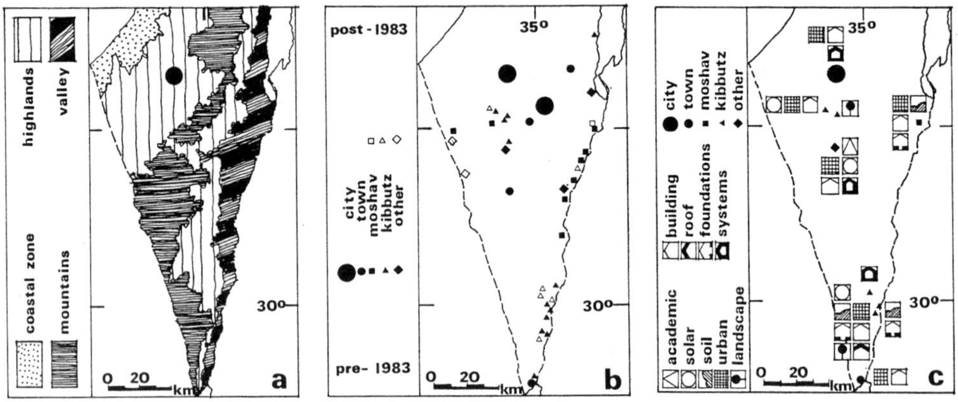

Fig.1. **a**. Settlement distribution in the Negev. **b**. Climatic subregions of the Negev. **c**. Locations and subjects of Workshop seminars and Workshop-stirred activities.

The participants were asked to classify by degree of importance ("vital", "very important", important") the different topics included in the seminars (case studies, theoretical background, field trips, open discussions). One of the most important conclusions derived from the questionnaire is that most of participants feel the need to investigate case studies: 84.3% of the participants classified the investigation of case studies as a "vital" part of the seminars. This conclusion conforms with Stern's (1990) conclusions on the importance of the "triability" and "observability" parameters affecting the innovation process. It seems that learning from an existing example is easier and more comprehensible than learning in a theoretical approach. It also seems that the existence of demonstration projects is vital to the promotion of innovative ideas and practices. To a great extent, the special nature of DAU, which is involved both in applicable research and in the design of pilot projects, plays a major role in the dissemination of methods and techniques in the field of appropriate design for the desert.

It is hard to quantitatively evaluate the success the *Workshop* has had so far in promoting ideas in the fields it treats. However, a few examples may enhance the process of acceptance of appropriate strategies and technologies introduced through it.

A *kibbutz* situated in the Arava valley, one of the hottest areas in the country, turned to DAU in an attempt to solve the problem of overheating of dwellings. A special seminar was held there, during which the concept of "floating roofs" shading the roofs of the buildings was proposed. Executed in an experimental scale in the beginning, the proposed solution caused the reduction of energy invested in air-conditioning to about half of that invested in the untreated units. As a result, the *kibbutz* decided to construct these shading devices on all the dwelling units of the type tested.

The activities of the *Workshop* brought the subject of appropriate design to the attention of a new *moshav* situated by the Dead Sea. (The area is very hot throughout most of the year, winds are strong and carry large amounts of dust, solar radiation rates are high and the saline and hot aquifer water reaches to about one meter below ground surface.) The settlers turned to DAU for consultation and evaluation of the plans the settling authorities had provided for them and, as a result, a number of meetings were held both at DAU and on the planned site.

The environmental aspects of urban design were treated in a number of seminars attended by representatives of the Ministry of Housing. Shortly after these seminars, DAU was asked to serve as consultant on environmental issues in the design of two new neighborhoods in the city of Elat (the southernmost settlement situated at the Red Sea Gulf). DAU was also invited to design prototypes for bio-climatic educational facilities for different climatic sub-regions of the Negev.

DAU was asked to investigate and monitor different projects in remote settlements, whose representatives participate in the seminars. As a result of the technical problems faced in several cases, but also in order to enable better, more thorough investigations in the future, the *Workshop* initiated the establishment of a mobile monitoring laboratory under the auspices of the Ministry of Energy and Infrastructure.

4. Conclusions

The real effect a framework such as the *Workshop* may have is yet to be seen. Nevertheless, activity around the *Workshop* shows a growing interest and awareness in the topics treated by it. This awareness seems to be reaching individuals attending the seminars, as well as the authorities.

It seems that the success of the *Workshop* so far is based on a combination of factors. The organizing body, DAU, is an academic institution (not for profit purposes). It combines theoretical research and pilot projects of varying scales and uses (devices and systems, buildings, complexes and clusters). Adoption of the methods and techniques advocated by the unit have obvious economic advantages and, therefore, acceptability of the innovation introduced becomes easier. The "informal" character of the framework, established through public funding rather than participation fees, as well as through the "nomadic" and informal character of the seminars, encourages people to participate in as many seminars as possible. It also encourages them to participate actively in both lectures and discussions, regardless of their professional background or the forum they address.

The flexibility of the *Workshop*, established through these means, enables it to tackle problems and issues of different scales and character, making *in situ* observations and studies in different locations regardless of distances.

It is hoped that the intervention of the *Workshop* in design and construction processes in the Negev will have a growing influence on decision makers, architects and users. The varied activity including seminars, consultancy, monitoring, pilot project construction and publications (so far spanning over more than seventy pages of seminar proceedings and a growing number of papers in professional journals) will strengthen the links between academe and professionals in the different fields. Thus, promotion of information dissemination will enable the adaptation of planning, design and construction to higher standards, toward the creation of better, more comfortable built environments. The author of this report believes that the success the *Workshop* has had so far may help in the creation of similar frameworks for the dissemination of information in remote and sparsely populated areas.

Acknowledgements

The *Negev Design and Construction Workshop* idea and activities have been promoted with the assistance and advice of Dr.S.Dover, Architects M.Cones, D.Shen, A.Kashtan, Mrss. I.Shoval, O.Barzilai, A.Elon, the DAU staff, and especially Dr.Y.Etzion, Unit Head. The financial support of the Municipalities, Local and Regional Councils in the Negev region, the Ministry of Housing - South District, and that of the J.Blaustein International Center for Desert Studies is kindly acknowledged.

References

Meir, I.A. (1989) "Climatic sub-regions and design contextualism", Building and Environment 3,245-251.

Meir, I.A., Etzion, Y., Faiman, D. (1990) Energy Aspects of Design in Arid Zones (in Hebrew), Desert Architecture Unit/Applied Solar Calculations Unit, J.Blaustein Institute for Desert Research, Sede Boqer Campus & Ministry of Energy and Infrastructure, Division of research and Development, Jerusalem.

State of Israel - Central Bureau of Statistics (1984) Localities : Population and Households, 1983 Census of Population and Housing Publications No.3, Jerusalem.

Stern, A. (1990) "Bridging the gap: building information - technology transfer" Building Research and Practice (CIB) 5, 309-313.

Modelling Wind/Building Interactions Using Computer Animated Graphics: An Evaluation of the Educational Effectiveness

K. McCARTNEY and P.J. RHODES
School of Architecture
Portsmouth Polytechnic
King Henry I Street
Portsmouth PO1 2DY
England

ABSTRACT. Results are presented from an experiment in which architectural students were required to predict the effects of various types of obstructions on wind speed and direction, after being presented with information in either an animated, or a conventional static graphic format. The superior performance of those students exposed to animated graphics lends support to the proposition that a wider range of representational techniques can improve the ability of designers in dealing with dynamic environmental problems.

Introduction

It can be argued that the difficulties which architects face in attempting to comprehend the nature and effects of many environmental phenomena, are compounded by the mismatch between the essentially dynamic nature of the phenomena, and the essentially static techniques used to represent them. The architect's traditional language of plans, cross-sections and perspectives is severely constraining when it comes to modelling the fourth dimension. This constraint may be responsible for many of the lost opportunities, and mistakes, which arise from a lack appreciation of, for example, the shifting path of the sun, the convective flow of heated air, or the acceleration of the wind.

The importance of the range of representational techniques available to the designer has been emphasised by Moran (1970) in his description of the "multilingual designer". Eastman (1970) analysed the actions and protocols of architects designing a bathroom layout. His conclusions supported the proposition that an architect who lacks the means, or the ability, to represent design-relevant characteristics, will ignore those characteristics.

In considering the specific problems of knowledge transfer between the fields of thermodynamics and architectural design, Yannas (1989) has suggested the "need to visualise time". The recent appearance of software packages enabling animated graphics to be produced on desk-top computers might make it considerably easier for designers to follow this advice, and to extend their repertoire of representational techniques in a significant new direction.

Air movement arising from the interaction between wind and buildings appears to be an obvious area in which to test these ideas. This paper describes work using animated graphics to represent basic concepts about the interaction between wind and obstructions. These concepts have been identified through empirical work, particularly observations of physical models in wind tunnels. Such work has been described by Penwarden and Wise (1975), Peterka et al (1985) and reviewed at an earlier PLEA Conference by Bowen (1983).

S. Alvarez et al. (eds.), Architecture and Urban Space, 799–804.

Although specialists now possess a great deal of information on the subject, the problem of transferring expert knowledge into practice remains. Wisse (1988) discussed this problem, and emphasised that the creation of the urban environment is a team effort in which effective communication has an important role to play. He referred to the need for shared knowledge and a shared image, to be formulated in understandable terms, and not, he stressed, "in a specialist's secret symbols". This paper describes an experiment to test a particular approach to the development of the "shared image".

The effectiveness of utilising dynamic representational techniques as an integral part of the design process is difficult to test. Our experiment was intended to test, in a pedagogical context, the potential of animated graphics to improve the ability of architectural students to understand, and predict, the effects of different types of obstruction on wind velocity and direction.

Experimental Materials

A briefing on some basic principles of wind interaction with obstructions such as walls, buildings and vegetation was prepared in two formats:

(A) STATIC FORMAT:
a series of nineteen illustrations in the form of cross-sections plans, and coordinate graphs (see a sample selection in fig (i))

(B) ANIMATED FORMAT:
a short sequence of animated graphics incorporating the nineteen static illustrations from (A)

As the intention was to identify the influence of the animation, attempts were made to minimise all other differences in the two briefings. The graphics used for the static format (A) were key frames selected from the animated format (B). A single explanatory text which could accompany both graphic formats was prepared.

All the graphics were prepared on an Apple Macintosh computer using Macromind Director for the animation.

Subjects and Procedures

Students in the second year of the BA(Hons) Architecture course at Portsmouth Polytechnic were randomly assigned to one of two groups. Group (a) received a briefing consisting of the (A) static format graphics, presented using an overhead projector, accompanied by an oral delivery of the explanatory text. Group (b) received a briefing consisting of the (B) animated format graphics, presented on a computer graphics monitor, accompanied by an oral delivery of the same explanatory text.

Immediately after the briefings, which lasted twenty minutes, the students were set two tasks which required them to indicate, by drawing and/or writing on supplied drawings, the influence of built artefacts on the wind.

The first task required an indication of the path of the wind on both sides of a perforated obstruction. The drawing supplied to the students was the same illustration which had been

Figure 1: Examples of the slides and commentary

Solid Obstructions

Slide-1

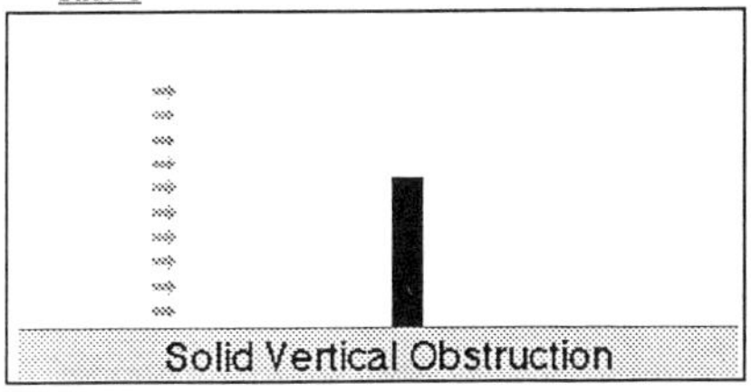

What happens when wind approaches an obstruction?

The arrows in this diagram represent the direction of the wind, which is at a right angles to the obstruction.

Slide-2

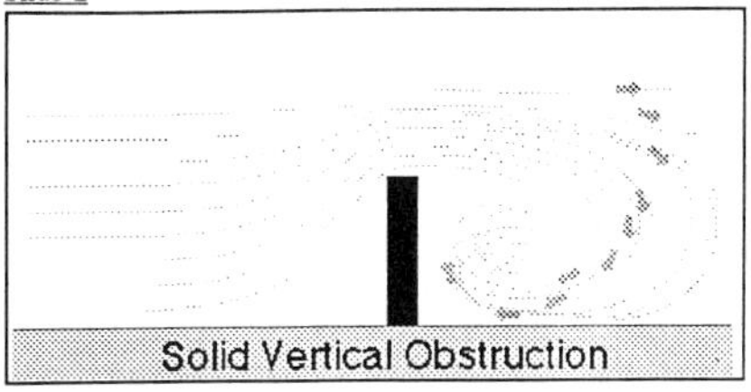

Wind rises, even before encountering the obstruction, then passes over the obstruction, and then curls down on the *leeward* (sheltered) side.

The path traced by the wind is shown diagramatically by the mean *streamlines* (lines drawn parallel to the direction of the flow).
The curling of the streamlines on the leeward side is described as an *eddying* current and is a sign of *turbulence*. (Turbulence is a rapid variation in both velocity and direction).
What causes the wind to change direction to create these eddies?

Slide-3

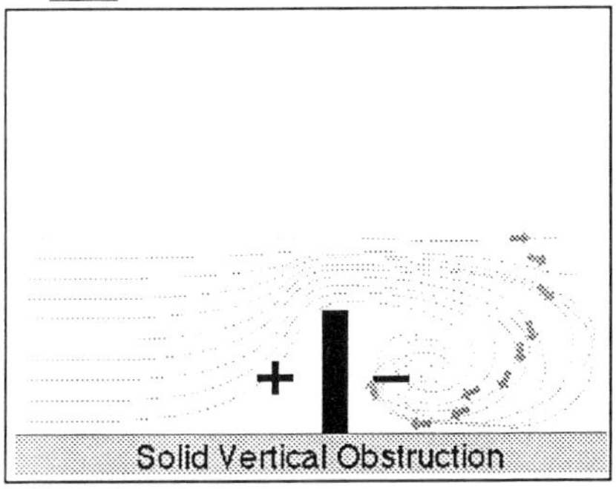

Changes in wind direction are caused by differential air pressures

The force of the wind creates a relatively high positive pressure on the windward side of the obstruction. This positive pressure deflects the wind over the wall. Negative pressure develops on the protected, leeward side. This negative pressure, or suction, draws the wind down beyond the obstruction.

Slide-4

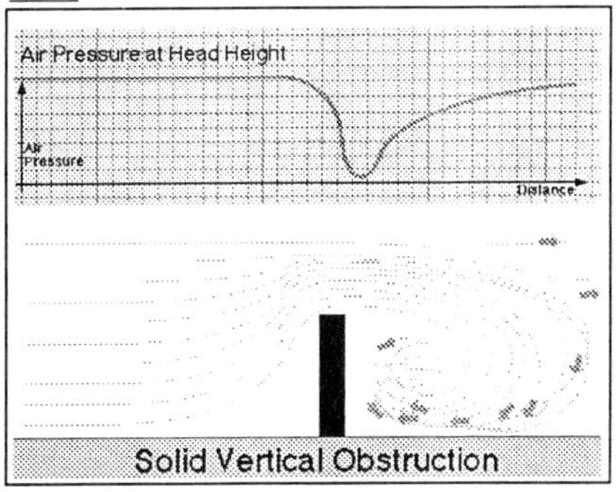

We can see these changing air pressures more clearly if we look at a graph.

This graph illustrates the air pressure, at about head height, at varying distances, upstream, and down stream of the wind.
Notice how the pressure drops suddenly and dramatically on the leeward side of the wall.
Notice also that the pressure begins to rise again a very short distance downwind of the wall.

used in the briefings (see the diagram in Table 1) with the wind streamlines removed. Performance on this task was therefore likely to be considerably influenced by short-term memory ability.

The second task involved a more complex arrangement of obstructions represented by a drawing which had not been presented in the briefing. (See the diagram in Table 2). This second task therefore required, to a greater degree, an understanding of the principles and an ability to apply the information in a new context.

The students performance was assessed by recording on a checklist those characteristic wind/obstruction interactions which had been correctly identified. (Identification required an indication of the locations at which one might expect occurrences of cavitation, turbulence, and also, less dramatic changes of direction and speed).

Hypothesis

The number of correct identifications of each characteristics in the two groups of students provided the basis for a limited testing of the hypothesis that *animated representations of wind/obstruction interactions will improve the ability of designers to predict changes in wind patterns.*

Results

The results are summarised in Tables (1) and (2). The significance of the differences recorded between the two groups of students was tested using the x^2 non-parametric test for two independent samples. None of the differences recorded for performance in the first task are significant.

However, in the second task, which required a greater degree of understanding and ability to apply concepts in a novel context, there are two examples of differences which are significant. These relate to the identification of the relatively sheltered area of cavitation which occurs on the windward side of an obstruction, and to the turbulence which occurs downstream of the corner of the building in plan view. In these cases the performance of the group of students who received the (B) animated format briefing was superior (the probability that this difference was due to random effects is less than 5%).

Conclusions

The experiment demonstrated that there was a significant difference in the ability of two groups of architectural students to identify areas of cavitation, and locations of turbulence when wind interacts with a complex three-dimensional obstruction.

If the significant superiority in the performance of the group (b) students can indeed be attributed to the animated graphic format of the briefing they received, this would support the working hypothesis and the more general hypothesis that access to animated graphic formats for representing design-related characteristics would enhance the ability of designers to understand and apply information about complex and dynamic environmental phenomena.

Table 1: Results for the Perforated Obstruction

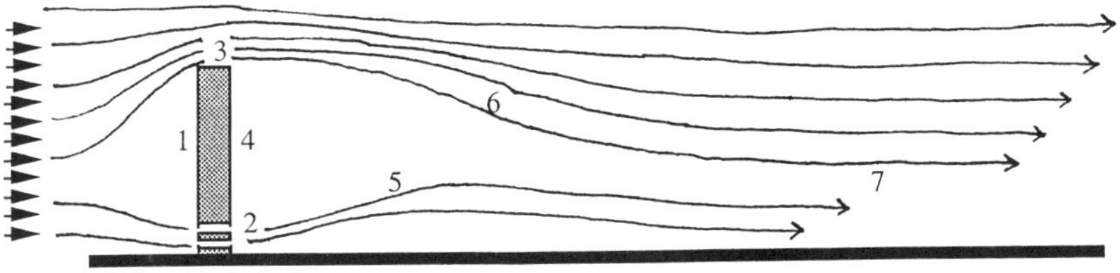

	Characteristic wind/obstruction interactions						
	1	2	3	4	5	6	7
Overhead Projection Slides (Total Sample 22)	81.82	100	100	90.91	72.73	95.45	13.64
Animation (Total Sample 25)	84	100	96	84	72	84	32

The diagram shows the characteristic key points where we might expect the wind to change direction or speed. The table indicates the percentage of students who correctly identified these characteristics.

Table 2: Results for the Multiple Obstructions

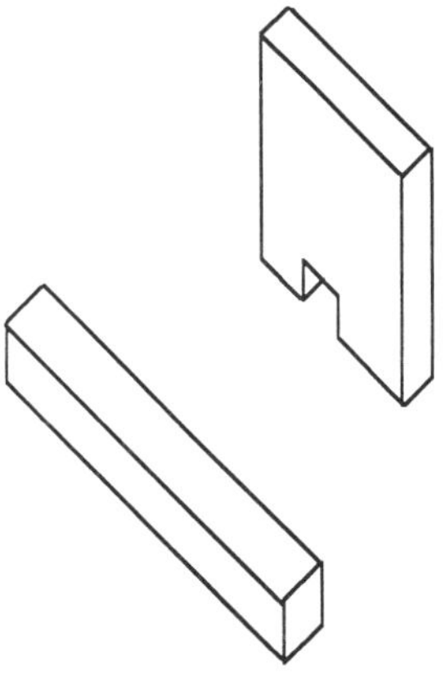

AXONOMETRIC

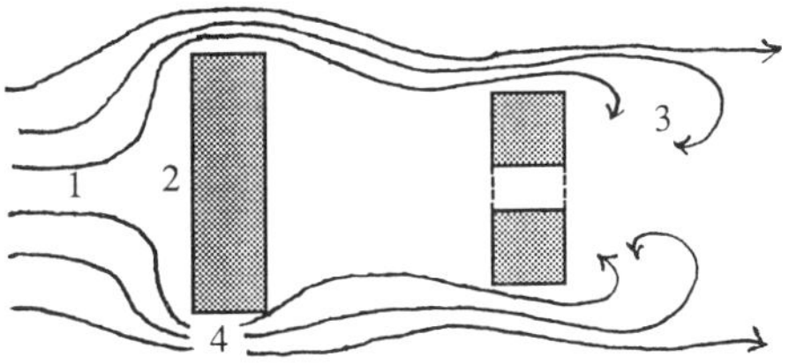

PLAN

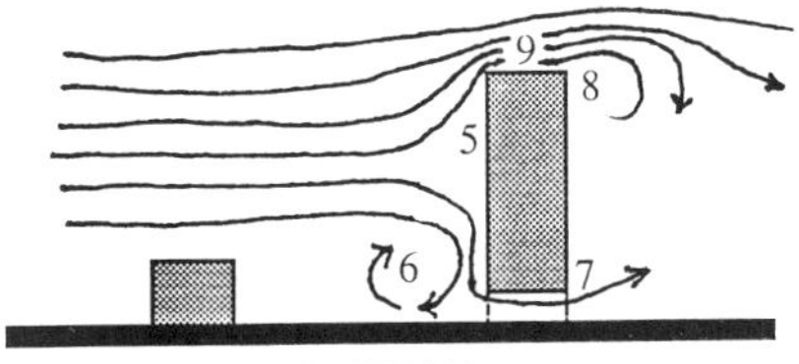

SECTION

	Characteristic wind/obstruction interactions								
	1	2	3	4	5	6	7	8	9
Overhead Projection Slides (Total Sample 22)	77.27	68.18	31.82	95.45	50	50	90.91	63.64	90.91
Animation (Total Sample 25)	96	96	68	96	56	72	88	72	96

The diagrams show characteristic key points where we might expect the wind to change direction or speed. The table indicates the percentage of students who correctly identified these characteristics.

Animation production has always been considered a very time-consuming, and hence expensive, graphic format. This is changing. The time may be well be ripe for animation to be added to the designer's repertoire of representational languages.

ACKNOWLEDGEMENT

The tasks used to assess students understanding were inspired by exercises devised by Antony Sealey, formerly of the School of Architecture at Birmingham Polytechnic and by Penwarden and Wise (1975).

References

Bowen, A. (1983) 'Design Guidelines on Lateral Airflow Through and Around Buildings', in S. Yannas, ed.(1983) 'Passive and Low Energy Architecture: The Second International PLEA Conference', Pergamon Press, Oxford.

Eastman, C. (1970) 'On the Analysis of Intuitive Design Processes', in G. Moore, ed. (1970) 'Emerging Methods in Environmental Design and Planning', MIT Press, Cambridge, Mass.

Moran, T. (1970) 'A Model of a Multilingual Designer', in G. Moore ed. (1970) 'Emerging Methods in Environmental Design and Planning', MIT Press, Cambridge, Mass.

Penwarden, A.D. and Wise, A.F.E. (1975) 'Wind Environment Around Buildings', HMSO, London.

Peterka, J.A., Merony, R.N. and Kothari, K.M. (1985) 'Wind Flow Patterns about Buildings', J. of Wind Engin. Indust. Aerodyn., 21, 21-38.

Wisse, J.A. (1988) 'A Philosophy for Teaching Wind in the Built Environment', Energy & Buildings, 11, 157-161.

Yannas, S. (1989) 'Physics and Architecture. Issues of Knowledge Transfer and Translation to Design', Solar and Wind Technology, 6, 4, 301-308.

Eco-Design-Aid

ir. Leo Gommans and ir. Kees Duijvestein
Delft University of Technology; Faculty of Architecture
Kabinet 11.04
Berlageweg 1
2628 CR Delft

Eco-Design-Aid: Computerprograms as tools for environmental-friendly designing of houses and urban layouts.

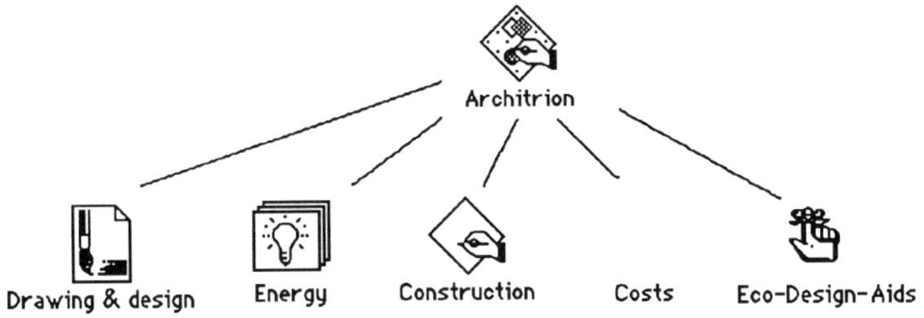

The Eco-Design-Aid (EDA) is part of the PEBI-project (Personal computers and Building Innovation) which is started in the EC COMETT-program. The first problem noticed by PEBI was the lack of education and training in the field of building innovation with microcomputers. Therefore several courses and computerprograms have been made, concerning the design of buildings. The Eco-Design-Aid is one of the courses.

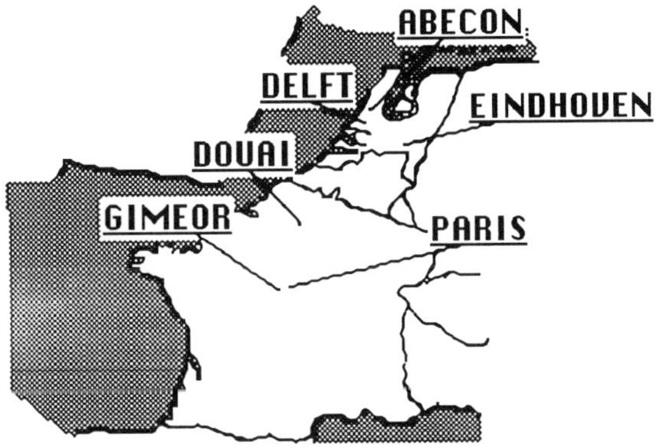

PEBI is a nascent foundation. Main participants are the Universities of Delft an Einhoven, the Ecoles de Mines in Paris and Douaie and the softwarehouses Gimeor and Abecon. A comment-group is formed in order to realize the objectives in a way that is useful for small and medium sized enterprices.

S. Alvarez et al. (eds.), Architecture and Urban Space, 805–808.

The state of the art (may '91)

Until now 12 modules have been developed. These modules have been given in one course, but recently this course is divided into two parts. The first part handles the ecology of the urban environment, the second part the ecology of the dwelling. Every part is devided into several modules, each concerning another environmental aspects.

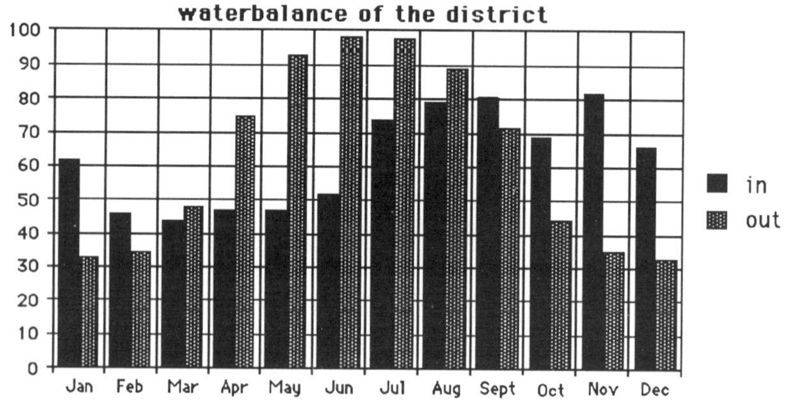

The modules are used and tested in student-courses on the Technical University in Delft. The last 3 years 10 courses were held for students (Architecture and Urban Design). At the end of each course a workshop was organized in which all environmental aspects of a course were handled. During the last years experiences have been used to improve the text-book and computer-programs used for these courses.

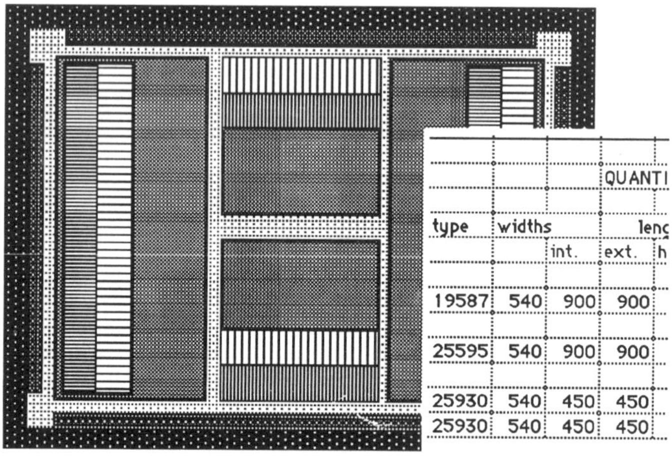

At the moment a Dutch course for architects and an experimental English course for students is prepared. In the Dutch course for architects Excel- and Hypercard-applications are use. The English course is a testcase in which we use some experiments in linkage between 3D-CAD (Architrion) and calculation in Hypercard (with external commands). On this part we do some further research. It is planned to convert all the Excel-application into Hypercard with a more grafical interface for the Technical University in Delft. Then the programs will be used in the new committed educational program for ecology and environmental design.

The course

The course is coördinated by the Technical University in Delft. They are given in day-parts (4 hours) in which several modules are handled. The course includes:

- A text-book about the subject
- A computer-program
- A manual for the computer-pogram
- A text-book with reference-data
- Excercises on the subject

The computer-programs can be used in the practise of an architect-office for making calculations and analyses for. In the course excersises will be made by the participants to:

- Have more feeling with the subject and problematic
- Get more skill in the use of the computer-programs
- Learn how an analystic approach of the problematic

The subjects for the dayparts of the Dutch course (the ecology of the dwelling) are:

1. Introduction comfort and references
2. Energy-, water-use and sewage
3. Trafficnoise, façades and buildingmaterials
4. Integration of the aspects and Effect-report

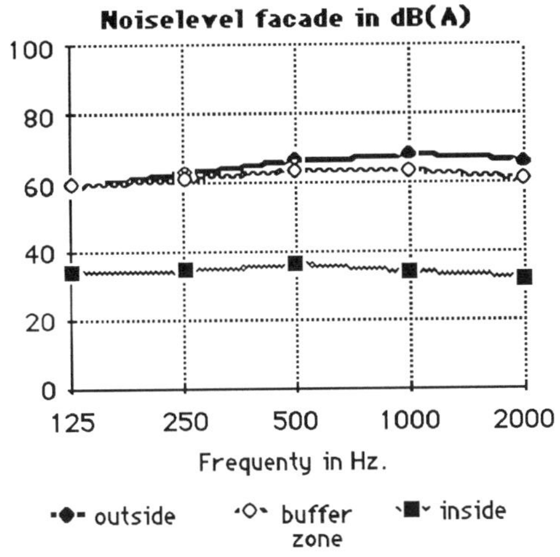

The English course will be prepared for the European Environment Campus which will take place in september 1991. The campus is organised by the General Directorate for the Environment (DG XI) of the European Commission and worked out in Delft by SOM/TUDelft.

The subjects for the dayparts of the English course are:

1. Introduction references and urban lay-outs
2. Sunshots and energy-use
3. Water-use sewage and urban water
4. Noise and polution by traffic and façades
5. Integration of the aspects and Effect-report

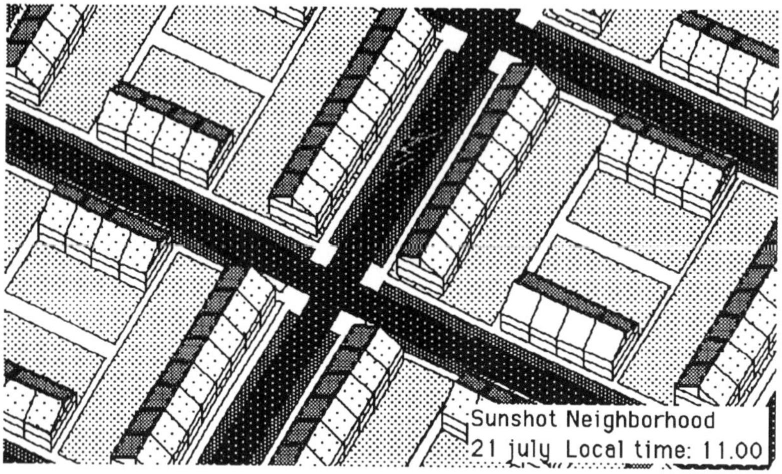

Leo Gommans
works in the SOM-group (Urban Design & Environment) Faculty of Architecture, Delft University of Technology, for BOOM, (Office for Environmental Research & Design) Delft, Holland and for MOSAIK, (Office for the Ecological Design) Maastricht, Holland

Kees Duijvestein
works in the SOM-group (Urban Design & Environment) Faculty of Architecture, Delft University of Technology and in BOOM, (Office for Environmental Research & Design) Delft, Holland

ASSESSING AND MEETING THE NEEDS IN ENERGY AND ENVIRONMENTAL EDUCATION IN THE 1990S: THE SITUATION IN BRITAIN

DR. SUSAN ROAF and MARY HANCOCK
School of Architecture
Oxford Polytechnic
Oxford OX3 0BP

ABSTRACT. In Britain there is a need to improve the quality of education for student and professional Architects in Energy and Environmental issues, methods and information. This paper outlines a number of problems encountered in teaching these subjects and some initiatives currently being developed to provide solutions to them, aimed ultimately at reducing energy consumption in buildings.

1. Reasons for Change

Through the spring of 1989 the national papers published a series of full page articles alerting their readers to the problems associated with global warming. This discussion of the problem ('The greenhouse gases causing global warming, principally carbon dioxide, are by-products of the industrial world's way of life, involving the burning of fossil fuels such as coal, oil and gas. To bring about the reduction (of harmful gases) will require energy strategies vastly different from those being pursued now ...' Time 26 April '89) and scientific assessments such as the Intergovernmental Panel on Climatic Change has helped to generate public awareness. The problem of over use of the earth's resources is moving from the sidelines, a peripheral issue, an interest of cranks, to a position of central importance for our generation and those that follow.

The architectural profession has recognised that it has a major role to play in the reduction of energy consumption in buildings. It is recognising that architects need continuing education if they are to respond appropriately to the new opportunities that are presented.

On the 13th April the Belfast Declaration was signed at the Conference of the Architects' Council of Europe (ACE) by delegates representing the 180,000 architects who serve the 320 million inhabitants of the European Community (EC). Paragraphs 6 and 9 of the 9 in the document state:

6) "Architectural Education as defined by the Architects' Directive 85/384 and continuing professional development must develop to provide architects with increasing environmental skills to a level at which environmental design will out-date assessment."

S. Alvarez et al. (eds.), Architecture and Urban Space, 809–816.

9) "The development of EC environmental design must include research feedback and the establishment of an EC wide environmental network utilising the schools of architecture defined by Architects' Directive 85/384."

The sooner we begin to discuss our common problems in the teaching of Energy and Environment to designers, the greater are our chances for their speedy solution. This is why we are opening up the subject at this conference with a review of the problems encountered in Britain in Energy Efficiency, Environment and Health in Buildings (EEEHB) education and some of the steps being taken to solve them there. We hope this may illicit useful feedback from our colleagues abroad.

We will first outline currently available and forthcoming programmes in Britain for EEEHB teachers sponsored by Government Departments, and then outline the current situation in schools of architecture and possible ways forward for the teaching of the subject in schools and in the profession.

2. Energy Education Support Programmes

In response to the recently acknowledged need for education support to architects, there have been a number of recent initiatives:

2.1 THE BEST PRACTICE PROGRAMME

BRECSU (Building Research Energy Conservation Support Unit) is a special support unit set up within the BRE, (Building Research Establishment) and funded by the EEO (Energy Efficiency Office) of the British Government. The best practice initiative is carefully targeted at improving efficiency in specific building sectors such as low income housing, commercial, health and school buildings. A best practice file is developed from a series of case studies by BRECSU, and explains in detail how savings can be made on particular aspects of the energy bill for the building type. The first best practice file sector file was launched in January 1991. Files on Schools and Low income housing are expected during 1991.

2.2 BICEPS

Building Industry Co-ordinated Packages. Teaching packages for Schools of Architecture and CPD (Continuing Professional Development). This is an EEO/BRECSU/ETSU (Energy Technology Support Unit) initiative to assist the education system by producing a series of teaching packages which will provide resource material for courses in energy efficiency. The course will provide validated, user-friendly packages for use in teaching undergraduates to MSC level and professionals. First packages to appear in 1991.

2.3 CPD ENERGY ROADSHOW

For designers in practice. EEO/BRECSU/RIBA initiative to organise a Roadshow of 4 annual CPD seminars for each of the 14 RIBA regions providing updates on market possibilities, energy resources information, thermal standards, risk and liability implications of energy efficiency design, housing using the BREDEM programmes, offices, schools, commercial buildings and design strategies. Start date c. September 1991.

2.4 DAS - DESIGN ADVICE SERVICE

EEO/BRECSU initiative based on the EADS scheme (Energy Advisory Design Service) in Scotland. This will provide advice on energy efficiency for particular buildings in the design stage for practices serviced by regional DAS offices. Start date 1991/1992.

2.5 THE ANNUAL ENERGY CONFERENCE

For teachers of EEEHB including engineers, architects, and surveyors to provide a forum for discussion of teaching methods, contents, new teaching support schemes such as BICEPS, and information updates for teachers. The first annual conference is to be held in July 1992. Organised by Oxford Polytechnic in conjunction with British Gas, Southern Region.

2.6 CLIENT GUIDE

BRECSU/EEO/RIBA initiative. This will provide a guide for clients on the principles and accepted standards for energy efficient design which should be sought in all newly commissioned and refurbished buildings. Should be available by summer 1992.

2.7 JOINT MSC PROGRAMME

The syllabus for a joint MSC course that has standardised contents and can be picked up in unit form in a number of schools is being experimented with - in discussion stage.

3. The Current Situation in EEEHB Education

This paper is based on four surveys and reports: two BICEPS (Building Industry Co-ordinated Education Packages) reports, one in 1988 on Energy Education and a second from a Workshop on Education for Building Professionals on Energy Efficiency, Environment and Health in May 1990. The third report was written by Robert and Brenda Vale for the Energy and Environment Committee of the Royal Institute of British Architects and the fourth survey was our own surveys of the teaching of thermal design in 17 of the 36 schools of architecture in Britain and within our own school. The paper covers the education of three different groups, teachers of environmental sciences, students, and professionals. Each group will be described in terms of current practice and problems, and a final section deals with possible solutions to these problems which are currently being explored by the Government and in the Schools of Architecture.

3.1 TEACHERS

Teachers of Energy and Environment in the schools of architecture are generally very committed to their subject. Problems include:

3.1.1 "The Teaching of Energy and Environment is not Treated with Urgency - Given the Scale of the Problems which Society as a Whole is Confronting" (RJD Vale, BICEPS, 1990, p.13). In the BICEPS report of 1988 75% of the Heads of Schools of Architecture and 82% of the specialist energy lecturers gave support to the idea that energy-efficiency was "a prime aim in building practice." Half the responding Heads of School said that energy related teaching was "very important at undergraduate, graduate and post-graduate level (BICEPS, 1988, p.30)." In 1990 the Vales pointed out that, with the exception of one school, and some others where energy focused units were offered, they did not find that "energy-related teaching was very important anywhere." They also hoped that ideally 100% of specialist energy lecturers would be committed to the idea (Vales, 1990, p.14). A difference in perception of the importance of this teaching area between Heads of Schools and specialist Lecturers appears to denote a lack of communication within schools.

The reasons for the lack of priority given to EEEHB teaching are not immediately clear. In a recent survey at the Oxford Polytechnic, the largest school of Architecture in Britain, staff were asked questions about their perceptions of global warming. In view of the problems expressed in the Vales report, regarding integration of environmental science principles into studio projects, and experience culled from our own survey of architectural schools, it was surprising that only 20% believed that the climate was not changing; the remaining 80% expressed the idea that the climate was changing and that increased action should be taken now to reduce the amount of non-renewable energy consumed in this country. We asked questions about anticipated changes attributable to the environment. 75% expected that significant changes to both lifestyle and buildings would be forced upon us in their working lifetimes, and about 33% anticipated that settlement patterns would also have to change. Why is this not expressed in studio work?

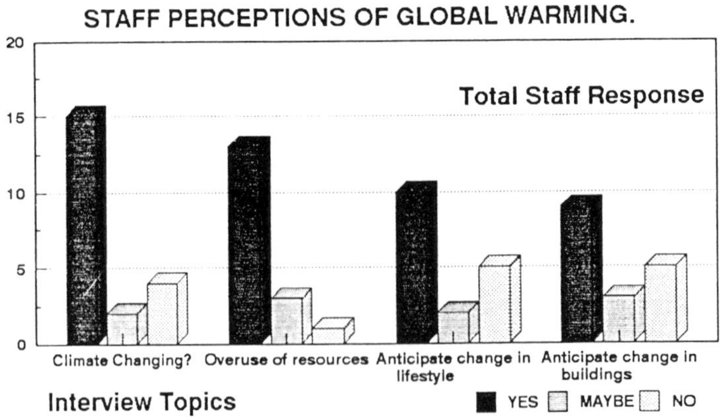

3.1.2 Teachers' Low Status in the Schools. Most schools have one person who is identified as being responsible for this subject area. Of 36 lecturers interviewed by the Vales, 6 at some point suggested that they were strongly marginalised, 7 more felt remote from the main stream of teaching. A further 7 said that they felt fully integrated into the main stream of teaching. The attitudes of the others were not gauged.

3.1.3 Part-Time Specialist Teachers. Teaching done by part-time staff only which meant that students could get the lecture information but had little time with part-time specialist staff to discuss their schemes. This results in unspecialised studio staff either a) dealing with the question themselves at the drawing board, with or without success, b) denigrating the importance of the specialist information he does not understand, c) referring the student back to the part-time tutor who has insufficient time to deal with the question or d) referring the student to published works which are often out of date. Such occurrences often results in a diminishment of the perceived importance of EEEHB teaching staff.

3.1.4 Specialist/Studio Staff Communication. Lack of communication between specialist EEEHB staff and studio staff. The former may be physicists or engineers who approach teaching from a scientific standpoint while the latter may have a more design or aesthetic approach and the two often do not mix in the studio at a strategic or technical level. This is exacerbated by the tendency for EEEHB teaching to be done in set lecture courses which are separate from studio projects. Projects which are set that concentrate on environmental issues at the board and are then assessed by "design" criteria (aesthetics/novelty/presentation) often result in the focused item, EEEHB, actually being devalued in the eyes of the students (Vales, 1990, p.6). Studio staff may have a low level of competence in EEEHB subjects.

3.1.5 Teaching Standards. There is a general lack of communication between schools on course content and form. The question of what are priorities in the teaching of the subject and what minimum standards should exist, and the range of methods that can be developed to convey the information.

3.1.6 Money. Schools lack funding to provide important facilities such as the hardware, software and specialist teaching for proper teaching of energy and environmental auditing programmes through the 5 years of student training.

3.2 STUDENTS IN SCHOOLS

It was generally found that schools do teach levels of heat, light and sound that equip undergraduates to cope with current requirements of building performance, generally to building regulation standards.

Our survey of 17 of the 29 Architectural schools indicates a wide variation in the amount of time given to EEEHB section of the syllabus. Top of the range were Cambridge, Sheffield, Bath and the Welsh School where teaching in the first year exceeded 20 hours. At the lower end of the range Polytechnic South West, Manchester Polytechnic and Oxford Polytechnic students received about 10 hours teaching per year on the topics defined as 'Thermal'. In a minority of the schools, assessment in these topics is solely by examination; most make some reference to project work but only Newcastle University and Manchester Polytechnic go so far as to declare percentages attributed to each section.

Generally the topics taught include heat transfer methods, climatic information, human comfort, passive solar design, heating and ventilating and air conditioning. Many schools do not cover the topics of energy auditing, energy management, energy strategy, global responsibilities or sick building syndrome. Very few of those questioned attempted any active solar techniques at all.

First year students were also questioned about their perceptions of global warming. 80% of the students agreed that global warming is occurring but half of those interviewed were uncertain if this would affect lifestyles or settlement patterns during their working lives. Staff perception of the problem had tended towards minimizing environmental impact "a carbon tax will be introduced." Student perceptions were towards accommodating any changes that might occur "as the earth gets hotter design will have to be adapted." Problems include:

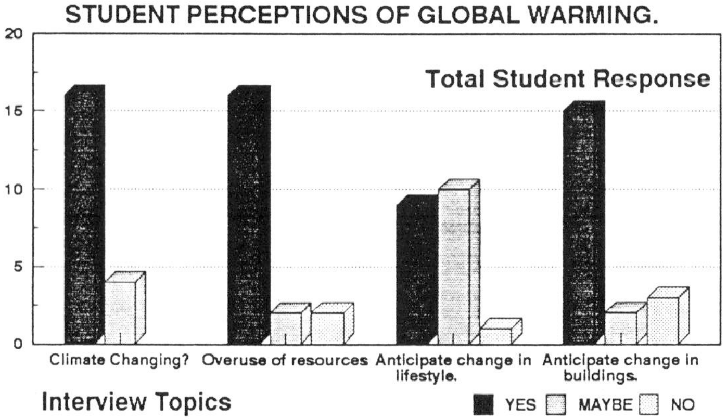

3.2.1 Application of Knowledge. The knowledge gained in lectures is often not applied in student own schemes. There appears to be an endemic deficiency in applying EEEHB skills in the studio. EEEHB teaching is often degraded by studio teachers who do not understand it.

3.2.2 Urgency. Not sufficient motivational teaching is input into courses to impress students with the necessity of becoming competent in EEEHB, for global reasons.

3.2.3 The Big Design Picture. The full role of energy efficiency in design is often not taught - from good briefing, good design to good building management. The centre is concentrated on and the peripheries neglected due often to lack of time.

3.2.4 Student Skills. There is a problem with levels of technical competence and numeracy skills of students and while some develop good strategic appreciation of problems, others may be stronger on technical issues. Some way of bridging skills gaps is required to ensure basic levels of competence are reached by all students.

3.3 THE PROFESSION - CPD - CONTINUING PROFESSIONAL DEVELOPMENT

Britain is divided into 14 separate CPD regions overseen by Regional Covenors who are seldom architects. Problems include:

3.3.1 The general standard of regional CPD is non-co-ordinated and often low. There is a need to create a standardised and high quality provision for teaching EEEHB to professional designers.

3.3.2 However high a standard of teaching is set for CPD this is no guarantee that designers will go to CPD events. At the moment CPD is not required in the architectural profession in Britain but in the beginning of 1993 it will become mandatory for members of the RIBA to attend 34 hours of CPD lectures and do 34 hours of study per annum. This will be done using a personal CPD plan and is to be discussed with colleagues, which means that it will be avoidable by those who do not want to do it.

4. Solutions

Raising the importance of EEEHB in building design with:

4.1 PROFESSIONALS. Because

4.1.1 Clients will demand it - the client guide may stimulate such a development.

4.1.2 The profile of liability for deficient design of EEEHB in buildings may rise and designers be forced to achieve minimum standards for which learning will become necessary. This point will be reinforced in CPD events.

4.1.3 Government Regulations for EEEHB standards in buildings will change forcing practitioners to CPD events to update their knowledge to keep up with legislation.

4.1.4 Better CPD/Advice/Information available for Professionals through BICEPS programmes, Roadshows, DAS, etc.

4.1.5 Growing awareness of the urgency of reducing energy consumption levels in Britain to comply with international legislation on CO2 emission reductions or increased awareness of the global responsibility of the profession.

4.2 STUDENTS

4.2.1 Making them understand that climate change is occurring and that it will change their world and that they have an important role to play in such changes.

4.2.2 BICEPS. It is hoped that by making available BICEPS packages to all teachers of EEEHB it will facilitate a) teaching of the wider view, b) provide easily understood material for most students to raise uptake rates of information reducing skill gaps, c) add prestige to lecture series lending weight to the importance of their application at studio level.

4.2.3 Practice Pressure. With increasing profile of EEEHB in offices, and client awareness.

4.2.4 Up-Grading Teaching Methods. EEEHB should be taken out of the sole realm of the lecture theatre and onto the studio drawing boards. This could be done either or by bringing EEEHB specialists in as studio teachers. This is a subject area we hope to discuss at the Educators in Energy Conference in July 1992.

4.2.5 All studio teachers should be brought up to a level of basic competence in applying EEEHB in design. Studio teachers should be enabled to learn, if necessary.

4.3 TEACHERS

4.3.1 Tackle Heads of Department. Raise importance of subject in school. Improve communication between Heads and EEEHB staff.

4.3.2 Increase quality of teaching using methods such as:
Up-grading of studio teachers' knowledge of strategic/technical subjects.
- Changing information input to project related teaching before schemes.
- Make marking of projects include a definite input on EEEHB.
- Raise priority of subject in studio teachers'/students' eyes.
- Provide updated information not just out of date text books, BICEPS.
- Innovative teaching methods, eg workshops/EEEHB teachers taking over studio responsibility.

4.3.3 Promote outside input on teaching through BICEPS and contact with other schools, eg through Annual Conference/Msc Programmes, etc.

4.3.4 Find alternative sources of funding for expensive teaching requirements.

Building 2000: Technology Transfer and the Design Process; Pilot Study of Five UK Projects

Sue Wolff, Mike Wilson, Low Energy Architecture Research Unit, Polytechnic of North London, Holloway Road, London N7 8DB

Abstract

Specialist energy advice is not often incorporated into architectural design projects although there is increasing interest in passive solar design. The CEC funded programme Building 2000 investigated the value of such advice by providing funding for energy consultancy and thermal modelling for selected European projects. This study consisted of an interview survey of a pilot group of five UK projects with the object of assessing the success of technology transfer to the design team and the conditions conducive to success. It concluded that broad advice conveyed in terms accessible to architects and with the involvement of the whole design team at the strategic planning stage was most helpful and identified several areas where further research and education is required.

1. Introduction

Building 2000 was a scheme by the Commission of the European Community Directorate General XII to provide expert assistance in passive solar technology to selected design projects involving a range of building types in EEC countries by funding the employment of energy and daylighting consultants by each project as well as group conferences in Noordvijk,Barcelona and Crete and local seminars in each region.

2. Methodology

The objectives of the pilot study were to:

1. Identify obstacles to the better transfer of technology by questioning whether conventional design practice needs to be changed.

2. Provide guidelines for better organisation of technology transfer.

3. Identify areas in which information is lacking or poorly communicated.

4. Identify other factors that could enable successful technology transfer.

S. Alvarez et al. (eds.), Architecture and Urban Space, 817–822.

The role of 'strategic' decisions

A central assumption of the pilot study was that 'strategic' decisions , described by L.Kealy(1989) [1] as 'soft or enabling decisions' were arrived at very early in the design process and determined the subsequent options investigated by detailed scientific study. Examples would include :

Construction heavy/lightweight.Super/well insulated
Solar orientation
Deep plan or shallow plan - Good daylight
Passive or active solar features Eg direct gain and/or sunspaces and whether integrated with heating/ventilation

A second consideration was to critically evaluate effective technology transfer by assessing the usefulness of scientific studies in fine tuning design choices and their subsequent implementation. This depended on the involvement of the design team in the learning process and interviews considered how the timing of individual's participation as well as their previous experience and background influenced this.

Based on the CEC Building 2000 published reports (from which the following illustrations are taken)[2] an initial appraisal was made by the author of the 'strategic' decisions which formed the basis for later interviews with the whole design team using a pro-forma intended to:

1. Elicit design decisions and to confirm or amend the initial evaluation of 'strategic' decisions.
2. Establish the time sequence in which ideas were developed.
3. Identify which people had been significant at each stage
4. Investigate methods used to convey ,illustrate or test ideas. 5. Investigate sources of information and training.
6. Evaluate the success of the exercise and what had been obstacles.

These were then evaluated in terms of :

1. The appropriateness of the initial 'strategic' decisions
2. The degree of success in making use of expert consultants

3. Observations on individual projects

3.1 METHILHILL PRIMARY SCHOOL Primary school for 464 pupils located on greenfield site with road access in windswept east coast of Scotland.

The preliminary design featured a courtyard planning principle with glazed courtyards using traditional heavyweight building materials. The initial selection of strategic features advised by the energy consultants included several non-starters : roofspace collectors and heat pumps which were thermally modelled. Subsequent advice from B2000 consultants allowed a more appropriate and simpler approach, using direct gain in the courtyards as preheat to a mechanical ventilation system, to be adopted in time to influence the sketch design. These were subsequently incorporated unchanged in the built design which now acts as an exemplar to other projects in the department and has been used as evidence to change government funding guidelines. The accessibility of rule of thumb advice combined with the continued involvement of the M&E consultant were identified as having assisted the success of technology transfer. However lack of involvement of both the external consultant and the thermal modelling consultants in the selection of features to be modelled was seen as undermining the value of the initial scientific study.

3.2. ROTHERHAM MAGISTRATES COURT Development of 10 courtrooms in Rotherham city centre, South Yorkshire (maritime climate) in restricted triangular site partially overshadowed in winter.

The initial scheme was a two storey building spread out over site in staggered blocks in roughly triangular shape around open courtyards with heavy weight masonry construction for sound insulation. Initial strategic decisions which included a two storey south facing direct gain space and high level windows into courtrooms were seen as appropriate and in the main incorporated in the final design. Detailed specification of glazing based on simulation was however amended when designed by the M&E engineer, highlighting the problem of compatibility of simulation methods with conventional CIBSE methods of calculation. This would indicate the need for the distribution of results in building services journals of wider comparisons and validation by monitoring.

3.3. NEWBURY CENTRAL LIBRARY Town library located in town centre in flat and sheltered built up area by major road in the provincial town of Newbury with deep plan heavy weight masonry construction.

A very detailed energy consultancy[3] including lighting and thermal simulation techniques could do little to alter the initial design which was derived primarily from planning considerations. The deep plan and narrow skylight were inappropriate strategic decisions which engineered lighting design attempted to improve. The late involvement of the energy consultant after fundamental design decisions had been made and lack of early M&E input contributed to this. Additionally this project raised queries about the effectiveness of light distribution methods in diffuse skies which indicated the need for further research and validation by monitoring.

3.4. GREEN PARK COMBINED SCHOOL. School for 360 children designed on open central courtyard principle using heavyweight masonry construction in Newport Pagnell on green field site close to new council estate and M1 motorway.

Further to the initial scheme, designed to give good daylighting, the use of energy consultancy developed the roofspaces as solar collectors to pre-heat air for the hot air heating system. Initial strategic decisions based on rule of thumb planning were confirmed by subsequent simulation which finetuned areas and location of glazing and provided design data on preheat available and ventilation required subsequently incorporated in the detailed M&E specification. A wide variety of technical sources and agencies were consulted by the architect resulting in well integrated considerations of fabric, construction and building services. Early involvement and commitment of the M&E engineer guaranteed subsequent implementation of energy advice. Experience gained through the energy consultancy was effectively transferred to the design team.

3.5. DOWNSVIEW COUNTY PRIMARY SCHOOL

Unlike the other projects, this involved a general review of possibilities for energy saving during the refurbishment of Kent schools with the aim of providing a refurbishment manual and specifically designed to transfer technology. Downsview School was a typical example of a school built in the 1960's as a single storey, flat roofed system built building with low levels of insulation, high ratios of single glazed windows, suffering from overheating in summer and being cold in winter. A well integrated review of both practical construction possibilities and the energy implications of fabric and building services upgrading combined with qualitative and quantitative lighting design resulted in a successful manual. Once initial simulations are used to define a range of refurbishment methods they are unnecessary in further applications to this particular type of buildings.

4. General Conclusions

Architects particularly found simple rules most useful which can be easily integrated into the initial sketch scheme. They identified closely with consultants who could give general advice and who were sympathetic to constructional and aesthetic issues. Their early involvement was therefore crucial. Time allocated to this sort of consultancy in the Building 2000 programme was generally very short , of the order of one to three days and therefore very good value for money if available from the early design stage. It was ideal to allow some of this time to be spent during the detailed design stage for the consultant to accommodate changes in design which may arise from other considerations.

Architects quoted their main source of information as their own college training and this is clearly a main avenue for energy education and technology transfer.

The early involvement of M&E and their confidence in the simulation methods used was necessary for their subsequent implementation of energy advice. The study identified the need for further validation of simulation methods against traditional calculation methods particularly in the case of overheating. Doubts about the effectiveness of methods of light distribution within a space such as reflectors and diffusers in obtaining satisfactory lighting require further study.

It is clear that M&E consultants are more reluctant to adopt passive techniques than architects. Though partly caused by lack of confidence in the newer prediction methods conventional approaches (including those that may be necessary in litigation) are not directed towards passive design. It may be the M&E consultant requires more education than the architect.

5. Recommendations

5.1. Energy advice cannot be incorporated into an existing scheme, it should be given at the sketch design stage, most effectively in terms of broad rules.

5.2. In the selection of energy consultants consideration should be given to their ability to integrate constructional and aesthetic aspects and effectively communicate general principles.

5.3. Sufficient experience in many types of building is now available to eliminate wasteful consideration of inappropriate technologies. Further promotion is required of the limits of effectiveness of particular solar features. Simulation is costly and should be selectively used after detailed consideration of the main strategic possibilities by the whole design team. In well researched categories of building it is probably unnecessary. Simplicity is almost always a guide to effectiveness.

5.4. M&E engineers should also be involved early and selected on the basis of their experience and commitment to passive solar technology.

5.5. The importance of architectural education as a means of transferring energy technology should be stressed.

5.6. Further research is required into the compatibility of simulation methods with traditional building services calculation techniques in the case of overheating with validation based on monitoring.

5.7. The energy aspects of any new building should be considered from the outset and can be accomplished at relatively little cost by the employment of a consultant if the architect or M&E engineer cannot provide that expertise.

1. L Kealy, Pilot Study of the Work Practices of Design Leaders in the Area of Passive and Low Energy Architecture, Solinfo, Energy Research Group, UNiversity College, Dublin, July 1989
2. Commission of the European Communities DGXII, Building 2000 Workshops: Brochures, September 1989
3. C.J.Hancock, J.Littler, Newbury Central Library - Bulding 2000 An Example of a Daylight Design Study, Polytechnic of Central London, Conference Proceedings CIBSE Daylighting Conference, 1990

Part 8

LATE PAPERS

ONE "MANZANA" ON THE WIND.
Aerodynamic comportment study of the exterior spaces in the colonial urban weaving, under a tropical climate in Colombia.

JORGE ALVARO RAMIREZ F.
Arquitectua & Bioclimática Ltda.
Calle 83 N° 5-57, T.B apto 2002.
Bogotá, Colombia.

ABSTRACT. The objetive of the investigation is to understand the aerodynamic comportment under tropical climate of the spanish colonial urban weaving and the effect of possible future interventions. Using a numeric heliodon and a wind tunnel, the results of simulations are destined to help urban designers to ameliorate comfort conditions in the exterior areas.

Introduction

During the spanish colonial period urban centres were developed based on an instinctive and empirical climatic principles. The historical downtown areas of Colombian cities are currently confronted to urban recuperation plans. In order to ameliorate life and confort conditions, future interventions must refer to climatic adaptation principles.

Within the scope of the C.E.A.A. "Energétique et Architecture Urbaine" in the "Ecole d'Architecture de Nantes", and the colaboration of the CSTB of Nantes, this work pretends to retake the climatic theory of 1573 and, using scientific simulation methods, produce a series of practical observations which may be useful for designers who work in colonial downtown urban projects. The study is focused on the natural ventilation of exterior spaces of downtown areas and its possible developments under tropical hot and humid climates. It deals specifically with the problem generated by vegetation used as a solar protection device and its effect on air movement.

1. The "manzana"

1.1. "LAS LEYES DE INDIAS", FROM 1573 TO NOWADAYS

The four main streets begin in the square, each one starts in the median point of each of the sides, and two other streets from each of the corners. The four corners must be oriented to the four cardinal points, therefore the streets which begin in the square are not directly exposed to the four main winds.(...) Streets are wide in cold regions and narrow in hot regions.(...) Extracted from "Las Leyes de Indias", 1573.

In 1573 the definitive rules for the foundation of Spanish colonies overseas were laid, and codifyed by Philipe II in "Las Leyes de Indias". The settlements were divided in equal parcels of land following an orthogonal site plan. The side dimention of each block of

S. Alvarez et al. (eds.), Architecture and Urban Space, 825–830.

houses or "manzana" varied betwen 80 and 100 meters. During the 270 years of colonial domination, Spain established more than 600 settlements following the chessboard layout.

Nowadays, the new projects are designed mantaining the old urban weaving. The "manzana" remains the regent element of the ensemble. The principle of central courtyard is generaly used in the "manzana" layout. Continuity of the urban plan may be modifided by a series of physical disruptions, therefore, the urban weaving becomes more flexible and articulate.

1.2. SOLAR COMPORTMENT IN THE COLONIAL DOWNTOWN AREAS

Using the numeric heliodon, designed by the C.E.R.M.A., we can simulate the solar comportment in the colonial downtown areas and its effect on urban future interventions. Simulations are run resembling the dry periods for a latitude of 5° North. The urban layout is oriented towards the cardinal points.

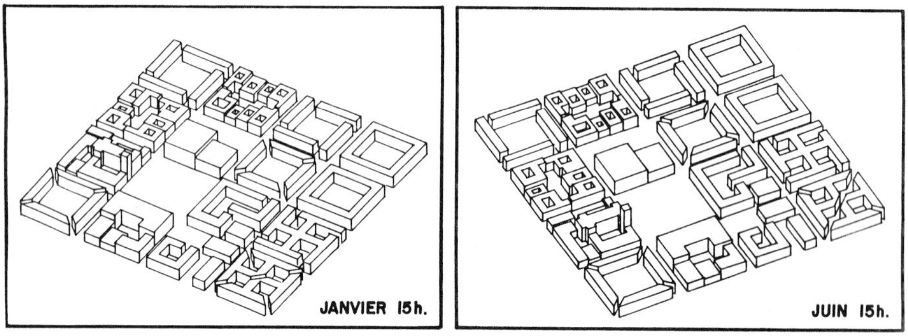

Figure 1. Heliodon view of the colonial nucleus at 15 hours, in January and June.

The comportment is rather regular. In January, the elevations facing South are insolated during the whole day, while North elevations show this behavior during the month of June. Elevations facing Est or West are exposed to the sun throughout the whole year. Narrow streets are less insolated than the wide ones. In the Est-West direction, one of the pavements remains in shadows. The big central courtyards are exposed while the small ones lie on the shadow casted by adjacent constructions. Central squares areas are insolated thoughout the whole year.

Arrangements of trees are currently used to protect exterior areas from solar radiatiation. This shadow search has an influence on the aerodynamic comportment of environment.

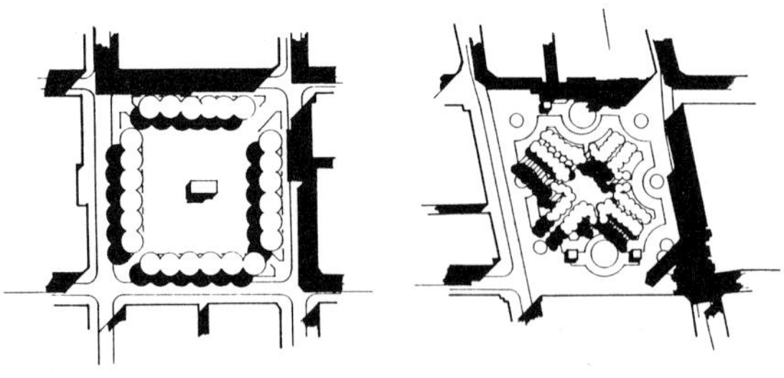

Figure 2. Tree arrangements on Bolivar's square in Pereira and Caycedo's square in Cali.

2. Wind and comfort

Comfort definition remains subjective, however, based on human sensations, we can attempt to establish some comfort values. For the dynamic comfort, a wind speed superior to 5 m/s will be considered uncomfortable. On the other hand, in a hot climate, a thermal comfort sensation is achieved when human metabolism produces a minimal amount of energy to dissipate the heat produced by the body.

In the Colombian hot tropical climate, annual average temperatures vary between 24°C and 30°C. Relative humidity values fluctuate between 65% and 80%. Average wind speed is rather low, the dynamic comfort limit of 5 m/s will be rarely surpassed.Wind helps in regulating thermal exchanges between the human beings and the environment. **In this way, natural ventilation is an effective means to ameliorate comfort conditions in exterior urban spaces of cities under a hot tropical climate.**

3. Experimental context

3.1. INVESTIGATION MEANS, SCALE MODEL AND EXPERIMENTAL WIND INCIDENCES

On site, experimental studies of this magnitude are rather difficult to conduct. Simulations are made using a scale model and the wind tunnel of the "Ecole d'Archiutecture de Nantes".

Tests are conducted for three case studies A, B and C which synthesize the heterogeneous characteristics of the colonial nucleus. These case studies have the following modifications: openings in the "manzanas", variation of the "manzana's" size, variation of the street profile, implantation of internal subdivisions in the bigger courtyards and modifications of tree arrangements in courtyards, streets and squares.

Wind incidence on the concerning cities is not regular, for simulations we use two theoretical wind directions: **"wind i=90°**, parallel to the urban weaving and **"wind i=45°**, 45° incidence with respect to the urban weaving.

3.2. MEASUREMENT TECHNIQUES AND RESULTS PRESENTATION

Measurements are carried out using an anemometric hot filament probe. Wind speed values are taken at 0.5 cm height, at equidistant points distributed in in the experimental "manzanas". Wind measurements taken are:

•U, the average wind speed measured in the variations of the cases A, B and C.
•U virgin, the average wind speed of the empty site occupied by preceding variations.
•U reference, the average wind speed measured out of the urban texture influence.

With the preceding measurements we can establish the following similitude conditions:

• **Coefficient** α:shows the tendencies generated by the implantation of urban components of the three cases in the empty site called "virgin".

• **Coefficient** β:represents the part of reference average wind speed registered in the different variations of the experimental model.

$$\alpha = \frac{\dfrac{\overline{U}}{\overline{U}\ \text{refere.}}}{\dfrac{\overline{U}\ \text{virgin}}{\overline{U}\ \text{refere.}}} \qquad \beta = \frac{\overline{U}}{\overline{U}\ \text{refere.}}$$

4. Simulations results

In order to visualize the comportment of the cases, coefficient α values of the "clear" arrangement and those of the most promising tree arrangement (for the cases A and B) are presented in an schematic layout.

4.1. "CASE A" OR THE MAIN SQUARE

Case A reproduces the spatial configuration of the city's main square, adjacent streets and representative buildings.

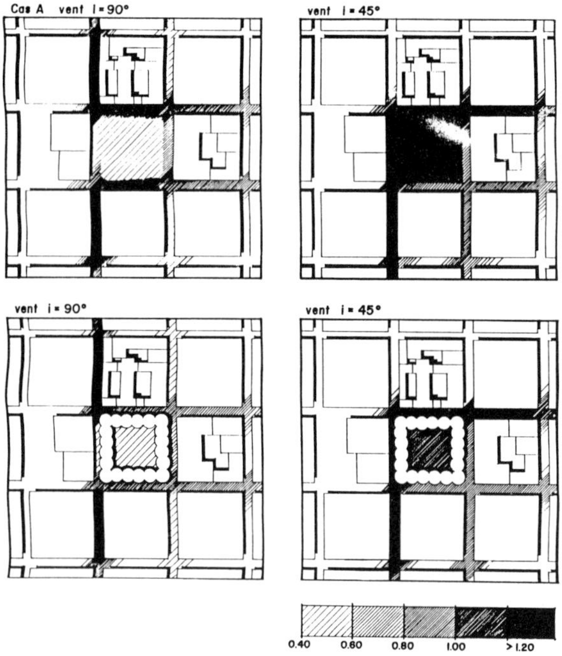

Figure 3. Case A. Tendencies (α) of the air movement for the "clear" and tree arrangements; under the experimental wind incidences: i=90° and i=45°.

The aerodynamic comportment of the model varies according to the wind incidence. With i=90°, wind flux tends to accelerate on the streets rather than on the square area. With i=45° incidence, the aerodynamic response becomes regular and homogeneous.

Tree arrangements cause a loss of wind speed (β) depending on the area, values fluctuate between 4% and 10%, in comparison to the "clear" arrangement.

4.2. "CASE B" OR NEW HOUSING PROJECTS

Case B represents new housing projects implanted in the colonial "manzanas". Small squares, streets and interior courtyards are created maintaining the colonial urban profile and historical buildings.

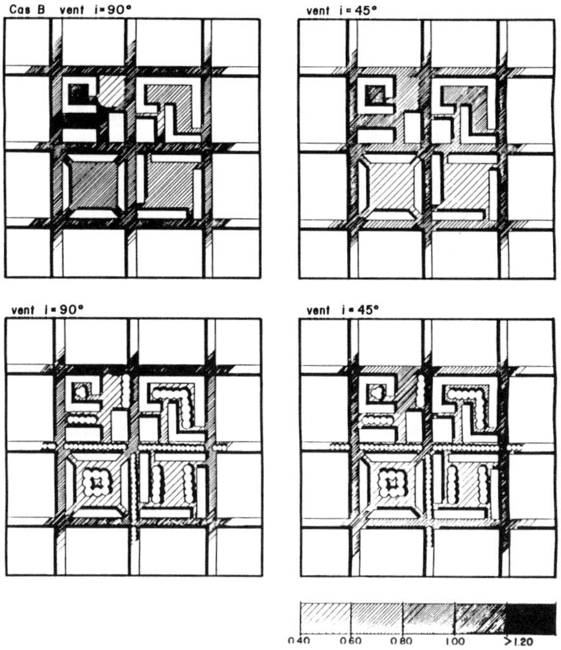

Figure 4. Case B. Tendencies (α) of the air movement for the "clear" and tree arrangements ; under the experimental wind incidences: i=90° and i=45°.

Urban components of the model respond differently. The ensemble of courtyards shows a tendency to reduce the wind flux speed. With i=90°, streets runing in the Est - West direction canalize and accelerate the wind producing favorable results. Perpendicular streets, with i=45°, produce similar results. Small squares' comportment varies depending on their location with respect to the incident wind.

The effect of tree arrangements differs depending on the shape of urban components. Loss of wind speed (β) fluctuates between 1% and 16% in comparison to the "clear" arrangement.

4.3. "CASE C" OR THE BIG DIAGONAL OPENING

The big diagonal free area simulates an avenue or a river course. New housing projects generate free exterior areas, streets and courtyards. The tree arrangement showed in the figure does not provide the best aerodynamic performance but is the most commonly used.

Wide streets always show the best ventilation conditions but remain being the most exposed ones to solar radiation. Courtyards conditions remain disadvantageous. Under a wind incidence of i=90°, the canalization effect in the zone parallel to the diagonal free area shows the highest value for the coefficient β obtained in the three experimental cases (β= 0,53). This beneficial effect disappears with i=45°. Wind speed decreases in the free areas adjacent to this parallel zone and accelerates when the wind flux reaches the streets.

Tree arrangements do not cause the same loss of wind speed as the case B. β values fluctuate between 2% and 22% in comparison to the "clear" arrangement.

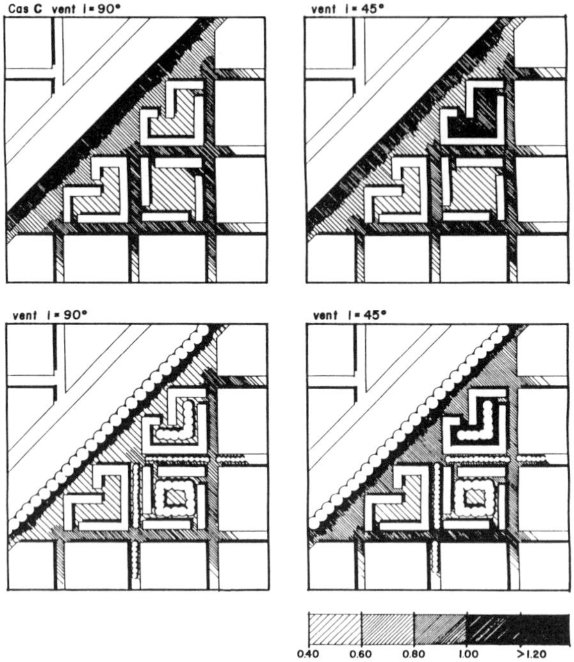

Figure 5. Case C. Tendencies (α) of the air movement for the "clear" and tree arrangements; under the experimental wind incidences: i=90° and i=45°.

Conclusion

In Colombian hot tropical climate, comfortable conditions may be obtained in exterior urban spaces with an harmonious equilibrium between natural ventilation and solar protection.

The main square presents the best ventilation conditions of all central exterior areas. It produces a remarkable gain of wind speed on adjacent streets. It can be protected from solar radiation using an adequate tree arrangement. Concatenation of squares could produce an amelioration of aerodynamic conditions on the urban orthogonal nucleus. Central courtyards present the worst conditions of the colonial downtown area. However, small central courtyards formed by the subdivision of the "manzana" are best ventilated and less insolated. Streets are the best ventilated spaces of all urban exterior areas. Aerodynamic performance of wide streets is superior than that of narrow ones only by 2% to 9%, but they remain excessively exposed to the sun. Therefore, intructions prescribed on "Las Leyes de Indias" are still valid.

Tree arrangements have a tendency to decrease air velocity and their influence is not uniform. The use of avantageous tree arrangements is highly recommended.

Building or vegetation volumes may produce "canalization" or "venturi" aerodynamic effects which generate a beneficial wind speed acceleration.

Buildings which exceed the height of the urban colonial profile, generate air movement in adjacent exterior areas.

ENERGY: A LETTER IN THE ALPHABET OF ARCHITECTURE

Prof.STANKO KRISTL
Faculty for Architecture
University of Ljubljana
61000 Ljubljana, Cojzova 5, Slovenia, YU

When talking about energetically economical construction or about solar architecture, architects do not consider this aspect as separated from the whole. Architecture is a complex notion, and being conscious of the problems concerning energy is just one of the letters in the alphabet of architecture.
When thinking about energy in housing construction (our research deals mainly with multi-storey housing construction), we were interested in reducing the consumption of energy, not only in active and passive exploitation of revewable sources, such as solar energy. We have tried to achieve this with a groundplan layout that changes the relation between the volume of an apartment and its facade to the benefit of the volume. We did not forget the quality of living, and have, therefore, introduced the possibility of fluid space, of every individual having his/her own room, etc.: all of this within the current housing standards. We have also introduced glasshouses and covered atria in the apartments, if possible, individual ones in all apartments, introducing a higher quality of living conditions.

We know that the living environment is not created only by buildings facing south. In a residential complex, apartment houses should also face east and west. Due to the demand that every apartment should be exposed to the sun at least 2 hours per day, south-oriented buildings are depth-limited (approximately 14 m), while in buildings facing east and west, and having an atrium, this depth is increased at least once. Dooing this research, we have found interesting parallel qualities and rationalisation possibilities regarding ecology, infrastructure and the use of construction sities (more green areas, reduded construction sities, shorter passageways, sewage systems, etc.).

S. Alvarez et al. (eds.), Architecture and Urban Space, 831–836.

Never before has our world been so professionally rationalistic and uniform, with such a small degree of emotional expression as it is today. On the other hand, there have never before been so many runinations about architecture and its theoretical aspects. The latter notwihstanding, we cannot claim to be more ingenious than our ancestors as regards our built environment, especially, if we consider our colleagues from the Classical and Medieval periods

Especially the latter were in this respect far more reserved, but their architectural creations were full of ingenious solutions as regards spiritual and material values, and answered the questions of identity, experiencing the ambiente, the contents, usefulness, economic aspects, etc.. There was no room for unilateral or formal pretentiousness, which in the later periods interrupted the flow of creative development with its selfsufficiency, nor for the blind infatuation with the almightiness of machines and technology, which undoubtedly did contribute their share in the form of widely varied offers and the rise of living standards, but simultaneously gravely endangered the natural balance of our planet.
Every architectural scene is - like any other being that wishes to survive - a part of whole, placed in time, here and now, based on yesterday's experiences for tomororrow's use, with built-in coordinates of life and with an interweaved network of economic, technical, technological and artistic aspects, the intended message, local characteristics, findings and knowledge of other sciences, however, it is not a sum of numerous functional, techonogical, technical and similar requirements, which in their selfsufficiency result in known ready-made models, but a reflection of experincing indomitable energy and vigour of mind which transforms these factors creatively into a unison of architectural expression. It is an architectural objectification of concentrated observation and an objectification of irrational, intuitive lines of force.

Passive solar architecture, is not a complete novelty in the architectural creation. However, it must be admitted that individual technical achievements, scientific discoveries and ascertainments have introduced new dimensions into architectural design. In the technological domain, we can talk about new possibilities for exploitation of renewable natural energy sources (in our case solar energy), while in the professional technical sphere, new interpretations of architecture evolve, above all in respect of semantic - semiotic dimensions. By freeing architecture from conventionalism, we make people to live in a completely different living environment. That involves our risk of their approval or refussal. The risk is reduced if definite basic principles are followed. One of them is that architecture can be particular regarding structural approach, technological and technical aspects, architectural forms, and if we want, even in its historical continuity, but it must remain legible. It must fulfill expectations ranging from economical and social trends upto the cultural messages. In architectural design, the elements of passive solar heating have no connection with the conventional approach. The main concern of our investigation has been, above all, organized housing construction, i.e. multi - storey buildings and larger housing aggregates. The research carried out on simulation models applies to the region of Ljubljana, the area with the continental climate.

Our starting point was the fact that organizing multi-storey construction requires, due to many reasons, different orientations. For that reason we cannot count only on building rows facing south, which are usually the subject of research dealing with passive solar energy exploitation. The question appears wheather it is possible to attain equivalent energy effects also with east-west oriented building rows.
The first critical point concerning the groundplan of a dwelling facing east-west is its facade strip: the longer it is, the greater are heat losses. These losses can be reduced by adequate measures (good insulation, glasshouses,..), however, in our case we have to do with a relative relation between south - oriented and eastwest - oriented building rows. In designing these layouts, we must, above all, try to reduce the external fasade strip, i.e. its surface area. The consequence is that the groundplan is to be developed into the depth.

Another energy waster is the exposure of a building to external thermal conditions (wind, frost, humidity,..). Here, we are facing the question of categorization of individual dwelling areas: e.g. the living room which is primarily linked to insolation, view, etc.. Dwelling areas are directly exposed to external effects and are therefore not to be treated in the same way as the areas which can do without these advantages (bedrooms, kitchens). The latter may be placed in a thermally more comfortable part of the building (leeside or facing a glazed atrium).

The third problem arising from the effort to provide the largest possible groundplan depth is sufficient daylighting, especially in the dwelling spaces which are further away from windows. We tried to solve this problem with reflective admission of light into the depth which gave no important results due to low floor heights of the dwelling spaces. By the introduction of duplex dwellings, where the influence of the angle of light incidence increases drastically due to the doubled floor height and by admitting daylight into the rooms, situated in the interior of the groundplan (e.g. kitchens, dining rooms) over light shafts (using the high degree of the skylight factor and new solutions. We have not apllied light transfer through glass fibres), we have come closer to the solution of the problem. (Fig.1, Fig.2)

According to our calculations, the eastwest- oriented models with increased groundplan depths and glasshouse as well as glazed central atria have 20% (B1), 25% (A2) up to 27% (K) lower energy consumption per sqm of a dwelling unit than classic ground plans with depths approx. 15 - 17m. Within all models, glasshouses have been taken into consideration (Fig.3.)

With south-oriented building rows, energy savings are higher, howewer, in the complete energy balance, they do not outweigh the energetic advantages of an eastwest-oriented builcing with a depth-stressed layout. New dwelling patterns developed by the use of the above-mentioned starting points do not prove only the same or even greater positive energy effects than the conventional solutions of south-oriented housing systems (which lose a considerable portion of the energy gained on the south side over their northern facades), but also their great advantage is evident in the structuring of the built space. Thus, a considerably greater number of dwelling units, or considerably larger dwelling surface areas, can be built on the same construction site, using the same heights and the same size of green areas. On the other hand, the lenght of utility lines required for this increased number of dwelling units is reduced, and the required infrastructure is considerably decreased.

The proposed layout represents our contribution to the reduction of site pollution. From the urbanistic point of view it revives the so-called extended living street with all its characteristics and advantages, taking into account that parking areas have been moved into the basement of the building core due to great ground-plan depths (25-30 m). The road becomes again the doamin of pedestrians, the groundfloor along the road a colorful activity area and the town space the subject of creative ambitions of architects. Thus, the problem of energy in architecture should not remain on the periphery, but should become a vivacious factor in the alphabet of its expression, intelligible and acceptable both to architects and users.

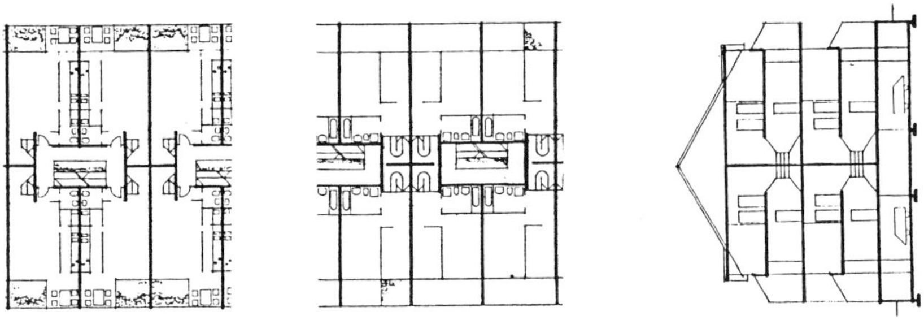

Figure 1. Duplex dwellings, designed during our research

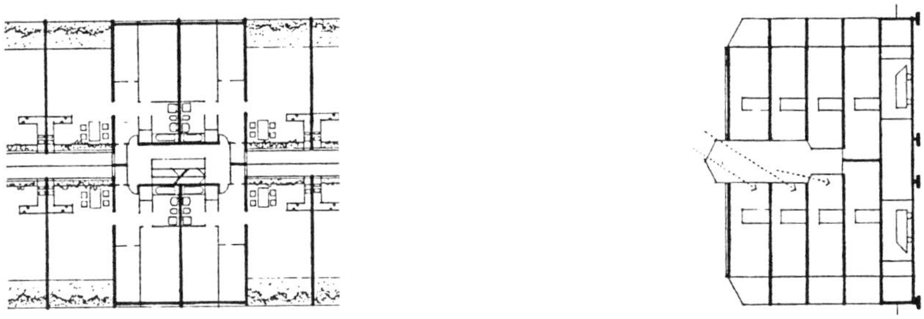

Figure 2. Residential building with skylight, designed during our research

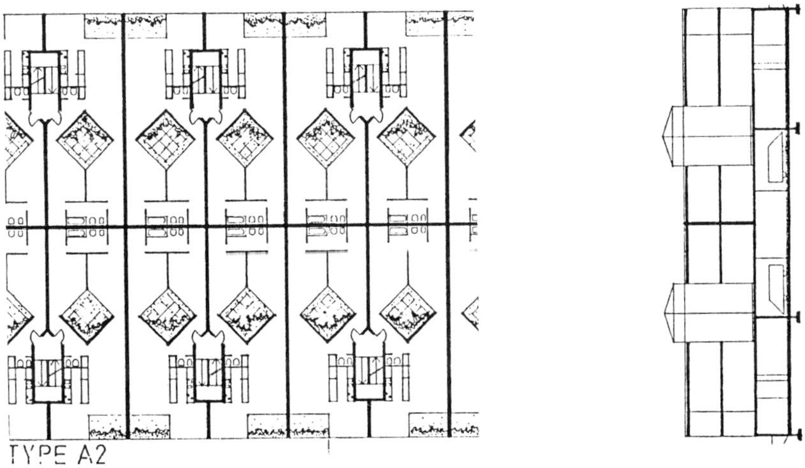

TYPE A2

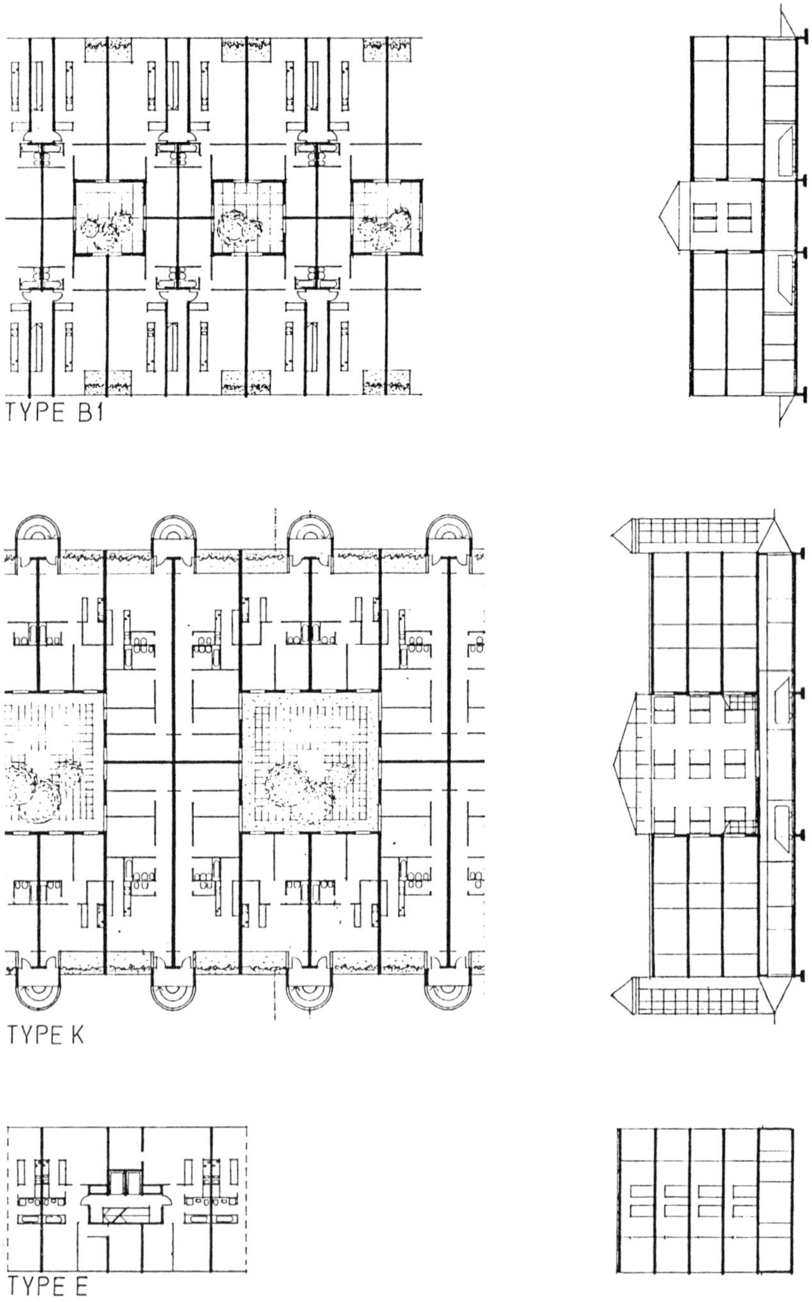

Figure 3. Residential buildings with increased ground plan depth, designed during our research

INFLUENCE OF SPECTRALLY SELECTIVE PAINT COATINGS ON BUILDING THERMAL RESPONSE

B. Orel, M. Klanjšek Gunde, A. Krainer*
Boris Kidrič Institute of Chemistry
P.O.Box 30, 61130 Ljubljana, Slovenia, YU
*University in Ljubljana, FAGG,
Jamova 2, 61000 Ljubljana, Slovenia, YU

ABSTRACT. Nocturnal temperature measurements of various infrared selective radiators performing radiative coolings have been carried out. Temperature differences between the radiators and ambient temperature achieved were about 14°C. Radiating surfaces have been prepared by applying different kind of paints on aluminium substrates. It was shown that addition of $BaSO_4$ extender in the paint dispersion increased the cooling performance of the paint radiators. Comparison between predicted cooling performances by calculations and measured ones are discussed in details.

1. Introduction

Passive cooling of dwellings and buildings takes place on account of the radiation emission from the surfaces facing the cold and dry atmosphere. Johnson |1| has shown for example that using simple infrared emitting radiator for a dwelling unit with 2000 ft^2 of radiating roof and exposed wall area would average a heat removal rate of nearly 7000 BtU/h during eight hours of night time cooling. This is equivalent already to a room air conditioner.

The basic physics of radiative cooling is simple: since the atmosphere is transparent for the thermal radiation having wave lengths in the spectral range 8-13 μm, the surface which is placed in thermal contact with the atmosphere is losing its heat by emitting the radiation in this spectral region. Emittance of the radiation from the surface is enhanced when the surface emittance for this spectral range is high ($e_{s2} \rightarrow 1$). At the same time, reflectance of the surface must be high for all the wavelengths outside the 8-13 μm window. In this way excessive heating of the surface on account to the absorptance of the atmosphere radiation impinging the surface is prevented. This kind of surface is performing the radiative cooling effects and is classified as an infrared selective radiator. Cooling power (W/m^2) which could be reached is up to 150 W/m^2. The ideally black surface with $e_T \cong 1$

S. Alvarez et al. (eds.), Architecture and Urban Space, 837–842.

exhibits the highest cooling power but the temperature drop of such of the surface exposed to the atmosphere temperature is not at its maximum. Granqvist |2| has shown that much higher temperature drops ($\Delta T = T_{atm}-T_{surface}$) could be expected when the surface is a selective radiator. The use of spectrally selective radiators instead of ideally black ones makes possible to achieve also a sizable cooling effect during daylight hours. However, to achieve this it is necessary to lower the average solar absorptance (a_s) of the solar radiation below 0.2. Nicklasson |3| has evaluated white (TiO_2) pigmented polyethylene plastic foil which performs as a selective radiator with shading efficiency. Shading i.e. low solar absorptance is achieved by intensive scattering of the solar radiation taking place by the pigment particles. The material has appropriate shading efficiency but it is not very effective as a selective radiator.

Nocturnal radiative cooling is much easier to achieve. The high reflectance for the thermal infrared region except for the 8-13 μm region is obtained by using the simple metal surface modified by applying onto it infrared transparent coatings which exhibit strong absorptions in the 8-13 μm region.

In this paper we are describing various paint coatings which act as infrared radiators having different selectivity. White pigmented paint was prepared and applied in thin films, with thicknesses from 3 to 70 μm, on highly reflecting metal substrate.

Our intention was to demonstrate how selective paint radiators perform in real conditions. In order to show these test cells (25x25x10 cm) equipped with radiators having different spectral selectivity have been exposed to the atmosphere and temperature drops of paint radiators were measured.

2. Experimental

The paints were prepared from the polymeric binder and various pigments. The binder was alkyd resin exhibiting strong absorption bands in the 8-13 μm (1250-770 cm^{-1}). The paint was pigmented with either pure TiO_2 pigment with the pigment-to-volume concentration (PVC) ratio 56 % (NO2 sample) or $TiO_2/BaSO_4$ (32:24) mixture having the same overall PVC ratio (NO1). The paints were applied on Al plates (AlMg2.5) in different thicknesses (Table I).

The applicator blade was used for the application and the thickness of the paint film was expressed in the number of grams of paint per m^2 (W/A ratio).

Test cells have been prepared from the polystyrene foams 5 cm thick with dimensions 25x25x5 cm. The polyester foil with thickness 50 μm was used as a convective cover. The paint samples (10x10 cm) were placed in the middle of the test cell and were equipped with termocouples (Cu/constant) firmly attached on the back side of samples. Data acquisition station HP 3852A was used for diurnal and nocturnal temperature measurements. The solar irradiation was measured by Solarimeter (Kuppen & Zonen).

Table I. Paint samples prepared as selective (NO1) and non-selective (NO2) radiators.

	Sample	W/A (g/m^2)	pigment(s)		Sample	W/A (g/m^2)	pigment
	1	2.68			9	3.16	
	3	5.18			2	6.12	
	4	13.04	$TiO_2/BaSO_4$		10	8.2	TiO_2
NO1	5	17.96		**NO2**	11	14.92	
	6	24.00	(32:24) PVC		12	18.00	56 % PVC
	7	44.28			13	26.28	
	8	60.92			14	26.36	

Infrared spectra were recorded on Fourier Transform Spectrometer FTS-80 Digilab. Paint films were measured in Diffuse Reflection Accessory respect to the backside of the painted Al plate. In this way, the surface roughness effects on the measured paint reflectance was eliminated.

3. Results

The results are divided in two parts: in the first one (i) paint radiators parameters which are relevant for the temperature drops and cooling powers are represented. In the second part (ii) the comparison among theoretically predicted properties and experimental data are given.

(i) Selectivity of the radiator performing radiative cooling has been determined by two parameters which describe the cooling power and the maximum temperature drop of the surface |2, 3| respect to the ambient temperature (T_a). These parameters are actually the integrated surface emittances, weighted by the atmosphere emittance |4| in the range 8-13 μm (e_{s2}) and in the whole spectral region (e_s). The third parameter ($\eta = e_{s2}/e_s$) governs the maximum achievable temperature drop, while the e_s governs the cooling power at ambient temperature. When the e_{s2} is high (ideally 1) the efficient cooling is expected and if the low temperatures are desired the large η values must be met at the same time. The maximum theoretical value η is about 3.32 at $T_a = 0°C$.

The e_{s2}, e_s and η parameters have been evaluated from the experimentally determined FT-IR reflectance spectra of paint coating (Table I). FT-IR reflectance spectra of only two samples (NO1 and NO2) differing in the type of the pigment contained are represented in Fig. 1a. The absorption bands characteristic for the resin binder and the pigment used are clearly observed indicating that the absorption of the paint NO1 is higher in the 8-13 μm (1250-770 cm^{-1}) spectral range due to the presence of ν SO2 stretching vibrations of the $BaSO_4$ extender.

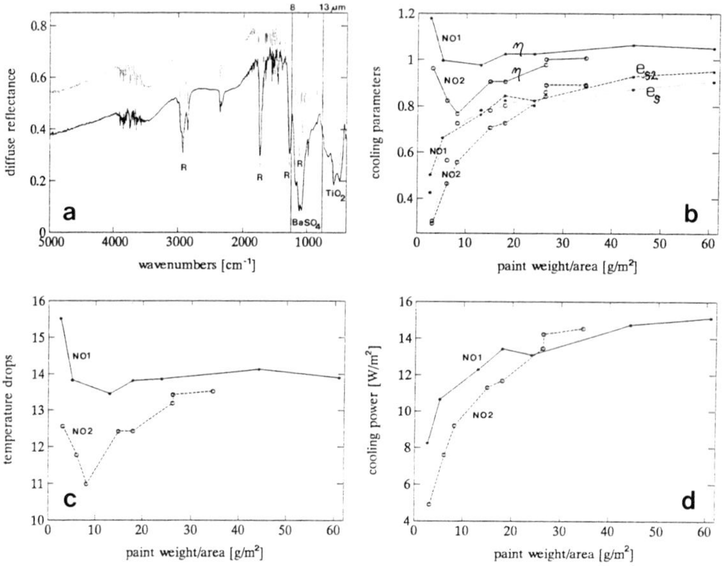

Fig 1. a - FT-IR spectra of radiators: NO1 (sample 3 - 5.18 g/m^2) full line, NO2 sample 2 - 6.12 g/m^2) dotted line. b - calculated cooling paramters of NO1 and NO2 radiators, c - calculated temperature drops (ΔT) of NO1 and NO2 samples, d - calculated cooling powers of NO1 and NO2 samples as a function of W/A and at cirrus covered sky atmosphere.

Cooling parameters e_s, e_{s2} and η for all tested samples are represented in Fig. 1b.
As expected the NO1 paint containing $BaSO_4$ exhibits higher e_{s2} and η values respect to the NO2 paint which contains TiO_2 pigment only (Fig. 1b). On the contrary, both paints have nearly equal e_s parameters indicating that the overall absorptance (=emittance) of both paint coatings in thermal infrared region is approximately the same.

Maximal temperature drop (ΔT) of various paint radiators have been determined from the condition (Fig. 1c) that emitted radiative flux from the surface is equal to the absorbed flux ($\Delta j = 0$) |2,3|. The $BaSO_4$ pigment containing paint (NO1) exhibits higher temperature drops.

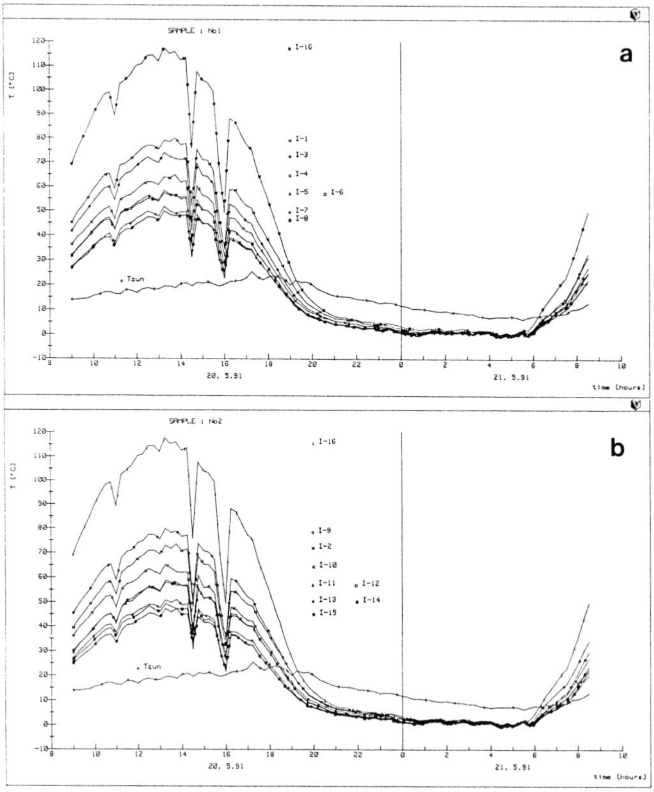

Fig. 2. Diurnal and nocturnal temperature responses of NO1 (a) and NO2 (b) sampels.

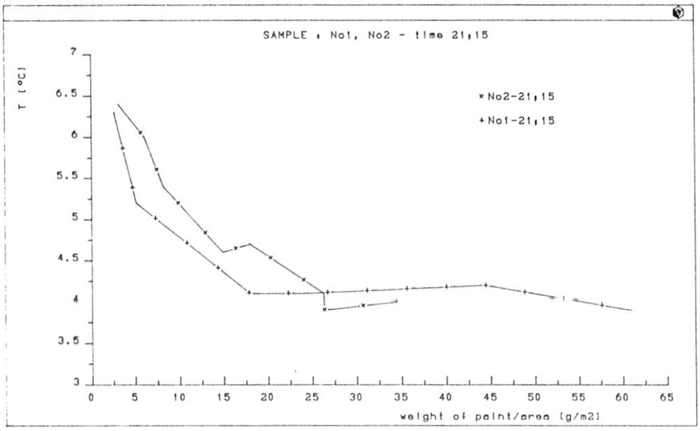

Fig. 3. Measured temperatures of various NO1 and NO2 samples as a function of W/A.

Similarly, the cooling powers of NO1 and NO2 paint radiators differ at low W/A ratios, being greater for NO1 respect to the NO2 paint (Fig. 1d). Similarly to ΔT values the differencies in the radiatively emitted fluxes become smaller with increasing paint film thicknesses and are reversed at W/A ratios greater than 25 g/m^2.

All the calculations have been performed for cirrus covered humid sky at the sea level. For other types of atmospheres the absolute values of cooling power and ΔT have been also determined, but are not represented here.

(ii) On average the temperature drops are up to 12°C below the ambient temperature (Fig. 2a,b). It is apparent that NO1 samples reach low temperatures more rapidly than NO2 samples.

At low W/A ratios - up to 15-20 g/m^2 - the NO1 paint samples cool more efficiently with respect to the NO2 samples while for greater W/A ratios (> 20 g/m^2) the NO2 samples override the first one. This behaviour is in agreement with the predictions based on e_{S2}, $e_T$and η the parameters for both type of paints (Fig.3).

4. Conclusions

At the present state the extensive practical tests of the paint radiators on the large scale have not been done yet. Our results has shown that concepts used give selective radiator, nevertheless the optimization of the actual pigment resin dispersion should still have to be made.

References

1 Johnson, T.E (1975) Radiation cooling of structures with infrared transparent wind screens, Solar Energy 17, 173-178.

2 Eriksson, T.S., Lushiku, E.M. and Granqvist, C.G. (1984) Materials for radiative cooling to low temperatures, Solar Energy Materials 11, 149-160.

3 Niklasson, G.A. and Eriksson, T.S (1988) Radiative cooling with pigmented polyethylene toils, SPIE 1016, 89-99.

4 Bell, E.E., Eisner, L., Young, J. and Oetjen, R.A. (1960) Spectral radiance of sky and terrain at wavelenghts between 1-20 microns. II. Sky measurements, J. Opt. Soc. of America 50, 1313-1320.

ENERGY ANALYSIS OF OFFICE BUILDING DESIGN FOR THE HOT ARID CLIMATE OF EGYPT

Dr.S.M.AZIZ
Dept. of Urban Planning
GOHBPR, P.O. Box 1770
Cairo, Egypt.

Prof. G.B.HANNA
Head, Dept. of Building Physics
GOHBPR, P.O. BOX 1770
Cairo, Egypt.

ABSTRACT. This article reports the impact of passive solar design techniques for an office building on heating and cooling conditionning demands. Analysis and discussions are applied on the Egyptian Building Research Center in Cairo, which is a typical governmental office building. Thermal analysis for the existing conditions are determined using a computer energy simulation program, then, energy savings opportunities are assessed through optimization analysis for fenestration, shading, reducing infiltration, walls and ceilings insulation and finishing. A set of inexpensive design enhancements succeeded in achieving a total reduction in the annual mechanical loads of about 36% compared to the initial existing conditions.

INTRODUCTION. Rapid increases in energy prices aroused strong interest in improving energy efficiency in all energy- consuming sectors. The annual electrical demand in Egypt reaches $45x\ 10^9$ Kwh against an annual production of about $60x10^9$ Kwh. Industry consumes 55% of the annual consumption, agriculture uses only 5%, Public & commercial buildings are second largest consumer of energy , 35% of the total consumption . Finally , the governmental buildings account for a share of 5%.

It is urgent to cut energy consumption through the implementation of technologies and policies for energy efficient buildings. For air conditionning, electric bills can be reduced through simple systems of thermal storage , insulation and shading which make it possible to move about 40 percent of electrical demand. In fact, it costs no more to constuct an energy-efficient office building than it does to construct an inefficient one.

Efficiently designed buildings will slash energy bills, liberate the capital investment and avoid the expense of constructing new power plants.

S. Alvarez et al. (eds.), Architecture and Urban Space, 843–848.

1. DESCRIPTION OF BASE CASE

The main office building is rectangular (fig. 1), elongated on a North -South axis. It is six storeys high. The East facade which is entirely occupied by offices has 2/3 of its area glazed while the West facade which contains offices and services, has 1/3 of its area glazed. Windows occupy the entire width of the office rooms. They are single glazed and externally shaded by a 0.6 m horizontal overhang on the West facade. East facade windows are shaded by a series of vertical fins, 10 cm away of the glass surface and having 0.6 m depth and a 0.9 m spacing in between.

External walls are made of red bricks , 0.25 m thick, plastered from inside and covered by 6 cm red brick tiles from outside, giving a thermal transmition of 1.7 W / m²°c . The roof of the last floor is a concrete slab , 0.16 m thick, covered by layers of water insulation, sand and cement tiles having an overall U-value of 1.12 W / m²°c.

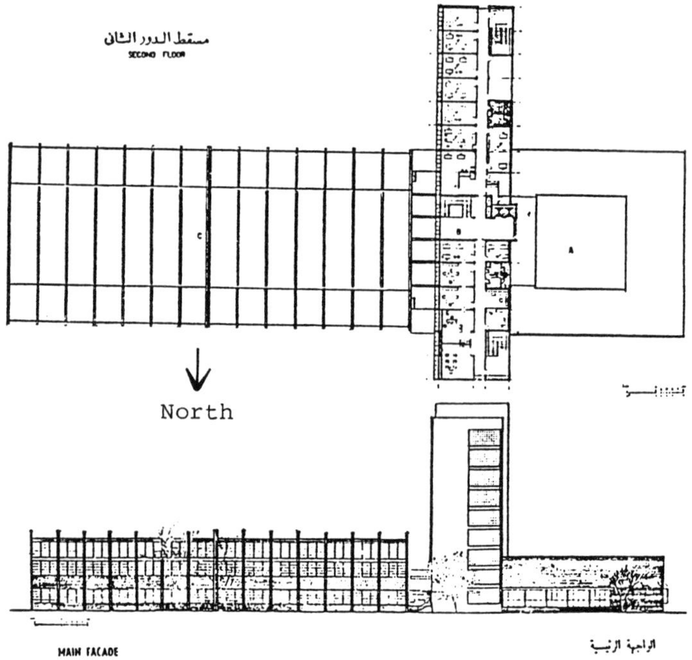

Fig. 1 : Typical floor and main facade of GOHBPR.

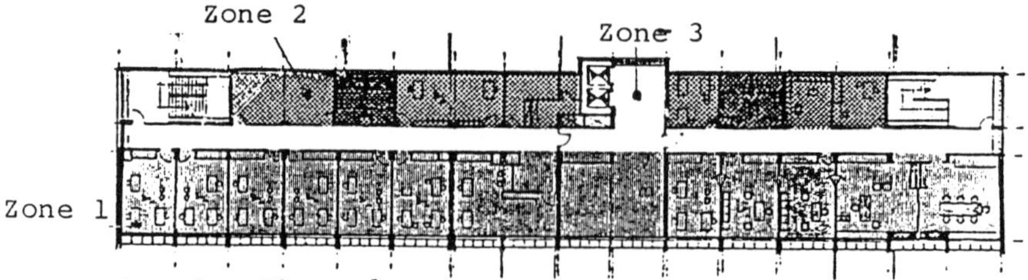

Fig. 2 : Thermal zones

2. THERMAL ANALYSIS FOR EXISTING CONDITIONS

The plan of the typical floor , Fig. 2, was devided into three thermal zones for evaluations of thermal loads and comparison:
Zone 1: a series of offices on the East facade.
Zone 2: offices on the west facade.
Zone 3: circulation area.

Table 1 shows the floor area for each zone, the area of the external walls and glazed areas according to orientation. Calculations of heating and cooling loads were performed using an energy simulation program CALPAS 3 (1), all year round from 7:00 AM to 3:00 PM in order to keep indoor temperatures at a minimum of 19.5 °C in winter and a maximum of 26.6 °C in summer.

Table 1. Characteristics of thermal zones(m^2).

zones	floor area m^2	walls N	walls E	walls S	walls W	windows N	windows E	windows S	windows W
zone 1	384	21	120	21	--	--	104	--	--
zone 2	112	--	--	--	77	--	--	--	21
zone 3	273	--	--	--	56	20	--	20	12

Table 2 shows the results of computer runs. Zone 1 requires more energy for both cooling and heating per unit floor area than zones 2 and 3. Figures 3 and 4 present a comparison of heating and cooling loads required per unit floor area for the office zones in a typical winter day (January 21) and a typical summer day (July 22) .West zone needs heating from 8:00 AM to 10:00 PM with rates 25% lower than east zones. This may be due to larger glazed area as well as fixed external shading devices.

Table 2. Comparison of mechanical loads per unit floor area for the three zones (KBtu/Ft^2)

zones	Cooling	Heating	Total	% of total load
Zone 1	35.97	6.31	42.28	57%
Zone 2	32.91	4.74	37.65	14.8%
Zone 3	25.7	3.8	29.5	28.2%

The cooling loads for East zone are much higher than for west zone untill noon. During afternoons, cooling loads for the east zone decreases while it gradually increases for the west zone and reaches a maximum at 3:00 PM, the end of the working day.

The difference in total annual loads between zones 1 and 2 is emphasized due to the larger floor area of zone 1 that is about 3.5 times the floor area of zone 2. Zone 1 is responsible for 57 % of the annual loads required per floor against 14.8% for zone 2.

3. IMPACT OF CHANGING ORIENTATION OF BUILDING

Calculations were done for heating and cooling loads in case of rotation of the existing design by 45°, 90°, and 135° to the East of South. The main office zone, (zone1) , is then facing North-East, North and North-West instead of East. Table 3 shows the results and we can deduce that the rotation causes a decrease in total loads for zone 1 and an increase in total loads for zones 2 and 3.

Table 3. Comparison of thermal loads for different orientations (KBtu/ Ft^2)

Zones	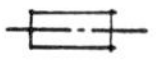			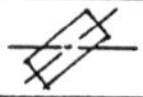			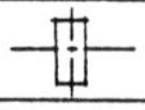			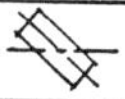		
	C	H	T	C	H	T	C	H	T	C	H	T
Zone1	35.9	6.3	42.2	29.4	7.5	36.9	24.2	7.6	31.8	25.4	7.6	33.0
Zone2	32.9	4.7	37.6	37.0	2.9	39.9	39.5	1.9	41.4	46.7	2.5	49.2
Zone3	25.7	3.8	29.5	27.6	3.6	31.2	28.9	3.4	32.3	29.3	3.5	32.8
Total Kbtu	306844			292292			276255			292090		
Saving	---			4.7 %			10 %			4.8 %		

A 90° rotation decreases the total annual loads of the whole building by 10 %.

4. OPTIMIZATION ANALYSIS FOR EXISTING CONDITIONS

There are many opportunities to improve the thermal performance of the existing building. The impact of the following design enhancements is considered:

1- External shading:
   * The horizontal overhang of the West facade windows is increased from 0.6 to 0.9 m.
   * The spacing of the vertical fins on the East facade is narrowed from 0.9 to 0.4 m by adding a new fin between each two.
2- Applying insulation of external envelope:
   * Wall insulation by using foam concrete spray (Celton), 1.5 cm thick (K=0.04 W/ m.°C) that will reduce the overall transmition factor from 1.7 W/m^2°C to 1 W/m^2.°C.
   * Roof insulation by adding a layer of expanded styrofoam 5 cm thick (K= 0.031 W/m.°C). This will decrease the U-value from 1.12 W/m^2.°C to 0.4 W/m^2.°C.
3- Reduction of infiltration from windows by changing steel frames by aluminium frames with rubber strips.
4- Light cream finishing colour for external walls surfaces.

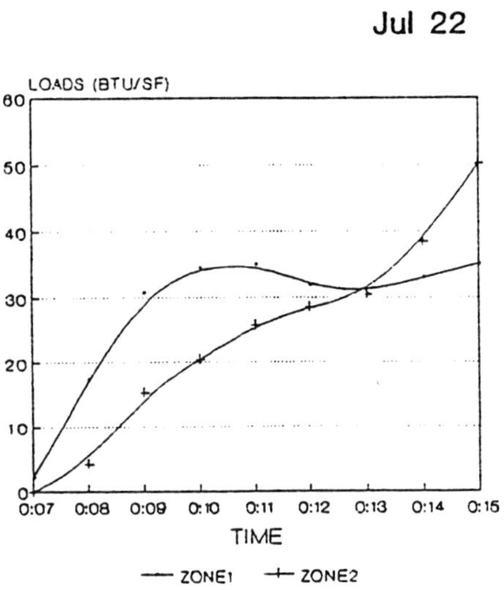

Fig. 3 : Cooling loads for office zones.

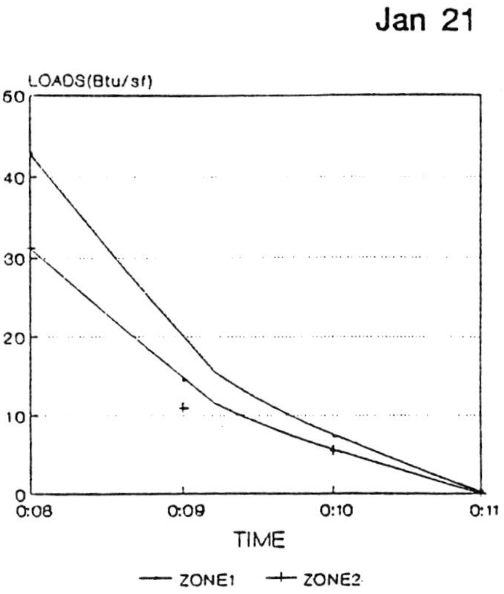

Fig. 4 : Heating loads for office zones

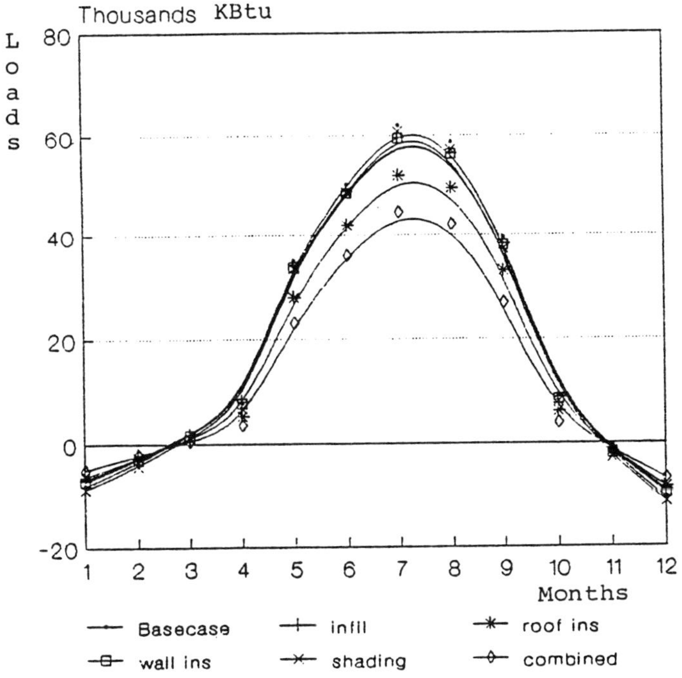

Fig. 5 : Thermal loads reduction by design enhancements for zone 1.

Table 4 shows the amount of savings achieved by each policy and the total saving that can be reached by combining them.

Table 4. Load reduction and saving opportunities of different design enhancement.

Policy	Loads in KBtu Zone 1	Zone 2	Zone 3	Total	Saving %
Base case	174768	45405	86671	306844	---
* Window shading.	140011	39645	86671	266327	13.2 %
* Wall insulation	166877	41894	84335	293106	4.5%
* Roof insulation	140833	35470	62624	238927	22.1%
*Reducing infiltra.	168094	43642	82449	294185	4.1%
*Light wall colour	172553	44269	96228	303050	1.5%
Combined	114549	28079	52292	194920	36.5 %

CONCLUSION

Most effective policies are roof insulation and optimizing the external shading of windows. A total saving of 36.5 % has been achieved in order to improve the existing conditions. On the other hand, thermal loads could have been minimized from the beginning by making the right design decisions as it has been proved by the analysis of the thermal impact of the orientation . Therefore, building energy analysis should be considered at the preliminary design phase, especially that with the advent of the personal computer energy analysis methods became accessible to a wider range of building professionals.

REFERENCES

1. CALPAS 3 Users Manual, Berkely Solar Group, California,(1984).
2. Energy Conservation Policies and Technologies for Building in Egypt. Seminar by Lawrence Berkely laboratory, Organization for Energy Planning, (1990).

AUTHOR INDEX